T0263746

TECHNIQUES AND INSTRUMENTATION IN ANALYTICAL CHEMISTRY — VOLUME 12

HAZARDOUS METALS IN THE ENVIRONMENT

TECHNIQUES AND INSTRUMENTATION IN ANALYTICAL CHEMISTRY

Volume 1 **Evaluation and Optimization of Laboratory Methods and Analytical Procedures. A Survey of Statistical and Mathemathical Techniques**
by D.L. Massart, A. Dijkstra and L. Kaufman

Volume 2 **Handbook of Laboratory Distillation**
by E. Krell

Volume 3 **Pyrolysis Mass Spectrometry of Recent and Fossil Biomaterials. Compendium and Atlas**
by H.L.C. Meuzelaar, J. Haverkamp and F.D. Hileman

Volume 4 **Evaluation of Analytical Methods in Biological Systems**
Part A. Analysis of Biogenic Amines
edited by G.B. Baker and R.T. Coutts
Part B. Hazardous Metals in Human Toxicology
edited by A. Vercruysse

Volume 5 **Atomic Absorption Spectrometry**
edited by J.E. Cantle

Volume 6 **Analysis of Neuropeptides by Liquid Chromatography and Mass Spectrometry**
by D.M. Desiderio

Volume 7 **Electroanalysis. Theory and Applications in Aqueous and Non-Aqueous Media and in Automated Chemical Control**
by E.A.M.F. Dahmen

Volume 8 **Nuclear Analytical Techniques in Medicine**
edited by R. Cesareo

Volume 9 **Automatic Methods of Analysis**
by M. Valcárcel and M.D. Luque de Castro

Volume 10 **Flow Injection Analysis – A Practical Guide**
by B. Karlberg and G.E. Pacey

Volume 11 **Biosensors**
by F. Scheller and F. Schubert

Volume 12 **Hazardous Metals in the Environment**
edited by M. Stoeppler

TECHNIQUES AND INSTRUMENTATION IN ANALYTICAL CHEMISTRY — VOLUME 12

HAZARDOUS METALS IN THE ENVIRONMENT

Edited by

M. Stoeppler
*Institut für Chemie, Forschungszentrum Jülich GmbH, Institut 4,
D-W-5170 Jülich, Germany*

ELSEVIER
Amsterdam — London — New York — Tokyo   1992

ELSEVIER SCIENCE PUBLISHERS B.V.
Sara Burgerhartstraat 25
P.O. Box 211, 10^0 AE Amsterdam, The Netherlands

ISBN 0-444-89078-5

© 1992 ELSEVIER SCIENCE PUBLISHERS B.V. All rights reserved.

No part of this publication may be reproduced, stored in a retrieval system or transmitted in any form or by any means, electronic, mechanical, photocopying, recording or otherwise, without the prior written permission of the publisher, Elsevier Science Publishers B.V., Copyright and Permissions Department, P.O. Box 521, 1000 AM Amsterdam, The Netherlands.

Special regulations for readers in the USA – This publication has been registered with the Copyright Clearance Center Inc. (CCC), Salem, Massachusetts. Information can be obtained from the CCC about conditions under which photocopies of parts of this publication may be made in the USA. All other copyright questions, including photocopying outside of the USA, should be referred to the publisher.

No responsibility is assumed by the Publisher for any injury and/or damage to persons or property as a matter of products liability, negligence or otherwise, or from any use or operation of any methods, products, instructions or ideas contained in the material herein.

Although all advertising material is expected to conform to ethical (medical) standards, inclusion in this publication does not constitute a guarantee or endorsement of the quality or value of such product or of the claims made of it by its manufacturer.

Transferred to digital printing 2005
Printed and bound by Antony Rowe Ltd, Eastbourne

CONTENTS

Preface
by Markus Stoeppler IX

INTRODUCTION

CHAPTER 1. METAL ANALYSIS – PROBLEM AREAS AND ANALYTICAL TASKS
by Markus Stoeppler
1.1. Introduction 1
1.2. Areas and tasks 2
1.3. Planning 6
References 6

SAMPLING AND SAMPLE TREATMENT

CHAPTER 2. SAMPLING AND SAMPLE STORAGE
by Markus Stoeppler
2.1. Introduction 9
2.2. Freshwater, seawater 10
2.3. Biological fluids 12
2.4. Solid biological materials 12
2.5. Environmental materials and materials of anthropogenic origin 13
2.6. Sampling for subsequent speciation analysis 14
2.7. Homogenization and storage 15
References 16

CHAPTER 3. ENVIRONMENTAL SPECIMEN BANKING
by Robert A. Lewis, Barbara Klein, Martin Paulus and Christoph Horras
3.1. Introduction and status 19
3.2. Historical aspects 20
3.3. The relationship of specimen banking to environmental monitoring, research and assessment 26
3.4. The role and application of ESB to the assessment of toxic and hazardous chemicals in the environment 27
3.5. Coordination, cooperation and technical assistance 31
3.6. The suitability of specimens 31
3.7. Methods 32
References 43

CHAPTER 4. WINE – AN ENOLOGICAL SPECIMEN BANK

by Heinz R. Eschnauer and Markus Stoeppler

4.1. Introduction 49
4.2. Enological specimen bank 50
4.3. Historical enological specimen bank 50
4.4. Results of the historical enological specimen bank 53
4.5. Some remarks on arsenic 70
References 70

CHAPTER 5. SAMPLE TREATMENT

by Peter Tschöpel

5.1. Introduction 73
5.2. Systematic errors and their avoidance 75
5.3. Sampling, sample storage and pretreatment 81
5.4. Decomposition procedures 81
5.5. Separation and pre-concentration methods 88
5.6. Conclusion 92
References 92

ELEMENTAL ANALYSIS

CHAPTER 6. ANALYTICAL METHODS AND INSTRUMENTATION – A SUMMARIZING OVERVIEW

by Markus Stoeppler

6.1. Introduction 97
6.2. Optical atomic spectrometry 98
6.3. Inductively coupled plasma–mass spectrometry (ICP–MS) 106
6.4. Electrochemical methods 107
6.5. X-Ray fluorescence spectrometry (XRF) 112
6.6. Nuclear methods 114
6.7. Chromatographic methods 117
6.8. Mass spectrometry (MS) 118
6.9. Spectrophotometry and related techniques 119
6.10. Future prospects 120
References 122

CHAPTER 7. CHEMICAL SPECIATION AND ENVIRONMENTAL MOBILITY OF HEAVY METALS IN SEDIMENTS AND SOILS

by Manfred Sager

7.1. General 134
7.2. Batch methods 140
7.3. Some experimental data from sequential leaching sequences 157
7.4. Speciation and availability to biota 164
7.5. Column methods 167
References 171

CHAPTER 8. CADMIUM

by Markus Stoeppler

8.1. Introduction 177

8.2. Environmental and biological levels of cadmium 178
8.3. Sampling and sample pretreatment 178
8.4. Enrichment and separation 181
8.5. Analytical methods 182
8.6. Speciation 210
8.7. Quality control and reference materials 213
8.8. Conclusion and prospects 218
References 219

CHAPTER 9. LEAD

by Steve J. Hill

9.1. Introduction 231
9.2. Analytical techniques for the determination of lead 232
9.3. Applications 244
9.4. Conclusion 251
References 252

CHAPTER 10. MERCURY

by Iver Drabæk and Åke Iverfeldt

10.1. Introduction 258
10.2. Sampling procedures 264
10.3. Determination procedures 273
10.4. Conclusions 281
References 282

CHAPTER 11. ARSENIC

by Kurt J. Irgolic

11.1. Introduction 288
11.2. Total arsenic versus arsenic compounds 290
11.3. Arsenic compounds in the environment 292
11.4. Total arsenic determinations 294
11.5. Determination of arsenic compounds 321
11.6. Standards for the determination of arsenic 340
References 340

CHAPTER 12. THALLIUM

by Manfred Sager

12.1. Introduction 352
12.2. Occurrence of thallium 352
12.3. Sample decomposition methods 354
12.4. Separation methods 357
12.5. Final determination methods 361
References 369

CHAPTER 13. CHROMIUM

by Nancy J. Miller-Ihli

13.1. Introduction 374
13.2. Sampling procedures 379
13.3. Determination procedures 387
References 400

CHAPTER 14. NICKEL AND COBALT

by Markus Stoeppler and Peter Ostapczuk

14.1. Introduction 405
14.2. Environmental levels of nickel and cobalt 406
14.3. Sampling and sample pretreatment 408
14.4. Enrichment and separation 408
14.5. Analytical methods 410
14.6. Speciation 432
14.7. Quality control and reference materials 434
14.8. Conclusions and prospects 440
References 442

CHAPTER 15. ALUMINIUM

by Wolfgang Frech and Anders Cedergren

15.1. General introduction 451
15.2. Environmental levels of aluminium 452
15.3. Sampling and pretreatment procedures 455
15.4. Methods for the determination of aluminium 463
15.5. Determination of aluminium by GFAAS 465
15.6. Standard reference materials 470
15.7. Need for future research 470
References 470

CHAPTER 16. SELENIUM

by Milan Ihnat

16.1. Introduction 475
16.2. Analytical methodology reviews 478
16.3. Analytical applications 483
16.4. Sampling, sample handling and storage 490
16.5. Sample treatment 495
16.6. Determination procedures and recommended methods 499
16.7. Conclusions 505
References 505

CHAPTER 17. QUALITY ASSURANCE AND VALIDATION OF RESULTS

by Bernard Griepink and Markus Stoeppler

17.1. Introduction 517
17.2. Relevant definitions (ISO) 519
17.3. Within laboratory quality measures 519
17.4. Intercomparisons 526
17.5. Certified reference materials (CRMs) 530
References 533

Subject Index 535

PREFACE

Metals and metalloids have long been mined, industrially processed and used in numerous applications. This has led, especially since the industrial revolution and during the 20th Century, to regional and global redistribution and for some – more or less hazardous – elements to a significant increase of their concentration in the upper part of the earth's crust. Therefore, in the plough-layer of soils, in plants, animals, lakes, rivers, estuaries and even in oceanic and arctic regions, in foodstuffs and human beings the levels of a variety of elements have substantially increased over time. Detailed studies on the fate and contents of these elements in various environmental compartments and human beings have become a major task in environmental research, and especially in analytical chemistry, and there is a continuous challenge to develop new methodology and optimize that already available. In this area remarkable progress has been observed during the last two decades, leading to new and improved information about the state of the environment.

This rapid progress in methodology and the information it has provided makes it desirable that achievements in this field with regard to actual knowledge of research strategies, environmental levels and methods are periodically reviewed and updated. This is the rationale behind this volume. It offers the reader a general introduction into the problem areas that have been identified and are currently being tackled, followed by chapters on sampling and sample preservation, strategies and applications of the archiving of selected representative specimens for long-term storage in environmental specimen banks. This is supplemented by the example of wine as a preserved – frequently already historical – specimen that clearly reflects technological changes over time. This is followed by a chapter on sample treatment, an overview of the most frequently and the at present most successfully applied trace analytical methods for metals and metal compounds, and an introduction into the increasingly important approaches to identify and quantify metal species in sediments and soils. It is not the total amount of a metal that is the most important, though necessary, information but its chemical form (species), which often is responsible for adverse toxic or ecotoxic effects.

The chapters in the second part provide detailed information on analytical methods for determining the levels of a number of selected toxicologically, ecotoxicologically and ecologically important elements in environmental and biological materials, including information on the separation and quantification of chemical and organometallic species. The elements treated are aluminium, arsenic, cadmium, chromium, cobalt, lead, mercury, nickel, selenium and thallium. The closing chapter treats quality assurance approaches and the paramount importance of appropriate reference materials to avoid incorrect data.

Since the chapters of this book have been contributed by acknowledged experts on each topic, it is hoped that the text and references will serve as a valuable aid for the many researchers involved in trace metal and species analysis in environmental and

X

biological materials in selecting the most promising method for a certain element in a certain matrix.

Finally I would like to thank my colleague Prof. Dr. Rokus A. de Zeeuw (Groningen), for the proposal to conceptualise and edit this book, and of course the authors without whose cooperation and engagement this volume would not have been possible.

Juelich, Germany   Markus Stoeppler

M. Stoeppler (Editor)/Hazardous Metals in the Environment
© 1992 Elsevier Science Publishers B.V. All rights reserved

Chapter 1

Metal analysis – problem areas and analytical tasks

Markus Stoeppler

Institute of Applied Physical Chemistry, Research Center Juelich, P.O. Box 1913, D-W-5170 Juelich (Germany)

CONTENTS

1.1. Introduction 1
1.2. Areas and tasks 2
    1.2.1. Without particular difficulties 2
    1.2.2. Moderate difficulties 3
    1.2.3. Slightly increased difficulties 4
    1.2.4. Great difficulties 5
    1.2.5. Extreme difficulties 5
1.3. Planning 6
References 6

1.1. INTRODUCTION

This book deals in its first part with the crucial steps of trace analytical procedures ranging from sample collection, sample treatment and instrumental techniques to the increasingly important approaches of chemical speciation and environmental mobility of metals and metal compounds in sediments and soils. The second part describes the application of analytical methods for ten selected metals and metalloids that are very frequently analyzed in numerous biological and environmental materials and at present considered as toxicologically and environmentally especially significant and often also hazardous.

These elements appear in man and in food and in various environmental compartments in a broad concentration range from natural ultratrace levels at the ng/kg or ng/l (ppt) level, sometimes even below, to the –often increased due to anthropogenic pollution– mg/kg level. For metal toxicology, in occupational medicine and ecotoxicology as well as in food control, accurate and precise trace analytical methods are very important. The application of these methods thus provides the basis for appropriate information about levels, fate, behavior and, if feasible, chemical form ("speciation") of metals and metalloids. This information is necessary for the observance of governmental rules and regulations and for decisions on the alteration of existing or the introduction of new measures for environmental conservation or protection [1]. Therefore it is evident that less reliable or in fact incorrect analytical results can lead to severe and often also expensive misinterpretations for toxicological, ecological and economic impacts of trace elements. Although remarkable methodological progress has been achieved, especially during the last two decades, the present situation in analytical chemistry is nonetheless far from being ideal. It has at its disposal –on average– fairly successful approaches for the quantification of total contents for estimates of true levels and dose-effect relationship as is obvious from the results of recent analytical interlaboratory comparisons [2-12].

The reasons for the frequent observation of disappointing results in total elemental as well as in speciation analysis are just as varied as the complex structure of trace analytical procedures. Despite remarkable progress in instrumental performance and automation, possibly indeed due to just this progress, multiple error sources already starting at the first steps, e.g. at the planning or the sampling stage, have obviously not lost importance.

If the present commercially available, instrumental trace analytical methods are treated this situation has to be mentioned to obtain a realistic view of their potential and applications. This book will therefore commence with definitions of different tasks in increasing order of difficulty and complexity for the determination of metals, metalloids and their organometallic compounds and valency states with a critical attitude, as far as necessary.

1.2. AREAS AND TASKS

In Fig. 1 the general structure of a trace analytical procedure from definition of tasks to data evaluation and interpretation is outlined.

These tasks belong to the categories discussed below that also require skill and equipment in increasing order.

1.2.1. Without particular difficulties

Rapid analyses are required in e.g. the case of an acute intoxication or heavy pollution for the identification and estimation of the concentration of the considered element Methods for this purpose have to be quick and specific. However, due to the clinical and sometimes legal importance of the results, false negative, sometimes also false positive, results must be minimized as far as the state of the art allows.

The same is the case for determinations aiming at the identification of significantly increased metal concentrations that exceed legal threshold limits, for instance for cadmium, lead, copper, chromium, mercury, nickel and zinc in sewage sludge [13-15]. This would preclude such a material from being used as a fertilizer. Analysis of this type can be performed with comparatively simple and less expensive or less sensitive methods like colorimetry (however, in decreasing frequency), flame atomic absorption spectrometry, solid sampling atomic absorption spectrometry, polarography, emission spectrometry with plasma (ICP and DCP) excitation and X-ray fluorescence (XRF). Because of concentration levels in the medium to high mg/kg (ppm) range sample preparation, at least for the

ANALYTICAL TASK
Degree of difficulty, frequency of sampling, number of samples, importance and urgency

PLANNING
Orientation of task to the potential of the included laboratory(ies), organizational structure
and time schedule

SAMPLING
Minimization of contamination and losses; if subsequent speciation is required, conservation of species is predominant

SAMPLE STORAGE
Short-term, medium-term or long-term storage under as far as feasible controlled conditions. In some cases long-term storage partly after homogenization, e.g. for
environmental specimen banking under cryogenic conditions (i.e. below - 80°C) for retrospective analysis

PRETREATMENT
Separation, preconcentration (liquids), homogenization, aliquotation, extraction or decomposition, matrix or compound separation, enrichment etc. (solids)

ANALYSIS
Application of a method selected according to the planning step for routine determination
of metals and/or their compounds (speciation)

QUALITY CONTROL
Use of independent or reference methods(s), application of control, reference or certified
reference materials (internal quality control), Interlaboratory comparison (external quality
control) for further validation of results, adherence to GLP (good laboratory practice)

DATA EVALUATION
Numerical values with precision data and estimations of accuracy

DATA INTERPRETATION
Toxicological, epidemiological, occupational, ecotoxicological, ecological, ecochemical,
geological, oceanographical etc.

Fig. 1.1. General diagram of a trace analytical procedure.

metals and metalloids treated in this book, is commonly simple and contamination risks
are not very pronounced.

1.2.2. Moderate difficulties

These tasks include the determination of elemental concentrations and, predominantly,
concentration ratios around the mg/kg (ppm) level. Here the most frequently performed
analyses aim at a differentiation between anthropogenically polluted and less or practically
non-polluted areas using bioindicators and bioaccumulators. Specimens to be analyzed
are different horizons of soils, sediments, a number of metal-accumulating terrestrial,
aquatic and marine plants, e.g. lichen, mosses, trees, and algae, and wild indicator

animals, e.g. earthworms, foxes, deer, birds (particularly bird's eggs, organs and feathers) mussels and fish (fillet and organs) [15,16]. This category also includes outdoor and indoor dust [17] and the determination of metal distributions –"fingerprints"– in many of the above-mentioned biological and environmental materials to arrive at sound conclusions about particular metal burdens.

Methods applied for these studies should attain an appreciable precision and also provide considerable element coverage, because of the main aim to compare similar species with different degrees of pollution. Thus, utmost accuracy is commonly not required, since the elemental levels predominantly to be determined are not significantly influenced by contamination. Rigid contamination controls, except in a few cases where metals with endogenously low levels (as e.g. mercury in some plants and terrestrial animals) have to be determined, are not required. The methods applied in this context are quite similar to those mentioned under Section 1.2.1, but should include graphite furnace AAS for some elements, occurring on the borderline of the detection power of the other methods.

1.2.3. Slightly increased difficulties

In cases of occupational and environmental exposure also comparative measurements in controls, i.e. the recognition of trace levels in materials with endogenously low elemental levels, is necessary [18]. The procedures applied for these purposes should thus be able to analyze accurately and precisely metals and metalloids at the higher $\mu g/kg$ or $\mu g/l$ up to the lower mg/kg level. It is, however, frequently sufficient that with these procedures a satisfactory differentiation from natural concentrations, providing determination or even detection limits in the lower $\mu g/l$ region, is possible with certainty. Examples are the determination of arsenic, cadmium, chromium, cobalt, lead, mercury, nickel, thallium etc. in body fluids and tissues of moderately to heavily exposed collectives and controls, tap and wastewater and effluents. This also includes a number of significant environmental species such as leaves, needles, grains, vegetables and many food products in order to ascertain that metals, metalloids or organometallic compounds are present significantly below or above certain values for risk assessment [19] threshold limits, legislative regulations [20] and also of importance for current food and feed control [21-25] and drinking water [26].

Methods recommendable for this task are graphite furnace, cold vapor and hydride AAS, several modes of voltammetry, inductively coupled plasma atomic emission spectrometry (frequently in conjunction with preconcentration), total reflection X-ray fluorescence spectrometry (TXRF) and, now increasingly applied, inductively coupled plasma mass spectrometry (ICP-MS) as a very potent but also expensive multielement technique with outstanding potential for rapid screening, particularly for natural waters.

1.2.4. Great difficulties

This area includes the very precise and accurate determination of average normal trace metal contents in numerous biological and environmental materials frequently still as baseline studies if formerly applied methods were not sensitive or not reliable enough to approach the true levels. Because it is nowadays evident from a number of careful studies performed in highly specialized and experienced laboratories that so-called normal levels of metals and metalloids, e.g. in body fluids, natural, river and sea water are much –sometimes up to several orders of magnitude– lower than assumed less than two decades ago [27-34]. This type of analysis also includes the determination of various chemical forms of metals in inland and seawater [36,37] and normal and somewhat elevated levels of trace metals in precipitation (rain and snow) [37]. Metal concentrations in these materials range from the lower μg/kg (ppb) down to the ng/kg level. It is obvious that at those levels high risks for contamination and losses during sampling as well as problems of detection power and precision of the applied methods are to be faced. Reliable data, therefore, can be only expected if the laboratory staff are already skilled in this matter and continuous efforts in analytical quality control are made. The methods applied in this task are among the most sensitive modes of atomic absorption spectrometry, for some elements already subsequent to various preconcentration procedures [38], voltammetry, usually in the anodic or cathodic stripping mode and, increasingly TXRF and ICP-MS. Voltammetry is very useful for speciation in aqueous samples [35] and both TXRF and ICP-MS usually require pretreatment and preconcentration [39-41].

1.2.5. Extreme difficulties

It is very difficult to determine trace metal contents at the ng/kg level in solid materials and below that level in aqueous matrices. Here, the task is to detect extremely low levels and concentration differences at low natural levels because of geochemical, behavioral and nutritional differences. Examples are the determination of extremely low levels of trace metals in open oceans and in snow of remote e.g. arctic areas [41-43]. The determination of ubiquitously distributed elements such as aluminium with extreme contamination problems from sampling to determination is of further importance [45,46]. Similar problems have to be faced for the determination of lead, cadmium and nickel in blood plasma or serum [47-49] as well as for the determination of cadmium and lead in human and retail milk [50-52]. Levels of some trace metals –e.g. mercury, cadmium, lead, cobalt, nickel– are very low in some foodstuffs with concentrations at the ng/kg level [53-55]. The same is the case for methyl mercury in various biological and environmental materials with levels significantly below 1 ng/l in rainwater, snow and seawater [56-60].

Tasks of this kind cannot be performed successfully without particular experience and staff training, clean-room techniques, use of the most powerful –frequently not yet commercially available– methods and continuous blank and quality control. This area of investigations will doubtless remain the domain of laboratories with highly skilled personnel and first-rate instrumental equipment.

References on p. 6

1.3. PLANNING

The planning phase for the performance of analyses within the context of medical, biochemical and environmental research or surveillance has to consider the particular tasks, its significance and hence urgency. The chosen approach has to consider the final requirements as far as data quality and frequency are concerned, as well as the laboratory equipment and the state of experience of the scientific and technical staff involved. If the level of experience is only fair this may preclude a successful execution of the whole program.

If the task is limited in time but urgently required, it is usually not advisable to introduce a new procedure in order to hasten sample throughput if the time for becoming acquainted with an approach not applied hitherto appears to be longer than the expected gain in time compared to the previous procedure. This is because one has to consider that the introduction of a new procedure may in addition be linked with unexpected disturbances and delays.

Similar but often still more severe problems might be faced if a completely new instrumental analytical method is to be introduced. In such a case, but also if only changes in the execution of the whole procedure are to be made, it is strongly recommended to still apply the earlier method (or procedure) in a transition phase, if it appears still useful as a reference method compared to the one newly introduced so that the staff can still use a method with which they are familiar and in which they have confidence. If the new method (procedure) performs better in direct comparison, which usually should be expected from theoretical insights into the potential of this method, this helps to convince and motivate them about the necessity of the methodological alteration. This, of course, is only absolutely necessary if another independent method is not already at hand to check the new one continuously and with all materials analyzed. This is regrettably very often not the case in many analytical laboratories. In this phase of change to a new method the increased use of reference or (internal) control materials and the at least occasional participation in external quality control programs is extremely valuable. If the new method, however, in addition is more sensitive, which logically is often the case, particular care must be taken in the adoption of this method. Thus extended internal and external quality control is mandatory.

REFERENCES

1 E. Merian, *Kem. Kemi,* 13 (1986) 959-969.
2 A. Scholz, *Staub, Reinhalt., Luft,* 41 (1981) 304-309.
3 J. Versieck, *Trace Elem. Med.,* 1 (1984) 2-12.
4 Analytical Quality Control Committee, *Analyst,* 110 (1985) 103-111.
5 A. Taylor and R.J. Briggs, *J. Anal. At. Spectrom.,* 1 (1986) 391-395.
6 H. Muntau, *Fresenius' Z. Anal. Chem.,* 324 (1986) 678-682.
7 L. Fishbein, *Int. J. Environ. Anal. Chem.,* 28 (1987) 21-69.
8 D.H. Loring and R.T.T. Rantala, *Mar. Chem.,* 24 (1988) 13-28.
9 H.S. Hertz, *Anal. Chem.,* 60 (1988) 75A-80A.
10 R. Wagemann and F.A.J. Armstrong, *Talanta,* 35 (1988) 545-551.

11 R. Schelenz, R.M. Parr, E. Zeiller and S. Clements, *Fresenius' Z. Anal. Chem.*, 333 (1989) 33-34.
12 R.F.M. Herber, M. Stoeppler and D.B. Tonks, *Fresenius' J. Anal. Chem.*, 338 (1990) 279-286.
13 A. Kloke, *Mitt. VDLUFA*, (1977) 32-38.
   A. Kloke, *Mitt. VDLUFA*, (1980) 1-3.
14 Klärschlammverordnung (Abf. Klär V), *Bundesgesetzblatt*, 1 (1982) 734-739.
15 J. Hertz, in E. Merian (Editor), *Metals and Their Compounds in the Environment*, VCH, Weinheim, 1991, pp. 221-231.
16 U. Arndt, W. Nobel and B. Schweizer, *Bioindikatoren*, Eugen Ulmer Verlag, Stuttgart, 1987.
17 L. Fishbein, in E. Merian (Editor), *Metals and Their Compounds in the Environment*, VCH, Weinheim, 1991, pp. 287-310.
18 U. Ewers and A. Brockhaus, in E. Merian (Editors), *Metals and Their Compounds in the Environment*, VCH, Weinheim, 1991, pp. 207-220.
19 R.L. Zielhuis, in E. Merian (Editor), *Metals and Their Compounds in the Environment*, VCH, Weinheim, 1991, pp. 651-686.
20 U. Ewers, in E. Merian (Editor), *Metals and Their Compounds in the Environment*, VCH, Weinheim, 1991, pp. 687-714.
21 P. Weigert, in E. Merian (Editor), *Metals and Their Compounds in the Environment*, VCH, Weinheim, 1991, pp. 449-468.
22 H.J. Hapke, in E. Merian (Editor), *Metals and Their Compounds in the Environment*, VCH, Weinheim, 1991, pp. 469-490.
23 *Bundesgesundheitsblatt*, 29 (1986) 22-23.
24 WHO/FAO Joint Committee on Food Additives, *Evaluation of Certain Food Additives and the Contaminants Mercury, Lead and Cadmium*, WHO, Geneva, Techn. Rep. Ser. 505, 1972.
25 WHO, *Trace Elements in Human Nutrition*, WHO, Geneva, Techn. Rep. Ser. 532, 1973.
26 WHO, *Guidelines for Drinking Water Quality*, Vols. 1-3, WHO, Geneva, 1984.
27 J. Versieck and R. Cornelis, *Anal. Chim. Acta*, 116 (1980) 217-254.
28 J. Versieck, *Trace Elem. Med.*, 1 (1984) 2-12.
29 J. Versieck, *CRC Crit. Rev. Clin. Lab. Sci.*, 22 (1985) 97-184.
30 L. Huynh-Ngoc, N.E. Whitehead and B. Oregioni, *Toxicol. Environ. Chem.*, 17 (1988) 223-246.
31 W.S. Dorten, F. Elbaz-Poulichet, L.R. Mart and J.M Martin, *Ambio*, 20 (1991) 2-6.
32 B.K. Schaule and C.C. Patterson, *Earth Plant Sci. Lett.*, 54 (1981) 97-116.
33 K.S. Wong, E. Boyle, K.W. Bruland and E.D. Goldberg (Editors), *Trace Metals in Sea Water*, Plenum Press, New York, London, 1983.
34 T.D. Jickels and J.D. Burton, *Mar. Chem.*, 23 (1988) 131-144.
35 T.M. Florence, *Analyst*, 111 (1986) 489-505.
36 G.M.P. Morrison, in G.E. Batley (Editor), *Trace Element Speciation: Analytical Methods and Problems*, CRC Press, Boca Raton, 1989, pp. 43-76.
37 H.W. Nürnberg, P. Valenta, V.D. Nguyen and M. Gödde, *Fresenius' Z. Anal. Chem.*, 317 (1984) 314-323.
38 K.W. Bruland, K.H. Cole and L. Mart, *Mar. Chem.*, 17 (1985) 233-245.
39 N.-S. Chong, M.L. Norton and J.L. Anderson, *Anal. Chem.*, 62 (1990) 1043-1050.
40 D. Beauchemin and S.S. Berman, *Anal. Chem.*, 61 (1989) 1857-1862.
41 A. Prange, *Spectrochim. Acta*, 44B (1989) 437-452.
42 L. Mart, *Tellus*, 35B (1983) 131-141.
43 C.F. Boutron, in J.O. Nriagu and C.I. Davidson (Editors), *Toxic Metals in the Atmosphere*, Wiley, New York, 1986, pp. 467-505.
44 J. Völkening and K.G. Heumann, *Fresenius' Z. Anal. Chem.*, 331 (1988) 174-181.
45 W. Slavin, *J. Anal. At. Spectrom.*, 1 (1986) 281-285.
46 A. Cedergren and W. Frech, *Pure Appl. Chem.*, 59 (1987) 221-228.
47 J. Everson and C.C. Patterson, *Clin. Chem.*, 26 (1980) 1603-1607.
48 M. Stoeppler, *Int. J. Environ. Anal. Chem.*, 27 (1986) 231-239.

49 F.W. Sunderman, A. Aitio, L.G. Morgan and T. Norseth, *Toxicol. Ind. Health*, 2 (1986) 17-78.
50 R.W. Dabeka, K.F. Kardinski, A.O. McKenzie and C.D. Bajdik, *Food Chem. Toxicol.*, 24 (1986) 913-921.
51 R.W. Dabeka and A.D. McKenzie, *J. Assoc. Off. Anal. Chem.*, 70 (1987) 753-757.
52 H.D. Narres, C. Mohl and M. Stoeppler, *Z. Lebensm. Unters. Forsch.*, 181 (1985) 111-116.
53 K. May and M. Stoeppler, in *Proc. Int. Conf. Heavy Metals in the Environment*, Vol. 1, CEP Consultants, Edinburgh, 1983, pp. 241-244.
54 H.D. Narres, P. Valenta and H.W. Nürnberg, *Z. Lebensm. Unters. Forsch.*, 179 (1984) 440-446.
55 G. Vos, H. Lammers and H. van Delft, *Z. Lebensm. Unters. Forsch.*, 187 (1988) 1-7.
56 M. Horvat, K. May, M. Stoeppler and A.R. Byrne, *Appl. Organomet. Chem.*, 2 (1988) 515-524.
57 M. Padberg and M. Stoeppler, *Curr. Top. Environ. Anal. Chem.*, 1991, in press.
58 N.S. Bloom and C.J. Watras, *Sci. Total Environ.*, 87/88 (1989) 199-207.
59 R.P. Mason and W.F. Fitzgerald, *Nature*, 347 (1990) 457-459.
60 Y.-H. Lee and H. Hultberg, *Environ. Toxicol. Chem.*, 9 (1990) 833-841.

M. Stoeppler (Editor)/*Hazardous Metals in the Environment*
© 1992 Elsevier Science Publishers B.V. All rights reserved

Chapter 2

Sampling and Sample Storage

Markus Stoeppler

Institute of Applied Physical Chemistry, Research Center Juelich, P.O. Box 1913, D-W-5170 Juelich (Germany)

CONTENTS

2.1. Introduction 9
2.2. Freshwater, seawater 10
2.3. Biological fluids 12
2.4. Solid biological materials 12
2.5. Environmental materials and materials of anthropogenic origin 13
2.6. Sampling for subsequent speciation analysis 14
2.7. Homogenization and storage 15
References 16

2.1 INTRODUCTION

Sampling is a crucial step –in many cases doubtless the most crucial– in trace and ultratrace analysis. This is because of severe contamination risks during collection from sampling tools and also from the persons performing this task. Hence all necessary precautions have to be taken and included in the sampling plan to ascertain that the collected samples are representative of the task as well. For this purpose the sampling strategy must be based on detailed –e.g. geological, ecological and biological– background information so that the first step can be executed cautiously and properly as a reliable prerequisite for subsequent analytical work. For practical performance a sampling protocol based on previously carefully evaluated standard operation procedures is mandatory. It must include all relevant information for final data evaluation and interpretation in agreement with the requirements of good laboratory practice and standardized analytical methods [1-3].

Because of distinct differences in sampling techniques between liquid and solid and between biological and environmental materials, these will be treated separately first from the view of total elemental analysis, then for speciation. Finally preparation for and storage of collected materials is discussed.

References on p. 16

2.2. FRESHWATER AND SEAWATER

Formerly, contamination during sampling was a severe and frequent source of error, indeed water collection is still difficult to perform properly and thus has to be executed with extreme care using scrupulously cleaned and prepared vessels and contamination precautions, e.g. by wearing clean gloves and other measures mentioned below, also for the person performing this task.

The collection of potable water should be performed by first taking a district volume (approx. 20 l) and discarding it in order to minimize contamination from the supply tubes [4]. In water works sampling can be done automatically and either continuously or sequentially analyzed with flow-through systems using a bypass to the main water line [5].

Wet precipitation is often collected together with dry deposition in so-called bulk samplers or just as wet deposition with systems equipped with suitable plastic (mainly polyolefine) vessels, that open –initiated by a humidity sensor– only during rain- or snow-fall. These samplers are called wet only samplers, they are frequently equipped with an electric heater to permit collection of snow as well. Because of problems with the proper collection of total deposition, although they provide limited information, from the present state of knowledge wet-only samplers appear to be generally more reproducible and thus more reliable compared with complex conditions of samplers for total deposition [6,7]. If precipitation has to be collected for the determination of mercury and methyl mercury, it is necessary to replace plastic by glass or quartz collection vessels. For methyl mercury collection, due to the sensitivity of this compound to UV radiation, these vessels have to be protected against light [8].

In polar and alpine areas snow and ice samples are collected manually or with sophisticated technical equipment using precleaned plastic containers and ultraclean sampling tools under utmost contamination precautions and in surface sampling in a direction opposite to the wind [9-11]. Determination of trace (ng/kg levels) metals in scrupulously prepared ice cores from arctic and antarctic regions can be performed directly at the sampling site in special containers with clean-room equipment [11] or after shipping in ultratrace laboratories [12]. Determination of trace metals in ice cores and snow from arctic and antarctic regions has provided scientifically invaluable information about retrospective and present pollution of these areas by heavy metals and thus about global transport mechanisms [13-15].

Sampling in rivers, lakes and in the sea depend on the problem to be investigated. If surface water has to be collected it is common practice to take samples manually or with bottles fixed on telescopic (e.g. fiberglas) bars, approx. 0.5 to 1.0 m below the water surface from a boat, the bank or the shore using scrupulously precleaned bottles from polyethylene, PTFE and other suitable materials.

For mercury, however, due to its particular properties, PTFE, glass or even quartz bottles are very recommendable. For total metal determinations acidification of the samples frequently after filtration from particulate matter is recommendable, depending on the subsequent analytical approach.

The British Standing Committee of Analysts has given a detailed description of suitable sample containers and storage recommendations for various aqueous samples including also the elements treated in this book. For the determination of total metal contents it is recommended to add either 20 ml of 5 M hydrochloric acid or 2 to 10 ml concentrated nitric acid per liter of sample and to store at 4°C, "but be guided by the conditions for the most sensitive metal likely to be present" [16].

As already mentioned for snow sampling, contamination from the person who performs the sampling and the sampling boat has to be avoided in that collection is performed opposite to the wind direction and far from any possible contamination source, such as the research vessel itself in oceanic operations. Descriptions of all necessary steps from purification of bottles and sample preparation to analysis under clean-room conditions are given by Mart and Mart et al. [9,15,17-20] (see also Fig. 1) and also addressed by Stoeppler [21] for freshwater in some detail and in a more general manner also considering subsequent chemical speciation by Batley [22].

ACID CLEANING OF SAMPLING BOTTLES
(Repeated acid leaching from outside and inside under clean-air conditions until blanks for analytes are negligible)

ENCLOSURE IN POLYETHYLENE BAGS

SAMPLING
(Manually or with sampling systems)

FILTRATION
(Clean-air conditions, 0.45 μm purified membrane filter)

SOLUTION	FILTER
(Aliquotation and acidification)	(Transport, intermediate storage)

| TRANSPORT AND INTERMEDIATE STORAGE | DECOMPOSITION |

| IRRADIATION (If necessary to destroy interfering matter) | ANALYSIS |

ANALYSIS

Fig. 2.1. Plan for preparation of sampling bottles, sampling, treatment and determination of metals in aqueous samples, after Mart [20].

References on p. 16

A comparatively new method for water collection in investigating ocean transects with modern research vessels is pumping water from a depth of 7 m depth by an all-PTFE pump at a rate of 1.2 m^3/h through polyethylene tubing from a towed stainless-steel fish suspended underneath the hull cruising at a speed of about 10 knots. If sampling is done for 5 min an integrated section of approx. one nautical mile is obtained [23].

For the collection of seawater with usually extremely low trace metal contents, with the exception of estuaries, besides collection close to the surface, sampling at different depths to obtain vertical concentration profiles is commonly important in marine geochemistry. For these tasks different, now very reliable samplers have been developed and successfully applied ranging from relatively simple GoFLO samplers or modifications of these for sampling down to approx. 200 ml [22,24-26] to complex and expensive systems for contamination-free seawater collection from different depths [27-29] or just very close to the bottom of the sea [30].

2.3. BIOLOGICAL FLUIDS

The most frequently analyzed biological fluids are whole blood, blood plasma or blood serum and urine, occasionally other fluids such as milk, gastric juice, sweat etc. are analyzed as well.

Since metal contents in body fluids are frequently very low and certainly much lower than formerly supposed [31-33], which is particularly the case for most of the metals treated in this book, it is now understood that sampling of body fluids can only lead to reliable results if utmost care is taken during sampling. Besides contamination that can be minimized by use of cleaned containers or those containing coagulants by careful contamination tests and special collection techniques, e.g. use of plastic cannulae instead of stainless-steel needles for the analysis of nickel, cobalt and chromium or discarding the first five ml of blood collected [34,35], also physiological and environmental sources of error have to be taken into account. Collection of blood should be performed at distinct times of the day and from a defined position of the proband's body [36-38]. Since the proband himself might often be a severe contamination source, particularly in occupational exposure, bathing or showering prior to urine voiding is mandatory. The most reliable specimens are 24 h urines samples, however, in many cases this is not practicable, hence spot urines well defined in time and normalization of metal contents to specific gravity, osmolality or creatinine will often lead to useful information [37-39]. Acidification of urines to approx. pH $\leq$ 2 is often necessary to avoid coprecipitation of metals with particulates.

2.4. SOLID BIOLOGICAL MATERIALS

Meaningful selection of solid biological specimens from human beings and animals (e.g. for purposes of specimen banking, see Chapter 3) requires knowledge of metabolic processes for the elements under consideration [36-41]. For autopsy materials post mortem influences can occur and hence cause changes in elemental concentrations due to autolysis, internal losses and contamination [42,43].

Metal contents in feces, organs (lung, kidney, liver), tissues, and bone are in most cases (except for e.g. lead and cadmium in muscle tissue) somewhat to significantly higher than in body fluids [44,45]. Dissection of materials with low endogenous levels such as fish fillet and muscle tissue should be performed with knives and pincers made of precleaned quartz [46] or plastic material and preferably under clean-room conditions.

Hair is a biopsy as well as an autopsy material. Because of severe difficulties in distinguishing between endogenous and exogenous contents epidemiological studies using hair as a bioindicator are doubtful for a number of metals of toxicological and environmental concern [47]. For retrospective studies in metal poisoning hair can be valuable. Longitudinal analysis of hair strands for instance can provide estimations about the time of poisoning and sometimes also about biological half-life [48,49]. Due to the comparatively high metal contents in hair, generally contamination during collection is insignificant. Nails behave similarly, however, due to the much smaller surface exogenous influences are less pronounced [50].

Teeth, are like hair, a biopsy as well as an autopsy material, they possess particular significance because of their composition with low moisture content. Contamination upon collection is also no problem with teeth.

2.5. ENVIRONMENTAL MATERIALS AND MATERIALS OF ANTHROPOGENIC ORIGIN

Compared with that of natural water, seawater, blood and urine, trace metal contents in many of the materials mentioned in this section are significantly higher so that contamination is usually of minor importance.

For the collection of environmental specimens the proper selection of sampling region, particular site and sample size is important for reliable information about the environmental and ecological situation if trace metals are concerned. Planning for tasks of this kind has to be performed in close cooperation between the analyst and ecologists, biologists and geologists for optimal results. A particular example of this is environmental specimen banking (see Chapter 3).

Sampling of soils and of sediments in undisturbed position is frequently performed in cores with tubes from plastics or metals to obtain retrospective information about geological or anthropogenic influences by stratified analysis [51,52]. The cores thus obtained can be stored complete directly after collection under cryogenic conditions or separated into slices and stored as well.

Some basic foods, such as flour, sugar, milk, dairy products, fruits and various beverages, contain very low levels of ecotoxic trace metals (see also Chapter 1, Section 1.2.5). Therefore sample collection for these materials must be performed very cautiously in order to minimize contamination by using the clean tools already mentioned under Section 2.4 and execution under dust-free working conditions. For materials in cans or boxes

References on p. 16

subsamples should also be taken from inside, if mixing is not feasible, since boundary surfaces often show higher trace metal levels due to adsorption leading to erroneously high average values.

2.6. SAMPLING FOR SUBSEQUENT SPECIATION ANALYSIS

In Sections 1 - 5 of this chapter collection approaches are cscussed that will lead to representative samples predominantly for total elemental analysis so that possible changes in valency state, complexation and deterioration of labile compounds are not especially prohibited.

If, however, the increasingly important investigation on occurrence and quantification of chemical forms is concerned special precautions are necessary.

Sampling procedures for water are in principle the same as outlined in Section 2.2 but subsequent treatment has to be minimized. The optimal approach is to perform any speciation analysis immediately following sampling [13], however, this is often not possible so that samples in most cases have to be stored for some time prior to analysis. Usually the first step, in most cases performed at the sampling site, is removal of particulate matter from the sample by filtration in scrupulously cleaned systems through 0.45 μm pore size membrane filters or –less frequently– to centrifuge the sample in order to avoid changes in the distribution of chemical species due to various influences of particulates.

If filtration cannot be performed immediately after sampling, the samples should be cooled down to approx. 4°C to prevent rapid changes, e.g. by bacteria. If storage of filtered samples prior to analysis is necessary and if species determination is required storage also at 4°C without acidification was shown to be effective for species preservation for some months [22]. If methyl mercury has to be determined, preservation of aqueous samples by acidification with hydrochloric acid, even in polyethylene bottles and storage in the dark was found to be effective [53,54]. Preservation of arsenic species was reported to be effective, if 0.2% sulfuric acid was added and samples stored in polyethylene [19]. For some other organometallic tin and arsenic compounds freezing was successful at least for a few weeks [22].

Speciation of metal-containing fractions by the use of various chromatographic methods in body fluids described for zinc in human erythrocytes [55], aluminium in blood serum [56] and chromium in erythrocytes [57] is performed directly in freshly prepared materials. For selenium speciation in urine, however, an alkaline denaturation step for proteins was necessary prior to chromatographic separation [58]. Various sample collection, sample preparation and sample preservation techniques for speciation of metals in biological materials were reviewed by Gardiner [59].

Freshly collected solid biological materials often have to be prepared immediately by dissection and separation from the parts not required. They can be used either fresh, stored at low temperatures or also frozen for speciation of metals and organometallic compounds. For fish, tissues, algae etc., frequently after homogenization, various extraction procedures followed by specific separation/determination steps are applied [60,61].

Environmental materials such as soils and sediments can be stored after collection preferably at 4°C. Freezing is sometimes also possible. For sediments wet sieving ($<20\,\mu$m, 63 μm) and/or centrifugation for separation of interstitial water before storage and suitable extraction/separation procedures is also applicable as is described in detail by e.g. Kersten and Förstner [62] and for organometallic compounds in general by Crompton [63].

For all steps in sample collection and direct preparation at sampling time or sampling site contamination precautions are often equally as important as for collection prior to total metal analysis [22,63].

2.7. HOMOGENIZATION AND STORAGE

Material of anthropogenic origin stemming from technological processes is often homogeneous. The same is also the case for dust filters [6] so that subsampling for analysis is easily possible. Other materials, e.g. fruits, vegetables, human and animal organs, coarse sausage, meat, bread, cake, sewage sludge etc., have to be first collected in larger amounts to start with a representative sample and then homogenized for subsampling.

For homogenization of materials for which drying and milling using conventional techniques is not advisable, cryogenic homogenization, preferably of fresh (wet) materials under cryogenic conditions, down to the temperature of liquid nitrogen (i.e. -196°C) can be performed with ball mills (brittle fracture technique) [64] or larger systems of different sizes recently developed for sample homogenization in specimen banking [65-67].

After collection and if necessary some preparatory steps, e.g. acidification of liquid and homogenization of solid materials, intermediate storage is often necessary. The reasons are manifold. Reference samples for forensic medicine, food control, environmental and basic research have to be kept as frequently surplus material is required as a reference for internal quality control. During this storage, predominantly at temperatures between $+4$ and -16°C, deterioration can occur and hence may negatively influence repetitive analysis.

In addition elemental losses by adsorption on the walls of storage vessels or contamination from these can be expected, if they are not carefully cleaned and checked prior to use. In the case of mercury, if vessels of plastic materials are used reduction to metallic mercury in solution and diffusion through the material as well as from outside can cause significant either positive or negative concentration changes as was shown with radiotracers [68]. If storage temperatures arc lowered diffusion becomes very slow and will be negligible as well as moisture loss from temperatures below -80°C. It has to be noted, however, that storage of whole blood at these temperatures will cause hemolysis. For the determination of chemical species in urine, however, deep frozen storage in the dark under cryogenic conditions is needed [69]. For materials containing easily decomposable compounds utmost care in evaluating suitable storage conditions is necessary if no detailed information is at hand.

References on p. 16

A straightforward approach for safe long-term storage of biological and environmental material is the preservation of carefully selected specimens at cryogenic temperatures in Environmental Specimen Banks in the U.S.A. and Germany (for details and background see Chapter 3).

After a preliminary phase with pilot studies in the U.S.A. [70] and Germany [71,72] permanent banking with chronological sampling campaigns started in 1985 in both countries. Storage temperatures are around −80°C for human specimens in Germany, stored in Münster, and below −150°C for specimens of marine, aquatic and terrestrial origin stored at the Research Center Juelich (KFA) [73]. Similar, however, less comprehensive programs are being performed in Japan, Canada, Sweden and recently also commenced in Denmark with a current exchange of data and ideas at international meetings between all scientists and institutions engaged in this task [74].

REFERENCES

1 D.L.M. Weller, *Anal. Proceed.*, 25 (1988) 199-200.
2 W.Y. Garner and M. Barge (Editors), *Good Laboratory Practices (ACS Symposium Series No. 369)*, American Chemical Society, Washington, DC, 1988.
3 W.D. Pocklington, *Pure Appl. Chem.*, 62 (1990) 149-162.
4 P. Klahre, P. Valenta and H.W. Nürnberg, *Vom Wasser*, 51 (1978) 199-219.
5 F.G. Bodewig and P. Valenta, *Gewässerschutz Wasser Abwasser*, 77 (1985) 173-179.
6 D. Klockow, *Fresenius' Z. Anal. Chem.*, 326 (1987) 5-24.
7 P. Valenta, V.D. Nguyen and H.W. Nürnberg, *Sci. Total Environ.*, 55 (1986) 311-320.
8 R. Ahmed, K. May and M. Stoeppler, *Sci. Total Environ.*, 60 (1987) 249-261.
9 L. Mart, H.W. Nürnberg and G. Gravenhorst, *Ber. Polarforsch.*, 6 (1982) 68-69.
10 C.F. Boutron, *Anal. Chim. Acta*, 106 (1979) 127-130.
11 C.F. Boutron, in J.O. Nriagu and C.I. Davidson (Editors), *Toxic Metals in the Atmosphere*, Wiley, New York, 1986, pp. 467-505.
12 C.F. Boutron, *Fresenius' Z. Anal. Chem.*, 337 (1990) 482-491.
13 J. Völkening and K.G. Heumann, *Fresenius' Z. Anal. Chem.*, 331 (1988) 174-181.
14 M. Murozumi, T.J. Chow and C.C. Patterson, *Geochim. Cosmochim. Acta*, 33 (1969) 1247-1294.
15 L. Mart, *Tellus*, 35B (1983) 131-141.
16 Standing Committee of Analysts, in *Methods for the Examination of Waters and Associated Materials*, Her Majesty's Stationary Office, London, 1980.
17 L. Mart, *Fresenius' Z. Anal. Chem.*, 296 (1979) 350-357.
18 L. Mart, *Fresenius' Z. Anal. Chem.*, 299 (1979) 97-102.
19 L. Mart, H.W. Nürnberg and P. Valenta, *Fresenius' Z. Anal. Chem.*, 300 (1980) 350-362.
20 L. Mart, *Talanta*, 29 (1982) 1035-1040.
21 M. Stoeppler, in A. Boudou and F. Ribeyre (Editors), *Aquatic Ecotoxicology: Fundamental Concepts and Methodologies*, Vol. I, CRC Press, Boca Raton, FL, 1989, pp. 77-96.
22 G.E. Batley, in G.E. Batley (Editor), *Trace Element Speciation Analytical Methods and Problems*, CRC Press, Boca Raton, FL, 1990, pp. 1-24.
23 K. Kremling and C. Pohl, *Mar. Chem.*, 27 (1989) 43-60.
24 K.W. Bruland, R.P. Franks, G.A. Knauer and J.H. Martin, *Anal. Chim. Acta*, 105 (1979) 233-245.
25 L. Mart, H.W. Nürnberg and D. Dyrssen, in C.S. Wong, E. Boyle, K.W. Bruland and E.D. Goldberg (Editors), *Trace Metals in Sea Water*, Plenum Press, New York, London, 1983, pp. 113-130.

26 P. Freimann, D. Schmidt and K. Schomaker, *Mar. Chem.*, 14 (1983) 43-48.
27 C.C. Patterson and D.M. Settle, *NBS Spec. Pub.*, U.S. Dept. of Commerce, Washington, DC, 1976, pp. 321-351.
28 H. Haas and L. Mart, *Ber. KFA Jülich*, Jül (1990) 1689.
29 L. Brügmann, E. Geyer and R. Kay, *Mar. Chem.*, 21 (1987) 91-99.
30 L. Sipos, H. Rützel and T.H.P. Thyssen, *Thalassia Jugosl.*, 16 (1980) 89-94.
31 J. Versieck and R. Cornelis, *Anal. Chim. Acta*, 116 (1980) 217-254.
32 J. Versieck, *CRC Crit. Rev. Clin. Lab. Sci.*, 22 (1985) 97-184.
33 T.W. Clarkson, L. Friberg, G.F. Nordberg and P.F. Sager (Editors), *Biological Monitoring of Toxic Metals*, Plenum Press, New York, London, 1988.
34 F.W. Sunderman, Jr., A. Aitio, L.G. Morgan and T. Norseth, *Toxicol. Ind. Health*, 2 (1986) 17-78.
35 S. Bro, P.J. Jørgensen, J.M. Christensen and M. Hørder, *J. Trace Elem. Electrolyt. Health. Dis.*, 2 (1988) 31-35.
36 D. Behne, *J. Clin. Chem. Clin. Biochem.*, 19 (1981) 115-120.
37 A. Aitio and J. Järvisalo, *Pure Appl. Chem.*, 56 (1984) 549-566.
38 A. Aitio, in T.W. Clarkson, L. Friberg, G.F. Nordberg and P.F. Sager (Editors), *Biological Monitoring of Toxic Metals*, Plenum Press, New York, London, 1986, pp. 75-83.
39 J. Angerer and K.H. Schaller, *Analyses of Hazardous Substances in Biological Materials*, Vol. 1, VCH, Weinheim, 1985.
40 B. Sansoni and V. Iyengar, *Ber. KFA Jülich*, Jül-Spez. (1978) 13.
41 G.V. Iyengar and W.E. Kollmer, *Trace Elem. Med.*, 3 (1986) 25-33.
42 G.V. Iyengar, *J. Pathol.*, 134 (1981) 173-180.
43 G.V. Iyengar, *J. Res. Natl. Bur. Stand.*, 91 (1986) 67-74.
44 C. Vanoeteren, R. Cornelis and E. Sabbioni, *EUR 104440 EN*, 1986.
45 H.D. Narres, P. Valenta and H.W. Nürnberg, *Z. Lebensm. Unters. Forsch.*, 179 (1984) 440-446.
46 M. Stoeppler and H.W. Nürnberg, *Ecotox. Environ. Safety*, 3 (1979) 335-351.
47 G. Chittleborough, *Sci. Total Environ.*, 14 (1980) 53-75.
48 H. Hagedorn-Götz, G. Küppers and M. Stoeppler, *Arch. Toxicol.*, 38 (1977) 275-285.
49 M. Yukawa, *Sci. Total Environ.*, 38 (1984) 41-54.
50 B. Gammelgaard and J.R. Andersen, *Analyst*, 110 (1985) 1197-1199.
51 U. Förstner and G.T.W. Wittmann, *Metal Pollution in the Aquatic Environment*, Springer, Berlin, 2nd revised ed., 1981.
52 MARC (Monitoring and Assessment Research Centre), *Historical Monitoring*, Report No. 31, University of London, 1985.
53 R. Ahmed and M. Stoeppler, *Anal. Chim. Acta*, 192 (1987) 109-113.
54 R. Ahmed, *Ber. KFA Jülich*, Jül-Spez. (1986) 349.
55 P.E. Gardiner, H. Gessner, P. Brätter, M., Stoeppler and H.W. Nürnberg, *J. Clin. Chem. Clin. Biochem.*, 22 (1984) 159-163.
56 P.E. Gardiner, M. Stoeppler and H.W. Nürnberg, in P. Brätter, P. Schramel (Editors), *Analytical Chemistry in Medicine and Biology*, Vol. 3, Walter de Gruyter, Berlin, New York, 1984, pp. 299-310.
57 B. Neidhart, S. Herwald, Ch. Lippmann and B. Straka- Emden, *Fresenius' J. Anal. Chem.*, 337 (1990) 853-859.
58 A.J. Blotcky, G.T. Hansen, N. Borkar, A. Ebrahim and E.P. Rach, *Anal. Chem.*, 59 (1987) 2063-2066.
59 P.H.E. Gardiner, in *Topics in Current Chemistry*, Vol. 141, 1987, pp. 145-174.
60 Y.K. Chau, *Sci. Total Environ.*, 71 (1988) 57-58.
61 A. Mazzucotelli, R. Frache, A. Viarengo and G. Martino, *Talanta*, 25 (1988) 693-696.
62 M. Kersten and U. Förstner, in G.E. Batley (Editor), *Trace Element Speciation: Analytical Methods and Problems*, CRC Press, Boca Raton, FL, 1990, pp. 245-317.
63 T.R. Crompton, *Environ. Int.*, 14 (1988) 417-463.
64 G.V. Iyengar and K. Kasperek, *J. Radioanal. Chem.*, 39 (1977) 301-316.

65 J.K. Langland, S.H. Harrison, B. Kratochvil and R. Zeisler, in R. Zeisler, S.H. Harrison and S.A. Wise (Editors), *The Pilot National Environmental Specimen Bank (NBS Spec. Pub. 656)*, U.S. Dept. of Commerce, Washington, DC, 1983, pp. 21-34.

66 U. Klussmann, D. Strupp and W. Ebing, *Fresenius' Z. Anal. Chem.*, 322 (1985) 456-461.

67 J.D. Schladot and F.W. Backhaus, in S.A. Wise, R. Zeisler and G.M. Goldstein (Editors), *Progress in Environmental Specimen Banking, (NBS Spec. Pub. 740)* U.S. Dept. of Commerce, Washington, DC, 1988, pp. 184-193.

68 K. May, K. Reisinger, R. Flucht and M. Stoeppler, *Vom Wasser*, 55 (1980) 63-76.

69 M. Stoeppler, in P. Brätter and P. Schramel (Editors), *Trace Element Analytical Chemistry in Medicine and Biology*, Vol. 2, Walter de Gruyter, Berlin, New York, 1983, pp. 909-928.

70 R. Zeisler, S.H. Harrison and S.A. Wise, *The Pilot National Environmental Specimen Bank (NBS Spec. Pub. 656)*, U.S. Dept. of Commerce, Washington, DC, 1983.

71 D. Kayser, U. Boehringer and F. Schmidt-Bleek, *Environ. Monit. Assess.*, 1 (1982) 241-255.

72 BMFT, *Umweltprobenbank-Bericht und Bewertung der Pilotphase*, Springer, Berlin, Heidelberg, New York, London, Paris, Tokyo, 1988.

73 U.R. Boehringer and W. Hertel, *Umwelt*, 1-2 (1987) 21-23.

74 S.A. Wise, R. Zeisler and G.M. Goldstein (Editors), *Progress in Environmental Specimen Banking (NBS Spec. Pub. 740)*, U.S. Dept. of Commerce, Washington, DC, 1988.

M. Stoeppler (Editor)/*Hazardous Metals in the Environment*
© 1992 Elsevier Science Publishers B.V. All rights reserved

Chapter 3

Environmental specimen banking

Robert A. Lewis, Barbara Klein, Martin Paulus and Christoph Horras

Department of Biogeography, University of Saarland, 6600 Saarbrücken (Germany)

CONTENTS

3.1. Introduction and status 19
3.2. Historical aspects 20
    3.2.1. The use of naturally-preserved materials 20
    3.2.2. Historic examples of the use of stored materials 23
    3.2.3. Problem-oriented ESB 25
3.3. The relationship of specimen banking to environmental monitoring, research and assessment 26
3.4. The role and application of ESB to the assessment of toxic and hazardous chemicals in the environment 27
    3.4.1. Role and application to chemical and environmental management 28
    3.4.2. Role and applications to the assessment of metals 30
3.5. Coordination, cooperation and technical assistance 31
3.6. The suitability of specimens 31
3.7. Methods 32
    3.7.1. Field and laboratory procedures 32
    3.7.2. Field sampling and statistical design 32
    3.7.3. Specimen collection, handling and transport 34
    3.7.4. Security 36
    3.7.5. Use of pooled (composite) specimens 37
    3.7.6. Chemical procedures and analyses 39
    3.7.7. Preservation and storage 40
    3.7.8. Quality assurance 42
    3.7.9. Financial aspects 43
References 43

3.1. INTRODUCTION AND STATUS

Environmental specimen banking (ESB) is the systematic collection and long-term storage of representative environmental specimens for deferred analysis and evaluation. During the past decade, the capacity of ESB to help to identify and to characterize the nature and extent of dispersion of potentially hazardous metals and other harmful chemicals in selected environments has been amply demonstrated [12]. National programs of ESB are being conducted in Sweden, Canada, the United States, and Germany. The long-standing Swedish program is conducted by the Swedish Museum of Natural History. The Canadian program is conducted by the Canadian Wildlife Service and has been active since the mid-sixties. Both the Swedish and Canadian programs concern wildlife. The West German program, which is administered by the Federal Environmental Agency, Berlin, on behalf of the Ministry of Environment, Nature Conservation and Reactor Safety, Bonn, is comprehensive, dealing with a range of ecological and human types of specimens on a national basis. The West German bank conducts a subprogram on inorganic analysis which includes the adaptation and development of methods, studies of long-term stability of the stored materials, and the production and characterization of control materials [3]. The U.S.A. program is conducted by the National Bureau of Standards. This program was originally oriented toward the testing, development and standardization of methods and pilot studies relating mainly to human health, but has expanded to include other types of materials. Institutes of France, Japan and the Netherlands have also contributed substantially to the development of ESB as a tool for the evaluation and management of potentially harmful environmental chemicals.

The U.S.A./Canada Great Lakes Water Quality Agreement of 1978 acknowledges the value of specimen banking for detecting new pollutants that enter the North American Great Lakes. The Great Lakes Water Quality Board [4] confirmed the banking of tissues and sediments as a "critically important component of monitoring and surveillance in the Great Lakes." The Board recommended that tissue and specimen banks should be maintained or enhanced wherever appropriate to permit retrospective analysis of previously unrecognized contaminants and to extend the long-term data record. Result to date have been impressive largely due to the initiative of Canadian agencies (e.g., refs. 5-8).

3.2. HISTORICAL ASPECTS

The preservation of environmental specimens, or the use of naturally-preserved materials are rather common activities occurring from ancient times. Such activities have sometimes intentionally and sometimes inadvertently served the purposes of chemical assessment. In biology and medicine for example, tissues are often preserved for future reference. Such specimens have established the time and geographic focus of onset of various communicable diseases and have solved any number of medical mysteries. A simple, but elegant example of ad hoc "specimen banking" occurred in the late 1960's [9], when authorities placed front-page ads in Swedish newspapers asking for old seal oil. Scientists at the Swedish Museum of Natural History were thereby enabled to determine trends in concentrations of DDT and PCB in seals from Swedish waters from about 1940 to 1970 and to correlate these trends with the occurrence of aborted seal pups (ibid.).

3.2.1. The use of naturally-preserved materials

Environmental materials that store chemical, biological and/or physical information over a number of years or that collectively represent periods of chemical uptake that span a number of years (e.g., pine needles, tree rings, lake sediments) and materials that can be stored without the use of special techniques (e.g., dry soil, egg shells) have often served in establishing baseline concentrations or records of past pollution. Even now, well-designed retrospective studies that employ such materials can provide valuable information on the origin, dispersion and status of environmentally stable elements and compounds [1,2,10-13]. A few examples are given below. Further examples are discussed by Lewis [1,2,11,12].

Annual rings of many temperate zone tree species can provide a concurrent record of the annual accumulation of elements and growth rate for decades or even centuries. The excellent studies of S.B. McLaughlin, C.F. Baes III and their colleagues (e.g., refs. 14 and 15) should be consulted. Working at two well-characterized sites, an experimental watershed at the Oak Ridge National Laboratory and a cove in the Great Smoky Mountains National Park, these scientists collected cores from shortleaf pine (*Pinus echinata*) and red oak (*Quercus borealis*) trees. Specimens exhibited strong temporal trends in concentrations of several metallic elements. Concentrations of Al, B, Cu, Fe, Ti, Zn, and Cr in the pine increased during the preceding 10-30 years as did Cu, Fe and Cr in red oak. Furthermore, the cambium and inner bark contained concentrations of Al exceeding that found in the most recent wood [14]. More recent data appears to document the advent of smelter operations at Copperhill, Tennessee in the 1870's [15].

At the beginning of the 1950's in Minamata, Japan, aquatic organisms died in vast numbers, and large number of humans, who mostly earned their living from fishing, also became seriously ill. Many people died. The diseases proved to be due to wastes containing mercury from acetaldehyde manufacture, which was willfully dumped into the bay by a chemical plant in Minamata. Methyl mercury formed by microorganisms in bottom sediments of the bay accumulated in fish and other aquatic organisms which were then consumed by humans. Neurological damage in man ranged from mild loss of sensation to irreversibly impaired vision or neuromuscular function and even death. The developing foetus was often severely affected, manifesting a cerebral palsy-like syndrome and postnatal loss of mentation. Because of inadequate knowledge of the biological effects of methyl mercury and its behavior in the aquatic environment, assessment was delayed. Scientists found heavy accumulations of methyl mercury in the hair of victims. Retrospective analyses of long female hair established the time of occurrence of the major introductions of mercury into the bay [16].

Feathers have frequently been employed in the assessment of mercury pollution. They can be easily and inexpensively collected and stored and can often be taken without disturbing the bird population. In Germany, the goshawk (*Accipiter gentilis*) is widely distributed and meets most of the criteria of a useful monitor of environmental chemicals [17] (see also refs. 18-21). The keratin of feathers - to which Hg is firmly bound - is not readily biodegradable [18] thus allowing retrospective analyses spanning long periods of time. Partly on the basis of feather analyses we know that the mercury contamination of piscivorous pelagic birds [22] of Minamata Bay persists today.

Concentrations of environmental chemicals in hair, feathers, and perhaps other integumental structures can be related not only to temporal and geographic gradients of exposure, but to nutritional deficiencies and disease states as well. Some precautions and

References on p. 43

disadvantages in the use of hair as a monitoring tissue have been noted [23]. Hair and feathers are employed in the Swedish Environmental Specimen Bank [9].

Concentrations of hazardous metals and other pollutants in bottom sediments may reflect trends in industrial and/or agricultural activities in the drainage area of the watershed in which they are situated. Actual concentrations of trace metals in fluvial sediments may, however, be greatly influenced by processes and factors such as a grain size and composition of the sediments; history of disturbances (e.g., via dredging); variations in normal flow-rate, rainfall and consequent variations in rates of erosion, leaching, and flooding or even purging. In the absence of metal pollution, the concentrations of trace metals in river sediments depend primarily upon the types of rocks and minerals that occur within the drainage system (e.g., ref. 24).

The history of Rhine River pollution was inferred from the analyses of banked specimens of dried sediment dating from 1870 [24,25]. By analyzing sediments that were collected more than 50 years ago, trends in trace metal concentrations of fluvial, marine and lagoon sediments in the Netherlands were established [24].

Mueller et al. [26] reconstructed the chronology of pollution in the Baltic Sea during the past by means of dated sediment cores. Concentrations of artificial radionucleides (^{137}Cs), nutrients (P and N), heavy metals (Pb and Zn) and chlorinated hydrocarbons were examined by means of three sediment cores from Geltinger Bay, Eckernfoerder Bay and the Luebecker Bay. Results confirmed the assumption that sediments can provide a useful record of serious environmental change. Following the observed maximum in radioactive precipitation (1962), ^{137}Cs reached peak concentrations in all three sediment cores. Increased P and N concentrations occurred in the layers from 1965 to 1980 in association with increasing eutrophication of the Western Baltic Sea by fertilizers and detergents. The author reported a parallel development of heavy metals (Pb and Zn) and industrial growth in the Baltic Sea, especially for the last 130 (±30) years, during which the use of coal having a high metal content increased.

About 11 000 years of sedimentation was surveyed within 3-m long sediment cores taken from the deepest point of Rachel Lake in the Bavarian Forest. Changes in concentration of organic substances, nutrients and metallic elements of the sediment layers clearly indicated climatic changes during this period In a region 20-30 cm deep, a general increase in heavy metal content was recognized. This resulted from the intrusion of heavy metals from the atmosphere beginning in the 14th century. Glassworks had been established in the region at this time [27,28].

Holden stated in 1975 that there are few precedents for a systematic program of sediment monitoring. This is no longer true. Worldwide interest in trace analysis of sediments as a way of determining the extent of environmental pollution dates back several decades [29]. By the mid-1970's, numerous toxic organic and inorganic compounds and elements had been demonstrated in sediments [e.g., 30-34]. More recently, at least one major program of sediment monitoring has made seminal contributions in Lake Ontario [35]. Analysis of the distribution of mercury, polychlorinated biphenyls, and Mirex in the surficial sediments of Lake Ontario have demonstrated the importance of the Niagara River as a major source of these pollutants. These analyses contributed also to an understanding of the net movements of the pollutants within the lake and their occurrence in various trophic compartments [36]. Investigations have further demonstrated that heavy metals and pesticides are actually transported to the lake from the Niagara River by suspended solids [37-39].

The value of glacial ice in assessing chemical burdening of the environment is unclear. Such ice may contain dust, trace elements and pollen from earlier centuries and may also reveal trends in atmospheric CO_2 and acidity of precipitation. Climate and certain major environmental trends may be assessed [40-41]. In warmer glaciers - perhaps those above -10°C - partial meltings may occur with some frequency and even cold glaciers may experience sublimation. Thus, not only are glacial ice layers eroded by melting, but translocation of chemicals might be expected to occur. Nevertheless, cold glaciers, at high latitudes or high altitudes, offer possibilities that have yet to be explored. Boutron et al. [40] reported on 22 atmospheric trace elements and compounds found in successive layers of a large block of blue ice of prehistoric age that was collected at a coastal ablation area in East Antartica. The concentrations of NH_4, Sc, Cr, Fe, Co, As, Se, for example, measured in the center probably represents the original, natural baseline level deposited more than 12 000 years ago. Enrichments of Se, Au, Cd, Zn, Cu probably resulted from natural processes. The results further indicate that the atmosphere in this remote polar region is still little affected by elemental pollution of human origin.

3.2.2. Historic examples of the use of stored materials

Human bones from medieval churches and ancient caves near Cracow and Kroczyce in Poland exhibited interesting Pb concentrations [42]. Specimens from the 3rd century A.D. that were found in limestone caves gave values from 1.81-3.40 $\mu g/g$ of dried bone. Samples from the 11th to 19th centuries showed a broad range of values; from 5.6 to 318.6 $\mu g/g$. Contemporary specimens yielded an average concentration of 3.9 $\mu g/g$. The authors attribute these differences to changes in dietary habits and lead concentrations in the food. They hypothesize that in the 11th to 19th centuries Polish people had access to lead-contaminated tin kitchen utensils, which modern Poles do not use. Drasch [43] compared lead content of human skeletal material from precolumbian Peru (500-1000 B.C.), prehistoric Bavaria, and from late Roman times to the middle ages in Augsburg with that of recent material from Bavaria. He established that the body burden in recent times is more than twice as high as in late Roman times and the middle ages. Drasch believes that this finding relates to the present-day ubiquitousness of pollution from industrial wastes and automobile exhausts.

Museum collections have frequently and sometimes effectively been employed in the assessment of past and present environmental pollutant burdens [1,2,11,44,45]. There are, however, significant drawbacks to the use of such collections [1,2]. For example, various treatments and preservatives are employed for most museum specimens to pre-vent decay or infestation by pests. Arsenic, formalin, ethanol and other organic solvents, boric acid, and paradichlorobenzene are frequently employed in the management of museum specimens. If the management practices of a given collection are well known,

References on p. 43

carefully planned retrospective studies employing these materials may provide useful results. Valuable information on the extent of historical pollution by trace contaminants such as lead, arsenic, cadmium, and mercury has been provided by collections of integumental materials such as hair, feathers or skin (e.g., ref. 10).

Hair is an effective monitor of exposure to heavy metals and is comparatively resistent to decay [46]. For example, Suzuki and Yamamoto [47] examined mercury levels in human hair with and without storage for eleven years. They found no loss of mercury with storage. A comparison of Pb concentration in old human hair from adults and children in 11 states of the U.S.A., using a collection of hair dating from 1871-1923, together with samples collected in 1971 from Pennsylvania and Michigan, was made by Weiss, Whitten and Leddy (cited in ref. 46). The Pb concentrations had decreased since the earlier period although ambient lead concentrations had meanwhile increased. The authors concluded that because of the use of Pb glazes in cook-ware, water containers, waterpipes, etc. people ingested a greater amount of lead during the earlier period.

During the 1950's, Swedish populations of the yellow-hammer (Emberiza citrinella), peregrine falcon (Falco peregrinus), sparrow hawk (Accipiter nisus) and a number of other bird species decreased dramatically [9,48-50]. Some of these populations recovered or began to recover in the mid to late 1960's [3,51]. Berg et al. [19] compared mercury content in feathers of various Swedish birds for a period of more than 100 years. They found a more or less constant concentration of mercury in each species from 1840 to 1940. During the following two and one-half decades, the concentration increased by a factor of 10-20 in several of the species studied. The chemical analytic data together with information on trends implicated alkyl mercury as an important cause of these population declines. Alkyl mercury had been used as a seed dressing in Sweden during the decades of decline. Its use was banned in 1966 [51]. Subsequently, the mercury content in the feathers of some species decreased [52-55].

Johnels et al. (cited in ref. 49) subsequently correlated mercury concentrations in feathers of the migratory fish-eating osprey (Pandian haliaetus) with local sources of mercury. Knowledge of the temporal pattern of malt in this species provided the key information to demonstrate that feathers grown in Sweden contained much higher levels of mercury than those grown in Africa. High mercury levels were also found in biota downstream from effluent outlets of pulp and paper mills that use phenyl mercury as a fungicide [56,57]. The levels of mercury in aquatic organisms and birds decreased only gradually, as much of the mercury remained in the water system [58]. If arsenic had been the problem standard museum collections could not have been employed as arsenic soap had long been used by curators to prevent infestations of bird skins [9].

Birds' eggs and egg shells have provided useful information regarding chemical exposure and effects. Huge collections of egg shells are maintained in museums throughout the world. They have been invaluable, for example, in establishing the hazards of DDT and other organochlorine pesticides residues by serving as monitors of accumulation and an associated thinning of the eggshell. We had to wait for ten years in the case of DDT- and PCB-research before a clear assessment of the Baltic Sea environment could be made. Today one can retrospectively study the appearance and increase of many toxic chemicals in the Baltic Sea from the material that has been collected and stored during the past 15 years [9].

Lichens and mosses are useful in establishing trends in heavy metals in the environment (e.g., ref. 13). These plants can be easily stored dry at room temperature. They possess a rough surface that readily retains particulate matter. Peat moss is especially useful for monitoring the deposition of atmospheric pollutants because it takes up nutrients only from atmospheric sources and precipitation [59]. Moss surveys have been widely employed in Western Europe and Scandinavia as an inexpensive means of acquiring information on national or regional distributions of atmospheric heavy metals and radionucleides [e.g., ref. 60].

Mosses are proven indicators of trends in Pb deposition when both herbarium and contemporary specimens from geologically similar sites are compared. Analyses of three species of woodland mosses collected from 1860 to recent times demonstrated a sharp upward trend in Pb content with the advent of the Industrial Revolution [61,62]. Kjellström [63] reported a significant increase in cadmium content of stored autumn wheat from "non-polluted" farms in Sweden from 1916 to 1972. Two separate sets of samples of the moss, *Hypnum cupressiforme*, that were collected near these wheat fields from 1917 to 1970 exhibited a 40% and a 250% increase in cadmium concentration respectively. For purposes of elemental analyses, both wheat and moss can be banked dry. Such specimens can simultaneously serve as taxonomic reference material.

3.2.3. Problem-oriented ESB

Herring gull (*Larus argentatus*) eggs, collected in 1980 throughout the Great Lakes, were analyzed by the Canadian Wildlife Service for tetrachorodibenzo-*p*-dioxins (TCDDs) [6]. Only the 2,3,7,8-TCDD isomer was found (in concentrations of 9 to 90 μg/kg). A retrospective analysis of herring gull eggs that had been collected from Scotch Bonnet Island in Lake Ontario almost annually since 1971 and stored at $-25°C$ to $-30°C$ was undertaken. Again, only the 2,3,7,8-TCDD isomer was found. The average concentration of TCDD in herring gull eggs from the early 1970's exceeded 1000 ppt. These results show a declining trend in 2,3,7,8-TCDD levels. The high levels of TCDD from the early 1970's may explain reproductive failures and an associated pathology in herring gull embryos that was observed during this period in Lake Ontario [6,64].

The widespread contamination of Lake Ontario by Mirex, a toxic insecticide and fire-retardant, was established by retrospective analyses of lake sediments and banked specimens of fish. Subsequent research revealed that Mirex is persistent, and accumulates through aquatic food chains [65,66].

An informal specimen bank has been in operation for some years at the Patuxen Wildlife Research Center (U.S. Fish and Wildlife Service). Bald eagles (*Haliaetus leucocaephalus*), and endangered species, found dead or dying in the wild are routinely sent to the Center for chemical analysis. The remaining portions of the carcass are stored at $-28°C$. In the mid-seventies, the state of Virginia became aware of a contamination problem involving Kepone in the James River. Bald eagle and osprey eggs collected in 1976

References on p. 43

from sites in the vicinity of the James River contained Kepone. Eagle carcass homogenate and liver samples from specimens collected in the vicinity of the James River from 1968 to 1975 were removed from storage and analyzed for Kepone. Results demonstrated the presence of Kepone in bald eagles at least since 1968 including specimens collected more than 130 miles from the primary source of contamination on the James River [67]. The chemical company that was established as the sole source of Kepone had claimed that at that time it was not discharging this chemical into the James River [68].

An examination of canned tuna fish demonstrated a significant relationship between mercury content and the geographic origin of the fish. From 12 countries in Southern Europe, Africa and Asia 306 samples were analyzed. Significant differences in contamination levels were associated with the country of origin. The highest mercury concentrations were found in samples from Southern Europe (Portugal, Spain, Italy) (median values: 0.26; 0.31; 0.33 μg Hg/g, wet weight). Specimens from the Solomon Islands contained extremely low concentrations (an average of 0.01 μg Hg/g). Samples from Malaysia, Thailand, the Philippines, Mauritius and Africa (Ivory Coast and Senegal) may represent – with an average concentration of 0.1 μg Hg/g – the global background [69].

3.3. THE RELATIONSHIP OF SPECIMEN BANKING TO ENVIRONMENTAL MONITORING, RESEARCH AND ASSESSMENT

There is considerable uncertainty associated with the assessment of environmental hazard from xenobiotic chemicals. This is largely due to inherent difficulties in the extrapolation of laboratory findings to the complex and varying conditions that obtain in the environment and to the fact that a large set of often incompatible and/or incomplete data sets must be employed in such assessments. For this reason alone, environmental monitoring (EM) is essential to the secure management of potentially harmful chemicals. However, environmental monitoring exhibits certain weaknesses. For example, monitoring programs cannot generally establish baseline information on anthropogenic chemicals that are already widely dispersed in the environment. Furthermore, it is impossible to monitor each of the numerous chemicals that may reach the environment by diverse mechanisms, nor especially to measure their transformation products. Indeed, only a small portion of those chemicals that are suspected of being potentially hazardous can be included in a monitoring program. Thus, even with the aid of EM new decisions (legislative or executive) will be based upon current, incomplete and non-reproducible data.

Specimen banking is thus a valuable complement to monitoring. Appropriately preserved specimens can be used, for example, as reference specimens for the determination of trends in chemical burdening of the environment, and especially for establishing the sources and times of intrusions of new chemicals into the environment. Long-term specimen banking is an essential complement to real time monitoring (see ref. 70) since, for example:

— total analysis of any sample to identify and quantify all anthropogenic chemicals is impractical because of the extreme expense alone (future analysis of specific compounds in stored samples, however, is always possible as needed) and
— future analytic improvements can be applied to representative samples of the past.

Unless an ESB project has very limited aims, it should be coordinated as well as possible with relevant environmental research and monitoring. Where such coordination is possible the value of a bank is greatly enhanced. Thus, coordinating with suitable local, regional or national networks that monitor environmental concentrations or effects of toxic metals or other pollutants is highly desirable. Coordination of EM and ESB is not usually difficult and simplifies many problems inherent in specimen selection and sampling designs.

Several efficiencies are achieved by the marriage of monitoring and banking (see also refs. 12 and 71):
— Criteria for specimen selection are similar for both.
— Sampling designs are similar for both and the combined sample set enhances statistical confidence for many applications.
— Criteria for the selection of study sites are similar.
— Temporal aspects of sampling are similar, although the frequency of sampling for specimen banking is necessarily considerably lower than for monitoring. In these respects, monitoring strongly supports and reinforces banking with respect to both environmental and statistical representativeness of samples.
— The monitoring data (biological, physical, and chemical) and assessments may serve as part of the sample characterization for banking.
— Monitoring provides essential baseline information that could not easily be gathered by means of a more narrowly-defined ESB program.
— On the other hand, specimen banking allows retrospective analyses that can provide baseline information to support monitoring results.

Research is an essential partner. EM or ESB may identify either changing patterns or adverse impacts in the environment. Nevertheless, EM and ESB when passively applied, are epidemiological in nature. That is to say, they deal only with contingencies and correlations and can not usually distinguish among the numerous possible causes of any impact unless data from appropriate field or laboratory research are available. In principle it is possible to test whether a particular environmental chemical contributes to a given impact. To do so, however, samples must have been taken from areas where the chemical is present and from those where it is absent but where all other factors are the same [72]. This constraint is too stringent to be strictly applied.

3.4. THE ROLE AND APPLICATION OF ESB TO THE ASSESSMENT OF TOXIC AND HAZARDOUS CHEMICALS IN THE ENVIRONMENT

The effective management of environmental chemicals requires timely, realistic assessment and interpretation of environmental processes. This should include the ability to

References on p. 43

distinguish between natural and anthropogenically induced changes in materials, media (soil, sediment, air, water), biotic populations, communities and ecosystems. Chemicals are important sources of such changes. Indeed, long-lived biologically available toxicants such as metals and metal compounds that accumulate in the environment are prime candidates for future incidents of human/animal/plant intoxication. Such chemicals pose serious, possibly irreversible, threats to environmental integrity and human health.

It may be impossible a priori to identify a particular chemical as a potentially detrimental environmental contaminant because, for example, reactions in the environment may transform essentially non-toxic substances to toxicants or conversely; unknown toxic properties of apparently non-toxic substances may be exhibited under natural conditions; a chemical that enters the environment in low concentrations may bioaccumulate or biomagnify; synergistic interactions of essentially non-toxic substances or low-level toxicants may result in toxicity. There is no compelling evidence that *currently-employed* toxicity tests and other predictive methods (e.g., structure-activity correlations) can be reliably applied to environmental systems (populations, communities, ecosystems) including human communities [1,2]. It is thus not surprising to note that health and environmental impacts of anthropogenic chemicals are generally recognized only when the damage has become obvious to everyone. Environmental protection measures are thus usually delayed while investigations into the probable causes of impact are conducted. This fact alone, provides a clear role for specimen banking.

3.4.1. Role and application to chemical and environmental management

All predictive methods depend upon extrapolation and therefore upon assumptions which may or may not be valid and which always include quantitative error and uncertainty and often qualitative uncertainty. With respect to chemicals, extrapolations are made from high dose to low dose, models to natural systems, one system to another (e.g., rat to man, laboratory to ecosystem), environmental concentrations to exposure concentrations, concentration to effects or impacts, effects of short-term exposure to long-term effects, etc. (e.g., refs. 73-75). EM and ESB can be employed to asses or to validate such extrapolations, assessments, and the consequences of resulting management or regulatory decisions, thereby,

— reducing costs where overprotection results from errors of assessment or of management;
— reducing impacts where predictions have led to underprotection;
— improving assessments in the future by helping to establish which of the numerous predictive methods are sensitive and effective under various circumstances.

ESB offers a means of providing reference specimens for the determination of trends in chemical burdening of the environment, and especially for establishing the sources and times of intrusions of new chemicals into the environment. The determination of concentrations of chemicals that were not recognized as hazardous at the time of collection and storage, or which could not be adequately monitored or accurately analyzed is possible.

EM and ESB can contribute indispensible evidence to form a reliable, cost-efficient basis for the protection of human health and well-being as well as that of terrestrial, limnic, and coastral marine ecosystems against intentional and inadvertent chemical pollution. From the administrative point of view, combined EM and ESB offer the possibility of detecting toxic chemicals before critical environmental concentrations are reached or before unacceptable or irreversible damage occurs. The information acquired by a system of monitoring and banking can be employed either to channel and to direct needed assessments and research related to the determination of sources, pathways, fates, and effects of hazardous metals or other persistent pollutants or to develop effective regulatory or management strategies. Information developed from EM/ESB programs can also serve functions such as the following (see also refs. 1,2,12,71):

— A local or regional record of trends in concentrations of pollution over a period of decades with respect to a variety of organic and inorganic pollutants can be provided.

— If the bank is designed in such a way that specimens are characterized with respect to properties that reflect human or environmental health or integrity, the biological or environmental significance of the chemical concentrations measured may be established or confirmed.

— Authentic, representative samples from the past, whose hazards were not recognized at the time of sample collection and storage, or for which sufficiently accurate analytical methods were not then available, can be characterized and evaluated.

— The effectiveness of predictive methods (e.g., risk assessment, hazard assessment; data from premanufacture notifications as predictors of hazards; toxicological tests) can be determined.

— Through routine fingerprint analyses of chemicals in the to-be-stored specimens, it is likely that some contaminants will be identified at an early stage of intrusion, whereas they might otherwise remain unsuspected until serious impacts are observed.

— Early detection of environmental increases in hazardous chemicals that were though to be under control is possible.

— The effectiveness of restrictions, regulations, or management practices that have been applied to a region or to the manufacture, distribution, disposal, or use of toxic chemicals can be assessed.

— Hazard evaluations and regulatory decisions generally suffer from a lock of quantitative data on pollutant burdens of earlier times or from inconsistencies or ambiguities among the available data. This can be offset through ESB.

— Data from monitoring and banking can help to determine assessment strategy. For example, if analysis shows that a particular harmful chemical occurs significantly only in soil, assessments and remedial actions can be appropriately focussed, saving time and money.

— Species, populations, communities and ecosystems that may be especially endangered by or are sensitive to toxic chemicals can be identified.

— Sources of environmental chemicals may be identified by means of specimen banking. Often, by the time a chemical is identified as an environmental problem (usually

References on p. 43

through accidental detection or by monitoring), it may be sufficiently widespread in occurrence to defy identification of major sources or pathways.

3.4.2. Role and applications to the assessment of metals

There has been a very rapid rise in the amount of "total elemental" data in the environmental sciences. Much more needs to be known, however, regarding the distribution in the environment of metal species. It is the form of the metal and not its concentration or total amount per se which determines its bioavailability, toxicity, mineralization rates, and modes of interaction with biological and other environmental components (e.g., refs. 1,2,76).

Unfortunately, except in a few cases such as mercury, elemental speciation data are sparse. It is often difficult to obtain data on the form of metals in complex environmental samples. Whereas recent developments in element-specific chromatographic detection pioneered by McCormack et al. [77] and Kolb et al. [78] reduce the degree of difficulty and bring such determinations within the capability of most analysts, many metal organic compounds are unstable under various circumstances and in various materials. Extreme care in collection, handling, preservation and storage are at least as important as sample preparation and analysis. Even the accuracy of total element analyses must often be called into question (e.g., refs. 1,2,7,79).

The mobilities, reactivities, bioavailability, and toxicities of chromium vary widely depending upon numerous properties of the local environment and the physical and chemical state of this element. As a consequence there is usually no clear relationship between the behavior or impacts of chromium and the concentration or amount of total Cr within a system; within compartments of a system; at one location versus another; or at one time versus another [80-83]. The situation is further complicated because the environmental distribution of Cr species is often highly inhomogeneous (e.g., ref. 80) and temporally variable (Lewis, unpublished data).

Improved knowledge of metal speciation in environmental materials should improve our ability to interpret and predict the environmental behavior of metals and the hazards to be expected in various environmental compartments under various environmental conditions. Such knowledge can in principle improve pollution control, medical practice, and land management, for example, through improved knowledge of rates and routes of transport; partitioning of species within the environment; fate; and bioavailability and toxicity of specific species, moieties, complexes and mixtures. Costs, lack of time and methodological problems in particular indicate the need for retrospective analyses of appropriately stabilized, banked specimens. If however, chemical speciation studies are to improve our ability to predict toxicities and other environmental effects, they must generally be conducted in concert with investigations of the environmental systems and the effects of interest.

3.5. COORDINATION, COOPERATION AND TECHNICAL ASSISTANCE

In general, ESB is a multidisciplinary activity. Unless a banking project is unusually constrained with respect to goals and scope of collection, ecologists, statisticians, analytical chemists, and perhaps others will be involved. Furthermore, cooperation among specimen banks, at least with respect to standards, conformability of methods and compatibility of data are highly desirable. New banking projects should generally be planned and implemented by a multidisciplinary team. Technical assistance may be required. Such is readily available from the U.S.A., Canadian, Swedish and West German ESB programs [1,2].

3.6. THE SUITABILITY OF SPECIMENS

The types of specimens that are ultimately of greatest value depend upon program objectives, the properties of as yet unknown chemicals, the probable route of entry into the environment, and other factors. A given type of specimen will accumulate or otherwise respond unpredictably to an unknown chemical regardless of how much we know about its responses to other chemicals. Individual species or types of specimens do not have characteristic responses to a broad spectrum of chemicals (e.g., refs. 84-89). Furthermore, little can be learned regarding the burdening of a study area as a whole by use of a single type of specimen. Thus, species "X" or specimen type "X" might behave in a completely idiosyncratic way with respect to any given chemical or might behave differently at different locations as a function of critical site factors. Butler [90], for example, reported that the uptake and retention of organochlorine residues by several species of estuarine shellfish varied unpredictably.

Other studies have demonstrated that the concentrations of specific environmental chemicals among the various compartments within an ecosystem or habitat can differ by several orders of magnitude (e.g., refs. 91-93).

Preliminary sampling may be required to determine the suitability of prospective specimens. Practical determinants of suitability will, for example, include the ability to collect a sufficient number of specimens to establish confidence limits and acceptable levels of variation. Only a few of the more important considerations with respect to the selection of individual types of specimens can be mentioned here. For an extended treatment of the selection of specimens, see Lewis [1,2].

Specimens selected for ESB should generally have an established information base with respect to those characteristics that affect their programmatic utility. Human tissues are suitable candidates for ESB. However, whereas the analysis of human tissues usually has direct implications regarding the health effects of exposure to environmental pollutants, the analysis of human tissues alone is generally insufficient because [94]:
— Not all chemicals to which human beings are exposed indirectly can be measured later in human tissues or fluids.

References on p. 43

— Some chemicals are metabolized by the body to other substances which could indicate initial exposure to more than one chemical compound.

— Chemicals that leave measurable residues in human specimens cannot always be attributed to a specific type of exposure or source.

— Residues in human specimens that are indicative of pesticides may, in fact, result from exposure to similar chemicals used for other purposes.

— It is important to identify sources and pathways of human exposure (e.g., through food products) before potentially harmful exposure occurs.

3.7. METHODS

3.7.1. Field and laboratory procedures

Field and laboratory procedures must be uniformly defined for the program and for each study site and each type of specimen to be processed and most include methods to prevent contamination of specimens (including cross-contamination from other specimens) until analysis. Some chemicals and specimens may deteriorate prior to storage unless special care is taken. Records must clearly indicate all relevant steps and procedures from collection through storage. Transportation of specimens to a storage facility should be the fastest practicable.

3.7.2. Field sampling and statistical design

Much of the literature on ESB fails to treat field sampling methods. This is unfortunate and indefensible. In such cases, one can not interpret, reproduce or verify the study. There is ample literature on statistics and field sampling design which can be consulted. Appropriate techniques are available and should be employed, and the significance of constraints upon such designs should be fully understood and acknowledged. Useful references include 72,95-119.

The object of sampling in the field is to acquire specimens that can provide unbiased and precise estimators or descriptors (e.g., chemical concentration) of some population or compartment of the environment about which inferences are to be made. Measurements performed on the sample population are estimates of the true population values. Biases may be introduced in collecting, handling, and measurement or by failure to apply appropriate statistical procedures. How well the sampling and collection procedures satisfy the assumptions of the estimation procedure must be assessed to determine the sources of error. Sources of bias, error and uncertainty abound. *The magnitude of natural variation compounded by field sampling error may be considerably greater than that of errors associated with chemical analysis.* This is true at all levels of organization. Even the true mean value of an environmental function may exhibit considerable temporal and geographic variation. With care, however, many sources of variation can be eliminated. Known or suspected sources of variability can be dealt with be: (1) randomizing the sampling protocol with respect to the variables of concern, (2) employing a procedure that further constrains or stratifies the population of interest, or (3) adjusting the choice of areas and specimen types. An appropriate field sampling design will reduce sampling error, will help to determine sources of variability and error, and can greatly improve the usefulness and credibility of conclusions. Where a type of variation results from a more or less general and systematic process (e.g., phenological stages, seasonal variation), knowledge of such can be used to help determine the timing, frequency, or location of sampling (e.g., refs. 9,49).

Sampling designs depend fundamentally upon the scope and objectives of the program and upon the sites and systems that are therefore selected. The designs and procedures employed depend also upon the abundance, behavior and distribution of the populations and/or materials to be sampled, and upon the extent and characteristics of the study area. The sampling design should take into account those factors that are most likely to importantly affect concentrations or burdens of chemical pollutants in the materials or populations to be sampled.

Preliminary surveys are essential to develop scientifically sound and efficient sampling designs and field methods. Such surveys are the only effective means to establish or determine: (1) distribution, abundance and availability of the specimens that are sought, (2) costs and ease of collection and processing, (3) a standard workable approach to collection, (4) ability to sex, age, and otherwise appropriately identify and characterize the population, (5) the number and distribution of sampling sites for each specimen type within the surveyed ecosystem, (6) sample size and the timing and frequency of sampling that are adequate for the objectives and required level of precision, (7) the extent, nature and frequency of variation.

Existing maps (e.g., topographic, soils, vegetation, land-use, air quality, water quality) and aerial photographs (especially infrared color) can be very valuable in designing and conducting a survey. Remote sensing imagery can be used in assessing or mapping land forms, major soil formations, plant communities, water bodies, historic patterns of land use, microclimate, distribution patterns of wildlife, key plant species, drainage patterns and erosion, and trends in ecosystem structure and dynamics. Representative environmental sampling generally requires:
— a clear, comprehensive and unambiguous description of the environmental compartment or the population to be sampled;
— application of sampling procedures that randomly draw specimens from the defined population or stratum;
— multiple samples in order to establish reproducibility of sampling;
— statistically sound estimates of central tendency and/or variation based upon the applied procedure.

Aspects of sampling design that may strongly influence statistical significance and power include sample size, sample locations, frequency and time of sampling, the methods and conditions under which sampling is performed, and the extent of pooling (compositing) of samples. An acceptable balance must be struck among the above features of design and other constraints. For example, if the sampling frequency is low,

References on p. 43

sample size must generally be larger. In the Swedish program, approximately 50 specimens each of several terrestrial species are banked at about five year intervals.

Sample size can greatly influence the error in estimates of chemical concentration or
other parameters. The number (and mass) of specimens comprising a sample must:
— be adequate to support sufficiently precise estimates of chemical concentrations;
— not incur unusual costs either for collection or for analyses;
— not require excessive storage space.

Satisfactory sampling in the field is usually achieved by a combination of deliberate
choice (e.g., to set boundaries on the population to be sampled) and a random or
stratified-random sampling approach. The frequency of sampling and the number of
specimens to be taken can be determined by pre-sampling or by use of prior knowledge
of population characteristics and the degree of reliability that is desired. The assessment of
systematic variation also requires prior knowledge of the specimen type and its natural environment. The essential descriptors to be represented by the sample population include measures
of central tendency (e.g., average, mode and median values) and appropriate measures
of variation including the types and degree of dispersion of values about the center.

Once the general approach has been determined, sample sizes, the timing and frequency of collection, sample strata, and the collection and treatment protocols must be
established. These processes are partly theoretical in nature and partly time-, site- and
cost-dependent. Prior estimates of sample variance as well s costs and effort of collection,
handling, storage, and analysis are needed to establish the desired sample size.

3.7.3. Specimen collection, handling and transport

Methods employed to collect and process specimens depend upon the type of specimen, location, accessibility, organizational and financial constraints, and other factors
including experience in collecting a particular type of specimen. Methods must thus be
developed on an ecosystem-specific and a specimen-specific basis. Specimen identification must be sure.

Collection, handling, transport and storage procedures must be uniform, reproducible
and specified in detail. Protocols and practice must assure that materials are collected
and processed according to program requirements, so that desired information sill be
retained. Procedures for inspection and quality assurance should be established.

Care must be taken to minimize gains or losses of chemical constituents via contamination, decomposition, volatilization, etc. Each type of specimen offers special problems.
For example, many organic chemicals decompose rapidly following death. Post mortem
cytolysis and other continuing biological activities may also produce chemical species not
present in the living specimen. Many chemicals are readily oxidized by sunlight. Thus, for
example, care must be exercised to avoid exposing subsurface soils to sunlight. The
collection of recently disturbed (eg., plowed) soils is contraindicated for this and other
reasons [1,2]. Koirtyohann and Hopps [120] provide an excellent treatment of procedures
and requirements of collection of human tissues for trace element analysis. This paper

also has relevance for field collection and handling of non-human materials. Collecting or sampling tools can be serious sources of contamination. Even the use of surgically-clean procedures does not protect specimens from metal contamination. Biopsy needles, and other surgical tools can introduce metals to the specimen [121]. Materials used in making equipment and the containers in which specimens are transported and/or stored should be selected so that the number and kinds of trace contaminants that may be introduced to the specimen will be limited. Surface contamination of instruments or containers that are improperly stored or maintained may also transfer contaminants to the specimens. Sample containers should consist of inert, non-absorptive materials such as Teflon. The collection protocol for human livers of the U.S.A. ESB program requires the use of non-contaminating materials such as dust- and talc-free vinyl gloves; pre-cleaned, dust free Teflon sheets and bags, high-purity water, and a specially designed titanium-bladed knife.

Records must clearly indicate all relevant steps, procedures and responsible persons from collection through storage. Information relevant to stability of specimens or the residues that ultimately come under analysis must be included (e.g., temperature, pH, time between collection and freezing). A matter of special concern is the speed with which specimens can be preserved. In specimens that are collected following death, varying and often rapid changes will have taken place. These include not only altered chemical concentrations, but changes in water and lipid content of whole body or of individual organs or tissues, biochemical functions, cell structure, and tissue structure. This problem might be resolved in the case of small intact organisms by transfer to the temperature of liquid nitrogen or dry ice immediately following collection. The immediate freezing of specimens, however, does not permit preliminary preparations to be made (e.g., washing or trimming, taking material for histological sections) and does not usually allow biometric or other characterizations to be made. Clearly, some time delay is unavoidable and will vary with the size and type of specimen (e.g., whole organisms vs. specific organs or tissues). For each specimen type, judgement must be exercised with regard to how much time can be permitted for measurement and other manipulations. The time from collection to preservation should be standardized as much as possible for each type of specimen and carefully recorded for each specimen collected. Weight determination and most other measurements must generally be made before preservation. Live trapping of animals and in situ measurements of plants prior to collection will sometimes be advantageous. Certain determinations such as dry weight or fat-free dry weight can be made for some types of specimens at the time of analysis if needed. Care must be taken to include measurements that may be needed by the chemical analyst. With animals that are collected by shooting, contamination by lead and other materials is a matter of concern. In spite of the results of several comprehensive studies (e.g., refs. 122-124), we do not yet know whether collection by shooting produces inadmissible contamination or alteration of the specimen [123].

Sampling for trend analysis in long-term programs must take place repeatedly at the same site. Sampling and associated activities must not be permitted to unduly alter the population from which samples are drawn. This requirement may sometimes limit the size of the sample and favors non-destructive methods when they are otherwise justified.

References on p. 43

Thus, one might collect eggs, individuals or molted feathers of the goshawk. Only in the last case it is possible to repeatedly and extensively sample the population without disturbing or altering it substantially. Such non-destructive methods of sampling sometimes also offer the possibility of acquiring repeated samples from the same individual. Repeated samples are correlated, however, thus raising questions concerning the statistical assumption of independence. There is an extensive literature on methods of collection for various types of environmental specimens which should be consulted [125]. Practical considerations for collecting include:

— The structure, form and chemical integrity of the specimen should be disturbed as little as possible.
— Disturbance of the area should be minimized.
— Care must be taken to minimize changes in chemical composition during collecting, handling or shipping.
— Field and laboratory methods of sample collection and processing should include appropriate quality assurance procedures.
— Once a frozen specimen has thawed, it should not be refrozen since this causes separation into phases and produces large statistical variation in results of analysis of subsamples [126].

Local and national laws and international conventions and treaties limit the collection and handling of various types of specimens (e.g., humans, rare and endangered species, migratory waterfowl, materials on private lands, materials on public lands, etc.) and the collection and processing of information. There are, for example, numerous local and national laws to protect privacy and the sanctity of the human body. Such laws may have serious implications for sampling designs and data processing. In the U.S.A. the principal law protecting humans in the Privacy Act of 1984, but where state laws are more stringent they may take precedence. In Europe, the Nüremberg Code of 1949 precludes any experiment on a human subject without his voluntary consent [71,127]. Knowledge and compliance with relevant laws, conventions and government regulations is necessary.

3.7.4. Security

Specimens, subsamples of specimens, or data derived from specimens may become the subject of public inquiry or controversy; used for legal purposes; or for the establishment of binding rules, regulations or restrictions. Specimens, samples and primary data must therefore be collected, transported, transferred, stored and analyzed under circumstances whereby their chemical and legal integrity are preserved. Thus the "chain of custody" from collector to transporter to archivist to analyst must be unbroken and verifiable. Each party in this chain must be able to guarantee that the materials and data are as represented with respect to their identity, amount and condition and that each specimen has been under supervision or otherwise secure at all times. This must be recorded in signed registers that identify the time, place, circumstances, method, etc. of receipt and custody of material. Transport of transfers between responsible parties should be direct or by tamperproof means.

Security, accountability, and a continuous record thereof, extend to the environment under which the specimens are transported and maintained and to the methods of processing. Facilities must be secure, access to materials and primary data must be restricted, and written accountability is required at all times. Preventive maintenance procedures to limit problems from equipment failures and human error must be developed and implemented. Secure storage of primary data (e.g., as logged in the field or laboratory) and computer tapes or discs, or microfilm copies of data sets is required. Film, data sets, tapes, etc. must be maintained in good condition for many years. This requires cool, dry, dark conditions in secure, fire-proof cabinets and occasional copying of tapes and perhaps film.

As a security measure and in order to conserve valuable specimens, it is generally wise to establish a policy as to who may have access to specimens, under what conditions and for what purposes. Authority to implement this policy must be delegated to an appropriate officer or committee. The value of stored, irreplacable specimens must be weighed against the value of the proposed, usually destructive, use.

3.7.5. Use of pooled (composite) specimens

The pooling or compositing of specimens following collection is used as a means of reducing the number of samples for analysis or in the case of specimen banking as a means of saving space. These are practical concerns, not scientific. To base decisions simply on such concerns is patently absurd. The key question is not so much whether the pooling and homogenization of specimens prior to storage is useful, but whether it is scientifically acceptable. Often the answer is no. The analyst should be aware that in a well-designed program, the field scientists will probably have collected the minimal set that is representative of the target population. To reduce this set further may dissolve any hope of making inferences about the population or environment of interest. In any event, the number of specimens in a sample from the field does not dictate the number that must be analyzed in a given case. The analyst can subsample from the set of specimens present, using a number that provides statistically valid data for the study in question. The number of specimens required for assay will vary depending on various factors (unknown at the time of storage) such as the sensitivity and accuracy of analysis, the variability in concentration and distribution of the determinant among specimens within a sample, and the costs and purposes of analysis. Often, the number of specimens to be analyzed can only be determined on the basis of preliminary analyses performed years after the specimens were placed in storage.

The pooling and homogenization of specimens prior to storage *may* be justified when: (1) individual specimens are too small for analysis, (2) the specimens are chemically inhomogeneous, (3) the costs of analysis are high, and/or (4) storage space is at a premium; *but only when* (5) the loss of information by pooling and homogenization is acceptable. Composite samples are especially useful in determining the accuracy of chemical analysis and in the development or improvement of analytical methods. Under the conditions of storage employed by the West German pilot bank using liquid nitrogen

References on p. 43

as the coolant, the space required by composite homogenized specimens is generally less than 20% of that required by whole specimens.

The use of pooling must be carefully weighed against the resulting loss of information and the consequent implications for inferential utility. Injudicious use of pooled samples can yield results that are erroneous and misleading or qualitative when applied to the population or area from which samples were drawn. For example, unless the type of frequency distribution of chemical concentration within the sampled population can be identified, true mean values may be substantially overestimated or underestimated and the magnitude and direction of error may remain unknown.

A single pooled sample may be useful for establishing the presence of a chemical, but the probability of false negatives can be high. Compositing may increase the number of samples that are below detection for a given chemical, depending upon the type of frequency distribution and the actual concentrations in relation to detection limits. Dilution due to the pooling of specimens can be partly compensated if the minimum detection limit (MDL) for a particular analytic procedure is known and the maximum acceptable level of contamination (MAL) can be specified. According to Skalski and Thomas [128], the maximum number of specimens (n_{max}) that can be grouped into a single pooled sample and still guarantee detection of the chemical is:

$$n_{max} \leq \frac{MAL}{MDL}$$

Whereas MAL is likely to be known in a monitoring program, it is never known for ESB at the time this decision must be made. Therefore, n_{max} *must be as small as practical.* Replicate sampling (as opposed to subsampling) is necessary to establish the degrees and types of variation in the target population.

If numerous samples are collected and stored, but only a few of these can be later analyzed, we have not only a wasteful situation, but we then need a subsampling design to determine which of the samples to analyze. If, on the other hand, sampling is too limited with respect to the number of locations or specimens sampled and subsequently analyzed, or if single pooled samples from each location are employed, analyses may fail to distinguish between regional pollution and local foci of pollution due to introduction from spills, plume strikes, waste dumps, applications, manufacture, or use. Such sampling may also fail to distinguish temporal trends. This is not a trivial point and such sampling can lead to rejection of the ESB results by decision makers or, if accepted, to costly and ineffective management activities. Thus, the field sampling design for a given type of specimen and the laboratory sampling and subsampling designs must be integrated.

There is very little experience in banking soils, but pooling and homogenization gener-ally offer no obvious advantages and the convenience and considerably greater information content of small, intact cores is highly desirable [1,2,129]. Ausmus and O'Neill [130] found, for example, that the leaching of homogenized soil columns removed more dissolved organic carbon and with greater variation than from intact soil cores. In another study, the leaching of organic carbon was greater than for intact columns [131,132].

There were no consistent differences between intact and homogenized columns. Many other types of specimens are useful only if stored whole. Clearly sediment cores should be stored intact, thereby preserving the structure of the column. It is possible thereby to infer the history of impacts associated with various stable chemicals [1,2].

3.7.6. Chemical procedures and analyses

Whereas environmental sampling and chemical analysis are sometimes regarded as distinct parts of ESB, they are in fact interdependent. Decisions in one activity affect those in the other. Certain decisions regarding analytic requirements must be made early in program planning so that sampling and statistical designs are appropriate and adequate. Chemical analytic aspects influence sample size and the methods to be employed in the collection, handling, transport, and storage of specimens.

Where accepted standard methods of subsampling, extraction and instrumental analysis exist, they should be employed. Rigorous quality assurance is essential. The identity of significant residues must be established to support executive decisions and perhaps legal actions. This may require two or more independent physico-chemical determinations of a residue's identity. The quantitative analytical methods employed should generally be sensitive and specific enough to identify the targetted chemicals at levels actually encountered in the environment. Qualitative analyses can be extremely important also. The detection of especially toxic chemicals can be very useful in establishing the need for quantitative analyses. Qualitative analyses may also provide means of early warning of the introduction of new chemicals into the environment and sometimes offer *prima facie* evidence of the pollution source or sources.

From the point of view of environmental chemistry it is useful to note that environmental burdens of chemical pollutants frequently do not follow normal or Gaussian distribution patterns but often exhibit a more or less positive (and sometimes a negative) skew which may be exponentially or log-normally distributed (e.g., refs. 7,127,128,133-135). This has important implications for sampling designs and evaluations. Sample size must be large, and the average or mean value together with a single measure of variability may not adequately describe the situation. Whereas such non-normally distributed data appear to be genuinely natural, one must guard against artifacts resulting from handling and analytical error. Skewed distributions may result for example, from error in values close to the limit of analytical detection. More research is needed on this point, but many cases of skewed distributions appear to represent true environmental variation. Errors in chemical analysis are of course unavoidable and, in EM and ESB programs, can be expected to derive not so much from limitations in the "state-of-the-art", but rather in the use of available instrumentation by diverse analysts of varying skill. Thus, a specimen bank that ultimately depends upon a variety of contractors and institutions for analysis will have a very substantial quality assurance problem. This is amply illustrated by Erdmann [136], who found that of 28 laboratories, no more than 5 performed acceptable chemical analyses of a standard reference water! Nevertheless, extreme accuracy of analysis is often

References on p. 43

wasted for the most part because of inadequate field sampling designs, and the stochastic behavior and heterogeneity of natural systems at *all* levels of organization.

3.7.7. Preservation and storage

Cryogenic storage. No single standard or optimal method of preservation and storage can be recommended for all types of chemicals and all types of specimens (e.g., refs. 1,2,9,127,137,138). There is, nevertheless, an emerging consensus that the best general method is deep-freezing at temperatures of –80°C of lower [1,2,12]. Such temperatures are employed by the West German and American programs. Temperatures at or above –40°C are generally not recommended. Thus, commercial freezers (ca. –10°C to –20°C) should not generally be employed for long-term storage. Studies of the effects of various storage conditions and temperatures on the form, stability, recovery, and migration of chemicals within stored specimens should continue, as should studies of collection and preparation for storage.

Deep freezing is not always essential to retain many stable elements and compounds including some classes of organic compounds (e.g., refs. 139-141). Because of the necessarily limited supply, deep frozen specimens should not be employed for initial surveys of compounds and elements or moities that are adequately stable under less intensive methods of storage. Nor should they be used for the testing or demonstration of new or unproven analytic methods or applications.

Of all methods that can be employed in the field, freezing is probably the most convenient and safest for metals [142]. A temperature approaching –80°C can be obtained with dry ice, but freezing may not be rapid where the mass of material introduced to the field chest is high relative to the amount and location of dry ice. For rapid deep freezing of specimens we recommend the use of liquid nitrogen (–196°C) when possible.

The problems of storing metals and metallic compounds for speciation analysis are poorly understood. Certainly, the use of preservatives such as mineral acids must be avoided if speciation rather than total concentration is to be evaluated. Conventional freezing of water or specimens with a high water content is also probably ill-advised where metal speciation studies are projected. Water may, for example, concentrate elements in the liquid fraction during freezing, sometimes producing irreversible reactions (for example, polymerization of iron hydrolysis products) as a result of the elevated concentrations (e.g., ref. 143).

Alternative methods. There are a number of methods of preservation and storage that, used for specified purposes or as an adjunct to cryogenic storage, can save space, optimize the use of valuable deep-frozen specimens, improve sampling designs and information content, and improve cost-effectiveness. Some of these methods offer the possibility of acquiring combined information on chemical burdening and environmental effects. For an extended treatment of alternative methods, see Lewis [1,2]. In most of these methods one may expect many chemicals to be lost or altered through volatilization, decomposition or other processes.

There are further questions to resolve regarding preservation and storage. Lundell [144] stored air-dried and frozen samples of humus and mineral soils for a few months prior to analysis for nutrients. In humus, concentrations of "relatively available" phosphorus and ammonium were 51% and 76% higher in frozen samples than in the air-dried samples. Conversely, concentrations of nitrate were 75% higher in air-dried than in frozen humus. In the mineral soil samples, concentrations of "relatively available" potassium were higher (21-64%) in the frozen samples and nitrate was higher (80-427%) in air-dried samples.

Dried specimens. Dry samples (e.g., soil, sediment, fly ash, bird feathers, seeds), if collected and stored according the standards and requirements of modern trace analysis, can be of great value. Dry specimens (maintained in a cool, dry environment) may find their greatest use in the assessment of toxic heavy metals and stable elements [12]. Results in the West German pilot bank indicate that lyophilization may be usefully applied to heavy metals. Precautions may be necessary in the case of metalloids and volatile metals [145]. Studies have shown that minimal losses of trace elements may occur for a period of a year in lyophilized specimens [137]. Freeze-dried bovine liver (NBS Liver-SRM 1577) that was certified for trace elements in 1972 remained unchanged with respect to trace element content for more than five years (ibid.) Freeze drying is employed to some extent in the Swedish Environmental Specimen Bank [9,146].

Plants such as mosses and lichens can be easily stored at room temperature for purposes of elemental analysis. These plants possess a rough and variable surfaces that readily retain particulate matter. In areas chronically affected by particulate air pollutants, the collection and analysis of mosses and lichens can provide useful information.

Concentrations of a wide range of chemical elements and inorganic compounds in bodies of water may at times exceed the solubility product of the parent compound and thus precipitate as minerals onto the surface of mud or sediment. Organic compounds can also be deposited onto sediments in various ways. These materials may be taken up by a wide variety of organisms. Chemical analyses of dried or frozen undisturbed sediment cores can provide valuable information on the normal elemental composition as well as the time of introduction and rate of accumulation of toxic chemicals.

With the help of more than 20 scientists, including several with experience in specimen banking, Britt and McFarland [147] evaluated the adequacy of existing methods for long-term storage of particulate matter. They were concerned primarily with the feasibility of storing combustion-related particulates for several months to several years for purposes of toxicity testing. The "best available" storage systems are described in their report, and recommendations for sample preparation, sample storage, and quality assurance are made. A number of sources cited in Britt and McFarland [147] indicate that fly ash and other particulates may be remarkably stable in many respects, even when stored dry at room temperature. A fly ash reference material is available from the U.S. National Bureau of Standards [148] it is certified for 14 trace elements and 6 of the more prevalent matrix elements. This material is stored at room temperature under conditions that NBS considers to be adequate to prevent changes in the certified chemical elements for a period

References on p. 43

of up to three years. The alteration of oxidation states during storage is a matter of concern of room temperature and even at conventional freezing temperatures, but becomes much less important at –80°C [149]. Even the mutagenicity of particulate material was observed in samples that were stored at room temperature for approximately 15 years [150]. Britt and McFarland [147] believe that (for purposes of toxicological research) storage of particulate material should be cold, preferably at –80°C.

Chemical preservation. The use of chemical preservatives is generally to be avoided. They can cause unwanted chemical changes, can introduce contaminants, and can be expected to stabilize some chemical constituents but not others. There is some indication that certain organic compounds (e.g., organochlorine pesticides) and some elements are stable under formalin or alcohol fixation. The use of chemical preservatives deserves further investigation because they also preserve structural information. The application of such methods for some purposes would simplify some of the technical and logistic problems (see also refs. 146,151-153). In the Swedish specimen bank preservatives are used only for future histological studies [9].

Barber et al. [154] and Miller et al. [155], who studied mercury concentration in recently-collected and historical museum specimens of marine fish, found no changes in mercury concentrations due to storage. Gibbs et al. [156], however, reported changes in concentrations of other heavy metals in fish specimens that they attributed to the preservatives (formalin and alcohol).

3.7.8. Quality assurance (QA)

Standards and QA methods must be established to assure an acceptable level of competence and performance. QA applies to *all phases* of ESB including chemical and biological procedures and analyses; sampling design; methods of collection, transport and storage; maintenance and calibration of equipment; security; and data processing and retrieval. Detailed written protocols are required for all procedures that may affect the quality or credibility of data. These protocols should be reviewed and as possible tested for methodological errors.

QA should include tests of randomness in field sampling and in the subsequent sub-sampling for chemical analysis. It is important to know whether systematic or random variations and error result mainly from analysis or from field sampling and collection. It is then possible to determine whether error/variation is best reduced through increased frequency of collections, number of sampling locations, larger sample size per location; or by means of improved analytic technique. Failure to recognize the major sources of error and variation may lead not only to poor results, but can greatly increase costs to no purpose.

Quality assurance procedures should include tests of sample stability during storage. A preliminary test of sample stabilisation is desirable, but may be impractical. The West German and U.S.A. specimen banks were fortunate to be able to conduct pilot studies of nearly a decade in duration that included such tests. Even today, however, no method of preservation and storage is known to assure the long-term stability of all species of hazardous metals or the vast majority of other chemicals that may be of concern environmentally.

The analyst who is concerned with environmental chemicals *must become familiar with the applicable statistical concepts.* An introduction to such concepts (and to the relevant literature) as they relate to environmental monitoring and specimen banking together with a discussion of some of the more important sources of variation, error, bias, and uncertainty is provided by Lewis [1,2]. The genus and species (if possible) of all biological specimens must be determined and verified by a competent taxonomist and some specimens should be appropriately prepared and maintained in a systematic collection.

One can not overemphasize the point that results of trace metal analyses cannot be regarded as accurate and certainly can not be regarded as accurately representing a portion of the environment unless all procedures are evaluated beginning with the field sampling design. Stringent quality assurance procedures for analysis are especially important because of the wide disparity of values reported by different investigators working on these materials and the undoubted inaccuracy of large amounts of analytical data [e.g., ref. 79]. *Accurate analysis does not,* however, *compensate for faulty sampling, collection, handling and storage.*

3.7.9. Financial aspects

The most difficult aspect of financial planning is probably that of allocating costs among field sampling, statistical analysis, storage and chemical analysis. Because ESB is a long-term activity, potential future outlays must be assessed. In our experience there is a strong tendency to underestimate the costs of field sampling with the result that inadequate sampling procedures may be employed. Space, special facilities and equipment may be required (e.g., refrigeration equipment, or cryogenic systems; air filtration, laminar airflow hoods). Such requirements will, of course, vary but might be as little as a single walk-in freezer (with back-up compressors and controls) together with a clean room for preparation of specimens. Special implements (e.g., Teflon containers, titanium knives, and a contaminant-free system for cryogenic homogenization) are required.

REFERENCES

1 R.A. Lewis, *Richtlinien für den Einsatz einer Umweltprobenbank in der Bundesrepublik Deutschland auf ökologischer Grundlage,* Umweltforschungsplan des Bundesministers des Innern, Univ. Saarland, Saarbrücken, Germany, 1985.
2 R.A. Lewis, *Guidelines for Environmental Specimen Banking with Special Reference to the Federal Republic of Germany. Ecological and Managerial Aspects.* U.S. MAB Report, No. 12, 1987.
3 M. Stoeppler, *Fresenius' Z. Anal. Chem.,* 317 (1984) 228-235.
4 *Report on Great Lakes Water Quality,* Great Lakes Water Quality Board, Kingston, Ontario, 1985.
5 A.M. Beeton and J.E. Gannon, in R.A. Lewis, N. Stein and C.W. Lewis (Editors), *Environmental Specimen Banking and Monitoring as Related to Banking,* Martinus Nijhoff, The Hague, 1984, pp. 143-163.

44

6 J.E. Elliott, in R.A. Lewis, N. Stein and C.W. Lewis (Editors), *Environmental Specimen Banking and Monitoring as Related to Banking*, Martinus Nijhoff, The Hague, 1984, pp. 45-66.

7 J.E. Elliott, in S.A. Wise and R. Zeisler (Editors), *International Review of Environmental Specimen Banking*, U.S. Dept. of Commerce, National Bureau of Standards, NBS Special Publication, No. 706, 1985, pp. 4-12.

8 J.E. Gannon, *J. Great Lakes Res.*, 8 (4) (1982) 591.

9 M. Olsson, *Memoranda Soc. Fauna Flora Fennica*, 59 (1983) 93-100.

10 R.C. Dorn, P.F. Philips, J.O. Pierce II and G.R. Chase, *Bull. Environ. Contam. Toxicol.*, 12 (1974) 626-632.

11 R.A. Lewis and B. Kaiser, *Der Einsatz von Umweltprobenbanken zur Bewertung von toxischen Umweltchemikalien bis in die heutige Zeit (Biogeographische Mitteilungen*, 16) Universität des Saarlandes, Saarbrücken, 1986.

12 R.A. Lewis, N. Stein and C.W. Lewis, *Environmental Specimen Banking and Monitoring as Related to Banking*, Martinus Nijhoff, The Hague, 1984.

13 M.H. Martin, P.J. Coughtrey, *Biological Monitoring of Heavy Metal Pollution*, Applied Science Publ., London, New York, 1982.

14 S.B. McLaughlin, C.F. Baes III, R.K. McConathy, L.L. Sigal and R.F. Walker, *Interactive Effects of Acid Rain and Gaseous Air Pollutants on Natural Terrestrial Vegetation*, EPA/DOE Interagency Acid Rain Review, Raleigh, NC, 1983.

15 C.F. Baes III and S.B. McLaughlin, *Science*, 224 (1984) 494-497.

16 P. Nuorteva, *Naturwiss. Rundschau*, 24 (6) (1971) 233-242.

17 H. Ellenberg and J. Dietrich, in R.E. Kenward and I.M. Lindsay (Editors), *Understanding the Goshawk*, International Assoc. for Falconry, Oxford Univ., 1982.

18 H. Applequist, S. Asbirk and J. Draebak, *Mar. Pollut. Bull.*, 15 (1984) 22-24.

19 W. Berg, A. Johnels, B.R. Sjordstrand and T. Westermark, *Oikos*, 17 (1966) 71-83.

20 A.A. Goede and M. De Bruin, *Environ. Pollut. Ser. B., Chem.-Phys.*, 8 (1984) 281-298.

21 P. Lindberg, T. Odsjo and M. Wilkman, *Ornis Fennica*, 60 (1982) 28-30.

22 R. Doi, H. Ohno and M. Harada, *Sci. Total Environ.*, 40 (1984) 155-167.

23 D. Jenkins, in A. Berlin, A.H. Wolff and Y. Hasegawa (Editors), *The Use of Biological Specimens for the Assessment of Human Exposure to Environmental Pollutants*, Martinus Nijhoff, The Hague, 1979.

24 W. Salomons and A.J. De Groot, in W.E. Krumbein (Editor), *Environmental Biochemistry and Geomicrobiology*, Ann Arbor Science Publ., Ann Arbor, MI, 1977.

25 U. Foerstner and W. Salomons, in E. Merian (Editor), *Metalle in der Umwelt*, Verlag Chemie, Weinheim, 1984.

26 G. Mueller, J. Dominik, R. Reuther, R. Malisch, E. Schulte, L. Acker and G. Irion, *Naturwissenschaften*, 67 (1980) 595-600.

27 H. Nirschl, *Untersuchungen zum Sedimentchemismus in Rachelsee*. Diplomarbeit der Fachhochschule München, Physikalische Technik, 1983.

28 J. Rosenberger, *Paläökologische Untersuchungen an Sedimentkernen des Rachelsees*. Diplomarbeit am Geographischen Institut der Ludwig-Maximilian Universität München, 1985.

29 H. Buchert, S. Bihler, P. Schott, H.P. Roper, H.J. Pachur and K. Ballschmiter, *Chemosphere*, 10 (1981) 945- 956.

30 M.F. Cox and H.W. Holm, in D.D. Hemphill (Editor), *Trace Substances in Environmental Health. IX. Symposium*, Univ. Missouri Press, Columbia, MO, 1975.

31 H.R. Feltz, W.T. Sayers and H.P. Nicholson, *Pestic. Monit. J.*, 5 (1) (1971) 54-62.

32 S.C. Harding and H.S. Brown, *Environ. Geol.*, 1(3) (1975) 181-191.

33 H.C. Mattraw, Jr., *Pestic. Monit. J.*, 9(2) (1975) 106.

34 W.L. Mauck, F.L. Mayer and D.D. Holz, *Bull. Environ. Contam. Toxicol.*, 16 (1976) 1-9.

35 *Report on Great Lakes Water Quality*, Great Lakes Water Quality Board, Kingston, Ontario, Vol. 9, 1985.

36 R.L. Thomas, *J. Great Lakes Res.*, 9 (1983) 118-124.

37 R. Frank, R.L. Thomas, M. Holdrinet, A.L.W. Kemp and H.E. Braun, *J. Great Lakes Res.*, 5 (1979) 18-27.

38 C.L. Haile, *Chlorinated Hydrocarbons in the Lake Ontario and Lake Michigan Ecosystems*. Dissertation, University of Wisconsin, Madison, WI, 1977.

39 M.V. Holdrinet, R. Frank, R.L. Thomas and L.J. Hetling, *J. Great Lakes Res., 4 (1978) 69-74.*

40 C. Boutron, M. Leclerc and N. Risler, *Atmos. Environ.*, 18 (9) (1984) 1947-1953.

41 E.W. Wolff and Peel, *Nature*, 313 (6003) (1985) 535-540.

42 Z. Jaworowski, *Nature*, 217 (1968) 152-153.

43 G. Drasch, *Die anthropogene Pb- und Cd-Belastung des Menschen. Untersuchungen an Skelett- und Organmaterial*, Habilitationsschrift, Universität, München, 1983.

44 D.O. Coleman and M. Hutton, *Museum Specimens*, in MARC, Technical Report, Historical Monitoring, London, 1985.

45 R.A. Lewis and C.W. Lewis, in N.P. Luepke (Editor), *Monitoring Environmental Materials and Specimen Banking*, Martinus Nijhoff, The Hague, 1979, pp. 369-391.

46 D.O. Coleman, *Human Remains*, in MARC, Technical Report, Historical Monitoring, London, 1985.

47 T. Suzuki and R. Yamamoto, *Bull. Environ. Contam. Toxicol.*, 28 (1982) 186-188.

48 E. Rosenberg, *Sveriges Natur*, 54 (1963) 75-76.

49 M. Olsson, *Mercury, DDT and PCB in Aquatic Test Organisms*, Statens naturvardsverk, National Swedish Protection Board, Stockholm, PM 900, 1977.

50 G. Otterlind and I. Lennerstedt, *Vor Fogelvarld*, 23 (1964) 364-415.

51 C. Bernes (Editor), *Monitor 1982. "Tungmetaller och organiska Miljogifter i svensk natur" (Heavy metals and organic environmental poisons in Swedish nature)*. Liber/Allmanna Forlaget, Stockholm, 1982.

52 S. Jensen, A.G. Johnels, M. Olsson and T. Werstermark, *The Avifauna of Sweden as Indicators of Environmental Contamination with Mercury and Chlorinated Hydrocarbons, in Proc. XV Int. Ornithol. Congr.*, E.J. Brill, Leiden, 1972.

53 T. Odsjo and V. Olsson, *Vor Fogelvarld*, 34 (1975) 117-124.

54 T. Westermark, T. Odsjo and A.G. Johnel, *Ambio*, 4 (1975) 87-92.

55 B. Broo and T. Odsjo, *Holarct. Ecol.*, 4 (1981) 270-277.

56 S. Johnel et al., 1968, cited in 49.

57 T. Westermark et al., 1965 cited in 49.

58 M. Olsson, *Ambio*, 5 (2) (1976) 73-76.

59 L. Steubing, in N.P. Luepke (Editor), *Monitoring Environmental Materials and Specimen Banking*, Martinus Nijhoff, The Hague, 1979, pp. 538-554.

60 H. Gydesen, K. Pilegaard, L. Rasmussen and A. Ruehling, *Moss Analyses used as Means of Surveying the Atmospheric Heavy Metal Deposition in Sweden, Denmark and Greenland in 1980, 1983.*

61 A. Ruehling and G. Tyler, *Botaniska Notiser*, 121 (1968) 312-342.

62 A. Ruehling and G. Tyler, *Oikos*, 21 (1970) 92-97.

63 T. Kjellström, *Environ. Health Perspect.*, 2 (1979) 169-197.

64 Gilbertson and Fox. *Chick Edema Disease and Porphyria in Lake Ontario herring Gull Embryos in the Early 1970's. Proc. Int. Symp. on Chlorinated Dioxins and Related Compounds*, Oct. 25-29, 1981, Washington, DC.

65 K.L.W. Kaiser, *Sci. Technol.*, 12 (1978) 520-528.

66 R.J. Norstom, D.H. Hallett, F.I. Onuska and M.E. Comba, *Environ. Sci. Technol.*, 14 (1980) 860-866.

67 C.J. Stafford, W.L. Reichel, D.M. Swineford, R.M. Prouty and M.L. Gay, *J. Assoc. Anal. Chem.*, 61 (1978) 8-14.

68 P.A. Butler, in N.P. Luepke (Editor), *Monitoring Environmental Materials and Specimen Banking*, Martinus Nijhoff, The Hague, 1979, pp. 156-164.

69 R. Kruse and K.E. Krueger, *Arch. Lebensmittelhyg.*, 35 (6) (1984) 128-131.

70 U. Boehringer (Editors), *Umweltprobenbank: Ergebnisse der Vorstudien*, Umweltbundesamt, Projektträger für den Bundesminister für Forschung und Technologie, (Teil 1, Februar 1981; Teil 2, Oktober 1981), 1981.

71 N.P. Luepke (Editor), *Monitoring Environmental Materials and Specimen Banking*, Martinus Nijhoff, The Hague, 1979.
72 R.H. Green, *Sampling Design and Statistical Methods for Environmental Biologists*, Wiley, New York, 1979.
73 L.W. Barnthouse, G.W. Suter and R.V. O'Neill, in H.G. Lund (Editor), *Proceedings of the International Conference on Renewable Resource Inventories for Monitoring Changes and Trends, 1983*.
74 J. Cairns, Jr., *Water Res.*, 15 (1981) 941- 952.
75 E.E. Kenaga, *Environ. Toxicol. Chem.*, 1 (1982) 69-79.
76 S.A. Dressing, R.P. Maas and C.M. Weiss, *Bull. Environ. Contam. Toxicol.*, 28 (1982) 172-180.
77 A.J. McCormack, S.C. Tong and W.D. Cooke, *Anal. Chem.*, 37 (1965) 1470.
78 B. Kolb, G. Kemmner, F.H. Schlesser and G. Wedeking, *Anal. Chem.*, 221 (1966) 116.
79 G.V. Iyengar, W.E. Kollmer and H.J.M. Bowen, *The Elemental Composition of Human Tissues and Body Tissues and Body Fluids*, Verlag Chemie, Weinheim, New York, 1978.
80 R. Eisler, *Chromium Hazards to Fish, Wildlife and Invertebrates: A Synoptic Review. Biological Report 85 (1.6)*, U.S. Fish and Wildl. Serv., U.S. Dept. of Interior, 1986.
81 B.R. James and R.J. Barlett, *J. Environ. Qual.*, 12 (1983) 169-172.
82 B.R. James and R.J. Barlett, *J. Environ. Qual.*, 12 (1983) 173-176.
83 J.D. Steven, L.J. Davies, E.K. Stanley, A. Abbott, M. Ihnat, L. Bidstrup and J.F. Jaworski, *Effects of Chromium in the Canadian Environment*, in Nat. Res. Counc. Can., NRCC No. 15017 (1976).
84 A. Brown, *Ecology of Pesticides*, Wiley, New York, 1978.
85 D.J. Jeffries, *J. Reprod. Fertil.*, 19 (1973) 337-352.
86 J.H. Koeman, in H. Geisbichler (Editor), *Advances in Pesticide Science*, Pergamon Press, Oxford, 1979.
87 F. Korte, in A. Berlin, A.H. Wolff and Y. Hasegawa (Editors), *The Use of Biological Specimens for the Assessment of Human Exposure to Environmental Pollutants*, Martinus Nijhoff, The Hague, 1979.
88 D.S. Miller, W.B. Kinter and D.B. Peakall, *Nature*, 259 (1976) 122-124.
89 W.H. Stickel, in A.D. McIntyre and C.F. Mills (Editors), *Ecological Toxicological Research*, Plenum Press, New York, 1975.
90 P.A. Butler, *Proceedings of Seminar on Methodology for Monitoring the Marine Environment*, U.S. Environmental Protection Agency Report No. 00/4-74-004, 1974.
91 C. Bernes, *The Environmental Monitoring Programme in Sweden, Bulletin 1*, SNV PM 1327, National Swedish Environment Protection Board, 1980.
92 T.W. Duke, J.I. Lowe and A.J. Wilson, *Bull. Environ. Contam. Toxicol.*, 5 (2) (1970) 171-180.
93 R.L. Meeks, *J. Wildl. Manage.*, 32 (1968) 376-398.
94 F.W. Kutz, *Residue Reviews*, 85 (1983) 277-292.
95 M.A. Buzas, *Ecology*, 51 (1970) 874-879.
96 J. Cairns, Jr. and K.L. Dickson, *J. Water Pollut. Control. Fed.*, 43 (1971) 755-772.
97 J. Cairns, Jr. and K.L. Dickson, *J. Test Eval.*, 6 (1978) 85-94.
98 J. Cairns, Jr. and J.R. Pratt, in B.G. Isom (Editor), *Developing a Sampling Strategy: Rationale for Sampling and Interpretation of Ecological Data in the Assessment of Freshwater Ecosystems*. ASTM STP 894, American Society for Testing and Materials, Philadelphia, PA, 1986, pp. 168-186.
99 W.G. Cochran, *Sampling Techniques*, Wiley, New York, 3rd ed., 1977.
100 W.W. Cooley and P.R. Lohnes, *Multivariate Data Analysis*, Wiley, New York, 1971.
101 J.C. Davis, *Statistics and Data Analysis in Geology*, Wiley, New York, 1973.
102 L.L. Eberhardt and J.M. Thomas, *Survey of Statistical and Sampling Needs for Environmental Monitoring of Commercial Low-level Radioactive Waste Disposal Facilities. A Progress Report in Response to Task 1*, Pacific Northwest Laboratory, Richland, WA, Report PNL-4804, 1983.

103 J.M. Elliott, *Some Methods for the Statistical Analysis of Samples of Benthic Invertebrates (Sci. Publ. No. 25)*, Freshwater Biol. Assoc., Ferry House, 1977.

104 G.V. Glass, P.D. Peckham and J.R. Sanders, *Rev. Educ. Res.*, 42 (1972) 237-288.

105 P. Greig-Smith, *Quantitative Plant Ecology (Studies in Ecology, Vol. 9)*, Blackwell, Oxford, 3rd ed., 1982, p. 359.

106 R.J. Livingston, in Cairns, Jr., K.L. Dickson and G.F. Westlake (Editors), *Biological Monitoring of Water and Effluent Quality*, ASTM (Amer. Soc. Testing Materials) STP 607, 1977, pp. 212-234.

107 O.P. Patil, *Statistical Ecology Series, Vol. 1- 13.* International Cooperative Publishing House, Fairland, MD, 1971-1979.

108 E.C. Pielou, *Population and Community Ecology*, Gordon and Breach, New York, 1974.

109 E.C. Pielou, *Mathematical Ecology*, Wiley, New York, 2nd ed., 1977.

110 E.C. Pielou, *The Interpretation of Ecological Data: A Primer on Classification and Ordination*, Wiley, New York, 1984, p. 263.

111 R.B. Platt and J.F. Griffiths, *Environmental Measurement and Interpretation*, Robert E. Krieger Publ., Huntington, NY, 1972.

112 D. Raj, *The Design of Sample Surveys*, McGraw-Hill, New York, 1972.

113 D. Rasch, *Biometrie. Einführung in die Biostatistik*, Verlag Harri Deutsch, Frankfurt/Mai, 1983.

114 M.R. Samford, *An Introduction to Sampling Theory*, Oliver and Boyd, Edinburgh, 1962.

115 G.W. Snedecor and W.G. Cochran, *Statistical Methods*, Iowa State University Press, Ames, IA, 6th ed., 1967.

116 R.R. Sokal and F.J. Rohlf, *Biometry*, Freeman, San Francisco, CA, 1969.

117 R.R. Sokal and F.J. Rohlf, *Introduction to Biostatistics*, Freeman, San Francisco, CA, 1973.

118 J.B. States, P.T. Haug, T.G. Shoemaker, L.W. Reed and E.B. Reed, *A Systems Approach to Ecological Baseline Studies*, U.S. Department of Interior Report FWS/OBS-78/21, 1978.

119 R.G.D. Steel and J.H. Torrie, *Principles and Procedures of Statistics with Special Reference to the Biological Sciences*, McGraw-Hill, New York, 1960.

120 S.R. Koirtyohann and H.C. Hopps, *Fed. Proceedings*, 40 (1981) 2143-2148.

121 J.M.J. Versieck and A.B.H. Speecke, *Nuclear Activation Techniques in the Life Sciences*, IAEA: SM 157/24, 1972, pp. 39-49.

122 J. Holm, *Fleischwirtsch.*, 62 (3) (1982) 1- 6.

123 J. Holm, *Fleischwirtsch.*, 64 (1984) 613- 619.

124 W. Schinner, *Untersuchungen über endogene und exogene Einflüsse auf den Blei-(Pb) und Cadmium-(Cd) Gehalt in Muskeln und Organen von Rehwild (Capreolus capreolus L.) und Wildkaninchen (Lepus cuniculus L.).*, Diss. Fachb. Veterinärmedizin und Tierzucht, Universität Giessen, 1981.

125 P. Mueller and G. Wagner, *Probenahme und genetische Vergleichbarkeit (Probendefinition) von repräsentativen Umweltproben in Rahmen des Umweltprobenbank-Pilotprojektes*, Forschungsbericht, Bundesministerium für Forschung und Technologie, Bonn, 1984.

126 C. Pries, W.C. De Kock and J.M. Marquenie, in R.A. Lewis, N. Stein and C.W. Lewis (Editors), *Environmental Specimen Banking and Monitoring as Related to Banking*, Martinus Nijhoff, The Hague, 1984, pp. 88-94.

127 F.H. Kemper and N.P. Luepke, in R.A. Lewis, N. Stein and C.W. Lewis (Editors), *Environmental Specimen Banking and Monitoring as Related to Banking*, Martinus Nijhoff, The Hague, 1984.

128 J.R. Skalski and J.M. Thomas, *Field Sampling Designs and Compositing Schemes for Cost-effective Detection of Spills and Migration*, Pacific Northwest Laboratory, Richland, WA, Report PNL-4935, 1984.

129 O. Fraenzle and G. Kuhnt, *Regional repräsentative Auswahl der Böden für eine Umweltprobenbank - Exemplarische Untersuchung am Beispiel der Bundesrepublik Deutschland*, Forschungsbericht 106 05 028, Umweltforschungsplan des Bundesministers des Inneren, 1983.

130 B.S. Ausmus and E.G. O'Neill, *Soil Biol. Biochem.*, 10 (1978) 425-429.

131 D.R. Jackson, B.S. Ausmus and M. Levine, *Water Air Soil Pollut.*, 11 (1979) 13-21.

132 D.R. Jackson and M. Levine, *Water Air Soil Pollut.*, 11 (1979) 3-12.

133 J.P. Giesy and J.G. Wiener, *Trans. Am. Fish Soc.*, 106 (1977) 393-403.

134 J. Holm, *Erkennung und Beurteilung von flächenhaften Schwermetall- und Pestizidkontaminationen beim Wild*, Abschlussbericht. Staatliches Veterinäruntersuchungsamt Braunschweig, 1982.

135 R.K. Murton, in N.P. Luepke (Editor), *Monitoring Environmental Materials and Specimen Banking*, Martinus Nijhoff, The Hague, 1979, pp. 450-469.

136 D.E. Erdmann, *Role of Reference Samples in the Selection of a Private Laboratory to Analyze Water Samples*, WRD Bulletin Oct.-Dec. 1976 - Jan.-May 1977, pp. 21-23.

137 T.E. Gills and H.L. Rook, in N.P. Luepke (Editor), *Monitoring Environmental Materials and Specimen Banking*, Martinus Nijhoff, The Hague, 1979.

138 G. Goldstein, in A. Berlin, A.H. Wolff and Y. Hasegawa (Editors), *The Use of Biological Specimens for the Assessment of Human Exposure to Environmental Pollutants*, Martinus Nijhoff, The Hague, 1979.

139 A. Berlin, A.H. Wolff and Y. Hasegawa (Editors), *The Use of Biological Specimens for the Assessment of Human Exposure to Environmental Pollutants*, Martinus Nijhoff, The Hague, 1979.

140 H. Egan, in N.P. Luepke (Editor), *Monitoring Environmental Materials and Specimen Banking*, Martinus Nijhoff, The Hague, 1979, pp. 230-246.

141 A.V. Holden, in N.P. Luepke (Editor), *Monitoring Environmental Materials and Specimen Banking*, Martinus Nijhoff, The Hague, 1979, pp. 320-341.

142 E.J. Maienthal, in T.E. Gills, H.L. Rook and R.A. Durst (Editors), *The National Environmental Specimen Bank Research Program*, EPA Report 600/1-79-017, 1979, pp. 1-13.

143 D.T.E. Hunt and A.L. Wilson, *The Chemical Analysis of Water*, The Royal Society of Chemistry, London, 2nd ed., 1986.

144 Y. Lundell, *Plant Soil*, 98 (1987) 363-375.

145 H.W. Nuernberg, in R.A. Lewis, N. Stein and C.W. Lewis (Editors), *Environmental Specimen Banking and Monitoring as Related to Banking*, Martinus Nijhoff, The Hague, 1984, pp. 95-107.

146 M. Olsson, in S.A. Wise and R. Zeisler (Editors), *International Review of Environmental Specimen Banking*, National Bureau of Standards, Washington, DC, 1985, pp. 26-33 .

147 D.L. Britt and A.R. McFarland, *Evaluation or Methods Used in the Long-Term Storage of Particulate Matter*, General Research Corporation, McLean, VA, Report No. 1353-01-82-CR, 1982.

148 U.S. National Bureau of Standards, cited in 147.

149 B.E. Richter, 1981, cited in 147.

150 R. Talcoot and E. Wei, 1977, cited in 147.

151 C.S. Klusek and M. Heit, *Bull. Environ. Contam. Toxicol.*, 28 (1982) 202-207.

152 C.J. Stafford and W.H. Stickel, *Formalin Preservation of Avian Blood for Organochlorine Analysis (Special Technical Publication*, No. 757) Am. Soc. for Testing and Materials, Philadelphia, PA, 1982.

153 W.H. Stickel, L.F. Stickel, R.A. Dyrland and D.L. Hughes, *Environ. Monit. Assess.*, 4 (1984) 113-118.

154 R.T. Barber, A. Vijayakumar and F.A. Cross, *Science*, 178 (1972) 636-639.

155 G.E. Miller, P.M. Grant, R. Kishore, F.J. Steinkruger, F.S. Rowland and V.P. Guinn, *Science*, 184 (1972) 1121-1122.

156 R.H. Gibbs, E. Jarosewich and H.L. Windom, *Science*, 184 (1974) 475-477.

M. Stoeppler (Editor)/*Hazardous Metals in the Environment*
© 1992 Elsevier Science Publishers B.V. All rights reserved

Chapter 4

Wine – an enological specimen bank

Heinz R. Eschnauer

Material Science Institute, Technical University of Aachen, Templergraben 55, D-W-5100 Aachen (Germany)

and

Markus Stoeppler

Institut 4, Angewandte Physikalische Chemie, Forschungszentrum Jülich GmbH, Postfach 1913, D-W-5170 Jülich (Germany)

CONTENTS

4.1. Introduction 49
4.2. Enological specimen bank 50
4.3. Historical enological specimen bank .50
4.4. Results of the historical enological specimen bank 53
      4.4.1. Cadmium 53
      4.4.2. Cobalt 56
      4.4.3. Copper 57
      4.4.4. Lead 58
      4.4.5. Mercury 64
      4.4.6. Nickel 67
      4.4.7. Zinc 67
4.5. Some remarks on arsenic 70
References 70

4.1. INTRODUCTION

On the basis of the *'definitionem vini'*, mineral components in wine are present in concentrations up to 10^3 mg/l, trace elements are around and below 1 mg/l, and ultra-traces around and below 1 μg/l.

The sum of mineral constituents (ash contents) in one liter of wine does not usually exceed 5 g. The total content of trace elements is below 50 mg (i.e. 1% of mineral constituents), and the total content of ultratrace elements is about 0.05 μg (i.e. 0.001% of mineral constituents).

Up to now eight mineral constituents, twenty-four trace elements, and approximately twenty ultratrace elements have been quantified in wine. Additionally, fourteen rare earths as well as some natural and artificial radioisotopes have been found.

4.2. ENOLOGICAL SPECIMEN BANK

Properly cultivated, correctly bottled and sealed natural wines constitute a unique environmental specimen bank, in this case an enological specimen bank. Since wine, unlike virtually any other foodstuff, is perfectly suited for long-term storage, it is a material from the past for the present and future which permits, for example, the determination and evaluation of selected trace metals over time. From these determinations, statements about trends over time, which are mainly about anthropogenic influences on wine production, are possible [1]. Based on this idea, it was possible to set threshold values relatively early, according to the wine directive for unwanted trace elements. To our knowledge, these values are among the first, or indeed probably are the first, threshold values of this kind introduced for food products.

The purity standards for wine are very old, possibly somewhat older and more effective than those frequently mentioned in Germany for beer. In this chapter this will be demonstrated for some trace metals.

4.3. HISTORICAL ENOLOGICAL SPECIMEN BANK

The total content of a trace or ultratrace element in commercial wines consists of a primary plus a secondary content [2-4].

The primary content of a trace or ultratrace element in wine is the natural content. It derives from vineyard soil, reaches the grapes via the roots, and represents the characteristic and major part of the total content. Frequently, the primary content varies widely because of significant geological differences between vineyards. A correct estimation of the primary content is quite difficult and thus has rarely been performed.

The secondary content of a trace or ultratrace element in wine is a contamination of geological or man-made (anthropogenic) origin. The genealogical tree for the various causes of this type of contamination is shown in Fig. 4.1. The term 'enological contamination' includes the influencing factors from viticulture (fertilizers, pesticides) and wine production (harvesting, treatment, storage, and shipping). These influences will be demonstrated in detail for the particular elements treated here.

Typical normal contents of inorganic constituents in contemporary wines, i.e. mineral constituents, trace and ultratrace elements, are listed in Table 4.1. Four of the eight mineral constituents are cations (potassium, magnesium, calcium, sodium) and four elements (carbon, phosphorus, sulfur, chlorine) are present in the anions of acids, i.e. as salts. The concentrations of trace elements are in the wide range of 6 mg/l to 0.001 mg/l. They can be divided into the following regions: 10 to 1 mg/l, 1 to 0.1 mg/l, 0.1 to 0.01 mg/l, and 0.01 to 0.001 mg/l. The contents of ultratrace elements are below 0.001 mg/l. They can be quantified down to the lowest detection limits of modern analytical methods. In addition, in three wines from Ingelheim (Germany) from the 1980 vintage the fourteen rare earths were also determined by radiochemical neutron activation analysis (RCNAA) [5]. The radioactive isotopes will not be treated here.

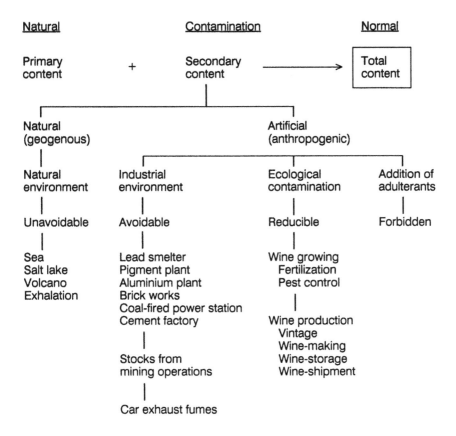

Fig. 4.1. Secondary content of trace elements in wine (contamination) [3].

Based on the stocktaking of inorganic constituents in contemporary wines, i.e. the enological specimen bank of today, the data from historical wines will be discussed from the standpoint of a historical enological specimen bank.

Due to fortunate circumstances (thanks to H. Rodenstock for the old wine samples) some series of old wines were made available and have so far been analyzed for seven elements. They were six wines from the 18th century, twelwe wines from the 19th century, and twenty-six wines from the first half of the 20th century. These wines were mainly of French origin, predominantly from the Bordeaux area. Only six old German wines have been included, since for different reasons these were not easily available. Of particular interest in these wines were the trace metals cadmium, cobalt, copper, lead, mercury, nickel and zinc. Data for arsenic, determined in a selection of old and numerous contemporary wines, will be discussed separately. These metals, and the metalloid arsenic, were determined using the same modern analytical methods which have been used to quantify trace element levels in many contemporary wines [6,7]. Thus, the results and the methodologies applied (in the case of the wines discussed here, different modes of

TABLE 4.1.

INORGANIC CONSTITUENTS (mg/l) IN WINE (TYPICAL NORMAL CONTENTS)

Mineral constituents	Trace elements				Ultratrace elements
1000-10	10-1	1-0.1	0.1-0.01	0.01-0.001	≤0.001
K 370 - 1120	B 5 - 2	Al 0.9 - 0.5	As 0.02 - 0.003	Co 0.02 - 0.001	Sb 0.006
Mg 60 - 140	Fe 1	F 0.5 - 0.05	Ba 0.3 - 0.04	Mo 0.01 - 0.001	Be 0.00008
Ca 70 - 140	Cu 0.5	I 0.6 - 0.1	Pb 0.1 - 0.03	Ag 0.02 - 0.005	Cd 0.001
Na 7 - 15	Mn 5 - 1.5	Rb 4.2 - 0.2	Br 0.7 - 0.01		Cs 0.0027
C 100 - 120	Si 6 - 1.5	Sr 3.5 - 0.2	Cr 0.06 - 0.03		Au 0.00006
P 130 - 230	Zn 3.5 - 0.5	Ti 0.3 - 0.04	Li 0.2 - 0.01		Hf 0.0007
S 5 - 10			Ni 0.05 - 0.03		Nb 0.001
Cl 20 - 80			V 0.26 - 0.06		Hg 0.00005
			Sn 0.7 - 0.01		Se 0.0006
					Ta 0.0005
					Tl 0.001
					Bi 0.00015
					W 0.003
					Rare earths
					Radioactive elements

atomic absorption spectrometry and voltammetry) are perfectly comparable. This is important insofar as the results of trace metal determinations, mainly at ultratrace levels, in previous papers may often be unreliable due to inadequate methodology. In addition it should be noted that wines of this age have never been investigated systematically before. Nevertheless, it should be mentioned that some uncertainties remain in assessing the results obtained, in view of the history of individual bottles over long periods of time, including their filling, the use of new cork stoppers, and the storage conditions which might also include transfers between the wine and the bottle itself. As far as trace metal contents are concerned, the results obtained are still valid.

4.4. RESULTS OF THE HISTORICAL ENOLOGICAL SPECIMEN BANK

The results of the studies referred to, except those for arsenic [8,9], are summarized in Table 4.2 for the above-mentioned wines from the 18th to the first half of the 20th century.

Evaluation of these data leads to Table 4.3, in which the metal contents of these wines are compared. Further, the results are associated with the results from wines from two distinct vineyards. For comparison, the average metal contents of contemporary wines, and the maximum permissible levels according to the German 'Weinverordnung' (wine ordinance) at present in force, are given as well. These results will be discussed in detail for single elements in the subsequent sections.

4.4.1. Cadmium

It is quite satisfactory that cadmium contents in German and foreign wines from the early 50's show a constant tendency towards lower values. In the first half of this century an average cadmium content of 0.003 mg/l appeared to be normal, whereas these levels dropped in more recent wines to around 0.001 mg/l, and often below. Thus, according to the 'definitionem vini', cadmium can be considered as an ultratrace element. The reasons for this decrease, and the current very low contents, are the extreme purity demands for wine-treatment chemicals and wine production equipment. Also, fertilizers and pesticides usually contain very little cadmium today. This precludes enological contamination of present commercial wines.

For several trace and ultratrace elements, such as toxic metals like cadmium, the vine root acts as a natural filter and prevents transfer of the metal from soil to vine. In addition the skin of grapes is very smooth and is covered by a waxy film so that no significant pollution from dust etc. can adhere to it, or it is easily removed by rain. Thus, the vine (*Vitis vinifera*) is practically free from environmental pollution not only by cadmium but also by other toxic trace metals, so that today one can nearly completely rule out secondary cadmium contents as well.

What is the significance of cadmium analyses in historical wines? Simply, and as may be expected, it can be said that cadmium contents are age-dependent. Wines from the first half of our century contain on average 0.0019 mg/l, wines from the 19th century 0.0039 mg/l, and those from the 18th century as much as 0.005 mg/l. There are seven

References on p. 70

TABLE 4.2.

TRACE ELEMENTS IN HISTORICAL WINES – ENOLOGICAL SPECIMEN BANK

Vintage	Content (µg/l)						
	Cd	Pb	Co	Cu	Ni	Hg	Zn
18th century							
1747	4.42	3460	8.22	260	34.7	0.08	1010
1748	3.67	4540	6.14	74	13.1	0.01	630
1750	7.00	5290	10.6	183	70.6	0.14	1220
1784	4.84	280	5.85	612	76.1	0.03	1060
1787	4.50	2280	6.30	338	57.4	0.04	2000
1787	5.58	243	3.77	359	12.9	0.30	362
n: 6 X̄:	5.00	2680	6.81	304	44.1	0.10	1047
19th century							
1811	9.09	3550	17.3	292	49.7	0.07	560
1847	5.83	11800	7.77	240	66.7	0.12	1180
1848	1.75	430	4.11	2	28.5	0.08	–
1858	10.2	5620	8.65	256	54.6	0.09	760
1865	1.96	490	7.60	63	65.0	0.30	–
1866	0.2	23100	77	1370	125	–	6400
1868	4.85	200	7.47	137	39.9	0.25	–
1869	1.86	3130	8.25	289	63.2	0.04	–
1884	0.86	2075	9.06	236	51.5	0.20	–
1890	3.53	1070	11.4	1	82.9	–	3830
1893	5.15	210	4.99	450	24.3	0.10	995
1895	1.30	180	3.83	3	13.4	0.07	–
n: 12 X̄:	3.88	4320	14.0	278	55.4	0.13	2290

20th century (1st half)

Year							
1900	0.83	190	7.12	42	19.5	—	2180
1908	1.13	2250	7.34	1530	41.5	0.22	920
1911	0.82	520	5.67	834	42.2	0.32	940
1917	1.48	150	10.9	107	44.2	—	1590
1921	3.69	2551	9.92	1005	65.0	—	—
1921	1.84	542	18.9	8950	66.2	0.55	—
1921	0.78	1350	6.06	322	19.8	0.09	—
1923	0.87	105	5.86	64	29.2	—	930
1924	1.10	155	8.84	427	80.5	—	2760
1926	1.01	850	8.35	395	44.2	0.05	750
1928	4.70	2605	12.1	643	71.7	0.05	—
1929	1.94	189	2.87	7	84.9	0.15	—
1929	5.58	242	6.94	2210	96.0	0.10	—
1934	0.83	970	5.1	240	48.0	—	1300
1934	1.24	136	5.50	126	23.7	0.08	—
1935	4.55	670	3.90	218	25.4	0.10	—
1937	0.90	451	9.70	160	66.8	0.05	699
1944	2.13	260	9.36	358	44.7	—	3390
1945	1.26	179	3.23	9	36.9	0.06	—
1946	1.03	390	8.89	682	40.0	—	1930
1947	2.28	240	10.2	682	36.0	—	1330
1947	1.79	65	10.7	225	143.0	0.05	—
1947	2.32	133	8.45	1848	39.1	0.18	—
1948	1.43	185	7.18	1498	65.9	0.53	—
1948	1.03	160	7.49	555	60.3	—	3500
1950	1.65	360	5.48	52	38.6	—	1150
n: 26 $\bar{x}$:	1.85	611	7.92	891	52.8	0.17	1667

References on p. 70

TABLE 4.3.

HISTORICAL ENOLOGICAL SPECIMEN BANK – SUMMARY

Century	Number	Trace element content (μg/l)								
		Cd			Pb			Co		
		Min.	Max.	$\bar{X}$	Min.	Max.	$\bar{X}$	Min.	Max.	$\bar{X}$
18th	6	3.67	7.00	5.00	243	5290	2680	3.77	10.6	6.81
19th	12	0.2	10.2	3.88	180	23100	4320	3.83	77	13.9
20th, 1st half	26	0.78	5.58	1.85	65	2605	611	2.87	18.9	7.92
Winery										
I	10	0.86	10.2	5.04	280	11800	3800	5.85	18.9	10.1
II	12	0.82	2.28	1.28	105	2250	470	548	10.9	7.96
$\bar{X}$ present wines		1.0			30-100			1-10		
present threshold (wine directive)		10			300					

wines with high contents above 0.005 mg/l, one of them containing more than 0.01 mg/l. This is inconceivable for contemporary wines.

When cadmium in wines from two wineries is evaluated, the typical trend with time, i.e. high levels in old wines, is obvious. This confirms that these levels were specific for their time, but not for distinct wine producers. However, it is interesting to note that there are mineral waters with much higher cadmium contents. For example, a mineral water from Ein Nun, Israel, contains around 1 mg/l cadmium [10].

4.4.2. Cobalt

The levels of cobalt in historical wines do not show unexpected figures. Cobalt in wines from the 18th century ranges from 0.0038 to 0.0106 mg/l, with an average 0.0068 mg/l, the values for the 19th century are 0.0038 to 0.077 mg/l, with average 0.014 mg/l. The respective data for the first half of this century are: range 0.0029 to 0.019 mg/l and an average of approx. 0.008 mg/l. These data do not show any trend, which is to be expected. The cobalt levels of old wines from two French wineries are, with averages of approx. 0.008 and 0.010 mg/l, within the expected limits. Out of the 44 old wines analyzed, nine contained more than 0.01 mg/l, only two of them 0.0173 and 0.0189 mg/l and one 0.08 mg/l. All these contents are well in accordance with those in contemporary wines, ranging from 0.001 to 0.02 mg/l cobalt in wine.

Vitamin B12 contains 4.35% cobalt and is thus an essential trace element. A glass of wine contributes significantly to the supply of nutritional requirements, possibly better than a multivitamin/multimineral preparation. Higher than average secondary cobalt contents in

Cu			Ni			Hg			Zn		
Min.	Max.	$\bar{x}$	Min.	Max.	$\bar{x}$	Min.	Max.	$\bar{x}$	Min.	Max.	$\bar{x}$
74	612	304	12.9	76.1	44.1	0.01	0.30	0.10	362	2000	1047
1	1370	278	13.4	125	55.4	0.04	0.25	0.13	560	6400	2290
7	8950	891	19.5	143.0	52.8	0.05	0.55	0.17	699	3500	1667
183	8950	1166	34.7	76.1	59.1	0.03	0.55	0.14	560	2000	1113
42	1530	477	19.5	80.5	43.7	0.05	0.32	0.16	750	3500	1790
	500		30-50			0.02-0.07				2000	
	5000									5000	

wine can occur after long-term storage in wine bottles made from blue glass that contains 0.02 to 0.05 weight percent cobalt. For example, the cobalt content in four 'old Burgundy' red wines of Ober-Ingelheim (Rheinhessen) that had been stored for twenty-five years in those blue bottles was on average 0.025 mg/l. In contrast, the content in the same wine stored in green bottles was on average only 0.010 mg/l. This indicates a migration of cobalt from the wine bottle into the liquid. A cobalt enrichment in the residue of these old red wines was not found. A level of 0.039 mg/l of cobalt was found in an old sherry. It was not possible to discover whether this might be due to storage in a blue glass bottle.

4.4.3. Copper

After turbidities due to iron, those from copper are the most frequent, and result in shortcomings in wine. High copper contents in wine could be caused by corrosion of copper vessels, even if these are tinned, or of alloys (brass or bronze) that are part of the wine processing equipment. Pesticides containing copper were previously applied to combat *Peronospora*. In dry years significant amounts of these pesticides remained on the grapes and were thus able to enter the wine via the must. Very high copper contents, sometimes up to 8.5 mg/l, are most probably due to the application of copper-containing pesticides previously in general use.

The analyzed wines were predominantly red Bordeaux wines, from which the high copper content had not been removed by a special treatment ('blue fining'). The origin of the extremely low contents, at the levels of a few microgram/liter, found in some wines, however, is still unknown since the history of these wines is not exactly known. Significant

References on p. 70

differences between samples from two wineries, with copper averages of 0.48 mg/l and 1.17 mg/l, are probably due to the increased use of vessels and tools made from copper and copper alloys in the second winery. The analyzed wines from the 18th and 19th century contain on average 0.30 and 0.28 mg/l, respectively, of copper, which is significantly lower than the wines from the first half of this century – whose average is 0.89 mg/l (see Table 4.2). This is probably caused by the already mentioned use of copper-containing pesticides.

4.4.4. Lead

Lead, lat. *plumbum*, is a metal *'sui generis enologicum'*. This metal was by far the most used material in classical antiquity, and it is appropriate to refer to the 'Roman Lead Age'. Lead was to the Romans what 'plastic' is to us. Therefore the Roman writer Marcus Varro was able to ask 'Do you not possess lead?' Lead was the raw material for various vessels, water pipes, tools, machinery etc., and in the form of numerous alloys. Thus it is self-evident that this material was also of central importance in a field where Roman culture played the leading role for two millenia, in wine cultivation. Roman lead capsules and lead caps etc., used for sealing small vessels which sometimes contained wine, are well known. So are Roman lead tablets (*'tesserae'*), labels in contemporary language, used to identify earthenware jars containing wine.

Marcus Cato recommended that, in the time before harvest, barrels (*'dolia'*) should be tightened with lead. Furthermore, to repair damaged earthenware barrels for wine storage one should apply an ingenious lead network to prevent them breaking into pieces. 'Roman cement', mixed with litharge (lead(II) oxide) for smoother handling, was used to seal earthenware jars. 'Roman mixtures', also containing lead(II) oxide as a significant ingredient, were used to coat iron hoops against rust and to impregnate the wooden staves of wine barrels to achieve better durability.

A unique and sophisticated procedure for the preparation of a sweet must concentrate ('sweet reserve') was invented by the Romans, but is certainly not worth imitating. To prepare this, the grape must (*'mustum'*) was concentrated in bronze boilers, and later in lead boilers, under exactly prescribed conditions, down to one third (*'carenum'*), half (*'sapa'*) or even two thirds (*'defrutum'*) of the initial volume. This sweet reserve was used for many purposes: for wine (*'vinum'*), for mixed wine (*'vinum confusum'*), double fermented wine (*'vinum recentatum'*), or sugar substitutes. It was also used to pickle fruits, as an ingredient of spicy sauces and cooking recipes, and for many other uses. Columelle and later Plinius recommended using lead vessels for boiling down. The Romans, who were connoisseurs of food and drink, quickly realized that a wine must concentrated from lead boilers was much sweeter and more palatable than that from copper boilers. The Romans also recognized the reason for this, namely the formation of lead salts. These salts can be seen partly as a white crust on the wall of the lead vessel upon concentration. It was only a short step from this observation to the direct addition of lead salt crusts to the must or even to the use of these or similar lead salts as sweeteners. In 1980 the Roman winery "Weilberg" in the vicinity of Bad Dürkheim-Ungstein, Germany, was exca-

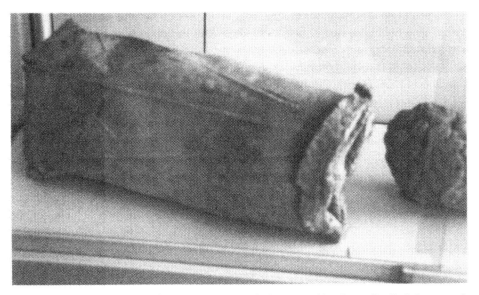

Fig. 4.2. Roman lead vessel for concentration of wine must (thanks to Dr. F. Schumann for the photograph).

vated and a Roman lead vessel found that had been used for the concentration of must as described above. This is one of the few lead vessels so far discovered that side of the Alps. Fig. 4.2 shows such a lead boiler for the concentration of must.

The Roman introduced their wine cultivation and production techniques to Germania superior. Since in the cooler northern regions the wines were frequently quite sour the production and use of the 'Roman sweet reserve' was very common. In the above mentioned lead vessel from Weilberg several hundred grape seeds were found. 'Mustum' was concentrated from grape skins, pulp and seeds. The determination of lead in the seeds from this vessel showed 250 mg/kg of lead. This is, compared to the average lead content of 0.5 mg/kg in grape seeds, a five-hundred-fold enrichment. It is also proof that this lead vessel had been used for the production of the 'Roman sweet reserve' (thanks to Dr. F. Schumann, Neustadt for the seeds and Ms Gemmer-Čolos, University of Mainz, for the determination of lead).

The Romans did not only have a perfect mastery of lead metallurgy for the ancient times, but they were also experienced chemists. Without doubt, no other element known at that time so had many compounds identified and applied for every-day use. The chemistry of lead thus attained a very high level, never previously achieved. Therefore it is obvious that the observations and experience with 'Roman sweet reserve' mentioned above would result in the use of various mineral lead compounds in the refinement of wines, and particularly for sweetening and de-acidifying wines from poor vintages, mainly from northern wine cultivation regions. At that time sugar was as expensive as gold and was – if at all – only imported via the silk road in very small amounts so that its price prohibited its use in wineries.

References on p. 70

The use of mineral lead compounds for wine refinement became an important part of Roman wine production techniques. A series of lead compounds was applied in distinct amounts according to detailed recipes. Initially various differently colored lead monoxides, red yellow gold litharge ('chrysitis'), pale yellow silver litharge ('argyritis') and lead grey litharge ('molybditis') were skilfully prepared. This was accomplished using red lead ('sandix' or 'minium'), i.e. Pb_3O_4 ($PbO \cdot Pb_2O_3$), a sort of 'burned lead white' as a mixture of marl and red lead, and basic lead carbonate, the real 'lead white' ('cerussa'), i.e. $Pb(OH)_2 \cdot 2PbCO_3$. Later the Romans learned to appreciate the properties of lead acetate ('plumbum aceticum') or 'sugar of lead' ($Pb[CH_3COO]_2 \cdot 3H_2O$) as a wine sweetener. There is, indeed, a wide assortment of Roman substances to sweeten sour wines. They remained in constant use from the 'Roman Lead Age' through the Middle Ages and until the beginning of modern times. This is true despite the fact that misuse of lead oxide and other compounds was already mentioned in a document from Esslingen in September 1495. In a book by the 'court physician' E. Gockelius, including many details, a recipe is given for the use of 'silver litharge' for sweetening sour wines with lead, which was already forbidden, in the years 1494, 1495 and 1496.

The colics ('Colica pictonum') occurring in 1572 in Poitou, France, were identified as being due to the treatment of wine with lead oxide. The standard work, 'Oenographia' by M.F. Helbachius, written in 1604, mentions the use of metallic lead for wine sweetening. In 1753 the same process is described in the wine book, 'Der curiose und offenherzige Wein/Arzt' ('The Curious and Outspoken Wine Doctor') [11]. Only in about 1750 did draconian laws succeed in stamping out these adulteration practices by punishing adulterators with the gallows.

For wine control and surveillance in the early days the "Würtemberg wine test" was the first available: then, soon after, the more reliable "Hahnemann wine test" (proof of lead by hydrogen sulphide). The latter was a very early success of wine analysis, which has always been the prerequisite for recognition and prevention of wine adulteration.

Even if the most severe abuses had been abolished due to the treat of draconian punishment, abuses due to the careless application of metallic lead in wine technology were still exposed from time to time. An example of this is a severe illness, called 'Morbus colicus dammoniorum', which occurred in 1767 and was caused by lead-coated cider presses in England. Around the turn of the 19th century lead balls and lead sheets were still put into wine barrels for de-acidification. Other sources of lead were lead slugs, used for bottle cleaning, and lead-rich solders for repairing winery machinery. However, as late as 1976, a French paper drew attention to a common practice in French wineries, of using lead sheets, fixed to the inner surface of damaged wooden staves by copper nails, to tighten the barrel [12].

The examples given from the history of 'enological plumbism' up to present times – a detailed work is in preparation [11] – illustrate the value of the results for lead which is an element 'Sui generis enologica' and thus of particular significance for the historical enological specimen bank.

From the results achieved so far it is obvious that the 'purity law' for lead in wines in the last two centuries was frequently ignored. Of the six wines from the 18th and the

TABLE 4.4.

TRACE ELEMENT VINOGRAM OF LEAD (CONTENT + BEHAVIOR + ORIGIN + DEPLETION)

	Average contents (μg/l)	Number of wines	Remarks
Total lead content			
approx. 1950-1960	300 - 500	9	Ingelheim/Rhein/Germany
approx. 1970-1980	100 - 150	200	mainly Germany white and red
mainly 1979	55	7	new winery, family owned ⎤
mainly 1979	124	13	older winery, family owned ⎬ Ingelheim white and red wines
1980	54	4	new winery, family owned ⎦
Primary lead contents	10 - 30	11	Ingelheim/Rhein/Germany
	63	18	California
Secondary lead contents			
Pesticides	1000		Pb H (AsO_4)
"Tin" covers	500 - 1000		$2PbCO_3 \cdot Pb(OH)_2$, $3PbCO_3 \cdot 2Pb(OH)_2 \cdot H_2O$ possibly others
Car exhaust	500		$Pb(C_6H_5)_4$, $PbClBr$, $xNH_4yPbClBr$
Emissions	500		$2PbCO_3 \cdot Pb(OH)_2$
Fertilizer			
Others			
Lead depletion			
Fermentation	30-65 % (80)		PbS (mainly)
Fining	10-35 % (50)		$Pb_2Fe(CN)_6$ (not stable?)
Turbidity			PbS, $PbSO_4$
Highest allowable lead content	300		
Geochemical abundance	15 ppm		

References on p. 70

twelve from the 19th century, only five had lead contents below the present German maximum allowable content of 0.3 mg/l. However, eleven wines contain more than 1 mg/l. These high levels are also reflected by the averages obtained (Table 4.2). The higher average of the wines from the 19th century is significantly influenced by one wine from 1866 with an extremely high lead level of 23.1 mg/l, and the comparatively small number of wines from the 18th century, and statistically valid conclusions cannot be drawn from the very low number of samples from former centuries.

Out of a number of Italian wines from the 19th and 20th century analyzed for lead, nine wines from the first half of the 19th century contained on average 0.6 mg/l [13]. About the same average (0.61 mg/l), with range 0.065 to 2.605 mg/l, was found for 26 wines of different origin from the first half of the 20th century (Table 4.2). However, there were only four wines above and one around 1 mg/l: the average of the remaining 21 samples is 0.29 mg/l, i.e. slightly below the present German maximum allowable content.

A drastic reduction in lead content can be seen for contemporary wines where, on the whole, only extremely low contents below 0.1 mg/l are common. This reduction is an outstanding example of the significant progress achieved in contemporary wine technology, so that the purity of wine now meets the requirement of offering consumers an optimal product. From the vinogram of lead in Table 4.4 it can be seen that today secondary lead contents can be nearly completely avoided. An exception, however, is

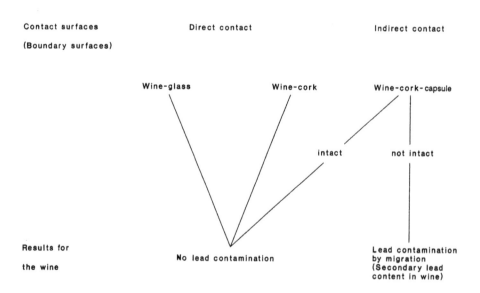

Fig. 4.3. Influence of the properties of boundary surfaces on contamination by lead.

63

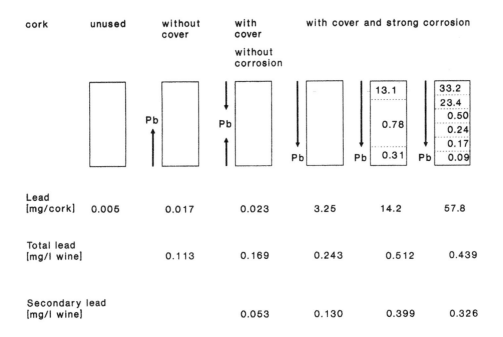

Fig. 4.4. Lead migration through the cork after storage for about 20 years (Alsheimer Silvaner from 1959).

TABLE 4.5.

CHEMICAL COMPOSITION OF METAL FOIL COVERS

Year	Pb (%)	Sn (%)	Cu (mg/kg)	Bi (mg/kg)	As (mg/kg)	Sb (mg/kg)
1790	a	4.1	190	170	100	1380
1811	0.01	a	300	75	250	–
1811	a	1.2	170	420	25	2130
1868	a	8.6	220	140	35	640
1868	a	7.6	50	<40	25	165
1870	a	3.4	20	120	5	100
1871	a	5.7	130	200	25	290
1880	a	4.7	70	220	40	265
1893	a	9.5	540	100	200	1800
1897	a	7.6	280	240	200	1600
1900	a	6.8	90	<40	120	250
1905	a	0.1	120	<100	10	30
1950	a	2.0	34	2900		1200
1975	a	1.0	3700	200		3900

a Main component (i.e. > 98%).

References on p. 70

contamination from metal foil covers for the decoration of wine bottles. If the system wine-cork-cover is not perfect (Fig. 4.3), secondary lead contents due to migration into the cork and wine cannot be excluded, as is shown by the examples in Fig. 4.4. These wines can attain lead levels above the present maximum allowable content. The different chemical composition of the alloy of metal foil covers for wines from 1790-1975 (Table 4.5) displays a continuous increase in the percentage of lead since 1811, when pure tin was used. Present foil covers contain nearly 100% lead. It is therefore apparent that in the event of corrosion there will be migration and uptake of lead by the wine.

4.4.5. Mercury

Mercury is a metal with unique properties and many industrial applications. The silvery sparkling metal is liquid at room temperature, has a high specific weight, and belongs to the toxic heavy metals.

The ultratrace element mercury undergoes to some extent an environmental circulation, promoted by its volatility. Extremely small amounts of the metal are volatilized into the atmosphere and return partly oxidized via wet and dry precipitation. The result is a ubiquitous distribution into all environmental categories [14]. Examples of anthropogenic influences are, on the one hand the increase of mercury from 1949 to 1966 in the feathers of goshawks, which prey on smaller birds that feed on grain. This resulted from the use of organomercurials as pesticides for the impregnation of seed corn, which was forbidden in 1966. On the other hand, from the turn of the century there may be noted a steady increase of mercury in the feathers of birds of prey, for example of fish hawks that feed on estuarine fish. The cause is mercury pollution of major rivers by electrolysis plants [15].

TABLE 4.6.

MERCURY CONTENT OF VINE PARTS AT TWO DIFFERENT MÜLLER-THURGAU VINE-YARDS FROM VEGETATION PERIODS 1975-82

Contents: soil (μg/kg dry weight), liquid (μg/l)

	Winery Durbach (granite soil)		Experimental winery Blankenhornsberg Ihringen (Loess clay)	
Hg content of soil	50		70	
Hg content in wine parts of different experimental variations	Leaves	Must	Leaves	Must
Control	27	0.11	252	1.63
Compost fertilizing	29	0.09	263	1.63
MK-fertilizing (75 t/ha/a)	28	0.10	253	1.54

Typical Hg contents: Basalt (Kaiserstuhl): 190; vineyard soils (Hessen): Average 90, n = 300 samples.

The mercury contents in some grape musts may be mainly due to natural, i.e. geogenic, origin. Depending on the growing area, vine leaves and grapes show distinct, but always very low, mercury contents at the μg/l level. It was shown, however, that the mercury content of grape must from a Müller-Thurgau vineyard (Kaiserstuhl) at an average of about 1.6 μg/l exceeded the values, common in other areas, of around 0.1 μg/l (Table 4.6). The reason for these Kaiserstuhl values lie in its volcanic geology with a significant mercury volatilization from soil.

Since small amounts of hydrogen sulfide originate during wine fermentation, mercury is transformed into the extremely insoluble mercury sulfide. This and some other sulfides are then nearly quantitatively coprecipitated with yeast and other turbid materials. In the final commercial product, therefore, only primary mercury contents remain, at levels of the same order as the determination limits of very sensitive modern analytical methods such as cold vapor AAS with preconcentration on gold traps [16].

Mercury, with its extremely low primary content in wine, is thus an ultratrace element and provides a good example of the value of mercury determinations in historical samples for providing information about secondary pollution from technology and the environment. Starting with wines from the 18th century, it was possible to determine the mercury content by modern analytical methods [6]. The results provide some basic information about the history of industrial production and the application of mercury and its compounds. Due to the circulation of mercury consequent on its extreme volatility, it was possible to recognize pollution and therefore an increase in the secondary mercury content in wines.

Out of the historical enological data bank six wines from the 18th century analyzed for mercury contained 0.01 to 0.30 μg/l, with an average of 0.1 μg/l: twelwe wines from the 19th century contained 0.04 to 0.25 μg/l, with an average of 0.3 μg/l, and twenty-six wines from the first half of the 20th century ranged from 0.05 to 0.55 μg/l with an average of 0.17 μg/l. The secondary mercury content shows a distinct elevation, probably due to the increased use of mercury during the industrial age. The highest mercury contents found in two wines from the first half of the 20th century amount to 0.53 and 0.55. In one sample from 1961 a mercury content of 2 μg/l was found (Table 4.7). These comparatively high secondary mercury levels cannot easily be explained. There could be some influence from bottle enamels containing cinnabar in the 19th and the outset of the 20th century. The less pure alloys used for winery machinery, and broken thermometers containing mercury could have caused some comparatively high secondary contents.

From approximately the middle of the 20th century there has been a significant reduction in the mercury levels in wine to today's value of 0.02 to 0.07 μg/l [16]. The reason for this is most probably the reduction in the industrial application of mercury and its compounds from the 70's onwards, as well as the extreme purity demands for all equipment used in wine production.

Mercury, like lead, is obviously enriched in the wine deposits (residues). For example, the deposits of two red Burgundy wines from 1747 and 1787 contained 1.0 μg and 0.3 μg mercury, compared to the mercury contents in the liquid phase of 0.08 μg/l and 0.04 μg/l

References on p. 70

TABLE 4.7.

VINOGRAM OF THE ULTRATRACE ELEMENT MERCURY

Year	Content (μg/l)	Average (μg/l)	Number of samples	Remarks
	Total mercury			
1980	<0.02 - 2.2	0.76	7	German grape must
1934	0.4 and 1.0	–	2	German white wines
1950-1961	<0.02-0.09	0.03	8	older wines from Ingelheim
1961	2.0	–	1	Ober-Ingelheimer Silvaner wine
1975-1979	<0.02 - 0.14	0.07	8	French white and red bottled wines
1976-1980	<0.02 - 0.25	0.05	10	younger wines from Ingelheim
1981	≤0.02 - 0.15	0.07	5	young wines from Ingelheim
1974-1980	<0.02 - 0.04	0.02	23	German white and red bottled wines
	Primary mercury			
	≤0.02 - (0.05)			
	Secondary mercury (contamination)			
1974	100 - 1700	670	12	Spanish wines (Hg-Fungicide)
1976	4 - 5	–	ca. 10	French wines
1979	3 - 8	5	3	Spanish barrel wines
	2	–	3	Spanish bottled wines
Typical secondary mercury contents	0.5 - (1) and significantly higher			
Depletion of mercury due to fermentation	80 - 100%			As HgS, Solubility product: 3 × 10^{-54}

mercury. All relevant results and recent knowledge about mercury are shown in the vinogram for the ultratrace element mercury (Table 4.7).

4.4.6. Nickel

The vinogram for nickel (Table 4.8) summarizes the most important data so far available for wine. First of all, the vinogram informs us about the total nickel content in wine, i.e. the average nickel content in contemporary commercial wines. The results indicate a slight, but not significant, increase in the average nickel contents in wine since the introduction of stainless (chromium-nickel) steels in wineries. However, some comparatively high nickel values found in old wines, sometimes up to around 0.14 mg/l (see also Table 4.2), may indicate nickel migration from the wine bottles, since these used to contain variable contents of nickel oxide [17]. Since, however, stainless (chromium-nickel) steels were not used in earlier periods, they cannot be the cause.

The reader's attention is also drawn to the vinogram of chromium (Table 4.9). It shows a significant increase in total chromium levels due to the introduction of stainless (chromium-nickel) steels in wineries. There was also a considerable migration of chromium from bottles of old French red and white wines, observed to reach up to 12.5 mg/l chromium. However, whether the yellow bottle varnishes containing zinc chromate, which were sometimes used on the outside of old wine bottles, contribute to the chromium contents has not yet been proven with any certainty. Regarding the vinograms of nickel and chromium, see also reference [18]. The element chromium was not considered further in the investigations of the historical enological specimen bank.

It was not possible to explain the surprisingly elevated nickel contents in two old sweet wines containing 0.1 and 0.12 mg/l, or the 0.094 mg/l in an old sherry; correspondingly elevated chromium contents were not observed in these cases.

4.4.7. Zinc

Trace elements like lead, or ultratrace elements like cadmium and mercury, belong to the so-called toxic metals and thus minimal contents in wine are permitted. Zinc acts differently in that it is a wine-sensitive metal which causes turbidities if the content exceeds distinct levels. To avoid this, a zinc threshold of 5 mg/l has now been set. This does not present serious problems since zinc-plated tools, vessels, and machinery parts in former use have been replaced by mainly stainless (chromium-nickel) alloys. Therefore the average content of zinc in contemporary wines is around 2 mg/l. In several of the investigated wines even lower zinc levels were found. As examples, an average of 1.05 mg/l was found in six wines from the 18th century, 2.3 mg/l in six wines from the 19th century, and 1.7 mg/l in fourteen wines from the first half of the 20th century. This can be only explained by the fact that these wines had almost exclusive contact with wooden surfaces and only minimal contact with materials that were zinc-plated or made from alloys containing zinc. This was also confirmed by comparatively low winery-specific zinc contents in wines from two wineries, with 1.1 mg/l ($n = 10$) and 1.8 mg/l ($n = 12$) respectively.

References on p. 70

TABLE 4.8.

TRACE ELEMENT VINOGRAM OF NICKEL

	Min.	Max.	$\bar{x}$	n	Remarks
Total nickel content (μg/l)	0.6	1.1	40	13	1947-1962 mainly Ober-Ingelheim
	9	170		153	Data from literature, with strong fluctuations
	10	500	54	158	1970-1977 young commercial wines
Primary content (μg/l)	1	3			
Secondary content (μg/l)					
Contact corrosion	154	622	437	3	'Sweet reserve', materials No. 1.4301/2.0945
Glass bottle-migration	25	125	58	78	1892-1976 French white and red wines
Depletion (%)					
Fermentation			55		No detailed information
Fining		40			No detailed information
Tap water (μg/l)			50		Threshold value
Food products (mg/kg)		0.01	2		Natural contents
Geochemical abundance (mg/kg)			44		
Vineyard soils (mg/kg)			23	17	Ingelheim, Germany
			27	300	Hessen, Germany

TABLE 4.9

TRACE ELEMENT VINOGRAM OF CHROMIUM

	Min.	Max.	$\bar{X}$	n	Remarks
Total chromium content (μg/l)	2.3	14.4	7	9	1947-1961 mainly Ober-Ingelheim, Germany
	10	30	17	6	1939-1943 mainly German wines
	10	152	61	68	1962-1981 mainly German wines
	1	450	71	88	1970-1977 young commercial wines
Primary content (μg/l)	1	25			
Secondary content (μg/l)					
Pin point corrosion			910		Experiments material 1.4301
Welding seam corrosion			520		Experiments material 1.4301
(Kieselguhr-)abrasion		109 and 290	2		Wine pump
Glass bottle-migration		12500	78		1892-1976 French white and red wines
Asbestos filter-migration	69	151	100	4	winery experiments
Depletion (%)					
Fermentation	30	45			No detailed information
Fining					No detailed information
Tap water (μg/l)	0.02	50	44		Threshold value
Food products (mg/kg)		2	28		Natural contents
Geochemical abundance (mg/kg)			27	17	Ingelheim, Germany
Vineyard soils (mg/kg)				300	Hessen, Germany

References on p. 70

4.5. SOME REMARKS ON ARSENIC

In the first half of this century arsenic-containing pesticides were used in vineyards [19], leading to comparatively high arsenic contents with values up to approx. 1 mg/l, in the range 0.06 to 0.92 mg/l. Of this, 60-70% of the total arsenic occurs as As(III) and only 30-40% as As(V) [20-25]. During fermentation of the must initial high arsenic contents are somewhat reduced, and part of the arsenic is precipitated as colloidal As_2S_3; however about 70-90% remains in solution [26]. The ban on arsenic in vine cultivation, in force since 1942 for Germany and now also in all other wine-producing countries, drastically reduced arsenic levels in commercial wines. Information on arsenic contents in contemporary wines is not very frequent, and so a sytematic investigation, including a total of 155 German and 36 foreign wines, was recently performed [27]. Five of the latter were wines from the 18th and 19th centuries which had already been analyzed for the metals mentioned above. From the results of this study, and investigations of wines from Spain [28] and Greece [29], a rather low arsenic level was found in old as well as contemporary wines. Average of 96 German wines from 1951 to 1984 vintages of about 0.01 mg/l, and a range 0.0017 to 0.04 mg/l. Only a few wines contained more than 0.02 mg/l arsenic. However, two German wines from 1937 and 1947 vintages had significantly elevated arsenic contents, 0.39 and 0.29 mg/l respectively, probably due to the use of arsenic-containing pesticides. Local differences in arsenic content are probably due to the geological background in vineyard soils [30], but may also be due to the influence of fertilizers with higher arsenic levels, for example those containing sewage sludges [31]. Another possible source of arsenic may be the metal foil covers already mentioned in Section 4.4.4, if corrosion and metal migration into the wine occur.

REFERENCES

1 H. Eschnauer and M. Stoeppler, in *Proc. Int. Conf. Heavy Metals in the Environment, Geneva*, Vol. 1, CEP Consultants, Edinburgh, 1989, pp. 393-397.
2 H. Eschnauer, *Spurenelemente in Wein und anderen Getränken*, Verlag Chemie, Weinheim, 1972.
3 H. Eschnauer, *Naturwiss.*, 73 (1986) 281-290.
4 H. Eschnauer, *Dtsch. Lebensm. Rundsch.*, 82 (1986) 320-325.
5 W. Schwellenbach, *Diploma Dissertation*, University of Cologne, 1982.
6 P. Ostapczuk, M. Apel, U. Bagschik, J. Golimowski, K. May, C. Mohl, M. Stoeppler, P. Valenta and H.W. Nürnberg, *Lebensm. Chem., Gerichtl. Chem.*, 39 (1985) 59-60.
7 H. Eschnauer and R. Neeb, in H.F. Linskens and J.F. Jackson (Editors), *Wine Analysis*, Springer, Berlin, 1988, pp. 67-92.
8 M. Stoeppler, M. Apel, U. Bagschik, J. Golimowski, K. May, C. Mohl, P. Ostapczuk, R. Enkelmann and H. Eschnauer, *Lebensm. Chem., Gerichtl. Chem.*, 39 (1985) 60-61.
9 P. Ostapczuk, M. Froning and H. Eschnauer, *Lebensm. Chem.*, 45 (1991) 44-45.
10 Y. Eckstein, *The Thermo Mineral Springs of Israel*, Health Resorts Authority, Jerusalem, 1985, p. 38.
11 H. Eschnauer, *Zur Reinheit des Weines seit 2000 Jahren – vinum et plumbum – Gesellschaft zur Geschichte des Weines*, Schriften zur Weingeschichte No 102, Wiesbaden, 1992, in preparation.

12 G. Hamelle, *Ann. Falsif. Expert. Chim.*, 69 No. 737 (1976) 101-105.
13 U. Pallotta and S. Galassi, *Atti. Accad. Ital. Vite Vino Siena*, 37 (1985) 449-465.
14 A. Stock and R. Cucuel, *Naturwiss.*, 22 (1934) 390-393.
15 R. Neeb, personal communication.
16 R. Enkelmann, H. Eschnauer, K. May and M. Stoeppler, *Fresenius' Z. Anal. Chem.*, 317 (1984) 478-480.
17 M. Medina and P. Sudraud, *Connaiss. Vigne Vin*, 14 (1980) 79-96.
18 H. Eschnauer, *Proc. 8th Int. Oenolog. Symposium, Kapstadt*, 27th-29th April, 1979, 281-290.
19 P. Claus, *Arsen zur Schädlingsbekämpfung im Weinbau 1904-1942*, Schriften zur Weingeschichte No. 58, Wiesbaden, 1981.
20 M. Ceribashi and K. Marga, *Bol. Shkencare Natyr. Univ. Sheteoror Tiranes*, 26 (1972) 85-90.
21 E.A. Crecelius, *Bull. Environ. Contam. Toxicol.*, 18 (1977) 227-230.
22 A.C. Noble, B.H. Orr, W.B. Cook and J.L. Campbell, *J. Agric. Food Chem.*, 24 (1976) 532-535.
23 M. Giaccio, *Bull. Lab. Prof.*, 1 (1975) 260-268.
24 F.S. Interesse, F. Lamparelli and V. Allogio, *Z. Lebensm. Unters. Forsch.*, 178 (1984) 271-278.
25 P.D. Handson, *J. Sci. Food Agric.*, 35 (1984) 215-218.
26 H.D. Mohr, *Weinberg und Keller*, 26 (1979) 277-288.
27 M. Burow, M. Stoeppler and H. Eschnauer, *Dtsch. Lebensm. Rundsch.*, 84 (1988) 16-19.
28 M.D. Garrido, M.L. Gil and C. Llaguno, *An. Bromatol.*, 26 (1974) 167-276.
29 A.P. Grimanis and A.G. Souliotis, *Analyst*, 92 (1967) 549-552.
30 H. Faath, R. Hindel, U. Siewers and J. Zinner, *Geochemischer Atlas der Bundesrepublik Deutschland*, E. Schweizer Barth'sche Verlagsbuchhandlung, Stuttgart, 1985.
31 H. Heinrichs, *Naturwiss.*, 69 (1982) 88.

Chapter 5

Sample treatment

Peter Tschöpel

Max-Planck-Institut für Metallforschung, Laboratorium für Reinststoffanalytik, Bunsen-Kirchhoffstrasse 13, D-W-4600 Dortmund (Germany)

CONTENTS

5.1. Introduction 73
    5.1.1. Definitions 75
5.2. Systematic errors and their avoidance 75
    5.2.1. Volatilization 76
    5.2.2. Adsorption 77
    5.2.3. Blanks from the vessel and the vessel material 77
    5.2.4. Blanks from the reagents 79
    5.2.5. Blanks due to dust from the air 79
    5.2.6. Contamination by analytical work 79
    5.2.7. Basic rules for the recognition and elimination of systematic errors . . 80
5.3. Sampling, sample storage and pretreatment 81
5.4. Decomposition procedures 81
    5.4.1. Requirements for decomposition procedures 82
    5.4.2. Examples of decomposition methods for extreme trace analysis . . . 84
5.5. Separation and pre-concentration methods 88
5.6. Conclusion 92
References 92

5.1. INTRODUCTION

The tasks of analytical chemistry, imposed by science, trade, industry, public administration and many other clients, are extremely varied. If the determination of trace organic compounds is excluded, however, they can be summarised thus: all naturally occurring elements have to be determined, within a very broad concentration range, in all imaginable organic and inorganic matrices. Although the concentrations of the elements in the bulk of the sample are still mainly of interest, data on their distribution on the surface of the sample, in micro-regions, or phase boundaries have become increasingly important, even at trace levels [1,2]. Further speciation analysis is also required more and more.

The main aim of analytical research is to improve the methods available to provide power of detection, reliability, and a high performance/cost ratio [3]. We want to have procedures available which are quick, simple and inexpensive. Moreover, they should be free of errors and allow the determination of absolute amounts of elements down to the picogram level in all matrices. These demands cannot all be fulfilled together.

In routine analysis there is a trend towards fully computer-controlled simultaneous or sequential multi-element methods. For the analytical characterization of high-purity materials, and in many areas of medicine, biochemistry, human nutrition and control of environmental pollution, analytical procedures are already required to allow the reliable determination of absolute quantities at the nanogram and picogram level. The realization of such low detection limits is, however, inconsistent with high analytical reliability and low-cost analysis. The most economical approach is the application of instrumental methods and direct sample excitation. But even when the sensitivity of the method is sufficient, the analytical output data can suffer from matrix interferences, and suitable standard reference materials must be used for the calibration [4]. Unfortunately, these materials are not yet available at extremely low concentration levels. Accordingly, each improvement in the power of detection again raises the question of the reliability of the data, and the economic aspects in their turn again depend on the degree of attainable reliability [5,6].

Therefore, the strategy of extreme trace analysis makes it necessary to use wet chemical, multistage combined procedures. These permit high levels of detection and analytical reliability to be obtained, provided that the element to be determined is completely separated from any concomitant sample components. Then the method can be calibrated using aqueous standard solutions.

Multistage combined procedures include a pretreatment and decomposition of the sample, a separation, and a preconcentration of the elements of interest into a volume, or onto a target surface, which is as small as possible, and finally the determination of the analytes [5-9]. These procedures have several advantages. First, matrix effects are avoided by the dissolution of the sample and the separation of the matrix. Further, the power of detection is improved by the preconcentration of the analytes, and an optimal analytical procedure can be chosen irrespective of the kind, shape and composition of the sample. Finally the most important advantage lies in the easy calibration of a wet chemical procedure.

Sample preparation is, however, inherently expensive and time consuming and is connected with very intensive work. Further, the procedures can be hampered by their own kinds of systematic errors [10-14], which are very complex and insidious. Various sources of error contribute, to different extents, at the various stages of an analytical procedure. Contamination, as well as losses of elements and compounds, introduce the main errors. Depending on the elements involved they may falsify the results by orders of magnitude. Most difficult is the determination of traces of the elements which are present in high abundance in the earth's crust [5,15] (e.g. Si, Al, Fe, Ca, Na, Mg (Table 5.1)) or which are introduced into our environment by man-made pollution (e.g. Cu, Ni, Co, Cd, Pb).

In spite of the enormous difficulties inherent in the multistage combined procedures, we have to apply this approach and strive to trace and eliminate the systematic errors. When the results of these procedures prove to be right, we can use them to produce reliable standard materials which will allow the direct instrumental methods to be calibrated.

TABLE 5.1.

ORDER OF ABUNDANCE OF THE ELEMENTS IN THE CONTINENTAL CRUST (AFTER TAYLOR, [15])

Element	%	% x 10^2		% x 10^3		% x 10^4		% x 10^5	
O	46.4	Ti	57	Ni	7	Th	9	J	5
Si	28.2	H	14	Zn	7	Sm	6	Tl	4.5
Al	8.2	Mn	10	Ce	6	U	2.7	Cd	2
Fe	5.6	P	10	Co	2.5	Sn	2	Hg	1.8
Ca	4.2	S	2.6	Li	2	Ta	2	Bi	1.7
Na	2.4	C	2.0	N	2	As	1.8	Ag	0.7
Mg	2.3	Zr	1.6	Pb	1.2	Mo	1.5	Se	0.5
K	2.1	Cl	1.3	B	1	W	1	Au	0.04

5.1.1. Definitions

In modern trace analysis, the term "sample preparation" cannot be exactly defined and covers a very broad field. It starts with mechanical pretreatment of the sample (e.g. turning, drilling, grinding, milling, sieving) prior to instrumental methods of analysis, and extends to chemical methods (e.g. etching, decomposition, separation, enrichment) required for a wet chemical procedure in which the solid, liquid or gaseous samples or targets are prepared and passed on to the real determination step.

Accurate sampling and mechanical pretreatments are essential prerequisites for reliable results from the analysis. These topics represent a specialised area of analytical chemistry with numerous problems, rules and procedures [16-20], which cannot be discussed at length within the limited frame of this chapter. We will therefore discuss mainly the chemical operations which precede the real determination step of the multistage combined procedure. The contribution mainly deals with modern trace analysis of elemental quantities in the ng/g range and below, and with problems and systematic errors inherent to this field [6-12].

5.2. SYSTEMATIC ERRORS AND THEIR AVOIDANCE

The most important problems in extreme trace analysis arise from systematic errors, which increase sharply with decreasing amounts or concentrations of the elements to be determined. These errors can exceed orders of magnitude, depending on the omnipresence and the distribution of the elements in the environment and in the laboratory (see Table 5.1).

References on p. 92

The most important sources of systematic errors [6] are:

(a) Inadequate sampling, sample handling and storage;

(b) contamination of the sample and/or the sample solution by tools, apparatus, vessels, reagents and airborne dust during the analytical procedure;

(c) adsorption and desorption effects at the surface of the vessels and phase boundaries (e.g. filters or precipitates);

(d) losses of elements (e.g. Hg, As, Se Cd, Zn) and compounds (e.g. oxides, halides, hydrides) due to volatilization;

(e) undesired or incomplete chemical reactions (e.g. change of valency, precipitation, ion exchange, formation of compounds and complexes);

(f) influences of the matrix on the generation of the analytical signals;

(g) incorrect calibration and evaluation as a result of incorrect standard materials, unstable standard solutions, false calibration functions or invalid extrapolation.

This section will be limited to a discussion of the particularly serious problems of element losses due to volatilization and adsorption, and to contamination from the three most important sources, vessels, reagents and dust [21-27].

5.2.1. Volatilization

Losses of elements by volatilization mainly occur at high temperatures. However, for very volatile elements such interferences can be remarkably high even at room temperature. Mercury, particularly, can be lost when an aqueous solution is stored in an open vessel. It could be proved, using the radioactive isotope ^{203}Hg that, at ng/ml-concentrations, up to 25% of the Hg was volatilized from an acidic solution within 6 hours [28]. During the dissolution of metal samples in non-oxidizing acids the hydrides of elements such as Se, Te, As, Sb or Bi escape.

The number of elements and compounds which can be lost increases with the temperature e.g., during the evaporation of solutions or in decomposition procedures (see Table 5.2). These systematic errors can be eliminated by using closed systems and by working at low temperatures. For the determination of elements which are relatively easily

TABLE 5.2.

ELEMENTS AND COMPOUNDS WHICH CAN BE SEPARATED BY VOLATILIZATION (20-1000°C) (AFTER BÄCHMANN AND RUDOLPH [113])

elements	gaseous elements, Te, Sn, Pb, Tl, P, As, Sb, S, Se, Br, I, Zn, Cd, Hg
oxides of	As, S, Se, Te, Re, Ru, Os, Zn, Cd, Hg
fluorides of	B, Si, Ge, Sn, P, As, Sb, Bi, S, Se, Te, Ti, Zr, Hf, V, Nb, Ta, Mo, W, Re, Ru, Os, Ir, Hg
chlorides of	Al, Ga, In, Tl, Ge, Sn, Pb, P, As, Sb, Bi, S, Se, Te, Ti, Zr, Hf, Ce, V, Nb, Ta, Mo, W, Mn, Fe, Ru, Os, Au, Zn, Cd, Hg
hydrides of	Si, Ge, Sn, Pb, P, As, Sb, Bi, S, Se, Te

volatilized (e.g. Hg, Se, As, Bi, Zn, Cd) from non-volatile matrices, we can make use of the very advantageous separation in the vapour phase (see Sections 5.4.2 and 5).

5.2.2. Adsorption

The concentrations of elements in very dilute solutions can change very quickly owing to adsorption and desorption [6,11,26,29]. Trace elements are bound onto the surface of the vessel as ions or molecules and may be released again when the composition of the solution changes. As a rule, such losses become considerable at concentrations below 10^{-6} M, and are of the order of 10^{-9} - 10^{-12} mol/cm^2 [30]. A general prediction of the error for a specific case is not possible. However, the losses can be monitored very easily with the aid of radioactive isotopes [31].

The amounts of the elements adsorbed depend on numerous factors. These all have to be specified, so as to enable the adsorption behaviour of an element to be estimated. The factors to be taken into account are:

(a) the element itself, its valence state and its concentration;

(b) the composition of the solution, the pH-value, the other elements present, their valency and concentration, as well as the salts and organic components contained in the sample;

(c) the vessel and its material, the composition and purity of the material, the dimensions and constitution of the surface, as well as the pretreatment and cleaning procedures applied;

(d) the time and temperature taken for the process.

In order to minimize losses of elements by adsorption the following conditions have to be fulfilled:

(a) The vessels should be of quartz, PTFE, or glassy carbon;

(b) the surface and the volume of the vessel should be as small as possible;

(c) the concentration of the elements to be determined should be as high as possible;

(d) the time of contact between the vessel and the solution should be short;

(e) acidic solutions should be used if possible because the losses are often lower than with neutral or alkaline solutions;

(g) the vessels should be pre-treated with acid vapour because this considerably reduces the adsorption.

5.2.3. Blanks from the vessel and the vessel material

No vessel material is totally resistant, even to water. Accordingly, each element present in the material of the vessel will be found at a more or less high level in the solution contained in the vessel [5-12,26,32-34]. Glass is a much more impure material than quartz or PTFE, and the losses of elements due to adsorption on it are very high. Therefore, glass vessels are not suitable for extreme trace analysis.

Quartz is available in different purity grades. It is definitely the purest material and should be preferred wherever possible. When acid solutions are stored in quartz vessels

References on p. 92

few impurities are introduced. Glassy carbon and PTFE are substantially less pure. After a pretreatment by steaming, glassy carbon introduces few impurities, because their diffusion from the bulk to the surface is inhibited by its structure. However, PTFE is permeable to many substances, for example, gases. Polypropylene, PTFE and especially, glassy carbon are used for solutions containing HF.

The cleaning procedure for the vessels is of very great importance [26,32], apart from the purity of their material. The usual way of cleaning laboratory glassware is rinsing and leaching it with acids and pure water. Sometimes the leaching can be enhanced by applying ultrasonic treatment [33]. It should be noted, however, that these procedures are not sufficiently effective to guarantee residual blanks in the lower pg/ml-region. Excellent cleaning of vessels can be achieved by a steaming procedure, with only few exceptions (e.g., traces of Fe in PTFE) [5-12]. For this, the vessels to be cleaned are placed upside down on top of quartz tubes in a glass apparatus (Fig. 5.1). Vapours are passed through the quartz tubes and mainly the inner surfaces of the vessels are washed continuously. For the first 4-6 h nitric or hydrochloric acid vapours are used, then water vapor for another hour. By this process the surfaces are not only cleaned but the adsorption of traces of elements during the subsequent procedure is considerably inhibited.

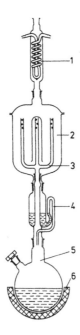

Fig. 5.1. Steaming apparatus for cleaning vessels in acid and water vapour [10-12]. (1) Reflux condensor, (2) steam chamber, (3) vapour tubes (quartz), (4) overflow, (5) flask, (6) heating device.

5.2.4. Blanks from the reagents

For reducing the blanks given by the reagents [6-14,24-27,35-41] the possibilities, unfortunately, are very limited. For most solid substances the procedures available are very laborious and sophisticated, and are effective for only a very few elements: the blanks of others might even be increased. Accordingly, in extreme trace analysis we often have to use only reagents which can easily be purified, such as gases and liquids.

The sub-boiling distillation technique, described by Kuehner et al. [6,35,36,38], is extremely efficient for most laboratory acids and for some organic solvents. With this technique boiling of the liquid is avoided and, accordingly, aerosols which can contaminate the distillate are not formed. Otherwise, no other single purification procedure is capable of removing all metallic or cationic impurities to such a low level. Ultra-purification of some other reagents, such as H_2O_2, hydrazine, and $AsCl_3$, can be performed with the aid of low-temperature sublimation [38,39].

5.2.5. Blanks due to dust from the air

The laboratory atmosphere carries particulate matter containing various kinds of inorganic and organic compounds which may contaminate the sample solution. In principle any element can be present, depending on the environment, the laboratory itself and its history. The dust may sometimes be excluded by very simple means, such as closed apparatus or glove boxes. Clean benches and clean rooms [6,12,31], whose working space is flushed with "dust-free" air are much more efficient and convenient. For such systems the air is filtered through high efficiency particulate air filters. Highest clean-liness can only be achieved with a laminar airflow, and can give a residual content of dust particles about 4-5 orders of magnitude lower (Class 100 of the US-Federal Standard 209b) than in the air of a conventional well cleaned analytical laboratory. For advanced technology, however, even higher purity demands exist. For example, the electronic industry claims a much lower dust content for the production of megabyte chips (Class 10 or 1 of the US-Federal Standard). The effort needed to achieve such a high air quality is enormous.

5.2.6. Contamination by analytical work

In spite of the very effective techniques for cleaning and purifying the equipment, the reagents, and the laboratory, further contamination of the sample solution during the analytical procedure cannot be totally excluded. At the ppb- and still more at the ppt-concentration level, very simple working steps such as pipetting, shaking, evaporating, or filtering, can increase the blanks very seriously [5,6,10-12,22,42,43].

During storage of high purity solutions the blank content increases with time, to an extent which depends on the pretreatment of the vessel, the composition of the solution, and the element [10,11,30,44]. Filters, mortars, and sieves may also become sources of serious contamination. Errors arising from blank contributions may be large. The most alarming aspect of this situation is not the value of the blank from a single step, but the fact that blanks occur even in very simple operations. Consequently, they accumulate during the whole analytical procedure to values lying in the ng/g-range, where numerous analyses are done every day. In particular, the scatter of the blanks, which in some cases can exceed several orders of magnitude, limits the possibilities for extreme trace analysis.

In principle, various sources of systematic errors are present in all steps of the analytical procedure: namely, sampling, transport, storage, sample pretreatment, decomposition and separation. Accordingly, the general statement can be made that, with decreasing absolute amounts of elements to be determined, systematic errors increase dramatically and become the main problem in extreme trace analysis.

5.2.7. Basic rules for the recognition and elimination of systematic errors

Particularly because of the lack of standard materials in extreme trace analysis there is no absolutely sure method to recognize and avoid systematic errors. However, at least some basic rules can be summarized which serve to minimize these problems [5-9,12]:

(a) The reproducibility of the results of an analytical procedure gives absolutely no information on its accuracy. A bad standard deviation might only give a hint that systematic errors are present. However, a low standard deviation does not prove the results to be reliable.

(b) The accuracy of a result must be confirmed by at least one or two independent procedures, differing in each analytical step such as decomposition, separation, preconcentration, and determination. A further way to confirm analytical results is to make use of inter-laboratory comparisons.

(c) All manipulations, and the number of working steps, have to be kept to a minimum.

(d) Monitoring of the different steps of a combined procedure can be done well with radioactive tracers.

(e) Microchemical techniques, which use small apparatus and vessels, should be preferred. It would be best if all steps of the procedure could be performed in one vessel (the single vessel principle).

(f) To avoid losses of elements by volatilization, closed systems should be used and the temperatures used should be as low as possible.

(g) To reduce blanks, and losses of elements by adsorption, all apparatus, vessels, and tools should be made of materials which are as pure and inert as possible. These requirements are met to a high degree by quartz, and to a lesser extent by glassy carbon, PTFE and polypropylene.

(h) For the cleaning of apparatus and vessels a treatment with vapours of nitric acid, then water, is most effective.

(i) Reagents have to be of high purity. If possible, liquids such as acids or organic solvents which can be purified by sub-boiling distillation are to be preferred.

(j) Contamination by the dust of the air can be excluded to a high degree by using clean benches and clean rooms.

(k) Perseverance and self-criticism of the analyst are most important prerequisites for obtaining reliable analytical data.

5.3. SAMPLING, SAMPLE STORAGE AND PRETREATMENT

Sampling, sample storage and pretreatment are most important and complex handling techniques (see Chapters 2-5). The problems inherent in these procedures cannot be considered here in detail. The theory and praxis of these topics must be well known by the analyst to avoid or minimize the inherent errors. General statements cannot be made since all the effects mentioned depend strongly on the matrix, on the respective elements, and on their concentrations.

Sampling, sample transport, storage, drying, and processing are associated with many kinds of systematic errors [16-20,45-53] when amounts of elements below about 1 μg are to be determined. In addition to the problems discussed in Section 5.2 we also have to consider lack of sample homogeneity and changes of the element content and/or the bonding of these elements by biological or photochemical reactions. One example can illustrate such biological interferences. It is well known that diluted sulfide solutions in the μg/ml-range are not stable, owing to oxidation by their oxygen content. Stabilization can be achieved by adding reducing agents, such as ascorbic acid (at pH > 12). However, this is only true for synthetic and sterile solutions and not for samples of natural water, such as river water or mineral spring water. In these samples we have to add Zn^{2+} up to a concentration of 20 mg/l (and pH 8) to avoid the disintegration of S^{2-} by bacteria [54]. Considerable losses of Hg may also occur as a result of volatilization [55] or of diffusion of elemental Hg through plastic containers. Further, red rubber tubing containing antimony sulphide binds Hg and other elements which form rather insoluble sulphides. It has also been reported that reactions may be induced by bacteria, which convert Hg^{2+} to elemental Hg or organic derivatives by a methylation process. For sea water the inherently low levels of trace metals are known to cause severe problems in sampling [48-51].

Several sources of error are inherent in sample treatment. When suspended matter is removed from liquid samples by filtration, traces of elements will be lost by adsorption, and blanks will be introduced. For solids, contamination as well as losses of elements arise during disintegration, homogenization or sieving. Therefore these techniques should be avoided whenever possible. Homogenization of food or biological matter, for example, can be well performed in a special mixing apparatus made of PTFE and cooled with liquid nitrogen, in order to avoid or minimize losses of trace elements by adsorption or cementation onto the metal of the cutting tool.

5.4. DECOMPOSITION PROCEDURES

In a multistage combined procedure the step which follows sampling and mechanical or chemical pretreatment, is dissolution or decomposition (or 'opening out') [56-60].

Dissolution is usually defined as the simple process of dissolving a substance in a suitable liquid at relatively low temperature, with or without a chemical reaction. The term decomposition denotes a more complex process which is usually performed at higher temperatures and/or at increased pressure, with the aid of reagents and of special apparatus. When the decomposition is achieved in a molten flux, the expression 'to open out' is used. A clear distinction between these terms cannot, however, be made.

The aim of dissolution and decomposition is to obtain a solution, or in some cases a gaseous sample, which contains all elements and compounds of interest from the sample, in unchanged amounts. In some cases homogenization or isoformation should also be achieved, as it is done by melting the sample substance with a flux for sample preparation in XRFA.

In analytical chemistry, an immense variety of different matrices ranging from waste water, sludge, cosmetics, and foodstuffs, to high-purity materials have to be analysed. For these tasks, the power of detection and the reproducibility of the analytical procedures have to meet very different requirements, and widely different numbers of elements have to be determined at different concentrations. It is evident that as large a number as possible of decomposition techniques has to be available. Table 5.3 [58] gives a short systematic survey of these decomposition methods.

5.4.1. Requirements for decomposition procedures

The conditions which must be fulfilled by a decomposition procedure often depend greatly on the analytical task. There are two main problems. The first is that the decomposition may be incomplete, i.e., parts of the sample or of the reaction products remain as an insoluble residue. This is of importance when very complex matrices, such as rocks, soils, or organic matter with a high content of inorganic material, must be analyzed. Very often, these types of samples can only be decomposed by two or more subsequent decomposition steps. The second problem is that the amounts of the elements or compounds in the decomposition product to be analysed do not correspond to those present in the sample, because of systematic errors.

For determinations of trace elements we have to consider the following points:

(a) The decomposition must be complete. Inorganic materials have to be converted completely into soluble compounds, and organic material has to be totally mineralized.

(b) Residues should be quantitatively soluble in a small volume of high-purity acid.

(c) The decomposition procedure has to be as simple as possible and should not require complicated apparatus.

(d) The decomposition must be adapted in an optimal manner to the whole analytical procedure.

(e) Preference should be given to procedures where decomposition and separation are achieved in one step.

(f) In order to minimize the systematic errors in the decomposition procedure, clean vessels made of an inert material, and the smallest amounts of high-purity reagents should be used, and dust should be excluded. Reaction chambers should be as small as possible. Precautions should be taken so as to minimize losses of elements due to adsorption and volatilization.

(g) The yield from the decomposition step should be checked by using radioactive tracers.

TABLE 5.3.

SCHEME OF DECOMPOSITION METHODS [58]

Decomposition technique	Procedure	Reagents/method	Samples
Fusion decomposition	acidic	sodium disulfate	inorganic
	alkaline	sodium carbonate	inorganic
	oxidizing	sodium carbonate + nitrate	inorganic
	reducing	metallic sodium	organic
	sulfonating	"Freiberger" decomposition potassium carbonate + sulphur	inorganic
Wet chemical decomposition	open system with acids		
	non oxidizing	HCl	inorganic
	oxidizing	HNO_3 / $KMnO_4$	inorganic/organic
	catalytic	H_2O_2 + Fe^{2+}	organic
	closed system		
	static	pressure bomb / Carius	inorganic/organic
	dynamic	Millon	inorganic/organic
Combustion	open system	dry ashing	organic
	closed system		
	static	oxygen flask (Hempel, Schoninger)	organic
	dynamic	Wickbold, Radmacher-Hoverath	organic
Miscellaneous			
electrolytic oxidation			metals/(organic)
pyrolysis/pyrohydrolysis	(heat / heat + H_2O)		inorganic/organic
photolysis	(UV + H_2O)		organic
decomposition with halogens	(e.g. F, Cl_2, $HCCl_3$)		inorganic/organic
decomposition by reduction	(e.g. H_2, C, NH_3, metals)		inorganic/organic
enzymatic decomposition			organic

References on p. 92

5.4.2. Examples of decomposition methods for extreme trace analysis

Only some of the most essential decomposition techniques, which are especially suitable for extreme trace analysis, can be treated in this section.

Decomposition by fusion with solid reagents is to be avoided [58]. Indeed, the excess of flux reagents which is normally required causes an immense contamination, and the impure crucible material is in any case attacked to a greater or lesser extent. Further, relatively high amounts of the elements can be lost by adsorption, by reaction with the crucible material, or by volatilization at the high temperatures required.

If volatile elements such as Se, Te, Bi, or Tl are to be determined in non-volatile matrices special use can be made of selective volatilization. For this, the sample is mixed with NaCl or V_2O_5 and heated in a small quartz tube, whose central part is extended to a capillary [6,9]. The quartz tube is heated in a crucible furnace up to about 1000°C (Fig. 5.2.). The traces of the volatile chlorides or oxides then escape and condense in the cold part of the capillary, outside the furnace. After the capillary has been cut off, the condensate can easily be taken up in some microlitres of hydrochloric acid.

Wet digestion procedures using oxidizing agents, such as HNO_3, $HClO_4$, or H_2O_2 in an open vessel are used widely-spread as methods for the decomposition of inorganic samples and for the mineralization of organic matter [56-59, 61-63]. They are suitable for both high sample weights and large numbers of samples. Indeed, they are easy to perform and cheap, as one only needs beakers and a hot-plate. However, they are not suitable when volatile elements or compounds are to be determined.

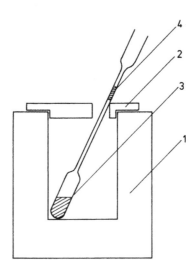

Fig. 5.2. Volatilization in a quartz capillary [6,9]. (1) Furnace (~1000°C), (2) cover, (3) quartz tube with sample and flux, (4) condensate of volatile compounds.

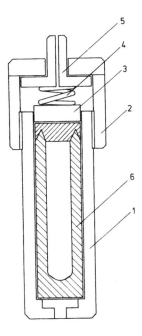

Fig. 5.3. Pressure decomposition device [64,65]. (1) Pressure bomb, (2) screw cap, (3) pressure plate, (4) pressure spring, (5) insert (1-5: stainless steel), (6) decomposition vessel and cover (PTFE or glassy carbon).

An automatic and continuous wet digestion device is described by Knapp et al. [59,62]. The temperature of the sample solution can be controlled by inserting the test tubes at different depths into the extension tubes of the heating block. By using a mechanical transport system, a high sample throughput is obtained. For organic matrices with a high fat content or stable organic compounds, however, complete decomposition can only be performed using a mixture of acids such as HNO_3 and H_2SO_4, under high pressure and temperatures (up to 350°C) in special apparatus (Fig. 5.3.). Decomposition at high pressure requires the use of only small volumes of acid, and prevents losses of volatile elements such as Hg, As, and Se. Various types of high-pressure decomposition systems are available [56-58,64-72]. They differ in construction, size, and container materials. For the latter, PTFE, glassy carbon or quartz are used. For ultra-trace analysis, volumes of only a few ml, and small amounts of sample (0.2-0.5 g) are preferred, in order to minimize systematic errors. A new technique for the wet ashing of biological samples, for example, which uses PTFE vessels, with heating in a microwave oven [73,75,75a] at moderate pressure, is of potential interest.

The quantitative mineralization of all organic matter cannot be achieved in high-pressure decomposition systems with PTFE-vessels under the usual conditions. This is, however, absolutely necessary, for example, for voltammetric determination methods

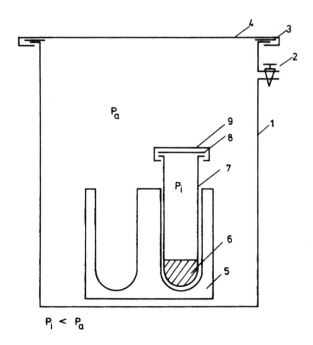

Fig. 5.4. High pressure decomposition autoclave [59,60]. (1) Autoclave, (2) pressure gas inlet, (3) O-ring, (4) cover, (5) heating block, (6) acid + sample, (7) quartz vessel, (8) seal (PTFE), (9) quartz cover.

[116-119]. With HNO_3, sufficiently high oxidation potentials are obtained only at decomposition temperatures $\geq 220°C$.

Using quartz vessels, temperatures up to 320°C and inside pressures up to 100 bar can be reached provided a comparable high pressure is applied outside the decomposition vessel so as to avoid bursting. This is possible in a microprocessor controlled autoclave, (Fig. 5.4) [59,60,76,77]. Using this method, up to 0.5 g of sample materials such as PVC, polypropylene, rubber, coal, sewage mud and others can be decomposed within 2-4 hours by only 2 ml of HNO_3.

For determinations of low concentrations of the more common elements in organic materials, combustion in pure oxygen is preferred over wet decomposition methods because the risks of contamination are lower. Dry ashing of the organic materials [56-58], however, should be avoided. Here, the sample is burned in an open crucible placed in a muffle furnace. This technique is impaired by losses of elements by volatilization, and contamination or reactions with the material of the crucible. There are two alternatives at our disposal, namely combustion in an oxygen plasma excited by a high frequency or ultra-high frequency field, or combustion in molecular oxygen either at slightly increased pressure in a glass flask or according to a quasi-static procedure (Trace-O-Mat: see p. 88).

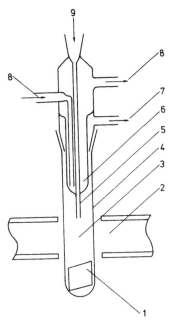

Fig. 5.5. Low temperature ashing by a microwave-induced oxygen plasma [79]. (1) Sample, (2) microwave field, (3) oxygen plasma, (4) test tube, (5) oxygen capillary, (6) cooling finger, (7) to vacuum pump, (8) water in and out, (9) oxygen in.

For the determination of volatile elements, such as Se, As, Sb, Cd, Zn, Tl, in organic materials, a very gentle treatment is required. For this a low-temperature ashing with excited oxygen at a pressure of 1-5 Torr is very suitable. The oxygen plasma can be produced either by a high-frequency electromagnetic field [59,60,77-79] or by microwave energy [79] (Fig. 5.5.). The sample must be placed close to the plasma zone, where the organic compounds are oxidized very slowly. Temperatures are usually below 200°C, depending largely on the electrical power, the oxygen pressure, the distance between the plasma and sample, and on the nature of the sample. Very stable materials, such as PTFE, graphite, or carbon black can be burned off. Many inorganic compounds, however, remain unchanged at these temperatures. If the samples are handled very carefully, it is even possible to retain their original shape.

The disadvantages of dry ashing methods can be avoided by performing the combustion with molecular oxygen in a closed flask. This technique, first described by Hempel in 1892, was improved by Schöniger [81] for the micro determination of the halogens, S, P, and other elements in organic materials. In order to avoid the high ratio of the volume of the flask to the sample weight, the sample can be burned in a stream of pure oxygen. This can be performed in a quartz apparatus (Trace-O-Mat: Fig. 5.6) [59,82,83], provided with a liquid nitrogen cooling unit and a combustion chamber having a volume of only 75 ml. The sample is ignited using an IR lamp and the volatilized elements, e.g., Hg, Se, Te, Tl, Bi, and Cd, together with other volatile combustion products, are frozen out completely. The trace elements can be taken up by refluxing in 1-2 ml of high purity HCl or HNO_3 and

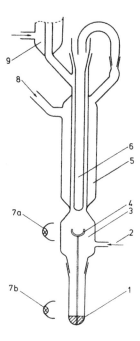

Fig. 5.6. Quartz apparatus for the combustion of organic materials in a stream of oxygen (Trace-O-mat) [59,82,83]. (1) Nitric acid, (2) O_2 in, (3) combustion chamber, (4) sample holder, (5) cooling jacket (liquid N_2), (6) cooling finger (liquid N_2), (7) infrared lamps: (a) for ignition, (b) for refluxing, (8) liquid N_2 in, (9) water-cooled condenser.

are then collected in the reagent vessel placed below. Non-volatile elements remain in the sample holder from which they can be recovered, together with the volatile ones, or separately. In this way the decomposition and separation are achieved in one step. After pulverization, inorganic samples can be mixed with high-purity cellulose powder and the mixtures pressed into pellets.

5.5. SEPARATION AND PRE-CONCENTRATION METHODS

The aims of the separation and pre-concentration methods are:
(a) To avoid matrix effects;
(b) to increase the sensitivity and power of detection;
(c) to enable aqueous standard solutions to be used for the calibration;
(d) to allow an optimal determination procedure to be chosen.
There are some disadvantages:
(a) high expenditure of work, time, manpower, and costs;
(b) additional systematic errors.

The requirements of a separation procedure are:

(a) that it must be complete (yield = 100%) or, if this is not possible, at least reproducible;

(b) it must be simple;

(c) only small amounts of high-purity reagents should be required;

(d) it should preferably use quartz vessels or, for example when HF solutions are needed, vessels made of PTFE, polypropylene, or glassy carbon;

(e) and it must be well adapted to the decomposition and determination steps.

The fields of the separation and pre-concentration of trace elements are very complex and do not permit one to provide a comprehensive review which will solve all kinds of problems (Table 5.4) [84-92]. Only some general points and some special procedures can be briefly discussed here. The requirements for separation and pre-concentration procedures which will be suitable for ultra-trace analysis are very similar to those we have met for the decomposition techniques and need not, therefore, be repeated. Of particular importance are the determination of the yield using radioactive isotopes, if possible, and the close coupling of the decomposition, separation and determination steps (using the single vessel principle).

In most cases, precipitation is not a suitable method because the formation of the

TABLE 5.4.

SURVEY OF SEPARATION METHODS

Precipitation
 direct precipitation
 precipitation exchange
 electrolysis, mercury cathode electrolysis, spontaneous electrochemical deposition
 co-precipitation (carrier precipitation)
Flotation
Liquid-liquid extraction
 batch extractions
 continuous extraction
Selective dissolution
Chromatographic separation (by physical sorption, ion exchange, complex formation and other chemical reactions)
 ion exchange (organic and inorganic exchanger)
 column-, paper-, thin-layer chromatography, ring oven
 separation with cellulose exchanger, activated carbon, organic reagents on a carrier
 sorption
 electrophoresis
 gas chromatography
 liquid chromatography (HPLC, ion chromatography)
Distillation, sublimation, volatilization
Freezing, zone melting

References on p. 92

precipitates is often incomplete. However, co-precipitation is used very often and numerous different procedures are at our disposal [84,87]. The same applies to ion-exchange techniques and to adsorption or sorption techniques, using charcoal or cellulose membranes impregnated, for example, with chelating agents or metal hydroxides [94-98]. However, large risks of contamination by impurities in the active material, and element losses by incomplete separation, are inherent in these procedures. It is possible, using chelate-GC, chelate-HPLC, or ion chromatography [99-101], to achieve simultaneous separation and determination of elements in the ng/g range in a large variety of analytical tasks.

Precipitation exchange on thin layers such as ZnS [102] should be briefly mentioned, as a special technique. By filtering a sample solution through a freshly-prepared ZnS-layer (-300 μg Zn), elements such as Ag, Cu, Pb, Bi, Cd, As, Sb, Sn, Se, and Te, whose sulfides have a solubility product lower than that of ZnS, will be exchanged very quickly against zinc. They can then be determined, for example by XRFA or GFAAS.

Extraction methods can be optimized by miniaturization and adaptation to the "single-vessel principle" [6]. The extraction is carried out directly in the decomposition vessel, e.g., in a quartz test tube having a volume of a few ml, and not in a separatory funnel. Pipettes are employed to separate the phases after centrifugation. Numerous variations of this separation method are available [103,104].

A further technique is the electrolytic pre-concentration of nanogram amounts of trace elements (e.g., Hg, Cu, Bi, Pb, Cd, Fe, Ni, Co, or Zn) from sample components which are not deposited electrolytically (e.g., Be, Zr, Nb, Ta, W) [105-111]. This technique can be used to quantitatively pre-concentrate the elements mentioned from a fairly large volume of electrolyte (about 50 ml) within a short time. This requires that the mass transport of the electrolyte is not effected by conventional stirring of the solution or by rotating the cathode: instead, a hydrodynamic flux system is used. Figure 5.7 [105,106] shows the principle of an apparatus, made of PTFE, in which the electrolyte is rapidly moved by a teflon pump through a small graphite tube, which serves as the cathode. The Pt/Ir-anode is situated along the axis of the cylinder. When the distance between cathode and anode is less than 1 mm current densities as high as 1 A/dm^2 can be applied. By this means, interferences caused by hydrogen evolution are avoided and yields higher than 98% are obtained. Metals such as Cu, Co, Fe, Zn, and Bi can be determined directly, after deposition inside the cathode tube. In GFAAS the tube can be directly used as furnace. However, optical emission spectrometry using an MIP can be applied instead [106].

Separation can easily be achieved for those elements and compounds that can be removed by volatilization from the mineral phase (Table 5.4). Examples are Hg, Se, Te, Tl, As, and Sb [85,113,114]. Here, the separation step can be performed simultaneously with the decomposition or combustion procedure. An appropriate dynamic ashing procedure using flowing oxygen is described in Section 5.4.2.

A simple method allows Se, for example, to be separated from metals and other materials. It can be volatilized as SeO$_2$ in an oxygen carrier gas stream [115]. The sample is placed in a quartz sample boat which is heated inside a quartz tube by a tubular furnace, (Fig. 5.8). The tube is directly connected to a cold trap (cooled with liquid

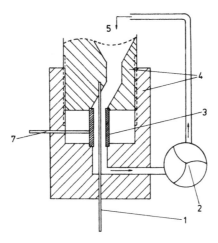

Fig. 5.7. Principle of the device for electrolytic pre-concentration of trace elements [105,106]. (1) Pt-Ir-anode, (2) PTFE-pump, (3) graphite tube cathode (i.d. = 2.8 mm, o.d. = 5 mm, l = 9 mm), (4) mounting device (PTFE) for anode and cathode, (5) storage vessel (PTFE, 70 ml) (omitted), (7) Pt-contact for the cathode.

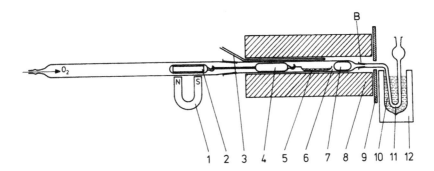

Fig. 5.8. Separation of nanogram-amounts of Se by volatilization [115]. (1) Magnet, (2) Fe-rod, covered with quartz, (3) thermocouple, (4) and (7) gas flow bodies, (5) sample, (6) quartz tube, (8) furnace, (9) shielding, (10) liquid N_2, (11) cooling trap, (12) dewar.

nitrogen), in which the SeO_2 and oxygen are condensed together. Then the liquid nitrogen is removed from the cold trap and the oxygen distils off. The remaining SeO_2 can be taken up with a few ml of diluted nitric acid and determined, for example by AAS.

The combustion of inorganic materials such as non-metals, alloys, steels, refractory metals, and ceramics, in elemental fluorine [112] instead of oxygen, leads to the volatile fluorides of more than 20 elements. This technique can be used for the separation of the volatile compounds of non-metals and some metals, and their simultaneous determination with the aid of, for example, a quadrupole mass spectrometer. With this method contamination is negligible.

Separation procedures related to those described can be performed by the combustion of organic matrices, by reduction with H_2 (for the determination of S in metals, for example), by chlorination with Cl_2 or CCl_4, or by hydrolysis or pyro-hydrolysis with steam (which has been applied, for example, for the determination of F).

5.6. CONCLUSION

With the help of the foregoing examples of ultra-trace analysis techniques, we have been able to review only some of the problems and possibilities. The role of the multistage procedures in modern elemental trace analysis cannot be generalized. Further progress in increasing the sensitivity and reliability of the procedures requires considerable effort: it is very time consuming and expensive, and can only be performed by laboratory teams having extensive knowledge and experience in this field. The importance of these procedures has diminished for analyses at the μg/g-concentration level, in favour of instrumental methods. But so far the multistage procedures are indispensible for the determination of ng/g, and particularly so for pg/g, contents. This will be true as long as standard reference materials at this concentration level remain unavailable for the calibration of the more economical and convenient instrumental methods.

When economic aspects have to be considered within the analytical strategy we have to keep in mind the facts that wrong analytical data are not only useless but can often lead to wrong conclusions and have unforeseeable consequences.

REFERENCES

1 G. Tölg, *Analyst*, 112 (1987) 365.
2 J.A.C. Broekaert and G. Tölg, *Fresenius' Z. Anal. Chem.*, 326 (1987) 495.
3 G. Tölg, Symposium der Gesellschaft f. Toxikologische und Forensische Chemie *"Anorganische Stoffe in der Toxikologie und Kriminalistik"*; Vlg. Dr. D. Helm, 6148 Heppenheim, 1983, pp. 109.
4 J.P. Cali, *Fresenius' Z. Anal. Chem.*, 297 (1979) 1.
5 G. Tölg and P. Tschöpel, *Anal. Sci.*, 3 (1987) 199.
6 P. Tschöpel and G. Tölg, *J. Trace Microprobe Techn.*, 1 (1982) 1.
7 G. Tölg, *Fresenius' Z. Anal. Chem.*, 283 (1977) 257.
8 G. Tölg, *Fresenius' Z. Anal. Chem.*, 294 (1979) 1.
9 G. Tölg, *Pure Appl. Chem.*, 50 (1978) 1075.
10 P. Tschöpel, L. Kotz, W. Schulz, M. Veber and G. Tölg, *Fresenius' Z. Anal. Chem.*, 302 (1980) 1.
11 K. Gretzinger, L. Kotz, P. Tschöpel and G. Tölg, *Talanta*, 29 (1982) 1011.
12 P. Tschöpel, *Pure Appl. Chem.*, 54 (1982) 913.
13 M. Zief, R. Speights (Editors), *Ultrapurity - Methods and Techniques*, Marcel Dekker, New York, 1972.
14 P.D. LaFleur (Editor), *Accuracy in Trace Analysis: Sampling, Sample Handling, Analysis*, NBS Special Publication 422, Vols. I and II, Washington, 1976.
15 R.S. Taylor, *Geochim. Cosmochim. Acta*, 28 (1964) 1273.
16 B. Kratochvil and J.K. Taylor, *Anal. Chem.*, 53 (1981) 924 A.
17 B. Kratochvil, *Anal. Chem.*, 56 (1984) 113R.

18 G. Brands, *Fresenius' Z. Anal. Chem.*, 314 (1983) 6.
19 P.M. Gy, *Analusis*, 11 (1983) 413.
20 J.M. Hungerford and G.D. Christian, *Anal. Chem.*, 58 (1986) 2567.
21 A. Mizuike and M. Pinta, Project Leaders, International Union of Pure and Applied Chemistry, *Pure Appl. Chem.*, 50 (1978) 1519.
22 M. Zief and J.W. Mitchel, in P.J. Elwing and I.M. Kolthoff (Editors), *Chemical Analysis*, John Wiley & Sons, New York, 1976.
23 B. Griepink, *Fresenius' Z. Anal. Chem.*, 317 (1984) 210.
24 J. Versieck, F. Barbier, R. Cornelis and J. Hoste, *Talanta*, 29 (1982) 973.
25 L. Kosta, *Talanta*, 29 (1982) 985.
26 K. Heydorn and E. Damsgaard, *Talanta*, 29 (1982) 1019.
27 L. Mart, *Talanta*, 29 (1982) 1035.
28 G. Kaiser, D. Götz, P. Schoch and G. Tölg, *Talanta*, 22 (1975) 889.
29 G. Tölg, *Talanta*, 19 (1972) 1489.
30 R. Massee, F.J.M.J. Maessen and J.J.M. De Gogeij, *Anal. Chim. Acta*, 127 (1981) 181.
31 V. Krivan, *Talanta*, 29 (1982) 1041.
32 H. Tamenori and J. Inoue, *Bunseki Kagaku*, 32 (1983) 337.
33 B. Kinsella and R.L. Willix, *Anal. Chem.*, 54 (1982) 2614.
34 A. Isasa, T. Yonemoto, I. Matsubara and K. Nakagawa, *Bunseki Kagaku*, 36 (1987) T113.
35 E.C. Kuehner, R. Alvarez, P.J. Paulsen and T.J. Murphy, *Anal. Chem.*, 44 (1972) 2050.
36 J.R. Moody and E.S. Beary, *Talanta*, 29 (1982) 1003.
37 H. Kunze, *Fresenius' Z. Anal. Chem.*, 316 (1983) 52.
38 J.W. Mitchell, *Talanta*, 29 (1982) 993.
39 J.W. Mitchell, *Anal. Chem.*, 50 (1978) 194.
40 J.W. Mitchell and C. McCrory, *Separation Purification Meth.*, 9 (1980) 165.
41 J.W. Mitchell, C. Herring and E. Bylina, *Appl. Spectroscopy*, 38 (1984) 653.
42 L. Mart, *Fresenius' Z. Anal. Chem.*, 296 (1979) 350.
43 L. Mart, H.W. Nürnberg and P. Valenta, *Fresenius' Z. Anal. Chem.*, 300 (1980) 350.
44 H. Scheuermann and H. Hartkamp, *Fresenius' Z. Anal. Chem.*, 315 (1983) 430.
45 S. Szymczyk and M. Cholewa, *Fresenius' Z. Anal. Chem.*, 326 (1987) 744.
46 P.M. Gy, *Anal. Chim. Acta*, 190 (1986) 13.
47 S.A. Katz, *Biotechnology Lab.*, June (1985) 10.
48 S.S. Berman and Ph.A. Yeats, *CRC Critical Reviews in Anal. Chem.*, 16 (1985) 1.
49 D. Schmidt, *Fresenius' Z. Anal. Chem.*, 316 (1983) 566.
50 A. Ashton and R. Chan, *Analyst*, 112 (1987) 841.
51 R. Sturgeon and S.S. Berman, *CRC Critical Reviews in Anal. Chem.*, 18 (1987) 209.
52 D. Klockow, *Fresenius' Z. Anal. Chem.*, 326 (1987) 5.
53 W. Ebing and G. Hoffmann, *Fresenius' Z. Anal. Chem.*, 275 (1975) 11.
54 P. Tschöpel, A. Disam, V. Krivan and G. Tölg, *Fresenius' Z. Anal. Chem.*, 271 (1974) 106.
55 G. Kaiser, D. Götz, P. Schoch and G. Tölg, *Talanta*, 22 (1975) 889.
56 R. Bock, *Aufschlussmethoden der anorganischen und organischen Chemie*, Verlag Chemie, Weinheim, 1972.
57 R. Bock, *A Handbook of Decomposition Methods in Analytical Chemistry*, International Textbook Co. Ltd., London, 1979.
58 P. Tschöpel, *Aufschlussmethoden*, in *Ullmann's Encyclopädie der Technischen Chemie*, Band 5, Verlag Chemie, Weinheim, 1980, p. 27.
59 G. Knapp, *ICP Information Newsletter*, 10 (1984) 91.
60 G. Knapp, *Fresenius' Z. Anal. Chem.*, 317 (1984) 213.
61 G. Knapp, B. Sadjadi and H. Spitzy, *Fresenius' Z. Anal. Chem.*, 274 (1975) 275.
62 G. Knapp, *Fresenius' Z. Anal. Chem.*, 274 (1975) 271.
63 R. Dumarey, M. Van Ryckeghem and R. Dams, *J. Trace Microprobe Techn.*, 5 (1987) 229.

94

64 L. Kotz, G. Kaiser, P. Tschöpel and G. Tölg, *Fresenius' Z. Anal. Chem.*, 260 (1972) 207.
65 L. Kotz, G. Henze, G. Kaiser, S. Pahlke, M. Veber and G. Tölg, *Talanta*, 26 (1979) 681.
66 K. Eustermann and D. Seifert, *Fresenius' Z. Anal. Chem.*, 285 (1977) 253.
67 B. Bernas, *Internat. Lab.*, 7/8 (1976) 29.
68 H.U. Kasper, *Fresenius' Z. Anal. Chem.*, 320 (1985) 55.
69 J.F. Woolley, *Analyst*, 100 (1975) 896.
70 M. Stoeppler and F. Backhaus, *Fresenius' Z. Anal. Chem.*, 291 (1978) 116.
71 P. Schramel, G. Lill and R. Seif, *Fresenius' Z. Anal. Chem.*, 326 (1987) 135.
72 M. Würfels, E. Jackwerth and M. Stoeppler, *Fresenius' Z. Anal. Chem.*, 329 (1987) 459.
73 M. Lachica, *Analusis*, 18 (1990) 331.
74 P. Aysola, P. Anderson and C.H. Langford, *Anal. Chem.*, 59 (1987) 1582.
75 H. Matusiewicz, R.E. Sturgeon and S.S. Bermann, *J. Anal. At. Spectrom.*, 4 (1989) 323.
75a G. Vermeir, C. Vandecasteele and R. Dams, *Anal. Chim. Acta*, 220 (1989) 257.
76 P. Schramel, S. Hasse and G. Knapp, *Fresenius' Z. Anal. Chem.*, 326 (1987) 142.
77 G. Knapp, *Intern. J. Environ. Anal. Chem.*, 22 (1985) 71.
78 C.E. Gleit and W.D. Holland, *Anal. Chem.*, 34 (1962) 1454.
79 S.E. Raptis, G. Knapp and A.P. Schalk, *Fresenius' Z. Anal. Chem.*, 316 (1983) 482.
80 G. Kaiser, P. Tschöpel and Tölg, *Fresenius' Z. Anal. Chem.*, 253 (1971) 177.
81 W. Schöniger, *Mikrochim. Acta*, (1955) 123.
82 G. Kaiser and G. Tölg, *Fresenius' Z. Anal. Chem.*, 325 (1986) 32.
83 G. Knapp, S.E. Raptis, G. Kaiser, G. Tölg, P. Schramel and P. Schreiber, *Fresenius' Z. Anal. Chem.*, 308 (1981) 97.
84 R. Bock, *Methoden der Analytischen Chemie*, Band 1: *Trennungsmethoden*, Verlag Chemie, Weinheim, 1974.
85 K. Bächmann, *CRC Crit. Rev. in Anal. Chem.*, 12 (1981) 1.
86 IUPAC, Anal. Chem. Division, *Commission on Microchem. Techniques in Trace Analysis*, Project Leaders: A. Mizuike, M. Pinta: *Pure Appl. Chem.*, 50 (1978) 1519.
87 A. Mizuike, *Enrichment Techniques for Inorganic Trace Analysis, Chemical Laboratory Practice*, Springer Verlag, Berlin, Heidelberg, New York, 1983.
88 A. Mizuike, Separation and preconcentration techniques, in R.F. Bunshah (Editor), *Modern Analytical Techniques for Metal and Alloys*, Interscience, New York, London, 1970.
89 A. Mizuike, *Fresenius' Z. Anal. Chem.*, 324 (1986) 672.
90 G. Schwedt, *Fresenius' Z. Anal. Chem.*, 320 (1985) 423.
91 J.C. Giddings, *Anal. Chem.*, 53 (1981) 943A.
92 E. Jackwerth et al., *Pure Appl. Chem.*, 51 (1979) 1195.
93 R. Van Grieken, *Anal. Chim. Acta*, 143 (1982) 3.
94 P. Burba and P.G. Willmer, *Fresenius' Z. Anal. Chem.*, 329 (1987) 539.
95 H. Berndt, U. Harms and M. Sonneborn, *Fresenius' Z. Anal. Chem.*, 322 (1985) 329.
96 R.E. Van Grieken, C.M. Bresseleers and B.M. Vanderborght, *Anal. Chem.*, 49 (1977) 1326.
97 K.H. Lieser, H.-M. Röber and P. Burba, *Fresenius' Z. Anal. Chem.*, 284 (1977) 361.
98 X.G. Yang and E. Jackwerth, *Fresenius' Z. Anal. Chem.*, 327 (1987) 179.
99 G. Schwedt, W. Fresenius, H. Günzler, W. Huber, J. Lüderwald and G. Tölg (Editors), *Analytiker- Taschenbuch*, Bd. 2, Springer-Verlag, Berlin, Heidelberg, New York, 1981, p. 161.
100 G. Schwedt, *Fresenius' Z. Anal. Chem.*, 320 (1985) 423.
101 J.S. Fritz, *Anal. Chem.*, 59 (1987) 335A.
102 A. Disam, P. Tschöpel and G. Tölg, *Fresenius' Z. Anal. Chem.*, 295 (1979) 97.
103 Yu.A. Zolotov, V.A. Bodnya and A.N. Zagruzina, *CRC Crit. Rev. Anal. Chem.*, 14 (1982) 93.
104 Yu.A. Zolotov, N.M. Kuz'min, O.M. Petrukhin and B.Ya. Spivakov, *Anal. Chim. Acta*, 180 (1986) 137.

105 G. Volland, P. Tschöpel and G. Tölg, *Anal. Chim. Acta,* 90 (1977) 15.
106 G. Volland, P. Tschöpel and G. Tölg, *Spectrochim. Acta,* 36B (1981) 901.
107 J.L. Anderson and R.E. Sioda, *Talanta,* 30 (1983) 627.
108 J. Golas and J. Osteryoung, *Anal. Chim. Acta,* 192 (1987) 225.
109 M. Veber, S. Gomiscek and V. Stresko, *Anal. Chim. Acta,* 193 (1987) 157.
110 W. Frenzel, *Anal. Chim. Acta,* 196 (1987) 141.
111 H. Matusiewicz, J. Fish and T. Malinski, *Anal. Chem.,* 59 (1987) 2264.
112 E. Jacob, *Fresenius' Z. Anal. Chem.,* 333 (1989) 761.
113 K. Bächmann and J. Rudolph, *J. Radioanal. Chem.,* 32 (1976) 243.
114 C. Spachidis and K. Bächmann, *Fresenius' Z. Anal. Chem.,* 300 (1980) 343.
115 G. Tölg, *Pure Appl. Chem.,* 44 (1975) 645.
116 M. Würfels, E. Jackwerth and M. Stoeppler, *Fresenius' Z. Anal. Chem.,* 330 (1988) 160.
117 M. Würfels, E. Jackwerth and M. Stoeppler, *Anal. Chim. Acta,* 226 (1989) 1.
118 M. Würfels, E. Jackwerth and M. Stoeppler, *Anal. Chim. Acta,* 226 (1989) 17.
119 M. Würfels, E. Jackwerth and M. Stoeppler, *Anal. Chim. Acta,* 226 (1989) 31.

M. Stoeppler (Editor)/*Hazardous Metals in the Environment*
© 1992 Elsevier Science Publishers B.V. All rights reserved

Chapter 6

Analytical methods and instrumentation — a summarizing overview

Markus Stoeppler

Institute of Applied Physical Chemistry, Research Center (KFA) Juelich, P.O. Box 1913, D-W-5170 Juelich (Germany)

CONTENTS

6.1. Introduction 97
6.2. Optical atomic spectrometry 98
     6.2.1. Atomic absorption spectrometry (AAS) 98
     6.2.2. Atomic fluorescence spectrometry (AFS) 103
     6.2.3. Atomic emission spectrometry (AES) with plasmas 103
6.3. Inductively coupled plasma–mass spectrometry (ICP–MS) 106
6.4. Electrochemical methods 107
6.5. X-Ray fluorescence spectrometry (XRF) 112
6.6. Nuclear methods 114
6.7. Chromatographic methods 117
6.8. Mass spectrometry (MS) 118
6.9. Spectrophotometry and related techniques 119
6.10. Future prospects 120
References 122

6.1. INTRODUCTION

There now exist a number of methods that can be successfully used for single and multielement analysis. These methods today cover in principle the whole range of trace element contents in biological and environmental materials, so that for each analytical task at least one suitable method can be found.

Hence, the intention for the conception of this chapter was to provide a condensed introduction to the principles and the most promising commercially available analytical techniques. For the elements treated in this book ample detailed information can be found in the elemental chapters.

General problems and prospects of trace analytical chemistry are discussed in recent work (e.g. refs. 1–15), references about particular methods are given in the following sections.

6.2. OPTICAL ATOMIC SPECTROMETRY

Free atoms, predominantly generated thermally, can absorb or emit radiation due to distinct transitions of the valence electrons of the outer shell of the atom. Hence element identification is feasible based on the atomic structure and ranges from wavelengths somewhat below 200 nm to around 850 nm. To achieve this, depending on the particular method, suitable systems for free atom generation, separation and quantification of radiation exist [15–18].

6.2.1. Atomic absorption spectrometry (AAS)

In AAS the sample is atomized in the light path of an appropriate radiation source. The extent of absorption of radiation is, theoretically, directly proportional to the number of atoms and hence to the elemental concentrations. This is mathematically formulated by the Lambert-Beer law:

$$A = \log \frac{I_o}{I_t}$$

where A = absorbance, I_o = initial radiation intensity, I_t = radiation intensity after passage through the sample.

This principle and the comparatively weak radiation from the applied light sources —initially mainly hollow cathode lamps [18]— limit the dynamic range. Hence calibration graphs for individual elements are only linear at lower concentrations. Absorption spectra are quite simple so that optical systems (gratings) do not require utmost resolution.

AAS was commercially introduced in 1955 with only rather cheap flame atomizers and cold vapor systems for mercury. Compared to methods previously applied such as colorimetry, classical atomic emission, X-ray fluorescence and neutron activation analysis it was quick —a single measurement could be completed in ten seconds— chemical separation was not required and the detection power was quite good at the trace element level. This soon made AAS the predominantly applied trace analysis method [19–23].

However, as with other analytical approaches, the method is not free from errors. These are due to instrumental-baseline drift interferences (physical-light loss by scattering on particles, absorption by molecules, emission from other elements) and chemical interferences (e.g. suppression of atomization).

Correction of these interferences was made possible by improvements to the optical systems (e.g. alternating current, double-beam systems with additional mirrors and lenses) that required more powerful radiation sources —electrodeless discharge and high intensity lamps. Correction of non-specific radiation losses was performed by measurement of a close non-absorbing element line and, with more universal application, by background correction using suitable continuum radiation (deuterium and hydrogen) sources [23].

In 1969, the commercial introduction of the graphite furnace broadened the application field of AAS further because of detection power improved by orders of magnitude for approximately 40 elements. From that time graphite furnace AAS became the preferred ultratrace analytical method in the biological and environmental sciences (e.g. refs. 19–32).

During the last two decades AAS has undergone continuous development and refinement. It can be regarded at present as a mature instrumental technique with various modes that range from flame AAS to the direct determination of metals and metalloids in solid samples.

The present state is characterized for *flame AAS* by quite effective atomization systems such as slotted quartz tubes [33] and the atom trapping technique [34,35] that enhance detection power and thus application range. A new field for flame AAS promising remarkable improvements in flexibility and detectability is to be expected if coupling with flow-injection techniques is performed [36–38]. Quick and successful progress is being made with this approach at present. It can be used with benefit for other AAS techniques such as hydride and cold vapor AAS as well [39].

With an element coverage of about 70 elements and quite good precision – under proper conditions a relative day-to-day standard deviation of around 2% is feasible – flame AAS still is a practicable method and one of the most reliable also allowing a high sample throughput.

Modern, fully equipped and computerized instruments for flame AAS, however, are not inexpensive and are at their limit if numerous elements have to be determined in the same analyte solution. Hence there is increasing competition at least from low – to medium – cost sequential plasma atomic emission spectroscopy (AES) systems. These attain for typical AAS elements at least the same detection power as flame AAS but are more sensitive for refractory ones with the further important advantage of being a multielement technique.

Graphite furnaces offer tremendous detection power, but are linked with a quite lengthy measuring cycle of on average around 100 s in an inert purging gas from sample injection, drying and thermal treatment to decomposing interfering matter (charring) to atomization and (thermal) cleaning. They were initially plagued by severe technical and analytical problems [e.g. 40–45]. Remarkable technical progress and the efforts of numerous researchers in close cooperation with manufacturers has meanwhile made this method a very promising and often the only practicable approach for the direct analysis of trace metals and metalloids in different materials.

To these developments belong, in conjunction with the early introduction of automated sample injection [23], the use of chemical modifiers. The latter alter the behavior of matrix and analyte in such a manner that the analyte signal can be delayed. Hence charring at higher temperatures in order to volatilize interfering matrix constituents prior to analyte atomization was made possible [46,47]. Further refinements were the observation that palladium and some of its compounds are universal modifiers [48–50], the application of graphite tubes coated with pyrocarbon, and the insertion of so-called L'vov platforms, completely made of pyrocarbon, into graphite tubes. The latter approach led to the "stabilized temperature platform furnace" (STPF) concept. In this technique the platform is predominantly heated by radiation from the tube wall. Thus vaporization occurs into an atmosphere that is at a higher temperature than the platform and in addition at a nearly constant temperature. These conditions are an important prerequisite for the minimization of interferences [51,52].

Another, similarly effective approach is the so-called probe technique which separates heating of the probe from heating of the tube [53,54].

An important step forward in graphite furnace AAS was the commercial introduction of Zeeman background correction. Here a magnetic field is applied either to the radiation source (direct Zeeman effect) or to the atomization system (inverse Zeeman effect). This effects a splitting of the emission lines so that the background radiation can be measured slightly off (direct Zeeman effect) or directly at the wavelength of the analyte (inverse Zeeman effect) [17,18,55–57]. This technique is much more effective than the formerly and nearly exclusively applied background correction with continuum light sources in that it is able to correct over the entire wavelength range. The potential of Zeeman effect background correction allows correction for optical background up to around 2 A and is particularly useful for a structured background. Another, less frequently applied, but in many cases similarly effective method is the Smith-Hieftje technique which also uses radiation from the same light source for background correction [17,18,25].

These techniques were combined with further improvements, e.g. maximum power heating and the choice of alternative gases such as hydrogen, freon and methane [58–61] for matrix modification and of oxygen for in-situ ashing of biological materials [62,63]. The latter was especially effective for materials with extremely low metal contents, e.g. cadmium and lead in human and retail milk and other materials difficult to analyze [64,65].

This progress permitted the use of integrated absorbance to evaluate "characteristic mass" data, which are signals equivalent to 0.0044 A•s, for various elements in different matrices. It was shown that under properly selected conditions for a number of metals this figure differed on average by only around 20% [66]. If this is compared with the large discrepancies reported in interlaboratory comparisons (see Chapter 1) this is remarkable progress and a first step towards absolute analysis in atomic absorption spectrometry [31]. Graphite furnace data for the elements treated in this book are listed in Table 6.1.

There were numerous attempts, due to problems with background correction often only partly successful, in previous years to analyze solid samples directly by AAS [67,68]. The availability of faster continuum source background correction and especially Zeeman background correction leads to the increasingly successful application of solid sampling and the analysis of aqueous slurries of finely ground solids by various instrumental concepts (e.g. refs.69–74). It has recently been possible to automate the slurry technique, particularly useful for the analysis of refractory elements [75,76], and it is now commercially available so that quick determinations are feasible with this technique.

An AAS instrument especially designed for solid sampling graphite furnace AAS, using the direct Zeeman effect, was also introduced commercially in the early 80's. It had a somewhat larger furnace than other systems and a sample boat that could take sample amounts up to at least 5 mg possessing properties similar to that of a L'vov platform

TABLE 6.1.

GRAPHITE FURNACE DATA FOR TEN ELEMENTS DUE TO STANDARD CONDITIONS
RECOMMENDED BY PERKIN-ELMER, 1990 (not applicable to 4100 ZL)

Notes: Tube indicates the type of tube, site the recommended sample position
The applicable charring temperature depends on the applied matrix modifier. The recommendations by Perkin-Elmer are relatively conservative and do not consider some progress achieved with palladium as a matrix modifier. Hence for some elements higher charring temperatures might be possible, leading to better working conditions. For the atomization step in general maximum power heating is recommended. For mercury the cold vapor technique is more sensitive, see text

Element	Wave-length (nm)	Tube/Site	Thermal pre-treatment (charring) (°C)	Atomization temp. (°C)	Characteristic mass (pg/0.0044A•s)	
Al	309.3	Pyro/Platf.	1700	2500	10	
As	193.7	Pyro/Platf.	1300	2300	17	
Cd	228.8	Pyro/Platf.	900	1600	0.35	
Co	242.5	Pyro/Platf.	1400	2500	7.0	
Cr	357.9	Pyro/Platf.	1650	2500	3.3	
Hg	253.7	Uncoated/Wall	150	2000	210	peak height
Ni	232.0	Pyro/Platf.	1400	2500	13.0[a]	
Pb	283.3	Pyro/Platf.	900	1800	12.0	
Se	196.0	Pyro/Platf.	900	2100	27.0	
Tl	276.8	Pyro/Platf.	600	1400	10	

[a] For nickel, wall atomization is somewhat more sensitive with a characteristic mass of 10 (pg/0.0044 A•s).

[77–81]. The instrument, particularly in its second improved version, showed in general a quite good performance in many duties from homogeneity studies to local analysis [81–84]. The basic instrument recently appeared in its third completely PC operated version [85]. In principle also automation of sample introduction is possible for solids, but at present not yet commercially available [86].

It was mentioned at the beginning of this section that commonly, i.e. in nearly all instruments available at present, the dynamic range is very limited. There is, however, a possibility to improve this by staircase modulation wave form AAS using a continuum radiation source that allows multielement analysis and covers 4–6 orders of magnitude in concentration (SIMAAC) [87]. The background correction was further improved recently [88] but the system is still not commercially available. Despite the limitation of the wavelength range there is a potential for the simultaneous analysis of up to sixteen elements. The limitation of the light source, however, might be overcome in the future by suitable laser techniques.

References on p.122

In 1987 the first commercial Zeeman-AAS system with graphite furnace was introduced that allowed the simultaneous measurement of four elements during one firing using normal element (hollow cathode) lamps. The data evaluation is somewhat similar to that described above for the SIMAAC system [89].

For mercury and hydride-forming metalloids special accessories to measure these elements after volatilization are commercially available from most manufacturers offering AAS instruments and others [90]. Mercury, being the first element analyzed by AAS [23], is commonly reduced after digestion of solid samples or directly in liquids to elemental mercury by the addition of stannous chloride or sodium tetrahydroborate and analyzed at ambient or slightly increased temperature with detection limits at the $\mu g/l$ level [23]. Since these reducing agents act differently, i.e. stannous chloride reduces only ionic mercury and sodium tetrahydroborate also transfers methyl mercury into elemental mercury, these reactions can be used for speciation. Other speciation approaches for mercury based on chromatographic techniques are described in Chapter 10.

If very low amounts of mercury have to be determined by AAS, preconcentration of the evolved elemental mercury on noble metals, preferably gold and silver traps,, and volatilization from these by rapid heating to around 500°C into quartz tubes is common. This produces sharp peaks in the recording system and attains detection limits at the low absolute picogram level (e.g. ref. 91–94). For preconcentration/subsequent heating the cold vapor technique was also successfully used in conjunction with gold- or platinum-lined graphite furnaces [95,96]. Also direct reduction/volatilization of mercury in solid samples by pyrolysis has been reported either with a special device and preconcentration on gold [97] or a nickel furnace as an accessory to the solid sampling Zeeman instrument described above [98]. The latter used air as a purge gas. This furnace has been redesigned recently and is now commercially available in an improved version [99].

Inorganic compounds of the hydride-forming metalloids arsenic and selenium can be transformed into volatile compounds by the addition of sodium tetrahydroborate to the analyte solution in appropriate, also commercially available, instruments and predominantly atomized in electrically heated quartz tubes [25]. Since organoarsenic compounds like arsenobetain are not transformed into volatile hydrides and different compounds behave differently depending on working conditions e.g. pH value, speciation is already possible by this method (see Chapter 11). Detection limits for arsenic and selenium can be significantly lowered to the picogram level by preconcentrating the hydrides at cryogenic temperatures [99–101].

The hydride method thus attains in its most sophisticated versions lower relative detection limits than the graphite furnace, but decomposition for total elemental levels prior to measurement has to be performed under proper conditions [102], which is also the case for the working conditions that have to be selected very carefully in order to achieve accurate results [100,103–106].

Furnace techniques can be used for the hydride method as well. Evolved selenium hydride e.g. could be trapped in a graphite tube at elevated temperatures and subsequently atomized with an absolute detection limit of 70 pg [107].

Although there is an increasing number of powerful competitive methods, it can be expected that due to the versatility and reliability of the various modes of AAS and steady progress [108,109] this approach will remain an important analytical tool at least for the foreseeable future.

6.2.2. Atomic fluorescence spectrometry (AFS)

In contrast to atomic absorption this method uses fluorescent radiation emitted from atoms which have initially absorbed radiation of an appropriate wavelength. Since the emitted radiation is not only proportional to the number of atoms present but also to the intensity of the radiation source, AFS combines the simplicity of AAS spectra with a wide dynamic range and an extraordinary detection power for a number of elements [18,110]. Despite the fact that the method introduced in 1964 [110] was not used commercially for years, numerous studies have been performed to evaluate the potential of the method by various conventional atomization techniques including also Zeeman background correction. Examples are cadmium in blood and urine [111,112], cadmium, mercury and zinc in environmental materials [113]. Cadmium, copper, iron, magnesium and manganese in water [114] as well as arsenic and selenium in coal [115].

Recently an atomic fluorescence spectrometer was described for the extremely sensitive determination of mercury and different mercury compounds — see also Chapter 10 — with species- dependent detection limits down to approx. 0.3 pg for elemental mercury [116]. Meanwhile it is commercially available and in successful use in a number of laboratories. With simple modifications it was improved so that a detection limit for elemental mercury below 0.1 pg has been achieved [117].

A further increase in detection power can be expected if vaporization in graphite tubes is combined with laser excitation [118] as was shown for lead [119] and other elements such as aluminium, cadmium, cobalt, copper, manganese, and thallium (e.g. refs. 119,120–123).

There are, however, only a few commercial instruments using the AFS principle. With the exception of the special AFS detector for mercury mentioned above at present only one commercial system is available that uses hollow cathode lamps for excitation and an inductively coupled plasma (ICP) as sample cell. The detection limits for 32 elements are in the order of that reported for flame AAS and ICP–AES with dynamic concentration ranges of up to five orders of magnitude. Spectral interferences and also matrix effects were negligible. However, detection limits for refractory elements were fair compared with ICP–AES [124,125]. Hence, the most promising version for future development because of superior sensitivity is undoubtedly laser-excited furnace AFS.

6.2.3. Atomic emission spectrometry (AES)

In AES — also called optical emission spectrometry (OES) — the radiation is generated by sparks, arcs, flames or plasmas producing numerous transitions. The spectra thus generated are, particularly for heavier elements, very complex and a number of lines depend on the excitation temperature. This multitude of lines can lead to interferences with lines of other elements, hence the quality of optical resolution systems must be very high. An advantage, especially in comparison to AAS, is the fact that radiation intensity is directly proportional to the atom concentration. Hence a dynamic range of several orders of magnitude can be achieved which makes AES a typical multielement method [18].

AES is a classical method and played a leading role in the past with excitation by sparks for solids and with flames for liquids. It was for some time, until the appearance of neutron activation analysis and X-ray fluorescence, the only commercially available multi-element method. With this increased competition its use stagnated or even decreased.

With the introduction of plasmas as excitation sources in the early 60's, reaching temperatures above 5000°C, the potential of the method was significantly widened [126]. After its commercial appearance in 1974 plasma, mainly ICP-, AES became a major analytical approach with a general potential for up to 70 elements. For about 40 elements detection power was drastically enhanced. Moreover, the high temperature of the plasmas led to a significant reduction of hitherto observed, mainly chemical, matrix interferences [127–131]. Very rapidly plasma–AES, predominantly ICP–AES but in some cases also with direct current (DCP) and microwave excitation (MIP), was introduced in numerous laboratories. This approach allowed sequential and simultaneous multielement determinations in numerous materials ranging from water [132,133], biological [134–136] and environmental [137–141] to technical samples [142].

If plasma AES is compared with flame AAS (see Table 6.2) its multielement potential and often superior detection limits make it frequently the more promising approach. Since often several elements have to be determined plasma–AES increasingly displaced flame AAS, since there are also several sequential instruments available at moderate prices. It has to be mentioned, however, that running costs for ICP–AES instruments are much higher because of the considerable argon consumption, so that in tasks in which only a few elements have to be determined and sensitivity is appropriate flame AAS still might be more economical. Precision, with typical values around 2%, is about the same for both methods. In comparison to neutron activation analysis and classical X-ray fluorescence plasma–AES offers much faster or simpler determinations if concentration levels are met.

Interferences of different origins that can affect accuracy as well as precision and detection power are still a problem for plasma–AES if complex matrices are to be analyzed [128,129,131,143,144]. There are different possibilities to overcome or at least to minimize these interferences, ranging from further optimization of nebulizers [145–147], preconcentration [148,149], coupling with chromatographic systems [150], flow injection [151–153] particularly if preconcentration is applied [154], internal standardization [155,156] to improved evaluation approaches such as simplex optimization [157] and Fourier transform techniques [158].

Analysis of solids is performed by spraying of aqueous suspensions of finely ground solids (slurries) into a conventional plasma torch by use of special nebulizers. Particle size must be $\leq 30\,\mu$m for proper operation (e.g. refs. 159–161). The analysis of slurries by the use of flow injection was also shown to be applicable to the multielement analysis of e.g. soil reference materials [162]. Despite somewhat poorer precision than for aqueous solu-

TABLE 6.2.

TYPICAL RELATIVE DETECTION LIMITS FOR 10 ELEMENTS FOR ICP–AES AND FLAME AAS FOR NON–INTERFERING AQUEOUS SOLUTIONS

Values are given in μg/l. The detection limit is defined as three times the standard deviation of the respective background noise or blank (3 s).

Element	ICP-AES[a]	Flame AAS
Al	4.2 (7)	45
As	17 (11)	30
Cd	0.7 (1)	0.7
Co	1.5 (3)	9
Cr	1.6 (1.5)	3
Hg	3 (7)	300
Ni	2 (4)	6
Pb	11 (15)	15
Se	15 (20)	150
Tl	0.5 (0.6)	15

[a] First set of data is based on published values by P.W.J.M. Boumans and J.J.A.M. Vrakking, *Spectrochim. Acta*, 42B (1987) 553–579; second set — in parentheses — are values reported 1988 by Jobin Yvon for the instruments JY 38 PLUS and JY 70 PLUS. Flame AAS data are based on a literature compilation [23] but corrected for 3 s.

tions, solid sampling techniques are of benefit for screening and homogeneity studies. Thermal vaporization prior to ICP–AES, including also solid sampling and future prospects such as the use of lasers were recently reviewed in great detail [163].

Since the detection power of plasma–AES is fair for elements such as arsenic, selenium and mercury, the hydride generation and cold vapor techniques can be combined with plasmas as well. For hydrides the direct current plasma combined with a suitable generator achieves detection limits in the order of ≤ 1 μg/l (e.g. refs. 164–166). The microwave-induced plasma is not applicable for the direct analysis of solutions but is very effective for dry analyte vapors, e.g. the extremely sensitive determination of mercury after preconcentration on gold with an absolute detection limit of 0.5 pg, corresponding to 0.01 ng/l [167].

Finally, carbon furnace AES should be mentioned, which is still not commercially available but might be of significance for the future development of simple multielement methods [168,169].

References on p.122

6.3. INDUCTIVELY COUPLED PLASMA–MASS SPECTROMETRY (ICP–MS)

The principle of ICP–MS is the combination of an inductively coupled plasma as ion source with a mass spectrometer as ion detector [18,170]. Most instruments available at present (ICP–MS was commercially introduced in 1983) are equipped with a quadrupole mass spectrometer. Recently more sophisticated systems have also become available that provide utmost mass resolution by a magnetic sector field mass spectrometer (cf. Section 6.8). This, however, is linked with appreciably higher costs, and ICP–MS is per se a very expensive technique even with the comparatively simple quadrupole MS detection.

The major advantages of ICP–MS are that it provides a nearly 100% ionization efficiency for most elements, relatively simple spectra, a wide dynamic range of about five orders of magnitude, a very low background level throughout a large section of the mass range, and high detection power, typically below 1 μg/l for most elements which is by far superior to that for ICP–AES and often approaches or even exceeds that of the graphite furnace, see Table 6.3. Moreover, there is also isotope ratio capability (cf. section 6.8) and the multielement potential allows high sample throughput.

There are of course also limitations. These are mainly due to the fact that the ICP operates at atmospheric pressure in the temperature range from 5000 to 9000 K and the mass spectrometer in a vacuum down to approx. 10^{-6} Torr and at ambient temperature. This interfacing often requires reduced-pressure sample introduction and is limited to

TABLE 6.3.

TYPICAL RELATIVE DETECTION LIMITS FOR 10 ELEMENTS FOR ICP–MS FOR NON–INTERFERING AQUEOUS SOLUTIONS

Values are given in μg/l. The detection limit is defined as three times the standard deviation of the respective background noise or blank (3 s).

Element	μg/l[a]
Al	0.16
As	0.04
Cd	0.03
Co	0.01
Cr	0.01
Hg	0.02
Ni	0.04
Pb	0.01
Se	1.0
Tl	(0.05)

[a] Most values are taken from a recent compilation [178], the remaining for Tl, given in parentheses, can be found in a table distributed by SCIEX 1988 for the elan ICP–MS elemental analysis system (10 s integration).

small quantities of sample. Also precision is commonly less than in ICP–AES if isotope dilutions are not applied. Moreover, also mass spectral interferences, mass overlap and interferences occur, e.g. from argon ions that are present at high concentrations in the plasma, and of molecular ions from acids like HCl, H_3PO_4 and H_2SO_4 used for sample treatment. From this, nitric acid is considered to be the most appropriate acid for decomposition prior to ICP–MS [18]. There are a number of detailed and critical reviews about the performance and potential of ICP–MS also addressing problems and their solution (e.g. refs. 170–179).

The field of applications of ICP–MS is growing rapidly with a steady increase of the relevant literature. Subsequently some examples will be given for applications of the method in various research areas.

ICP–MS is excellently suited for rapid screening of a large number of elements in freshwater [180–182] and in ultrapure acids [183]. Since signal suppression occurs due to higher amounts of alkali salts, e.g. in seawater, this can be corrected by internal standardization [184] or by on-line sample pretreatment to separate interfering alkali and alkaline earth elements and anions from the metals to be determined by adsorption on suitable chelating columns and elution prior to ICP–MS [185]. The same technique was also applied for the determination of trace metals in seawater and in an open ocean water reference material [186,187].

ICP–MS was applied after appropriate decomposition procedures for the determination of trace metals in various organic materials [188], in milk and blood, including isotope ratios [189–191], in food [172,192] environmental and reference materials [191,193], in soils [194] and marine sediments [195]. Methyl mercury and trace metals were determined in marine biological reference materials and seawater by the application of flow injection, standard addition, and isotope dilution for isotope ratios also to cope with chloride interference [189,196–198].

It was demonstrated with certified reference materials and industrial catalysts that the introduction of solid samples as slurries is practicable as well [199]. The detection power could be further increased for a number of metals in different materials if ICP–MS was coupled with electrothermal pre- volatilization [200,201]. Pneumatic nebulization and continuous hydride generation were investigated as sample introduction methods for isotopic analysis of stable isotopes of selenium in human metabolic experiments [202]. Further examples of ICP–MS can be found in the elemental chapters.

6.4. ELECTROCHEMICAL METHODS

The basic principle of polarography is the following electrode process:

$$M^{n+} + ne^- \underset{\text{oxidation}}{\overset{\text{reduction}}{\rightleftharpoons}} M$$

References on p. 122

Sometimes, however, the electrode process consists only of a change in valence state, as is e.g. the case during the reduction of Cr(VI) to Cr(III).

The electrolysis is performed at a microelectrode consisting of an inert conductive material such as mercury, graphite, graphite paste, gold etc. in an aqueous or non-aqueous solution. The solution is made conductive by the addition of a supporting electrolyte. The type of electrode and supporting electrolyte determines the potential range in which reduction or oxidation of a distinct metal takes place.

Polarography was introduced in the early 20's by Heyrovsky, hence electrochemical methods can be considered as belonging to the classical trace analytical methods. From its early days this analytical approach has undergone steady progress, but it appears that owing to new developments and fast growing computer techniques this progress is now in an expansive phase.

Here the terminology recommended by IUPAC (International Union of Pure and Applied Chemistry) will be used for this method [203]. According to this terminology the original method that applied a dropping mercury electrode is called *polarography* whereas *voltammetry* includes all techniques that evaluate current–potential relations, i.e. polarography as well. Hence, subsequently only the term voltammetry will be used for these techniques.

The history and progress of electrochemical methods is reviewed in many books and papers (e.g. refs. 204–213).

On the basis of Faraday's law, 1 mol of a compound transformed in an electrode process is equivalent to the high electric charge of $n \times 96500$ Coulomb (where n is the number of transferred electrons). Hence detection power and dynamic range are commonly very high in voltammetry.

In contrast to the methods treated above with thermal procedural steps, for voltammetric analysis the metals to be determined must be present completely dissolved in a non-interfering analyte solution, so that decomposition of biological and environmental materials prior to voltammetric analysis must be of utmost effectivity for accurate data as is the case for pressurized decomposition at $\geq 300°C$ with nitric acid [214] (see also Chapter 5).

Since each metal has a certain redox potential at a certain working electrode, frequently several metals can be sequentially determined at the same electrode in the analyte solution. An example is the sequential determination of copper, lead, cadmium, zinc and selenium(IV) at the mercury electrode. The gold electrode can be used for the determination of mercury, copper, chromium(VI) and arsenic(III) [210,213]. Since dissolved oxygen interferes, deaeration of the analyte solution with ultrapure nitrogen is necessary prior to measurement. During the electrode process always the potential E is given and the current i, which is proportional to the concentration of the analyte in the solution, measured. Material consumption is negligible, the process occurs only very close to the working electrode.

At present, *differential pulse techniques* have gained importance due to a favorable signal to noise ratio since E consists of a series of rectangular pulses superimposed on a voltage ramp. This allows determinations down to extremely low levels around 10^{-12} mol/l (0.1 ng/l) [210,213].

If the analyte concentration is above 50 μg/l, commonly direct determination with the dropping mercury electrode (DME) in the pulse mode (termed differential pulse polarography, DPP) is performed.

For lower metal concentrations an *in-situ enrichment* is necessary by using the stripping technique [210]. Metals that form an amalgam with mercury, e.g. cadmium, copper, lead, zinc etc., are accumulated by cathodic deposition in a mercury electrode while the solution is agitated frequently by rotating the electrodes [203]. Subsequently the amalgam is anodically oxidized in the determination step. The mercury drop electrode is applicable down to approx. 0.1 μg/l. If the metal concentration is even lower, the mercury film electrode (MFE) has to be applied. This electrode is electrolytically formed after addition of mercury(II) nitrate to the analyte solution by simultaneous *in-situ* deposition of a very thin mercury film during the cathodic enrichment step. Mercury, bismuth and arsenic can be deposited as an elemental monomolecular film at a gold or graphite electrode. In both cases the film is anodically oxidized as well. This procedure is called anodic stripping voltammetry (ASV). Metals forming insoluble compounds with mercury such as selenium are accumulated by anodic oxidation and the compound is subsequently cathodically reduced, which is called cathodic stripping voltammetry (CSV); see also adsorptive preconcentration techniques below.

Progress in computer techniques permitted the commercial implementation of square wave voltammetry (SWV). The principle of this technique was first proposed in 1952 [215]. It consists of the combination of a large amplitude square wave modulation with a staircase wave form. The voltammograms obtained show excellent sensitivity and rejection of background currents (see examples in Chapters 8 and 14). The main advantage of this technique compared to conventional stripping techniques, however, is the lowering of analysis time without losing the advantages of differential pulse stripping voltammetry (DPSV) [212,216–218].

Another very promising technique, applied with success during the last couple of years for an increasing number of elements, is voltammetry with adsorptive preconcentration [203,219,220]. In this technique, metals able to form a complex with large organic ligands are accumulated by adsorption of this complex at the surface of mainly a hanging mercury drop electrode maximally forming a monomolecular layer. Subsequently the complex is cathodically reduced in a similar way to that already described above for CSV [210,219–221].

This approach was initially applied using dimethylglyoxime as the chelating agent for nickel and cobalt [222–224]. Somewhat later nioxime was found to be even more effective for the determination of cobalt in seawater [225]. Up to now many investigations have been performed to evaluate optimal ligands and working conditions for a number of elements. At present 25 elements are known to be accessible to the adsorptive preconcentration technique [219,220,226]. Of the elements treated in this book except nickel and cobalt, aluminium [210,227], cadmium [228], chromium [229,230] and lead [228] can be determined with cathodic stripping after adsorptive preconcentration with low detection limits. Since several of these are not or only poorly accessible to classical voltammetric techniques, this is a significant extension of voltammetric methods at least for trace analysis in natural waters since up to now only a few of these procedures (e.g. for nickel and cobalt) can be used for metal determinations in biological and environmental materials after an appropriate decomposition [214]. Surface adsorption on the electrode, however, significantly limits the dynamic range so that for higher analyte concentrations either reduction of adsorption time or dilution is required [220].

There are other possibilities in voltammetry to achieve higher sensitivity. For instance, selenium(IV) can be determined by cathodic stripping voltammetry after the formation of piazselenol and its extractive preconcentration from natural waters down to the lowest natural levels [231].

The combination of different electrochemical techniques frequently permits oligoelement determinations. If, e.g. classical polarography, different normal stripping modes, and adsorptive preconcentration are combined up to eight elements can be sequentially determined in the same analyte solution with an average analysis time of 25 min for each element, the necessary calibration — mainly standard addition — steps and deaeration included [232]. If the square wave mode already mentioned above is applied, the average analysis time can be significantly reduced further. By technical optimization such as use of parallel cells that allow deaeration prior to the start of the analytical cycle, sample throughput is comparable to that attainable with single element graphite furnace AAS, often with a superior detection power. This is particularly the case if aqueous samples with extremely low metal levels are concerned, since the above mentioned in-situ enrichment directly from the analyte solution has — if carefully precleaned working cells from PTFE are used — practically negligible blanks in contrast to enrichment techniques applied for other trace analytical methods. Hence voltammetry has proved to be an ideal method for the direct, accurate determination of metal contents at the low ng/l level in seawater, inland water and precipitation [225,228,233–235].

For many water types, however, UV digestion prior to voltammetric determination is necessary in order to decompose interfering organic matter. If it is performed in carefully precleaned quartz tubes only spurious contamination is introduced [236].

Table 6.4 summarizes the present potential and working conditions of electroanalytical methods in non-interfering analyte solutions for the ten subsequently treated elements.

Since voltammetry is specific for distinct chemical compounds it is in addition a valuable tool for the speciation of metals in fresh- and seawater [237–240].

The extremely low detection limits for trace metals in fresh- and seawater due to often negligible blanks cannot be achieved if materials have to be analyzed that require a complete decomposition. In this case obtainable blanks are typically in the range of ≤ 1 to a few ng per digestion vessel and element. Since intakes commonly vary from approx. 100 to 400 mg dry weight [214] practically attainable detection limits are in the older of μg/kg. Compared to the graphite furnace that frequently permits the direct analysis in liquid and solid samples this is somewhat disappointing [218]. The potential range over which most metals are oxidized (or reduced) is comparatively narrow. Hence in materials with relatively high metal contents it can be difficult to resolve the responses of metals that

TABLE 6.4.

TYPICAL WORKING CONDITIONS AND PRACTICALLY ATTAINABLE DETECTION
LIMITS OF TEN ELEMENTS WITH VOLTAMMETRIC METHODS IN A NON-INTERFERING
AQUEOUS SOLUTION [5,203,210,219–221,227–230]

Element	Method	Electrode	Chelate	Detection limit [ng/l] approx.
Al	DPAV	HMDE	DASA	50
			SVRS	2000
As	DPASV	Au		200
Cd	DPASV	HMDE		50
		MFE		0.5
	DPAV	HMDE	Q-8-O	10
Co	DPAV	HMDE	DMG/NO	1
Cr	DPAV	HMDE	DTPA	20
Hg	DPASV	Au		50
Ni	DPAV	HMDE	DMG	1
Pb	DPASV	HMDE		50
		MFE		0.5
	DPAV	HMDE	Q-8-O	50
Se	DPCSV	HMDE		200
Tl	DPASV	MFE		10

Method: DPASV = differential pulse anodic stripping voltammetry; DPCSV = differential pulse cathodic stripping voltammetry; DPAV = differential pulse voltammetry with adsorptive preconcentration [203] also termed adsorptive stripping [219] or adsorptive cathodic stripping voltammetry [221].
Electrode: Au = rotating gold electrode; HMDE = hanging mercury drop electrode; MFE = rotating mercury film electrode.
Chelates: DASA = 1.2-dihydroxyanthraquinone-3-sulphonic acid; DMG = Dimethylglyoxime; DTPA = diethylenetriaminepentaacetic acid; NO = Nioxime; Q-8-O = quinolin-8-ol; SVRS = Solochrome violet RS.

are very close so that overlap is probable. This depends, however, also on the choice of electrodes and other measures, e.g. chemical (masking, separation of interfering element) mathematical and instrumental that can be used to solve the problem [210].

A comparatively new method is potentiometric stripping (PSA). It is based on the potentiostatic reduction and amalgamation of metals. Subsequently the potential–time curve is registered when the reduced metals are reoxidized with the aid of mercury(II) ions. At the same time a mercury film electrode is formed by deposition of metallic mercury on glassy carbon or fiber electrodes of different materials. The reoxidation reaction employed for the determination of the preconcentrated metals occurs by disconnecting the potentiostatic circuitry and hence without an external current. Electrochemical reoxidation is performed with the nobler Hg^{2+} ions, the oxygen dissolved in the analyte solution and other oxidizing substances and the potential–time relation recorded. The metals deposited as amalgams were sequentially and selectively reoxidized based upon their individual electrochemical redox potential, the reoxidation time being proportional to the individual concentration in the analyte solution.

The advantages of this technique are that in contrast to other electrochemical techniques deaeration frequently is not necessary. Moreover, the method is less sensitive to traces of organic substances that severely interfere in voltammetry [241,242]. This technique does not provide the detection power of voltammetry but has, besides the properties already explained, the advantage of a simple and relatively inexpensive instrumentation. Increasing interest in the continuous monitoring of heavy metals has resulted in the adaptation of PSA to flow-through measurements which offer several advantages, e.g. ease of automation, high sample throughput and medium exchange [243]. The recent introduction of the −however more expensive− computerized PSA and improved, mainly fiber, electrodes [244,245] has significantly extended the working range down to the low μg/l level with the possibility for the determination of e.g. cadmium and lead in whole blood [246], various metals in milk and milk powder [247], mercury in tapwater [248] and arsenic in seawater and urine [249]. The already mentioned properties also make PSA well suited for coupling with flow injection systems [250,251]. Hence PSA can be considered as a promising extension of electrochemical methods in that it has led to a remarkable gain in simplicity, versatility and speed.

6.5. X-RAY FLUORESCENCE SPECTROMETRY (XRF)

If atoms are subjected to radiation of appropriate energy, electrons from the inner (K, L, M) atomic shells are eliminated. Due to this elimination, electrons from the outer shells drop, following distinct laws, into the free positions. This produces a comparatively simple electromagnetic radiation ranging from approx. 0.6 to approx. 120 keV, which corresponds to wavelengths from approx. 2 to approx. 0.01 nm. The radiation thus generated is unique and can be precisely computed for each element, and the frequency of the emitted lines is proportional to the square of the nuclear charge number (Z) of the respective element. Hence, this radiation is used after wavelength or energy-dispersive registration for qualitative detection or quantification of all elements with $Z > 6$. These rays were called X-rays by their discoverer, Röntgen.

The generation of X-rays is performed by X-ray tubes, direct irradiation with X-rays, gamma radiation from radionuclide sources but also by particles such as electrons, protons, alpha and even heavier particles [252–255]. The induction of X-rays by particles is termed PIXE. This technique is especially useful for thin targets and elements with lower nuclear charge numbers. It has considerable detection power down to concentrations in the order of μg/kg [256–259].

The wavelength-dispersive −recently reviewed for its sequential mode and progress achieved up to now [260]− as well as the energy-dispersive X-ray technique are perfectly suited for multielement determinations with a comparatively high sample throughput. However, typical absolute detection limits are in the order of mg/kg. Hence classical XRF methods are only applicable to the determination of higher contents from approx. 5 mg/kg [261] i.e. for fingerprint studies and the analysis of e.g. polluted biota, sewage sludge, soils and dust filters. For the determination of lower contents, preconcentration techniques are commonly required. If concentrations and sample properties are suitable the attainable precision can reach 2% or less. Hence the method can be used successfully for homogeneity estimations, e.g. during preparation of reference materials.

The matrix dependence of each analytical line from each element present in the sample to be analyzed with $Z > 20$ can cause irregular stray effects of the exciting radiation and thus lead to inaccurate results [254,262]. Hence XRF requires a careful and appropriate sample preparation by superfine milling or fusion and the use of certified reference materials with a matrix composition that is as close as possible to the unknown sample. If such materials are not at hand, the assessment of accuracy by comparative analyses with independent analytical methods is mandatory if reliable quantitative results have to be obtained. For the analysis of samples with very high elemental levels, the dilution of a matrix with low Z such as cellulose leads to a minimization of those effects and can be favorably applied for the preparation of reference samples.

Collimation of the exciting radiation by special arrangements e.g. mirrors on very small areas, is applied in microprobe techniques. By using this method, absolute detection limits in the ng range can be achieved with e.g. PIXE technique [256,259]. If the X-ray fraction of synchrotron radiation (SYXRF) is used, absolute quantities down to the pg-level can be measured with spot sizes of the beam of less than 10 x 10 μm for the elements from $Z = 30$ (Zn) to $Z = 19$ (K) [263,264]. This is a very effective technique for metal distribution studies in biological matter such as tissue, hair, plants, and sediments.

A comparatively new methodological approach, total-reflection X-ray fluorescence (TXRF) analysis, has found improved and successful application during the last decade. The principle is the use of the total reflection of the exciting beam from conventional radiation sources at a flat support. Due to a remarkable improvement of the signal-to background ratio absolute detection limits can be attained that are 2–3 orders of magnitude lower than that of conventional X-ray techniques [264,265]. The method, however, requires proper sample preparation techniques that include preconcentration, digestion and in some cases also separation from interfering matrix constituents [266]. The method has been increasingly and successfully applied for multielement analysis in environmental and biological materials ranging from rain, fresh and seawater to a number of solid materials [264,267–269]. It was demonstrated that the technique can be applied to the direct analysis of solid samples as well [270].

The analytical potential as far as TXRF detection limits in non- interfering aqueous analyte solutions are concerned is shown for the elements treated in this book in Table 6.5.

XRF also has a potential for *in vivo* elemental analysis for concentration assessment for lead in bones. However, it can be used for cadmium contents in kidney, mercury and some other elements such as strontium and platinum as well [272,273].

References on p. 122

TABLE 6.5.

DETECTION LIMITS FOR 10 ELEMENTS USING TOTAL REFLECTION X-RAY FLUORES-
CENCE IN A NON-INTERFERING AQUEOUS ANALYTE SOLUTION

Based on a sample volume of 50 μl [271].

Element	Detection limit (μg/l)
Al	—
As	0.2
Cd	0.4
Co	0.1
Cr	0.4
Hg	0.2
Ni	0.1
Pb	0.2
Se	0.2
Tl	0.2

6.6. NUCLEAR METHODS

Nuclear methods in analytical chemistry consist of activation analysis and the use of radiotracers. The principle of *activation analysis* is the alteration of the atomic nucleus by bombardment with particles. These are e.g. neutrons, protons, helium nuclei and high-energy photons generated in facilities such as nuclear reactors and different accelerators. The reaction of these particles with the atomic nucleus leads to an excitation either by energy supply or the absorption of a particle. If a particle is absorbed, initially either an isotope of the target element (neutron absorption) or of another element (proton, helium nucleus) is formed [e.g. 274–281].

The number of excited atoms depends on the intensity of radiation, the irradiation time, and on the probability of absorption of radiation by the nucleus, termed cross section, and is directly proportional to the number of atoms present.

The excited atomic nucleus is predominantly labile. This means that it tends to reach an energetically lower level, which is always accompanied by emission of alpha- or beta-particles and/or gamma-quanta. This is called radioactive decay. The emitted gamma-radiation is characteristic of each isotope and each energy level and shows sharp, nearly monoenergetic, lines in the electromagnetic spectrum.

Since the intensity of radiation only depends on the initial number of excited nuclei, radioactive decay follows a 1st order reaction. Thus the excited nucleus is characterized by the decay constant $\lambda = \ln 2/T$ and the energy of the emitted radiation, where T is the half-life, i.e., the time after which half of the excited atoms have "decayed", i.e. have reached a lower energy level. Half-lives range from fractions of seconds up to years.

Identification and quantification of the isotope representing the element to be determined is performed by measuring the energy that typically ranges from a few keV up to some MeV, the intensity of radiation expressed as decays per unit of time and in some cases also the half-life. For many elements large cross sections, high intensity of the exciting radiation and a favorable half-life lead to low absolute detection limits (see Table 6.6). The very high number of lines and half-lives, especially in complex environmental and biological materials sometimes make simultaneous multielement analysis difficult, even if the modern germanium (lithium)-detectors with excellent energy resolution and PC-based data evaluation are applied [281,282]. In these cases and also in order to use the whole methodological potential, radiochemical separations for single elements but also groups of elements are common. This version is termed radiochemical activation analysis in contrast to instrumental activation analysis [278,280,283].

Introduced after the 2nd World War, activation analysis, mainly with neutrons (NAA), was for more than two decades for many elements the most sensitive method. This was in addition accompanied by an extremely high reliability due to its specifity and the important fact that commonly the introduction of errors (e.g. contamination and losses) during the minimal sample pretreatment was much less probable compared to other contemporary methods.

The invaluable merit of all modes of activation analysis is the gain achieved for insights into bioinorganic elemental relations in medicine, biology and ecology. This was due to significantly improved data of normal levels of trace metals in numerous materials [284–289].

Despite the eminent progress in instrumental analysis in general, activation analysis, mainly NAA, is still a very useful approach. It can be regarded, if applied with expertise and modern equipment, as a potentially definitive method, if all sources of uncertainty can be accounted for [281,282]. The latter is often possible, hence it is still of great importance for quality assessment in comparison to and complementation of other methods and particularly in the certification of reference materials [281,290–300].

An example of a relatively new version of activation analysis is the *in vivo* elemental analysis of cadmium by fast neutrons from a neutron source in order to assess the level of this element in organs of exposed persons (see Chapter 8). The technique is also applicable for the analysis of some other elements [301]. Another example is the use of "cold" neutrons for the generation of short-lived isotopes to determine elements that are less accessible under regular reactor conditions (prompt gamma NAA). Since only isotopes with extremely short half-lives are generated the irradiated samples can be used after measurements for other trace analysis techniques [302].

As there are only a few research reactors or other irradiation facilities such as linear accelerators, cyclotrons etc. a broad routine application of activation analysis techniques is not achievable. Moreover, the comparatively long "cooling", i.e. storage, times often necessary, expensive security precautions and the need for specially trained staff limit their application to a couple of specialized and highly experienced laboratories.

Radiotracer techniques are very promising for methodological development and studies on the behavior and reaction mechanisms of metals and metalloids in biochem-

TABLE 6.6.

TYPICAL RELATIVE DETECTION LIMITS FOR 10 ELEMENTS FOR INSTRUMENTAL NEUTRON ACTIVATION ANALYSIS AND RADIOISOTOPES FOR RADIOTRACER STUDIES

The detection limit is defined as three times the standard deviation of the respective background noise or blank (3 s)

Element	Radioisotope for INAA[a]	Half life	Det. limit (μg/kg)	Radioisotope for tracer studies (commercially available)	Decay mode[b]	Half life
Al	^{28}Al	2.25 min	4	^{26}Al	β^+,γ	$7.2 \cdot 10^5$ a
As	^{76}As	26.4 h	0.05	^{76}As	β^-,γ	26.4 h
Cd	^{111m}Cd	49 min	1.5	^{109}Cd	ε	453 d
				^{115}Cd	β^-,γ	44.8 d
Co	^{60m}Co	10.5 min	0.03	^{60}Co	β^-,γ	5.27 a
Cr	^{51}Cr	27.7 d	20	^{51}Cr	β^+,γ	27.7 d
Hg	^{197}Hg	64.1 h	0.03	^{197}Hg	ε,γ	64.1 h
				^{203}Hg	β^-,γ	46.6 h
Ni	^{65}Ni	2.52 h	15	^{63}Ni	β^-	100 a
Pb	^{207m}Pb	0.8 s	3000	^{210}Pb	β^-,γ	22.3 a
Se	^{77m}Se	17.5 s	1.1	^{75}Se	γ	120 d
Tl	^{204}Tl	3.78 a	40	^{204}Tl	β^-,ε	3.8 a

[a] INAA data are based on a sample weight of 500 mg (non-interfering matrix) and the following irradiation and counting conditions. Thermal neutron flux $1 \cdot 10^{13}$ n$\cdot$cm$^{-2}\cdot$s^{-1}, irradiation time 5 h (maximum), counting with a 40 cm^3 Ge(Li)detector, with a sample-to-detector distance of 2 cm; zero decay time before start and maximum counting time = 100 min.

[b] Decay modes: β^+ = positron decay; ε = electron capture; β^- = negatron decay; γ = emission of gamma quanta.

istry as well as analytical and environmental chemistry [303–310], but require appropriate equipment and experience for handling and measurement.

6.7. CHROMATOGRAPHIC METHODS

Chromatographic systems are able to separate different compounds by using their behavior to travel under certain conditions at different speeds through columns that consist of a stationary and a mobile phase. This compound-specific separation technique is commonly combined with a compound- or element-specific detection system [311]. These properties make all chromatographic techniques well suited for chemical speciation [312–314].

In *gas chromatography* (GC) the mobile phase is an inert gas (nitrogen, hydrogen, argon). The compound to be separated and determined is usually volatilized by heat, hence there are sometimes problems with thermally unstable compounds. Present GC instrumentation has reached a high technical level with computer-controlled pressure and temperature and automated sample changers. Detectors commonly applied are e.g. conductivity (CD), flame ionization (FID) and electron-capture (ECD) detectors. For metal determinations atomic spectrometric detectors are directly coupled with gas chromatographs [315].

Inorganic compounds are not directly accessible to GC techniques. Hence they have to be transformed into appropriate organometallic compounds. Numerous elements are transformed into di(trifluoroethyl)dithiocarbamato chelates for determination by capillary GC with FID or ECD detection allowing sensitive multielement determinations (ref. 316, see also Chapter 14).

Aluminium in seawater has been determined by transformation into the 1,1,1-trifluoro-2,4-pentanedione by GC–ECD down to 0.6 nmol/l [317]. Transformation into volatile hydrides was applied for As, Sb and Sn and butyltin species in water [318,319]. Picogram levels of methyl mercury and ionic mercury were transformed into the respective ethyl-methyl and diethyl compounds, separated by cryogenic GC and determined by cold vapor atomic fluorescence [320].

In *liquid chromatography*, predominantly applied as high-performance liquid chromatography (HPLC), the mobile phase is always a liquid. Hence, the working temperatures are comparatively low, which makes the determination of thermally less stable compounds possible [321,322]. Since this technique requires elevated pressure it is more expensive than GC. Detection is performed spectrophotometrically, electrochemically [323] and especially for metals by atomic spectrometric methods [324,325]. Examples are the use of dithiocarbamate complexes for Cu, Ni and Pb in urine [323] of bis(ethoxyethyl)dithiocarbamates for Cd, Co, Cu, Hg and Ni in water [326] and of Cu, Be, Al, Ga, Pd and Fe after solvent extraction as actylacetonates [327].

Ion chromatography using various chelating cation-exchange resins has significantly gained in importance since its introduction in 1975 [328–331]. Usually no pre-column derivatization is necessary and separations are performed as in anion exchange by elution with appropriate solutions. In some cases also post-chromatographic derivatization is performed prior to quantification. Using various detectors such as UV–VIS, conductivity and pulsed amperometric the multielement potential and the achievable detection limits are outstanding [332–334]. Ion chromatography can be regarded as being complementary to atomic spectrometry and voltammetry since it combines total metal and metal species detection and quantification. Moreover, it allows — at least in aqueous samples — the simultaneous determination of cations and anions [335]. Here the reader's attention should be drawn to a recent comprehensive handbook of ion-exchange resins and their analytical application [336]. *Thin-layer chromatography* can be used for the separation and determination of metal chelates [337].

As for the other trace and ultratrace methods, treated in this chapter, contamination precautions are mandatory to attain meaningful results.

6.8. MASS SPECTROMETRY (MS)

MS analysis of metals starts with the thermal generation of gaseous ions by a spark of high potential, by an electron- or an ion current. The generated ions or molecular fragments are then separated by energy/mass focusing in strong magnetic fields. The detection or quantification was formerly performed by use of photoplates [338]. This has now been replaced by direct-reading electronic systems. A particular advantage is that all elements of the Periodic Table are accessible to this method with absolute detection limits down to 10^{-12} g absolute [339–341].

Spark source MS (SSMS) is predominantly applied for simultaneous multielement analysis. The method requires transformation of the sample into a conducting target if not initially present as a conducting material. The preparation of the conducting target is performed by mixing the sample with ultrapure graphite. Hence the attainable detection limits depend mainly on the purity of this graphite. Precision is around 10% under optimal conditions, whereas accuracy generally is $\leq$ 30% if calibration is applied or within one order of magnitude without calibration [342]. These properties and the expensive instrumentation have decreased the use of this method because it is not flexible enough for the analysis of biological and environmental materials.

If, however, isotope dilution MS (IDMS) is applied, excellent precision and accuracy can be achieved under optimal conditions. The method, however, is only applicable with success if the sample is completely decomposed and if an appropriate isotope of the element to be determined is added to the sample and this isotope attains exactly the same chemical form as the ions of the element to be determined in the final analyte solution. Since only the mass ratio is measured for quantification no exactly known sample amount has to be transferred into the mass spectrometer. The method is only applicable to elements with at least two stable isotopes or those that posses radioisotopes with half-lives of $> 16^6$ years. Therefore, of the elements treated in this book hence aluminium, arsenic and cobalt cannot be analyzed by IDMS.

IDMS is applicable with most ionization modes, e.g. thermal ionization [343], spark source [344,345] electron impact [346] and field desorption [346,347]. The method needs in most cases chemical pretreatment, i.e. single element or group separation and preconcentration procedures prior to determination [346]. For reliable application down to the μg/kg or even ng/kg level utmost contamination precautions and practical experience as well as technically well equipped laboratories are required (e.g. refs. 348–351).

Due to the expensive instrumentation and laboratory equipment and the time-consuming sample preparation IDMS is not a routine method. However, it is the method of choice when extreme requirements for accuracy and precision, e.g. for certification of reference materials and if control of other less reliable methods is required [351–355] and is also considered as a potentially definitive method. Since isotope ratios can be determined with very high precision, MS is an invaluable approach for the identification of e.g. anthropogenic and geogenic lead sources in the environment [356–358] and for metabolic studies with stable isotopes in man and experimental animals [359–361].

Trace element distributions on surfaces of solid biological materials and in very small samples can be determined by mass spectrometers coupled with ion microprobes, termed *secondary ion mass spectrometry* (SIMS) [340,362] or by the use of laser beams for ionization (LAMMA) [363,364].

In these techniques extremely low absolute detection limits can be attained. The instrumentation is, however, very expensive and usually does not allow exact quantification because of problems with calibration.

There are additionally some also very sensitive and comparatively new mass spectrometric techniques. Examples are fast atom bombardment MS [365], and resonance ionization MS based on thermal ionization and lasers [366].

Glow discharge MS, already in use 50 years ago, is now again being increasingly used for the direct bulk analysis of solids [367]. The possibilities of various modes of MS, also including the glow discharge approach, for inorganic mass spectrometry of solid samples has been reviewed recently [368].

6.9. SPECTROPHOTOMETRY AND RELATED TECHNIQUES

UV–VIS-spectrophotometry is, in contrast to *atomic* spectroscopy, the *optical spectroscopy of molecules*. Absorption and emission processes of molecules due to splitting of electron transfer into numerous oscillation and rotation terms, however, are much more complex than that of atoms. This leads to series of lines in the gas phase −which might cause interferences in atomic absorption and atomic fluorescence spectrometry− and to absorption continua with relatively broad absorption or fluorescence maxima in solutions [369–372]. A distinct colored metal compound cannot be accurately determined colorimetrically in a complex sample that contains other metals. Hence colorimetric methods require selective separation from interfering ions which is commonly performed by various methods such as ion exchange, chelate formation and solvent-extraction. Colored chelates can also be used for determination after separation from the matrix.

References on p.122

Spectrophotometric instruments are similar to those for AAS, however, the commonly occurring broadband spectra require optics of less high quality. The modern double-beam spectrophotometers predominant available offer high stability, low noise and often the advantage of computerized background correction, area integration, sample supply, as well as derivative spectroscopy [373], diode array [374] and Fourier transform techniques, etc. with on average a precision below 5% [373,375,376]. Highly sensitive spectrophotometric methods are feasible if a catalytic reaction is performed prior to determination [377]. Spectrophotometers are applied with an increasing tendency in biochemistry, organic and clinical chemistry and as detectors for e.g. HPLC [378]. Sensitive multielement determinations are possible with chemiluminescence using spectrofluorimeters after ion chromatographic separation [379]. For total metal determination in environmental chemistry there is increasing competition from faster methods with multielement potential. Thus spectrophotometry has been replaced in many cases by atomic spectroscopic and electrochemical methods. With the exception of special applications mentioned above it still serves as a reliable routine or reference method if higher elemental contents have to be analyzed [380].

6.10. FUTURE PROSPECTS

Though forecasts of future developments are difficult, in this final section a few, in the opinion of the author, possible trends and prospects shall be mentioned in the order of the methodological sections.

For *atomic spectrometry* in general on the one hand some decrease in the application of single element techniques is probable [381]. This is especially the case for flame AAS that is competing with decreasing success with plasma emission instruments which offer quick multielement determinations at moderate costs (see Section 6.2.3). On the other hand, it is obvious that in the near future the graphite furnace will remain a technique permitting direct, accurate and sensitive analysis of many metals in complex matrices including solids with appreciable precision. Further improvements and new applications, e.g. in hyphenated techniques such as AES and ICP–MS can be expected as has been recently reviewed [382,383]. Hence the forecast of a still frequent, even slightly increasing, use of furnaces seems to be justified. A promising technique, now also commercially available, is certainly all versions of atomic spectroscopy and ICP–MS flow injection analysis with a remarkable gain in separation/preconcentration [39,384]. Promising developments during the last decade might in addition lead in the near future to e.g. the commercial introduction of an ICP as an atomizer for atomic absorption in multielement mode [385,386], the multielement methods *coherent forward scattering* [387] and *furnace atomic non-thermal excitation spectrometry* (FANES) [388–390]. Another possibility for *multielement AAS* could be the use of a modulated Grimm-type glow discharge plasma as primary light source [391]. Laser techniques are promising in all atomic spectrometric methods, but also for graphite furnace AAS [392].

Atomic fluorescence spectrometry with laser excitation achieves extremely low detection limits down to the femtogram range [122,393–395]. Despite this promising potential some doubts might be expressed about a commercial introduction of instruments in the near future, with affordable prices necessary for broader use.

Atomic emission spectrometry will certainly benefit from progress in flow injection techniques as already mentioned above as well as from further progress in solid sample analysis, e.g. by vaporization into the plasma source, possibly combined with use of lasers [396]. Since Fourier transform spectrometers attain a much higher resolution and complete spectral record compared to scanning monochromators, improvements of commercial instruments in this direction can be predicted [397,398].

In *ICP–MS* further progress is to be expected with the appearance of additional instruments from competing manufacturers. The introduction of instruments with improved resolution (see Section 6.3) should lead to a further minimization of ion interferences and a gain in accuracy even for difficult samples.

Electrochemical methods have gained importance in the last decade due to permanent methodological and instrumental progress. Commercial availability of computerized potentiometric stripping systems will stimulate further advances in direct analysis of a number of trace elements in numerous liquid samples ranging from natural waters and beverages to biological fluids (see Section 6.4). The problems observed hitherto in sample decomposition have now been completely overcome by the HPA decomposition technique. It is to be expected that this will lead to a further extension of voltammetric methods in various analytical applications.

In *X-ray fluorescence spectrometry* the progress of the more sensitive modes, such as e.g. TXRF will continue, probably also supported by the HPA decomposition method that can further minimize interfering residues.

Neutron activation analysis and *radiotracer applications* will still be of benefit. NAA will remain of importance as an invaluable reference method, possibly with more emphasis on radiochemical methods and for some important elements of anthropogenic or geogenic origin less accessible to other methods. The continuing application of radiotracers for many studies in method development and elucidation of mechanisms and the commercial production of some new isotopes hitherto not available or only with difficulty would be of increasing importance particularly in environmental chemistry.

Chromatographic methods, especially in conjunction with most speciation approaches, will certainly find increased use in many fields. Here, HPLC will probably be more generally applicable than GC to speciation studies, particularly if compounds are to be detected that are sensitive to higher temperatures or to any derivatization technique [399].

For *mass spectrometry* improvements towards simpler and less expensive, but for many analytical purposes still sufficient, commercial quadrupole mass spectrometers allow higher sample throughput with still excellent accuracy and satisfactory precision, comparable to that of voltammetry and the graphite furnace. These systems will most probably increase the use of this very important technique [400].

References on p.122

Spectrophotometry, mainly in combination with chromatographic and FIA systems, could provide extremely low detection limits for some applications, particularly in speciation studies, if the potential of laser excitation, chemiluminescence [401,402] split-beam optics [403] and diode array technique [404] is used.

REFERENCES

1 G. Tölg, *Analyst*, 112 (1978) 365–376.
2 IAEA (International Atomic Energy Agency), *Elemental Analysis of Biological Materials. Current Problems and Techniques with Special Reference to Trace Elements (Tech. Rep. Ser.,* No. 197), IAEA, Vienna, 1980.
3 I.L. Marr and M.S. Cresser, *Environmental Chemical Analysis,* Int. Textbook Comp., Glasgow, London, 1983.
4 M. Stoeppler and H.W. Nürnberg, in A. Vercruysse (Editors), *Hazardous Metals in Human Toxicology,* Elsevier Science Publishers, Amsterdam, 1984, pp. 95–149.
5 M. Stoeppler and H.W. Nürnberg, in E. Merian (Editor), *Metalle in der Umwelt,* Verlag Chemie, Weinheim, Deerfield Beach, FL, Basel, 1984, pp. 45–104.
6 W.R. Wolf and J.R. Harnly, in R.D. King (Editor), *Developments in Food Analysis Techniques,* Vol. 3, Elsevier, London-New York, 1984, pp. 69–97.
7 G. Knapp, *Fresenius' Z. Anal. Chem.,* 317 (1984) 213–219.
8 B. Sansoni (Editor), *Instrumental Multielement Analysis.* VCH, Weinheim, Deerfield Bech, FL, Basel, 1985.
9 L. Fishbein, *Int. J. Environ. Anal. Chem.,* 28 (1987) 21–69.
10 G. Tölg, *Fresenius' Z. Anal. Chem.,* 329 (1988) 735–736.
11 G.V. Iyengar, *Elemental Analysis of Biological Systems, Vol. 1: Biological, Medical, Environmental, Compositional, and Methodological Aspects,* CRC Press, Boca Raton, FL, 1989.
12 C.S. Creaser and A.M.C. Davies, *Analytical Applications of Spectroscopy,* Royal Society of Chemistry, Cambridge, 1989.
13 W. Fresenius and I. Lüderwald (Editors), Presentations at the *11th Int. Symposium on Microchem. Techniques, Wiesbaden, Fresenius Z. Anal. Chem.,* 334 (1989) 601–721.
14 G. Tölg, *Fresenius' Z. Anal. Chem.,* 329 (1988) 735–736.
15 E. Merian (Editor), *Metals and Their Compounds in the Environment − Occurence, Analysis and Biological Relevance,* VCH, Weinheim, New York, Basel, Cambridge, 1991.
16 W. Slavin, *Anal. Chem.,* 58 (1986) 589A–597A.
17 E. Metcalf, *Atomic Absorption and Emission Spectroscopy − Analytical Chemistry by Open Leraning,* Wiley, Chichester, New York, Brisbane, Toronto, Singapore, 1987.
18 J.W. Robinson, *Atomic Spectroscopy,* Marcel Dekker, New York, Basel, 1990.
19 G.F. Kirkbright, in: *Elemental Analysis of Biological Material, (Tech. Rep. Ser.,* No. 197), IAEA, Vienna, 1980, pp. 141–165.
20 J.E. Cantle (Editor), *Atomic Absorption Spectrometry (Techniques and Instrumentation in Analytical Chemistry,* Vol. 5) Elsevier, Amsterdam, Oxford, New York, 1982.
21 D.L. Tsalev and Z.K. Zaprianov, *Atomic Absorption Spectrometry in Occupational and Environmental Health Practice, Vol. I, Analytical Aspects and Health Significance,* CRC Press, Boca Raton, FL, 1983.
22 D.L. Tsalev, *Atomic Absorption Spectrometry in Occupational and Environmental Health Practice, Vol. II, Determination of Individual Elements,* CRC Press, Boca Raton, FL, 1984.
23 B. Welz, *Atomic Absorption Spectrometry,* VCH, Weinheim, Deerfield Beach, FL, Basel, 2nd ed., 1985.

24 M. Suzuki and K. Ohta, *Prog. Anal. At. Spectrosc.,* 6 (1983) 49–223.
25 W. Slavin, *Graphite Furnace AAS – A Source Book,* Perkin-Elmer Corp. Norwalk, CT, 1984.
26 B. Welz (Editor), *Selected Topics from Graphite Furnace and Hydride Generation AAS, Fresenius' Z. Anal. Chem.,* 323 (1986) 95–101.
27 K.S. Subramanian, *Prog. Analyt. At. Spectrosc.,* 9 (1986) 237–334.
28 S. Greenfield, G.M. Hieftje, N. Omenetto, A. Scheeline and W. Slavin, *Anal. Chim. Acta,* 180 (1986) 69–98.
29 R. Sturgeon, *Fresenius' Z. Anal. Chem.,* 324 (1986) 807–818.
30 K.S. Subramanian, *Prog. Analyt. At. Spectrom.,* 3 (1988) 9–12.
32 W. Slavin, D.C. Manning and G.R. Carnrick, *J. Anal. At. Spectrom.,* 3 (1988) 13–19.
33 A.A. Brown and A. Taylor, *Analyst,* 110 (1985) 579–582.
34 A.A. Brown, D.J. Roberts and K.V. Kahokola, *J. Anal. At. Spectrom.,* 2 (1987) 201–204.
35 T.S. West, *Anal. Proc.,* 25 (1988) 240–244.
36 Z. Fang and B. Welz, *J. Anal. At. Spectrom.,* 4 (1989) 83–89.
37 Z. Fang, B. Welz and G. Schlemmer, *J. Anal. At. Spectrom.,* 4 (1989) 91–95.
38 M. Sperling, Z. Fang and B. Welz, *Anal. Chem.,* 63 (1991) 151–159.
39 J.L. Burguera, *Flow Injection Atomic Spectroscopy,* Marcel Dekker, New York, Basel, 1989.
40 H. Massmann and S. Gücer, *Spectrochim. Acta,* 39B (1974) 283–300.
41 J.M. Ottaway, *Proc. Anal. Div. Chem. Soc.,* 13 (1976) 185–192.
42 G. Volland, G. Kölblin, P. Tschöpel and G. Tölg, *Fresenius' Z. Anal. Chem.,* 284 (1977) 1–12.
43 R.E. Sturgeon and C.L. Chakrabarti, *Prog. Anal. At. Spectrosc.,* 1 (1978) 5–199.
44 J.P. Matousek, *Prog. Anal. At. Spectrosc.,* 4 (1981) 247–310.
45 W. Slavin and D.C. Manning, *Prog. Anal. At. Spectrosc.,* 5 (1982) 243–340.
46 R.E. Ediger, *At. Absorpt, Newslett.,* 14 (1975) 127–130.
47 S. Callio, *Atom. Spectrosc.,* 1 (1980) 80–81.
48 G. Schlemmer and B. Welz, *Spectrochim. Acta,* 41B (1986) 1157–1165.
49 L.M. Voth-Beach and D.E. Schrader, *J. Anal. At. Spectom.,* 3 (1988) 52–64.
50 K.B. Knowles and K.G. Brodie, *J. Anal. At. Spectrom.,* 3 (1988) 62–64.
51 E.J. Hinderberger, M.L. Kaiser and S.R. Kortyohann, *At. Spectosc.,* 1 (1981) 1–7.
52 D.C. Manning and W. Slavin, *Appl. Spectrosc.,* 37 (1983) 1–11.
53 J. Marshall, D.C. Baxter, J. Carroll, S. Cook, S.P. Corr, S.K. Girie, D. Durie, D. Littlejohn, J.M. Ottaway, S.C. Stephen and S. Wright, *Anal. Proc.,* 22 (1985) 371–373.
54 A.A. Brown and U. Riedel, *Int. Lab.,* 17 (1987) 42–50.
55 R. Stephens, *Crit. Rev. Anal. Chem.,* 9 (1980) 167–195.
56 K. Yasuda, H. Koizumi, K. Ohishi and T. Noda, *Prog. Anal. At. Spectrosc.,* 3 (1980) 299–368.
57 W. Slavin, G.R. Carnrick, D.C. Manning and E. Pruszkowska, *Atom. Spectrosc.,* 4 (1983) 69–86.
58 W. Frech and A. Cedergren, *Anal. Chim. Acta,* 82 (1976) 93–102.
59 L. Novak and M. Stoeppler, *Fresenius' Z. Anal. Chem.,* 323 (1986) 737–741.
60 B. Welz and G. Schlemmer, *At. Spectrom.,* 9 (1988) 76–80.
61 B. Welz and G. Schlemmer, *At. Spectrom.,* 9 (1988) 81–83.
62 M. Beaty, W. Barnett and Z. Grobenski, *Atom. Spectrosc.,* 1 (1980) 72–77.
63 D.K. Eaton and J.A. Holcombe, *Anal. Chem.,* 55 (1983) 946–950.
64 H.D. Narres, C. Mohl and M. Stoeppler, *Z. Lebensm. Unters. Forsch.,* 181 (1985) 111–116.
65 C. Mohl, H.D. Narres and M. Stoeppler, in: B. Welz (Editor), *Fortschritte in der Atomspektrometrischen Spurenanalytik, Band 2,* VCH, Weinheim, 1986, pp. 439–446.
66 W. Slavin and G.R. Carnrick, *Spectrochim. Acta,* 35B (1984) 271–282.
67 B.V. L'vov, *Talanta,* 23 (1976) 109–118.
68 T.J. Langmyhr, *Analyst,* 104 (1979) 993–1016.
69 I. Atsuya and K. Itoh, *Spectrochim. Acta,* 38B (1983) 1259–1264.

70 U. Völlkopf, Z. Grobenski, R. Tamm and B. Welz, *Analyst,* 110 (1985) 573–577.
71 K.O. Olayinka, S.J. Haswell and R. Greeskowiak, *J. Anal. At. Spectrom.,* 1 (1986) 297–300.
72 G. Schlemmer and B. Welz, *Fresenius' Z. Anal. Chem.,* 328 (1987) 405–409.
73 E. Lücker, A. Rosopulo, S. Koberstein and W. Kreuzer, *Fresenius' Z. Anal. Chem.,* 329 (1987) 31–34.
74 I. Lindberg, E. Lundberg, P. Arkhammer and P.O. Berggen, *J. Anal. At. Spectrom.,* 3 (1988) 497–501.
75 M.S. Epstein, G.R. Carnrick, G.R. Slavin and N.J. Miller-Ihli, *Anal. Chem.,* 61 (1988) 1414-1419.
76 N.J. Miller-Ihli, *Fresenius' J. Anal. Chem.,* 337 (1990) 271–274.
77 U. Kurfürst, *Fresenius' Z. Anal. Chem.,* 315 (1983) 304–320.
78 U. Kurfürst, *Fresenius' Z. Anal. Chem.,* 316 (1983) 1–7.
79 U. Kurfürst, *Doctoral Thesis,* Univ. Bremen, 1984.
80 A. Rosopulo, K.H. Grobecker and U. Kurfürst, *Fresenius' Z. Anal. Chem.,* 319 (1984) 540–546.
81 K.H. Grobecker, C. Mohl and M. Stoeppler, in S.E. Lindberg and T.C. Hutchinson (Editors), *Proc. Int. Conf. Heavy Metals in the Environ., Vol. 1,* CEP Consultants, Edinburgh, 1987, pp. 486–488.
82 C. Mohl, K.H. Grobecker and M. Stoeppler, *Fresenius' Z. Anal. Chem.,* 328 (1987) 413–418.
83 P.S. Low, G.J. Hsu, *Fresenius' J. Anal. Chem.,* 337 (1990) 299–305.
84 E. Hahn, K. Hahn, C. Mohl and M. Stoeppler, *Fresenius' J. Anal. Chem.,* 337 (1990) 306–309.
85 *Grün SM 30 Zeeman-Atomic-Absorption-Spectrometer,* Grün Optik, Wetzlar, 1990.
86 U. Kurfürst, M. Kempeneer, M. Stoeppler and O. Schuirer, *Fresenius' J. Anal. Chem.,* 337 (1990) 248–252.
87 J.M. Harnly, *Anal. Chem.,* 58 (1986) 933A–943A.
88 T.C. O'Haver, J. Carroll, R. Nichol and D. Littlejohn, *J. Anal. At. Spectrom.,* 3 (1988) 155–157.
89 M. Retzik and D. Bass, *Int. Lab.,* 18/8 (1988) 49–56.
90 M. Stoeppler, *Sonderheft Spektroskopie, Nachr. Chem. Tech. Lab,* 37 (1989) 29–55.
91 C.E. Oda and J.D. Ingle, *Anal. Chem.,* 53 (1981) 2030–2033.
92 N.S. Bloom and E.A. Crecelius, *Mar. Chem.,* 14 (1983) 49–59.
93 B. Welz and M. Melcher, *Atom. Spectrosc.,* 5 (1984) 59–61.
94 M. Schintu, T. Kauri and A. Kudo, *Water Res.,* 23 (1989) 699–704.
95 S.H. Lee, K.-H. Jung and D.S. Lee, *Talanta,* 36 (1989) 999–1003.
96 D.C. Baxter and W. Frech, *Anal. Chim. Acta,* 225 (1989) 175–183.
97 R. Dumarey, R. Heindrickx and R. Dams, *Anal. Chim. Acta,* 118 (1980) 381–383.
98 K.H. Tobies and W. Grossmann, in B. Welz (Editor), *5. Colloquium Atomspektrometrische Spurenanalytik,* Bodenseewerk Perkin-Elmer, Überlingen, 1989, pp. 513–521.
99 V. Arenas, M. Stoeppler and G. Bergerhoff, *Fresenius' Z. Anal. Chem.,* 332 (1988) 447–452.
100 J. Piwonka, G. Kaiser and G. Tölg, *Fresenius' Z. Anal. Chem.,* 321 (1985) 225–234.
101 F. Alt, J. Messerschnidt and G. Tölg, *Fresenius' Z. Anal. Chem.,* 327 (1987) 233–234.
102 B. Welz, M.S. Wolynetz and M. Verlinden, *Pure Appl. Chem.,* 59 (1987) 927–936.
103 B. Welz and M. Melcher, *Analyst,* 108 (1983) 213–224.
104 B. Welz and M. Melcher, *Analyst,* 109 (1984) 577–579.
105 B. Welz and M. Schubert-Jacobs, *J. Anal. At. Spectrom.,* 1 (1986) 81–83.
106 K. Itoh, M. Chikuma and H. Tanaka, *Fresenius' Z. Anal. Chem.,* 330 (1988) 600–604.
107 S.N. Willie, R.E. Sturgeon and S.S. Berman, *Anal. Chem.,* 58 (1986) 1140–1143.
108 N.J. Miller Ihli, *Spectrosc. Int.,* 3/3 (1991) 26–30.
109 M. Sperling, Z. Fang and B. Welz, *Anal. Chem.,* 63 (1991) 151–159.
110 J.C.C. van Loon, *Anal. Chem.,* 53 (1981) 332A–361A.
111 R.G. Michel, M.L. Hall, J.M. Ottaway and G.S. Fell, *Analyst,* 104 (1979) 491–504.

112 E.J. Ekanem, C.L.R. Barnard and J.M. Ottaway, *J. Anal. At. Spectrom.*, 1 (1986) 349–353.
113 D.A. Naranjit, B.H. Radziuk and J.C. van Loon, *Spectrochim. Acta*, 38B (1984) 969–977.
114 G. Tieh-Zheng and R. Stephens, *J. Anal. At. Spectrom.*, 1 (1986) 355–358.
115 L. Ebdon and J.R. Wilkinson, *Anal. Chim. Acta*, 914 (1987) 177–187.
116 N. Bloom and W.F. Fitzgerald, *Anal. Chim. Acta*, 208 (1988) 151–161.
117 Å. Iverfeldt, personal communication, 1991.
118 N. Omenetto and H.G.C. Human, *Spectrochim. Acta*, 39B (1984) 1333–1343.
119 N. Omenetto, P. Cavalli, P. Broglia, P. Qi and G. Rossi, *J. Anal. At. Spectrom.*, 3 (1988) 231–235.
120 K. Dittrich and H.J. Stärk, *J. Anal. At. Spectrom.*, 2 (1987) 63–66.
121 L.M. Garden, D. Littlejohn, K. Dittrich and H.J. Stärk, *Anal. Proc.*, 25 (1988) 230–232.
122 J.P. Dougherty, F.R. Preli and R.G. Michel, *J. Anal. At. Spectrom.*, 4 (1989) 429–434.
123 R.L. Irwin, D.J. Butcher, J. Takahashi, G.-T. Wei, R.G. Michel, *J. Anal. At. Spectrom.*, 5 (1990) 603–610.
124 D.R. Demers and E.B.M. Jansen, in B. Sansoni (Editor), *Instrumentelle Multielementanalyse*, VCH, Weinheim, Deerfield Beach, FL, Basel, 1985 pp. 397–410.
125 R.F. Sanzolone, *J. Anal. At. Spectrom.*, 1 (1986) 343–347.
126 S. Greenfield, I.L. Jones and C.T. Berry, *Analyst*, 89 (1964) 713–720.
127 V.A. Fassel, *Fresenius' Z. Anal. Chem.*, 324 (1986) 511–518.
128 P.W.J.M. Boumans (Editor), *Inductively Coupled Plasma Emission Spectroscopy, Part 1, Methodology Instrumentation and Performance*, Wiley, New York, Chichester, Brisbane, Toronto, Singapore, 1987.
129 P.W.J.M. Boumans (Editor), *Inductively Coupled Plasma Emission Spectroscopy, Part 2, Applications and Fundamentals*, Wiley, New York, Chichester, Brisbane, Toronto, Singapore, 1987.
130 J.A.C. Broekaert, *Anal. Chim. Acta*, 196 (1987) 1–21.
131 M. Thomson and J.N. Walsh, *Handbook of Inductively Coupled Plasma Spectrometry*, Blackie, Glasgow, 2nd ed., 1988.
132 I.T. Urasa, *Anal. Chem.*, 56 (1984) 904–908.
133 R. Speer, P. Hoffmann and K.H. Lieser, *Fresenius' Z. Anal. Chem.*, 325 (1986) 558–560.
134 K.S. Subramanian and J.C. Méranger, *Sci. Total Environ.*, 24 (1982) 147–157.
135 A.L. Molinero, J.R. Castillo and A. de Vega, *Fresenius' Z. Anal. Chem.*, 331 (1988) 721–724.
136 F.D. Bamiro, D. Littlejohn and J. Marshall, *J. Anal. At. Spectrom.*, 3 (1988) 379–384.
137 P. Schramel, X. Li-Qiang, A. Wolf and S. Hasse, *Fresenius' Z. Anal. Chem.*, 313 (1982) 213–216.
138 P. Schramel and X. Li-Qiang, *Fresenius' Z. Anal. Chem.*, 314 (1983) 671–677.
139 S.S. Que Hee and J.R. Boyle, *Anal. Chem.*, 60 (1988) 1033–1042.
140 Y. Kanda and M. Taira, *Anal. Chim. Acta*, 207 (1988) 269–281.
141 C. Mohl and M. Stoeppler, in B. Welz (Editor), *5. Coll. Atomspektrometr. Spurenanalytik*, Bodenseewerk Perkin-Elmer, Überlingen, 1989, pp. 729–736.
142 K.H. Karstensen and W. Lund, *Sci. Total Environ.*, 79 (1989) 179–189.
143 M.H. Ramsay and M. Thomspon, *J. Anal. At. Spectrom.*, 1 (1986) 185–193.
144 M.H. Ramsay, M. Thompson and E.K. Banerjee, *Anal. Proc.*, 24 (1987) 260–265.
145 D.E. Nixon and G.E. Smith, *Anal. Chem.*, 58 (1986) 2886–2888.
146 D.R. Luffer and E.D. Salin, *Anal. Chem.*, 58 (1986) 654–656.
147 M.W. Routh, J.E. Goulter, D.B. Tasker and S.D. Arellano, *Int. Lab.*, 17/3 (1987) 54–61.
148 T. Kumamaru, Y. Okamoto, Y. Yamamoto, F. Nakata, Y. Nitta and H. Matsuo, *Fresenius' Z. Anal. Chem.*, 327 (1987) 777–781.
149 F. Buhl and W. Galas, *Fresenius' Z. Anal. Chem.*, 332 (1988) 366–367.
150 J.A. Tielroy, P.H.M. Vleeschhouwer, J.C. Kraak and F.S.M.J. Maessen, *Anal. Chim. Acta*, 207 (1988) 149–159.

151 M.F. Gineé, F.J. Krug, H.G. Filho, B.F. dos Reis and E.A.G. Zagatto, *J. Anal. At. Spectrom.*, 3 (1988) 673–678.

152 J. Ruzicka, *Fresenius' Z. Anal. Chem.*, 324 (1986) 745–749.

153 J. Ruzicka and A. Arndal, *Anal. Chim. Acta*, 216 (1989) 243–255.

154 D.R. Anderson and C.W. McLeod, *Anal. Proc.*, 25 (1988) 67–69.

155 G.J. Schmidt and W. Slavin, *Anal. Chem.*, 54 (1982) 2491–1395.

156 M.H. Ramsay and M. Thompson, *J. Anal. At. Spectrom.*, 2 (1987) 497–502.

157 L. Ebdon and R. Carpenter, *Anal. Chim. Acta*, 209 (1988) 135–145.

158 R.D. Snook, *Anal. Proc.*, 25 (1988) 354–355.

159 S.T. Sparkes and L. Ebdon, *Anal. Proc.*, 23 (1986) 410–412.

160 L. Ebdon and J.R. Wilkinson, *J. Anal. At. Spectrom.*, 2 (1987) 39–44.

161 A.A. Verbeek and I.B. Brenner, *J. Anal. At. Spectrom.*, 4 (1989) 23–26.

162 A.J. Ambrose, L. Ebdon, M.E. Foulkes and P. Jones, *J. Anal. At. Spectrom.*, 4 (1989) 219–222.

163 H. Matusiewicz, *J. Anal. At. Spectrom.*, 1 (1986) 171–184.

164 P. Ek and S.-G. Hulden, *Talanta*, 34 (1987) 495–502.

165 R.J. Watling and A.R. Collier, *Analyst*, 113 (1988) 345–346.

166 Z. Li, S. Xiao-Quan and N. Zhe-Ming, *Fresenius' Z. Anal. Chem.*, 332 (1988) 764–768.

167 Y. Nogiri, A. Otsuki and K. Fuwa, *Anal. Chem.*, 58 (1986) 544–547.

168 J. Marshall, D. Littlejohn, J.M Ottaway, J.M. Harnly, N.J. Miller-Ihli and T.C. O'Haver, *Analyst*, 108 (1983) 178–188.

169 D. Littlejohn, *Anal. Proc.*, 25 (1988) 217–220.

170 D.J. Douglas and R.S. Houk, *Prog. Anal. At. Spectrosc.*, 8 (1985) 1–18.

171 C.J. Pickford and R.M. Brown, *Spectrochim. Acta*, 41B (1986) 183–187.

172 S. Munro, L. Ebdon and D. McWeeny, *J. Anal. At. Spectrom.*, 1 (1986) 211–219.

173 D.J. Douglas and J.B. French, *Spectrochin. Acta*, 41B (1986) 197–204.

174 M. Selby and G.M. Hieftje, *Int. Lab.*, 17/8 (1987) 28–38.

175 J. Luck and U. Siewers, *Fresenius' Z. Anal. Chem.*, 331 (1988) 129–132.

176 G.R. Gillson, D.J. Douglas, J.E. Fulford, K.W. Halligan and S.D. Tanner, *Anal. Chem.*, 60 (1988) 1472–1474.

177 A.R. Date and A.L. Gray (Editors), *Applications of Inductively Coupled Plasma Mass Spectrometry*, Blackie, Glasgow, 1989.

178 G.M. Hieftje and G.H. Vickers, *Anal. Chim. Acta*, 216 (1989) 1–24.

179 K.E. Jarvis, A.L. Gray, I. Jarvis and J. Williams (Editors), *Plasma Source Mass Spectrometry (Spec. Pub. No. 85)*, Royal Society of Chemistry, Cambridge, 1990.

180 F. Fietz, *Fresenius' Z. Anal. Chem.*, 324 (1986) 212–223.

181 B. Sansoni, W. Brunner, G. Wolff, H. Rupert and R. Dittrich, *Fresenius' Z. Anal. Chem.*, 331 (1988) 154–169.

182 J.M. Henshaw, E.M. Heithmar and T.A. Hinners, *Anal. Chem.*, 61 (1989) 335–342.

183 P.J. Paulsen, E.S. Beary, D.S. Bushee and J.R. Moddy, *Anal. Chem.*, 60 (1988) 971–775?

184 C. Vandecasteele, M. Nagels, H. Vanhoe and R. Dams, *Anal. Chim. Acta*, 211 (1988) 91–98.

185 M.R. Plantz, J.R. Fritz, F.G. Smith and R.S. Houk, *Anal. Chem.*, 61 (1989) 149–153.

186 J.W. McLaren, A.P. Mykytiuk, S.N. Willie and S.S. Berman, *Anal. Chem.*, 57 (1985) 2907–2911.

187 D. Beauchemin, J.W. McLaren, A.P. Mykytiuk and S.S. Berman, *J. Anal. At. Spectrom.*, 3 (1988) 305–308.

188 R.C. Hutton, *J. Anal. At. Spectrom.*, 1 (1986) 259–263.

189 J.R. Dean, L. Ebdon and R. Massey, *J. Anal. At. Spectrom.*, 2 (1987) 369–374.

190 H.T. Delves and M.G. Campbell, *J. Anal. At. Spectrom.*, 3 (1988) 343–348.

191 M.J. Campbell and H.T. Delves, *J. Anal. At. Spectrom.*, 4 (1989) 235–236.

192 R.D. Satzger, *Anal. Chem.*, 60 (1988) 2500–2504.

193 N.I. Ward, in: *Proc. Int. Conf. Heavy Metals Environ., New Orleans*, Vol. 2, CEP Consultants, Edinburg, 1987, pp. 23–28.

194 R. Dolan, J. van Loon, D. Templeton and A. Paudyn, *Fresenius' J. Anal. Chem.*, 336 (1990) 99–105.
195 J.W. McLaren, D. Beauchemin and S.S. Berman, *Anal. Chem.*, 59 (1987) 610–613.
196 D. Beauchemin, J.W. McLaren, S.N. Willie and S.S. Berman, *Anal. Chem.*, 60 (1988) 687–691.
197 D. Beauchemin, K.W.M. Siu and S.S. Berman, *Anal. Chem.*, 60 (1988) 2587–2590.
198 P.S. Ridout, H.R. Jones and J.G. Williams, *Analyst*, 113 (1988) 1383–1386.
199 J.G. Williams, A.L. Gray, P. Norman and L. Ebdon, *J. Anal. At. Spectrom.*, 2 (1987) 469–472.
200 C.J. Park, J.C. van Loon, P. Arrowsmith and J.B. French, *Anal. Chem.*, 59 (1987) 2191–2196.
201 D.C. Gregoire, *J. Anal. At. Spectrom.*, 3 (1988) 309–314.
202 M. Janghorbani and B.T.G. Ting, *Anal. Chem.*, 61 (1989) 701–708.
203 P. Valenta, *GIT Fachz. Lab.*, 32 (1988) 312–320.
204 J. Heyrovsky and P. Zuman, *Einfürung in die praktische Polarographie*, Verlag Technik, Berlin, 1959.
205 J. Heyrovsky and J. Kuta, *Grundlagen der Polarographie*, Akademie-Verlag, Berlin, 1965.
206 L. Meites, *Polarographic Techniques*, Wiley Interscience, New York, 2nd ed., 1965.
207 H.W. Nürnberg and B. Kastening, in F. Korte (Editor), *Methodicum Chimicum*, Vol. I, Part A, Academic Press, New York, San Francisco, London, 1974, pp. 548–607.
208 K.Z. Brainina, *Stripping Voltammetry in Chemical Analysis*, Wiley, New York, 1974.
209 A.M. Bond, *Modern Polarographic Methods in Analytical Chemistry*, Marcel Dekker, New York, 1980.
210 J. Wang, *Stripping Analysis, Principles, Instrumentation and Applications*, VCH, Weinheim, Deerfield Beach, FL, Basel, 1985.
211 J. Wang, *Electroanalytical Techniques in Clinical Chemistry and Laboratory Medicine*, VCH, Weinheim, Deerfield Beach, FL, Basel, 1988.
212 J.G. Osteryoung and R.A. Osteryoung, *Anal. Chem.*, 58 (1985) 101A–110A.
213 G. Henze and R. Neeb, *Electrochemische Analytik*, Springer, Berlin, Heidelberg, New York, Tokyo, 1986.
214 M. Würfels, *Mar. Chem.*, 28 (1989) 259–264.
215 G.C. Barker and I.L. Jenkins, *Analyst*, 77 (1952) 685–496.
216 E.G. Buchanan and D.D. Soleta, *Talanta*, 30 (1983) 459–464.
217 P. Ostapczuk, P. Valenta and H.W. Nürnberg, *J. Electroanal. Chem.*, 214 (1986) 51–64.
218 P. Ostapczuk, M. Stoeppler and H.W. Dürbeck, *Fresenius' Z. Anal. Chem.*, 322 (1988) 662–665.
219 J. Wang, *Int. Lab.*, 16 (11/12) (1986) 50–59.
220 C.M.G. van den Berg, *Sci. Total Environ.*, 49 (1986) 89–99.
221 C.M.G. van den Berg, *Analyst*, 114 (1989) 1527–1530.
222 J. Golimowski, H.W. Nürnberg and P. Valenta, *Lebensm. Chem. Gerichtl. Chem.*, 34 (1980) 116–120.
223 C.J. Flora and E. Nieboer, *Anal. Chem.*, 52 (1980) 1013–1020.
224 B. Pihlar, P. Valenta and H.W. Nürnberg, *Fresenius' Z. Anal. Chem.*, 307 (1981) 337–346.
225 J.R. Donat and K.W. Bruland, *Anal. Chem.*, 60 (1988) 240–233.
226 J. Wang, *Fresenius' J. Anal. Chem.*, 337 (1990) 508–511.
227 C.M.G. van den Berg, K. Murphy and J.P. Riley, *Anal. Chim. Acta*, 188 (1986) 177–185.
228 C.M.G. van den Berg, *Electroanal. Chem.*, 215 (1986) 111–121.
229 J. Golimowski, P. Valenta and H.W. Nürbberg, *Fresenius' Z. Anal. Chem.*, 322 (1985) 315–322.
230 K. Torrance and C. Gatford, *Talanta*, 35 (1987) 939–944.
231 Ph. Breyer and B.P. Gilbert, *Anal. Chim. Acta*, 201 (1987) 33–41.
232 S.B. Adeloju, A.M. Bond and M.H. Briggs, *Anal. Chem.*, 57 (1985) 1386–1390.
233 L. Mart, *Talanta*, 29 (1982) 1035–1040.

128

234 L. Mart, *Tellus*, 35B (1983) 131–141.
235 L. Mart, H.W. Nürnberg and D. Dyrssen, in C.S. Wong, E. Boyle, K.W. Bruland and E.D. Goldberg (Editors), *Trace Metals in Sea Water*, Plenum Press, New York, London, 1983, pp. 113–130.
236 W. Dorten, P. Valenta and H.W. Nürnberg, *Fresenius' Z. Anal. Chem.*, 317 (1984) 367–379.
237 P. Valenta, in G.G. Leppard (Editor), *Trace Element Speciation in Surface Waters and its Ecological Implications*, Plenum Press, New York, 1983, pp. 49–69.
238 H.W. Nürnberg, in: G.G. Leppard (Editor), *Trace Element Speciation in Surface Waters and its Ecological Implications*, Plenum Press, New York, 1983, pp. 211–230.
239 T.M. Florence, *Analyst*, 111 (1986) 489–505.
240 T.M. Florence, in G.E. Batley (Editor), *Trace Element Speciation: Analytical Methods and Problems*, CRC Press, Boca Raton, FL, 1990, pp. 77–116.
241 D. Jagner and A. Graneli, *Anal. Chim. Acta*, 83 (1976) 19–26.
242 D. Jagner, *Anal. Chem.*, 50 (1978) 1924–1929.
243 P. Ostapczuk, M. Stoeppler and H.W. Dürbeck, *Toxicol. Environ. Chem.*, 27 (1990) 49–53.
244 A.S. Baranski and H. Quon, *Anal. Chem.*, 58 (1986) 407–412.
245 H. Huiliang, C. Hua, D. Janger and L. Renman, *Anal. Chim. Acta*, 193 (1987) 61–69.
246 L. Almestrand, D. Jagner and R. Renman, *Anal. Chim. Acta*, 193 (1987) 71–79.
247 L. Almestrand, D. Jagner and L. Renman, *Talanta*, 33 (1986) 991–995.
248 H. Huiliang, D. Jagner and L. Renman, *Anal. Chim. Acta*, 201 (1987) 1–9.
249 H. Huiliang, D. Jagner and L. Renman, *Anal. Chim. Acta*, 207 (1988) 37–46.
250 L. Almestrand, M. Betti, C. Hua, D. Jagner and L. Renman, *Anal. Chim. Acta*, 209 (1988) 329–334.
251 G. Schulze, E. Han and W. Frenzel, *Fresenius' Z. Anal. Chem.*, 332 (1989) 844–848.
252 E.P. Bertin, *Principles and Practices of X-Ray Spectrometric Analysis*, Plenum Press, New York, 1971.
253 R. Jenkins, *An Introduction to X-Ray Spectrometry*, Heyden & Sons, London, 1975.
254 H.W. Liebhafsky, E.A. Schweikert and E.A. Meyers, in P.I. Elving (Editors), *Treatise on Analytical Chemistry*, Part I, Vol. 8, John Wiley & Sons, New York, 2nd ed., 1986, pp. 209–309.
255 C. Whiston, *X-Ray Methods*, Wiley, Chichester, 1987.
256 B. Gonsior and M. Roth, *Talanta*, 30 (1983) 385–400.
257 W. Maenhaut, *Anal. Chim. Acta*, 195 (1987) 125–140.
258 R. Klockenkämper, B. Raith, S. Divoux, B. Gonsior, B. Brüggerhof and E. Jackwerth, *Fresenius' Z. Anal. Chem.*, 326 (1987) 105–117.
259 S.A.E. Johansson and J.L. Campbell, *PIXE: A Novel Technique for Elemental Analysis*, John Wiley & Sons, Chichester, New York, Brisbane, Toronto, Singapore, 1988.
260 B.W. Adamson, *Int. Lab.*, 21/4 (1991) 50–53.
261 V. Talbot and W.J. Chang, *Sci. Total Environ.*, 60 (1987) 213–223.
262 J.A. Helsen and B.A.R. Vrebos, *Int. Lab.*, 16/10 (1986) 66–71.
263 R.D. Giauque, A.C. Thompson, J.H. Underwood, Y. Wu, K.W. Jones and M.L. Rivers, *Anal. Chem.*, 60 (1988) 855–858.
264 A. Knöchel, *Fresenius' J. Anal. Chem.*, 337 (1990) 614–621.
265 A. Prange, *Spectrochim. Acta*, 44B (1989) 437–452.
266 A. Prange and H. Schwenke, *Adv. X-Rat-Anal.*, 32 (1989) 209–218.
267 W. Michaelis and A. Prange, *Int. J. Radiat. Appl. Instrum.*, E2 (1988) 231–245.
268 A. Prange, R. Niedergesäss and C. Schnier, in W. Michaelis (Editor), *Estuarine Water Quality Management (Coastal and Estuarine Studies, Vol. 36)*, Springer, Berlin-Heidelberg-New York, 1990, pp. 426–436.
269 S. Mukhtar, S.J. Haswell, A.T. Ellis and D.T. Hawke, *Analyst*, 116 (1991) 333–338.
270 A.V. Bohlen, R. Eller, R. Klockenkämper and G. Tölg, *Anal. Chem.*, 59 (1987) 2551–2555.
271 W. Michaelis, private communication, 1990.

272 L. Wielopolski, J.F. Rosen, D.N. Slatkin, D. Varsky, K.J. Ellis and S.H. Cohn, *Med. Phys.,* 10 (1983) 248–251.
273 M.C. Scott and O.R. Chettle, *Scand. J. Work. Environ. Health,* 12 (1986) 81–96.
272 V.P. Guinn and J. Hoste, in: *Elemental Analysis of Biological Materials (Tech. Rep. Ser.,* Vol. 197), IAEA, Vienna, 1980, pp. 105–140.
273 M.C. Scott and O.R. Chettle, *Scand. J. Work. Environ. Health,* 12 (1986) 81–96.
274 V.P. Guinn and J. Hoste, in: IAEA, *Elemental Analysis of Biological Materials (Tech. Rep. Ser.,* Vol. 197), Vienna, 1980, pp. 105–140.
275 K. Heydorn, *Neutron Activation Analysis for Clinical Trace Element Research,* 2 Volumes, CRC Press, Boca Raton, FL, 1984.
276 V. Krivan, *Analytiker Taschenbuch, Vol. 5,* Springer, Heildelberg, 1985, pp. 35–68.
277 G. Erdtmann and H. Petri, in P.J. Elving (Editor), *Treatise on Analytical Chemistry,* Vol. 14, Part 1, Wiley, New York, 2nd ed., 1986, Ch. 7, pp. 419–643.
278 J. Hoste, in: P.J. Elving (Editor), *Treatise on Analytical Chemistry,* Vol. 14, Part 1, John Wiley & Sons, New York, 1986, 2nd ed., 1986, pp. 645–777.
279 IAEA, *Nuclear Techniques for Analysis of Environmental Samples,* IAEA/RL/135, Vienna, 1986.
280 L. Kosta, *Fresenius' Z. Anal. Chem.,* 324 (1986) 649–654.
281 K. Heydorn, *Fresenius' J. Anal. Chem.,* 337 (1990) 498–502.
282 R. Cornelis, J. Hoste and J. Versieck, *Talanta,* 29 (1982) 1029–1034.
283 L. Xilei, D. van Renterghem, R. Cornelis and L. Mees, *Anal. Chim. Acta,* 211 (1988) 231–241.
284 H.J.M. Bowen, *Trace Elements in Biochemistry,* Academic Press, London, 1966.
285 J. Versieck and R. Cornelis, *Anal. Chim. Acta,* 116 (1980) 217–254.
286 R. Cornelis, *J. Trace Microprobe Tech.,* 2 (1985) 237–265.
287 J. Versieck, *Trace Elem. Med.,* 1 (1984) 2–12.
288 J. Versieck, *CRC Crit. Rev. Clin. Lab. Sci.,* 22 (1985) 97–184.
289 J. Versieck, L. Vanballenberghe, A. de Kesel, J. Hoste, B. Wallaeys, J. Vandenhaute, N. Baeck, H. Steyaert, A.R. Byrne and F.W. Sunderman, Jr., *Anal. Chim. Acta,* 204 (1988) 63–75.
290 M. Esprit, C. Vandecasteele and J. Hoste, *Anal. Chim. Acta,* 175 (1986) 79–88.
291 J.J.M. De Goeij, *Trans. Am. Nucl. Soc.,* 56 (1988) 194–195.
292 R. Zeisler, S.H. Harrison and S.A. Wise, *The Pilot National Environmental Specimen Bank - Analysis of Human Liver Specimen (NBS Spec. Pub.* No. 656), U.S. Dept. of Commerce, Washington, DC, 1983.
293 R. Zeisler, S.F. Stone and R.W. Sanders, *Anal. Chem.,* 60 (1988) 2760–2765.
294 G.E.M. Hall, G.F. Bonham-Carter, A.I. MacLaurin and S.B. Ballantyne, *Talanta,* 37 (1990) 135–155.
295 R. Dams, *Fresenius' J. Anal. Chem.,* 337 (1990) 492–497.
296 A. Chatt, R.R. Rao, C.K. Jayawickreme and L.S. McDowell, *Fresenius' J. Anal. Chem.,* 338 (1990) 399–407.
297 R. Cornelis, S. Dyg, B. Griepink and R. Dams, *Fresenius' J. Anal. Chem.,* 338 (1990) 414–418.
298 M. Petra, S. Landsberger and G. Swift, *Fresenius' J. Anal. Chem.,* 338 (1990) 567–568.
299 P.M. Lindstrom, A.R. Byrne, D.A. Becker, B. Smodis and K.M. Garrity, *Fresenius' J. Anal. Chem.,* 338 (1990) 569–571.
300 J.R.W. Woittiez, *Fresenius' J. Anal. Chem.,* 338 (1990) 575–579.
301 M.C. Scott and O.R. Chettle, *Scand. J. Work Environ. Health,* 12 (1986) 81–96.
302 M. Rossbach, O. Schärpf, W. Kaiser, W. Graf, W. Schirmer, A. Faber, W. Duppich and R. Zeisler, *Nucl. Inst. Meth. Phys. Res.,* B35 (1988) 181–190.
303 J.W. Mitchell, *Int. Lab.,* 11/1 + 2 (1982) 12–25.
304 L. Maggi, V. Caramella-Crespi, N. Genova and S. Meloni, *Int. J. Appl. Radiat. Isot.,* 33 (1982) 217–221.
305 J.F. Klaverkamp, M.A. Turner, S.E. Harrison and R.H. Hesslein, *Sci. Total Environ.,* 28 (1983) 119–128.
306 M. Maggi, M.T. Ganzerli Valentini and R. Stella, *Analyst,* 112 (1987) 1617–1618.

130

307 V. Krivan, in: P.J. Elving (Editor), *Treatise on Analytical Chemistry*, Vol. 13, Part I, Wiley, New York, 2nd ed., 1986, pp. 339–417.
308 V. Krivan, *Sci. Total Environ.*, 64 (1987) 12–25.
309 V. Krivan and S. Arpadjan, *Fresenius' J. Anal. Chem.*, 335 (1989) 743–747.
310 V. Krivan and K. Petrick, *Fresenius' J. Anal. Chem.*, 336 (1990) 480–483.
311 Z. Deyl, K. Macek and J. Janak (Editors), *Liquid Column Chromatography (J. Chromatogr. Library*, Vol. 3), Elsevier, Amsterdam, 1975.
312 G.E. Batley and G.K.-C. Low, in G.E. Batley (Editor), *Trace Element Speciation: Analytical Methods and Problems*, CRC Press, Boca Raton, 1990, pp. 185–218.
313 Y.K. Chau and P.T.S. Wong, in G.E. Batley (Editor), *Trace Element Speciation: Analytical Methods and Problems*, CRC Press, Boca Raton, 1990, pp. 219–244.
314 P.H.E. Gardiner, *Top Curr. Chem.*, 141 (1987) 145–174.
315 L. Ebdon, S. Hill and R.W. Ward, *Analyst*, 111 (1986) 1113–1138.
316 H. Schaller and R. Neeb, *Fresenius' Z. Anal. Chem.*, 323 (1986) 473–476.
317 C.I. Measures and J.M. Edmond, *Anal. Chem.*, 61 (1989) 544–547.
318 S. Clark, S. Ashby and P.J. Craig, *Analyst*, 112 (1987) 1781–1782.
319 S. Clark and P.J. Craig, *Appl. Organomet. Chem.*, 2 (1988) 33–46.
320 N. Bloom, *Can. J. Fish. Aquat. Sci.*, 46 (1989) 1131–1140.
321 J.H. Knox, J.N. Done, M.T. Gilbert, A. Pryde and R.A. Wall, *High Performance Liquid Chromatography*, University Press, Edinburgh, 1978.
322 J.W. Dolan, L.R. Snyder and M.A. Quarry, *Int. Lab.*, 17/8 (1987) 66–71.
323 A.M. Bond and N.M. McLachlan, *Anal. Chem.*, 54 (1986) 756–758.
324 L. Ebdon, S. Hill and R.W. Ward, *Analyst*, 112 (1987) 1–16.
325 O. Nygren, C.A. Nilsson and W. Frech, *Anal. Chem.*, 60 (1988) 2204–2208.
326 A. Munder and K. Ballschmiter, *Fresenius' Z. Anal. Chem.*, 323 (1986) 869–874.
327 S. Ichinoki, N. Hongo and M. Yamazaki, *Anal. Chem.*, 60 (1988) 2099–2104.
328 H. Small, T.S. Stevens and W.C. Bauman, *Anal. Chem.*, 47 (1975) 1801–1809.
329 R.W. Frei (Editor), *Proc. 5th Sils-Maria IAEAC Symposium on Ion-Chromatography, J. Chromatogr.*, 439 (1988) 1–170.
330 R.W. Frei (Editor), *Proc. 6th Sils-Maria IAEAC Symposium on Ion-Chromatography, J. Chromatogr.*, 483 (1990) 263–425.
331 D. Chambaz and W. Haerdi, *J. Chromatogr.*, 482 (1989) 335–342.
332 R.B. Rubin and S.S. Heberling, *Int. Lab.*, 17/9 (1987) 54–60.
333 J. Weiss, *Fresenius' Z. Anal. Chem.*, 327 (1987) 451–455.
334 D. Yan and G. Schwedt, *Fresenius' Z. Anal. Chem.*, 327 (1987) 503–508.
335 V.K. Jones and J.G. Tarter, *Int. Lab.*, 15/9 (1985) 36–39.
336 J. Korkisch, *CRC Handbook of Ion Exchange Resins: Their Application to Inorganic Analytical Chemistry*, 6 Vols, CRC Press, Boca Raton, 1988.
337 M. Schuster, *Fresenius' Z. Anal. Chem.*, 324 (1986) 127–129.
338 A.J. Ahearn (Editor), *Trace Analysis by Mass Spectrometry*, Academic Press, New York, 1972.
339 S. Facchetti (Editor), *Applications of Mass Spectrometry to Trace Analysis*, Elsevier, Amsterdam-Oxford-New York, 1982.
340 F. Adams, R. Gijbels and R. van Grieken (Editors), *Inorganic Mass Spectrometry*, Wiley, New York, 1988.
341 K.G. Heumann (Editor), *Element Trace Analysis by Mass Spectrometry, Fresenius' Z. Anal. Chem.*, 331 (1988) 103–222.
342 J.R. Bacon and A.M. Ure, *Analyst*, 109 (1984) 1229–1254.
343 K.G. Heumann, *Toxicol. Environ. Chem. Rev.*, 3 (1980) 111–129.
344 J.R. Moody and P.J. Paulsen, *Analyst*, 113 (1988) 923–927.
345 K.P. Jochum, H.M. Seufert, S. Midinet-Best, E. Rettmann, K. Schönberger and M. Zimmer, *Fresenius' Z. Anal. Chem.*, 331 (1988) 104–110.
346 K.G. Heumann, in: F. Adams, R. Gijbels and R. van Grieken (Editors), *Inorganic Mass Spectrometry*, Wiley, New York, 1988, pp. 301–376.
347 H.-R. Schulten, U. Bahr and R. Palavinskas, *Fresenius' Z. Anal. Chem.*, 317 (1984) 497–511.
348 B.K. Schaule and C.C. Patterson, *Earth Planet. Sci. Lett.*, 54 (1981) 97–116.

131

349 V.J. Stukas and C.S. Wong, in C.S. Wong, E. Boyle, K.W. Bruland, J.D. Burton and E.D. Goldberg (Editors), *Trace Metals in Sea Water*, Plenum, New York, 1983, pp. 513–536.
350 E. Michiels and P. de Bièvre, in I.K. O'Neill, P. Schuller and L. Fishbein, (Editors) *Environmental Carcinogens, Selected Methods of Analysis (IARC Scientific Publication*, No. 71), IARC, Lyon, 1986, pp. 443–450.
351 J. Völkening and K.G. Heumann, *Fresenius' Z. Anal. Chem.*, 331 (1988) 174–181.
352 P. de Bièvre, *Fresenius' J. Anal. Chem.*, 337 (1990) 766–771.
353 P. de Bièvre, J. Savory, A. Lamberty and G. Savory, *Fresenius' Z. Anal. Chem.*, 332 (1988) 718–721.
354 A. Götz and G. Geumann, *Fresenius' Z. Anal. Chem.*, 332 (1988) 640–644.
355 S.K. Aggerwal, M. Kinter, M.R. Wills, J. Savory and D.A. Herold, *Anal. Chem.*, 62 (1989) 111–115.
356 J.D. Schladot, K. Hilpert and H.W. Nürnberg, *Adv. Mass Spectrom.*, 8 (1980) 325–329.
357 V.J. Stukas and C.S. Wong, *Science*, 211 (1981) 1424–1427.
358 O. Tera, D.W. Schwarzman and T.R. Watkins, *Arch. Environ. Health*, 40 (1985) 120–123.
359 M.B. Rabinowitz, G.W. Wetherill and J.D. Kopple, *Science*, 182 (1973) 725–727.
360 M.B. Rabinowitz, G.W. Wetherill and J.D. Kopple, *Arch. Environ. Health*, 30 (1976) 220–223.
361 J.R. Turnlund, *Sci. Total Environ.*, 28 (1983) 385–392.
362 A. Benninghoven, F.R. Rudenauer and H.W. Werner, *Secondary Ion Mass Spectrometry: Basic Concepts, Instrumental Aspects, Applications and Trends*, Wiley, New York, 1987.
363 P.K.D. Feigl, F.R. Krueger and B. Schueler, *Microchim. Acta*, II (1984) 85–96.
364 K.P. Jochum, L. Matus and H.M. Seufert, *Fresenius' Z. Anal. Chem.*, 331 (1988) 104–110.
365 R. Self, J. Eagles, S.J. Fariweather-Tait and D.E. Portwood, *Anal. Proc.*, 24 (1987) 366–367.
366 L.J. Moore, J.D. Fassett and J.C. Travis, *Anal. Chem.*, 56 (1984) 2270-2775.
367 N. Jakubowski, D. Stuewer and W. Vieth, *Fresenius' Z. Anal. Chem.*, 331 (1988) 145–149.
368 F. Adams and A. Vertes, *Fresenius' J. Anal. Chem.*, 337 (1990) 638–647.
369 A. Knowles and C. Burgess (Editors), *Practical Absorption Spectrometry*, Chapman & Hall, London, 1984.
370 A.B.P. Lever, *Inorganic Electronic Spectroscopy*, Elsevier, Amsterdam, 1984.
371 T. Nowicki-Jankowska, K. Gorgzynska, A. Michalik and E. Wietecka, in G. Svehla (Editor), *Comprehensive Analytical Chemistry, Vol. 19*, Elsevier, Amsterdam, 1986.
372 C. Burgess and K.D. Mielenz (Editors), *Advances in Standards and Methodology in Spectrophotometry*, Elsevier, Amsterdam, 1987.
373 A.G. Melgarejo, A.G. Céspedes and J.M.C. Pavon, *Analyst*, 114 (1989) 109–111.
374 J. Sanz, F. Gallarta, J. Galban and J.R. Castillo, *Analyst*, 113 (1988) 1387–1391.
375 M. Komata and I.-I. Itoh, *Talanta*, 35 (1988) 723–724.
376 J.L. Garcia, P. Navarro and H. Cordoba, *Talanta*, 35 (1988) 885–889.
377 H. Müller, *Crit. Rev. Anal. Chem.*, 13 (1982) 313–372.
378 T.L. Threlfall, *Eur. Spectrosc. News.*, 78 (1988) 8–17.
379 P. Jones, T. Williams and L. Ebdon, *Anal. Chim. Acta*, 237 (1990) 291–298.
380 S. Koelling, J. Kunze and C. Tauber, *Fresenius' Z. Anal. Chem.*, 332 (1988) 776–789.
381 G.M. Hieftje, *J. Anal. At. Spectrom.*, 4 (1989) 117–122.
382 R.E. Sturgeon, *Fresenius' J. Anal. Chem.*, 337 (1990) 538–545.
383 P. Hulmston and R.C. Hutton, *Spectrosc. Int.*, 3/1 (1991) 35–38.
384 M. Valcarcel and M.D. Luque de Castro, *Fresenius' J. Anal. Chem.*, 337 (1990) 662–666.
385 D.C. Liang and M.W. Blades, *Anal. Chem.*, 60 (1988) 27–31.
386 M.A. Mignardi, B.T. Jones, B.W. Smith and J.D. Winefordner, *Anal. Chim. Acta*, 227 (1989) 331–342.

387 G.H. Hermann, *CRC Crit. Rev. Anal. Chem.*, 19 (1988) 323–377.
388 H. Falk, E. Hoffmann and C. Ludke, *Spectrochim. Acta*, 39B (1984) 283–294.
389 H. Falk and J. Tilch, *J. Anal. At. Spectrom.*, 2 (1987) 527–531.
390 J.M. Harnly, D.L. Styris and N.E. Ballou, *J. Anal. At. Spectrom.*, 5 (1990) 139–144.
391 K. Ohls, J. Flock and H. Loepp, *Fresenius' Z. Anal. Chem.*, 332 (1988) 456–463.
392 K. Niemax, *Fresenius' J. Anal. Chem.*, 337 (1990) 551-556.
393 D. Goforth and J.D. Winefordner, *Anal. Chem.*, 58 (1986) 2598–2602.
394 M.B. Leong, J. Vera, B.W. Smith, N. Omenetto and J.D. Winefordner, *Anal. Chem.*, 60 (1988) 1605–1610.
395 J.A. Vera, M.B. Leong, C.L. Stenson, G. Petrucci and J.D. Winefordner, *Talanta*, 36 (1989) 1291–1293.
396 K. Dittrich, I. Mohamad, H.Th. Nguyen, K. Niebergall, M. Pfeifer and R. Wennrich, *Fresenius' J. Anal. Chem.*, 337 (1990) 546–550.
397 A. Thorne, *J. Anal. At. Spectrom.*, 2 (1987) 227–232.
398 L.M. Faires, *J. Anal. At. Spectrom.*, 2 (1987) 585–590.
399 W. Lund, *Fresenius' J. Anal. Chem.*, 337 (1990) 557–564.
400 K.G. Heumann, W. Schindlmeier, H. Zeininger and M. Schmidt, *Fresenius' J. Anal. Chem.*, 320 (1985) 457–462.
401 E.A. Boyle, B. Hardy and A. VanGeen, *Anal. Chem.*, 59 (1987) 1499–1503.
402 C.M. Sakamoto-Arnold and K.S. Johnson, *Anal. Chem.*, 59 (1987) 1789–1794.
403 J. Bishop, *Int. Lab.*, 19/4 (1989) 58–63.
404 M. Kendall-Tobias, *Int. Lab.*, 19/4 (1989) 64–69.

M. Stoeppler (Editor)/*Hazardous Metals in the Environment*
© 1992 Elsevier Science Publishers .B.V. All rights reserved

Chapter 7

Chemical speciation and environmental mobility of heavy metals in sediments and soils

Manfred Sager

Geotechnical Institute, BVFA-Arsenal, A-1030 Vienna (Austria)

CONTENTS

7.1. General 134
    7.1.1. The experimental approach towards the mobility and availability of
        trace elements 134
    7.1.2. Brief about general speciation and transformation reactions in
        sediments and soils 136
        7.1.2.1. Sample storage 136
        7.1.2.2. Grain size 137
        7.1.2.3. Carbonates 137
        7.1.2.4. Weathering of silicates 137
        7.1.2.5. Fe/Mn hydrous oxides 138
        7.1.2.6. Organics 138
        7.1.2.7. Interstitial waters 139
        7.1.2.8. Sediment reduction 139
7.2. Batch methods 140
    7.2.1. Adsorption on the solid 140
        7.2.1.1. Model substances 140
            7.2.1.1.1. Organics and active carbon 140
            7.2.1.1.2. Al hydroxide 141
            7.2.1.1.3. Mn oxides and hydroxides 141
            7.2.1.1.4. Fe oxides 142
            7.2.1.1.5. Clay minerals 143
            7.2.1.1.6. Carbonates 144
            7.2.1.1.7. Sulfides 144
        7.2.1.2. Adsorption on real samples 144
    7.2.2. Desorption and dissolution experiments 145
        7.2.2.1. Salt and buffer solutions in the neutral pH range 145
        7.2.2.2. Weak acids without dominant complexation or redox re-
            actions 147
        7.2.2.3. Strong acids 148
        7.2.2.4. Leaching with alkaline solutions 148
        7.2.2.5. Effect of reductants 150
            7.2.2.5.1. Hydroxylamine 150
            7.2.2.5.2. Others 152

134

       7.2.2.6. Effect of oxidants 153
       7.2.2.7. Action of complexants 154
       7.2.2.8. Leaching with organic solvents 156
7.3. Some experimental data from sequential leaching sequences 157
    7.3.1. Arsenic 157
    7.3.2. Cadmium 157
    7.3.3. Chromium 158
    7.3.4. Copper 159
    7.3.5. Cobalt 161
    7.3.6. Nickel 162
    7.3.7. Lead 163
    7.3.8. Zinc 163
7.4. Speciation and availability to biota 164
    7.4.1. Aquatic systems 165
    7.4.2. Terrestrical systems 166
7.5. Column methods 167
    7.5.1. Experiments with uniform solids 167
    7.5.2. Experiments with cored soil profiles 168
       7.5.2.1. General 169
       7.5.2.2. Organics 169
       7.5.2.3. Saturation with respect to inorganics 170
       7.5.2.4. Variations of pH 170
       7.5.2.5. Microbial activity 171
References 171

7.1. GENERAL

7.1.1. The experimental approach towards the mobility and availability of trace elements

In the ecological cycling, a significant part of elements is bound to solids: few elements are enriched in the aqueous phase or in living biota. Mobility means either the movement of solutes, or the passage from the solid into the hydrosphere or the biosphere. For the development of appropriate methods, as well as the estimation of their applicability and limitations, mere calculations with solubility data from pure phases are insufficient, and experiments with model substances or model solutions have to be considered. Calculations start from pure solids and ideal mixtures, and try to approach reality by the introduction of correction factors. In sediments and soils, however, we have to expect non-ideal mixtures of an unknown number of components; some dissolve incongruently and some are colloidal, with unknown surface energies. Sorbed amounts are very important for the cycling of minor elements, whose concentrations are too low to make a phase of their own. Numerous speciation forms are possible, but it is reasonable to cut the number of gained speciations of each element to, say, 4 to 6. These should be accessible to experiment in a single sample, which can be performed by routine methods on a larger number of samples.

In batch methods, a solid sample is shaken with the reagent solution, until the concentration of dissolved species does not measurably increase - that is, presumably, until solubility equilibrium is established. The solid and solution are finally separated by centrifugation or filtration. In sequential leaching, the solid residue is used as sample for the next reagent.

Another experimental approach is to pack the solid material, either unchanged, or homogenized, into a column. A test solution is continuously applied and the eluted fractions are either collected, or there is a feed-back to the solid until it is saturated ("column method"). When the solid material is substantially dissolved the penetration velocity suddenly increases, because of gaps: the contact time ceases to be reproducible, and the experiment has to be stopped. Therefore, in the column, the solid has to be in large excess over the reagent.

In batch methods, high liquid–solid ratios are often used to avoid exhaustion of the reagent. Finally, there is the approach to equilibrium, which may be relevant for the uptake in organisms and plant in the long term [1]. In column experiments, however, low liquid–solid ratios appear, and the concentration gradients in the column and equilibria can differ from layer to layer, which may be relevant for the migration in the pore water, and for the short term uptake by plant roots or the benthic and soil fauna.

Batch methods resemble the dissolution of sediment particles upon re-suspension in a chemically different environment, whereas column experiments act as models for the dissolution and movement of dissolved species in the undisturbed solid. Column methods are therefore well suited for investigations of the transport of solutes in sediment layers, which are similar to the profiles for solutes in the pore water, or of the transport by groundwater of solutes in aquifers. The direct compatibility between the results of mobilization by column or by batch methods is thus only low [1]. For example, the retardation capacity for Zn of a clay from the Austrian Molasse [2], as calculated from batch tests, was far higher than the effective value measured in percolation tests in the form of retardation factors.

In the batch design, it is possible to solubilize materials as soluble complexes, depending on the equilibrium conditions. But not all soluble complexes move through the column, because sorption of the reagent itself creates additional sorption sites for the heavy metals in the column design. However, the solubility of humic complexes is increased by delivering additional H^+ to the slower moving humates [3]. Of most importance to man, however, has to be considered the passage of toxic materials and also nutrients (such as phosphorus) from residual solids into the food chain.

For availability studies, it is necessary to have data for biota, and species analyses of the surrounding solid and liquid phases, to give a better understanding of the mechanisms of uptake. The data may be rather specific for the site, also. Biota from natural environments, or plants grown in greenhouses, have been considered. The results are influenced, however, by the plant species, supply of nutrients and other micro-elements, genetic adaptation to the site, the season and climate, etc., so it seems more reasonable to predict availabilities from a more general level, in terms such as mobility and chemical speciation.

References on p. 171

7.1.2. Brief about general speciation and transformation reactions in sediments and soils

The mobility and cycling of trace metals depend on the properties of the trace metals themselves, such as their solubility, reactivity for complexation or adsorption, on the solution characteristics, and on the solid biotic and abiotic surfaces encountered [4]. The experimental results are strongly influenced by the sampling methods (e.g., digging, coring, freeze-coring), the selection of grain size to be investigated, and the sample storage. As speciation shifts occur during storage (see below), the samples should be processed as soon as possible.

The sediment derives from the settling of suspended matter, either from surface run-off or riverine input, as well as precipitation reactions in the water body itself, and settling organic detritus. Some contributions emanate directly from the soil fauna, like shells or corals. The influence of the terrestrial ecosystem is much greater for lakes than for ocean basins, owing to the inputs of terrestrial organic and mineral particles [5].

Some particular problems have been treated in specific reviews: these include the surface chemistry of calcium carbonate minerals in natural waters [6], the chemistry of weathering [7], and the availability to biota of enhanced total trace-metal levels [8]. This material will not be repeated here in detail.

7.1.2.1. Sample storage

Because one aims at a picture of the fractionation at the time of taking the samples, possible changes during storage have to be considered.

Wet storage of oxidized estuarine sediments at 4 or 25°C generally changes the phase pattern of sequential leaching techniques, especially for organic substances and Fe oxides, due to microbial activity. Only the leach with dilute HCl and, in the case of Cu and Zn, the leach with acetic acid, are not affected [9]. When the biomass is starved, there is significant release of Zn and other elements into the pore water. Similarly, freeze-thaw cycles in soils have a significant effect upon the chemical composition of stream-water. Substantial initial increases in leachate concentrations of Ca, Mg, Na, K, Fe, Al, Si, and total organic carbon, occur after a freezing period. This can be interpreted as the result of a breaking-open of decomposing plant cells and soil micro-fauna as the intracellular fluid expands. During snow-melt, and at autumn flashes, the pH of riverwaters falls, due to organic acids originating from decomposition products, if there is insufficient buffering by carbonates [10].

Upon drying, destruction of microbial cells and dehydration of hydrous phases occurs, leading to an increase in exchangeable Cu and Zn [9]. Even the dehydration caused by shaking with ethanol shifts the leaching patterns of Fe, Mn, and Al from the hydrous oxides to the crystalline forms [11]. Grinding greatly affects the amount of Fe which is extractable from soil, but has less effect on Zn and Mn, but none on Cu. Similarly, the shaker speed affects the extractability, except for Cu [12]. Oxidation of anoxic sediments in the air, even during freeze drying, changes the leaching pattern. Iron carbonate is converted into hydroxides, and Cd- and Zn sulfides to carbonates, in exchangeable and Fe-bound forms [13]. The oxidation of sulfides generates acids, resulting in additional mobilization of most of the trace elements [14].

7.1.2.2. Grain size

The total contents as well as the speciation of a given sample depends on the grain size selected. Surface phases, like hydrous oxides, and clay minerals, which contain most of the trace elements, are of course enriched on small grains, whereas the products of primary abrasion are rather coarse [11] (See Section 7.4). However, the grain-size fraction below 0.2 mm does not contain maximum element contents in any case. Elements which are preferentially bound to heavy minerals, like cassiterite, scheelite, tourmaline, and zircon, are enriched at 1-5 mm [15].

7.1.2.3. Carbonates

The formation of carbonates is significantly influenced by the Mg/Ca ratio in the liquid phase. At Mg/Ca <2, a low-Mg calcite is precipitated. A ratio of Mg/Ca in the range 2-12 results in the formation of a high-Mg calcite, which can undergo further dolomitization. At Mg/Ca $>$ 12, aragonite is formed together with hydrous Mg carbonates, from which no dolomitization can occur [16]. In natural waters, complexes of transition metal ions, hydroxycarboxylic acids and carbonate act as nucleation centers for carbonate precipitation. The kind of precipitated carbonate depends in some way on the metal cation of the crystallization nucleus as well as the Mg/Ca ratio [17].

The surface of carbonate minerals is a micro-zone of increased pH. This can lead to new surface phases like apatite or rhodochrosite, or the deposition of hydroxides. Also, organic compounds adsorb strongly onto carbonate mineral surfaces [6,18].

The preservation of precipitated or biogenic $CaCO_3$ in sediments depends on the input of readily metabolised organic matter and the immediate establishment of anoxic/sulfidic conditions. Aerobic respiration leads to the acidification by CO_2, with the subsequent dissolution of carbonates [19].

7.1.2.4. Weathering of silicates

The chemical weathering of aluminosilicates leads either to the formation of non-crystalline Al and Fe hydroxides and silicic acid, or the dissolution is accompanied by the formation of secondary minerals, such as kaolinite and other clay minerals. The weathering products often occur as patches on the surface of the primary minerals [20]. Weathered biotite grains are heavily coated with Fe/Al oxides. The weathering leads to a low K and a high water content [21]. In general, clay minerals have alternating layers of hydrated alumina and silica, with both tetrahedral and octahedral sites. For charge balance, cations like Li, Na, K, Ca, Mg, and Fe are usually present between the silica and alumina layers [22].

References on p. 171

7.1.2.5. Fe/Mn hydrous oxides

The composition of iron deposits depends much more on the redox potential than on the pH. Hematite is stable under oxidizing conditions: the transition to reducing conditions is marked by the occurrence of magnetite, while siderite and sulfide correspond to feebly and strongly reducing environments [23].

Hydrous Fe/Mn oxides are formed where anoxic groundwater is discharged to the surface, where neutralization of acidic waters takes place, and where riverwater is mixed with seawater [18]. Iron oxides and oxyhydroxides occur as surface coatings and as small particles, and contribute significantly to the total available surface area. Organic-rich groundwaters contain high amounts of Fe and Mn, deriving from the reduction of the oxide coatings by the phenolic reducing groups of degraded organic matter [24]. If Mn and Fe oxides precipitate separately, the Mn phase is much lower in humic substances. Electron microscopy and micro-probe analyses reveal that the Mn phase consists of crumpled thin sheets which must have a considerable surface area [25]. Hydrous oxides have a much greater adsorption capacity than crystalline surfaces of goethite, hematite or magnetite. Divalent cations can enhance the adsorption of humic substances on goethite, hematite, and oxides of Mn [26].

The transformation of ferric hydroxide to goethite is strongly retarded by the presence of silicic acid [20]. Under anoxic conditions, the amount of soluble Mn is controlled by the solubility of $MnCO_3$ [25].

7.1.2.6. Organics

There are three possibilities for metal interaction with organic substances: the action of organisms, the decomposition of plant and animal material, and the sorption of lower molecular weight organic matter onto clay or metal oxides [18]. Complex-forming ligands (oxalate, citrate and others) from micro-organisms and plants increase the rate of dissolution and extend the domain of congruent dissolution of silicate minerals such as feldspar [27,28].

Micro-organisms such as bacteria and fungi can have marked bio-accumulation capacities towards some specific metals. It is not possible, however, to predict the uptake of this biomass, because it is influenced largely by genetic adaptation to the site. Thus, adapted organisms take up less Be, Cr, Co, Zn, and Cd, whereas no general trend has been observed for Mn, Fe, Pb, Cu, and Mg [29].

Roots of higher plants influence the soil by respiration, by releasing cells and cell-wall material, and by excreting organic compounds like amino acids and carboxylic acids. In the rhizosphere, bacteria, fungi, actinomycetes, and protozoa are enriched. In a moderate climate, the excretion by roots changes within the year's cycle and passes a maximum at maximum vegetative growth. Healthy roots can change the pH of their surroundings towards their optimum, by the excretion of bicarbonate or organic acids. They also influence the redox potential [30]. Humic substances are polyanionic materials of molecular weight up to 1 000 000, having carboxyl and phenolic hydroxyl groups [26]. They comprise up to 70% of the organic matter in sediments. As a minor fraction, lipids can be found [31]. Marine humates are formed in marine environments but there is negligible contribution from the neighbouring land masses [32]. The marine humates contain more aliphatic and heterocyclic structures than do soil humates, and more S and N. They are formed by degradation of planktonic material.

Terrestrial humic substances extracted from a sludge-treatment and composting process with 0.5 M NaOH show 3 peaks in the ^{13}C NMR, due to 3 different types of carbon structure. The dominant signal is from aliphatic carbons, mainly from methylene and methyl groups. As in the aquatic humics, aromatic carbons are found. Methoxy groups, probably derived from lignin-like structures, are, however, typical for terrestrial residual organic material [33]. When sludge is applied to soil, the organic matter largely influences the transfer of heavy metals. Sludge solutions generally increase the mobility due to complexation, high ionic strength, and high background. Some organic compounds are easily degraded by bacteria and the action of light, with subsequent variation of the leachability of the metals [34-36].

7.1.2.7. Interstitial waters

In the interstitial waters, high-molecular-weight polymers are dissolved, which are assumed to be precursors of sedimentary humates. Major parts of Cu, Fe, Ni, and Zn but no Ca and Mn are thus bound to dissolved organic carbon compounds [32]. Vertical trends in a core through oxidized surface sediment into reducing sediment are controlled by the redox interface. The As, Cu, Zn, Mn and Fe in the pore-water are controlled by the solubility of Fe and Mn oxyhydroxides or of metal sulfides [37].

Arsenic dissolved in the pore water of sediment-cores increases with depth, whereas the dissolved Fe is at a maximum near the oxic/anoxic boundary. The depth-distribution of dissolved P and dissolved As show similarities [38]. The Fe(II) and Mn(II), generated by reductive dissolution in the anoxic zone, can diffuse upward, be re-oxidized in the oxic layer, and form new Fe or Mn hydroxides. These can be collected *in situ*, best on a PTFE sheet [39].

Concerning the mobility in soils, for Fe, Mn, Zn, Co, and Cd, the water-solubility increases with decreasing soil pH; Mo as an anion behaves in the opposite way. No clear relationship between water-solubility and soil pH could be found for Cu, Ni, Cr, and Pb [40]. Thus, acidification provokes the mobilization of many trace elements (see below). Similarly, in column experiments, addition of acid leads to mobilization of cations (e.g. Na > K > Ca > Mg > Mn > Fe > Al), in exchange for H^+ [41].

7.1.2.8. Sediment reduction

Most of the important reactions in a water-body occur at or near geochemical boundaries. At the sediment-water interface, the gradients of density, pH, pE, ligand concentrations, biota abundance, etc., are at their greatest. The bacterial degradation of organic carbon regulates the redox potential, and thus the 'redox-sensitive' elements, such as Fe, Mn, Cr, Se, V, and others [5]. Under reducing conditions, organic complexes are not able to compete with the bisulfide for the complexation of trace metals. The pH is governed by $CaCO_3$ and by the conversion of goethite to siderite. Under oxidizing conditions, the mobility of trace metals is generally higher (except for Fe, Mn, P) [14].

7.2. BATCH METHODS

7.2.1. Adsorption on the solid

When effluents are poured into the watershed, contaminants can be either transported as solutes, or adsorbed on solid particles and thus sedimented further on. In order to predict the adsorption, possible adsorption sites, transformation reactions, and release from known components in the solid, many test experiments with model substrates have been performed. These lead also to the design and improvement of the selectivity of reagents for use in leaching procedures (see Section 7.2.2.).

7.2.1.1. Model substances

7.2.1.1.1. Organics and active carbon

The surfaces of algal cells (*Chlamidomonas reinhardii*) have a high affinity for Cu and Cd, even in the presence of Ca. For Cu, this is greater than for colloidal iron hydroxide. The functional group ligands can compete with soluble complex-formers which are typically present in natural waters. The adsorption can be interpreted in terms of equilibria of surface complex formation. The tendency to form complexes decreases with increasing load. A fast initial adsorption is followed by a slow diffusion-controlled step. The formation of surface complexes at comparable pH values is of the same order of magnitude for bacterial cell suspensions (*Klebsiella pneumonia*) [42]. Any Cr(III) is removed from the water phase faster than Cr(VI). Chromate, however, is rapidly reduced by dissolved organic compounds, suspended organic matter, or organics in the sediment, leading to rapid depletion of Cr in the liquid of watersheds of high organic load [5,43,44].

Because of the lack of further experiments with naturally occurring organic substances, results obtained with active carbon are presented here, also. Experiments with humic substances packed in columns are mentioned in Section 7.3, and the behaviour of mixtures of natural organic substances with Fe/Mn oxides will be treated in 7.2.1.1.3.

Upon hydration of active carbon, hydroxo groups are developed, which behave amphoterically. The surface can thus be made positively or negatively charged, or neutral, just by changing the pH. The amount of metal removed from solution increases abruptly at a specific pH value. The sorption capacity (for 14 samples of activated C tested) decreased in the order Pb > Cu > Zn > Ni > Co > Cd. The metal-removal capacities are as good as, or even better than Al hydroxide of Fe hydroxide gels. The carbon can be regenerated only by strong acid [45].The sorption of dissolved amino-acids on activated carbon is nearly quantitative if they contain aromatic or heterocyclic structures (tryptophan, phenylalanine, phenylglycine, histidine), but very low for aliphatic chains (glycine, serine, cystine, alanine, valine). The S-containing cysteine and methionine are about half adsorbed. Amino acid chelates of Ni and Cu are sorbed like the pure amino acids (at pH 10) [46].

7.2.1.1.2. Al hydroxide

If, in the course of weathering of alumosilicates, more Si is removed than Al, the formation of hydrous oxides of Al (bayerite and pseudoboehmite) on the primary minerals is observed [47]. This is the case for weathering by inorganic agents, in an environment low in organics. Because of sorption of H^+ or OH^-, the surface of Al-hydroxide is positively charged in acid, and negatively charged in alkaline solutions. The point of zero charge for free $Al(OH)_3$ gel lies at a pH > 9, but the adsorption of carbonate conveys additional negative charge to the surface, thus shifting the point of zero charge down to pH 6.3 [48]. Therefore, the surface adsorption characteristics for Al hydroxide coatings may be different for carbonaceous and non-carbonaceous sediments [11,49].

The sorption of Pb, Cu, and Cd onto Al hydroxide from acetate buffer is weaker than for similar Fe and Mn compounds, but stronger than for clay minerals or Fe ore minerals (for example, hematite). It is quantitative at pH > 6, but negligible at pH < 5 [50]. A significant fraction of the Al hydroxide surface can be covered by organic matter [51], as mentioned for carbonate. This changes the behaviour towards the adsorption of cations, which are complexed more by the organics. Increasing pH leads to an increase in the adsorption of trace-metal cations, but to a decrease in the adsorption of (probably negatively charged) organic matter. Copper, which bonds very strongly to organics, changes its adsorptive behaviour from metal-like to a completely organic-matter-like adsorption, similar to that of naturally occurring fulvic acid. In contrast, Cd adsorption shows only slight effects, and Zn is not affected [52]. In the presence of dissolved organic carbon, the maximum of Cu adsorption occurs at pH 5-6, with desorption in alkaline solution at pH > 6.5. On pure alumina, however, quantitative adsorption has been reached at pH > 7.5. At a level of 10 mg/l DOC, adsorption of Cu starts at pH 3.7, instead of the pH 4.5 found in purely inorganic conditions.

Microcrystalline gibbsite is less reactive than non-crystalline alumina and boehmite of similar surface area. The Cu adsorption on gibbsite as a function of pH is largely reversible. At pH > 5, the gibbsite appeared to promote the hydrolysis and polymerisation of Cu, with further adsorption at the surface. Less Cu was adsorbed at a given pH in a short time than over 1 day [53,54].

7.2.1.1.3. Mn oxides and hydroxides

Iron and Mn in natural waters have been called "scavenging-type" elements, together with Co, Cr, Pb, Pu, Sn, Pa, and Th, because their concentration in liquids is largely controlled by precipitation and co-precipitation reactions [5].

The properties of Fe and Mn oxides vary greatly, according to their broad oxidational status and crystallinity. Thus, freshly precipitating Mn and Fe oxides are much better sorbents for the trace-elements Pb, Cu, and Cd than are hydrous alumina or clay minerals; ore minerals, like hematite or goethite, are rather inert [50]. Under anoxic conditions, soluble Mn is controlled by the solubility of $MnCO_3$. Oxygenation of Mn-rich water samples, and maybe also bacterial action, cause the precipitation of hydrous Mn oxide in the pH range 6-8. This is rather amorphous, with some reflections due to MnOOH, MnO, and vernadite (MnO_2) in the X-ray diffractogram. Electron microscopy and micro-probe analysis reveal that the Mn phase consists of crumpled thin sheets which must have a considerable surface area. Organic matter Fe, Ca, and trace elements are co-precipitated [25]. The release of Mn from MnO_2 is inversely related to pH. At pH 1-3, up to 90 % of the metal goes into solution, without dissolution of the oxides themselves. At pH 3, the sorption decreases in the order Fe = Pb = Mn > Cu > Cd > Ba > Zn > Ca [55].

The binding to Mn oxides is usually very strong. On the surface, Co(II) is oxidized to Co(III) [5]. The uptake of Pb, Cu, and Cd by freshly prepared hydrous Mn(IV) oxide, from acetate buffer, was nearly complete, over the whole pH range 3–9. On more crystalline MnO_2, however, Cd and Cu uptake was low below pH 5 [50].

Iron and Mn were precipitated when an ammoniacal ammonium sulfate leachate of manganese deep-sea nodules was allowed to age, co-precipitating some Co, Cu and Ni [56]. The surface of Mn oxides is preferably negatively charged. The isoelectric point of β-MnOOH is at pH 2.8, and of the more crystalline Mn_3O_4 at pH 5.4. Humic substances adsorb more strongly to the oxide with the higher zero-point of charge, while the effect for Ca is opposite. The adsorption of humics on manganese oxides is enhanced by $CaCl_2$ as is found for Fe and Al hydroxides [57]. Mixtures of synthetic MnO_2 and naturally occurring montmorillonite from Wyoming at pH 4.2 adsorbed 77% of Co, 67% of Cu, 69% of Ni, but only 28% of Ca onto the manganese phase: the rest was taken by the clay. This somehow reflects the geochemical preferences. Electron microscopy and leaching with hydroxylamine yielded the same results. Oxides of Mn need a higher redox potential to maintain stability than do Fe-hydroxides. Thus, birnessite, $Mn_7O_{13} \cdot 5H_2O$ is reduced by Fe^{2+} to yield soluble Mn^{2+} at pH < 4: in the range pH 4–6, in a slow surface-controlled reaction, FeOOH is precipitated, adsorbing further Fe^{2+} [58]. No noticeable dissolution of Mn oxides, however, results from the action of humics [57].

7.2.1.1.4. Fe oxides

The co-precipitation of trace elements with $Fe(OH)_3$ or $FePO_4$ is used for the clean-up of effluents. Any Cu, Zn, As, and Pb are nearly quantitatively co-precipitated with $Fe(OH)_3$ from 0.1 M NaCl, and Ag and Cd are partially. Carbonate and humic substances increase the co-precipitation rate. Similarly, $FePO_4$ takes Cu, Zn, Cd, and Pb, quantitatively, and 72% of As [59]. Thus, when Fe- oxides are precipitated from natural waters, phosphate sorption has no significant effect on the co-precipitation of many trace elements.

At the isoelectric point, K^+, Cl^-, and NO^{3+} were not adsorbed on goethite or hematite. Drying of hematite at 110°C greatly reduced H^+ adsorption from solution, while the phosphate adsorption was only slightly depressed. Below pH 3.5, no reliable adsorption equilibrium could be established, because dissolution occurred. At pH 11, however, there was no evidence for appreciable dissolution, and only a small proportion of available dissolution sites became changed [60].

Copper and Pb are strongly sorbed on hydrated Fe(III) oxide and on kaolin, moderately on anhydrous Fe oxide, and only slightly on quartz [61]. The adsorption of Pb onto hydrous Fe oxide is unaffected by the ionic strength: the Cd adsorption resembles those on marsh-sediments [62]. The sorption of Cd on goethite ($FeOOH$) was quantitative at pH > 7.5, but not below pH 6. The sorption of Pb on goethite was quantitative at pH > 6, and about half complete at pH 4.5 [50].

Humic acid is sorbed strongly on hydrated Fe(III) hydroxide, but only slightly on kaolin, anhydrous Fe(III) oxide and quartz [61]. The adsorption of humic substances on synthetic goethite declines towards alkaline pH; sulfate does not compete [63]. Divalent cations can enhance the adsorption of humic substances on goethite, hematite and oxides of Mn. On the other hand, the uptake of humic substances changes the adsorption characteristics of Fe oxides, and, in particular, the surface charge. In the absence of adsorbed humics, goethite particles carry a positive charge, whereas the coating of adsorbed humics renders them negative [26]. Copper, is tightly bound to humics: at pH < 6 more Cu is adsorbed by mixtures of goethite and humics than by the components independently, while at pH > 6 it is less [63]. The adsorption of Cu on goethite is negligible below pH 5.5, but quantitative above pH 6.5. Addition of humics increases Cu in the particulate phase, down to pH 4. When goethite and total Cu are kept constant, the ratio of sorbed Cu to humics passes through a maximum [26].

7.2.1.1.5. Clay minerals

The sorption capacity of clay minerals for cations (for example, Cs and Sr) increases with increasing pH and increasing surface. The pH dependence, however, is less than on hydroxides or carbonates. The adsorption of Cs at pH 3.5 is about half of the value at pH 6.5, but at pH 10.5 it is only slightly more. Most of the sorption is completed after 2 min, but then a considerable increase of sorbed amounts occurs after long times [64].

Treatment of clay minerals with aqueous extracts of humified clover changes the net surface charge to more negative values. The change was largest at low Al contents relative to Si. The amount of organic matter retained by the aluminosilicates is not only determined by the specific surface areas; it increases with increasing content of poorly ordered materials, as indicated by fluoride reactivity [47]. Substitution of interlayer-cations, like Na, K, Ca, Mg, or Fe, by quaternary ammonium cations or other organic cationic species causes the clay to become organophilic. Penetration by water is prevented and lipids can be readily adsorbed, as can transition metals *via* chelation mechanisms [22].

The adsorption capability of the various clay minerals for cations varies widely. Illite and montmorillonite retain Pb at pH > 6, but kaolinite retains only half of the Pb at pH > 8 [50]. In mixtures of montmorillonite and synthetic MnO_2, the major part of the Cu, Co and Ni moves to the Mn: only Ca prefers the clay mineral [65]. For Cs and Sr, the highest sorption was found for a bentonite with a high content of smectite and high specific surface area, lower for illite, and lowest for kaolinite, both in batch and column experiments [64].

Sequential leaching of kaolinite with hydroxylamine, oxalate, and citrate/dithionite did not significantly alter the exchange capacity or the surface area. Higher temperature and vigorous shaking stabilize the colloids of kaolinite, leading to an increase of the adsorbable metal in the liquid phase [66].

7.2.1.1.6. Carbonates

Carbonates are not the best adsorbents from solution, but the quantity of carbonate sedimentation, especially in the oceans, is appreciable. It has been estimated that carbonate sedimentation removes 28% of the sedimented P, 11% of the sedimented F, and 4.3% of the sedimented B from seawater [67]. The surface of the carbonates is alkaline, but rather smooth; the carbonates are not known to be amorphous. They can scavenge many cations, but their adsorption capacity is usually lower than that of hydrous oxides or clay minerals [6]. The deposition of mixed carbonates on calcite from solutions containing Mn, Mg, Na, and Sr is a function of the precipitation rate and the concentrations in the liquid; equilibrium conditions are not established. The incorporation of Mn in calcite increases with a decrease in the precipitation rate. The amount of Mg incorporated in the deposits decreases with increasing $MnCO_3^-$ content. In marine sediments, $MnCO_3$ has been found to occur in solid solution up to 50 mole % below the redox boundary [68].

In the sedimentation of carbonates, the smaller Mg atoms enter the lattice of calcite, while the larger Sr is only able to enter aragonite. Aragonite and strontianite are isomorphous, as are calcite and magnesite [23]. The sorption of Cd on calcite takes place at pH > 6.5. In a first fast step, completed within 24 h, Cd diffuses into a surface layer of hydrated $CaCO_3$ that overlies crystalline calcite. In a second step, needing 7 days, a solid solution is formed in new crystalline. The sorption is reversible and Cd can be partially desorbed with EDTA [69].

7.2.1.1.7. Sulfides

The adsorption on sulfides is strongly pH dependent, the dependence being related to the hydrolysis of the metal ions. Hydrolyzed species are adsorbed directly onto the sulfide groups. On pyrite, pyrrhotite, galena and sphalerite, the order of sorption decreases from Hg > Pb > Zn > Cd. Although Hg is sorbed much more than on Fe or Mn oxides, it is not influenced by CO_2. Pyrite takes up Pb quantitatively at pH > 6.5, and sphalerite takes up Cd quantitatively only at pH > 10 [70].

7.2.1.2. Adsorption on real samples

In naturally occurring sediments, metal sorption on organic substance can happen either via the action of the organisms themselves, by adsorption on decomposing plant and animal tissues, or the sorption of organic matter on clay minerals and hydrous oxides [18]. Generally, the heavy-metal load from the aqueous phase moves into the more mobile fractions during sequential leaching procedures [18,49]. The adsorption capacity of oxidized estuarine sediments is greatly influenced by organic matter. After exhaustive extraction of humic materials from the sediment in 0.1 M NaOH, and oxidation of the sediment at room temperature, the organic carbon was negligible, and the adsorption capacity for Cd decreased to a quarter. However, the Cd adsorption characteristics were unaffected by any other chemical selective extraction. To the contrary, the adsorption characteristics of the same sample for Pb were unaffected by the removal of organics [62].

Among 26 elements investigated, F, B, and Tc have been found to be the least adsorbed on two soil samples of different pH but nearly equal in organic carbon content [36]. Manganese, Sr and Sb were only mobile in the soil of pH 5, whereas the mobility of Mo and P was higher in the soil of pH 8. The sorption capacity of an Orthic Luvisol for Cd, Pb, and Cu individually in the Ah horizon (pH 5.5; 1.7% organic carbon; 14% < 2 μm) from 0.01 M NaNO$_3$ was quite high: 6 mg/g for Pb, 2.5 mg/g for Cu, and 3.0 mg/g for Cd. Competition by the other metals reduced the saturation level: Cu was least affected by competition [71]. The sorption of boron by soils depends on the organic carbon content, the pH, and the amounts of the clay-size fractions. Dihydroxy-organic compounds are believed to be responsible for the B retention by humus. The total contents of Fe and Al, however, do not significantly influence the sorption of boron [72].

The sorption of As on soils with high sesquioxide content (lateritic) decreased with increasing pH. Phosphate substantially suppressed the sorption of arsenate, but not in proportion with the phosphate–arsenate molar ratio. Chloride, nitrate, and sulfate did not compete with arsenate and phosphate for the sorption sites [73]. Measurements on power on-plant ash from brown coal showed that the sorption increases with rising pH, increasing solution concentration, and specific surface. Cobalt was quantitatively sorbed at pH > 8, Mn at pH > 10, but the quantitative sorption of varied between the samples pH > 5 to pH > 8. The results can be interpreted in terms of the pH-dependent occurrence of the hydrolysis products of the metal ions in solution and their reaction with the ash [74].

7.2.2. Desorption and dissolution experiments

7.2.2.1. Salt and buffer solutions in the neutral pH range

The treatment of sediments and soils with water or solutions of inorganic salts and buffers at a neutral pH results in the leaching of easily mobile amounts, often termed "exchangeable". The simple addition of water leads to the elution of soluble phases, as does an increase of the pore-water volume. Changes of the ionic strength without changes in pH (e.g. from adding MgCl$_2$, or NaNO$_3$) alters the colloidal status of organic macromolecules in the pores, kills the soil or sediment fauna by changing the osmotic pressure, and leads to competition between sorbed and the excess, so-called "inert" ions, for the adsorption sites. Finally, salts of a weak acid or a weak base (e.g. ammonium acetate), have buffering capacities superior to the solid in most cases: this results in a change of pH relative to the original pore-water, in addition to salt effects. In the latter case, the exchangeable amounts are not determined for the pH actually present in each individual pore-water or soil solution, but at a uniform pH for a whole series of samples, which is easier for practical work and inter-comparison.

References on p. 171

In surface sediments of Lake Macquarie, partially polluted by a Pb/Zn smelter, the pore-water data did not parallel the ammonium acetate extraction data [75]: the same applied to the sediments of the reservoir at Altenwörth of the River Danube [11]. In 40 lakes from Quebec and Ontario, in Canada, the pH of the pore-water ranges from 4.0 to 8.4 [4]: thus, a somewhat broader range than strictly neutral has to be considered for solubility and mobilization. In solid urban waste and compost the quantity of metals depends on the actual pH of the leachate and is quite low, although most of the metal load is desorbable by complexants like EDTA. With finer grain sizes, the concentrations in the leachate with distilled water and 1 M KNO_3 increase [76]. On the other hand, about 15% of the organic carbon of urban sludges, together with major amounts of N, have been found to be soluble in distilled water but, obviously, the load of heavy metals in this soluble fraction is low [35,77].

When salt solutions instead of distilled water are used for the determination of exchangeable amounts, one must be aware of certain differences among various possible salts. Thus, the amounts of Pb, Cu, and Cd desorbable from hydrous oxides differ when magnesium chloride or ammonium acetate are used. This is due to different affinities for the sorbed cation on the one hand, and to the solid source on the other [50]. Similarly, in samples of the River Danube and the River Gurk, different amounts were found to be exchangeable with ammonium acetate at pH 7 and with ammonium chloride–NaOH at pH 7 [11]. It is thus important to know which salt has been used to define the "exchangeable" fraction. In ammonium acetate at pH 7, there is some dissolution of calcite. Acetate buffer, pH 8.2, has been suggested for the exchangeable fraction of carbonaceous samples [78] but has hardly been used so far. The use of 1 M $MgCl_2$ did not affect the composition of Fe hydroxide coatings, determined with citrate–dithionite [79].

In the reducing marine sediments of Saanich Inlet (Canada), high-molecular-weight polymers were found, which were assumed to be precursors of sedimentary humates. They carried aliphatic, benzenecarboxylic, and phenolic acid groups. From the trace elements in the interstitial water, 100% of Zn, 80–100% of Cu, 50–100% of Fe, 48–80% of Ni, 30–50% of Co, and 50% of Sr, but no Ca and Mn were bound to dissolved organic polymers [32]. In a model experiment, humic substances loaded with Cd, Zn, Cu, and Pb released appreciable amounts (30–70%) with neutral salts: the lowest exchange was for Zn and Cd with $MgCl_2$ [97]. In carbonate-rich soils, the high pH of the soil solution (pH > 7.7) determines the water-soluble and matrix-bound forms of the micro-elements (e.g. Co, Cu, Zn). Ammonium acetate of pH 7 was used to gain the trace elements sorbed on carbonate surfaces (calcite, dolomite) [80]. In carbonaceous forest-soil, containing about 10% of organic carbon, the proportion of metals exchangeable with 1 M NH_4Cl increases for Zn at a soil pH < 6.5, for Mn at a soil pH < 5.5, for Pb at a soil pH < 4.5, and for Fe at a soil pH < 3.5 [81]. Because in the test forest, in central Germany, the soil pH near the stems of the beech trees has been lowered below pH 3 by the action of acid rain, the mobilization of soil constituents is of some practical importance for the plant growth. It was found that 40% of $PbSO_4$ added to an uncontaminated soil, as a model for Pb contamination in roadside soils, was "exchangeable" with 1 M $MgCl_2$, and a further 20% as carbonate (weak acid buffer) [82].

In clay-minerals, interlayer K is not directly exchangeable with ammonium acetate of pH 7, but is slowly released when 1 M HCl is applied, or if the pore-water concentration of K falls below 50–400 μmol/l for the trioctaedric sites, and below 2.5 μmol/l for the dioctaedric sites [83].

7.2.2.2. Weak acids without dominant complexation or redox reactions

Experiments with weak acids or weakly acidic buffers without pronounced complexing, reducing, or oxidizing capabilities may help the understanding of natural processes which liberate acids, and of buffering capacities. Dissolution with weak acids is mainly used to render the carbonates and trace metals exchangeable with acid. The importance of the leaching behaviour of a substratum towards weak acids has been recognized by the US Environmental Protection Agency, which set regulatory levels for the classification of hazardous waste by ordering a leaching test with dilute acetic acid at pH 5 [84].

Aerobic respiration leads to the release of CO_2 into the pore-water, which is sufficiently acidic to dissolve calcite. Thus, calcareous shells are only completely preserved when they are deposited in anaerobic environments. In moderate climates, oxygen consumption is lower in winter, and aerobic respiration reaches lower sediment depths than in summer, whereas in tropical and sub-tropical climates, this borderline is fairly constant [19]. Another natural source of acid is the oxidation of sulfides, by either oxygen or Fe(III), yielding sulfate [14,85].

Huntite $CaMg_3(CO_3)_4$ is considerably more soluble than the corresponding mechanical mixtures of calcite or dolomite, and magnesite. It dissolves congruently [86]. The solubility of mixed Ca and Mn carbonates such as $CaMn(CO_3)_2$, Kutnahorite, is lower than pure $MnCO_3$ (rhodochrosite), but higher than for calcite [68]. In ammonium acetate buffer of pH 5, synthesized $CaCO_3$, $CdCO_3$, $MnCO_3$, and $PbCO_3$ are more than 90% dissolved [87]. However, all of the Ca from a calcite doped with Pb has been found in acetate buffer of pH 5, but only 40% of the Pb [88]. The $CaCO_3$ dissolved reaches equilibrium in less than 5 hours [89]. Acetate buffer of pH 5 partially dissolves zinc oxide, zinc phosphate, and lead carbonate, but not CuO and PbO [50].

Carbonaceous soils contain primary carbonates from erosion of rocks, and secondary carbonates which have been precipitated in the soil, derived from soil fauna etc. The latter contain micro-elements, e.g., Co, Cu, Zn, Mn, which are important for nutrition, and can be re-mobilized more easily than from the primary carbonates. Acetate buffer of pH 5 selectively dissolves the carbonates (calcite) which have been produced in the soil itself, whereas for primary abrasion, a pH below 3.5 is needed [80]. Treatment for one hour with ammonium acetate at pH 4.4 did not change the X-ray-diffractograms of gibbsite, goethite, palygorskite, kaolinite, or hematite [90]. The dissolution of muscovite in acid and alkaline solutions (pH 1.4–11.8) is slightly faster than at pH 5–7, but negligibly slow compared with carbonates, etc., and even slower than feldspar and quartz. It leads to free toxic Al^{3+} [91] and silicic acid [92]. With 0.5 M acetic acid, Pb, Cu, and Cd are partially desorbed from goethite, hydrous Al hydroxide, and MnOOH [50]. Treatment with 25% acetic acid overnight dissolves all carbonates, but will not destroy clay mineral lattices

[93]. Similarly, 25% acetic acid dissolves the oxides and phosphates of Zn, Cu, and partially those of Pb [94], whereas their sulfides are negligibly attacked.

7.2.2.3. Strong acids

Strong acids exert more aggressive, and sometimes also oxidative, dissolution behaviour towards samples than the weak acids discussed above. Complete dissolution of many types of samples is achieved by decomposition with $HF/HNO_3/HClO_4$, but this is not wanted in the present context of selective dissolution techniques.

Sometimes, heterotrophic and chemolithotrophic micro-organisms synthesize strong mineral acids, such as nitric acid or sulfuric acid, under natural conditions to derive micro-nutrients from silicates and other poorly soluble material [95]. It is worthwhile also to consider these reagents in the natural cycling of elements. From acid rain, sulfuric acid is applied to surface soils in variable amounts.

Dilute (0.3 M) nitric acid, with boiling for 30 min, is a suitable reagent for the distinction of various Mn and Fe minerals. It dissolves manganocalcite, rhodochrosite, 85% of biotite, and 80% of chlorite: however, pyrolusite MnO_2, manganite $MnOOH \cdot H_2O$, psilomelan, magnetite, and hematite are not attacked. The method has been successfully applied to ore prospecting, if sulfides are absent. Pyrite and pyrrhotin are about half attacked [96]. 0.5 M HCl or 1 M HNO_3 nearly quantitatively release Cu, Pb, Cd, and Zn from technical grade commercially available humic acids, after adsorption of these metals from aqueous solution [97]. They also displace most of the sorbed ions from hydrous oxides, like those of Pb, Cu, and Cd from FeOOH (goethite), MnOOH, and $Al(OH)_3$ [50]. From estuarine sediments of South-West England, the Cu extracted with 1 M HCl was just the sum of the Cu extractable with 25 % acetic acid and that with 1 M NH_4OH, which could be interpreted as the sum of an inorganic and an organic available form of Cu [98]. Boiling 1 M nitric acid dissolved the same amounts of K from sandy, feldspathic soils of Eastern Nebraska as did the complexant tetraphenylboron: this was called "slowly available", according to experiments with alfalva. Considerably more K was leached from micas than from the feldspars [99]. In the leaching of carbonaceous rocks, 22% HCl always dissolves more than 25% acetic acid, because it also attacks the minor fraction of clay-minerals [93].

Boiling in the typically non-oxidizing 6 M HCl allows hydrogen sulfide to be evolved quantitatively from pyrrhotite, sphalerite, and galena, which can thus be clearly distinguished from pyrite, which is not reactive in 6 M HCl. Simultaneously, the leach solution gains "acid soluble sulfate", which is every sulfate except barite [100].

7.2.2.4. Leaching with alkaline solutions

On various surfaces, ion exchange of anions versus OH^- occurs, leading to general desorption and mobilization of anions and the largely negatively charged organic matter. The solubility of cations in alkali is limited, because of precipitation of sparingly soluble hydroxides. Only complexes which are more stable than the respective hydroxides, or amphoteric elements like Al, can be expected in an alkaline leach. Tissues and residual organic substances are more easily dissolved in alkaline than in acid solutions. Irreversible saponification of esters of carboxylic acids occurs. Esters of phosphoric acid, however, are rather stable. Oxidation by oxygen from the air is more pronounced than in acid media. The organic matrix carries carboxylic and phenolic groups which are uncharged in neutral and acidic solution, but get a negative charge in alkali, rendering them more hydrophilic. Some complexes of metal ions with groups of moderate acidity are favoured in alkaline solution, if the respective hydroxy complexes of the cations involved exert no strong competition. On the other hand, anionic hydroxy and other complexes can have interactions with amino-groups and other basic sites on the organic matrix.

Humic substances can be expected to be the main organic matter in sediments. Marine humates are formed from a marine environment, mainly by degradation of planktonic material, whereas limnic and riverine humates are partly detritus from the neighbouring land masses. As well as trace elements, the humics also collect major constituents of the earth's crust, like Si, Al, Fe, Ca, and Mg, but their proportions in the humics do not correspond to any silicate mineral species [32].

If humics or Pb and Cu humic/fulvic complexes are sorbed on kaolin covered with Fe hydroxide, which has been found to be a major part of the suspended matter in a Japanese river, they can be desorbed with 0.1 M NaOH. Although organically-bound Pb and Cu are thus desorbable from Fe hydroxide, the desorbed Fe itself does not necessarily correspond to the humic complexes, because it is sparingly soluble [61]. Moreover, Fe and Mn are precipitated from ammonia–ammonium sulfate leachates of manganese nodules, when the solution is allowed to age, and co-precipitate some dissolved Co, Cu, and Ni again. Thus it is essential to separate or to complex Fe and Mn in alkaline leaches as soon as possible [56].

Similar quantities of humic substances and trace metals were removed by 1 M ammonia and 1 M NaOH; the range was 1-23% of the total organic matter in the area studied [101]. A high proportion of Cu and Ag was found in this fraction, relative to other metals. The Fe correlated strongly with the humic substance, but probably not all of the complexed Fe was leached out. Pretreatment with oxalate enhanced the alkali- soluble fraction of Fe and Cu, indicating a certain cover of Fe hydroxides on the humics: the humic substance itself was only weakly associated with the oxalate-leachable amount [101]. In oxidized estuarine sediments, the Cu associated with extractable organic matter showed some agreement with the proportion of Cu extracted with aqueous ammonia. Substantially more Cu was extracted by ammonia than by acetic acid [98].

On the whole, NaOH does not markedly react with the organic compounds in sediments, soils, or sludge. Only some nitrogen is lost as ammonia. Thus, the treatment of urban sludge with cold 0.1 M NaOH does not change the IR spectra of organic compounds subsequently extracted with various organic solvents, such as diethyl ether, benzene, acetone, or ethanol. Only in the extract with tetrahydrofuran, do some aromatic-type compounds appear after the NaOH treatment [34]. Furthermore, NaOH and organic solvents tend to dissolve different compounds from environmental matrices. Refluxing with 1 M NaOH, and Soxhlet-extraction with a benzene–dichloromethane–methanol (1:1:1) mixture, dissolve different fractions of the organic matter from riverine and estuarine sediments: the contents of the trace elements Cu, Zn, Pb, Ni, Cr, and Cd do not correlate, and show no trends at all. This means that the trace metals can be bound to hydrophilic and hydrophobic sites, which are rather independent of one another. The amounts of Fe and Mn are significantly lower in the NaOH extract, owing to their poor solubility in alkaline solutions [102].

The effect of weakly alkaline solutions on substances other than organic detritus is slight. Thus, treatment overnight with 5% carbonate solution of gibbsite, goethite, palygorskite, kaolinite, and hematite had negligible effect [90]. Similarly, 0.1 M Na_3PO_4 and 0.1 M $Na_4P_2O_7$ do not desorb appreciable amounts of Pb, Cu, or Cd from goethite FeOOH and manganite MnOOH [50]. Strong alkali also attacks some silicates. Boiling 5 M NaOH was used to remove gibbsite and kaolinite from lateritic bauxite samples, derived from weathering of basalts. Goethite, however, is not attacked [90]. Diatoms, such as *Nitzschia linearis* and *Thalassiosira nana*, gradually dissolve in alkaline solution with rather constant speed. *Thal. nana* dissolves quantitatively in 37 days at pH 10, and 17% of at *Nitzschia* pH 10, although they are stable at pH < 8 [103]. The dissolution in alkali of micas, which are relatively inert, yields soluble aluminate and soluble silicate: the dissolution rate increases linearly with pH [92]. Carbonates are not known to be attacked or changed by alkaline treatment.

7.2.2.5. Effect of reductants

It is important to consider the mobilization of elements from the solid under reducing conditions for the understanding of their natural cycling. Bacterial degradation of organic carbon regulates the redox potential, which can go down enough to allow the formation of sulphide and methane. Redox-sensitive elements, like Mn, Fe, and Co, together with the trace elements sorbed on their surface, are generally mobilized by reduction (compare ref. 5), and may subsequently be depleted as their sulfides. The vertical trends in a core through oxidized surface sediment into reducing sediment are controlled by the redox interface. In particular, the As, Cu, Zn, Mn, and Fe in the pore-water are controlled by the solubility of Fe and Mn oxyhydroxides or metal sulfides. Under very reducing conditions, organic complexes are not able to compete with the bisulfide for the complexation of trace metals. The pH is governed by $CaCO_3$, which is completely preserved, and by the conversion of goethite to siderite [14].

7.2.2.5.1. Hydroxylamine

Hydroxylamine is very often used as a mildly reducing agent. It does not introduce additional cations to the sample, it has no complexation capabilities, and it allows the discrimination of different Fe and Mn phases. Carbonates have to be removed before the hydroxylamine leach if it is used in acid solution, where it is most effective.

The oxygenation of Mn-rich waters at pH 6-8 causes the precipitation of a Mn oxide, resembling vernadite (delta MnO_2). This rather amorphous solid contained 10% Ca and many other trace elements (Mg, Si, P, S, Cl, K, Ba, Fe), and 20-30% organic matter. It could be completely dissolved with 1% hydroxylamine–1 M HCl. The co-precipitated matter is humic material, which can be isolated from the extract with *n*-butanol, and stripped with 0.1 *M* NaOH [25]. A mixture of MnO_2 and ferrihydrite completely dissolved in 5% NH_2OH–5% oxalate. For the dissolution of goethite, acidification of 5% NH_2OH to 5 *M* in HCl was necessary [55]. It is possible to use reduction by 5% hydroxylamine at pH 3 to selectively desorb chromate from particles and colloids like humics, kaolin, Fe hydroxides, and silica: Cr(III) associated with any kind of negatively charged colloids is not affected [104]. Although amorphous or poorly crystalline phases are completely dissolved by hydroxylamine in weakly acidic solution, the attack on crystalline solids is not so easy to document, because weathered surfaces can be amorphous as well, and are not easy to detect by independent methods. Among the tested sulfides, 90% of Cu was released from FeS and about half of Cr from PbS. The tested vivianite released its Mn, and partially Fe. Main- and trace-elements leached with acid hydroxylamine from hematite, goethite, and chlorite were slight [87]. Cobalt sorbed on MnO_2 mixed with clay-minerals (montmorillonite) could be selectively leached with hydroxylamine, which was in agreement with electron microscopic measurements on solid MnO_2 particles [65].

The reducing action of hydroxylamine is quite dependent on the pH of the reagent. It is much more effective at pH 1 than at pH 3 [94]. A 3% aqueous hydroxylamine hydrochloride solution was used to remove hydrated Fe oxides from oxidized ferruginous quartzite samples. Hematite and magnetite, carbonates, and silicates were not attacked [105]. The amounts of Cu and Mn in the extract from hydroxylamine–HNO_3 at pH 2 were substantially higher when pyrophosphate, used for the removal of organics, was applied after, rather than before the extraction. From this, the authors conclude that pyrophosphate partially dissolve the Mn oxides [106], but it seems much more probable that phosphates built up in the pyrophosphate leach cannot be attacked by hydroxylamine any more. Humics co-precipitated with amorphous Mn oxide are found in the hydroxylamine leach as well [25]. Leaching with 0.1 *M* hydroxylamine–0.01 *M* HNO_3 is extremely sensitive to the presence in sediments of carbonate, which easily raises the pH and thus lowers the reducing capability of the reagent [101]. In this case, the pH has to be controlled and readjusted to pH 2 by the addition of further dilute nitric acid. For carbonaceous samples, hydroxylamine in 25% acetic acid is preferred, because of its high buffer capacity. However, when it is used only low levels of Si and Al are found in the extracts: there is no chlorite transformation, but there is a decrease in the XRD peak for smectites [89]. Hydroxylamine–25% acetic acid is reasonably selective in dissolving amorphous Fe and Mn hydroxides, provided that carbonates have been previously removed. It does not attack goethite, FeOOH [4], but does quantitatively dissolve the Mn-rich layer of deep-sea Fe/Mn nodules, leaving the Fe-rich phase intact [87]. The fraction of sediments which can be leached by hydroxylamine–25% acetic acid dominates in the upper layers, and decreases with depth. Upon reduction of the deeper layers, the hydroxylamine-leachable fraction becomes richer in sulfides, but not uniformly. Manganese has been found to be dominated by the hydroxylamine-leachable fraction in the sulfidic layers also; this could be explained by a lack of transformation into MnS and incorporation into other sulfide phases, or by the dissolution of MnS in 25% acetic acid [37].

References on p. 171

The selectivity of 1 M NH_2OH–25% acetic acid has been tested, using the sulfides, oxides, and phosphates of Zn, Cu, and Pb as model substances [94]. Although 51% of CuO and 39% of PbO are dissolved, the amounts of the others are negligible. As these substances are hardly ever met in natural environments, the selectivity is reasonable. However, care has to be taken in the prospecting for ores, where rather unusual solids can occur. Of the Zn incorporated into Fe sulfides 70% was found using hydroxylamine–25% acetic acid. This may be due to the solubility of ZnS in acid, rather than the reductive action of hydroxylamine [88]. The extracts from surface sediments, polluted with Zn and Pb, with hydroxylamine–HNO_3 at pH 2, and with 0.05 M EDTA, both from the original sample, correlated well for Zn ($r=0.984$), moderately for Fe ($r=713$), but not for Cu and Pb [75]. As the carbonates and "exchangeables" were not removed beforehand, they are included in these cases, and thus relations between reducing- and complexing-agents cannot be inferred.

7.2.2.5.2. Others

Like hydroxylamine, oxalate is often termed a reductant for Fe oxide layers. Because its action is dominated by its complexation capabilities, however, it is treated among the complexants in the present context.

Dithionite is a strong reductant, applicable to Fe oxides, close to neutral pH. In dithionite solution, the reduced Fe is sparingly soluble, and mere desorption from surface sites occurs [11,107,108]. To keep the Fe in solution, dithionite is often used in citrate buffer at pH 7 (citrate–dithionite–bicarbonate reagent, abbreviated CDB), but the complexing activities of citrate severely increase the desorbable amounts. Thus, more than twice as much P is dissolved by CDB from limnic sediments, than by dithionite buffered with carbonate alone [109]. The CDB leach does not in any case correlate with hydroxylamine or oxalate leaches; in some cases, it can be smaller than that from hydroxylamine–acetic acid [89]. Citrate–dithionite at pH 7, and acetate–dithionite at pH 3.4, dissolved some goethite from the more deeply weathered samples of lateritic bauxites deriving from weathered basalts [90]. Citrate–dithionite at pH 7 dissolved negligible amounts of transition metals from natural organic material, chlorite, and montmorillonite [79].

Phenolic reductants, either natural or anthropogenic, reduce amorphous Fe oxides and also some goethite to soluble Fe(II) in the absence of oxygen. Terrestrial phenolic substituents are generated through microbial processes and the degradation of lignins. Trihydroxybenzenes, such as pyrogallol and gallic acid, are related to tannins found in natural waters, and are strong reductants of goethite. Organic-rich ground-waters contain high amounts of dissolved Fe and Mn owing to the reduction of Fe/Mn surface coatings by these degradational phenolic groups. Hematite, however, is quite inert: its rate of reductive dissolution by hydroquinone was extremely small in the pH range 2.0 to 3.8. At pH 2, the dissolution rates for hematite and goethite are similar. At pH 3.4, however, goethite undergoes reductive dissolution two orders of magnitude more quickly than hematite. No change of the reduction rate by illumination was observed, so the process is not photoinduced. Sulfate at a level of 10^{-4} M had no effect on the reductive dissolution of goethite. At high levels, it decreased the reaction rate because of competition with the reductant for the adsorption sites [24].

The reduction of chromate by dissolved organic matter in enclosed tanks was not affected by an increase in the living biomass and the particle flux such as occurs in rivers. A possible role for dissolved organic carbon is that it takes part in a photoreaction of the weakly adsorbing chromate to cationic Cr(III) [5]. Similarly, freshly precipitated Mn oxide (average oxidation state 3.5) can undergo humic-mediated photoreduction in oxic lake waters: this has been proposed as a possible mechanism for supplying Mn(II) to phytoplankton [110].

The reduction of urban sludge with SO_2 results in solubilization of Cu and Zn, possibly by its acid action, and in a 50% increase of the solubility of organics, especially N compounds, in organic solvents like tetrahydrofuran or ethanol [35].

Although Fe(II) is an unusual reductant in artificial chemical leaching methods it is a very important intermediate in the processes in the oxic/anoxic boundary in sediments. It diffuses upwards from the anoxic zone and selectively reduces the Mn oxides, e.g. birnessite $Mn_7O_{13} \cdot 5H_2O$, which need a higher redox potential to be stable, than do the Fe(III) oxides [58].

Pyrite is the most abundant sulfur compound in anoxic sediments. Labeling with [^{35}S]sulfate has shown that the pyrite quantitatively derives from sulfate reduction, whereas organic sulfur compounds are reduced to elemental sulfur. Pyrite and elemental sulfur can be oxidized to sulfate with aqua regia, but it is more specific and more sensitive to reduce them to H_2S, which can be trapped as ZnS or Ag_2S. This reduction can be done either with Cr(II) in 4 M HCl [111], or with $LiAlH_4$ [100]. Unfortunately, the reduction methods for pyrite and elemental sulfur were not used for the determination of trace-elements bound to these phases. The reduction of sulfate from solid $BaSO_4$ to hydrogen sulfide is possible either with a mixture of $HI–H_3PO_2–HCl$, or with Sn(II) in phosphoric acid [100].

It should be kept in mind that Fe oxides can precipitate from reductant solutions on prolonged standing in air, scavenging trace elements. After the separation from residual solid, the Fe should be stabilized by addition of a complexant, like EDTA (unless citrate is already present, as in CDB) [109].

7.2.2.6. Effect of oxidants

In sediments and soils, organic matter and sulfides are the main targets of oxidation reactions. In the presence of moisture, oxygen, and Fe-oxidizing bacteria (*Thiobacillus ferrooxidans*), pyrite oxidation yields sulfate and H^+ ions. The sulfate may form gypsum, and the excess acid dissolves heavy metals, whose migration is controlled by the buffering capacity of the sample [85]. *Thiobacillus ferrooxidans* transforms sulfide ores to soluble sulfates, at an optimum pH 2.0 - 2.4 [95].

During storage in air, oxidation processes provoke severe speciation shifts in anoxic sediments. Cadmium changes from sulfidic to exchangeable and carbonate forms, and Fe changes from carbonate to hydroxides (oxalate-leachable). Therefore it is necessary to process anoxic sediments in an inert atmosphere, for example N_2, if speciation studies are made [13]. The trace elements, Cr, S, Se, U, and V are potentially mobilized by oxidation [5]. For example, the aeration of anaerobically digested sewage sludge results in a mobilisation of Co, Se, and Cu, but fixation of Sn [36]. In marine ecosystems, Cr(III) can be oxidized rapidly to soluble chromate [5]. It is thus necessary, when characterizing the mobility of the constituents of a given sample, to take possible oxidational changes into consideration. Alkaline hypochlorite, hydrogen peroxide, or concentrated nitric acid have been used as oxidants for leaching reactions. As these reactions are quite aggressive, a leaching sequence should be designed so that labile phases are removed beforehand. For trace analysis it should be mentioned that the blanks of nitric acid, and of hydrogen peroxide are much more easily controlled and minimised than those of alkaline hypochlorite. Sodium hypochlorite at pH 8.5 removed 95% of the organic carbon from natural organic material, whereas hydrogen peroxide removed only 65-80% [79]. Salim and Shaikh [112] used NaOCl at pH 9.5 to oxidise organic matter in loess-derived alkaline calcareous soils. Pyrite, together with sorbed As, could be quantitatively dissolved with hot HNO_3 [113]. Double $KClO_3$/HCl extraction has been found to dissolve 86–100% of galena, chalcopyrite, cinnabar, stibnite, sphalerite, and tetrahedrite, and also 71–86% of pyrite [37]. The resistance of sulfides towards the attack of hydrogen peroxide, with subsequent leaching by ammonium acetate, as proposed in the sequence of Tessier et al. [89], is variable. From PbS, the Cu, Pb, Cd, and Zn were found quantitatively in the resulting solution: from FeS, however, the recoveries were 90% of Zn, 70% of Pb and Cd, 60% of Fe, and 50% of Cr, Mn and Ni [87]; compare [88].

7.2.2.7. Action of complexants

Roots of higher plants, as well as micro-organisms, are known to excrete organic acids with complexation capabilities, possibly to get some nutrients or micro-nutrients from the surrounding solid. Complexants are therefore frequently used to simulate so-called "available" fractions (see below). In chemical terms, complexation reactions can affect all types of species in the solid to various extents. All reagents have an optimum pH for this action, lying between their protonation, and the hydrolysis of their cations. Sorption of the complexant on active sites of the solid, and thus its fixation, competes with the transport of soluble complexes in the pore-water. Reactive forms of main elements, like hydrous oxides of Fe, Al, and Mn, compete with the trace-elements for the complexants.

The presence of water-soluble anions of organic acids, such as malic, malonic, oxalic, acetic, succinic, tartaric, vanillic, and p-hydroxybenzoic acids, derived from microorganisms and plants has been demonstrated in top soils at a level of 10^{-5} M to 10^{-4} M. Oxalate was the most abundant [27]. The clay minerals dissolve incongruently in pure water, yielding more Si than Al. Organic acids, however, dissolve more Al than Si, if they are complexing like citric, tartaric, salicylic and tannic acid, whereas the non-complexing aspartic acid yields about the same proportion of dissolution of Si/Al as does water. Dissolution by complexing acids is favoured in the order illite > montmorillonite > kaolinite. The dissolution behaviour of Fe parallels that of Al [114].

Various complexones, like EDTA, are synthetic compounds. Certainly they cannot match the conditions typical of a natural substance, but they are not degraded during longer leaching periods. Application of a complexant to an untreated sample clearly desorbs the "exchangeable" amounts such as other salt solutions, and the carbonates if the solution contains sufficient acid. A solution of 0.05 M EDTA quantitatively dissolves ZnO, PbO, $ZnCO_3$, $PbCO_3$, Zn and Cu phosphate in an optimum range of pH 5–7: CuO and $Pb_3(PO_4)_2$ are only moderately attacked [94]. Cadmium sorbed on calcite surfaces can be partially desorbed with EDTA at pH 8. On the other hand, the Cd-EDTA complex is not sorbed by calcite like the free ion [69]. Interaction of goethite FeOOH, manganite MnOOH and hydrous $Al(OH)_3$ with complexing agents, like citrate or salicylate, leads to displacement of surface-adsorbed Pb, Cd and Cu, but to minimal attack on the sample component structure [50].

Oxalate, salicylate, and F^- enhance the dissolution rate of delta-Al_2O_3 and various Fe hydroxides in weakly acid solution [27]. Whereas the extent of dissolution of crystalline Fe-oxides is only about 1% after 300 h contact with naturally occurring complexants as well as complexones, the non-crystalline oxide was completely dissolved at pH 3.5 in the presence of oxalic, malonic, malic, citric, or tartaric acids, or by EDTA, DTPA or NTA. Acetic, lactic, salicylic, succinic, phthalic and fulvic acids were much less reactive. Dissolution at pH 5.5 was markedly slower than at pH 3.5 [115]. This proves that oxalate buffer at pH 3, within the sequence of Tessier et al. [89], selectively dissolves non-crystalline hydrous oxides. Their main components are Fe, Al, Mn and: they carry a great amount of trace-metals, and are strongly surface-correlated [11].

Complexation reactions with carboxylic and phenolic groups within the framework of humic substances is sufficient for the adsorption to hydrous oxides of Fe, Mn, and Al. No noticeable dissolution of Mn_3O_4 and MnOOH, due to possible reduction to soluble Mn(II), has been observed [57]. Humics are strongly sorbed on hydrated Fe oxides, but only slightly on kaolin, anhydrous Fe_2O_3, and quartz [61]. Sodium citrate, EDTA, and DTPA quantitatively released Cd, Zn, Cu, and Pb, previously sorbed to solid humics [97].

Heterotrophic micro-organisms produce organic complexing acids, like oxalic, citric, gluconic, and amino-acids for the dissolution of silicate minerals, provided sufficient nutrients are available (phosphate, sulfate, ammonium, Ca, K, CO_2). Strikingly, they get their energy from the oxidation of Fe(II). Biotite, orthoclase, and even zircon could be dissolved this way with the aid of micro-organisms, at an optimum temperature of 32°C, and an optimum pH of 2.0 – 2.4. This can be taken as a model for the natural weathering processes of silicates [95]. The dissolution of quartz in 0.02 M citrate at pH 7 was 8 to 10 times faster than in pure water. Salicylic, oxalic, and humic acids also accelerated the dissolution of quartz, but to a minor extent; acetate was not effective. In comparison with other solid phases, the dissolution rate is, however, negligible [116]. Lead sulfide and Ag_2S are only partially attacked by citric acid at pH 1.3, but very efficiently by ammoniacal citrate at pH 9.5. Addition of hydrogen peroxide accelerates the dissolution, so that 1 h in the cold is sufficient [117].

In carbonaceous forest-soils containing 7.4–11.9% organic carbon, the Fe and Pb in the EDTA extract correlate with organic carbon, but Zn and Mn do not. When the soil pH

References on p. 171

is down at pH 3.0 the EDTA-extractable Zn, Mn, and Pb are very low, because it has already been washed into deeper soil layers; the Fe-EDTA in acid soils, however is quite high [81].

7.2.2.8. Leaching with organic solvents

Generally, organic compounds occurring in sediments and soils, either within living biota, or degradation products, carry many hydrophilic groups, like hydroxyls and carboxyls. The only lipophile species encountered are fat, and chlorophyll and other porphyrins, together with their degradation products. Thus, only a small amount of organic matter, and metal ions bound to organic matter, can be extracted into organic solvents. The applied solvent has to have a certain miscibility with water, to be able to displace the water in the capillaries of the solid sample, gradually, and ensure overall contact.

In reverine and estuarine sediments, the extract achieved with 1 M NaOH did not correlate with that from Soxhlet-extraction with benzene–dichloromethane–methanol, and no trends were observable. Only Fe and Mn were significantly lower in NaOH, because of low solubility in alkali. The Soxhlet-extract was eluated with benzene, and subsequently with methanol; the main amounts of Cu, Zn, Pb, Ni, Cr, and Cd, however, which remained in the residue in the Soxhlet-extractor, were called "asphaltenes" [102]. For sediments of the river Danube, the ethanol extract was significantly weaker than the NaOH extract, and exceeded 1% of total amount only for Cu; the content of other main and trace-elements was very low [11]. Similarly, from a purification sludge, less than 1% of Zn and Pb could be desorbed with ethyl acetate, chloroform, or methanol; only Cu reached the range 1–2%. Compared with the amounts exchangeable with ammonium acetate at pH 7, or acetate buffer at pH 5.5, only up to 1/5 of Fe and Cu, and less than 1% of Zn and Pb could be dissolved with organic solvents [118].

The extraction of labeled benzo[a]pyrene and anthracene from contaminated soil samples gave higher recoveries with Soxhlet extraction (hexane–acetone, 1:1) than with with high-velocity mixing (with acetone), but was not quantitative in either case [119]. "Solvent-extractable" organic sulfur has been isolated from sediment samples with benzene–methanol–acetone (70:15:15). Inorganic sulfides and sulfates are not attacked but a residue of sulfur-containing organics remained in the residue, and could be determined after decomposition with H_2O_2. No examples for the respective sulfur compounds or other elements in the extracts are given [100].

From urban sludge, organic solvents (diethyl ether, benzene, acetone, tetrahydrofuran, ethanol) extract only a minor part (15%) of the organic compounds; the water-soluble and NaOH-soluble parts are usually larger. Organic solvents preferentially extract aliphatic compounds, and no aromatic bands could be detected in the IR spectra of the extracts. Although water-soluble organics form sludges contained large amounts of N, no nitrogenous compounds were soluble in organic solvents. Of the metals, only Cu appeared to be soluble in solvent, and about 10% of total, largely in the tetrahydrofuran fraction. It is suggested, from the spectra, that Cu interacts with conjugated ketones, which are extractable in tetrahydrofuran, and also in NaOH [34]. The UV-irradiation of urban sludge increases aliphatic compounds and the number of alcoholic, olefinic, and carboxylic groups, but does not strongly affect the leachability of metals [77]. Treatment with sulfur dioxide increases the solubility in organic solvents of residual organic compounds, but the influence on metal mobility is small [35].

7.3. SOME EXPERIMENTAL DATA FROM SEQUENTIAL LEACHING SEQUENCES

7.3.1. Arsenic

Only a few data from leaching methods are available for arsenic, probably because the final determination in the leachates, which is chiefly done by hydride-AAS, is rather laborious.

Concentration and partitioning of As is largely controlled by the redox status. The amounts exchangeable with neutral salts or weak acid buffers are generally poor. No carbonates are formed, and arsenates are generally more soluble than phosphates. Under oxidizing conditions, adsorption on hydrated ferric oxide surfaces occurs over a wide pH range [120]. In contaminated oxic sediments of the Clark River, in Western Montana, the hydroxylamine–acetic acid fraction was therefore found to be dominant [37]. In the background pelagic sediments east of the Sahara the As is preferentially bound to crystallized Fe phases, or is mainly exchangeable with OH^-; major amounts are leached with NaOH–dithionite, or with HNO_3 after hydroxylamine–acetic acid [120]. In the deeper anoxic layers of the contaminated sediments of the Clark River [37], however, As is largely present as the sulfide, which is gained by oxidation with $KClO_3$–HCl, and proved by analyses of total sulfur and sulfate. Organically bound As is low in either case. Arsenic from clay minerals and micas has been found mainly in the oxalate or the dithionite fractions [108], but as the overall content was low, this could be due to impurities on the weathered surface. In studies of the same samples, the release of As, in a sequence originally developed for the speciation of phosphorus, is negligible in neutral ammonium chloride, but is appreciable in dithionite and NaOH, applied subsequently, where the silicate lattice is not attacked.

In the sediments of the River Danube at Altenwörth, whose total contents are close to the background values [11], arsenic is the only investigated element, all of whose fractions have been surface-correlated. Using the sequence devised by Tessier et al. the exchangeable As and its carbonate are found to be very low. The major fractions have been extractable by oxalate and dithionite, which means a close connection with iron hydroxides. A significant part has been leachable with NaOH, which can be interpreted as due to its being organically bound, or exchangeable against OH^-. A minor part, dissolved in hot HNO_3 or in hot NaOH, might be attributed to sulfide.

7.3.2. Cadmium

The speciation of Cd varies widely according to the oxidation status of the sediment. Under anoxic conditions Cd is highly immobile, which means it can be released only after strong oxidation with hydrogen peroxide. During short-term aeration a transformation occurs to hydroxylamine-reducible phases, and drying causes a transformation to carbonate and exchangeable fractions [121,122]. To preserve the original status in anoxic samples, their transport and leaching under N_2 are strongly recommended.

Exchangeable Cd is quite variable. It has been found to be negligible in reducing sediments (Eh < −100 mV) in the North Sea, in the European River sediments of Ems, the Rhine in Germany, the Deule in France, the Aliakmon in Greece [18,123,124], and in the sediments of the saline Lake Macquarie in Australia, which is heavily polluted by a smelter [75]. In North Sea sediment, an increase in the redox potential from −100 mV to +300 mV resulted in an increase of the exchangeable fraction from negligible to 25% of the total [125]. Thus, in river sediments which are probably oxic, from the Somme, Seine, Gironde, and Garonne in France [18], as well as from the Yamaska and St. Francois Rivers in Canada [78], more than 25% has been found to be exchangeable. The highest exchangeable proportion of Cd has been reported from the non-contaminated sediments of Lake Biwa in Japan, in the range of 40–50%. In the top layers of Lake Biwa sediment, an increase in Cd contents is due only to exchangeable and hydroxylamine-reducible fractions [126].

Cadmium which is leachable by weak acids, and is likely to be bound to carbonate, has not been determined in any case. It exceeds and parallels exchangeable amounts in most cases. It is about 1% under reducing conditions, and can rise to 40% upon oxidation. Hydroxylamine-reducible Cd is likely to be the major fraction in moderately and heavily polluted sediments, unless geogenic sulfides prevail. The highest value has been reported from the heavily contaminated River Deule in France (up to 1040 μg/g, with no mining activity!) with an average of 92% of total in this leach. Polluted European rivers, as well as Hamilton Harbour [127], yield 40–60% of their total with hydroxylamine. Lower amounts appear, of course, if much has already been released in an exchangeable or carbonate form, or is highly residual as a sulfide in mining areas. After hydroxylamine treatment, the release by oxalate is generally low, but only few data are available.

Oxidation with hydrogen peroxide yields only a minor fraction, except for the saline sediments of Lake Macquarie in Australia, where it rises to 60–90% [75], probably due to geogenic sulfides. After oxidation, the residual amounts are often small. Concerning other extractants, 0.5 M HCl takes up approximately the sum of Cd in its exchangeable, carbonate, and hydroxylamine-reducible forms [124]. The leachability in cold NaOH, together with humics, is low, but measurable (at 3.5–9.1%; [124]).

7.3.3. Chromium

In general, the fraction exchangeable with neutral salts is about 1% or less, as found in the sediments of the River Danube or the River Gurk. Even in the transport phase of the Amazonas and Yukon, less than 4% has been found to be exchangeable with $MgCl_2$ [79]. The highest mobility has been found in sediments from the Karst region of Yugoslavia, with 10–39% exchangeable with ammonium acetate at pH 7, and 12–36% in the carbonate fraction which is leachable with weak acid [128]. Chromium(III) adsorbed from the water phase is not exchangeable against ammonium acetate at pH 7, or ammonium chloride buffer of pH 7, whereas Cr(VI) can be partially recovered with ammonium chloride, depending on the organic carbon content [43]. It is thus possible to discriminate between the Cr(III) and Cr(VI) inputs into the sediments, which partially move to different adsorption sites. The part of Cr which is leachable with NaOH is only about 1–2%, in all cases found in the literature [11,49,124], but approximately 10% of adsorbable Cr(III) or Cr(VI) can move to this fraction. Weak acid buffers release marginal amounts, which are correlated with Fe containing phases, but probably only because of uncertainties in the leaching procedures.

The amount of Cr which is leached out by hydroxylamine 25% acetic acid seems to be very indicative of the pollutional status of the sediment resulting from uptake from the water phase. More than 50% of Cr(III) and Cr(VI) adsorbed to sediments of the River Gurk (1.3% Ca, 4.8% Fe, 7.7% Al) have been recovered with this reagent [49]. This is due to the strong affinity of an Fe-containing Mn phase which provides adsorbing sites for Cr. In the Danube sediments, only 5–10% was found to be leachable with hydroxylamine–acetic acid; in the Greek rivers Axios and Aliakmon there was about 3–9%; in the top sediment of the polluted River Deule in Northern France 20%; in the sediment of the Rhine, 86%; but in the unpolluted regions of the River Gurk, only 1% [11,49,123,124,129]. On the whole, the percentage of hydroxylamine-leachable Cr does not depend on the grain size, unlike that which is oxalate-leachable. In the sediments of the River Danube which were investigated, and also in the non-polluted region of the River Gurk, oxalate leached more Cr than did hydroxylamine/acetic acid: the relative amount increased with smaller grain sizes. The Cr from municipal effluents in a sediment which was poor in carbonates did not move preferably into the hydroxylamine–acetic acid extractable, but to the oxalate extractable fraction [108]. Any H_2O_2 oxidizable fractions are small and of minor importance, if they are obtained without the addition of concentrated HNO_3.

A citrate/dithionite mixture, dissolved about the same amount of Cr as did dilute HCl from a number of different sediment samples from the US [130]. In comparison with total decomposition, the amount of Cr dissolved in HCl is about $1/3 - 1/2$ of total, except at sites of input of municipal or mine waste [130]. With cold dilute HCl, usually only 10–20%, and in the Gurk sediments [49] only 6%, of the total Cr was found [11,124,131]. Any HCl leachable Cr is bound to Fe, and surely does not derive from carbonates. In the sediments of the Danube from the Altenwörth Reservoir, and in the sediments of the River Gurk, the main fraction is a residue, and could not be dissolved without HF. The Cr is possibly present in chromites and heavy minerals, which are not attacked in the sequences applied. The HNO_3-leachable fraction may derive from micas or chlorite and is thus also highly immobile and geogenic. Any Cr(III) and Cr(VI) adsorbed to the sediments from solution did not increase the HNO_3-leachable fraction in the Tessier sequence.

7.3.4. Copper

The amount of copper which is exchangeable using neutral ammonium acetate can be as low as 0.5%, or less, of the total weight of material. This is found for sediments in

Wisconsin Lakes [132], or from the Gaspé Peninsula in Eastern Quebec [133,134], which are low both in Ca and in Fe. In the Yamaska and St. Francois River in Quebec, the exchangeable Cu is about 1% of total Cu but it is also present in in appreciable amounts in soluble forms [78]. In contrast, the exchangeable Cu can rise to about 10% of the total in carbonaceous (Krka estuary, [135]) and in heavy loaded samples such as from the Elbe estuary [18]. In the sediments of the reservoir at Altenwörth of the River Danube, in Lower Austria, as well as in the small River Gurk in Carinthia, the relative amounts of exchangeable copper (2–10%) show a pronounced increase towards larger grain sizes, which is unlike all other elements investigated [11,49]. In the transport phase of the Amazonas and the Yukon, 5.2% and 2.4%, respectively, have been found exchangeable with $MgCl_2$ [79], and in the sediments of Lake Biwa in Japan about 4% [126].

It can be concluded from leaching with slightly acidic buffers that, in most cases, the carbonates do not take up significant amounts of copper (commonly 1–6%), except in the almost exclusively carbonaceous sediments of the Krka estuary in the Karst [135]. Even in the highly polluted sediments of the Neckar, and Rotterdam Harbour [136], the so-called "carbonate fraction" is below the detection limit. Quite high values for the carbonate fraction have been reported from the Yamaska and St. Francois River in Quebec (8% and 14%) [78]. On the other hand, the solubility in NaOH can be very high, and even exceed the solubility in dilute HCl, as found for the urban sludge of the wastewater treatment plant east of Rome [3]. In reducing marine sediments, up to 79% of the total Cu is reported to be soluble in 1 M NaOH [32]. From the available data it seems that the NaOH-leachable fraction is much larger in samples from moderate climates than from hot climates, but this has to be investigated further. Thus, any Cu leached with alkaline Na-pyrophosphate from Rivers around Jamshedpur in India was found to be below detection limits [137]. These effects can be explained by the formation of strong complexes with slowly degradable organic material, and have been repeatedly found in the water phase, also. A further proof of the dominant affinity of copper for organics is the high release usually found after oxidation with H_2O_2 of the sediment after extraction of the iron-bound parts with oxalate: this yielded 10–30% of the total from sediments of the Danube, the Yellow River, and the Gurk [11,49,129].

The affinity of Cu towards Fe-dominated phases can be monitored by the action of hydroxylamine, oxalate, or dithionite. The extractability of Cu with dithionite, even in the presence of citrate, is strikingly low, so that the formation of a precipitate with this reagent has been assumed [130,138]. In some cases, weakly acidified hydroxylamine dissolves the main fraction, as from the Krka [135], Ijssel, Maas and Scheldt [18], which is probably a fraction bound to an iron-containing Mn hydroxide phase. For sediments of the Gaspé Peninsula in Eastern Quebec, it could be shown that Cu-pollution preferentially moved to the fine-grain sizes of the fraction which was leachable by hydroxylamine–25% acetic acid [133,134]. Strong affinity to Fe has been found in the sediments of the Neckar and in Rotterdam Harbour, by leaching 72% and 81% of the total with oxalate buffer after decalcification [136]. Similarly, in the non-carbonaceous samples of the River Gurk in Carinthia [49], oxalate applied after hydroxylamine–25% acetic acid released the main amounts of Cu from the sediment (40–60%), with more for the finer grain sizes.

Leaching of a fresh sample with dilute HCl is always more effective than with 25% acetic acid [11,124,131], which can be explained either by the dissolution of Fe phases, or sulfides, or by release from clay minerals (illite, montmorillonite; [108]). Data about residual Cu, only found by the action of HF, and probably silicate bound, vary greatly. This is irrespective of the extent of the load, from 50% in the sediment in Rotterdam Harbour [1], and the Axios in Greece [124], down to zero in that of the Wisconsin lakes [132]. In moderately polluted sediments, Cu from municipal effluents passes into the refractory organic fraction, which is not soluble in acetic acid, but is found after oxidation with H_2O_2 [108]. From the literature data it seems that, in the more heavily polluted sediments, the oxalate extractable Fe phase is filled up when the sorption capacity of the organics is exhausted. In the sediments of the Danube at Altenwörth, much less Cu is soluble in dithionite than in oxalate. The NaOH-leachable Cu, which means the organically-bound part, is less than the fraction liberated after oxidation with H_2O_2, in contrast to the situation for Zn and Pb [11].

7.3.5. Cobalt

Although Co has rarely been an environmental problem in the past it can be concluded from the few data available that it behaves like nickel. About 10% has been found to be exchangeable against ammonium acetate in sediments of the Danube (a third in carbonate), but only 1.5% in the Gurk (a crystalline rock area), and in sediments of Lake Biwa in Japan about 6% was exchangeable versus $MgCl_2$ [11,49,126]. The release into NaOH is low, and near the detection limit [32]. Up to 10% has been found to be released into weakly acidic buffers, but this does not necessarily reflect binding to carbonates. The similarity of leaching patterns for model pure minerals led to the conclusion that Co and Ni in natural occurring limestone and dolomite are bound to traces of clay minerals, which are not detectable with present methods of mineral analysis [108]. In the carbonaceous samples of the Krka in the Karst, only about 20% is released into the carbonate fraction [128].

For Co, the hydroxylamine–acetic acid leach is rather important. Its value is maximal in the top layers of Lake Geneva and Lake Biwa, and reaches a constant value at a depth of a few centimeters [126,139], as found for Mn. It has been reported to be the main fraction in the Yamaska and St. Francois River in Quebec [78]. When oxalate is applied after hydroxylamine, the release of Fe-bound parts into the oxalate has been found to be lower for the sediments of the Danube, but higher for the sediments of the Gurk, which might be due to the different geochemical characteristics of the sample groups. For the rather oxic sediments of the Gurk and the Danube, the regain after oxidation with H_2O_2 (organic/sulfidic) was small, but it could reach 20% in non-surface layers of Lake Biwa [126]. The amount dissolved by dilute hydrochloric acid is always more than with citrate-dithionite [130] or hydroxylamine–acetic acid [131]. Although for the sediments of the Danube and the Gurk most of the Co could be attributed to Fe-bound fractions, in Lake Ontario sediments the Co was found to be highly refractory, with only 18% leachable without the action of HF [131]. In Lake Biwa, about 40%, and in Lake Geneva, 40–60% of Co could only be released from the sediments with strong hot acids [126,139]. Grain-size effects are generally low.

7.3.6. Nickel

In the sequence described by Tessier et al. [89], Ni tends to be rather uniformly distributed between the carbonate, Fe hydroxide, organic, and silicate phases, whereas in other sequences it is markedly influenced by its low solubility in alkaline solution. The exchangeability towards neutral ammonium acetate or $MgCl_2$ has generally been found at a level of 1–2% of the total, in sediments from the Danube and the Gurk in Austria, the Yamaska and St. Francois River in Canada, and Lake Biwa in Japan), and is near the detection limit in unpolluted samples. The percentage of the exchangeable fraction does not rise significantly at fine grain sizes, or in polluted areas. Approximately 4–15% can be expected to be leachable with weakly acidic buffers: some affinity to carbonates, however, is not proven. Below 20% has been found in the carbonate fraction of the highly carbonaceous sediment of the Krka estuary in the Karst region [128]. In three lakes of Wisconsin, poor in Ca, Mg and Fe, only 1.5–3.5% of Ni has been found as carbonates [132].

The hydroxylamine–acetic acid-leachable fraction has been found to range from below the detection limit in Wisconsin Lakes [132] to 24% in the St. Francois River in Quebec [78]. Pollution with Ni does not significantly alter the relative distribution between exchangeable, weak-acid-leachable, and hydroxylamine-reducible amounts in the Yamaska and St. Francois River in Quebec, or in the River Gurk in Austria [49,78]. In sediment cores of Lake Biwa in Japan, however, a significant decline of Mn from the top to deeper layers is accompanied by a decrease of Ni in the hydroxylamine–acetic acid fraction, indicating the importance of Mn as sorption site [126]. Oxalate releases about twice as much as hydroxylamine, and the oxalate-leach is grain-size, and thus surface, dependent [11,49].

For the sediments from the River Danube at Altenwörth, and the Gurk in Carinthia, dithionite at pH 7 releases much less Ni than do the acid iron-attacking reductants hydroxylamine or oxalate. One striking difference lies in the fact that dithionite reduces Fe on surfaces, but does not dissolve much Fe itself. The amount of HCl-extractable Ni is always greater than the gain using citrate–dithionite for a variety of sediment samples: the difference could be due to additional amounts bound to carbonate or sulfide [130], or exchangeable material from clay minerals [108]. The release into 1 M NaOH of Ni from sediments from the Danube and the Gurk in Austria was below 10% of the total: this was also true for the polluted samples. Nickel can be chiefly bound to the silicate lattice (60–75%), as in the Schelde, Gironde and Garonne [18], or in a sediment from Lake Ontario [131]. In the sediments from Altenwörth, about a quarter was unattacked by both leaching sequences applied, and was thus present in the residual silicates [11]. The HNO_3 fraction in the Tessier sequence seems to be found as HCl fraction in the sequence according to Psenner et al., run in parallel [107,109]. Grain size effects have not been pronounced.

7.3.7. Lead

In many samples, the exchangeable part has been reported to be below 1%. This is found for the Yamaska and St. Francois Rivers in Quebec, sediments at metalliferous deposit sites on the Gaspé Peninsula in Quebec, the River Deule in France, and the reducing sediments of Rotterdam Harbour [1,78,123,134]. In case of the oxidizing and carbonaceous sampling sites, the action of ammonium acetate at pH 7 already requires an acidification, leading to extra "exchangeable" amounts. In the sediments of the Danube at Altenwörth in Lower Austria, about 2% has been found to be exchangeable, and in some samples from the carbonaceous Krka estuary the exchangeable part even became the main fraction [128]. Similarly, the weak-acid-leachable "carbonate" fraction can be as low as 2% in the Rotterdam Harbour [1] or the metalliferous deposit sites in Canada [134]. In the sediments of the River Danube which were investigated, the weak-acid-leachable Pb was in the range 20–30%. Strong increases towards the smaller grain-sizes, and correlations with carbonate minerals have been found [11]. In many cases, hydroxyl-amine–acetic acid released the major part, as from the Yamaska River and St. Francois River in Canada, or from sediments of metalliferous deposits in Canada [78,89], but also from the Deule River in France and the Danube in Austria. In samples from the Elbe, Ems, Rhine, Ijssel, Scheldt, and Somme, hydroxylamine released the largest part of Pb as well, but no weak-acid-leachable fraction had been isolated before, so the carbonates are included [18]. In sediments from Lake Biwa in Japan, the highest absolute and relative amounts of Pb leachable with hydroxylamine–25% acetic acid are enriched in the top few centimeters, in parallel with Mn as adsorption sites [126]. After hydroxylamine treatment, oxalate extracted only small amounts of Pb from the sediments of the Danube at Alten-wörth in Lower Austria. Oxidation with peroxide, but without strong acids, yields only low amounts.

From the sediments of the Danube and other rivers, about 20% of the total is finally found in boiling nitric acid: in Rotterdam Harbour the figure is even 38%. The fraction which only dissolves with HF, which means silicates, ranges from zero in Wisconsin Lakes [132] to almost 30% in the Greek Rivers Axios and Aliakmon [124] and in sediment from Lake Ontario [131], and even up to 70% in non-enriched samples from the Gaspé Penin-sula in Canada [134]. This means that fundamental geochemical conditions significantly influence the leaching pattern achieved. The largely unknown behaviour of clay minerals, and the different stability of Pb in sulfides [108], complicate the situation. In the sediments of the Danube at Altenwörth there was found to be much more NaOH-extractable- than H_2O_2-degradable Pb; Zn shows a similar behaviour. An explanation for this might be an exchange versus OH^-, or release of humic-bound Pb into acid buffers. The Pb can be highly mobile upon acidification, but not upon reduction.

7.3.8. Zinc

The leaching patterns found for Zn exhibit a variable shape, dependent upon the geochemical matrix and the loading status. Thus, in polluted European rivers like the

References on p. 171

Danube, Rhine, Elbe, Ijssel, Seine, Meuse, Scheldt, Gironde, and Garonne the ammonium acetate-exchangeable Zn is about 10% of total [18]. However, $MgCl_2$ seems to release significantly less Zn- for example, less than 1% from sediments of the Po [140], and from the polluted rivers Yamaska in Canada, and Deule in France [78,123].

In the carbonaceous samples of the Krka estuary [128], and for grain sizes below 20 μm in samples from the River Danube at Altenwörth, the weak-acid-leachable Zn, which means the carbonate-bound part, is the largest fraction. In the non-carbonaceous sediments of the Gurk River in Carinthia, it is of minor importance, at 10–15%. In the hydroxylamine–acetic acid-extractable fraction, which represents material bound to Fe/Mn hydroxide or to dolomite, there can also be the release of large amounts of Zn, as in the Yamaska and St. Francois Rivers in Quebec (40%), the River Po (28–43%), and Rotterdam Harbour sediment (56%), and up to 72% from highly polluted sediments of the small River Deule in Northern France, and the Elbe, Ems, Ijssel, Meuse and Scheldt. In sediment from Lake Biwa in Japan, an increase in the concentration of Zn from a steady value at depths greater than 5 cm, up to the top, is mainly due to an increase of the hydroxylamine-reducible fraction, which correlates well with the Mn contents of the sediment [126].

Oxalate does not release major Zn fractions: about 10% comes from samples from the Danube, and 16% from the samples form the Gurk, mentioned above. Peroxide-degradable Zn, obtained without the action of strong acids, is sparse: in cored bottom-sediments from Lake Geneva it was found to be below 1% [139].

If anoxic sediments, for example from Hamburg Harbour, are preserved under nitrogen during shaking [121], the major part is dissolved in the last step of the leaching sequence, by oxidation with nitric acid–hydrogen peroxide. Similarly, hot nitric acid leached the main fractions from sediments of the Gurk [49], and from Lake Biwa [126]. In samples from the Yellow River [129], and the Gironde and Garonne [18], a major part has been attributed to residual silicates. Zinc can be appreciably soluble in NaOH in reducing marine sediments [32], or in urban sludge [34], which implies an organically-bound fraction. The amounts which are extractable with dilute HCl are generally much greater than those which are soluble in citrate–dithionite at pH 7 [130], or in hydroxylamine–acetic acid [131].

In a test sediment with a low carbonate content, a municipal effluent of Zn loaded onto both the carbonate and the iron-bound fractions [108]. In the sediments of the River Danube at Altenwörth, about 20% of Zn, together with the humic substances, was leachable with NaOH, whereas in the Tessier sequence the H_2O_2-degradable fraction was much smaller. A possible explanation lies either in the dissolution of Zn-bearing organics in preceding steps of the Tessier sequence, or an additional liberation of Zn from clay minerals by the NaOH.

7.4. SPECIATION AND AVAILABILITY TO BIOTA

One of the goals of speciation studies of nutrients and trace-elements in the water phase, in sediments, and soils is to obtain a more precise connection between the element-contents in the substratum and in the living biomass. Many experiments show that an enhanced level in the water, the sediment, the soil, or the food leads to an increase in the living biomass also. In this context, only those availability studies are reviewed in which growth experiments are coupled with the analysis of chemical speciation. "Available" means that the living cell-membrane can be passed. It depends on the speciation: for example, the size and charge of the ion or molecule, the actual need in the case of active uptake, and also on competition with similar species. The effect on living biota is due not only to the availability, but also to the pathways of excretion or depletion, the genetic adaptation to enhanced levels of toxicants, and the competition with similar species. One cannot, therefore expect to design a general reagent to deliver an "available" fraction, which is valid for any plant and animal at any sampling site; but it is interesting to search for connections between chemical forms and uptake.

7.4.1. Aquatic systems

Both the total and the EDTA-extractable metal concentrations in shallow marine sediments, which had been polluted by a Pb/Zn smelter in Southern Australia, correlated well with metal concentrations in the leaves of *Posidonia* (seagrass). Whereas the leaves get their load by absorption from the sea water, the roots take it up from the sediment, but the two figures correlated strongly in the test area. The metal content varied seasonally [141]. In Lake Macquarie, which was polluted by a discharge from a Pb/Zn smelter, the Zn and Pb in seagrass (*Zostera capricorni*) correlated well with the concentrations in 0.05 M EDTA, hydroxylamine–HNO_3 at pH 2, and peroxide extracts (after hydroxylamine) of the sediments. The correlations with pore-water and surface-water, however, were poor. In contrast to Zn and Pb, the Cd and Cu in the plants were more closely related to their contents in the water [75]. For the Yellow Water Lily (*Nuphar variegatum*) growing on sediments downstream of a Cu/Zn mining/smelting facility, the distribution of metals within a given plant was far from homogenous. The sediment was characterized according to the sequential leaching method of Tessier et al. [89]. The Zn in the stems correlated strongly with that in all the sediment fractions: this was interpreted as an uptake from the water column, rather than from certain phases of the solid. The Cu in the rhizome and stems, and Zn in rhizome were not correlated with any of the sediment fractions. However, Cu in the rhizome correlated with Cu/Fe in the $MgCl_2$ extract, and with (Cu in $MgCl_2$)/(Fe in hydroxylamine–25% acetic acid). This indicates a competition for Cu uptake between the plants and the Fe-hydroxides [142]. For phytoplankton, Fe was available not only from soluble salts, but also from co-precipitates of Fe oxides with humic substances, from Fe-EDTA, and from fresh ferrihydrite. Strikingly, all these forms "available" to phytoplankton could be reduced by light, whereas "non-available" forms could not [63].

Ahlf measured the uptake of trace elements by green algae (*Ankistrodesmus bibraianus Korshikov*) from the polluted freshwater sediment of the river Elbe in Northern Germany, using a filter chamber device [143]. The algae were separated from the sediment by a 0.45 μm pore-diameter membrane. The sediments were characterized by sequential leaching according to Tessier et al. [89]. No statistically significant correlations between the results of sequential leaching and the uptake by green algae were found.

References on p. 171

Bio-accumulation decreased with increasing pH for Cu, Pb, Ni, and Zn. The sorption of As was at a minimum at neutral pH, and Cd at its maximum. The toxicity of the sediment significantly declined with increasing pH [143].

In estuaries, many metals in deposit-feeders are dominated by the metal uptake from sediments. For the deposit-feeding bivalve, *Scrobicularia plana*, living on estuarine surface-sediments in south-west England and northern France, the Pb content correlated strongly with the Pb/Fe ratio in the 1 *M* HCl extracts of the respective sediments [144]. For the bivalve *Macoma balthica*, kept in a filter chamber device, the solute uptake contributed most of the 14-day total body burdens of Zn and Cd, whereas the Co uptake largely resulted from the ingestion of isotope-laden bacteria [145].

7.4.2. Terrestrial systems

In agriculture, the availability of K from soils is of some concern, because it is an important nutrient. The mechanism of uptake of K from different soils by the roots of ryegrass (*Lolium perenne*) could be successfully modelled using leaching reagents. In the direct vicinity of the root, at first the K depletes which is exchangeable with ammonium acetate and can be regarded as rapidly available. Within longer periods (1 day), the K fraction which is leachable with 1 *M* HCl is supplied to the plant, mobilized from the interlayers of the clay minerals. The clay minerals are not attacked directly, but gradually release their K into the pore-water, from the trioctaedric sites if the dissolved concentration of K falls below 2 mg/l, and from the dioctaedric sites if it falls below 0.1 mg/l [83].

The uptake of Cu by *Eichornia crassipes*, grown by a greenhouse solution-culture technique, was due not only to the free Cu ions, but some other chelates (with glycine, tartrate, or citrate) were partially taken up also. Whether the Cu really moves into the cell as a complex, or dissociation at the surface takes place, could not be distinguished [146].

Treatment of soils with sewage sludge raises the level of many heavy metals, but simultaneously organic C, N, P, Ca, Mg, etc., are introduced into the system, which means an increase of complexed species. In pot experiments with maize, rape, and peas, the soil pH had a great influence on the availability of the added heavy metals. Thus, Zn was accumulated largely independent from soil pH, whereas changes in the Cd contents of the plants were only slight for pH changes in both directions, and increases in the Mo uptake by rape and by maize were found from acid soils [147,148].

The amounts of heavy metals in their exchangeable and carbonate forms controlled their uptake from cabbage, in pot experiments, much more than their total content. The rate of uptake of the total exchangeable and carbonate forms (according to the Tessier sequence) was the same for Cd, Zn and Pb. The correlation of Pb in cabbage with the leached fractions was best: for Cd it was worst [149].

The Zn uptake by wheat plants from loess-derived alkaline calcareous soils with a previous Zn deficiency could be correlated well with the fractions of Zn in the soil which were exchangeable with ammonium acetate at pH 7 and desorbed using DTPA. The lime content, organic matter, and amounts of EDTA-desorbable and HCl-leachable Zn were not important [112].

7.5. COLUMN METHODS

The investigation of the genesis of groundwater and the impact of contaminants is important for the protection of aquifers from environmental hazards, and for the saving of clean water reserves for future generations. Transport phenomena derive from the input of dissolved species into the liquid, and the composition and speciation of solids, especially if the solid constituents are accessible for mobilization. When sites have to be selected for the disposal of hazardous waste it is essential to have a profound knowledge of the hydraulic conductivity of the underlying material, as well as possible changes due to the action and migration of micro-organisms, salts, acids, alkalies, and organic compounds. Research in this field is just at its beginning. The experiments last for a long time, and many data have to be obtained from one column. Detailed investigations are rather expensive, and only a few results are available. Column methods should help one to find conditions for the mobilization or retardation of contaminants in soils and aquifers, and to find simpler and cheaper parameters in the future. Last, but not least, column methods should help the design and testing of technical measures for groundwater protection at hazardous waste-disposal sites, or to predict infiltration of contaminants from heavily polluted surface waters.

7.5.1. Experiments with uniform solids

For uniform solids, a model of multi-component chromatography has been successfully used to simulate the leaching of major cations during steady, continuous infiltration of electrolyte solutions. Cations with higher affinities for exchange sites move more slowly through the column than species with lower affinities. The breakthrough curve in the effluent follows the order of increasing preference. The average velocity for each ion will vary with time and with all the other concentrations in the solution phase. Ion exchange is assumed to occur instantaneously for all of the partitioned exchanged sites [150]. The chromatographic model is, however, only valid if local chemical equilibrium is achieved, and no precipitation/dissolution reactions occur on the solid. Precipitation leads to irreversible adsorption, and dissolution leads to an increase of dissolved species competing with the original input at sites lower in the column. If transport is possible only after complexation, the metals compete for transport sites in the mobile phase, too [151]. In carbonate rock aquifers the velocity is usually high, and local equilibrium is hardly obtained [152].

The hydraulic conductivity is almost totally determined by the flow-rates through the largest flow channels. The fabric of the solid is not necessarily constant, but is influenced by the water content (from the swelling of colloids and clay minerals, see below), the method and effort of compaction, the hydraulic gradient, physical and chemical properties of the permeant, and the extent of biological processes which may lead to pore-clogging and to changes in speciation [153]. Clay-minerals, hydrous oxides, organics, sulfides, and carbonates are possible adsorption sites on the solid. The mobility of nearly all inorganic ions, except F and Cl, thus depends on the actual redox potential as well as the pH. Carbonate strongly buffers this [152].

Increasing salt concentrations at a constant pH and redox potential lead to flocculation of dissolved humics in the pore-water, and to reduced swelling of the clay-minerals. From peat, less organic carbon is eluted with salt solutions than with distilled water, but no correlation with the Cd and Hg contents of the respective leachates was found [154]. Ion exchange with Ca and Na on the peat can lead to strong acidification if no carbonate is present [154]. Reduced swelling of the clay minerals leads to increased hydraulic conductivity for solids with high clay contents [153]. Input of dissolved NaCl (brine contamination) has no discernible effect on grain-size distribution and density, but does lead to shrinkage of the double-layers of the clay particles. The Na and Cl can be regarded as non-retarded species [153]. Weak organic acids lead to the dissolution of carbonates and, partially, Fe oxides. Additionally, alkali-metals and alkaline earths are liberated by ion exchange with H^+. When the effluent fractions are graphed against time, each element's concentration increases to a maximum, and then decreases, owing to the exhaustion of leachable amounts from the solid phase. Complexing acids, like oxalic or tartaric acid, mobilize significantly more Fe, Al, and Cr than does the weakly complexing acetic acid [41]. The hydraulic conductivity, however, is not necessarily increased, due to pore-clogging by migration of small particles of the insoluble residue of the dissolution reaction.

Among the strong acids, 5% HCl increased the hydraulic conductivity of a soil-bentonite backfill, but sulfuric acid did not, owing to the precipitation of gypsum in the pores [153]. Alkali input primarily leads to a decrease of pore-water concentrations because of the precipitation of hydroxides of, e.g., Ca and Mg, with possible clogging of the pores [41]. No dissolution of clay minerals by organic bases has been observed; only aniline increased the hydraulic conductivity [153].

Water-soluble organic compounds, such as simple alcohols and ketones, have no effect on the hydraulic conductivity at contents below 80%. Replacement of water molecules in clay minerals, and flocculation of humics, increase hydraulic conductivity [153]. Complexants dissolve cations, and especially Fe and Al from the solid, and they mask complexed metals in the liquid with respect to adsorption. However, the retardation of heavy metals from the mobile phase is not changed much by an excess of simultaneously applied complexants, like oxalate, because these are adsorbed by the solid itself, thus increasing the retention capacity [41].

7.5.2. Experiments with cored soil profiles

For studies of the effect of the input of liquids and dissolved species on the vertical migration in a soil profile, soils from the test area are cored as far as the C horizon, and put into a column with rigid walls, called 'lysimeters'. From the analysis of dissolved species in various layers and the eluate, the conditions of the fluxes are deduced. It is possible to keep natural vegetation on top.

7.5.2.1. General

The first step of groundwater formation is the infiltration of precipitation into the soils [155]. A minor part derives from the infiltration of riverine and limnic waters through the sediments, which can exert filtration behaviour, but also release contaminants [123]. Serious problems can emerge if polluted riverwater infiltrates into aquifers used for drinking-water supply. Fortunately, tight sediments act as 'geochemical filters'. In a retention zone at the top of the fine sediments of the river, particles are physically trapped and pollutants are sorbed by Fe- and Mn hydroxides. In the lower biodegradation zone, however, the generation of acid as a by-product of microbial oxidation with oxygen, and from decomposition of organic ligands, leads to re-mobilization of metals which can finally be precipitated as sulfides in a reduction zone [123,156].

The aim of lysimeter studies with undisturbed soil profiles is to find the parameters mainly responsible for heavy metal mobility between the different soil horizons, and the influence of environmental impacts such as acid rain [157]. Atmospheric precipitates mainly contain Na, Ca, Mg, NH_4, Cl, NO_3 and SO_4. Normal rainwater is saturated with CO_2, and has pH about 5.6 - 6.0. CO_2 is largely neutralized by the dissolution of carbamate particles deriving from the abrasion of the earthcrust. If the pH of the rainwater is lower (more acidic), chiefly because of excess sulfate, it is called 'acid rain', which is quite common nowadays in central Europe.

Groundwater serves not only as a source of water, but also as a host and transporting medium for contaminants. Advection, hydrodynamic dispersion, and dilution, affect the spreading of contaminants in groundwater. The mobility of nearly all inorganic ions depends on the redox potential, with the exception of Cl and F [152], as well as on physical processes, geochemical reactions like solution and precipitation, acid–base reactions, complexation, adsorption and desorption reactions, and biochemical processes [151]. The development of the chemical properties of the pore-water is completed below the zone of intense weathering [151].

7.5.2.2. Organics

Dissolved organics in the top-soil are high-molecular-weight polymers of humic type. With increasing depth in the soil profile, they are degraded to acids of lower molecular weight. In podsols, a stable group of molecular weight 250–300 remains in solution; no organics of lower molecular weight are formed. In contrast, in sandy soils the humics are degraded to compounds of molecular weight below 200 and are sorbed on layered silicates and oxides [155].

Dissolved organic carbon in the pore-water decreases with increasing depth and increasing duration of acid rain [157]. In the B horizon, dissolved high-molecular-weight organics are quite sparse, and in the C horizon they are not found any more [157]. The amino-acid content in the pore-waters of podsol lysimeters passes a maximum in the A horizon: the level is low, but still detectable, in the C horizon as well [158].

References on p. 171

The Pb, Fe, and Al in pore-waters have strong affinities to high- and low-molecular-weight organic substances. The Ca, Mg, Mn, and Zn are present mainly as ions in solution, and are complexed by low-molecular-weight humics at low levels. Therefore, in the A horizon of the water-unsaturated zone of a podsol, Fe, Cr, Al, Pb, V, and partially Cu, move in organically bound forms, whereas the complexation of Cd, Mn, Ni has been found to be below 1% [157]. This leads to the dissolution or weathering of the inorganic matrix. In the B horizon, where degradation of the humics begins, the maximum dissolved Fe can be found, about half of it bound to organics. The Cr, V, Pb, and Cu, however, are precipitated in the B horizon, and only small amounts are transported further down. In the B and C horizons, a strong increase of inorganic dissolved Al can occur, but in deeper layers Fe and Al are sorbed or precipitated again [155,159]. Mercury is especially bound to high-molecular-weight humic compounds, which are the principal carriers in the transport of Hg from terrestrial to aquatic ecosystems [154]. Significant amounts of Cd exist as hydrophilic organic complexes [160].

The concentrations of heavy metals in the pore-water decrease strongly with increasing depth [158]. In the water-unsaturated region of a podsol, no significant influence could be observed of the pH of the rain upon the tendency of the heavy metals to bind to the organics [158].

7.5.2.3. Saturation with respect to inorganics

In many cases, concentrations of dissolved ions in the pore-water do not approach the saturation levels, as calculated from the solubilities of the pure components. This implies the existence of unknown colloids or complexes, or non-equilibrium conditions. In the top layers of limnic sediments, over-saturation often occurs with CO_2, derived from aerobic respiration. The pore-water in a quaternary outwash sediment was found to be under-saturated with respect to amorphous $Fe(OH)_3$ down to 12 m, but over-saturated with respect to goethite [161]. The pore-water was also under-saturated with respect to calcite and dolomite down to a depth of 16 m. Similarly, over-saturation with respect to kaolinite was found down to 25 m, but slight under-saturation with respect to albite and anorthite [161]. In carbonaceous B horizons, there is often over-saturation with respect to $CaCO_3$ [155]. Similarly, the retardation of Cd with respect to the precipitation of crystalline otavite has been too optimistic. This means that over-saturation is reached with respect to the crystalline mineral; possibly colloids are formed [162].

7.5.2.4. Variations of pH

In sediments and soils, various compounds can buffer the pH. Carbonates and silicates act mainly in alkaline and neutral media, exchangeables in weak acid, and Al and Fe mainly at pH 3–4, where their hydroxides begin to dissolve. Shifts in pH occur from the direct input of bases or acids such as acid rain, from the dumping of waste or the application of fertilizers of different pH to the soil, and as a result of microbial processes. Microbial oxidation of ammonia with oxygen to nitrate, and of (undissociated) hydrogen sulfide to (dissociated) sulfate, directly produce acids. Oxidation of Fe(II) to Fe(III), as well as the degradation of Fe and Al organic complexes at neutral pH, lead to precipitation of the hydroxides, which consumes hydroxyl ions and shifts the pH to more acid values. The addition of previously aerated waste-deposit leachate to a luvisol significantly lowered the pH, due to the intense microbial oxidation of the input ammonium to nitrate, and of sulfide to sulfate in the soil [163].

As long as there is aerobic respiration in the soil or the sediment, CO_2 is formed, which has acidic properties. In forest soils, the pH of the pore-water increases with depth towards the C horizon: there is a simultaneous increase of HCO_3^- and Ca [161]. The leaching of Mg, Ca, Mn, Cd, Zn, and Ni in forest soils mainly depends on the acidity of the soil, whereas for Fe, Cu, Pb, Cr, and V, it is more associated with the leaching of organic matter [159]. Non-calcareous quartz sand does not contain many exchange sites, and adsorbs Cd poorly. At the more alkaline pH values, achieved by the addition of carbonate, adsorption and precipitation reactions take place. The transport velocities of the main breakthrough curves and dispersivities of Cd and Cl are in the same range in acidic quartz sand, but the addition of $CaCO_3$ can retard, and even stop, the movement of Cd through the column, by irreversible adsorption [162]. Addition of a neutral salt solution (Ca or Na nitrate or sulfate) to peat and sand in Finland caused strong acidification, due to ion exchange of salts of the humics, and thus to the liberation of H^+ [154]. Reduction of the temperature resulted in acidification of the leachates, as well.

7.5.2.5. Microbial activity

Microbial activity is governed by the availability of nutrients and substrates, of oxygen, and the temperature, water content, and pH. Acid soils tend to have lower microbial activity than neutral and alkaline ones [163]. In acidic sandy loam soil (pH 4, 2.3% organic carbon), the retention of applied soluble Cd was the same for sterile and non-sterile conditions. Addition of nutrients, however, led to a significant rise under non-sterile conditions, after an induction period of 9 days, owing to the activation of micro-organisms which excrete acid and organic complexants [160].

REFERENCES

1 R.C.H. Steneker, H.A. Van der Sloot and H.A. Das, *Sci. Total Environm.*, 68 (1988) 11-23.
2 J.F. Wagner, *Oberrhein. Geol. Abh.*, 35 (1989) 187-195.
3 L. Campanella, E. Cardanelli, T. Ferri, B.M. Petronio and A. Pupella, *Proc. Int. Conf. Heavy Met. Environm. Athens, 1985*, pp. 336-338.
4 A. Tessier, R. Carignan, B. Dubbreuil, F. Rapin, *Geochim. Cosmochim. Acta*, 53 (1989) 1511-1522.
5 P.H. Santschi, *Limnol. Oceanogr.*, 22 (1988) 848-866.
6 J.W. Morse, *Mar. Chem.*, 20 (1986) 91-112.
7 J.I. Drever, *The Chemistry of Weathering*, Reidel, Dordrecht/Boston/Lancaster, 1985.
8 P.G.C. Campbell, A.G. Lewis, P.M. Chapman, A.A. Crowder, W.K. Fletcher, B. Imber, S.N. Luoma, P.M. Stokes and M. Winfrey, *Nat. Res. Council Canada*, (1988) Publ. 27694.

172

9 E.A. Thomson, S.N. Luoma, D.J. Cain and C. Johansson, *Water, Air Soil Poll.*, 14 (1980) 215.
10 A.C. Edwards, J. Ceasey and M.S. Cresser, *Water Res.*, 20 (1986) 831-834.
11 M. Sager, R. Pucsko and R. Belocky, *Arch. Hydrobiol.*, Suppl. 84 (1989) 37-72.
12 P.N. Soltanpour, A. Khan and W.L. Lindsay, *Commun. Soil Sci. Plant Anal.*, 7 (1976) 797-821.
13 U. Förstner, W. Ahlf, W. Calmano and M. Kersten, *Proc. Int. Conf. Heavy Met. Environm., Athens, 1985*, pp. 34-36.
14 W. Salomons, *Proc. Int. Conf. Heavy Met. Environm., Athens, 1985*, p. 23.
15 M. Birke and W. Schulze, *Z. Angew. Geol.*, 34 (1988) 135-139.
16 G. Müller, G. Irion and U. Förstner, *Naturwiss.*, 59 (1972) 158-164.
17 I.A. Mirsal and H. Zankl, *Geol. Rundschau*, 74 (1985) 367-377.
18 W. Calmano and U. Förstner, *Sci. Total Environ.*, 28 (1983) 77-90.
19 C.M. Reaves, *J. Sed. Petrol.*, 56 (1986) 486-494.
20 R. Giovanoli, J.L. Schnoor, L. Sigg, W. Stumm and J. Zobrist, *Clays Clay Minerals*, 36 (1988) 521-529.
21 S.K. Ghabru, A.R. Mermut and R.J.S. Arnaud, *Geoderma*, 40 (1987) 65-82.
22 R. Soundararajan and J.J. Gibbons, *Pers. commun.*, 1988.
23 K. Krejci-Graf, *Proc. Yorkshire Geol. Soc.*, 34 (1964) 469-521.
24 J. Lakind and A.T. Stone, *Geochim. Cosmochim. Acta*, 53 (1989) 961-971.
25 E. Tipping, D.W. Thompson and W. Davison, *Chem. Geol.*, 44 (1984) 359-383.
26 E. Tipping, J.R. Griffith and J. Hilton, *Croat. Chem. Acta*, 56 (1983) 613-621.
27 W. Stumm, G. Furrer, E. Wieland and B. Zinder, in J.I. Drever (Editor), *The Chemistry of Weathering*, Reidel Dordrecht, 1985.
28 M.A. Naga, I. Hossny, S.I. Abdel-Aal and R.R. Shahin, *Egypt. J. Soil Sci.*, 17 (1977) 237-250.
29 A.J. Drapeau, R.A. Laurence, P.S. Harbec, G. Saint-Germain and N.B. Lambert, *Sci. Techn. Eau*, 16 (1983) 359-363.
30 J. Tauchnitz, R. Schnabel, G. Kiesel, D. Rehorek, G. Knobloch and H. Hennig, *Z. ges. Hyg.*, 28 (1982) 718.
31 H. Goossens, R.R. Düren, J.W. De Leeuw and P.A. Schenck, *Org. Geochem.*, 14 (1989) 27-41.
32 A. Nissenbaum and D.J. Swaine, *Geochim. Cosmochim. Acta*, 40 (1976) 809-816.
33 W.V. Gerasimowicz and D.M. Byler, *Soil Sci.*, 139 (1985) 270-278.
34 B.M. Petronio, L. Campanella, E. Cardarelli, T. Ferri and A. Pupella, *Ann. Chim.*, 77 (1987) 721-733.
35 B.M. Petronio, L. Campanella, T. Ferri and A. Pupella, *Ann. Chim.*, 79 (1989).
36 R.G. Gerritse, R. Vriesema, J.W. Dalenberg and H.P. De Roos, *J. Environ. Qual.*, 11 (1982) 359-364.
37 J.N. Moore, W.H. Ficklin and C. Johns, *Environ. Sci. Technol.*, 22 (1988) 432-437.
38 N. Belzile, *Geochim. Cosmochim. Acta*, 52 (1988) 2293-2302.
39 N. Belzile, R. De Vitre and A. Tessier, *Nature*, 340 (1989) 376-377.
40 O. Horak and R. Rebler, *ÖFSZ-Bericht*, No. 4175 (1982).
41 M. Sager, P. Hacker, F. Kappel, E. Schwab and J. Ullrich, manuscript in preparation.
42 Han Bin Xue, W. Stumm and L. Sigg., *Water Res.*, 22 (1988) 917-926.
43 M. Sager, *Mikrochimica Acta*, 1991, in press.
44 A.C. Harzdorf, Intern. *J. Environ. Anal. Chem.*, 29 (1987) 249-261.
45 C.P. Huang, *Proc. Int. Conf. Heavy Metals Environ., Athens, 1985*, pp. 286-287.
46 E. Piperaki, H. Berndt and E. Jackwerth, *Anal. Chim. Acta*, 100 (1978) 589-596.
47 K.W. Perrott, *Aust. J. Soil Res.*, 16 (1978) 327-339.
48 E.C. Scholtz, J.R. Feldkamp, J.L. White and S.L. Hem., *J. Pharm. Sci.*, 74 (1985) 478-481.
49 M. Sager and W. Vogel, *Acta Hydrochim. Hydrobiol.*, (1992) in press.
50 T.U. Aualiitia and W.F. Pickering, *Water Res.*, 1986, 171-186.
51 J.A. Davis, *Geochim. Cosmochim. Acta*, 48 (1984) 679-691.
52 A.C.M. Bourg and P.W. Schindler, *Proc. Int. Conf. Heavy Met. Environm., Athens, 1985*, pp. 97-99.

53 M.B. McBride, *Clays Clay Min.,* 33 (1985) 397-402.
54 M.B. McBride, A.R. Fraser and W.J.I. McHardy, *Clays Clay Min.,* 32 (1984) 12-18.
55 E. Tipping, D.W. Thompson, M. Ohnstad and N.B. Hetherington, *Environ. Technol. Lett.,* 7 (1986) 109-114.
56 K.C. Nathsarma and P.V.P. Bhaskara Sarma, *Erzmetall,* 42 (1989) 259-262.
57 E. Tipping and M.J. Heaton, *Geochim. Cosmochim. Acta,* 47 (1983) 1393-1397.
58 D. Postma, *Geochim. Cosmochim. Acta,* 49 (1985) 1023-1033.
59 W. Dyck and K.H. Lieser, *Vom Wasser,* 56 (1981) 183-190.
60 R.J. Atkinson, A.M. Posner and J.P. Quirk, *J. Phys. Chem.,* 71 (1967) 650-558.
61 M. Hiraide, Y. Arima and A. Mizuike, *Mikrochim. Acta,* 1988, III, 231-238.
62 L.W. Llon, R.S. Altmann and J.O. Leckie, *Environm. Sci. Technol.,* 16 (1982) 660-666.
63 E. Tipping, *Mar. Chem.,* 18 (1986) 161-160.
64 K.A. Czurda, A. Rashidchi and J.F. Wagner, *Appl. Clay Sci.,* 2 (1987) 129-143.
65 S.J. Traina and H.E. Doner, *Clay Clay Min.,* 33 (1985) 118-122.
66 K.H. Lieser, S. Peschke and B. Gleitsmann, *Fresenius' Z. Anal. Chem.,* 321 (1985) 119-123.
67 M. Okumura, Y. Kitano and M. Idogaki, *Geochem. J.,* 17 (1983) 105-110.
68 A. Mucci, *Geochim. Cosmochim. Acta,* 52 (1988) 1859-1868.
69 J.A. Davis, C.C. Fuller and A.D. Cook, *Geochim. Cosmochim. Acta,* 1990.
70 G.E. Jean and G.M. Bancroft, *Geochim. Cosmochim. Acta,* 50 (1986) 1455-1463.
71 H.W. Schmitt and H. Sticher, *Z. Pflanzenern. Bdk.,* 149 (1986) 157-171.
72 M.A. Elrashidi and G.A. O'Connor, *Soil Sci. Soc. Am. J.,* 46 (1982) 27-31.
73 G.A. Manful, M. Verloo and F. De Spiegeleer, *Pedologie,* 39 (1989) 55-68.
74 V. Ender, J. Bosholm and S. Erb, *Acta Hydrochim. Hydrobiol.,* 16 (1988) 197-203.
75 G.E. Batley, *Aust. J. Mar. Freshw. Res.,* 38 (1987) 591-606.
76 G. Petruzzelli, I. Szymura, L Lubrano and B. Pezzarossa, *Environ. Technol. Lett.,* 10 (1989) 521-526.
77 L. Campanella, E. Cardarelli, T. Ferri, B.M. Petronio and A. Pupella, *Sci. Total Environ.,* 76 (1988) 41.
78 A. Tessier, P.G.C. Campbell and M. Bisson, *Can. J. Earth Sci.,* 17 (1980) 90-105.
79 R.J. Gibbs, *Geol. Soc. Am. Bull.,* 88 (1977) 829-843.
80 E.K. Kruglova and T.T. Turaev, *Agrokhim.,* 6 (1981) 106-110.
81 H. Neite, *Z. Pflanzenern. Bdk.,* 152 (1989) 441-445.
82 R.M. Harrison, D.P.H. Laxen and S.J. Wilson, *Environ. Sci. Technol.,* 15 (1981) 1378-1383.
83 T. Kong and D. Steffens, *Z. Pflanzenern. Bdk.,* 152 (1989) 347-343.
84 M. Berg, *US Envir. Prot. Agency,* Method 1310, 1987.
85 N.K. Dave, T.P. Lim and N.R. Cloutier, *Proc. Int. Conf. Heavy Met. Environm., Athens, 1985, pp. 24-26.*
86 E. Königsberger and H. Gamsjäger, *Ber. Bunsenges Phys. Chem.,* 91 (1987) 785-790.
87 F. Rapin and U. Förstner, *Proc. Int. Conf. Heavy Metals Environm., Heidelberg, 1983,* pp. 1074-1077.
88 C. Kheboian and C.F. Bauer, *Anal. Chem.,* 59 (1987) 1417-1423.
89 A. Tessier, P.G.C. Campbell and M. Bisson, *Anal. Chem.,* 51 (1979) 844-851.
90 J.J. McAlister, G. Svehla and W.B. Whalley, *Microchem. J.,* 38 (1988) 211-231.
91 M. Sager, *Schweiz. Z. Hydrol.,* 48 (1986) 71-103.
92 K.G. Knauss and T.J. Wolery, *Geochim. Cosmochim. Acta,* 53 (1989) 1493-1501.
93 D.M. Hirst and G.D. Nicholls, *J. Sed. Petrol.,* 28 (1958) 468-481.
94 K. Itoh, *Bunseki Kagaku,* 31 (1982) 657-662.
95 S. Becker, M. Bullmann, H.J. Dietze and U. Iske, *Fresenius' Z. Anal. Chem.,* 324 (1986) 37-42.
96 M.I. Popova, *Zh. Anal. Khim.,* 41 (1986) 1590-1595.
97 J. Slavek, J. Wold and W.F. Pickering, *Talanta,* 29 (1982) 743-749.
98 S.N. Luoma, *Mar. Chem.,* 20 (1986) 45-59.
99 D.L. McCallister, *Soil Sci.,* 144 (1987) 274-280.

174

100 G.E.M. Hall, J.C. Pelchat and J. Loop, *Chem. Geol.*, 67 (1988) 35-45.
101 S.N. Luoma and G.W. Bryan, *Sci. Total Environ.*, 17 (1981) 165-196.
102 B.S. Cooper and R.C. Harris, *Mar. Pollut. Bull.*, 5 (1974) 24-26.
103 E.G. Jorgensen, *Phys. Plant.*, 8 (1955) 846-851.
104 M. Hiraide and A. Mizuike, *Fresenius' Z. Anal. Chem.*, 335 (1989) 924-926.
105 M.I. Popova, *Zav. Lab.*, 54 (1988) 19-23.
106 W.P. Miller, D.C. Martens and L.W. Zelazny, *Soil Sci. Soc. Am. J.*, 50 (1986) 598-601.
107 M. Sager, *Arch. Hydrobiol. Beih.*, 30 (1988) 71-81.
108 M. Sager, R. Belocky and R. Pucsko, *Acta Hydrochim. Hydrobiol.*, 18 (1990) 157-173.
109 R. Psenner, R. Pucsko and M. Sager, *Arch. Hydrobiol.*, Suppl. 70 (1984) 111-155.
110 E. Tipping, J.G. Jones and C. Woof, *Arch. Hydrobiol.*, 105 (1985) 161-175.
111 R.W. Howarth and S. Merkel, *Limnol. Oceanogr.*, 29 (1984) 598-608.
112 M.S. Rahmatullah and B.Z. Shaikh, *Z. Pflanzenern. Bdk.*, 151 (1988) 385-389.
113 N. Belzile and J. Lebel, *Chem. Geol.*, 54 (1986) 279-281.
114 W.H. Huang and W.D. Keller, *Am. Minerol.*, 56 (1971) 1082-1095.
115 W.P. Miller, L.W. Zelazny and D.C. Martens, *Geoderma*, 37 (1986) 1-13.
116 P.C. Bennett, M.E. Melcer, D.I. Siegel and J.P. Hassett, *Geochim. Cosmochim. Acta*, 52 (1986) 1521-1530.
117 A.A. Antipina, M.I. Timerbulatova and D.P. Shcherbov, *Zh. Anal. Khim.*, 37 (1982) 2182-2185.
118 W. Schlösser and G. Schwedt, *Fresenius' Z. Anal. Chem.*, 321 (1985) 136-140.
119 P.J.A. Fowlie and T.L. Bulman, *Anal. Chem.*, 58 (1986) 721-723.
120 W.A. Maher, *Chem. Geol.*, 47 (1984) 333-345.
121 M. Kersten, U. Förstner, W. Calmano and W. Ahlf, *Vom Wasser*, 65 (1985) 21-35.
122 U. Förstner, *Münchn. Beitr. Abwasser-, Fisch. Flussbiol.*, 34 (1982) 249-271.
123 A.C.M. Bourg, D. Darmendrail and J. Ricour, *Geoderma*, 44 (1989) 229-244.
124 V. Samanidou and K. Fytianos, *Sci. Total Environ.*, 67 (1987) 279-285.
125 M. Kerner, H. Kausch and M. Kersten, *Arch. Hydrobiol. Suppl.*, 75 (1986) 118-131.
126 S. Nakashima, *Jpn. J. Limnol.*, 43 (1982) 67-80.
127 K.R. Lum and D.G. Edgar, *Analyst*, 108 (1983) 918-924.
128 E. Prohic and G. Kniewald, *Mar. Chem.*, 22 (1987) 279-297.
129 Y.T. Hong and U. Förstner, *Proc. Int. Conf. Heavy Met. Environm., Heidelberg, 1983*, pp. 872-875.
130 B.A. Malo, *Environ. Sci. Technol.*, 11 (1977) 277-282.
131 H. Agemian adn A.S.Y. Chau, *Analyst*, 101 (1976) 761-767.
132 J.M. Henshaw and E.M. Heithmar, *Water Res.*, 1988.
133 A. Tessier, F. Rapin and R. Carignan, *Geochim. Cosmochim. Acta*, 49 (1985) 183-194.
134 A. Tessier, P.G.C. Campbell and M. Bisson, *J. Geochem. Explor.*, 16 (1982) 77-104.
135 M. Juračič, E. Prohič and V. Pravdič, *Proc. VII Workshop Mar. Poll. Medit., Lucerne, 1984*, pp. 151-157.
136 U. Förstner and W. Calmano, *Vom Wasser*, 50 (1982) 83-92.
137 N.N. Roy and N.P. Upadhyaya, *Toxicol. Environm. Chem.*, 10 (1985) 285-298.
138 G. Pfeiffer, U. Förstner and P. Stoffers, *Senckenberg. Mar.*, 14 (1982) 23-38.
139 Z.J. Wang, H. El Ghobary, F. Giovanoli and P.Y. Favarger, *Schweiz. Z. Hydrol.*, 48 (1986) 1-17.
140 A. Barbanti and G.P. Sighinolfi, *Environ. Technol. Lett.*, 9 (1988) 127-134.
141 K.G. Tiller, R.H. Merry, B.A. Zarcinas and T.J. Ward, *Est. Coast. Shelf Sci.*, 28 (1989) 473-493.
142 P.G.C. Campbell, A. Tessier, M. Bisson and R. Bougie, *Can. J. Fish. Aquat. Sci.*, 42 (1985) 23-32.
143 W. Ahlf, *Vom Wasser*, 65 (1985) 183-188.
144 S.N. Luoma and G.W. Bryan, *J. Mar. Biol. Ass. U.K.*, 58 (1978) 793-803.
145 R.W. Harvey and S.N. Luoma, *Hydrobiologia*, 121 (1985) 97-102.
146 Y.M. Nor, *Proc. Int. Conf. Heavy Met. Environm., Athens, 1985*, pp. 283-285.
147 O. Horak, J. Zvara, R. Rebler and P. Herger, *Die Bodenkultur*, 33 (1982) 298-303.
148 O. Horak, *ÖFSZ-Ber.*, 4141 (1981).

149 X. Xian, *Plant and Soil*, 113 (1989) 257-264.
150 R.S. Mansell, S.A. Bloom, H.M. Selim and R.D. Rhue, *Geoderma,* 38 (1986) 61-75.
151 A. Gruhn, *Ber. Geol. Pal. Inst. Univ. Kiel,* Nr. 14 (1986).
152 B.A. Memon and E. Prohič, *Environ. Geol. Water Sci.,* 13 (1989) 3-13.
153 R.D. Woods, *Geotechnical Spec. publ.* No. 13, Am. Soc. Civil Engineers, 1987.
154 M. Lodenius and S. Autio, *Arch. Envir. Contam. Toxicol.,* 18 (1989) 261-267.
155 H. Holthusen, *Meyniana,* 34 (1982) 29-94.
156 M. Sager, *Proc. 4th Int. Conf. Conserv. Manag. Lakes, Hangzhou, 1990.*
157 A. Gruhn, G. Matthess, A. Pekdeger and A. Scholtis, *Z. Dt. Geol. Ges.,* 136 (1985) 417-427.
158 A. Scholtis, *Bericht Geol. Pal. Inst. Univ. Kiel,* Nr 13 (1986).
159 B. Bergkvist, *Report SNV-PM-1686* (1983); from *Chem. Abs.,* 102 (1985) 148065v.
160 P. Chanmugathas and J.M. Bollag, *Arch. Environm. Contam. Toxicol.,* 17 (1988) 229-237.
161 W. Ohse, G. Matthess and A. Pekdeger, *Z. Dt. Geol. Ges.,* 134 (1983) 345-361.
162 M. Isenbeck, J. Schröter, T. Taylor, M. Fic, A. Pekdeger and G. Matthess, *Meyniana,* 39 (1986) 7.
163 S. Roth-Kleyer and B.M. Wilke, *Wasser und Boden,* 41 (1989) 599-603.

M. Stoeppler (Editor)/*Hazardous Metals in the Environment*
© 1992 Elsevier Science Publishers B.V. All rights reserved

Chapter 8

Cadmium

Markus Stoeppler

Institute of Applied Physical Chemistry, Research Center (KFA) Juelich, P.O. Box 1913, D-W-5170 Juelich (Germany)

CONTENTS

8.1. Introduction 177
8.2. Environmental and biological levels of cadmium 178
8.3. Sampling and sample pretreatment 178
8.4. Enrichment and separation 181
8.5. Analytical methods 182
      8.5.1. Overview 182
      8.5.2. Atomic absorption spectrometry (AAS) 184
              8.5.2.1. Flame AAS 184
              8.5.2.2. Graphite furnace AAS 186
                        8.5.2.2.1. Liquid samples 186
                        8.5.2.2.2. Solid samples 188
                        8.5.2.2.3. Direct GFAAS analysis of solids 193
      8.5.3. Plasma atomic emission spectrometry (AES) 195
      8.5.4. Inductively coupled plasma-mass spectrometry (ICP-MS) 197
      8.5.5. Electrochemical methods 200
              8.5.5.1. Aqueous samples 200
              8.5.5.2. Other liquid and solid samples 202
      8.5.6. X-Ray methods 204
      8.5.7. Nuclear methods 206
      8.5.8. Chromatography 207
      8.5.9. Mass spectrometry 208
      8.5.10. Spectrophotometry and related techniques 210
8.6. Speciation 210
8.7. Quality control and reference materials 213
8.8. Conclusion and prospects 218
References 219

8.1. INTRODUCTION

Cadmium (Cd) is the sixty-seventh most abundant element in the earth's crust. Its atomic number is 48; atomic weight 112.4; density 8.64 g/cm^3; melting point 320.9°C; boiling point 767°C. In its metallic form it is silver-white, lustrous and ductile. Cadmium is relatively soft, this can, however, be improved by alloying, e.g. with zinc. It belongs, along with zinc and mercury, to the 2nd subgroup of the Periodic Table; its oxidation state is +2 in all compounds. With a normal electrochemical potential of −0.40 relative to the hydrogen electrode, it is slightly more noble than zinc.

Cadmium metal is readily soluble in nitric acid, but only slowly in hydrochloric and sulfuric acids and insoluble in basic solution. With regard to speciation one has to differentiate between bivalent cadmium ions (e.g. $CdCl_2$), chlorocadmium complexes, cadmium bound to proteins, cadmium bound to colloidal substances, and acid cadmium complexes. Some cadmium compounds are colored (yellow, red, brown), others are colorless. Salts of cadmium with strong acids are readily soluble in water. Less soluble compounds are the sulfide, the carbonate, the fluoride and the hydroxide. The latter, however, is readily soluble under complex formation in ammonium hydroxide. More details on cadmium and its compounds are given elsewhere [1,2].

Cadmium is according to present knowledge not essential for plants, animals and man but there are also papers claiming the essentiality of cadmium for animals. Higher doses of cadmium from ingestion and inhalation can cause toxic effects for humans. Acute effects can be seen in the respiratory and digestive tracts. Cadmium accumulates predominantly in the kidneys with a biological half-life of between ten and twenty years. Heavy long-term cadmium exposure can cause irreversible adverse renal effects. Health risks and adverse environmental influences due to cadmium and its slowly rising levels in the biosphere have been extensively discussed but are also controversial. There are several comprehensive reviews about cadmium chemistry, bio- and geochemistry (e.g. refs. 1–5) as well as toxicology and biological monitoring (e.g. refs. 6–14).

8.2. ENVIRONMENTAL AND BIOLOGICAL LEVELS OF CADMIUM

Typical prevalent contents for cadmium in air, precipitation, natural waters, sediments, limnic, marine and terrestrial materials, the food basket and human samples ranging from the ng/l or ng/kg to the mg/kg level based on numerous original papers, reports and compilations [5,11,14–52] are listed in Table 8.1. This demonstrates that in a number of materials very low levels occur requiring utmost skill on the part of the analytical staff.

8.3. SAMPLING AND SAMPLE PRETREATMENT

Collection of materials with cadmium contents above approx. 100 μg/kg normally poses no particular problems as far as contamination is concerned. However, if reliable results have to be achieved a proper sampling strategy also requires environmental, ecological, biological, toxicological and statistical considerations in relation to the respective analytical problem [5,11,39] as already discussed in Chapter 1 of this book.

For cadmium levels below 100 μg/kg, occurring in many environmental and biological materials, contamination precautions and proper sampling techniques have to be applied that are mentioned in some detail in Chapter 2. For work in the laboratory often the use of dust-free benches (clean benches) and, if samples with extremely low metal contents −e.g. rain and seawater, arctic snow− have to be treated, especially designed clean

TABLE 8.1.

TYPICAL LEVELS OF CADMIUM IN ENVIRONMENTAL AND BIOLOGICAL MATERIALS

Material	Cadmium content	Unit	Ref.
Air and deposition			
Airborne (worldwide)			15
Industrial areas	20 – 300	ng/m^3	
Urban areas	0.1 – 50	ng/m^3	
Remote/rural areas	0.003 – 4	ng/m^3	
Deposition rates (worldwide)			
Industrial areas	100 – 3000	ng/cm^2 month	15,16
Urban areas	3 – 70	ng/cm^2 month	15,16
Rural areas	0.05 – 12	ng/cm^2 month	15,16
Wet precipitation (Germany)			
Industrial areas	1.6 – 20	$\mu g/m^2/d$	17
	0.7 – 10	$\mu g/l$	17
Urban areas	0.6 – 2.5	$\mu g/m^2/d$	17
	0.25 – 0.9	$\mu g/l$	17
Rural areas	0.35 – 0.90	$\mu g/m^2/d$	17
	≤0.05 – 0.30	$\mu g/l$	6,17
Arctic precipitation	<0.0002 – 0.005	$\mu g/l$	18
Natural waters and sediments			
River water			
dissolved and particulate bound	<0.05 – 0.2	$\mu g/l$	19–20
Estuarine water	<0.04 – 2	$\mu g/l$	23–25
Seawater, open sea surface	≤0.001 – 0.05	$\mu g/l$	26–30
Depths below 1000 m	≤0.15	$\mu g/l$	
River sediments polluted	30 – >800 (DW)	mg/kg	20,31,32
Non-polluted	0.04 – 0.8 (DW)	mg/kg	
Marine sediments	<0.1 – 8 (DW)	mg/kg	33–36
Limnic and marine materials			
Oysters	<1 – 12 (DW)	mg/kg	37,38
Mussels (Mytilus), marine	<1 – 20 (DM)	mg/kg	37–41
Algae, marine	<0.2 – 2 (DW)	mg/kg	39–41
Fish muscle (fillet)	<0.01 (DW)	mg/kg	39
Fish organs	2 – 20 (DW)	mg/kg	39
Herring gull eggs	<0.002 (DW)	mg/kg	41
Terrestrial materials			
Soils, polluted	≤0.2 – ≥50 (DW)	mg/kg	5,41–43
Soils, non-polluted	≤0.01 – 0.5 (DW)	mg/kg	5,41,44

(Continued on p. 180)

TABLE 8.1. (continued)

Material	Cadmium content		Unit	Ref.
Sewage sludge (9 plants) (threshold in Germany: 20 mg/kg)	2 – 50	(DW)	mg/kg	45
Fossil fuels	< 0.03 – 400		µg/kg	5
Materials of the German Environmental Specimen Bank (Differently polluted areas)				
Mussels, freshwater	≤3	(DW)	mg/kg	46
Earthworm	5 – 20	(DW)	mg/kg	41
Grass	≤1	(DW)	mg/kg	46
Spruce shoots	< 0.01 – 0.7	(DW)	mg/kg	41
Poplar leaves	< 0.5 – 2	(DW)	mg/kg	41
Beech leaves	< 0.01 – 0.2	(DW)	mg/kg	41
Bird's feathers (various species)	≤0.1 – 15	(DW)	mg/kg	47,48
Various invertebrates in a non-polluted forest	< 0.01 – 8	(DW)	mg/kg	49

Food basket etc.

Material	Cadmium content		Unit	Ref.
Vegetables from polluted soils (parsley, celery, carrots, red beets)	< 0.2 – 20	(DW)	mg/kg	42
some mushrooms, dark chocolate, blue poppy seeds	> 200	(FW)	µg/kg	5
Kidneys and livers, most mushrooms, some rice brands	≤0.2	(FW)	mg/kg	5,50
Wheat flour, wheat bread, bran, potatoes root and foliage vegetables, rice	≤0.04	(FW)	mg/kg	5,50
Meat from various animals, wine, beer, fruit juices	≤0.005	(FW)	mg/kg	5
Tap water, milk and dairy products	≤0.001	(FW)	mg/kg	5
Cigarettes	< 0.5 – 3	(DW)	mg/kg	5,51

Human materials   5,11,14,52

Material	Cadmium content		Unit	Ref.
Kidneys, exposed persons	up to 500	FW	mg/kg	
Kidneys, smokers	≤6	FW	mg/kg	
Kidneys, nonsmokers	≤3	FW	mg/kg	
Liver	0.1 – 3	FW	mg/kg	
Lung tissue, unexposed	≤0.1	FW	mg/kg	
Bones	< 0.01 – 0.3	FW	mg/kg	
Whole blood, exposed persons	up to 0.2	FW	mg/kg	
Whole blood, smokers	≤0.0002 – 0.006	FW	mg/kg	

TABLE 8.1. (continued)

Material	Cadmium content	Unit	Ref.
Whole blood, nonsmokers	<0.0002 – 0.002 FW	mg/kg	
Urine, exposed persons	up to 0.2 FW	mg/kg	
Urine, non-exposed persons	<0.0001 – 0.003 FW	mg/kg	
Milk	<0.0001 – 0.003 FW	mg/kg	

laboratories are mandatory for accurate results [53]. Needless to say, blank control of all vessels and tools used in cadmium analysis has to be performed permanently and with utmost care if lower levels are to be determined [54,55].

Decomposition of all types of materials can be performed by common techniques using nitric acid in open and closed (pressurized) systems for optical spectroscopy ICP–MS and some other methods. Decomposition vessels should consist of quartz, for lowest levels of extremely pure material, and of PTFE or similar plastics. For details see Chapter 5.

If voltammetry is used the interfering matrix has to be completely decomposed. This was performed earlier by wet decomposition with a mixture of nitric, sulfuric and perchloric acids but is lengthy, and cumbersome, and not complete in all cases. Much better performance is achieved if a recently introduced decomposition method is employed, high pressure ashing with nitric acid in quartz vessels up to ≥ 300°C. The method achieves in all hitherto studied materials a complete decomposition of interfering substances [56,57] but the system is quite expensive.

8.4. ENRICHMENT AND SEPARATION

Analyses of trace metals at very low levels in sea and freshwater, body fluids and in complex environmental and biological materials often require preconcentration and/or selective separation from interfering matter. The reasons are either a lack of detection power of the applied method(s) or matrix interferences. Nevertheless, there are very sensitive methods available for cadmium. The concentration of this element in a number of environmental materials, particularly in some seawater and arctic (snow) samples, is frequently at the ng/l or ng/kg level so that direct analysis is only feasible with electroana-lytical methods and for the other methods preconcentration is necessary. For the analysis of several elements including cadmium in seawater by graphite furnace AAS (GFAAS) a solvent extraction procedure applying various dithiocarbamates and heavy solvents (chloroform, Freon, carbon tetrachloride) was introduced in the 70's [58,59]. This approach was and still is successfully used in marine analytical chemistry for determinations down to ng/l levels with the graphite furnace (e.g. refs. 23,30,60–64) but recently also for determination by ICP–AES [65]. This technique was also applied in flow systems [66,67].

Cadmium determinations in biological fluids are usually performed directly. There have, however, also been some successful attempts for chelate extraction of cadmium with hexamethylene ammonium–hexamethylene dithiocarbamidate (HMA–HMDC) from urine prior to GFAAS in order to achieve accurate results even at the lowest cadmium levels [68,69]. This approach is still used as a reference method for cadmium in urine introduced by the German Commission for the Investigation of Health Hazards of Chemical Compounds in the Work Area [70].

The application of chelating resins for the separation/preconcentration of a number of trace metals including cadmium from natural waters is a frequently applied approach prior to, mainly, atomic spectrometric methods and ICP–MS. Examples are batch treatment with Chelex-100 for subsequent ICP–AES [71] and GFAAS [72] determination. Also solvent extraction in combination with a Chelex-100 column for ion exchange was applied prior to GFAAS determination [73]. The same material was used in columns for trace metal preconcentration from seawater prior to flame AAS [74], and ICP–AES [75], and on board of a research ship achieving a concentration factor of 250 [76]. Other column materials used for trace metal preconcentration, including cadmium, are a poly(dithiocarbamate) resin prior to ICP–AES [65], Kelex 100, a commercially available alkylated oxine derivative [77], and CAD-4 resins impregnated with 7-dodecenyl-8-quinolinol (DDQ) [78] and sodium bis(2-hydroxyethyl)dithiocarbamate in an automated device [79] all for subsequent GFAAS determination. Cellulose collectors were also applied for sorption of metal dithiocarbamate chelates using either batch or column procedures. Analysis is performed by flame AAS, GFAAS and ICP–AES [80]. Oxine-silica gel columns were used for preconcentration prior to the GFAAS determination of a number of metals, including cadmium [28,81]. The same technique applying silica-immobilized 8-hydroxylchinoline was used for subsequent multielement analysis with ICP–MS in an open ocean water reference material [82] and with GFAAS in seawater in a flow system [83].

If a suitable collector element is used a number of elements, included cadmium, can be preconcentrated with recoveries $\geq$ 95%. An example is the precipitation of a small amount of iron with HMA–HMDC as collector for numerous trace elements from soils for subsequent AAS and XRF determination [84]. By co-precipitation of a number of trace metals, cadmium included, with magnesium hydroxide at pH 10 from seawater a 40-fold preconcentration for subsequent GFAAS determination can be achieved [85]. A procedure using co-flotation of a number of trace metals including cadmium on iron hydroxide was described recently [86].

Electrochemical preconcentration using mercury film electrodes was also applied for subsequent multielement determination including cadmium by ICP–AES after electrothermal vaporization in different solid materials and aqueous solutions [87].

8.5. ANALYTICAL METHODS

8.5.1. Overview

There are a variety of methods available that can be applied for the determination of cadmium in a broad selection of environmental and biological materials with appreciable

TABLE 8.2.

TYPICAL RELATIVE LIMITS OF DETECTION (LOD) FOR CADMIUM FOR ATOMIC SPEC-
TROMETRIC METHODS, INDUCTIVELY COUPLED PLASMA MASS SPECTROMETRY,
VOLTAMMETRY, TOTAL REFLECTION X-RAY FLUORESCENCE AND INSTRUMENTAL
NEUTRON ACTIVATION ANALYSIS

Conditions: Non-interfering aqueous analyte solutions or non-interfering solid matrix
(INAA). Values are given in $\mu g/l$ or $\mu g/kg$. The LOD is defined as three times the standard
deviation of the background noise or blank (3 s) [92]. See Section 5.1 for definition and
realistic estimation of limit of quantification (LOQ).

Methods:	ICP-AES[a]	Flame AAS	GFAAS[b]	ICP-MS[c]	Voltammetry	TXRF[d]	INAA[e]	Radio-nuclide
	0.7 [1]	0.7	0.01	0.03	≤ 0.0005	0.4	1.5	^{111m}Cd

[a] Data for ICP–AES are based on determined values for the most sensitive lines with a 50
MHZ conventional argon ICP at 15 pm bandwidth. Data in [brackets] are values reported by Jobin Yvon for their instruments JY 38 PLUS and JY 70 PLUS.

[b] For graphite tube furnace an injected volume of 30 μl on the L'vov platform is assumed;
characteristic mass for cadmium is about 0.35 (pg/0.0044 A•s) which is practically an
absolute detection limit, wavelenght 228.8 nm.

[c] The given value is for single element optimization, if multielement optimization is performed, the detection limit is approx. one order of magnitude higher.

[d] TXRF (Total Reflection X-Ray Fluorescence) data are based on a sample volume of 50
μl.

[e] INAA data are based on a sample weight of 500 mg non-interfering matrix; irradiation
and counting with a 40 cm³ Ge(Li-)detector, with a sample-to-detector distance of 2 cm;
zero decay time before start of count and maximum counting time = 10 min.

detection power [7,11,88–93]. Due to the favorable detection limits of several techniques
for cadmium frequently direct analysis is feasible either in liquids or digests of solid
samples and even in solids without any further treatment. Determination methods for
cadmium with examples of their application ranging from atomic spectrometry to spectrophotometry are subsequently discussed.

In Table 8.2 typical limits of detection (LOD) for cadmium using commercially available
instrumentation based on a recent compilation [93] are listed for the most frequently
applied methods. It has to be mentioned in this context that the limit of quantification
(LOQ), the lowest concentration for which a quantitative figure can be given with an
agreed degree of confidence − often ≥ 20%, is at least a factor of three higher. But this is
always method- and matrix-dependent. Hence it can be assumed generally that the determination limit is roughly equal to a concentration of ten times the standard deviation of the
background noise or blank [94]. This, of course, only applies to a direct determination
without dilution or decomposition of the matrix. In many cases, however, either dilution or

decomposition is unavoidable. Therefore practically attainable limits of quantification are approximately one order of magnitude, often far more, higher than LOQ data. This has always to be considered for the realistic judgement of the potential of a certain method in a certain matrix.

8.5.2. Atomic absorption spectrometry (AAS)

AAS with flame and furnace atomization is, in many research areas, the predominantly applied method for cadmium. With flame it is still very useful if single or oligoelement analysis is required due to its reliability and speed, with graphite furnace due to its versatility and detection power [11,88,89,91,92,94–97]. The determination of cadmium is commonly performed at the most sensitive 228.8 nm resonance line (see Table 8.2). There is also another much less sensitive line at 326.1 nm useful for the determination of higher concentrations,, without excessive dilution of the analyte solution or a solid matrix.

Because of the high volatility of cadmium that initially in some matrices only permitted thermal pretreatment (charring) up to about 300 to 400°C its determination in the graphite furnace was often difficult due to interferences from matrix constituents during atomization [98]. The progress in AAS methodology in the last two decades, however, has improved this remarkably. One of the most important developments for trace metal determinations in the furnace was the observation that the addition of certain chemical compounds to the sample led to matrix and analyte modifications during the whole thermal process so that higher pretreatment temperatures and hence a reduction of matrix influences could be achieved. This was first reported for the addition of ammonium nitrate and other ammonium salts to reduce cadmium volatility in the furnace [99]. Even simple acidification with nitric acid was quite effective [100]. Further improvements and refinements were achieved by the introduction of alkaline earth nitrates and phosphates as chemical modifiers in combination with the L'vov platform technique [101,102] leading to the so-called stabilized temperature platform (STPF) concept with charring temperatures up to 900°C for cadmium. Further improvement was attained with the introduction of palladium in combination with magnesium and ammonium nitrates as a universal modifier for a number of elements [103–106]. Recently also the usefulness of vanadium(V) was reported for chemical modification [107]. This progress was supported further by the introduction of Zeeman background correction, important for the analysis of solid samples, additional program refinements for furnaces and the introduction of PC operation etc. (cf. Chapter 6).

8.5.2.1. Flame AAS

Cadmium determination with flame AAS can be performed with appreciable sensitivity and accuracy using the air–acetylene flame down to a few μg/l analyte solution (see Table 8.2). Due to its reliability and freedom from interferences it is still the recommended approach in some official analytical methods. Examples are the AOAC methods for cadmium and lead in cookware [108] and earthenware [109] after leaching with acetic acid, for cadmium in food after digestion with nitric acid, sulfuric acid and hydrogen peroxide, extraction at pH 9 with dithizone–chloroform, and stripping the chloroform solution with dilute hydrochloric acid [110] and for a number of elements, cadmium included, in water. Cadmium is analyzed in this method either subsequent to evaporation of the sample and uptake of the residue with nitric acid or extraction with ammonium pyrrolidine dithiocarbamate (APDC) into methyl isobutyl ketone (MIBK) in the organic phase [111]. Cadmium in sewage sludges was hitherto determined after a prescribed aqua regia digestion – the threshold level for cadmium in sewage sludge is set e.g. for Germany at 20 mg/kg (DW) [112] with flame AAS [45]. Further examples of the use of flame AAS are given below.

Cadmium in more or less polluted river sediments was determined after freeze drying, sieving and digestion of 0.1–1 g subsamples of the particle fraction < 200 μm under pressure in PTFE vessels by adding 4–8 ml nitric acid–hydrochloric acid (3:1, inverse aqua regia) and making to volume, by flame AAS if the final cadmium content was ≥ 0.05 mg/l in the analyte solution [113].

Amounts of 5 g liver and kidney samples of wild Swedish birds and mammals were automatically wet ashed in 15 ml of a mixture of 65% nitric and 70% perchloric acids (7:3, v/v). The residue was diluted with water and adjusted to a final volume of 15 ml for flame AAS. The median levels found ranged from 0.08 to 1.8 mg/kg (wet weight) for liver and from 0.2 to 8.8 mg/kg in kidney [114].

Baseline data on heavy metal accumulation, including cadmium in organs and tissues and their variations with age, sex and habitat were determined in Japanese serows (Capricornis crispus). Homogenized tissue samples (1–10 g wet weight) and feces (0.05–1 g dry weight) were wet digested with various acid mixtures, depending on the matrix, and the resultant solutions diluted to a known volume. Some elements were directly analyzed by flame AAS, some others occurring in lower concentrations, including cadmium, were determined by AAS after extraction with diethyl dithiocarbamate–MIBK. Cadmium contents were at the μg/kg level [115].

Cadmium and lead contents of suspended particulate matter from the stack of a large refuse incinerator in the U.K. containing markedly elevated levels of both elements were determined by flame AAS after wet digestion of appropriate subsamples with concentrated nitric acid. A cadmium content of about 1600 mg/kg was found [116].

Amounts of 500 mg dried and ground (60 mesh) soil samples from experiments with sludge-amended soils were digested in 4 ml of a mixture of nitric acid–perchloric acid (2:1, v/v), brought to a 50 ml volume and analyzed for cadmium depending upon concentration either by flame or the graphite furnace. The same was the case for plant material with subsamples of 250 mg, which were digested in 2 ml of the same acid mix [117].

Sensitivity and detection limits of flame AAS for cadmium can be easily improved for routine use up to a factor of two to three by relatively simple devices that reduce the removal rate of the atoms from the optical path. These devices are called slotted quartz tube [118] or atom concentrator tube [119] but work only with the cooler air–acetylene flame, because the hotter nitrous oxide–acetylene flame would destroy these tubes.

If the atoms are trapped physically on the surface of a narrow diameter water-cooled silica tube placed just above the cone of the flame and heated up after a suitable collection period (by stopping the flow and removing the water) a transient signal is obtained enhancing the signal by about one order of magnitude. This was reported for the determination of cadmium in calcium chloride extracts of soils [120]. However, practical difficulties limit the application of this technique [119].

A combination of the slotted quartz tube and cadmium preconcentration by the use of an on-line alumina microcolumn was shown to increase the sensitivity for the measurement of cadmium in urine nearly thirtyfold. Cadmium in a subsample of 2.5 ml was trapped on the basic alumina and rapidly eluted with 8% v/v nitric acid. Results obtained with ten urine specimens compared well with those obtained by the graphite furnace technique in the concentration range $\leq$ to 15 μg/l [121]. In a similar approach a small glass column, packed with silanized glass beads, conditioned by tri-n-octylamine in isoamyl alcohol was used for preconcentration at pH below 3.0 from a sample volume between 1 and 50 ml. The eluent in this case was 0.2–0.3 ml of an aqueous solution of 0.05 M EDTA–0.3 M NH$_4$OH. In this analysis of tapwater using an aliquot of 20 ml a characteristic cadmium concentration of 0.23 μg/l was achieved [122].

Another promising technique for sensitivity enhancement in flame AAS certainly is flow injection [123]. An efficient flow injection system with on-line ion exchange pre-concentration for the determination of heavy metals including cadmium by flame AAS was recently investigated. With a sample consumption of only 1.6 ml per determination and a sampling frequency of 120 per hour the performance is comparable to that of conventional flame AAS with enrichment factors around 30. The ion-exchanger column contained quinoline-8-ol azo-immobilized on controlled pore glass [124].

8.5.2.2. Graphite furnace AAS

For the analysis of cadmium down to the lowest natural levels in liquid and solid materials the graphite furnace approach can now be considered as a mature and reliable technique often indispensable if extremely low contents have to be determined. If is also of particular value as a reliable checking method during the development of new analytical procedures and the improvement of already existing ones. Hence, examples of GFAAS applications in liquids, in solids after decomposition and directly in solids are treated separately below. It should be noted, however, that the temperatures given in the text below for thermal pretreatment and atomization are in most cases rather *temperature settings* than exactly determined ones. There might be large differences between settings and the true temperature of different instruments. Hence some differences in reported temperatures should not be taken too seriously. An experienced analyst should always adjust the temperature setting to optimal analyte/background signals as indicated on a recorder or monitor.

8.5.2.2.1. Liquid samples

Lowest cadmium levels in open ocean waters and arctic precipitation in the order of < 1 to a few ng/l are, despite its extraordinary detection power, not accessible to direct GFAAS measurements due to severe, not completely overcome, salt interferences. Thus extraction/preconcentration, already described in Section 8.4 [23,30,58–64], is unavoidable. Recently the tedious initial extraction procedure based on a heavy solvent has been simplified by replacing it by *n*-hexane following a re-extraction into and GFAAS determination in nitric acid [125]. Using the stabilized temperature platform furnace and Zeeman background correction, it could be shown, however, that it is possible in principle to analyze cadmium directly in seawater down to approximately 20 ng/l with and without modifiers [126–128] and in estuarine waters with a detection limit of 100 ng/l applying the stabilized temperature furnace, Zeeman background correction and ammonium nitrate as chemical modifier [129]. Recently the STPF concept, Zeeman background correction and mixed palladium and magnesium nitrates as chemical modifiers were studied for the direct determination of a couple of elements, cadmium included, in highly mineralized waters used for medicinal purposes. The authors reported a limit of determination of 50 ng/l for cadmium in undiluted samples [130].

For the direct determination of cadmium in freshwater and rainwater different figures were given for the limit of detection by some authors using the STPF concept and Zeeman background correction but different instruments: either 2 ng/l and 3 ng/l for a 50 μl sample [128,131] or 20 ng/l for a 20 μg/l sample [102,132]. In Table 8.2, for a 30 μl sample the detection limit, based on measurements in the author's laboratory, was given as 10 ng/l, which is equal to a limit of determination of about 33 ng/l if a 30 μl sample is used. For a 20 μl sample the determination limit would be about 50 ng/l which is in good agreement with data given in refs. 102 and 132.

The determination of cadmium in urine and whole blood has been increasingly performed in most laboratories from the early 80's by GFAAS. A frequently and successfully applied approach as far as accuracy of the data is concerned was the simple acidification of urine with nitric acid and protein precipitation prior to determination by the same acid for whole blood. This permitted for urine as well as for blood a temperature program for proper separation of background and cadmium signal with typical detection limits predominantly below 0.2 μg/l, allowing the reliable determination of prevalent levels depending on instrumentation and applied temperature program. At least for urine the use of Zeeman background correction was reported by some authors as being advantageous compared to the continuum source background correction often still sufficient for cadmium (e.g. refs. 100,133–142). Other workers, achieving comparable performance used mainly alkaline earth phosphates, nitric acid and nitrates and recently also palladium as chemical modifiers [143–149].

A few authors performed the determination in urine and blood without the use of modifiers but applying the STPF concept [150] and additionally Zeeman background correction with optimized temperature programs [151,152]. A performance similar to that for the other procedures applying chemical modifiers was reported.

Latest versions of the procedures using nitric acid as a chemical modifier for whole blood and urine reach, under carefully adjusted STPF conditions, detection limits < 0.08 μg/l for whole blood [141] and ≤ 0.02 μg/l for urine [142].

For the determination of cadmium in serum with Zeeman GFAAS under STPF conditions, palladium nitrate was used as a chemical modifier and the thermal treatment performed with hydrogen as an alternate gas (cf. Chapter 6). If 20 μl of analyte solution is injected a detection limit of 0.05 μg/l was achieved [153].

The duration of the furnace temperature program requires on average about 100 s in body fluids. Thus it is highly desirable from the view of decreasing running costs and an improved sample throughput to shorten the time of the furnace program considerably. As for other matrices (see Section 8.5.2.2.2) this has been investigated recently for the direct analysis of liquids by dispensing the sample into a pre-heated graphite tube. Hence, the majority of the sample volume evaporates on contact with the tube. A temperature program of less than 20 s using an ammonium oxalate modifier and deuterium background correction was achieved for the analysis of cadmium in a urine reference material. The duration of the complete analysis, including sampling time, was reported to be less than 60 s [154]. The value found, 4.9 ± 0.1 μg/l, was in good agreement with the recommended value (6 ± 2 μg/l) of the analyzed material. There was, however, no information about the attainable detection limit of the method, since prevalent cadmium levels in urine are well below the content of the reference material.

For the direct analysis of more difficult matrices as e.g. mother's and retail milk, just the addition of chemical modifiers is, however, not sufficient for the required sensitivity [149], because cadmium levels in these materials are frequently below 0.1 μg/l as confirmed by a study in which conventional methods, i.e. wet decomposition solvent extraction determination of cadmium in the organic phase by the graphite furnace, were applied under utmost contamination precautions [155].

Progress in this matter, however, could be achieved if an oxidizing alternate gas, i.e. air or oxygen, was introduced during the thermal pretreatment step as was demonstrated in a few previous studies (e.g. refs. 156–158). Applying the STPF concept, Zeeman GFAAS and only the addition of some Triton X-100 and, however, rather lengthy, charring at 600°C even extremely low cadmium contents < 0.05 μg/l in numerous retail milk and a few mother's milk samples could be reliably analyzed directly (cf. Table 8.1). With a 20 μl injection the detection limit (3 s) was about 0.02 μg/l [159]. Similar procedures were applied for the routine determination of cadmium in mother's milk in Germany [160] and of chromium, lead and cadmium in Danish dairy products and cheese [161].

8.5.2.2.2. Solid samples

The graphite furnace was used in numerous materials for cadmium analysis in solid materials in the μg/kg to mg/kg range. A number of examples from the literature of the last decade are given below followed by the discussion of the reliability of the method and some investigations to arrive at shorter temperature programs.

Cadmium was determined in a number of fly ash samples from U.K. coal-fired power plants and a municipal refuse incinerator. The samples were dried at room temperature, thoroughly mixed and 1 g subsamples open-digested at 150 °C with 10 ml concentrated nitric acid at 150°C to near dryness, subsequently made to volume and analyzed for cadmium with optimized GFAAS programs [162,163].

In various extraction solutions, some containing chelates from a study about the mobilization behavior of cadmium in lake sediments, cadmium was determined by GFAAS [164].

Cadmium in freshwater mussels (*Dreissena polymorpha*) used in a biomonitoring program close to an industrial area of the River Rhine was determined in soft parts of depurated specimens. A sample consisted of five homogenized individual mussels. Subsamples of 0.5 fresh weight were open-digested with 3 ml nitric acid and 1 ml sulfuric acid and determined under STPF conditions with GFAAS [165].

Cadmium was determined in bags of aquatic mosses used to monitor heavy metal pollution in rivers. Subsamples of 2-cm tips were digested after drying to constant weight at 105°C in 2 M HNO_3 and determined by GFAAS with deuterium background correction. Calibration was performed by acid matrix-matched solutions [166,167].

Mushroom samples were freeze-dried and finely ground; 10–100 mg subsamples were wet ashed with 3 ml of 65% nitric acid for 20 min at 80°C, made to volume with distilled water and determined by GFAAS according to the recommendations of the manufacturer (Perkin-Elmer) [168].

Heavy metals including cadmium were determined in vegetables growing above ground and some fruits after pressurized decomposition at 120–130°C of sample amounts of 6 g (fresh weight) with 2 ml of 35% nitric acid and 5 ml of 30% hydrogen peroxide for approx. 1 h. After cooling down the sample was made to volume (20 ml) by the addition of bidistilled water and subsequent determination by GFAAS with deuterium background correction. The contents usually were far below 0.1 mg/kg [169].

Cadmium in various marine biota was determined in deep-frozen subsamples of freeze-dried or wet materials. 0.5–2.5 wet or 0.1 to 0.5 g freeze-dried subsamples were then decomposed in PTFE crucibles of 35 ml volume after addition of 2–8 ml conc. nitric acid up to 160°C [170] and subsequently either directly analyzed by GFAAS or by solvent extraction (APDC, pH 3, NaDDC, pH 5) if levels were very low. Evaluation was performed against matrix-matched calibration graphs or by standard (analyte) addition. Accuracy was tested by independent analysis with e.g. voltammetry [39].

Cadmium in seaweed (*Enteromorpha*) was determined after drying and treatment of a 0.5 g subsample with 2 M nitric acid for approx. 1 h on a heating block, evaporated to near dryness then resuspended in 10 ml 2 M nitric acid and analyzed by graphite furnace AAS [171,172].

The analysis of various, also marine, materials for cadmium e.g. seaweed, freshwater and seawater mussels, earthworm, spruce shoots, poplar and beach leaves of the German environmental specimen bank is performed by a generalized standard operation procedure. It commences with the pressure decomposition of up to 0.2 g freeze-dried materials plus 2 ml concentrated nitric acid of appropriate purity in a decomposition system with four positions, equipped with extremely durable 35 ml PTFE vessels with smooth surfaces to avoid contamination and metal carry-over [173]. The resulting acidic analyte solution is appropriately diluted for each material so that the final analyte concentration lies in the optimal range of the Zeeman instruments used. The GFAAS measurements are performed in pyrolytically coated graphite tubes with L'vov platform at

STPF conditions using a mixed matrix modifier consisting of palladium and magnesium nitrates and a sample volume of 10 μl. Thermal pretreatment is at approx. 500°C, atomization (with maximum power) at approx. 1500°C (see remarks under 8.5.2.2). Data evaluation is performed by peak area, calibration either by matrix-matched reference solutions or matrix-identical reference materials [173,174]. Using standard conditions recommended by the manufacturer of the AAS instruments, including a mixed modifier consisting of 200 μg $NH_4H_2PO_4$ and 10 μg $Mg(NO_3)_2$ per firing under routine conditions, a characteristic mass of approx. 0.4 pg/0.0044 A•s was obtained on average for several matrices [175].

The determination of cadmium in plant materials was performed after pressure decomposition in a microwave oven by the use of delayed atomization cuvettes, Smith-Hieftje background correction and matrix/analyte modification using $(NH_4)_2HPO_4$ and $Mg(NO_3)_2$. The authors, however, extended a warning that care should be taken to restrict the amount of modifier used as too high amounts may lead to problems with both tube life and over-correction by the Smith-Hieftje background correction system [176].

The content of cadmium and selenium in horse kidneys from Jutland, Denmark, was determined in relation to age, local geographical variation and possible relationship between the two elements. For analysis the kidney cortex was separated from the medulla. Amounts of 3–5 g were taken and homogenized and subsamples of about 20 mg decomposed with 0.5 ml of 16% nitric acid prepared from ultrapure acid in a PTFE bomb for 1 h at 150°C. The measurement was performed by GFAAS with deuterium background correction [177].

The liver and kidneys of 30 roe deer and 30 hares from five selected areas in Germany were examined for lead, cadmium, thallium and mercury. The organs were homogenized, freeze-dried and milled. Subsamples were wet digested with purified nitric acid and the cadmium content determined by GFAAS. In comparison to already published contents of the soils of the selected areas a quite good correlation between cadmium in soils and cadmium in organs was found [178].

Low levels of cadmium were and still are determined after decomposition by extraction procedures prior to GFAAS as was reported for milk and dairy products [155], for fish muscle [39] and cereals, for which, however, also direct determination in samples from low-temperature ashing which introduced only spurious contamination were performed [178].

Cadmium in blue poppy seeds and in products containing poppy seed was determined after homogenization, pressure decomposition of appropriate subsamples with nitric acid, and determination in the analyte solution by GFAAS under STPF-conditions and with matrix modification. Maximal contents were around 2 mg/kg in poppy seeds [179].

The determination of trace elements, including cadmium in finished products in the chocolate industry, is quite difficult because of the often very low contents and samples with high fat content. Thus a procedure was proposed that started with high temperature/high pressure ashing of representative samples of about 0.35 g in the HPA system [56] at 270°C for two hours. For cadmium determination by GFAAS the method of analyte (standard) addition is used. The measurements are performed with 20 μl injection in a graphite tube with L'vov platform. In order to avoid a rapid degradation from the nitric acid solutions, the graphite tubes were coated with tantalum carbide. Thermal pretreatment was performed at 400°C and atomization at 2000°C. Cadmium contents in some cocoa products range from approx. 8 to approx. 215 μg/kg; accuracy was tested by analysis of certified reference materials with similar cadmium levels and spiked samples [180].

Recently a rapid routine analytical procedure for the determination of lead and cadmium in muscle and offal of slaughter animals after open wet decomposition was described. The material is freshly homogenized. Decomposition is performed with subsamples of 1 g in an automated open wet digestion system (VAO) in 40 ml glass tubes by the addition of 8 ml of 65% suprapure nitric acid with a maximal temperature of 250°C. The whole decomposition program requires 5–6 h for 20 tubes. After completion of the decomposition procedure the analyte solution is made to 25 ml with deionized water. The measurement is performed with a Zeeman GFAAS system applying STPF conditions: L'vov platform, thermal pretreatment at approx. 650°C, atomization (max. power) at 1400°C, matrix modification with a mixed modifier consisting of 0.5% $NH_4H_2PO_4$ and 0.2% conc. (65%) HNO_3. Injected volume is 20 μl sample, 10 μl modifier, calibration by standard addition, evaluation, however, in contrast to the manufacturer's recommendation peak height. Current quality control is carried out by analyzing an appropriate certified reference material (NIST, Bovine Liver 1577a). The detection limit of the method is approx. 0.01 mg/kg fresh weight and it was thus not possible to detect cadmium in animal muscle [181].

Cadmium in cigarettes from 331 packets out of a total of 314 brands produced in various areas of the world was determined after vacuum drying of five cigarettes from each packet by GFAAS after wet ashing in Kjeldahl flasks with an acid mixture consisting of nitric, sulfuric and perchloric acids. The cadmium contents of the investigated cigarettes ranged from 0.25 to 2.7 mg/kg (dry weight) with an average of 1.45 ± 0.6 mg/kg [182].

The determination of cadmium in bones is not simple because of the high content of calcium phosphate in the matrix and its quite low cadmium content so that an appropriate temperature program is necessary. In one case bone fragments of approx. 0.07 to 0.6 g are decomposed under pressure in PTFE bombs by addition of 4 ml 65% suprapure nitric acid and heating up to 130°C, made to volume with ultrapure water (15 ml) and analyzed by GFAAS at relatively low thermal pretreatment and atomization temperatures of 300–350°C and 1200°C respectively. Calibration was performed by the method of standard addition, quality control by analyses with voltammetry and an average content found of approx. 25 μg/kg, range 4–64 μg/kg [183,184]. In another study methodological aspects of the determination of cadmium, lead and manganese in rib bones of fetuses and children up to 2 years of age are investigated. The procedure started with a homogenization and fat extraction with diethyl ether. Hereafter subsamples from 0.05 to 0.140 g were decomposed under pressure by addition of 1 ml purified concentrated nitric acid up to 160°C. Finally the analyte solution was diluted with 5–10 ml bidistilled water and analyzed by Zeeman GFAAS under STPF conditions, injected volume 10 μl. Due the high phosphate content of the matrix no addition of a chemical modifier was necessary. Thermal pretreatment was at approx. 600°C and atomization at approx. 1700°C. Calibration was by standard addition, evaluation by peak area. Quality control was performed by analyzing bone powder H5 from IAEA (see Table 8.3). Contamination was continuously monitored by a Monetit ($CaHPO_4$) prepared in-house with very low cadmium and lead contents. The characteristic masses for cadmium were also determined in each of the eight series of analyses and over the range of approx. 0.45 to 0.7 pg/0.0044 A•s fluctuating results with an average obtained at approx. 0.55 pg/0.0044 A•s demonstrating the difficulty of these analyses. Results given for 66 samples ranged from 7 to 115 μg/g [185] in good agreement with former studies performed with GFAAS as well [183,184,186].

Eighty-nine autopsy samples of human and also fetal kidney cortex and liver were collected at random from three hospitals in London, Ontario. For metal analysis (Cd, Zn, Cu) 0.2 g of tissue was digested with nitric acid overnight and adjusted to 2 ml with deionized distilled water and analyzed by GFAAS, in some cases also by flame AAS. Detection limits for cadmium in tissue were 0.1 mg/kg. Cadmium in fetal samples was below the detection limit, and low in liver with mean values from 0.7 to 2.3 mg (fresh weight). Mean cadmium in kidney samples ranged from 5.4 to 41.8 mg/kg (fresh weight) [187].

Cadmium content in the kidney cortex of 388 deceased persons from some German cities and surrounding areas was determined by the following procedure. From each kidney two cortex samples of approx. 0.5 g were taken and open wet digest by 1.8 ml of a mixture of H_2SO_4–HNO_3–$HClO_4$ (2:2:1, v/v) up to 350°C. The obtained residue is diluted with 0.5% nitric acid to 25 ml and an aliquot of 300 μl of this solution diluted further to 25 ml. Determination is performed by Zeeman GFAAS against aqueous reference solutions. The atomization temperature was 1300°C. Quality assurance was performed by comparative analyses with another laboratory and the use of internal control samples. The obtained results showed a large scatter with a range from 1.7 to 94.3 mg/kg (wet weight) [188].

Cadmium in hair of 474 pre-school children living in an industrial and a rural area was determined. Two strands of hair were clipped with stainless-steel scissors as close as possible to the scalp from symmetrical occipital regions. Two cm of the procimal ends were marked by knotting a thread around them and used for analysis. Washed and dried hair samples (20–40 mg) were decomposed with 3 ml of a mixture of HNO_3 (65%), $HClO_4$ (70%), H_2SO_4 (96%) and bidistilled water (5:2:2:2, v/v) in 10 ml quartz flasks and heated in an aluminium block up to 240°C. Hereafter the analyte solution was diluted to 10 ml with 1.5% HCl. All operations were performed with suprapure acids and unter strict contamination precautions and control. Measurement was performed with Zeeman GFAAS and pyrocoated graphite tubes using $(NH_4)_2HPO_4$ as a chemical modifier. Thermal treatment was at 700°C, atomization at 1700°C. Calibration was against aqueous reference samples, quality control by analyzing certified reference material. Values for all samples ranged from 10 to 4280 μg/kg with a geometric mean of 89.5 μg/kg [189].

The progress in GFAAS technology and the proper use of chemical modifiers has considerably improved the reliability of this analytical approach. Even if one might have different opinions about GFAAS as a potentially absolute method, it is obvious from the recent literature that often a surprisingly good agreement can be observed in basic data, e.g. in published values for the characteristic mass. As far as cadmium is concerned for a number of matrices, e.g. various aqueous samples, urine, fish etc., the range of characteristic masses reported in a paper about "the possibility of standardless furnace atomic absorption spectroscopy" was relatively narrow with values from 0.31 to 0.45 pg/0.0044 A•s [190]. Another study confirmed these data under properly controlled STPF conditions using the already mentioned mixed modifier [$NH_4H_2PO_4$ + $Mg(NO_3)_2$] with characteristic masses of 0.39, 0.43, 0.43, 0.40 and 0.45 for acid digests of spruce needles, spruce shoots, brown seaweed and rainwater respectively [175]. The fact that there are, however, some matrices that do not behave similarly, i.e. that show larger fluctuations in characteristic masses — an example is cadmium in human bones (ref. 185 see above) — demonstrates that a perfect program and chemical modifier still has not been found. The matrix composition of bones with its extremely high calcium phosphate content is probably not favorable for optimal results, while for more organic materials one can expect a much better agreement.

Successful attempts were reported recently in adapting stabilized temperature platform furnace (STPF) methods to reduce the analytical time to less than 1 min per sample with no loss of analytical precision or accuracy. The authors showed that the analytical time could be reduced to about 30 s if certain changes in instrumentation are implemented, especially in the software and firmware that control the autosampler. These changes were sample uptake for the autosampler during the cooldown of the tube from the previous determination and deposition of the sample onto a heated platform. Further, the thermal pretreatment step and, in most cases, the use of a chemical modifier was omitted. The larger background signals thus produced could be handled with Zeeman systems. For cadmium, digests of five certified reference materials were analyzed using a 30 s protocol obtaining good agreement with certified values. The authors, however, stated that their paper "is not primarily intended to provide routine and reliable methods, it is intended to test the feasibility of these fast methods" (ref. 191, see also ref. 154).

8.5.2.2.3. Direct GFAAS analysis of solids

The analysis of solid samples has been an important objective for atomic spectroscopy for a long time since very frequently the analyst receives a sample in the solid state rather than as a liquid. With the introduction of optimized background correction systems, especially based on the Zeeman effect, this could be realized more reliably by GFAAS for a number of elements in many materials [192–194]. Since the analysis of more volatile elements is common in this AAS mode, there are numerous determinations of cadmium reported employing solid sampling either by direct introduction of solids employing appropriate devices for commercial instruments and those that are especially designed for solid samples [194] or of slurried samples [195] into the furnace. From the numerous papers already published dealing with the analysis of cadmium a few examples are given below.

Freeze-dried samples of pancreatic tissue were placed directly into the graphite furnace. Possible interference from other metals was evaluated by measuring NIST Bovine Liver. Thermal pretreatment was at 300°C, atomization at 1190°C. For higher contents the

References on p. 219

326.1 nm resonance line was used [196].

A graphite cup inserted into a graphite tube of a Zeeman GFAAS instrument was investigated for the direct analysis of solid samples and a number of elements, cadmium included, determined in reference and other materials and plastics. Atomization from the cup delayed sample volatilization until the tube had stabilized in temperature so that nearly the same effect was obtained as with a L'vov platform. For cadmium determination the thermal pretreatment temperature was about 600 – 650°C and atomization temperature about 1800 to 2000°C. The reproducibility of the results was typically around 10% [197–200]. Using the same system relatively low cadmium levels in solids could be determined directly with intakes up to 10 mg using oxygen ashing, thermal pretreatment at 600°C, atomization at 1500°C. Evaluation was done against a certified reference material. A determination limit of < 0.3 μg/kg was reported [201].

Using an instrument especially designed for solid sampling that allows the analysis of solids without oxygen ashing in up to at least 10 mg amounts, trace metals, including cadmium in various foodstuffs of marine origin [202], in inorganic solids [203], in environmental samples [204], in the caryopsis of wheat [205], in whole wheat [206], in various polymers [207,208] and in the placenta [209] were determined. An automatic sample introduction system has also been presented recently for this instrument [210].

Using the so-called autoprobe (50 μl volume) and improved deuterium background correction the reliable determination of trace metals, including cadmium, in various materials, mainly of biological origin, was demonstrated. Thermal pretreatment was at 750°C and atomization at 2100°C [211].

Local analysis was reported in wheat [205] and for the study of lead and cadmium profiles in bird's feathers [212].

The solid sampling technique is further particularly suited for the rapid estimation of the distribution of cadmium, i.e. its homogeneity, in numerous materials [213].

Another quite effective approach for solid sample analysis is the Zeeman inner miniature cup technique, with thermal pretreatment for cadmium at 240°C and atomization at 2010°C [214].

Since it is evident now that the most reliable calibration approach is the use of reference materials for the solid sampling technique and there are still gaps to be filled, a synthetic reference material prepared by coprecipitation of eight metal ions, including cadmium, with magnesium 8-quinolinate was recently introduced and successfully applied also for the inner miniature cup approach [215].

The practical advantages of the slurry technique, in which finely ground solid materials are suspended in an aqueous solution, is that it can be calibrated by aqueous reference solutions, chemical modifiers applied and in principle used with practically all conventional GFAAS systems [195] and that now an automated ultrasonic agitator is commercially available so that solids can be analyzed as easily and rapidly as liquids, sometimes with cycle times of less than a minute [216].

There are already numerous papers describing applications of this technique, e.g. for cadmium in soil [217], in sediments and suspended matter collected in natural waters [218] and in various foods [219,220]. Thermal pretreatment and atomization temperatures varied from approx. 700–800°C to 1300–2100°C respectively.

Recently also the use of this technique was described for cadmium in biological material with a molybdenum tube atomizer and sulfur as a chemical modifier. The attainable absolute detection limit was reported to be 0.31 pg [221].

8.5.3. Plasma atomic emission spectrometry (AES)

AES has gained remarkably in importance with the commercial introduction of plasmas as excitation sources in 1974. The dominant plasma-AES systems at present are equipped with an inductively coupled plasma (ICP) and provide sequential as well as simultaneous multielement potential for approx. seventy elements (refs. 222–226, cf. Chapter 4). The plasma technique also can be combined with solid sampling. Commercial instruments have reached maturity as far as optical, technical and computer performance is concerned. Quite promising detection limits also for more volatile elements like cadmium compete favorably with flame AAS. Hence ICP–AES instruments, despite high running costs, have begun to replace flame AAS rather quickly if multielement determination is required. From Table 8.2 it is obvious that the attainable detection limits for cadmium with ICP–AES and flame AAS are comparable.

In the recent comprehensive work [223] already mentioned applications of ICP–AES in various research areas, e.g. geology, agriculture, food surveillance, clinical chemistry etc., solid sampling included, are reviewed in detail.

Multielement determination in samples with comparatively low elemental levels requires preconcentration procedures prior to ICP–AES determination. A few examples of these procedures in seawater [65,75], also applying a flow system [83], in natural water [71,80] and in general by using an electrochemical preconcentration technique [87] have already been mentioned in Section 8.4.

Using a sequential ICP–AES instrument as a detector in a flow injection system, eight elements were determined by one working operation in potable water. Since lead and cadmium are commonly present in concentrations below the detection limit of ICP–AES one-line enrichment was necessary. This was achieved by inserting Chelex-100 micro ion-exchange columns into the injection valve instead of the sample loop and both elements were determined after preconcentration for 2 minutes, the other elements without enrichment. Enrichment is possible by a factor of 25 to 60. Dual detection based on potentiometric stripping analysis (PSA), see Section 8.5.5) and ICP–AES was used to control accuracy and to examine the influence of Fe(III) on the enrichment process. Cadmium contents in a number of potable water samples ranged from 0.5 to 1.3 μg/l [227].

An example of decomposition-selective extraction/preconcentration is multielement determination, cadmium included, in Japanese ribs. The material was decomposed with nitric acid under pressure in PTFE vessels and subsequent extraction of some trace metals with APDC (ammonium pyrrolidine dithiocarbamate) into chloroform. The solvent is evaporated and the residue treated with nitric acid to obtain a suitable analyte solution for ICP–AES [228].

The direct introduction of organic solvents into an ICP is desirable after solvent extraction e.g. for ICP–AES–HPLC coupling. This was recently studied in detail, also for cadmium for different nebulizers. It was observed that cooling the spray chamber, besides improving the limiting aspiration rate, produces a much more stable plasma and improves detection limits [229].

If liquid samples are introduced into an ICP by direct insertion on a wire loop absolute detection limits can be attained that are in the absolute picogram to subpicogram range. The investigation of matrix effects with Ca as matrix element showed that peak heights vary considerably with Ca concentration but peak area was more constant. For cadmium an absolute detection limit of approx. 4 pg and a relative detection limit of approx. 0.4 μg/l was reported [230].

Due to the frequent limitation of sample sizes the analysis of microsamples is increasingly important. Thus different sample introduction systems were investigated for amounts of 1 ml or less. Using a fritted disk sample introduction system with a consumption of 0.25 ml/min it was possible to determine cadmium and other metals in EPA 481 trace elements in water reference material with a certified concentration of 39 μg/l (found 37 μg/l) with a detection limit of 20 μg/l. It was stated, however, that further investigations are necessary before the system can be used routinely [231].

The domain of ICP–AES, however, still is the sequential or simultaneous analysis of higher contents in biological, environmental and technical materials.

NIST certified reference material SRM 1571, Orchard Leaves, was used as a multielement standard to determine 14 elements, including cadmium, at different concentration levels in some BCR botanical reference materials simultaneously by ICP–AES after pressurized decomposition with nitric acid. Cadmium contents found in the BCR materials ranged from < 0.2 to 2.3 mg/kg with a typical precision $\leq$ 10% [232].

ICP–AES was applied to the determination of metals and metal compounds, cadmium included, in samples collected in the workplace for the assessment of occupational exposure of workers. The materials analyzed after hot-plate and microwave-assisted decomposition using different acids and acid mixtures were air sampling filters, dusts, ashes and paints. Quality control was performed by analyzing appropriate certified reference materials, the application of independent methods and participation in interlaboratory comparisons. A practical detection limit of 0.7 mg/kg dry weight was reported for cadmium in dusts and ashes, assuming a sample mass of 50 mg and a final volume of 50 ml [233].

Samples from different horizons of natural podzolic soils from all over Norway were analyzed for the nitric acid extractable part of 27 elements cadmium included, by leaching with hot 1:1 nitric acid by ICP–AES. A regional trend with on average higher cadmium contents in the A_0 horizons of the southern part of Norway confirmed suggestions about long-range atmospheric transport of cadmium (and also lead) with contents up to approx. 10 mg/kg [234].

Following nitric acid decomposition in a PTFE bomb ICP–AES was applied for the determination of cadmium, chromium, copper, nickel, lead, zinc, arsenic and mercury in sewage sludge reference materials and other sludge-based samples. The results were compared with those obtained using an AAS standard flame method after sample decomposition with boiling nitric acid. The results obtained for cadmium in the range from 3.4 to 31.5 mg/kg showed some discrepancies between AAS and ICP–AES. However, in five of seven out of twelve instances the ICP–AES result was closer to the certified value than the AAS result. In general it was concluded that the obtained results compared favorably between both methods [235].

The application of an integrated direct-reading double polychromator – sequential scanning monochromator system for precise and accurate ICP–AES multielement analysis of rocks and waters, including brines, was described recently. The authors used an internal reference technique by synthetic multielement calibration solutions. Detection limits [35] reported for cadmium were 1 μg/l analyte solution for the 214.430 nm spectral line and $\leq$ 0.8 μg/l for the 226.5 nm line [236].

Sequential ICP–AES was used to determine 28 elements, cadmium included, in heavy slag and electrofilter ash from a large-scale city waste incineration plant. Pretreatment of samples was performed either by pressurized decomposition with nitric and hydrofluoric acids in PTFE vessels, lithium tetraborate fusion subsequent to hydrofluoric acid treatment or just leaching with water. Cadmium was measured at the less sensitive 214.438 nm spectral line. The cadmium contents ranged from 100 to 800 mg/kg in ash and slag, while the concentrations in the aqueous leachate were below the detection limit of the method [237,238].

Using an ICP–AES with a computer-controlled rapid-scanning echelle monochromator the major, minor and trace elements were determined in reference sediments and soils. The concentration of 17 elements in five reference materials, cadmium due to its low concentration in only detected in one material, could be determined due to the high resolving power of the echelle spectrometer by using a single set of analytical lines without any corrections for line overlap interferences. In river sediments the best agreement for cadmium (certified content 10.2 μg/kg) was found in the sequential mode [239].

It is also possible in principle to introduce powdered samples as aerosols into the ICP (for AES and MS, see also Section 8.5.4) using a powder dispenser. Recent studies have demonstrated that this method and device have potential for at least qualitative analysis of solid powdered samples [240].

8.5.4. Inductively coupled plasma–mass spectrometry (ICP–MS)

ICP–MS is a relatively new method. Since its commercial introduction in 1984 [241,242], though investment costs are high, it is, although there are still some limitations to be overcome (refs. 242–244, cf. Chapter 4), already applied routinely in an increasing number of laboratories. This is because of its comprehensive element coverage and detection power that compete for a number of elements with that of the graphite furnace while being superior for many others. For cadmium the detection limit under conditions optimized for single-element determination is very close to that of the graphite furnace. If multielement optimization is performed it can still compete with the detection limit of total reflection X-ray spectrometry (cf. Table 8.2). This makes ICP–MS one of the most sensitive and versatile commercially available instrumental analytical techniques in the medical, biological and environmental sciences. From this potential a further steady gain in many research fields, particularly for cadmium analyses in materials with cadmium levels up to a few mg/kg, can be expected.

Since ICP–MS cannot be used for liquids with comparatively high salt content, e.g. seawater, if dilution is not applicable separation/preconcentration is necessary.

The analysis of an open ocean water reference material was performed after 50-fold preconcentration of several trace metals, including cadmium by selective adsorption onto silica-immobilized 8-hydroxy-chinoline (I-8-HOQ) at pH 8. After elution with 1 M HCl–0.1 M HNO$_3$ the eluate was brought to dryness and taken up several times by the addition of nitric acid to remove as much HCl as possible to minimize spectroscopic interferences from molecular chlorine (cf. Chapter 4). After the addition of appropriate spikes for isotope dilution, the ICP–MS isotopic ratio determination of the trace metals was performed. The cadmium content in the reference material was found to be 27.2 ng/l [82].

In a riverine water reference material fifteen elements were determined directly. Cadmium and some other elements (Co, Ni and Pb), however, required the preconcentration procedure described above. Accurate results were obtained by external calibration, standard additions or isotope dilution. The latter approach gave the most accurate and precise results. The cadmium content in this material was found to be 14 ng/l [245].

The procedures outlined above were substantially simplified by the implementation of an on-line preconcentration technique using a miniature column packed with I-8-HOQ for trace metal determination in riverine and open ocean water. The procedure required only 10 ml aliquots and a preconcentration factor of 10 was sufficient. Another important improvement compared to the previous procedure was that the time needed could be reduced for the analysis of three samples and three column blanks from at least 8 h to 45 min [246].

Another approach for the determination of ultratrace levels of metals in seawater was electrochemical preconcentration for ten elements, cadmium included. After dissolving the deposit ICP–MS determination of Mn, Ni, Cd, Pb and Hg is easily possible down to levels between 13–86 ng/l [247].

Since the concentration of interfering elements is low, ICP–MS was an ideal method for reliable multielement determinations, including cadmium in wet precipitation [248] and in surface waters from lakes in the Eastern USA. However, since cadmium concentrations in lake waters are very low it could only be determined by ICP–MS in just 7% of the investigated lakes with a maximum measured concentration of 0.8 μg/l [249].

The determination of trace elements, including cadmium, in a marine reference material of lobster hepatopancreas was performed after open wet decomposition of 0.2 g subsamples with nitric acid up to 160°C and making to volume (10 ml) by ICP–MS with a determination limit (10 s) calculated from measured detection limits of 1.9 μg/l. The obtained value was in good agreement with the certified concentration [250].

ICP–MS was applied to the determination of 11 trace elements, included cadmium in two marine sediment reference materials (MESS-1 and BCSS-1) by using isotope dilution after initial decomposition of a 0.5 g sample in a mixture of 3 ml of hydrofluoric acid, 3 ml of nitric acid and 2 ml of perchloric acid under pressure in a PTFE vessel and in a boiling water bath. After some additional treatments and transformation into nitrates the isotopic spikes were added and isotope ratios determined [251].

The potential of ICP–MS for the analysis of trace elements in foods was investigated aiming at evaluating the effects of the major elements on the response of Al, Cr, Zn, Mo, Cd and Pb. Results for dry ashed NIST reference materials compared well with the certified data. It was stated, however, that "a higher resolution quadrupole or a narrower ion energy distribution would improve the accuracy of analyte elements whose m/z is adjacent to major background or matrix ions" [252].

ICP–MS was recently used to establish accurate values for 34 elements, including cadmium in cerebrospinal fluid or individuals with neurological fluids and controls. A large number of the elements determined at or below the μg/l level had not been reported before in this body fluid. Except for copper for which significantly lower values were found in motor neurone disease and multiple sclerosis and iron with significantly lower levels in Parkinson's disease no statistically significant differences were seen [253].

ICP–MS was applied in the semi-quantitative mode as well as with external calibration for the analysis of up to 19 elements, cadmium included, in three decomposed certified biological reference materials. For most elements good agreement with certified values and appreciable short-term and long-term precision was found. It was stated that major advantages for routine use are the wide dynamic range (5 to 8 decades depending on the applied mode) and the scan speed of analysis with 15 to 20 elements per minute [254].

The potential of semi-quantitative analysis by ICP–MS was recently investigated with water and more complex biological and botanical NIST certified reference materials. Analytical results of 23 elements, including cadmium, indicated that the elemental concentrations were within 30% of the certified values [255].

The suitability of ICP–MS after pressurized decomposition of polluted and control soil samples in PTFE vessels with HNO_3, $HClO_4$ and HF was studied for the elements Pb, Sb, Cd, As, Mn, Mo and Cr. NIST certified reference materials were run to permit an assessment of precision and accuracy. It was concluded that detection power and accuracy is acceptable for routine environmental soil work [256].

Isobaric molecular ion interferences resulting from the formation of species such as MO^+ and HCl^+ are a significant shortcoming of ICP–MS. In a recent study, in which low levels of cadmium were determined in the presence of Zr, Mo, Ru, In and Sn, this problem was circumvented by the use of two multivariate calibration methods–multiple linear regression and principal components regression. It was concluded that these as well as other multivariate approaches merit further study [257].

Powdered materials were introduced into an ICP–MS by using a spark dispersion-merging sample introduction technique for the direct multielement analysis of more than

40 elements including the rare earth elements (REE) and cadmium in four geological standard rocks. Optimization of the operational settings is possible by using an analyte solution. It was concluded that differences from reported values were better than ± 30% for the majority of the elements, including the REE. An absolute detection limit of 0.1 ng was obtained for most elements [258].

8.5.5. Electrochemical methods

Since their introduction in around 1920, electrochemical methods have undergone a steady, sometimes less pronounced, but during the last decade obvious progress (refs. 259–264, cf. Chapter 4). This was due to significant improvements by the introduction of new and more powerful modes that enhanced sensitivity such as differential pulse anodic and cathodic stripping and square wave voltammetry [265,266] as well as potentiometric stripping [267,268]. Hence, at present electroanalysis provides the lowest detection limits for some metals for *in-situ* determinations of all commercially available analytical methods in conjunction with appropriate computerization [269].

Cadmium belongs to the few elements that can be determined electrochemically with an extraordinary detection power. If a rotating mercury film electrode on a glassy carbon support is used a detection limit of at least 0.5 ng/l depending on the electrolysis time can be attained in an analyte solution (cf. Table 8.2) and even with the rather simple mercury drop electrode the detection limit of 50 ng/l is sufficient for many tasks in trace and ultratrace analysis particularly if the square wave mode is applied (refs. 264–266,270, cf. Chapter 4). Moreover, potentiometric stripping can be favorably used for direct cadmium analysis in matrices that were hitherto only accessible after decomposition [271].

Though not very frequently used for practical analysis, it has to be mentioned that cadmium also forms a chelate with 8-hydroxychinoline that is accessible to cathodic stripping after adsorptive accumulation at the hanging mercury drop electrode with an appreciable detection limit around 10 ng/l analyte solution, which might be used if interferences from other elements prohibit the use of anodic stripping techniques [272].

Electrochemical methods are excellently suited for studies on cadmium speciation in fresh and marine waters because of their ability to discriminate between different chemical forms of an element [273].

8.5.5.1. Aqueous samples

Direct determination of the extremely low levels of cadmium in natural waters is only feasible with differential pulse anodic stripping voltammetry (DPASV) using the rotating mercury film electrode (MFE) because of its extraordinary detection power. The pretreatment often consists, due to appreciable salt contents, particularly in sea and river water, just in an acidification to pH 2–3 by the addition of suprapure hydrochloric acid. If the sample to be analyzed contains interfering surfactants and other organic substances UV-irradiation in scrupulously precleaned quartz tubes in the presence of 0.01 M HCl and

0.03% H_2O_2 is necessary (e.g. ref. 274). Samples that contain particulate matter have to be filtered through 0.45-μm pore size membrane filters for differentiation from the dissolved metal (cf. Chapter 2).

Seasonal variations of Cd, Pb, Cu and Ni levels in snow from the eastern Arctic Ocean were determined by voltammetric methods, for cadmium with the MFE, during the Swedish Arctic expedition "Ymer 80". Snow samples analyzed consist of freshly fallen snow and snow from the previous fall/winter season collected from ice floes. Pb, Cd and Cu were analyzed under clean-room conditions on board ship in one run. Cadmium levels in freshly fallen snow were in the range of 0.2–0.6 ng/kg. Snow a few days old and snow from the 1979/1980 season contained cadmium levels in the range from 0.5 to 6 ng/kg [18]. Similarly low cadmium contents were found in arctic seawater [275].

Similar values have been reported recently for antarctic snow by using DPASV under clean room conditions during a German Antarctic expedition in winter 1987. In addition to voltammetric analysis at the sampling site, determinations with isotope-dilution mass spectrometry were performed after the expedition in the home laboratory and good agreement found with DPASV. The detection limit for DPASV was 0.5 ng/kg, for IDMS 0.2 ng/kg, cadmium contents in surface snow ranged from < 0.2 to approx. 4 ng/kg [276].

The same methodological and sampling procedures already mentioned in Chapter 2 were applied in comparative studies on cadmium levels in the North Sea, Norwegian Sea, Barents Sea and the (already mentioned above) eastern Arctic Ocean. This work included waters from polluted coastal zones usually rich in particulate matter as well. Hence filtration and UV irradiation of the filtrate was often necessary. The UV irradiation was performed with a 150 W UV-lamp in the spectral range of 238–580 nm. This decomposition was effective within 1–2 h if the sample was heated up close to boiling temperature during irradiation [277]. The results of these extended studies to test instrumentation and sampling devices, initially performed in Lake Constance and in Jade Bay, North Sea, yielded extremely low trace metal levels that were at that time in sharp contrast to literature data. Dissolved cadmium in Lake Constance ranged from 4 to 12 ng/l, in Jade Bay from 13 to 21 ng/l, in some open seawater samples from the Atlantic the contents were around 20 ng/l and in the open Baltic around 30 ng/l. Depth profiles in the Norwegian Sea and the Arctic Ocean are not very pronounced, whereas cadmium contents up to 40 ng/l were found at depths around 1000 m in the North Atlantic Ocean and up to 100 ng/l 120 km off the coast of California at 1000 m depth. Dissolved cadmium in the River Elbe estuary ranged from approx. 30 to approx. 200 ng/l. At Cuxhaven there was a significant difference between total cadmium content — determined after decomposition of the filter by a low temperature ashing method — with up to 200 ng/l. The dissolved part ranged from 40 to 60 ng/l [278]. Based on these and other studies using voltammetry it was possible to compare typical average trace metal levels in various oceanic regions showing lowest levels for dissolved cadmium in the Pacific Ocean (around 2 ng/l) with some higher values (around 25 ng/l) in the Weddell Sea ("Upwelling") and highest ones along the North Sea coastline from Belgium to Germany (40–50 ng/l) [279].

Workers from Sweden applied the above described procedures in comparison to a MIBK extraction and a freon extraction procedure with subsequent GFAAS determination.

They also applied UV-irradiation prior to DPASV but found the voltammetric method preferable because it was faster and provided a better detection limit. Concentrations down to 0.5 ng/l were determined with a relative standard deviation (RSD) of 20% [280].

DPASV with the MFE using the methodology described above was used in further studies on levels of heavy metals in the tidal Elbe as well as aiming at defining anthropogenic background levels (35 ng/l for cadmium) [19] and in the German Bight. The data of the latter study indicate that the Weser is a distinct source of cadmium with levels of dissolved cadmium up to approx. 400 ng/l in the tidal part. These values were somewhat higher than those of the tidal Elbe [281].

Applying the same method, dissolved cadmium in the Rhine and some other European, notably Italian, rivers was also studied. The anthropogenic baseline level for the Rhine was approximately the same as for the River Elbe. During a long-term investigation in the lower course of the Rhine from Cologne to the Dutch–German border a distinct decrease in the concentration of dissolved cadmium with time was observed [20].

An intercomparison of the analysis of seawater for dissolved cadmium, copper and lead between the voltammetric approach using the MFE and metal preconcentration either by solvent extraction with dithiocarbamates or by a Chelex-100 column prior to GFAAS was performed at that time as well. The results obtained by two very experienced teams for the three trace metals studied for a wide range of natural seawater samples displayed similar narrow limits (15%). It was mentioned that an advantage of the voltammetric approach is its amenability to real-time shipboard analysis [282].

Based on the determination of a number of dissolved metals (Cd, Pb, Cu and Zn) with voltammetry (DPASV with MFE) in the major rivers (Rhine, Ebro, Po, Arno, Krka, Nile and Tiber) draining into the Mediterranean and in the Mediterranean, a reassessment of the riverine input of these metals and Hg was recently made. The results obtained indicate that river inputs of trace metals to the Mediterranean are lower than previously published data suggest [22]. The values found for some Italian rivers agree quite well with those of a previous study also with DPASV [20].

Wet deposition of acid and of heavy metals (Cd, Pb, Cu and Zn) by rain and snow in Germany has been followed since 1980 with a network of automated wet-only samplers in rural and industrial areas in order to investigate differences in areas and trends over time. The above mentioned metals are sequentially determined from the same solution by DPASV at the hanging mercury drop electrode (HMDE). After acidification to pH 2 the sample is irradiated by UV. The detection limit of 50 ng/l and precision (RSD < 5% at cadmium contents $\geq 0.5 \mu g/l$) is sufficient for cadmium determination with typical contents of around 0.05 to several $\mu g/l$ [17,283].

8.5.5.2. Other liquid and solid samples

If liquids other than water and solid materials have to be analyzed for cadmium by voltammetric methods, exceptions are potentiometric stripping approaches (see below), the applied decomposition procedures have to be complete and blanks should be minimized. However, there are often significant blanks that amount to one or several nanograms per digestion vessel. If typical subsample weights between 0.1 and 1.0 g and fluctuating blanks with a rather high RSD are taken into account the blank level, corresponding to the sample weight, is typically somewhat below or around one μg/kg, sometimes lower but sometimes also higher. Therefore it is obvious that after decomposition and if the dilution factor is considered, the practically attainable detection limit in the analyzed material is often comparable to that for atomic absorption. In materials that can be analyzed directly by GFAAS the detection power of DPASV might be even inferior if the blank level is not favorably low or very constant.

For the determination of Cd, Pb and Cu in milk powder the sample was dry ashed under a carefully controlled temperature of 550°C with the addition of H_2O_2. The ash was dissolved with dilute hydrochloric acid. Determination was with DPASV and the HMDE. Using intakes of 5–10 g the cadmium content found was 1.1 μg/kg [284].

DPASV at the HMDE was used in an official AOAC method of analysis based on 10-g samples and dry ashing with K_2SO_4 as an ashing aid at 500°C. Due to the obtained blanks, the estimated limit of quantification for cadmium was 5 μg/kg, for lead 10 μg/kg, verified by a collaborative study [285,286].

Heavy metals (Zn, Cd, Pb, Cu, Ni and Co) were determined in meat and organs of slaughter cattle in Germany with DPASV/MFE for Zn, Cd, Pb and Cu after wet decomposition of up to 0.5 g samples (dry weight). Decomposition is performed in two steps with suprapure acids in a quartz crucible. In the first step the majority of the matrix is oxidized with 2 ml nitric acid, whereas in the second step the remaining organic matter is destroyed with the aid of 1 ml perchloric acid up to max. 200°C. Due to the low blank (0.2 ng cadmium/crucible) cadmium levels down to less than 2 μg/kg (FW) were determined [287]. A similar decomposition procedure in combination with DPASV/MFE was applied for the same elements in various biological and environmental materials of the German pilot environmental specimen bank [288] and in a broad collection of food types, including vegetables and plants of an industrially polluted area, mushrooms and tobacco. The determination limit was approx. 0.2 μg/kg and certified reference materials were analyzed at the same time to control accuracy [289].

In a recent investigation, several trace elements were determined in meat, livers and kidneys of sheep slaughtered in the Netherlands. Sample decomposition was by dry ashing. Lead and cadmium were determined by DPASV/HMDE, with a detection limit of 1 μg/kg (fresh weight) for cadmium. The reported cadmium values are quite similar to those reported earlier for cattle in Germany [290].

A multielement approach for the sequential determination of up to eight elements in digests of biological samples by different voltammetric modes started with a nitric acid–sulfuric acid decomposition up to 290°C either of 5 ml of urine or 0.3 g of solid material. It was reported that the determination of all eight elements (Se, Cd, Pb, Cu, Zn, Ni and Co) in the digested samples took 3 h or approx. 25 min for each element per sample. It was noted that decomposition had to be complete for reliable values and that sometimes problems exist with low cadmium levels so that single element approaches may be necessary in these cases. Reliability of the method was tested by the analysis of certified reference materials [291].

References on p. 219

The application of square wave voltammetry increases sensitivity for the determination of Zn, Ni and Co (cf. Chapter 14) in comparison with the normal ASV or CSV mode. For Cd and Pb, however, no significant improvement in sensitivity can be seen. The advantage of this approach is that the determination time is shortened to half the value by DPASV and hence attains the time needed for a single element determination by GFAAS [265].

For the analysis of metals such as Zn, Cd, Pb and Cu by voltammetry the frequently applied nitric acid pressure decomposition up to maximal 180°C is not complete in many matrices. Some remaining compounds often lead to severe interference effects [292], hence by the addition of perchloric or/and sulfuric acid a much more effective decomposition is attainable due to the higher oxidation potential and temperatures up to 300°C. Decomposition procedures with acid mixtures, however, are time consuming and require experience, and the use of perchloric acid may be hazardous. With the example of some marine materials it was shown that the oxidation of these materials using concentrated nitric acid in closed quartz vessels under pressure at 320°C completely destroys interfering organic compounds, so that voltammetric determinations free of interferences are possible (ref. 293, see also refs. 56,57).

Potentiometric stripping analysis (PSA) in combination with a very capable computer now commercially available facilitates direct determination in liquids that have to be decomposed for DPASV. Oxygen does not interfere, in contrast dissolved oxygen supports stripping as an oxidant, and allows direct measurement in various liquids and in digests from the nitric acid treatment of biological and environmental materials [269,271].

Computerized flow PSA, after a simple dilution with suprapure hydrochloric acid and with the use of glassy-carbon electrodes pre-coated with a film of mercury that can be employed for several analytical runs, permits the determination of cadmium and lead in e.g. milk and milk powder and whole blood. For milk and milk powder the sample is diluted fivefold with a suprapure hydrochloric acid and is electrolyzed for 0.4–4 min prior to stripping. The analytical results agreed satisfactorily with the certified values for three certified milk powder reference materials. After a pre-electrolysis time of 4 min the detection limit for cadmium is 0.8 μg/l [294]. Cadmium and lead in whole blood were determined by the same approach using disposable carbon fiber electrodes. Two sample aliquots of 0.2–0.4 ml are diluted (1 + 19) with 0.5 M hydrochloric acid, with a standard addition to one of the aliquots, prior to injection into the flow system. The computer-controlled flow system executes a pre-programmed number of cycles on each sample pair before presenting the final result. The method was verified for certified whole blood reference samples and by comparison with results obtained by GFAAS. A detection limit below 1 μg/l was reported [295].

8.5.6. X-Ray methods

The different modes of X-ray fluorescence including X-ray tubes, direct X-ray irradiation, gamma radiation from radionuclide sources, but also particles such as electrons, protons, α- and heavier particles, and also the comparatively new approach of total-reflection X-ray fluorescence are described in some recent books and compilations (e.g. refs. 296–298, cf. Chapter 4).

The application of classical X-ray fluorescence is, without preconcentration, rather limited for the analysis of cadmium, because of relatively high detection limits, and the fact that cadmium appears in many biological and environmental materials below this level. An example is the determination of Cd, Ag, As, Pb, Tl, Zn, Cu, Ni and Cr, using a sequential X-ray spectrometer in various materials. After drying, milling and disk preparation the metals mentioned above were determined in foods and feeds of animal and vegetable origin, soil extracts, sewage sludges, partly after decomposition, and barn floor coatings. Detection limits for cadmium varied from approx. 2 to approx. 5 mg/kg depending on the matrix and the pretreatment procedure [299,300].

Because of a very effective suppression of background radiation, total reflection X-ray fluorescence (TXRF) allows a remarkable improvement of detection power [297,298]. For the use of the potential of this method various procedures have been evaluated for liquid and solid samples.

In *liquids* after aliquotation and addition of an internal standard, which is frequently cobalt, several techniques were applied:

(i) direct measurement, after evaporation of an appropriate aliquot, on the sample carrier (which is often a quartz plate);

(ii) freeze drying, leaching of the residue with dilute nitric acid and then proceeding as described under (i);

(iii) freeze drying of the liquid and digestion of the material with concentrated nitric acid (under pressure or in open systems), followed by metal chelation or direct chelation from solution. Subsequently the chelate is adsorbed on a chromatographic column. Hereafter the metal chelates are eluted with small amounts of appropriate organic solvents [301].

In *solids*, appropriate subsamples can be either directly analyzed, e.g. air dust collected on the sample carrier, or prepared after digestion, dilution, addition of standard solution or even preconcentration as described above for liquids.

TXRF was applied, using the different pretreatment versions mentioned above, for the analysis of 26 elements, including cadmium in urban air dust [302], for the determination of up to 20 elements, including cadmium in rainwater with typical detection limits down to 5–20 ng/l for heavy metal traces [303], for up to 25 elements, including cadmium in seawater [304] and in various stages of the water cycle — rainwater, air borne particulates, riverwater, particulate matter, river sediment, seawater and mussel homogenate [305,306]. In the latter and in two further papers [307,308] also comparison with other analytical methods and values found in certified reference materials were reported. Using preconcentration techniques detection limits of 20 ng/l for cadmium in rainwater and approx. the same also for seawater were recorded.

In vivo X-ray fluorescence analysis of cadmium in the kidney cortex of man is performed using plane polarized photons for excitation as recently reviewed [309]. The polarized photons are produced by scattering the radiation from an X-ray tube in a polymethacrylate disc at a 90° angle. The minimum detectable concentration for a counting time of 1800 s and a skin-kidney distance of 30 mm is 8 mg/kg. The method was used to determine the cadmium concentration of six occupationally exposed persons and the results varied between 15 and 170 mg/kg [310].

8.5.7. Nuclear methods

Activation analysis, mainly with reactor neutrons, still remains a useful multielement approach for fingerprint studies in environmental and biological materials and as an independent method if analytical accuracy has to be confirmed [311,312]. As can be seen from Table 8.2, the detection power of instrumental neutron activation analysis (NAA) for cadmium is in the order of that of ICP–AES and flame AAS, however, it can be improved if radiochemical separations are applied as is shown below. For the development and control of analytical procedures and other e.g. biological and environmental distribution studies the use of radioactive isotopes of cadmium ([109]Cd and [115m]Cd with half-lives of 453 d and 44.8 d respectively, cf. Chapter 4) is often a considerable help [313,314].

Multielement analysis, including cadmium, of airborne dust and aerosols was performed by instrumental NAA in comparison to other techniques. The cadmium content in aerosols of cities was in the order of a few ng/m^3, whereas dust contained approx. one order of magnitude higher cadmium levels [315,316].

A sequence of instrumental methods was employed for the determination of 44 elements in marine bivalve tissue. The techniques applied were XRF, prompt gamma activation analysis (PGAA), and instrumental NAA with short and long irradiation time. Cadmium was determined with PGAA and long irradiation; contents ranged from 0.5 to 8 mg/kg [317].

Fast NAA with neutrons obtained by irradiation of a thick beryllium target with 14.5 MeV deuterons producing [11m]Cd was applied for the determination of cadmium in different reference materials. The formed [111m]Cd is separated by liquid–liquid extraction with zinc diethyldithiocarbamate in chloroform. For lower contents, cadmium is precipitated as cadmium ammonium phosphate after extraction. The cadmium contents of the analyzed materials varied from 3 to 500 mg/kg. The authors stated that the method though less sensitive than thermal neutron activation analysis, had the advantage of being much faster [318].

An analytical procedure has been developed for the determination of mercury and cadmium simultaneously by radiochemical NAA. After irradiation of subsamples of 0.2–0.5 g for 1–4 h and a cooling (decay) time of 3 days, the samples are decomposed after addition of appropriate mercury and cadmium carriers and holdback carriers. This is followed by solvent extraction with metal diethyldithiocarbamates. Nickel(II) diethyldithiocarbamate is used to isolate mercury and zinc compounds in chloroform, followed by a back extraction into 2 M HCl to isolate cadmium. The absolute detection limit is about 1 ng if care is taken to optimize irradiation, decay and counting parameters. The method was verified by analyzing certified reference materials [319].

A similar procedure was developed, adding cadmium carrier, [109]Cd for yield determination, and copper holdback carrier to the decomposed irradiated sample, followed by separation of cadmium by two extraction steps using antimony–diethyldithiocarbamate and a back-extraction into dilute hydrochloric acid. The method was applied to the determination of cadmium in serum and packed blood cells. In serum cadmium levels of < 0.2 μg/l were found [320].

Using an automated radiochemical group separation procedure multielement determinations, including cadmium, were performed on tissues of deceased smelter workers and of rural controls. Typical median values for cadmium in lung tissue of smelter workers and in rural controls were 166 μg/kg and 39 μg/kg, respectively [321].

The simple and rapid simultaneous determination of cadmium, indium and tin in reference materials has been carried out by sub-stoichiometric extraction of indium radionuclides ^{115m}In, ^{114m}In and ^{113m}In. The analytical results were in good agreement with the certified values. An absolute detection limit of 10 ng was reported for cadmium [321].

In vivo NAA is very useful for the assessment of cadmium accumulation in the kidney and liver of occupationally exposed persons as well as those of smokers and non-smokers. In this method the liver or kidney is irradiated with a shielded collimated neutron beam from e.g. a ^{238}Pu-Be neutron source. The isotope ^{113}Cd, with an abundance of 12.29%, reacts with slow neutrons to form ^{114}Cd which promptly decays to the ground state in less than 10^{-14} seconds. Detection limits are approx. 1.6 mg (absolute) for kidney and 1 mg/kg for liver [309,322,323].

8.5.8. Chromatography

Chromatographic methods combine a compound specific separation technique with a compound (or element) specific detection system. This approach is therefore also well suited for speciation. As metals can not directly be determined by chromatographic techniques they have to be transformed into appropriate metal chelates. Since the chelate used commonly reacts with several elements multielement determinations are possible for low metal contents. Some examples are given to demonstrate this.

Cadmium, nickel, lead, zinc, cobalt, copper and bismuth were determined at mg/kg levels in certified NIST reference materials Bovine Liver and Oyster Tissue. Subsamples of approx. 0.25 g were ashed in a muffle furnace, the ash extracted with hydrochloric acid, the metals transformed into their hexamethylenedithiocarbamates and extracted into chloroform. After separation on a reversed-phase column, determination of the metals was performed spectrophotometrically at 260 nm in a flow cell. All metals, except cobalt and bismuth, could be accurately determined in the range from 0.5–850 mg/kg with relative standard deviations of approx. 7% [324].

Several trace metals were extracted from aqueous solutions with di(trifluoroethyl)dithiocarbamic acid. This was performed by coupling of columns and improvement of the injection technique as well as the (flame ionization and electron-capture) detectors. This procedure allowed the simultaneous determination of chromium(VI), cobalt, copper, nickel, zinc and cadmium at the lower μg/kg or μg/l level [324].

References on p. 219

The behavior of bis(ethoxyethyl)dithiocarbamates of 14 metals, including cadmium was investigated. After preconcentration from rivers and lakes with an on-line system the chelates were separated by C_{18} reversed-phase liquid chromatography. A quaternary solvent mixture with admixture of a surfactant was used as eluent. UV-detection was performed at 254 nm. The detection limit was 5 ng for cadmium [325].

Further examples for the application of chromatographic techniques to cadmium speciation in various matrices will be given in Section 8.6.

For gas and liquid chromatography the metals have to be determined as chelates or similar organometallic compounds. Ion chromatography (IC), however, initially nearly exclusively used for the analysis of inorganic anions, is now also successfully applied for the direct separation of inorganic cations [326,327].

The relatively simple preconcentration and separation of trace metals from matrix elements, e.g. the ion chromatographic separation and determination of minute amounts of lead, copper, cadmium and cobalt at $\mu g/l$ level from high mg/l levels of nickel, is important to remove virtually any interfering metals from the element of interest [328].

Eleven heavy metals (iron, copper, lead, zinc, nickel, cobalt, cadmium, manganese, calcium, magnesium, strontium and barium) could be separated on a 200 mm long strong acid silica-based cation-exchanger column, in 24 min. Detection using a post-column derivatization technique for Pb, Cd, Mn etc. is performed by a PAR C4-(2-pyridylazo)resorcinol-Zn-EDTA solution which enhances sensitivity for these metals with detection limits ≤ 2 $\mu g/l$ [329].

Recently cadmium, cobalt, copper, iron, manganese, nickel, lead and zinc were separated/preconcentrated from complex matrices by pumping the analyte solution (e.g. 10 ml) through a chelation resin column, which selectively retains transition metals but not alkali or alkaline earth metals or common anions. The transition metals concentrated on the column are then automatically eluted and determined, e.g. by ICP–AES. Detection limits can be enhanced by a factor of 100 compared to direct nebulization, i.e. to approx. 0.1 $\mu g/l$ for cadmium and most other metals [330].

8.5.9. Mass spectrometry

Cadmium can be determined by classical and modern versions of mass spectrometry (MS) with detection limits down to the pg, in some cases even the fg level [331,332].

Since cadmium has eight naturally occuring stable isotopes it is accessible to the extremely sensitive, accurate and precise isotope dilution technique that could significantly contribute to the determination of a number of environmentally and toxicologically important metals and also to the correction of erroneously high data [333,334].

Iron, cadmium, zinc, copper, nickel, lead and uranium have been determined in seawater after preconcentration on Chelex-100 by stable isotope dilution spark source MS. The results are compared with those obtained by GFAAS and ICP–AES. The values found for cadmium in seawater were significantly below the 1 $\mu g/l$ level [335].

Electrodeposition on an amalgamated gold wire was used to preconcentrate cadmium from water of Lake Michigan for subsequent determination by isotope dilution MS (IDMS) or GFAAS. It was reported that the overall blank was in the order of 3 ng/l and the instrumental limit of detection < 0.1 ng/l. The mass spectrometer used was a thermal-ionization instrument. Cadmium concentrations measured in samples from 1978 ranged from 31 to 41 ng/l [336].

Seawater samples from the Indian Ocean 20 km west of Fremantle, Australia, were analyzed for silver, cadmium, lead, zinc and palladium by thermal-ionization IDMS. Preconcentration of lead and cadmium from water samples, to which appropriate spikes were previously added, was performed on the Chelex-100 resin. The cadmium content found was around 2 ng/l with similarly low values for the other elements and the authors stated that these levels were "lower than some reported previously" and that palladium in seawater was reported for the first time [337].

The same IDMS technique was applied to coastal seawater after preconcentration by solvent extraction of dissolved copper, cadmium, lead, zinc, nickel, iron and chromium. For the total concentration a rigorous aqua regia digestion of seawater was performed. The values found for total and dissolved cadmium did not vary significantly and were around 70–80 ng/l [338].

An elaborate and scrupulous procedure in a clean laboratory for the determination of cadmium in whole blood by IDMS was published and the authors mentioned that "the method is applicable to samples with cadmium content ranging from less than 1 ng/g up to 100 mg/g (10%) or more. The method used 2-g samples that are decomposed with a mixture of acids (nitric, hydrochloric and perchloric acid), purified by subboiling distillation and an anion exchanger for separation from interfering elements. The authors reported that the analytical blank for cadmium can be reduced to less than 1 ng and that total system blanks of 100 ± 50 pg were achieved. Five analysis of one blood sample were performed and an average mass fraction of 7.05 ng cadmium/g with a coefficient of variation of 0.28% was obtained [339].

With a newly introduced compact thermal ionization quadrupole instrument IDMS measurements were performed after electrolytical separation/preconcentration. Lead, cadmium, thallium, copper and zinc were determined in sewage sludges, soils, a sediment, a phosphate rock and a city waste incineration ash. The detection limit for cadmium was reported to be 6 μg/kg [340].

The same instrument and method was used for the determination of lead, cadmium, copper, zinc and iron in a number of certified food reference materials. Excellent agreement with the certified values, ranging from 0.013 to 7.51 mg/kg was found. The authors reported a detection limit of 0.8 μg/kg for cadmium based on a sample weight of 1 g and 3 s of the blank of the whole procedure [341].

Using the same analytical approach for lead, cadmium, copper, zinc and chromium in three sedimentary reference materials the total content and the aqua regia soluble portion were determined. For cadmium only in one material a measurable difference was found. Concentrations ranged from 0.43 to 11.5 mg/kg [342].

References on p. 219

The IDMS method with thermal ionization was also applied to the in Section 5.5 mentioned determination of heavy metals in Antarctic snow. Sample preparation was performed by addition of thallium, copper, cadmium and zinc isotopic spikes to 200–300 ml of the molten snow samples. Hereafter, the solutions were concentrated to about 10 ml by evaporation under rigid contamination control. After this concentration process the metals were cathodically deposited on Pt electrodes and after dissolution used for the IDMS measurement. Cadmium blanks at 1 ng/l concentrations amounted to approx. 16% of this value, if a subsample of 250 ml was used. The detection limit was found to be 0.2 ng/l [276].

8.5.10. Spectrophotometry and related techniques

The total elemental analysis for cadmium by UV/VIS spectrophotometry has now been superseded by more sensitive and selective quantification procedures such as AAS, ICP–AES, voltammetry etc. Therefore spectrophotometry is applied with decreasing frequency in conventional analysis and for the control of occupational exposure. However, in official analytical procedures spectrophotometric reference methods are still prescribed, e.g. for the analysis of cadmium in food after wet digestion of samples of 5–10 g and selective extraction after some purification steps with dithizone [343].

A list of chromophoric reagents that have been exploited for the spectrophotometric determination of cadmium together with wavelengths of maximum absorption and corresponding molar absorptivities are listed in a recent review and literature compilation of toxicology and analysis of cadmium [92].

There are, however also extremely sensitive reagents that can be used for e.g. determination of metals after selective preconcentration with HPLC and IC. Recently the use of a new and highly sensitive reagent $\alpha,\beta,\gamma,\delta$-Tetrakis (4-N-trimethylaminophenyl)porphine was studied. It forms a chelate with cadmium in alkaline solution with the very high molar absorptivity of $5.77 \cdot 10^5$ l•mol^{-1}•cm^{-1}. The calibration curve was linear for up to 9 μg of cadmium. By use of a spectrophotometer with an expanded full-scale absorbance range of 0.1 a cadmium concentration of 10 μg/l can be directly determined [344].

Another sensitive and rapid spectrophotometric method, based on the formation of a blue complex at pH 4 between the anionic iodide complex of cadmium(II) and Malachite Green (molar absorptivity $61 \cdot 10^4$ l•mol^{-1}•cm^{-1}), was adapted for flow injection analysis (FIA). The peak height was proportional to the cadmium concentration over the range 0.1–3 mg/l [345].

8.6. SPECIATION

In seawater, cadmium is calculated to exist for 92% as the $CdCl^+$ and $CdCl_2^0$ complexes, while in riverwater, the dominant inorganic forms are Cd^{2+} and $CdCO_3$, depending on pH. A high proportion (> 70%) of cadmium is ASV (anodic stripping voltammetry) labile in both seawater and fresh waters. Because cadmium ions are adsorbed on colloidal particles at only relatively high pH values, very little cadmium is present as pseudocolloids. In anoxic waters, cadmium may exist as non-labile CdHS [273]. Speciation studies for cadmium in liquids and solids have to consider the differentiation between the liquid phase — ionic species at different valency states, dissolved complexes — and the metal adsorbed on colloidal particles and on distinct solid phases in soils as well as sediments determining bioavailability (see also Chapter 3). The metals bound to distinct protein fractions in biological matter have to be considered also. A few examples from this steadily growing field will be given subsequently.

The speciation of Cu(II), Cd(II) and Zn(II) in natural water samples was determined rapidly with an automated two-column ion-exchange system. Two fractions of dissolved trace metal species are directly determined by on-line flame AAS after preconcentration on sequential columns of Chelex-100 chelating resin and an AG MP-1 macroporous anion resin. A third fraction was determined by standard addition with a 10 ml sample loop, 6 samples were analyzed per hour with a detection limit of 0.08 μg/l for Cd(II) [346].

Organically-associated trace metals (Ni, Cu, Mo, Mn, Cd and Pb) were determined in estuarine seawater by chloroform extraction at pH 8 and pH 3. While nickel, copper, molybdenum and manganese where more or less extracted, this was not the case with cadmium and lead, i.e. these metals were not associated with the dissolved organic matter (DOM) [347].

A flow-injection/Donnan dialysis/DPASV system was developed for the determination of free cadmium concentrations [Cd^{2+}] in soil solutions containing organically complexed Cd(II). In solutions containing 1.0 μM Cd(II) and 200 mg fulvic acid/l, the inorganic fraction (Cd^{2+} + $CdNO_3^+$) decreased from 57% to 10% when the pH increased from 4.04 to 5.51. In a soil solution from an orthic podzol, having a high concentration of dissolved organic carbon (22.4 mM), the inorganic fraction contained 53% of the total cadmium(II) concentration [348].

An analytical procedure was developed involving sequential chemical extractions for the partitioning of particulate trace metals (Cd, Co, Cu, Ni, Pb, Zn, Fe and Mn) into five fractions exchangeably bound to carbonates, to Fe-Mn oxides, to organic matter and residual. The accuracy of this approach was verified and the scheme applied to river sediments. The results indicated no detectable cadmium contents in fractions 1 and 2 and low levels in fractions 3 and 4. Fraction 5 (residual metals) contained more than 50% of the total metal concentration [349].

The same extraction procedure was applied to anthropogenically polluted sediments in two rivers which show a large difference in the partitioning range of heavy metals (Pb, Cd, Cu, Fe, Mn, Zn and Cr) in the fractions, indicating a different origin and transport media of the metals [350]. In another paper problems due to some operationally defined fraction categories are outlined and discussed [351].

Ion-exchange resins of different types (held in porous cages) were equilibrated with aqueous suspensions of the sediments studied for Cu, Pb, Zn and Cd. It was reported that this technique provides an alternative means to the sequential extraction scheme of subdividing the metal content of sediments into different "labile" or "available" fractions, and that the transfer mechanism may resemble the action of plant roots more closely than chemical extractant processes do [352].

A sequential extraction scheme, acidification to pH 4.2 and 0.5 and repetitive $CaCl_2$ (0.005 M) extractions, was used to evaluate the forms of cadmium, copper, nickel, lead and zinc in seven sewage sludges. There was more variation in the fractionation profiles of different metals in the same sludge than of the same metal in different sludges. The largest cadmium and nickel fractions were extractable in 0.1 M sodium EDTA (pH 6.5) corresponding to the carbonates. Progressive acidification of liquid sludges mobilized significant quantities of zinc at pH 4 and of cadmium and lead at pH 2 [353].

A procedure was developed for the determination of cadmium species in solid waste leachates employing a cation-exchange resin (Chelex-100) in a batch-column-batch sequence. The procedure allowed for determination of free divalent cadmium complexes, labile cadmium complexes, slowly labile cadmium complexes and stable cadmium complexes [354]. The method was applied to ten leachate samples representing five different types of solid waste: refuse compost, fly ash from coal combustion, sewage sludge, refuse incineration residues and landfill municipal waste. In all leachates, the labile and slowly labile cadmium complexes were predominant, accounting for 69-100% of the leachate cadmium concentrations. Amounts of free divalent cadmium and stable complexes varied highly, < 0.1-33% and < 0.2-25%, respectively and both did not exceed the 10% level in most cases. The authors mentioned that the obtained results stress the importance of considering cadmium species and not only total cadmium concentrations, when the fate of cadmium in the environment is studied [355].

Various polluted soils were selectively extracted by water, $BaCl_2$ solution, acetate buffer and EDTA and nitric acid to investigate the distribution of cadmium, zinc, copper and lead. The extraction solutions were directly injected into an ICP-AES instrument. This permitted modifications of the extraction procedures for each investigated sample. The authors found good agreement of the sum of the elemental contents in the fractions with the total metal content [356].

Cadmium speciation in biological systems is performed by coupling of chromatography with appropriate detection methods [357]. A few recent examples are mentioned below.

A two-stage procedure was applied comprising simulated gastric and intestinal juices. All of the cadmium in a simulated gastric digest of canned crab meat was associated with soluble low-molecular-weight species of less than 1000 daltons. Cadmium solubility resulted from the acidic conditions of the digest rather than enzymatic solubilization of binding proteins. In the digest 90% of the cadmium became rebound to the insoluble fraction on adjustment to pH 7. Subsequent simulated intestinal digestion increased the amount of soluble cadmium to 20%; 1-1.5% was associated with soluble species in the range of 26 000 to 37 000 daltons and 15-20% with soluble species of less than 3000 daltons [358].

Directly coupled flame AAS-HPLC was used for the determination of metals associated with various proteins. The potential application of this technique to clinical analysis was demonstrated by determinations of zinc, copper and cadmium in various proteins and zinc in blood plasma [359].

In leaves from unpolluted rape of two different growth stages the cytoplasmic fraction, containing 37% of the whole cadmium, was subjected to gel filtration. This revealed cadmium species of high (MW > 80 000 g/mol) and low molecular weight (MW = 4400 g/mol). No free ionic species have been detected. Analysis was performed by GFAAS [360].

Using size-exclusion chromatography (SEC) coupled directly to an ICP–MS system speciation of cadmium in retail pig kidney has been studied. Approx. 35% of the cadmium from uncooked kidney could be extracted with water at pH 8. SEC–ICP–MS revealed three peaks with retention times corresponding to relative molecular masses of approx. $1.2 \cdot 10^6$, $7 \cdot 10^4$ and $6 \cdot 10^3 - 9 \cdot 10^3$. In the cooked kidney, 35% of the cadmium was soluble and was associated with a peak of a relative molecular mass (M_r) of $6 \cdot 10^3 - 9 \cdot 10^3$. Following simulated gastric and intestinal digestion, it was observed that the majority of soluble cadmium in retail pig kidney was associated with a protein similar to metallothionein that survived both cooking and simulated *in vitro* gastro-intestinal digestion [361].

A model gut system was used to perform studies of cadmium bioavailability in crab meat. The use of a biological membrane indicates that although soluble cadmium forms may exist in the gut, only protein-complexed forms will pass through the membrane. Chromatographic separation by HPLC and GFAAS determination indicated that though cadmium-metallothioneine type complexes are present at the pH values of both the stomach and the intestine, their solubility is significantly reduced in the intestine [362].

Analysis of local distribution patterns is also important in speciation studies. This can be performed in various different specimens by solid sampling AAS (cf. Section 8.5.2.2.3) and various beam techniques [363]. For those studies in biological materials laser micro-probe mass analyzers (LAMMA combination of an optical microscope with a laser source and a time-of-flight mass spectrometer) are excellently suited as was shown for studies on the localization of cadmium in renal cortex of experimental animals [364].

8.7. QUALITY CONTROL AND REFERENCE MATERIALS

From interlaboratory comparisons regarding the determination of trace metals in biological and environmental materials it is evident that despite of frequently achieved acceptable precision, many of the participating laboratories were not in a position to produce sufficiently accurate results for trace elements [365]. This was, also and still is, the case for cadmium. Hence current quality control from sampling to analytical determination is of paramount importance in an analytical laboratory to maintain or to arrive at meaningful data. This can be achieved via different, commonly simultaneously applied approaches: scrupulous contamination control for sampling stages, use of different and reference methods in intro- and interlaboratory comparisons, and the use of appropriate —if available certified— reference materials.

Major advances in methodology (IDMS, GFAAS, DPASV) have made possible the reliable determination of trace elements in seawater down to the ng/l level and even lower

TABLE 8.3.

REFERENCE MATERIALS WITH CERTIFIED RESPECTIVELY INFORMATION OR SIMI-
LAR VALUES FOR CADMIUM IN DECREASING ORDER OF THE ELEMENTAL
CONCENTRATION

Concentrations are given on a dry weight basis and in mg/kg unless noted "liquid", than
mg/l.

Material	Code	Conc.	Error (confidence interval) (%)	
Biological materials, certified				
Lobster Hepatopancreas	NRCC-TORT-1	26.3	8.0	
Dogfisch Liver	NRCC-DOLT-1	4.18	6.7	
Oyster Tissue	NIST-SERM-1566	3.5	11	
Pig Kidney	BCR-CRM-186	2.71	5.5	
Aquatic Plant	BCR-CRM-060	2.20	4.5	
Rice Flour	NIES-CRM-10c	1.82	3.3	
Aquatic Plant	BCR-CRM-061	1.07	7.5	
Copepoda	IAEA-MA-A-1/TM	0.75	4.0	
Bovine Liver	CZIM-LIVER	0.48	6.3	
Bovine Liver	NIST-SRM-1577a	0.44	14	
Mussel Tissue	BCR-CRM-278	0.34	5.9	
Rice Flour	NIES-CRM-10B	0.32	6.3	
Sea Lettuce	BCR-CRM-279	0.274	8.0	
Fish	EPA-FISH	0.16	50	
Sargasso	NIES-CRM-09	0.15	13	
Skim Milk powder	BCR-CRM-151	0.101	7.9	
Olive Leaves	BCR-CRM-O62	0.10	20	
Human Hair	SHINR-HH	0.095	12	
Urine (Spiked)	NIST-SRM-2670	0.088	3.4	liquid
Dogfish Muscle	NRCC-CORM-1	0.086	14	
Wholemeal Flour	BCR-CRM-189	0.0713	4.2	
Fish Flesh	IAEA-MA-A-2/TM	0.066	6.1	
Mixed Diet	NIST-RM-8431a	0.042	26	
Wheat Flour	ARC/CL-WP	0.039	10.3	
Potato Powder	ARC/CL-PP	0.035	5.7	
Single Cell Protein	BCR-CRM-274	0.030	6.7	
Tea Leaves	NIES-CRM-07	0.030	10	
Hay Powder	IAEA-V-10	0.03	50	
Urine	KL-140-II	0.03	20	liquid
Citrus Leaves	NIST-SRM-1572	0.03	33	
Rice Flour	NIST-SRM-1568	0.029	14	
Brown Bread	BCR-CRM-191	0.0284	4.9	
Wheat Flour	NIST-SRM-1567a	0.026	7.7	
Rice Flour	NIES-CRM-10A	0.023	13	
Animal Muscle (pork)	ARC/CL-AM	0.022	18.6	

TABLE 8.3. (continued)

Material	Code	Conc.	Error (confidence interval) (%)	
Skim Milk powder	BCR-CRM-150	0.0218	6.4	
Total Diet	ARC/CL-TD	0.021	14.3	
Bovine Muscle	BCR-CRM-184	0.013	15	
Bovine Blood	BCR-CRM-196	0.0124	4.0	liquid
Urine	KL-140-I	0.01	25	liquid
Bovine Blood	BCR-CRM-195	0.00537	4.5	liquid
Skim Milk Powder	BCR-CRM-063	0.0029	41	
Human Serum	GHENT SERUM	0.0020	20	
Milk Powder	IAEA-A-11	0.0017	11	
Bovine Blood	BCR-CRM-194	0.0005	20	liquid
Milk Powder	NIST-SRM-1539	0.0005	40	
Biological materials, values only informative				
Milk Powder	ARC/CL-MP	< 2		
Kale	Bowen's Kale	0.899	28	
Pine Needles	NIST-SRM-1575	< 0.5		
Spruce Shoots	ESB-RM-F2	0.37		
Chlorella	NIES-CRM-03	0.026		
Spruce Shoots	ESB-RM-F1	0.020		
Whole Blood	NYCO-906	0.019		liquid
Rye Flour	IAEA-V-8	0.017		
Whole Blood	NYCO-905	0.0124		liquid
Urine	NYCO-108	0.0062		liquid
Whole Blood	NYCO-904	0.0056		liquid
Cotton Cellulose	IAEA-V-9	0.002		
Urine (Normal)	NIST-SRM-2670	0.0004		
Environmental materials, certified				
Incinerated Sludge	FISHER-SRS012	3582	1.9	
City Waste	BCR-CRM-176	470	1.9	
Copper Plant Flue Dust	IRANT-KHK	442	6.3	
Sewage Sludge	BCR-CRM-146	77.7	3.3	
Urban Particulates	NIST-SRM-1648	75	9.3	
Steel Plant Flue Dust	IRANT-OK	37.9	5.3	
Soil	BCR-CRM-143	31.1	3.9	
Municipal Sludge	EPA-SLUDGE	19.1	23	
Sewage Sludge	BCR-CRM-145	18.0	6.7	
Estuarine Sediment	BCR-CRM-277	11.9	3.4	
Sewage Sludge	BCR-CRM-144	4.82	20	
Fly Ash	BCR-CRM-038	4.6	6.5	
River Sediment	NIST-SRM-2704	3.45	6.4	
Diatomaceous Filter	FISHER-SRS004	3.1	29	
Coal Fly Ash	IRANT-ECH	3.06	36	

(Continued on p. 216)

TABLE 8.3. (continued)

Material	Code	Conc.	Error (confidence interval) (%)	
Coal Fly Ash	IRANT-EOP	2.94	51	
Marine Sediment	NRCC-PACS	2.38	8.4	
Coal Fly Ash	IRANT-ECO	1.99	60	
Lake Sediment	BCR-CRM-280	1.6	6.3	
Vehicle Exhaust	NIES-CRM-08	1.1	9.1	
Coal Fly Ash	NIST-SRM-1633a	1.0	15	
Pond Sediment	NIES-CRM-02	0.82	7.3	
Marine Sediment	NRCC-MESS-1	0.59	16	
River Sediment	BCR-CRM-320	0.533	4.9	
Soil	BCR-CRM-141	0.36	28	
Estuarine Sediment	NIST-SRM-1646	0.3	26	
Lake Sediment	IAEA-SL-1	0.26	19	
Soil	BCR-CRM-142	0.25	36	
Marine Sediment	NRCC-BCSS-1	0.25	16	
Gas Coal	BCR-CRM-180	0.212	5.2	
Coal	BCR-CRM-040	0.11	18	
Coal (Bituminous)	NIST-SRM-1632b	0.0573	4.7	
Steam Coal	BCR-CRM-182	0.057	7.0	
Coking Coal	BCR-CRM-181	0.051	5.9	
Coal	NIST-SRM-1635	0.03	33	
Water	NIST-SRM-1643b	0.02	5	liquid
Sea Water	NRCC-NASS-2	0.000029	13	liquid
Sea Water	NRCC-CASS-2	0.000019	21	liquid
Estuarine Water	NRCC-SLEW-1	0.000018	10	liquid
Environmental materials, values				
Coal Fly Ash	ICHTJ-CTA-FFA-1	2.8		
Soil	IAEA-SOIL-7	1.3		

in the late 70's. This was linked with a revolution in the knowledge of the distribution of these elements, particularly of metals like cadmium, lead, copper, nickel, cobalt and mercury in the oceans and therefore a first-order understanding of the geochemical basis of this distribution.

In order to arrive at comparable data between laboratories working in the field of marine analytical and geochemical chemistry the performance of intercalibrations of techniques for measuring trace metals in marine samples was increasingly performed on a national as well as an international basis. A series of intercalibrations for a number of elements, including cadmium, in seawater and estuarine water was undertaken by the International Council for the Exploration of the Sea (ICES) (e.g. refs. 366–369) and by

TABLE 8.4.

ADRESSES AND REFERENCES OF SUPPLIERS FOR THE CODES GIVEN IN TABLE 8.3

Most of these RMs can be purchased either directly from the suppliers or from Pro-mochem, P.O. Box 1246, D-W-4230 Wesel, Federal Republic of Germany

ARC/CL	Kumpulainen and Paaki [374]
BCR	Community Bureau of Reference, Belgium
BOWEN'S KALE	Bowen [387]
CZIM-LIVER	Kucera et al. [388]
ESB-RM-F	Environmental Specimen Bank, KFA Juelich, P.O. Box 1913, D-W-5170 Juelich, Germany, freshly homogenized and cryogenically stored material
FISHER	Fisher Scientific, Springfield, NJ, USA
GHENT	Versieck et al. [382]
IAEA	International Atomic Energy Agency, P.O. Box 100, A-1400 Vienna, Austria
ICHTJ-CTA	Commission of Trace Analysis of the Committee for Analytical Chemistry of the Polish Academy of Sciences and Institute of Nuclear Chemistry and Technology, Warsaw, Poland
IRANT	Institute of Radioecology and Applied Nuclear Techniques, Kosice
KL	Kaulson Laboratories Inc, West Caldwell, NJ 07006, USA
NIES	National Institute for Environmental Studies, Tsukuba, Yatabe-machi, Ibaraki, 305 Japan
NIST	National Institute of Science and Technology (formerly National Bureau of Standards, NBS) Washington, DC, USA
NRCC	National Research Council, Canada, Ottawa
NYCO	Nycomed Pharma, Oslo, Norway
SHINR	Shanghai Institute of Nuclear Research, Academia Sinica, Shanghai, People's Republic of China

others [370] with, usually, a distinct scatter of results but also indicating some progress in accuracy over time. Examples of the internal use of different methodological approaches for the analysis of the same sample, are: the use of DPASV and AAS for Baltic Sea Water [280], the already amply discussed determination of cadmium and other elements in antarctic snow samples [276], and analytical approaches for environmental specimen banking [46,173] which is also important to achieve more reliable results.

The intercomparison of low-level ocean waters in independent laboratories in this field and the excellent agreement even at lowest natural levels was a further proof of the reliability that had already been achieved at these elemental levels in some laboratories with appropriate expertise [278,282].

Reliability in marine analytical chemistry was further significantly improved by the intro-duction of certified seawater reference materials with realistic trace metal levels [ref. 371, see Table 8.3].

References on p. 219

Similar intercomparisons were performed in marine plants and tissues (e.g. refs. 372,373), for trace elements in foods (e.g. refs. 374,375) and for a number of human biological materials for current improvement and quality assessment of routine analytical procedures in biological monitoring (e.g. refs. 376–379) but also for the evaluation of standardized methods as performed by the Cadmium Subcommittee of IUPAC Commission on Toxicology for cadmium in whole blood and urine [141,142].

Most of these intercomparisons were supported and improved by certified, candidate and internal reference or control materials. Progress in these studies had often not been possible without the use of appropriate reference materials [380].

Cadmium was certified or characterized in a broad range of reference materials and some previous data were improved by more precise and sensitive methods, particularly recently for lower levels in animal muscle for eleven difficult elements including cadmium [381]. Further progress in this field is to note if matrix identical or 2nd generation reference materials are concerned. These materials are either prepared under scrupulously controlled clean room conditions, as e.g. human serum reference materials in order to maintain even lowest normal trace element levels [382], or prepared as matrix identical materials for specimen banking or general quality control purposes and subsequently stored fresh at extremely low (cryogenic) temperatures [383–385].

Table 8.3 summarizes, based on a recent compilation [386], reference materials at present (mainly commercially) available from different suppliers with either certified or informative values for cadmium. Table 8.4 lists addresses and references of suppliers.

8.8 CONCLUSIONS AND PROSPECTS

In this chapter it was shown by a number of selected examples from the recent literature that the instrumental analytical methods commercially available at present are excellent tools for the analysis of cadmium in environmental and biological materials. Especially the graphite furnace (for solid materials, some liquids and biological fluids), voltammetry (for aqueous matrices but also for solids), inductively coupled plasma mass spectrometry and for elevated levels inductively coupled plasma atomic emission spectrometry are well suited for routine applications. Since also total reflection X-ray fluorescence, mass spectrometry and in some cases instrumental or radiochemical activation analysis provide an appreciable potential for nearly every task at least one additional reference method is available.

Preconcentration and selective preconcentration/separation approaches such as the various solvent and column chelate extraction and ion chromatographic methods, recently also coupled with flow-injection, have further enhanced the potential of the graphite furnace and ICP–MS which are extremely prone to element or anion interferences (e.g. in seawater).

Further improvements of the electrochemical approach in terms of reliability, speed and detection power are the introduction of high-pressure or, more correctly, high-temperature ashing voltammetry, the square wave mode, and recent remarkable progress in computerized potentiometric stripping with a detection limit for cadmium of about 1 ng/l [389].

As far as validation of methods is concerned, the impressive list of biological and environmental reference materials, already available with certified or informative cadmium values from the lowest to the highest natural contents in a broad range of matrices, is an excellent prerequisite for further work [385]. Moreover, due to the efforts of many laboratories and suppliers around the world it is to be expected that some still existing gaps in materials and/or contents will be bridged in the near future, especially by the introduction of more second generation reference materials completely identical with natural materials. Since these materials can be prepared and stored at liquid nitrogen vapor temperatures (i.e. $< -150°C$) practically indefinitely long shelf lives can be achieved. Materials of this kind were recently introduced for quality assessment and trend recognition in specimen banking [383–385], which is amply discussed in Chapter 3.

Despite the fact that some of the methods presently available reach absolute detection limits at the pg level or even lower, it can be foreseen that the need for extremely small samples, investigation of tiny surface areas, as well as speciation in combination with chromatographic techniques will demand even lower routinely attainable detection limits. As this will certainly be in combination with simultaneous or sequential fingerprint analysis for a probably increasing number of essential and non-essential elements the further growth and improvement of multielement techniques, like total reflection and the still more sensitive Synchrotron excited X-ray modes and multielement mass spectrometric techniques making use of laser excitation, can be expected [390]. Of the latter techniques certainly atomic absorption and atomic fluorescence might benefit from further increase of detection power and reliability even at femtogram levels [391,392]. There might also be some potential for further development, possibly including the multielement analysis of solid samples, in furnace-atomization plasma emission coupling [393,394].

Finally furnace atomic non-thermal excitation spectrometry (FANES) a method currently being evaluated in detail should be mentioned as it might also be a useful technique in the near future for multielement analysis including cadmium at detection limits comparable to that of graphite furnace AAS [395].

REFERENCES

1 J.O. Nriagu (Editor), *Cadmium in the Environment, Part I,* Wiley, New York, 1980.
2 K.H. Schulte-Schrepping and M. Piscator, in: *Ullmanns Encyclopädie der technischen Chemie, Vol. A4,* VCH, Weinheim, Deerfield Beach, FL, Basel, 4th ed., 1985, pp. 499–514.
3 H. Mislin and O. Ravera (Editors), *Cadmium in the Environment,* Birkhäuser, Basel, Boston, Stuttgart, 1986.
4 J.O. Nriagu and J.B. Sprague (Editors), *Cadmium in the Aquatic Environment,* Wiley, Somerset, NJ, 1987.
5 M. Stoeppler, in E. Merian (Editor), *Metals and Their Compounds in the Environment,* VCH, Weinheim, New York, Basel, Cambridge, 1991, pp. 803–851.
6 K. Tsuchiya, *Cadmium Studies in Japan — A Review,* Kodansha, Tokyo, North Holland Biomedical Press, Amsterdam, New York, Oxford, 1978.
7 J.H. Mennear, *Cadmium Toxicity,* Marcel Dekker, New York, 1979.

8 M. Piscator, *Environ. Health Perspect.*, 40 (1981) 107–120.
9 L. Alessio, P. Odone, G. Bertelli, V. Foà, in L. Alessio, A. Berlin, R. Roi, M. Boni (Editors), *Human Biological Monitoring of Industrial Chemical Series*, EUR 8476 EN, JRC Ispra, 1983, pp. 25–44.
10 L. Friberg, M. Piscator, G.F. Nordberg and T. Kjellström, *Cadmium in the Environment*, CRC-Press, Cleveland, OH, 2nd ed., 1984.
11 L. Friberg, G.-C. Elinder, T. Kjellström and G.F. Nordberg (Editors), *Cadmium and Health, A Toxicological and Epidemiological Appraisal, Volume I. Exposure, Dose and Metbolism*, CRC Press, Boca Raton, FL, 1985.
12 L. Friberg, G.-C. Elinder, T. Kjellström and G.F. Nordberg (Editors), *Cadmium and Health: A Toxicological and Epidemiological Appraisal, Volume IV. Effects and Response*, CRC Press, Boca Raton, FL, 1986.
13 G. Kazantzis, *J. Toxicol. Environ. Chem.*, 15 (1987) 83–100.
14 U. Ewers, A. Brockhaus, J. Freier, E. Jermann and R. Dolgner, in O. Hutzinger and S.H. Safe (Editors), *Environemntal Toxins, Vol. 2, Cadmium* (M. Stoeppler and M. Piscator, Vol. eds.), Springer, Berlin, Heidelberg, New York, 1988, pp. 92–113.
15 C.R. Williams and R.M. Harrison, in H. Mislin and O. Ravera (Editors), *Cadmium in the Environment*, Birkhäuser, Basel, Boston, Stuttgart, 1986, pp. 17–23.
16 D. Jost, *Staub-Reinhalt Luft*, 44 (1984) 137–138.
17 V.D. Nguyen, A.G.A. Merks and P. Valenta, *Sci. Total Environ.*, 99 (1990) 77–91.
18 L. Mart, *Tellus*, 53B (1983) 131–141.
19 L. Mart, H.W. Nürnberg and H. Rützel, *Sci. Total Environ.*, 44 (1985) 25–49.
20 R. Breder, in: O. Hutzinger and S.H. Safe (Editors), *Environmental Toxins, Vol. 2, Cadmium* (M. Stoeppler and M. Piscator, Vol. eds.), Springer, Berlin, Heidelberg, New York, 1988, pp. 159–169.
21 L. Huynh-Ngoc, N.E. Whitehead and B. Oregioni, *Wat. Res.*, 22 (1988) 571–576.
22 W.S. Dorten, F. Elbaz-Poulichet, L.R. Mart and J.-M. Martin, *Ambio*, 20 (1991) 2–6.
23 K.R. Sperling, *Vom Wasser*, 58 (1982) 113–142.
24 L. Mart, H. Rützel, P. Klahre, L. Sipos, U. Platzek, P. Valenta and H.W. Nürnberg, *Sci. Total Environ.*, 26 (1982) 1–17.
25 P. Valenta, E.K. Duursma, A.G.A. Merks, H. Rützel and H.W. Nürnberg, *Sci. Total Environ.*, 53 (1986) 41–76.
26 K.W. Bruland, in J.R. Riley and R. Chester (Editors), *Chemical Oceanography, Vol. 8*, Academic Press, London, 1983, Ch. 45.
27 L. Mart and H.W. Nürnberg, in H. Mislin and O. Ravera (Editors), *Cadmium in the Environment*, Birkhäuser, Basel, 1986, pp. 28–39.
28 P.A. Yates, *Sci. Total Environ.*, 72 (1988) 131–149.
29 H.J. Brumsack, *Naturwiss.*, 76 (1989) 99–106.
30 K. Kremling and C. Pohl, *Mar. Chem.*, 27 (1989) 43–60.
31 U. Förstner, in J.O. Nriagu (Editor), *Cadmium in the Environment, Part I*, Wiley, New York, 1980, pp. 306–363.
32 U. Förstner, in H. Mislin and O. Ravera (Editors), *Cadmium in the Environment*, Birkhäuser, Basel, 1986, pp. 40–46.
33 S. Noriki, N. Ishimori, K. Harada and S. Tsunogai, *Mar. Chem.*, 17 (1985) 75–89.
34 L. Brügmann, *Beitr. Meeresk.*, 55 (1986) 3–18.
35 D. Martincic, Z. Kwokal, M. Stoeppler and M. Branica, *Sci. Total Environ.*, 84 (1989) 135–147.
36 T.F. Pedersen, R.D. Waters and R.W. McDonald, *Sci. Total Environ.*, 79 (1989) 125–139.
37 *The International Mussel Watch*, National Academy of Sciences, Wash. D.C., 1980.
38 E.D. Goldberg, M. Koide, V. Hodge, A.R. Flegal and J. Martin, *Coastal Shelf Sci.*, 16 (1983) 69–93.
39 M. Stoeppler and H.W. Nürnberg, *Ecotox. Environ. Safety*, 3 (1979) 335–351.
40 M. Stoeppler, F. Backhaus, M. Burow, K. May and C. Mohl, in S.A. Wise, R. Zeisler, G.M. Goldstein (Editors), *Progress in Environmental Specimen Banking*, (NBS Spec. Pub., No. 740), Washington, DC, 1988, pp. 53–61.

41 *Data from current analysis of materials of the German Environmental Specimen Bank,* to be published.
42 J. Gzyl, *Sci. Total Environ.,* 96 (1990) 199–209.
43 J. Einax, K. Oswald and K. Danzer, *Fresenius' J. Anal. Chem.,* 336 (1990) 394–399.
44 J. Thornton, in M. Mislin and O. Ravera (Editors), *Cadmium in the Environment,* Birkhäuser, Basel, Boston, Stuttgart, 1985, pp. 7–12.
45 U. Osberghaus, *Comparative Study for Trace Metal in Nine Seawage Sludge Plants,* (1991) to be published.
46 M. Stoeppler, H.W. Dürbeck, J.D. Schladot and H.W. Nürnberg, in BMFT (Editor), *Umweltprobenbank, Bericht und Bewertung der Pilotphase,* Springer, Berlin, Heidelberg, New York, London, Paris, Tokyo, 1988, pp. 59–76.
47 E. Hahn, K. Hahn and M. Stoeppler, *J. Ornitol.,* 138 (1989) 303–309.
48 E. Hahn, P. Ostapczuk, M. Stoeppler and H. Ellenberg, *Ecol. Birds,* 11 (1989) 265–281.
49 R. Knutti, P. Bucher, M. Stengl, M. Stolz, J. Tremp, M. Ulrich and C. Schlatter, in: O. Hutzinger and S.H. Safe (Editors), *Environmental Toxins, Vol. 2, Cadmium* (M. Stoeppler and M. Piscator, Vol. eds.), Springer, Berlin, Heidelberg, New York, 1988, pp. 171–191.
50 T. Watanabe, H. Nakatsuka and M. Ikeda, *Sci. Total. Environ.,* 80 (1989) 175–184.
51 T. Watanabe, M. Kasahara, M. Nakatsuka and M. Ikeda, *Sci. Total Environ.,* 66 (1987) 29–37.
52 R. Zeisler, R.R. Greenberg and S.F. Stone, in S.A. Wise, R. Zeisler and G.M. Goldstein (Editors), *Progress in Environmental Specimen Banking,* (NBS Spec. Pub., No. 740), Washington, DC, 1988, pp. 82–90.
53 C.F. Boutron, *Fresenius' J. Anal. Chem.,* 337 (1990) 482–491.
54 M. Zief and R. Speights, *Ultrapurity-Methods and Techniques,* Marcel Dekker, New York, 1972.
55 P.D. LaFleur (Editor), *Accuracy in Trace Analysis: Sampling, Sample Handling, Analysis (NBS Spec. Pub.* No. 422), Washington, DC, 1976.
56 G. Knapp, *Fresenius' Z. Anal. Chem.,* 317 (1984) 213–219.
57 M. Würfels, E. Jackwerth and M. Stoeppler, *Fresenius' Z. Anal. Chem.,* 329 (1987) 459–461.
58 J.D. Kinrade and J.C. Van Loon, *Anal. Chem.,* 46 (1974) 1894.
59 L.G. Danielsson, B. Magnusson and S. Westerlund, *Anal. Chim. Acta,* 98 (1978) 47–57.
60 K.W. Bruland, R.P. Franks, G.A. Knauer and J.H. Martin, *Anal. Chim. Acta,* 105 (1979) 233–245.
61 P.J. Statham, *Anal. Chim. Acta,* 169 (1985) 149–159.
62 R.F. Nolting, *Mar. Poll. Bull.,* 17 (1986) 113–117.
63 D. Chakraborti, F. Adams, W. van Mol and K.J. Irgolic, *Anal. Chim. Acta,* 196 (1987) 23–31.
64 S.C. Apte and A.M. Gunn, *Anal. Chim. Acta,* 193 (1987) 147–156.
65 E. Mentasi, A. Nicolotti, V. Porta and C. Sarzanini, *Analyst,* 114 (1989) 1113–1117.
66 M. Bengtsson and G. Johansson, *Anal. Chim. Acta,* 158 (1984) 147–156.
67 K. Bäckström and L.-G. Danielsson, *Mar. Chem.,* 29 (1990) 33–46.
68 A. Dornemann and H. Kleist, *Analyst,* 104 (1979) 1030–1036.
69 R. Heinrich and J. Angerer, *Fresenius' Z. Anal. Chem.,* 315 (1983) 528–533.
70 J. Angerer, in J. Angerer and K.H. Schaller (Editors), *Analyses of Hazardous Substances in Biological Materials, Vol. 2,* VCH, Weinheim, 1988, pp. 85–96.
71 C.J. Cheng, T. Akagi and H. Haraguchi, *Anal. Chim. Acta,* 198 (1987) 173–181.
72 F. Baffi and A. Cardinale, *Int. J. Environ. Anal. Chem.,* 41 (1990) 15–20.
73 R.E. Sturgeon, S.S. Berman, A. Desaulniers and D.S. Russell, *Talanta,* 27 (1980) 85–94.
74 G.R.W. Denton and C. Burdon-Jones, *Mar. Poll. Bull.,* 17 (1986) 96–98.
75 K. Vermeiren, C. Vandecasteele and R. Dams, *Analyst,* 115 (1990) 17–22.
76 S.C. Pai, T.-H. Fang, C.-T.A. Chen and K.-L. Jeng, *Mar. Chem.,* 29 (1990) 295–306.
77 V. Pravski, A. Corsini and S. Landsberger, *Talanta,* 36 (1989) 367–372.

78 K. Isshiki, F. Tsuji, T. Kuwamoto and E. Nakayama, *Anal. Chem.*, 59 (1987) 2491–2495.
79 A. van Geen and E. Boule, *Anal. Chem.*, 62 (1990) 1705–1709.
80 P. Burba and P.G. Willmer, *Fresenius' Z. Anal. Chem.*, 329 (1987) 539–545.
81 R.E. Sturgeon, S.S. Berman, S.N. Willie and J.A.H. Desaulniers, *Anal. Chem.*, 53 (1981) 2337–2340.
82 D. Beauchemin, J.W. McLaren, A.P. Mykytiuk and S.S. Berman, *J. Anal. At. Spectrom.*, 3 (1988) 305–308.
83 S. Nakashima, R.E. Sturgeon, S.N. Willie and S.S. Berman, *Fresenius' Z. Anal. Chem.*, 330 (1988) 592–595.
84 R. Eidecker and E. Jackwerth, *Fresenius' Z. Anal. Chem.*, 328 (1987) 469–474.
85 M. Hartmann, H. Lass and D. Puteanus, in B. Weltz (Editor), *Proc. 5th CAS*, Bodenseewerk, Perkin-Elmer, Überlingen, 1989, pp. 703–709.
86 J.M. Diaz, M. Caballero, J.A. Pérez-Bustamante and R. Cela, *Analyst*, 115 (1990) 1201–1205.
87 H. Matusiewicz, J. Fish and T. Malinski, *Anal. Chem.*, 59 (1987) 2264–2269.
88 M. Stoeppler, in A. Vercruysse (Editor), *Hazardous Metals in Human Toxicology, Techniques and Instrumentation in Analytical Chemistry - Vol. 4. Evaluation of Analytical Methods in Biological Systems*, Elsevier, 1989, pp. 199–223.
89 M. Stoeppler, in *Analytiker-Taschenbuch, Vol. 5*, Springer, Berlin, Heidelberg, New York, Tokyo, 1985, pp. 199–216.
90 M. Astruc, in M. Mislin and O. Ravera (Editors), *Cadmium in the Environment*, Birkhäuser, Basel, Boston, Stuttgart, 1986, pp. 12–17.
91 M. Stoeppler, in D. Wilson and R. Volpe (Editors), *Proc. 5th Int. Cadmium Conference*, Cadmium Assoc., London, Cadmium Council, New York, ILZRO, Research Triangle Park, 1988, pp. 149–152.
92 K. Robards and P. Worsfeld, *Analyst*, 116 (1991) 549–568.
93 M. Stoeppler, in E. Merian (Editor), *Metals and Their Compounds in the Environment*, VCH, Weinheim, New York, Basel, Cambridge, 1991, pp. 105–206.
94 L.H. Keith, W. Crummet, J. Deegan, Jr., R.A. Libby, J.K. Taylor and G. Wentler, *Anal. Chem.*, 55 (1983) 2210–2218.
95 W. Slavin, *Graphite Furnace AAS — A Source Book*, Perkin-Elmer, Norwalk, CT, 1984.
96 B. Welz, *Atomic Absorption Spectrometry*, Weinheim, Deerfield Beach, FL, Basel, 2nd ed., 1985.
97 J.E. Cantle (Editor), *Atomic Absorption Spectrometry — Techniques and Instrumentation in Analytical Chemistry, Volume 5*, Elsevier, Amsterdam, Oxford, New York, 1982.
98 B. Welz, *Atomic Absorption Spectroscopy*, Verlag Chemie, Weinheim, New York, 1976.
99 R.E. Ediger, *At. Absorpt. Newsl.*, 14 (1975) 127–130.
100 M. Stoeppler, K. Brandt, *Fresenius' Z. Anal. Chem.*, 300 (1980) 372–380.
101 E.J. Hinderberger, M.L. Kaiser and S.R. Koirtyohann, *At. Spectrosc.*, 2 (1981) 1–7.
102 D.C. Manning and W. Slavin, *Appl. Spectrosc.*, 37 (1983) 1–11.
103 G. Schlemmer and B. Welz, *Spectrochim. Acta*, 41B (1986) 1157–1165.
104 X. Yin, G. Schlemmer and B. Welz, *Anal. Chem.*, 59 (1987) 1462–1466.
105 B. Welz, G. Schlemmer and J.R. Mudakavi, *J. Anal. At. Spectrom.*, 3 (1988) 93–95.
106 B. Welz, G. Schlemmer and J.R. Mudakavi, *J. Anal. At. Spectrom.*, 3 (1988) 695–701.
107 D.L. Tsalev, T.A. Dimitrov and P.B. Mandjukov, *J. Anal. At. Spectrom.*, 5 (1990) 189–194.
108 K. Hilrich (Editor), *Official Methods of Analysis of the AOAC*, Association of Official Analytical Chemists, Arlington, VA, 15th ed., 1990, p. 241.
109 K. Hilrich (Editor), *Official Methods of Analysis of the AOAC*, Association of Official Analytical Chemists, Arlington, VA, 15th ed., 1990, pp. 241–242.
110 K. Hilrich (Editor), *Official Methods of Analysis of the AOAC*, Association of Official Analytical Chemists, Arlington, VA, 15th ed., 1990, pp. 247–248.

111 K. Hilrich (Editor), *Official Methods of Analysis of the AOAC,* Association of Official Analytical Chemists, Arlington, VA, 15th ed., 1990, pp. 324–325.

112 *Klärschlammverordnung* (Abf. Klär V) Bundesgesetzblatt 1, (1982) 734–739.

113 R. Breder, H.W. Nürnberg, J. Golimowski and M. Stoeppler, in: H.W. Nürnberg (Editor), *Pollutants and Their Ecological Significance,* Wiley, New York, 1985, pp. 205–225.

114 A. Frank, *Sci. Total Environ.,* 57 (1986) 57–65.

115 K. Honda, H. Ishihashi and R. Tatsukawa, *Arch. Environ. Contam. Toxicol.,* 16 (1987) 551–561.

116 A. Wadge and M. Hutton, *Sci. Total Environ.,* 67 (1987) 91–95.

117 R.J. Mahler, J.A. Ryan and T. Reed, *Sci. Total Environ.,* 67 (1987) 117–131.

118 A.A. Brown and A. Taylor, *Analyst,* 110 (1985) 579–582.

119 J. Moffet, *Spectroscopy,* 5/6 (1990) 41–44.

120 S.M. Fraser, A.M. Ure, M.C. Mitchell and T.S. West, *J. Anal. At. Spectrom.,* 1 (1986) 19–21.

121 A. Karakaya and A. Taylor, *J. Anal. At. Spectrom.,* 4 (1989) 261–263.

122 D.L. Tsalev, Li Chotjin and M.D. Ivanova, in B. Welz (Editor), *5th Coll. Atmospektrometr. Spurenanalytik,* Bodenseewerk Perkin-Elmer, Überlingen, 1989, pp. 465–474.

123 J.L. Burguera, *Flow Injection Atomic Spectroscopy,* Marcel Dekker, New York, Basel, 1989.

124 Z. Fang and B. Welz, *J. Anal. At. Spectrom.,* 4 (1989) 543–546.

125 K.-R. Sperling and B. Bahr, in B. Walz, *5th Colloquium Atmospektrometr. Spurenanalytik,* Bodenseewerk Perkin-Elmer, Überlingen, 1989, pp. 715–727.

126 E. Pruszkowska, G.R. Carnrick and W. Slavin, *Anal. Chem.,* 55 (1983) 182–186.

127 Z. Grobenski, R. Lehmann, B. Radzik and U. Völlkopf, *At. Spectrosc.,* 5 (1984) 87–90.

128 K.R. Lum and M. Callaghan, *Anal. Chim. Acta,* 187 (1986) 157–162.

129 W. Calmano, W. Ahlf and T. Schilling, *Fresenius' Z. Anal. Chem.,* 323 (1986) 865–868.

130 G. Bozsai, G. Schlemmer and Z. Grobenski, *Talanta,* 37 (1990) 545–553.

131 K.R. Lum, E.A. Kokotich and W.H. Schroeder, *Sci. Total Environ.,* 63 (1987) 161–173.

132 B. Welz, G. Schlemmer and J.R. Mudakavi, *J. Anal. At. Spectrom.,* 3 (1988) 695–701.

133 O. Vesterberg and K. Wrangskogh, *Clin. Chem.,* 24 (1978) 681–685.

134 A. Brockhaus, I. Freier, U. Ewers, E. Jermann and R. Dolgner, *Int. Arch. Occup. Environ. Health,* 52 (1983) 167–175.

135 H. Jessen, H. Kruse and I. Piechotowski, *Int. Arch. Occup. Environ. Health,* 554 (1984) 45–54.

136 M. Stoeppler, J. Angerer, M. Fleischer and K.H. Schaller, in J. Angerer and K.H. Schaller (Editors), *Analyses of Hazardous Substances in Biological Materials, Vol. 1,* VCH, Weinheim, 1985, pp. 79–91.

137 M. Hoenig, *Int. Clin. Prod. Rev.,* 6/1 (1987) 38–41.

138 D.J. Halls, M.M. Black, G.S. Fell and J.M. Ottaway, *J. Anal. At. Spectrom.,* 2 (1987) 305–309.

139 O. Vesterberg and A. Engqvist, in S. Safe and O. Hutzinger (Editors), *Environmental Toxin Series, Vol. 2, Cadmium,* (M. Stoeppler and M. Piscator, Vol. eds.), Springer, Berlin, Heidelberg, New York, London, Paris, Tokyo, 1988, pp. 195–204.

140 P. Dube, C. Krause and L. Windmüller, *Analyst,* 114 (1989) 1249–1253.

141 R.F.M. Herber, M. Stoeppler and D.B. Tonks, *Fresenius' J. Anal. Chem.,* 338 (1990) 269–278.

142 R.F.M. Herber, M. Stoeppler and D.B. Tonks, *Fresenius' J. Anal. Chem.,* 338 (1990) 279–286.

143 H.T. Delves and J. Woodward, *At. Spectrosc.,* 2 (1981) 65–67.

144 E. Pruszkowska, G.R. Carnrick and W. Slavin, *Clin. Chem.,* 29 (1983) 477–480.

145 K.S. Subramanian, J.-C. Meranger and J.E. MacKeen, *Anal. Chem.,* 55 (1984) 1064–1067.

224

146 K.G. Feitsma, J.P. Franke and R.A. de Zeeuw, *Analyst*, 109 (1984) 789–791.
147 X. Ying, G. Schlemmer and B. Welz, *Anal. Chem.*, 59 (1987) 1462–1466.
148 Z.A. de Benzo, R. Fraile and N. Carrion, *Anal. Chim. Acta*, 231 (1990) 283–288.
149 J. Smeyers-Verbeke, Q. Yang, W. Penninckx and F. Vandervoort, *J. Anal. At. Spectrom.*, 5 (1990) 393–398.
150 J.J. McAughey and N.J. Smith, *Anal. Chim. Acta*, 156 (1984) 129–137.
151 F. Claeys-Thoreay, *Atomic Spectrosc.*, 3 (1982) 188–191.
152 K.R. Lum and .G. Edgar, *Int. J. Environ. Anal. Chem.*, 33 (1988) 13–21.
153 C. Mohl, L. Novak, P. Ostapczuk and M. Stoeppler, in B. Welz (Editor), *4. Colloquium Atmospektrometr. Spurenanalytik*, Bodenseewerk Perkin-Elmer, Überlingen, 1987, pp. 453–460.
154 M.B. Knowles, *J. Anal. At. Spectrom.*, 4 (1989) 257–260.
155 R.W. Dabeka and A.D. McKenzie, *Can. J. Spectrosc.*, 31/2 (1986) 45–52.
156 R.D. Beaty and M.M. Cooksey, *At. Abs. Newsl.*, 17 (1978) 53–58.
157 M. Beaty, W. Barnett and Z. Grobenski, *Atom. Spectrosc.*, 1 (1980) 72–76.
158 D.K. Eaton and J.A. Holcombe, *Anal. Chem.*, 55 (1983) 946–950.
159 H.-D. Narres, C. Mohl and M. Stoeppler, *Z. Lebensm. Unters. Forsch.*, 181 (1985) 111–116.
160 C. Müller, *Trace Elem. Med.*, 4 (1987) 4–7.
161 E.H. Larsen and L. Rasmussen, *Z. Lebensm. Unters. Forsch.*, 192 (1991) 136–141.
162 A. Wadge, M. Hutton and P.J. Peterson, *Sci. Total Environ.*, 54 (1986) 13–27.
163 S.C. Wadge, *Ph.D. Thesis*, Univ. of London, 1985.
164 K. Günther, W. Henze and F. Umland, *Fresenius' Z. Anal. Chem.*, 327 (1987) 301–303.
165 D. Putzer and L. Matter, *VDI-Ber.*, 609 (1987) 177–187.
166 M.G. Kelly, G. Girton and B.A. Whitton, *Water Res.*, 21 (1987) 1429–1435.
167 J.D. Wetzer and B. Whitton, *Hydrobiologia*, 100 (1983) 261–284.
168 R. Seeger, *Z. Lebensm. Unters. Forsch.*, 166 (1978) 23–34.
169 W. Barudi and H.J. Bielig, *Z. Lebensm. Unters. Forsch.*, 170 (1980) 254–257.
170 M. Stoeppler and F. Backhaus, *Fresenius' Z. Anal. Chem.*, 291 (1978) 116–120.
171 P.J. Say, I.G. Burrows and B.A. Whitton, *Enteromorpha as a Monitor of Heavy Metals in Estuarine and Coastal Interfidal Waters*, Northern Environmental Conasultants, Durham, 1986.
172 Standing Committee of Analysts, *Atomic Absorption Spectrophotometry 1979 Version: Methods for the Examination of Waters and Associated Materials*, Her Majesty's Stationery Office, London, 1979.
173 M. Stoeppler, in: *Analytiker Taschenbuch, Band 10*, Springer, Berlin, Heidelberg, New York, Tokyo, 1991, pp. 53–84.
174 C. Mohl and M. Stoeppler, *Standard Operation Procedures for the Environmental Specimen Bank of the Federal Republic of Germany, Metals, Atomic Absorption Spectrometry*, Juelich, August, 1991.
175 M. Stoeppler, C. Mohl, L. Novak and P.E. Gardiner, in B. Welz (Editor), *Forschritte in der Atmospektrometrischen Spurenanalytik*, VCH, Weinheim, 1986, pp. 419–427.
176 A.P. Jackson and B.J. Alloway, *Int. J. Environ. Anal. Chem.*, 41 (1990) 119–131.
177 A.M. Teilmann and J.C. Hansen, *Nord. Vet.-Met.*, 36 (1984) 49–56.
178 H. Lorenz, H.-D. Ocker, J. Brüggemann, P. Weigert and M. Sonneborn, *Z. Lebensm. Unters. Forsch.*, 183 (1986) 402–405.
179 J. Hoffmann and P. Blasenbrei, *Z. Lebensm. Unters. Forsch.*, 182 (1986) 121–122.
180 J.C. Ciurea, Y.F. Lipka and B.E. Humbart, *Mitt. Gebiete Lebensm. Hyg.*, 77 (1986) 509–519.
181 A. Stelz, *Lebensmittelchem. Gerichtl. Chem.*, 41 (1987) 110–112.
182 T. Watanabe, A. Koizumi, H. Fujimoto, A. Ishimori and M. Ikeda, *Tohoku J. Exp. Med.*, 138 (1982) 443–444.
183 J. Simon and T. Liese, *Fresenius' Z. Anal. Chem.*, 309 (1981) 383–385.
184 J. Simon and T. Liese, *Fresenius' Z. Anal. Chem.*, 314 (19) 483–486.

185 M. Hedrich, U. Rösick, P. Brätter, R.L. Bergmann and K.E. Bergmann, in: B. Welz (Editor), *4. Colloquium Atomspektrometr. Spurenanalytik,* Bodenseewerk Perkin-Elmer, Überlingen, 1987, pp. 461–468.
186 T.L.M. Syversen and G.B. Syversen, *Bull. Environ. Contam. Toxicol.,* 13 (1975) 97–100.
187 J. Chung, N.O. Narteg and M.G. Cherian, *Arch. Environ. Health,* 41 (1986) 319–323.
188 R. Hahn, U. Ewers, E. Jermann, I. Freier, A. Brockhaus and H.W. Schlipköter, *Int. Arch. Occup. Environ. Health,* 59 (1987) 165–176.
189 M. Wilhelm, D. Hafner, I. Lombeck and F.K. Ohnesorge, *Int. Arch. Occup. Environ. Health,* 60 (1988) 43–50.
190 W. Slavin and G.R. Carnrick, *Spektrochim. Acta,* 39B (1984) 271–282.
191 W. Slavin, D.C. Manning and G.R. Carnrick, *Spectrochim. Acta,* 44B (1989) 1237–1243.
192 F.J. Langmyhr and G. Wibetoe, *Prog. Analyt. Atom. Spectrosc.,* 8 (1985) 193–256.
193 F.J. Langmyhr, *Fresenius' Z. Anal. Chem.,* 322 (1985) 654–656.
194 U. Kurfürst, in: *Analytiker-Taschenbuch, Band 10,* Springer, Berlin, Heidelberg, New York, Tokyo, 1991, pp. 189–248.
195 W. Slavin, N.J. Miller-Ihli and G.R. Carnrick, *Am. Lab.,* Oct. (1990)
196 T. Nilsson and P.O. Berggren, *Anal. Chim. Acta,* 159 (1984) 381–385.
197 U. Völlkopf, Z. Grobenski, R. Tamm and B. Welz, *Analyst,* 110 (1985) 573–577.
198 W. Scholl, *Fresenius' Z. Anal. Chem.,* 322 (1985) 681–684.
199 G. Schlemmer and B. Welz, *Fresenius' Z. Anal. Chem.,* 328 (1987) 405–409.
200 U. Völlkopf, R. Lehmann and D. Weber, *J. Anal. At. Spectrom.,* 2 (1987) 455–458.
201 C. Mohl, H.D. Narres and M. Stoeppler, in B. Welz (Editor), *Forschritte in der Atomspektrom. Spurenanalytik,* VCH, Weinheim, 1986, pp. 439–446.
202 K.H. Grobecker and B. Klüssendorf, *Fresenius' Z. Anal. Chem.,* 322 (1985) 673–676.
203 P. Esser, *Fresenius' Z. Anal. Chem.,* 322 (1985) 677–680.
204 L. Steubing and K.H. Grobecker, *Fresenius' Z. Anal. Chem.,* 322 (1985) 692–696.
205 K. Piczonka and A. Rosopulo, *Fresenius' Z. Anal. Chem.,* 322 (1985) 697–699.
206 L. Horner and U. Kurfürst, *Fresenius' Z. Anal. Chem.,* 328 (1987) 386–387.
207 W.J. Rühl, *Fresenius' Z. Anal. Chem.,* 322 (1985) 710–712.
208 A. Janssen, B. Brückner, K.H. Grobecker and U. Kurfürst, *Fresnius' Z. Anal. Chem.,* 322 (1985) 713–716.
209 R.F.M. Herber, A.M. Roelofsen, W. Hazelhoff Roelfzema and J.H.J. Copius Peereboom-Stegemann, *Fresenius' Z. Anal. Chem.,* 322 (1985) 743–746.
210 U. Kurfürst, M. Kempeneer, M. Stoeppler and O. Schuirer, *Fresnius' J. Anal. Chem.,* 337 (1990) 248–252.
211 E. Lücker, W. Kreuzer and C. Busche, *Fresenius' Z. Anal. Chem.,* 335 (1989) 176–188.
212 E. Hahn, K. Hahn, C. Mohl and M. Stoeppler, *Fresenius; J. Anal. Chem.,* 337 (1990) 306–309.
213 C. Mohl, K.H. Grobecker and M. Stoeppler, *Fresenius' Z. Anal. Chem.,* 328 (1987) 413–418.
214 I. Atsuya, K. Itoh and K. Akatsuka, *Fresenius' Z. Anal. Chem.,* 328 (1987) 338–341.
215 K. Akatsuka and I. Atsuya, *Anal. Chem.,* 61 (1989) 216–220.
216 G.R. Carnrick, G. Daley and A. Potinopoulos, *At. Spectrosc.,* 10 (1989) 170–174.
217 G. Rygh and K.W. Jackson, *J. Anal. At. Spectrom.,* 2 (1987) 3973400.
218 M. Hoenig, P. Regnier and R. Wollast, *J. Anal. At. Spectrom.,* 4 (1989) 631–634.
219 K.O. Olayinka, S.J. Haswell and R. Grzeskowiak, *J. Anal. At. Spectrom.,* 1 (1986) 297–300.
220 S. Lynch and D. Littlejohn, *Talanta,* 37 (1990) 825–830.
221 K. Ohta, W. Aoki and T. Mizuno, *Mikrochim. Acta,* I (1990) 81–86.
222 V.A. Fassel, *Fresenius' Z. Anal. Chem.,* 324 (1986) 511–518.
223 P.W.J.M. Bouman, *Inductively Coupled Plasma Emission Spectroscopy (2 Volumes)* Wiley, New York, Chichester, Brisbane, Toronto, Singapore, 1987.
224 J.W. Robinson, *Atomic Spectroscopy,* Marcel Dekker, New York, Basel, 1990.
225 J.A.C. Broekaert, *Anal. Chim. Acta,* 196 (1987) 1–21.

226 P. Schramel, in: M. Stoeppler and R.F.M. Herber (Editors), *Trace Metal Analysis in Biological Systems,* Year Book Medical Publishers, Chicago, IL, in press.

227 G. Schulze, O. Elsholz and W. Genthe, in B. Welz (Editor), *4. Colloquium Atomspektrometr. Spurenanalytik,* Bodenseewerk Perkin-Elmer, Überlingen, 1987, pp. 187–196.

228 J. Yoshinaga, T. Suzuki and M. Morita, *Sci. Total Environ.,* 79 (1989) 209–221.

229 T. Brotherton, B. Barnes, N. Vela and J. Caruso, *J. Anal. At. Spectrom.,* 2 (1987) 389–396.

230 R.L.A. Sing and E.D. Salin, *Anal. Chem.,* 61 (1989) 163–169.

231 M.W. Routh, J.E. Goulter, D.B. Tasker and S.D. Arellano, *Int. Lab.,* 17/2 (1987) 54–61.

232 P. Schramel and X. Li-Qiang, *Fresenius' Z. Anal. Chem.,* 314 (1983) 671–677.

233 A.M. Paudyn and R.G. Smith, *J. Anal. At. Spectrom.,* 5 (1990) 523–529.

234 B. Bølviken and E. Steinnes, in S.E. Lindberg and T.C. Hutchinson (Editors), *Proc. Heavy Metals in the Environment, New Orleans, Sept. 1987, Vol. 1,* CEP Consultants, Edinburgh, pp. 291–293.

235 D.J. Hawke and A. Lloyd, *Analyst,* 113 (1988) 413–417.

236 J.B. Brenner, Y. Lang, A. LeMarchand and P. Gorsdaillon, *Int. Lab.,* 17/12 (1987) 18–33.

237 K.H. Karstensen and W. Lund, *J. Anal. At. Spectrom.,* 4 (1989) 357–359.

238 K.H. Karstensen and W. Lund, *Sci. Total Environ.,* 79 (1989) 179–189.

239 Y. Kanda and M. Taira, *Anal. Chim. Acta,* 207 (1988) 269–281.

240 P.E. Pfannerstill, J.T. Creed, T.M. Davidson, J.A. Caruso and K. Willeke, *J. Anal. At. Spectrom.,* 5 (1990) 285–291.

241 G.M. Hieftje and G.H. Vickers, *Anal. Chim. Acta,* 216 (1989) 1–24.

242 J.A.C. Broekaert, in W. Fresenius, H. Günzler, W. Huber, J. Lüderwald, G. Tölg and H. Wisser (Editors), *Analytiker-Taschenbuch, Band 9,* Springer, Berlin, Heidelberg, New York, Tokyo, 1990, pp. 127–163.

243 G.H. Gillson, D.J. Douglas, J.E. Fulford, K.W. Halligan and S.D. Thanner, *Anal. Chem.,* 60 (1988) 1472–1474.

244 J.G. Williams and A.L. Gray, *Anal. Proc.,* 25 (1988) 385–388.

245 D. Beauchemin, J.W. McLaren, A.P. Mykitiuk and S.S. Berman, *Anal. Chem.,* 59 (1987) 778–783.

246 D. Beauchemin and S.S. Berman, *Anal. Chem.,* 61 (1989) 1857–1862.

247 N.-S. Chong, M.L. Norton and J.L. Anderson, *Anal. Chem.,* 62 (1990) 1043–1050.

248 N.W. Reid and M.A. Lusis, in S.E. Lindberg and T.C. Hutchinson (Editors), *Proc. Int. Conf. Heavy Metals in the Environment, Vol. 1,* CEP Consultants, Edinburgh, 1987, pp. 281–283.

249 J.M. Henshaw, E.M. Heithmar and T.A. Hinners, *Anal. Chem.,* 61 (1989) 335–342.

250 P.S. Ridout, H.R. Jones and J.G. Williams, *Analyst,* 113 (1988) 1383–1386.

251 J.W. McLaren, D. Beauchemin and S.S. Berman, *Anal. Chem.,* 59 (1987) 610–613.

252 R.D. Satzger, *Anal. Chem.,* 60 (1988) 2500–2504.

253 N.I. Ward, N. Walker, A.E. Ward and M.A. Hall, in P. Brätter, P. Schramel (Editors), *Trace Element Analytical Chemistry in Medicine and Biology, Vol. 5,* Walter de Gruyter, Berlin, New York, 1988, pp. 513–520.

254 U. Völlkopf, K. Prosen and M. Paul, in B. Welz (Editor), *Proc. 5. Colloquium Atomspektrometr. Spurenanalytik,* Bodenseewerk Perkin-Elmer, Überlingen, 1989, pp. 31–44.

255 C.J. Amarasiriwardena, B. Gercken, M.D. Argantine and R.M. Barnes, *J. Anal. At. Spectrom.,* 5 (1990) 457–462.

256 R. Dolan, J. van Loon, D. Templeton and A. Paudy, *Fresenius' J. Anal. Chem.,* 336 (1990) 99–105.

257 M.E. Ketterer, J.J. Reschl and M.J. Peters, *Anal. Chem.,* 61 (1989) 2031–2040.

258 T. Hirata, T. Akagi and A. Masuda, *Analyst,* 115 (1990) 1329–1333.

259 A.M. Bond, *Modern Polarographic Methods in Analytical Chemistry,* Marcel Dekker, New York, 1980.

260 H.W. Nürnberg, *Pure Appl. Chem.,* 54 (1982) 853–878.

261 J. Wang, *Stripping Analysis*, VCH, Weinheim, Deerfield Beach, FL, Basel, 1985.
262 G. Henze and R. Neeb, *Elektrochemische Analytik*, Springer, Berlin, Heidelberg, New York, Tokyo, 1986.
263 J. Wang, *Electroanalytical Techniques in Clinical Chemistry and Laboratory Medicine*, VCH, Weinheim, Deerfield, Beach, FL, Basel, 1988.
264 P. Valenta, *GIT Fachz. Lab.*, 32 (1988) 312–320.
265 J.G. Osteryoung and R.A. Osteryoung, *Anal. Chem.*, 57 (1985) 101A–110A.
266 P. Ostpaczuk, P. Valenta and H.W. Nürnberg, *J. Electroanal. Chem.*, 214 (1986) 51–64.
267 D. Jagner and A. Graneli, *Anal. Chim. Acta*, 83 (1976) 19–26.
268 D. Jagner, *Anal. Chem.*, 50 (1978) 1924–1929.
269 P. Ostapczuk and M. Stoeppler, *Chem. Ind. (Duesseldorf)*, 4 (1990) 76–78.
270 P. Ostapczuk and M. Stoeppler, in S. Safe and O. Hutzinger (Editors), *Environmental Toxins, Vol. 2, Cadmium*, M. Stoeppler and M. Piscator (Vol. eds.) Springer, Heidelberg, New York, London, Paris, Tokyo, 1988, pp. 213–226.
271 P. Ostapczuk, M. Stoeppler and H.W. Dürbeck, *Toxicol. Environ. Chem.*, 27 (1990) 49–53.
272 C.M.G. van den Berg, *J. Electroanal. Chem.*, 215 (1986) 111–121.
273 T.M. Florence, in G.E. Batley (Editor), *Trace Element Speciation: Analytical Methods and Problems*, CRC Press, Boca Raton, FL, 1990, pp. 77–116.
274 W. Dorten, P. Valenta and H.W. Nürnberg, *Fresenius' Z. Anal. Chem.*, 317 (1984) 367–371.
275 L. Mart, H.W. Nürnberg and D. Dyrssen, in C.S. Wong, E. Boyle, K.W. Bruland, J.D. Burton and E.D. Goldberg (Editors), *Trace Metals in Sea Water*, Plenum Press, New York, London, 1983, pp. 113–130.
276 J. Völkening and K.G. Heumann, *Fresenius' Z. Anal. Chem.*, 331 (1988) 174–181.
277 L. Mart, H.W. Nürnberg and P. Valenta, *Fresenius' Z. Anal. Chem.*, 300 (1980) 350–362.
278 L. Mart, H.W. Nürnberg and H. Rützel, *Fresenius' Z. Anal. Chem.*, 317 (1984) 201–209.
279 L. Mart and H.W. Nürnberg, *Verteilung und Schicksal ökotoxischer Schwermetalle in den Ozeanen und Küstengewässern, Umweltforschung*, 1985, KFA Jülich, 1985, pp. 41–51.
280 J. Gustavsson and L. Hansson, *Int. J. Environ. Anal. Chem.*, 17 (1984) 57–72.
281 L. Mart and H.W. Nürnberg, *Mar. Chem.*, 18 (1986) 197–213.
282 K.W. Bruland, K.H. Coale and L. Mart, *Mar. Chem.*, 18 (1985) 285–300.
283 P. Valenta, V.D. Nguyen and H.W. Nürnberg, *Sci. Total. Environ.*, 55 (1986) 311–320.
284 M. Fariwar-Mohseni and R. Neeb, *Fresenius' Z. Anal. Chem.*, 296 (1979) 156–158.
285 R.J. Gajan, S.G. Capar, C.A. Subjoc and M. Sanders, *J. Assoc. Off. Anal. Chem.*, 65 (1982) 970–977.
286 S.G. Capar, R.J. Gajan, E. Madzsar, R.H. Albert, M. Sanders and J. Zyren, *J. Assoc. Off. Anal. Chem.*, 65 (1982) 978–986.
287 H.-D. Narres, P. Valenta and H.W. Nürnberg, *Z. Lebensm. Unters. Forsch.*, 179 (1984) 440–446.
288 P. Ostapczuk, M. Goedde, M. Stoeppler, H.W. Nürnberg, *Fresenius' Z. Anal. Chem.*, 317 (1984) 252–256.
289 P. Ostapczuk, P. Valenta, H. Rützel and H.W. Nürnberg, *Sci. Total Environ.*, 60 (1987) 1–16.
290 G. Vos, H. Lammers and W. van Delft, *Z. Lebensm. Unters. Forsch.*, 187 (1988) 1–7.
291 S.B. Adeloju, A.M. Bond and M.H. Briggs, *Anal. Chem.*, 57 (1985) 1386–1390.
292 M. Würfels, E. Jackwerth and M. Stoeppler, *Anal. Chim. Acta*, 226 (1989) 31–41.
293 M. Würfels, *Mar. Chem.*, 28 (1989) 259–264.
294 L. Almestrand, D. Jagner and L. Renman, *Talanta*, 33 (1986) 991–995.
295 L. Almestrand, D. Jagner and L. Renman, *Anal. Chim. Acta*, 193 (1987) 71–79.
296 C. Whiston, *X-ray Methods*, Wiley, Chichester, 1987.
297 R. Klockenkämper, B. Raith, S. Divoux, B. Gonsior, S. Brüggerhoff and E. Jackwerth, *Fresenius' Z. Anal. Chem.*, 326 (1987) 105–117.

228

298 A. Prange, *Spectrochim. Acta,* 44B (1989) 437–452.
299 H. Rethfeld, *Fresenius' Z. Anal. Chem.,* 310 (1982) 127–130.
300 H. Rethfeld, *Fresenius' Z. Anal. Chem.,* 324 (1986) 720–727.
301 A. Prange and H. Schwenke, *Adv. X-Ray-Anal.,* 32 (1989) 209–218.
302 W. Michaelis, H. Böddeker, J. Knoth and H. Schwenke, *Fast and economical multielement analysis of urban air dust using total reflection XRF (TXRF), GKSS 83/E/33,* GKSS-Forschungszentrum Geesthacht GmbH, 1983.
303 R.-P. Stössel and A. Prange, *Anal. Chem.,* 57 (1985) 2880–2885.
304 A. Prange, A. Knöchel and W. Michaelis, *Anal. Chim. Acta,* 172 (1985) 79–100.
305 A. Prange, J. Knoth, R.-P. Stössel, H. Böddeker and H. Kramer, *Anal. Chim. Acta,* 195 (1987) 275–287.
306 W. Michaelis and A. Prange, *Nucl. Geophys.,* 2 (1988) 231–245.
307 W. Michaelis, *Multielement Analysis of Environmental Samples by Total-reflection X-Ray Fluorescence Spectrometry, Neutron Activation Analysis and Inductively Coupled Plasma Optical Emission Spectroscopy, GKSS 85/E/65,* GKSS-Forschungszentrum Geesthacht, 1986.
308 A. Prange, R. Niedergesäss and C. Schnier, in W. Michaelis (Editor), *Estuarine Water Quality Management; Coastal and Estuarine Studies, Vol. 36,* Springer, Berlin, Heidelbert, New York, 1990, pp. 429–436.
309 M.C. Scott and D.R. Chettle, *Scand. J. Work Environ. Health,* 12 (1986) 81–96.
310 J.O. Christoffersen and S. Mattson, *Phys. Med. Biol.,* 28 (1983) 1135–1144.
311 G. Erdtmann and H. Petri, in P.J. Elving (Editor), *Treatise on Analytical Chemistry,* Part 1, Vol. 14, Wiley, New York, pp. 419–643.
312 J. Hoste, in P.J. Elving (Editor), *Treatise on Analytical Chemistry,* Part I, Vol. 14, Wiley, New York, 2nd ed., 1986, pp. 645–777.
313 V. Krivan, in P.J. Elving (Editor), *Treatise on Analytical Chemistry,* Part I, Vol. 14, Wiley, New York, 2nd ed., 1986, pp, 339–417.
314 V. Krivan, *Sci. Total Environ.,* 64 (1987) 21–40.
315 V. Krivan and K.P. Egger, *Fresenius' Z. Anal. Chem.,* 325 (1986) 41–49.
316 V. Krivan, M. Franek, H. Baumann and J. Pavel, *Fresenius' J. Anal. Chem.,* 338 (1990) 583–587.
317 R. Zeisler, S.F. Stone and R.W. Sanders, *Anal. Chem.,* 60 (1988) 2760–2765.
318 M. Esprit, C. Vandecasteele and J. Hoste, *Anal. Chim. Acta,* 185 (1986) 307–313.
319 R.R. Greenberg, *Anal. Chem.,* 52 (1980) 676–679.
320 J. Versieck, L. Vanballenberghe and A. DeKesel, in S. Safe and O. Hutzinger (Editors), *Environmental Toxin Series, Vol. 2, Cadmium,* (M. Stoeppler and M. Piscator, Vol. eds.), Springer, Berlin, Heidelberg, New York, London, Paris, Tokyo, 1988, pp. 205–212.
321 K. Kobayashi, *Analyst,* 113 (1988) 1121–1123.
322 K.J. Ellis, D. Vartsky, J. Zanzi, S.H. Cohn and S. Yasamura, *Science,* 205 (1979) 323–325.
323 D. Chettle, T.C. Harvey, I.K. Al Haddad, R. Lauweyrs, J.P. Buchet, H. Roels and A. Bernard, *Proc. 2nd Int. Cadmium Conf. Cannes, 1979,* Metal Bulletin, London, 1980, pp. 161–163.
324 S. Ichinoki, T. Morita and M. Yamazaki, *J. Liquid Chromatogr.,* 7 (1984) 2467–2482.
325 H. Schaller and R. Neeb, *Fresenius' Z. Anal. Chem.,* 327 (1987) 170–174.
326 A. Munder and K. Ballschmiter, *Fresenius' Z. Anal. Chem.,* 323 (1986) 869–874.
327 J. Weiss, *Fresenius' Z. Anal. Chem.,* 327 (1987) 451–455.
328 R.B. Rubin and S.S. Heberling, *Int. Lab.,* 17/9 (1987) 54–60.
329 D. Yan and G. Schwedt, *Fresenius' Z. Anal. Chem.,* 328 (1987) 503–508.
330 R.J. Joyce and A. Schein, *Int. Lab.,* 20 Jan./Feb. (1990) 48–53.
331 F. Adams, R. Gijbels and R. van Grieken (Editors), *Inorganic Mass Spectrometry,* Wiley, New York, 1988.
332 K.G. Heumann (Editor), *Element Trace Analysis by Mass Spectrometry, Fresenius' Z. Anal. Chem.,* 60 (1988) 2070–2075.
333 K.G. Heumann, *Toxicol. Environ. Chem. Rev.,* 3 (1980) 111–129.
334 P. De Bièvre, *Fresenius' J. Anal. Chem.,* 337 (1990) 766–771.

335 A.P. Mykitiuk, D.S. Russell and R.E. Sturgeon, *Anal. Chem.*, 52 (1980) 1281–1283.
336 J. Mühlbaler, C. Stevens, D. Graczyk and T. Tisue, *Anal. Chem.*, 54 (1982) 496–499.
337 K.J.R. Rosman, J.R. de Laeter and A. Chegwidden, *Talanta*, 29 (1982) 279–283.
338 V.J. Stukas and C.S. Wong, *Mar. Chem.*, 12 (1983) 133–146.
339 E. Michiels and P. De Bièvre, in I.K. O'Neill, P. Schuller and L. Fishbein (Editors), *Environmental Carcinogens, Selected Methods of Analysis, Volume 8 – Some Metals: As, Be, Cd, Cr, Ni, Pb, Se, Zn (IARC Scientific Publications* No. 71, Int. Agency for Research on Cancer, Lyon, 1986, pp. 443–450.
340 A. Götz and K.G. Heumann, *Fresenius' Z. Anal. Chem.*, 325 (1986) 24–31.
341 A. Götz and K.G. Heumann, *Fresenius' Z. Anal. Chem.*, 326 (1987) 118–122.
342 A. Götz and K.G. Heumann, *Fresenius' Z. Anal. Chem.*, 332 (1988) 640–644.
343 K. Hilrich (Editor), *Official Methods of Analysis of the AOAC*, Assoc. of Off. Anal. Chemists, Arlington, VA, 15th ed., 1990, pp. 246–247.
344 M. Komata and J.-I. Itoh, *Talanta*, 35 (1988) 723–724.
345 I. Lopez Garcia, P. Navarro and M.H. Cordoba, *Talanta*, 35 (1988) 885–889.
346 Y. Liu and J.D. Ingle, Jr., *Anal. Chem.*, 61 (1989) 525–529.
347 H. Hayase, K. Shitashima and H. Tsubota, *Talanta*, 33 (1986) 754–756.
348 D. Berggren, *Int. J. Environ. Anal. Chem.*, 41 (1990) 133–148.
349 A. Tessier, P.G.C. Campbell and M. Bisson, *Anal. Chem.*, 51 (1979) 844–851.
350 V. Samanidou and K. Fytianos, *Sci. Total Environ.*, 67 (1987) 279–285.
351 T.U. Aualitia and W.F. Pickering, *Talanta*, 35 (1988) 559–566.
352 A. Beveridge, P. Waller and W.F. Pickering, *Talanta*, 36 (1989) 535–542.
353 T. Rudd, D.L. Lake, I. Mehrotra, R.M. Sterrit, P.W.W. Kirk, J.A. Campbell and J.N. Lester, *Sci. Total Environ.*, 74 (1988) 149–175.
354 T.H. Christensen and X.Z. Lun, *Water Res.*, 23 (1989) 73–80.
355 X.Z. Lun and T.H. Christensen, *Water Res.*, 23 (1989) 81–84.
356 P.O. Scokart, K. Meens-Verdinne and R. de Berger, *Int. J. Anal. Chem.*, 29 (1987) 305–315.
357 G.E. Batley (Editor), *Trace Element Speciation: Analytical Methods and Problems*, CRC Press, Boca Raton, FL, 1990.
358 R.C. Massey, J.A. Burrell, D.J. McWeeny and H. Crews, *Toxicol. Environ. Chem.*, 13 (1986) 85–93.
359 L. Ebdon, S. Hill and P. Jones, *Analyst*, 112 (1987) 437–440.
360 K. Günther and F. Umland, *Fresenius' Z. Anal. Chem.*, 331 (1988) 302–309.
361 H.M. Crews, J.R. Dean, L. Ebdon and R.C. Massey, *Analyst*, 114 (1989) 895–899.
362 K.O. Olayinka, S.J. Haswell and R. Grzeskowiak, *J. Anal. At. Spectrom.*, 4 (1989) 171–173.
363 M. Grasserbauer, *Anal. Chim. Acta*, 195 (1987) 1–32.
364 P.F. Schmidt, R. Barckhaus and W. Kleimeir, *Trace Elem. Med.*, 3 (1986) 19–24.
365 H. Muntau, *Fresenius' Z. Anal. Chem.*, 324 (1986) 678–682.
366 P.G.W. Jones, *A Preliminary Report on the ICES Intercalibration of Sea Water Samples for the Analysis of Trace Metals*, Int. Council Explor. Sea, Sep. No. CM 1977/E:16.
367 J.M. Bewers, J. Dalziel, P.A. Yeats and J.K. Barron, *Mar. Chem.*, 10 (1981) 173–193.
368 P.G.W. Jones, *Anal. Proc.*, 19 (1982) 565–566.
369 S.S. Berman and V.J. Boyko, *ICES Sixth Round Intercalibration for Trace Metals in Estuarine Water, Cooperative Research Report No. 152*, ICES, Copenhagen, 1988.
370 L. Brügmann, L.G. Danielsson, B. Magnusson and S. Westerlund, *Mar. Chem.*, 13 (1983) 327–339.
371 S.S. Bermann, R.E. Sturgeon, J.A.H. Desaulniers and A.P. Mykytiuk, *Mar. Poll. Bull.*, 14 (1983) 69–73.
372 H. Brix, J.E. Lyngby and H.-E. Schierup, *Mar. Chem.*, 12 (1983) 69–85.
373 G. Schlemmer and B. Welz, *Fresenius' Z. Anal. Chem.*, 320 (1985) 648–649.
374 J. Kumpulainen and M. Paaki, *Fresenius' Z. Anal. Chem.*, 326 (1987) 684–689.
375 H. Schauenburg and P. Weigert, *Fresenius' J. Anal. Chem.*, 338 (1990) 449–452.
376 B.J. Starkey, A. Taylor and A.W. Alker, *J. Anal. At. Spectrom.*, 1 (1986) 397–400.

377 K.H. Schaller, J. Angerer, G. Lehnert, H. Valentin and D. Weltle, *Fresenius' Z. Anal. Chem.*, 326 (1987) 643–646.
378 B. Lind, C.-G. Elinder, L. Friberg, B. Nilsson, M. Svartengren and M. Vahter, *Fresenius' Z. Anal. Chem.*, 326 (1987) 647–655.
379 M. Vahter and L. Friberg, *Fresenius' Z. Anal. Chem.*, 332 (1988) 726–731.
380 E.A. Maier, *Anal. Proc.*, 27 (1990) 269–270.
381 A.R. Byrne, C. Camara-Rica, R. Cornelis, J.J.M. de Goeij, G.V. Iyengar, G. Kirkbright, G. Knapp, R.M. Parr and M. Stoeppler, *Fresenius' Z. Anal. Chem.*, 326 (1987) 723–729.
382 J. Versieck, L. Vanballenberghe, A. de Kesel, J. Hoste, B. Wallaeys, J. Vandenhaute, N. Baeck, H. Steuaert, A.R. Byrne and F.W. Sunderman, Jr., *Anal. Chim. Acta*, 204 (1988) 63–75.
383 M. Stoeppler, F. Backhaus, J.D. Schladot, N. Commerscheidt, *Fresenius' Z. Anal. Chem.*, 326 (1987) 707–711.
384 J.D. Schladot, F.W. Backhaus, in S.S. Wise, R. Zeisler, G.M. Goldstein (Editors), *NBS Spec. Pub.* No. 740) *Progress in Environmental Specimen Banking*, U.S. Dept. of Commrce, National Bureau of Standards, Washington, DC, 1988, pp. 184–193.
385 A.R. Byrne et al., *Sci. Total Environ.*, in preparation.
386 E.C. Toro, R.M. Parr, S.A. Clements, *Biological and Environmental Reference Materials for Trace Elements, Nuclides and Organic Microcontaminants*, IAEA/RL/128 (Rev. 1) IAEA, Vienna, 1990.
387 H.J.M. Bowen, in W.R. Wolf (Editor), *Biological Reference Materials*, Wiley Interscience, New York, 1985, pp. 3–18.
388 J. Kucera, P. Mader, D. Mikolova, J. Cibulka, M. Polakowa, D. Kordik, *Report on the Interlaboratory Comparison of the Determination of the Contents of Trace Elements in Bovine Liver 12-02-01*, Czechoslovak Institute of Metrology, Bratislava, November, 1989.
389 O. Ostapczuk, *Labor Praxis*, 15 (1991) 460–468.
390 G. Tölg, *Fresenius' Z. Anal. Chem.*, 331 (1988) 226–235.
391 E.J. Ekanem, C.L.R. Barnard, J.M. Ottaway and G.S. Fell, *J. Anal. At. Spectrom.*, 1 (1986) 349–353.
392 N. Omenetto, P. Cavalli, M. Broglia, P. Qi and G. Rossi, *J. Anal. At. Spectrom.*, 3 (1988) 231–235.
393 R.E. Sturgeon, S.N. Willie, V. Luong and S.S. Berman, *Anal. Chem.*, 62 (1990) 2370–2376.
394 R.E. Sturgeon, S.N. Willie, V.T. Luong and S.S. Berman, *J. Anal. At. Spectrom.*, 5 (1990) 635–638.
395 S. Geiss, J. Eina, J. Mohr and K. Danzer, *Fresenius' Z. Anal. Chem.*, 338 (1990) 602–605.

M. Stoeppler (Editor)/*Hazardous Metals in the Environment*
© 1992 Elsevier Science Publishers B.V. All rights reserved

Chapter 9

Lead

Steve J. Hill

Department of Environmental Sciences, Polytechnic South West, Plymouth (UK)

CONTENTS

9.1. Introduction 231
9.2. Analytical techniques for the determination of lead 232
    9.2.1. Atomic spectroscopy 232
    9.2.2. Colorimetry 235
    9.2.3. Titration 236
    9.2.4. Electroanalytical methods 237
    9.2.5. X-ray methods 237
    9.2.6. Inductively coupled plasma–mass spectrometry 238
    9.2.7. Coupled (or hyphenated) techniques 239
9.3. Applications 244
    9.3.1. Determination of total lead 244
    9.3.2. Alkyllead compounds in petroleum 245
    9.3.3. Determination of lead in air 246
    9.3.4. Determination of lead in water 248
    9.3.5. Determination of lead in sediments, soils and biological materials 250
9.4. Conclusion 251
References 252

9.1. INTRODUCTION

Although lead (Pb) rarely occurs naturally in its elemental state, it has been used by man throughout history. One reason for this is that its principal ore, lead(II) sulphide known as galena, is easily recognised and readily reduced by heating with charcoal. It has been suggested that the discovery of metallic lead may have resulted from the accidental dropping of galena into a campfire, but whatever its origins, the use of lead by the Greeks and Romans is well documented, and certainly it has been used throughout Europe for centuries [1]. The occurence and properties of lead are discussed in more detail else-where [2], however, in the early years of the industrial revolution, much lead was added to the environment due to inefficient smelting processes and the poor design of early fur-naces. Atmospheric fallout of such inputs now have important implications when considering levels of lead in the environment, particularly with regard to soil levels and food crops presently grown in such areas. Today, however, although lead continues to be used in a wide range of products, the greatest threat from lead in the environment is due to pollution from man made organolead compounds, especially the use of tetraalkylleads (R_4Pb) in petrol. Such compounds have been added to petrol since 1923 to improve the octane ratings for fuels used in high compression internal combustion engines. The maximum lead content in petrol advised by the European Economic Community is 0.15 g l^{-1}, although in many countries the lead content is much higher than this. Although there is currently a move away from using such compounds in petrol, the detection of these lead compounds is certain to remain an important feature of environmental monitoring in the foreseeable future.

9.2. ANALYTICAL TECHNIQUES FOR THE DETERMINATION OF LEAD

There are a considerable number of analytical techniques which may be used for inorganic lead analysis. These include atomic absorption spectrometry (AAS), colorimetry, electrochemical techniques such as anodic stripping voltammetry (ASV), X-ray fluorescence spectroscopy (XRF), atomic emission spectrometry (AES), mass spectrometry (MS), radioactivation methods, and titration methods. Of these, the first three techniques may be identified as being of the most widely used for the determination of inorganic lead, although the choice of analytical method will obviously depend on a number of factors including the availability of instrumentation. For environmental work, however, the detection limits which may be achieved by a particular technique may be a prime consideration, although the possibility of interference must always be borne in mind when selecting the most appropriate technique. In addition, care should be taken to avoid contamination during sampling prior to analysis. For example, borosilicate glass and not soda-glass should be used for storage of samples, as should polyethylene instead of polypropylene. In many cases PTFE is an ideal material for containers. Pre-cleaning is essential for trace work (e.g. soaking in 10–50% HNO_3 for 24 hours prior to use). For ultra-trace work clean room facilities with filtered air may be necessary. In all cases blanks should be used throughout the analysis.

In the case of organoleads most of the analytical methods available are either not sensitive or not specific enough for many applications including environmental work. To overcome this problem so called 'coupled techniques' are now widely used for such applications. These techniques couple the separatory powers of chromatography to the sensitive element specific detection of atomic spectroscopy. The various combinations used for the determination of organoleads are summarised in Table 9.1 and discussed below in Section 9.2.7, although the subject has also been fully reviewed in the literature [3,4]. Typical detection limits for lead using a range of techniques are shown in Table 9.2.

9.2.1. Atomic spectroscopy

Lead is usually determined using AAS employing an air–acetylene flame although other flames may be used without large scale interference. Minor interferences from aluminium,

TABLE 9.1

CHROMATOGRAPHIC–ATOMIC SPECTROMETRY TECHNIQUES FOR LEAD SPECIA-TION

Coupled GC techniques	Coupled HPLC techniques
GC–FAAS	HPLC–FAAS
GC–ETA–AAS	HPLC–ETA–AAS
GC–FAFS	HPLC–ICP–AES
GC–MIP–AES	HPLC–ICP–MS
GC–ICP–AES	

Abbreviations:
GC gas chromatography
HPLC high-performance liquid chromatography
FAAS flame atomic absorption spectroscopy
ETA–AAS electrothermal atomisation atomic absorption spectroscopy
MIP–AES microwave induced plasma–atomic emission spectroscopy
ICP–AES inductively coupled plasma–atomic emission spectroscopy
ICP–MS inductively coupled plasma–mass spectrometry

TABLE 9.2

TYPICAL DETECTION LIMITS FOR LEAD ANALYSIS

Technique		Detection limit
Atomic absorption	- flame	$10 \ \mu g \ dm^{-3}$
	- graphite furnace	$0.02 \ \mu g \ dm^{-3}$
	- hydride generation	$10 \ \mu g \ dm^{-3}$
Direct current plasma–AES		$10 \ \mu g \ dm^{-3}$
Inductaively coupled plasma–AES		$20 \ \mu g \ dm^{-3}$
Differential pulse anodic stripping voltammetry		$0.01 \ \mu g \ dm^{-3}$
Colorimetry – using dithizone		$20 \ \mu g \ dm^{-3}$
Ion selective electrodes		$1.5 \ \mu g \ dm^{-3}$
X-ray fluorescence		$0.2 \ \mu g \ cm^{-2}$ [a]
Neutron activation		$10^3\text{-}10^4 \ \mu g \ dm^{-3}$
Coupled techniques: GC–AAS		17 pg [b]
HPLC–AAS		160 pg [b]
ICP–MS		$0.01 \ \mu g \ dm^{-3}$

[a] For analysis of the surface of an air filter by the thin film technique [116].

[b] Absolute detection limit for lead. Value will vary depending on organolead species.

References on p. 252

beryllium, zirconium and anions such as phosphate and sulphate can be largely eliminated by adding EDTA [5]. However, large excesses of other ions in the sample may
promote errors. For example, 10,000 ppm of iron has been shown to enhance the response from 5 ppm of lead by 35% in a lean air–acetylene flame [6]. Detection limits in the
order of 0.01 mg l^{-1} may be achieved using flames, although the sensitivity may be
improved using the boat technique [7] or the Delves cup [8]. Several resonance lines may
be used for lead, the most sensitive of which is the 217.0 nm line, however in practice the
283.3 nm line is often used. This is due to the poor signal-to-noise ratio, and the greater
background attenuation effects when using the 217.0 nm line. The principal resonance
lines for lead are shown in Table 9.3.

To determine lead at low concentrations, atomic absorption using electrothermal
atomisation is probably the most widely adopted method. The use of this technique is not
totally without problems, however, and a complex pattern of interference effects have
been identified. The origins of these interferences have been ascribed to the formation of
volatile lead species like lead oxide [9], lead sulphide [10], lead chloride [11], and the
presence of the chlorides of the alkaline earth metals [12]. Large excesses of many
elements as nitrates do not cause interference. Various methods to improve the analytical
performance have been investigated such as surface modification with oxygen ashing
[13] or metal impregnation [14], addition of hydrogen to the purge gas [15,16] and
chemical matrix modification. Using matrix modifiers such as ammonium phosphate or
magnesium nitrate and atomising from a L'vov platform, thermal pretreatment at temperatures of around 1000°C may be used with atomisation temperatures of up to 1900°C.
Normally, however, ashing of samples for lead determination should be carried out below
700°C to avoid vaporisation losses.

Lead also forms a volatile hydride, although it has only recently been determined using
this technique because of the difficulties encountered in lead hydride formation, i.e. low
yield and poor stability. To increase the efficiency of the lead hydride generation, an
oxidising agent is usually added prior to the reduction step. However, the reported detection limits are often little better than for flame techniques. The addition of tartaric acid and
potassium dichromate have been shown to increase the sensitivity although interferences
from iron [17], copper and nickel [18] may be severe. The effects of other soluble

TABLE 9.3

PRINCIPLE RESONANCE LINES FOR LEAD

Wavelength (nm)	Characteristic concentration (mg/l)
217.0	0.08
283.3	0.2
261.4	5
368.4	17
364.0	40

oxidants such as potassium dichromate–malic acid [18], hydrogen peroxide–nitric acid or hydrochloric acid [19–21,23] and nitric acid–ammonium persulphate [19], on the efficiency of lead hydride generation have also been reported. The increase in yield may be attributed to the formation of metastable Pb(IV) compounds prior to the formation of lead hydride using tetrahydroborate(III) [17]. More recently lead hydride generation has been used for isotope analysis by inductively coupled plasma–mass spectrometry (ICP–MS) (see Section 9.2.6).

Various emission techniques may also be used to determine total lead. For flame emission a nitrous oxide flame is recommended and the most sensitive emission wavelength is 405.8 nm. Alternate emission wavelengths for lead are 368.4 nm, 283.3 nm, and 261.4 nm respectively. An air–acetylene flame may be used, but with reduced sensitivity. More recently the ICP technique has been used although the sensitivity for lead is disappointing. Specialist techniques such as ICP–atomic fluorescence spectroscopy have also been used, for example, to determine lead in a variety of geological materials. Samples in this case may be decomposed with nitric, hydrofluoric and hydrochloric acids, and the residue dissolved in hydrochloric acid and diluted to volume prior to analysis [19]. The precision and accuracy of the technique are reported to compare well with both AAS and ICP–AES. Atomic fluorescence using an air–hydrogen flame at 405.8 nm may also be used. In an early study using this technique [20] the atomic fluorescence characteristics of lead in air–acetylene, nitrous oxide–hydrogen, and argon–oxygen–hydrogen flames were investigated. Using an electrodeless discharge tube as the source of excitation a detection limit of 0.01 μg ml^{-1} was obtained in the argon–oxygen–hydrogen flame. The effects of 30 cations and anions were also examined and only aluminium interfered significantly. However, the study also revealed that the analytical performance of the technique was highly dependent on the nature of the optics used in the experimental arrangement. The determination of three lead isotopes 206, 207 and 208 by AAS and the construction of isotope hollow cathode lamps have also been reported [21,22] (see also Section 9.2.6).

9.2.2. Colorimetry

The use of dithizone (diphenylthiocarbazone) as a colorimetric reagent for the determination of lead was widespread for many years. However today, except in cases where instrumental techniques are not available, it is little used due to its rather poor sensitivity and the possibility of interferences from other metals. Dithizone forms coloured complexes with 17 metals including thallium and cadmium which are co-extracted with lead. These two metals are not masked by cyanide like many other interferents, although there are many modifications of the dithiozone method for lead analysis in the literature [24–27]. In one such procedure the lead is extracted with sodium diethyldithiocarbamate and then extracted with a mixture of equal volumes of pentanol and 'sulphur-free' toluene. The organic layer is treated with dilute hydrochloric acid, so that the lead complex passes into the aqueous layer. The latter is mixed with ammoniacal dithizone solution and the lead dithizone extracted with carbontetrachloride, and the absorbance measured at 515.0 nm

References on p. 252

[27]. Such methods are, however, somewhat time consuming and this coupled to the need for experience to obtain reliable results may discourage many potential users.

When sulphide is added to a solution containing lead ions a brown colour due to the formation of colloidal lead sulphate is observed. This may be used to detect lead at low levels (0.005–0.25 mg) although in many samples the method is of limited use due to interference from natural salts such as tartrates, citrates and ammonium chloride, and metals such as copper, iron, bismuth and aluminium. The effects of copper and iron may be overcome by the addition of a few drops of a 10% aqueous solution of potassium cyanide, and aluminium may be retained in ammoniacal solution by the addition of ammonium citrate, a corresponding amount of the latter being added to the standards [27].

Colorimetric methods may also be used for the determination of microgram quantities of organolead compounds in a solution [28,29]. Here the classical separation using dithizone is used, with the addition of EDTA to complex any inorganic lead thus removing possible interferences. The tetraalkyllead compounds are trapped in a solution of iodine monochloride–hydrochloric acid in which they are quantitatively converted into dialkyllead halides. The lead is then preferentially extracted into carbon tetrachloride as the dithizonate. Finally the lead is extracted with nitric acid–hydrogen peroxide solution and an aliquot analysed using electrothermal atomic absorption (see Section 9.2.1). This method offers a detection limit of 7 nanograms of lead.

9.2.3. Titration

The lead concentration of an aqueous solution may be determined by amperometric titration with a standard potassium dichromate solution. The reaction is:

$$2Pb^{2+} + Cr_2O_7^{2-} + H_2O \rightarrow 2\ PbCrO_4(s) + 2H^+$$

The titration can be performed with a dropping electrode that is maintained at either 0 or -1.0 V (versus SCE). At 0 V, the current remains near zero until equivalence; then it rises rapidly as a consequence of reduction of the excess dichromate ion. At -1.0 V both dichromate and lead ions are reduced. Thus the current decreases to a minimum and then rises as the equivalence point is passed. In principle, the titration error should be less with the V-shaped curve. The advantage of the titration at 0 V is that oxygen does not have to be removed. Full details of this method have been described by Skoog and West [30].

Alkyllead compounds in petrol may be determined by the titrimetric–iodimetric chromate method [31], but the procedure is time consuming and tedious. Better is the method based on the reaction and extraction of lead with iodine mono–chloride followed by complexometric measurement [32]. Lead in minerals may also be determined by titration techniques. Here the lead ore may be dissolved in an hydrochloric–nitric acid mixture [33]. The lead is separated by extraction with a CHCl$_3$ solution of sodium diethyl dithiocarbamate alkaline cyanide solution to mask interferences, and titrated in an ammonia–

ammonium chloride medium at pH 10 using eriochrome black T. Other possible indicators for lead include pyrocatechol violet and xylenol orange.

9.2.4. Electroanalytical methods

The isolation of lead by electro-deposition of the metal from a solution of biological materials was suggested as early as 1941 [34]. Iron and copper were found to interfere but their effects could be minimised by the addition of potassium cyanide. However until the introduction of hanging mercury electrodes, electrodeposition techniques found little application in the analysis of environmental materials. In 1961 Kublik used this method for the measurement of lead in waters [35], and a number of other applications followed [36]. The polarographic behaviour of lead in patrol–water emulsions with various surfactants has been investigated [37]. In this case the polarographic process was found to be reversible and controlled by diffusion, and the adsorption on the dropping-mercury electrode was higher in the emulsion medium than in aqueous surfactant media. Anodic stripping voltammetry (ASV) and cyclic voltammetry have been used to study lead in organic free seawater [38]. Near seawater pH values (8.5) most of the lead was present as $PbOH^+$ (88%) although $PbCO_3$ (10%), and $PbCl^+$, Pb^{2+} and $PbSO_4$ (2% total) were also found. Applications of ASV to the analysis of lead in clinical samples has also been reported [39–41]. The possibility of producing lead ion-selective electrodes (ISE), based on the tetraphenylborate (TPB) salts of non-ionic surfactant polyalkoxylates such as polyethylene glycol (PEG 1540), nonylphenoxypoly(ethyleneoxy) ethanol (Antarox CO-880) and polypropylene glycol (PPG 425), has been reported [42]. Since the radius of the Pb^{2+} ion (0.12 nm) is close to that of barium (0.135 nm) for which a very good polyethoxylate-based ISE exists [43] the possibility of producing a selective lead electrode seemed high. Of the electrodes studied, the PVC matrix electrode based on Pb Antarox CO-880.TPB system, using 2-nitrophenyl phenyl ether (NPPE) as the plasticising solvent mediator, offered the highest selectivity towards Pb^{2+} ions in the presence of a large number of interferent metal ions, and had a Nernstian response between 10^{-1} and 10^{-5} mol dm^{-3} of Pb^{2+}.

9.2.5. X-ray methods

X-ray fluorescence (XRF) has been used particularly for the analysis of particulate lead in air. The technique has the advantage of avoiding sample pretreatment, since once the particles have been collected on a filter, the filter may be inserted directly into the instrument. In most cases standards are prepared by drying lead containing solutions, or filtering suspensions of lead salts onto a filter surface [44]. XRF may also be used for the analysis of geological samples, e.g. lead in zinc ores, although the complex matrix requires the use of a correction method. To determine lead in such ores it is possible to use the Lb and Ly1 lines, although the most useful Pb Lx line suffers from spectral interferences due to the close proximity of the As Kx line. Thus the Pb Ly1 line has been used [45] because the mass absorption coefficient of the specimens often used as standards is less influenced by the Zn concentration at this wavelength than at the Pb Lb line. It is also recommended that PbS is used to prepare standards in order to improve the accuracy of this method [45], as when PbO is used differences are obtained in the measured XRF intensities [46]. The major drawback in using XRF to determine lead is, however, the poor sensitivity of the technique for many applications, and the lengthy sampling times that may be required to collect sufficient lead for analysis. X-ray diffraction (XRD) may also be used to determine lead species, however the technique is dependent on the presence of crystalline forms susceptible to XRD analysis. Application of this technique to the determination of lead compounds in soils are given in Section 9.3.5.

9.2.6. Inductively coupled plasma–mass spectrometry

There is little doubt that the use of ICP–MS in environmental analysis is going to increase. The multielement capabilities, low detection limits (0.01 μg dm^{-3} for lead in aqueous solutions), rapid throughput of samples, potential for small sample analysis, the expanding range of sample introduction techniques, and the ability to readily determine isotopic ratios for trace and speciation studies, makes this a particularly versatile technique.

For the determination of lead, the use of ICP–MS is particularly well suited due to the large ion, freedom from interferences, and the fact that lead has four isotopes, the ratios of which can be used to investigate the sources of lead in environmental samples. This approach has already been used to study concentrations of lead along British motorways [47]. In addition, ICP–MS has been used to determine lead in geological materials, waters, food-stuffs, petroleum products and biological materials (see Section 9.3).

As indicated above, ICP–MS may be used to obtain isotope ratios on directly aspirated solutions. This ability is exploited using isotope dilution analysis which facilitates measurement of total elemental concentrations. Using this technique, the natural ratio of the element is altered by the addition of a stable non-radioactive trace of an artificially enriched minor abundance isotope. Thus by measurement of the isotope ratio in the fortified sample solution, corrected for measurement bias, the elemental concentration can be determined. The accuracy and precision of lead isotope ratio measurements at low concentration have been described for a number of materials [48,49]. The formula employed by Dean et al. [49] for isotope dilution analysis calculation for lead has been described as:

$$A = \frac{[(xB_2M_1/M_2) - B_1]}{(Z - Zx/y)}$$

where: A = the number of grams of the element in the sample; x = measured isotope ratio (208/206 Pb) in fortified sample; B_1 = the number of grams of 208 Pb; B_2 = the number of grams of 206 Pb; M_1 = the atomic weight of 208 Pb; M_2 = the atomic weight of 206 Pb; Z = the fractional abundance of 208 Pb; and y = the isotope ration (208/206 Pb) in the unfortified sample.

In addition, lead hydride generation may be employed for isotope analysis by ICP–MS. Wang and Barnes [50,51] have reported optimised conditions for lead hydride generation in a hydrogen peroxide–hydrochloric acid system using flow injection ICP–AES, and later modified this system for ICP–MS [52].

9.2.7. Coupled (or hyphenated) techniques

The use of directly coupled systems which utilise the separatory powers of chromatography and the sensitive yet, selective, detection of atomic spectroscopy are now widely used in speciation studies. In the case of lead, over twelve alkyllead compounds have been tentatively identified in chromatograms [53], although not all of these are likely to be found in the environment. However the environmental methylation of inorganic lead has been reported [54], and evidence has been put forward for the existence of a biogeo-chemical cycle of organic lead [55]. The tetraalkyllead (R_4Pb) compounds in petrol (see Section 9.3.2) are also reported to decompose to inorganic lead via fairly persistent intermediates such as trialkyllead (R_3Pb^+) and dialkyllead (R_2Pb^{2+}) compounds [56]. Thus the need for techniques which can unequivocally determine individual species in environmental media is obvious, and coupled techniques often provide the most suitable analytical approach available in most laboratories. In addition, this approach is often attractive since it requires little specialised equipment, the interfaces are relatively cheap to construct, and the instruments employed may be readily decoupled after use. The applications of coupled techniques generally have been fully reviewed elsewhere [3,4], although the principal combinations used for lead speciation studies are summarised in Table 9.1 and an overview of recent literature shown in Table 9.4.

The most successful combination for lead speciation is that of gas chromatography (GC)–AAS. The approach adopted by many laboratories is based on the interface developed by Ebdon et al. [57] where the GC effluent is mixed with hydrogen and introduced through a hole in the side of a ceramic tube suspended in an air–acetylene flame in the optical path of the instrument. The hydrogen is burnt as a small diffusion flame within the tube, and it is this that atomises the sample, the air–acetylene flame serving only to keep the hydrogen flame alight and to heat the ceramic tube to prevent condensation reactions. The ceramic tube helps increase the residence time of the atoms once formed. The system is shown in detail in Fig. 9.1, and a typical chromatogram obtained using this technique shown in Fig. 9.2. The operating conditions employed at the Department of Environmental Sciences in Plymouth are given in Table 9.5. More recently other studies [58] have confirmed that a hydrogen diffusion flame burnt inside the tube in this way improves the atomisation efficiency. In the absence of hydrogen, much of the lead will not atomise and will be deposited on the walls of the tube. Forsyth and Marshall [59] have suggested that the hydrogen radicals produced result in the formation of lead hydride which helps the lead to atomise. A number of modifications to this system may be used, for example, using custom-built electrically heated cells in place of the flame, although all work on a similar principal. The detection limit of 17 pg of lead reported for the above flame system [57] is still one of the lowest reported for a directly coupled system.

References on p. 252

TABLE 9.4

OVERVIEW OF COUPLED CHROMATOGRAPHY–ATOMIC SPECTROSCOPIC APPLICA-
TIONS FOR LEAD

Application	Reference
Directly coupled GC–MIP	
Tetraalkyllead compounds in the atmosphere.	64
Trialkyllead compounds in water samples.	60,63
Tetraalkyllead compounds in petrol.	127,128
Directly coupled GC–ICP–AES	
Lead alkyls in petrol.	71,129
Directly coupled GC–AAS	
Lead alkyls in petrol.	57,88,104,105,130,132, 134,135,136
Tetraalkyllead compounds in the atmosphere.	88,106,107,137,141
Methylation of lead salts.	130
Tetraalkyllead compounds in water samples.	75,113,124,139,140
Tetraalkyllead compounds in sediments.	75,138
Tetraalkyllead compounds in biological materials.	75,124
Directly coupled GC–AFS	
Tetraalkyllead compounds.	142
Directly coupled HPLC–AAS	
Tetraalkyllead compounds in petrol.	69,70
Tetraphenyllead.	143
Separation of inorganic and organolead species.	68

GC may also be coupled to various plasma atomic emission instruments (see Table 9.1). The high excitation temperature of the microwave induced plasma (MIP) compared with its gas temperature would seem to make it well suited to gas analysis and coupling to a gas chromatograph. Unfortunately, however, interfacing is not altogether straightforward since venting of the solvent from the plasma is required to prevent the plasma being extinguished. Carbon deposition on the walls of the atom cell may also be a problem. Estes et al. [60] have described a method for the measurement of triethyl- and trimethyl-lead chloride in tapwater using a fused-silica capillary column gas chromatograph with microwave excited helium plasma lead detection. The method involved an initial extraction of trialkyllead ions from water [61,62] into benzene, which was then vacuum reduced to further concentrate the compounds. However, the method proved time-consuming and

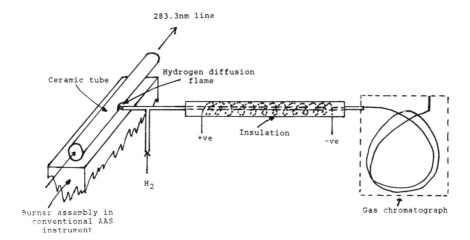

Fig. 9.1. Coupled GC–AAS system.

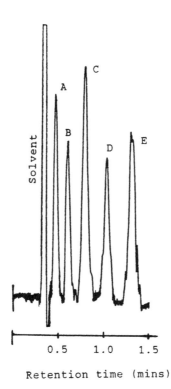

Retention time (mins)

Fig. 9.2. Typical GC–AAS chromatogram showing alkyllead compounds present in petrol. Peaks, with retention times in minutes in brackets: (A) tetramethyllead (0.052); (B) tri-methylethyllead (0.68); (C) dimethyldiethyllead (0.90); (D) triethylmethyllead (1.18); (E)

TABLE 9.5

OPERATING CONDITIONS FOR LEAD USING COUPLED GC–AAS

GC	AAS
Injector temperature 150°C	Lamp HCL(Pb) 6 mA
Column temperature 50°C at 10°C min^{-1} to 250°C	Wavelength 283.3 nm
Interface line 140°C	Bandpass 0.5 nm
N_2 carrier gas 40 ml min^{-1}	Air flow 4.5 l min^{-1}
H_2 auxiliary gas 105 ml min^{-1}	Acetylene flow 0.6 l min^{-1}
	Deuterium arc background correction applied

only semi-quantitative. The same authors later reported the n-butyl Grignard derivization of the trialkyllead ions extracted as the chlorides from spiked tapwater and industrial effluent [63]. Here the analyte compounds are extracted into benzene from an aqueous solution saturated with sodium chloride before being quantatively converted into the n-butyltrialkyllead derivatives. Precolumn Tenax trap enrichment of the derived trialkylbutylleads then enables the determination at low ppb levels to be carried out. In a further study looking at R_4Pb compounds in petrol and air samples [64], hydrogen (ca. 1%) was added to the argon carrier gas to prevent formation of lead deposits on the walls of the silica capillary.

Other forms of plasma may also be coupled to GC although such couplings are not often utilised. ICP has for example been little used as a GC detector, despite its wide adoption as a spectroanalytical emission source. However, this type of plasma has the advantage of accepting organic solvents more readily than does MIP, due to its higher gas temperature, and thus ICP has been more widely used in liquid chromatography couplings.

The coupling of high-performance liquid chromatography (HPLC) to AAS generally has not been widely used for organolead analysis, since the properties of many organolead compounds make them particularly well suited to GC separation. If liquid chromatography is to be used, simple interfaces such as the direct connection of the HPLC column to the nebuliser of the AAS instrument may lead to problems. This is due in part to the mismatch in flow-rates between the mobile phase of the HPLC and the uptake of the nebuliser if both instruments are to operate under optimal conditions. The difficulties in interfacing HPLC to AAS have been described in detail elsewhere [4,65], although many of the problems may be overcome and a number of successful interface designs have been constructed at our department [66–68]. These interfaces readily lend themselves to the separation of a range of lead species, and have the advantage of being able to separate a number of species which are not thermally stable and therefore not readily amenable to coupled GC systems. For example, as shown in Fig. 9.3, inorganic lead, tetraethyllead,

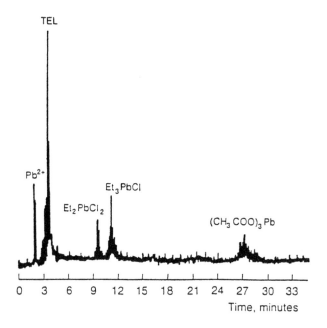

Fig. 9.3. Chromatogram showing the separation of inorganic and organolead compounds by directly coupled HPLC–AAS (TEL = tetraethyllead).

diethyllead dichloride, triethyllead chloride and triacetyllead chloride may be separated using a directly coupled HPLC–AAS system [68]. Here the sensitivity for some species is not as good as with coupled GC–AAS, although species which are not sufficiently volatile to be separated by GC, or indeed thermally unstable, may be readily determined.

The coupling of HPLC to electrothermal atomisers for AAS (ETA–AAS) is more problematical due to the discrete mode of operation of most furnace designs. Thus interface systems need to operate in a stop-flow fashion, or some form of fraction collector is required. For this reason there have been few reliable published methods for lead using this technique, although the use of post-column digestion [69] and high-temperature furnaces [70] have been reported. Once again full details of possible interface designs and their applications have been described elsewhere [4].

As mentioned above, the separation and detection of various lead species is also possible using HPLC coupled to ICP. However, once again nebulisation presents something of a problem since nebuliser uptake rates are again low, and in addition the plasma often exhibits a low tolerance to many organic solvents. To overcome these problems, various nebuliser and mobile phases have been examined [71,72] and a number of successful applications for lead have been reported including the separation of diethyllead dichloride, triethyllead chloride, trimethyllead acetate and triphenyllead.

9.3. APPLICATIONS

9.3.1. Determination of total lead

Until fairly recently, much of the information published on the concentration of lead species in the environment was unreliable due to contamination of the samples during sampling and analysis. In addition, the importance of speciation data was not recognised for many applications. However, a number of techniques have been developed to remove lead form organic and inorganic materials usually by variations of an acid digestion [73–75]. For fish, vegetation and sediments, mixtures of nitric acid, perchloric acid, and sulphuric acid may be used in the ratios of 4:1:0.5, 4:1:1, and 4:1:0 respectively [75]. Samples are best soaked overnight in the acid mixture followed by a low heat digestion for 1 hour and then increasing the temperature for another 2–3 hours until the solution evaporates. Subsequent determination of the lead may then be accomplished by diluting the sample and then using one of the variety of instrumental techniques described in Section 9.2. Hexane extractable lead may also be determined by adding 1 g of sample and 2 ml of hexane into a screw top container [76], the sample is then placed in an

TABLE 9.6

TYPICAL CONCENTRATION OF LEAD IN ENVIRONMENTAL SAMPLES

Sample	Lead concentration
Urban air (UK) (particulate)	0.1–10 μg m^{-3}
Rural air (UK) (particulate)	0.01–2.25 μg m^{-3}
Groundwater	<0.01 mg dm^{-3}
Rainwater	<0.4 mg dm^{-3} [a]
Inland waters and rivers (dissolved)	0.001–0.010 mg dm^{-3} [b]
Coastal waters	<0.008 mg dm^{-3} [c]
Ocean waters	0.004–0.065 μg dm^{-3}
Tapwater	<0.3 mg dm^{-3}
Soil	10–10 000 mg kg^{-1} [d]
Vegetation	0.1–1000 mg kg^{-1} [d]

[a] Highest levels for urban areas.
[b] Dominating tendency for lead to be bound to suspended matter, levels depend on anthropogenic imputs.
[c] 0.004 μg dm^{-3} reported for deep ocean water (see ref. 116).
[d] Higher levels found in polluted or mineralised areas.

ultrasonic bath before leaving to settle. In the case of water samples, a 200-ml aliquot may be shaken with 5 ml of hexane prior to leaving the phases to separate. In both cases final analysis may be achieved using ETA–AAS with inorganic lead standards for calibration. Typical lead concentrations for a range of environmental samples are shown in Table 9.6.

Other less common techniques such as thermal ionisation mass spectrometry [77] may also be used for the determination of total lead, although often these techniques have been developed for specific applications. One example is a procedure developed for determining lead in lubricating oils by means of electrodeposition and secondary ion mass spectrometry using an ion microprobe mass analyser (IMMA) [78–80]. This particular technique consists of three stages. First, the organic lead is treated to release the metal to an aqueous solution. Second the aqueous lead ions are electroplated onto platinium electrodes. Third the surfaces of the electrode are analysed with the IMMA, which provides data on the relative amount of each mass number component in the sample. Quantitative analysis can be obtained by isotope dilution, adding a known amount of a standard lead compound to the sample. Obviously the standard should have an isotopic composition which is distinct from those expected from the sample and from that of natural lead, so that it may be distinguished from those other components of the sample.

9.3.2. Alkyllead compounds in petroleum

The major organolead species present in the environment are the tetraalkyllead compounds and their di- and trialkyl decomposition products [80]. The presence of these species may be derived from two possible sources, either (i) anthropogenic petroleum inputs and/or (ii) environmental methylation of natural lead compounds. The latter of these possibilities is discussed in Section 9.3.5, however, since the determination of alkyllead compounds derived from petrol has received so much attention, and indeed is of importance to the lead chemistry of many environmental samples, it is worth discussing some of the analytical techniques that have been used to determine the lead, and more specifically the speciation of lead in petrols in some detail.

It is generally accepted that the first synthesized organolead compound was made in 1853 by Lowig [81]. Lowig's product was made by reacting a sodium–lead alloy with ethyl iodide to give either tetraethyllead or hexaethyldilead. From this early start organolead chemistry has become one of the largest branches of organometallic chemistry. Although antiknock additives are the major use of organolead compounds, there are other commercial applications, such as their use in the manufacture of organomercury fungicides by alkylation [82], their use in marine antifouling paints, lubricant additives, as catalysts in the production of polyurethan foams [83], stabilizers for PVC, in flame retardants, and rodent repellants.

A number of methods may be used for the determination of tetraethyl- or total lead in petrol. The most common methods comprise the use of various wet processes [84,85], polarography [86], GC [87–89], HPLC [90], atomic absorption [91], and coupled techniques as described in Section 9.2.7. Wet methods are often tedious since they require conversion of tetraethyllead to lead salts which are then determined gravimetrically, titri-metrically, or spectrophotometrically. Polarography requires the same sample pretreat-ment and so may be classified as a wet method. The determination of tetraethyllead in petrol by HPLC techniques as described in ASTM methods is generally confined to high levels of lead (0.2–5.0 g Pb/US gal range in most cases), often requires large amounts of sample, i.e. between 5 and 50 ml, and reproducilities may also be poor [90]. In the case of GC the use of conventional detectors is complicated because of peak overlaps and special columns must be used for retarding the aromatic hydrocarbons [89]. Thus coupled GC–AAS systems are now usually employed to utilise the benefits of specific element detection. This technique offers a dection limit of 17 pg for alkyllead and is thus suitable for environmental work [52]. In addition the technique may also be used in forensic studies [92], and at our department we have successfully been able to identify traces of petrol on hand swabs taken up to 7 hours after the original contact with the petrol.

9.3.3. Determination of lead in air

Airborne lead compounds are found in both gaseous and particulate matter. Although there are many organolead compounds, the environmental chemistry is dominated by a relatively small number of tetraalkyllead compounds, their salts and decomposition prod-ucts. As trialkyllead compounds are probably more chemically stable than tetraalkyllead compounds [93], their existence in the atmosphere is also plausible. Most lead com-pounds enter the environment as small aerosol particles of less than 1 μm. Such fine particles may travel long distances by atmospheric transportation processes prior to removal, however, larger particles are deposited by fallout or wet washout much closer to the source. Early methods proposed for the determination of volatile organolead com-pounds often relied on the removal on any particulate lead compounds present in the sample by a filter positioned before the absorption solution. However, it has been sug-gested that a significant proportion of atmospheric particulate lead may escape collection by conventional filters [94]. In such cases any lead that passes the filter would be ab-sorbed and measured as tetraalkyllead compounds. A further complication with any non-specific chemical method is that, because lead is naturally ubiquitous, blank levels tend to be relatively high. The de-leading of reagents minimises the problem but unfortu-nately some compounds are extremely difficult to de-lead by conventional procedures and a residual blank is usually present.

Several of the early methods for the determination of tetraakyllead compounds in air were based on dry-sampling procedures. In one such method, suitable for field use, iodine crystals were used as a collector [95], and a standard volume of sample taken at a fixed sampling rate. This may be modified [96] to collect the tetraalkylleads on activated carbon over a period of up to 4 days, but it is necessary to carry out a lengthy extraction procedure with nitric acid-perchloric acid prior to the lead determination. Other work has also been reported using a dry sampling system with sampling periods of up to 8 hours [97].

Collection System.

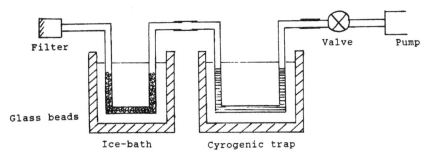

Filter

Valve   Pump

Glass beads

Ice-bath   Cyrogenic trap

Desorption / Detection System.

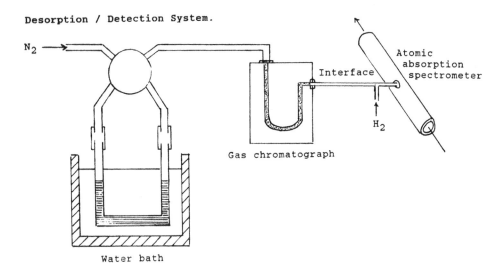

N₂ →

Atomic
absorption
Interface   spectrometer

H₂

Gas chromatograph

Water bath

Fig. 9.4. Apparatus for collection of tetraalkyllead compounds.

Air samples may be passed through a glass-fibre-ionised carbon filter on which the tetraalkyllead compounds can be collected [98]. In this case the lead is extracted from the filter and converted into the inorganic state by treatment with iodine solution followed by colorimetric determination as lead dithizone using a commercially available test kit. In an alternative procedure involving the use of nitric acid-bromide regent for extraction, the lead is determined by ETA–AAS.

Laveskog was the first to determine individual alkyllead compounds in street air [99], although his method involved GC–MS instrumentation which may not be available to all workers. Cantuti and Cartoni were the first to describe a method for the direct collection of tetraethyllead from polluted air and its chromatographic determination at ppm levels by using electron-capture detection (ECD) [100]. However, in this case although ECD offers good sensitivity, it lacks specificity especially for environmental samples which may contain a variety of compounds with high electron affinity. Flame ionisation detection (FID) has rather poor responses for tetraalkylleads compounds, however, flame emission detectors [101] and hydrogen-rich FID detection [102] may be used for the analysis of tetraethyllead with reasonable sensitivity. Today, however, the use of AAS detection as the final stage of the method is now utilised by most workers [88,103–106].

To facilitate sampling using this technique, the lead compounds are usually collected on an adsorbent held in a U-shaped glass tube at low temperatures. Although GC adsorbents are often used [107], glass beads have also been used to good effect [49,106,108]. Gas-phase and particulate material are separated by employing an on-line filter. One problem associated with this approach is the build-up of moisture in the trap, and so the collection tube is usually preceded by some form of condenser. An empty U-tube at $-78°C$ (dry-ice–methanol mixture) may be used [59,109], although such an approach may not remove all the water if large volumes of air are to be sampled, and so the tube may be packed with glass beads to increase the drying efficiency. Once collected, thermal desorption is used to release the organolead compounds to the detection system. The complete system is shown schematically in Fig. 9.4. The use of coupled GC–atomic spectroscopy for the analysis of organolead compounds in air has been reviewed in detail elsewhere [48].

The determination of tetramethyllead and tetraethyllead in the atmosphere has also been achieved using a two-step enrichment method and GC–MS isotope dilution analysis [110]. Using this technique the tetraalkyllead compounds are collected on Porapak QS or N at ambient temperature, desorbed, re-collected at $-80°C$ on a small column containing 4% Apiezon M on Chromosorb PAW-DCMS, and then analysed by GC–MS with ion monitoring. The isotope dilution technique is used by adding known amounts of [12-^{2}H]tetramethyllead and [20-^{2}H]tetraethyllead to the sampling columns in advance, thus making it possible to correct for decomposition during the sampling and/or analysis. This method has proved successful in determining tetramethyllead in air with levels down to 20 pg m^{-3}. It is also suitable for field measurements since tetramethyllead is collected at ambient temperatures and because it is possible to correct for any decomposition of tetramethyllead taking place during sampling, and the collected samples may be transported over long distances.

9.3.4. Determination of lead in water

Lead may exist as both dissolved and suspended species in water. In many cases it is important to explore the possible biomethylation of lead and the distribution of organolead compounds from man made sources in the environment. Lead in aqueous environments is most likely to be found in the +2 oxidation state. Only small quantities are likely to reach water via air deposition, drainage water or runoff from agricultural land, the most important source being industrial effluents. It is important to remember that the high-molecular-weight humic acid content of some waters may also play an important role in the mobilising and transportation of many heavy metals including lead. Galena is an important source of lead in surface waters, since although the mineral itself is insoluble, it may be slowly oxidised by air to give the rather more soluble sulphate:

$$PbS + 2O_2 \rightarrow PbSO_4$$

Species such as $Pb(OH)_2$, $Pb(OH)_3^-$ are likely to be found when the lead concentrations are low, although polymeric lead ions such as $Pb_2(OH)_3^+$ and $Pb_4(OH)_4^{4+}$ may also be present, in addition to insoluble compounds such as PbO, $PbCO_3$, and $PbSO_4$.

As implied above, the presence of suspended matter may markedly reduce the recovery of lead compounds from aqueous samples [49]. Thus any suspended matter must be separated from the water body prior to analysis. This tendancy of lead to be bound in suspension is also reflected in the capacity of riverwater to reduce its lead burden via sedimentation resulting in only small variations in the levels of dissolved lead within natural water bodies [111]. The association of lead with suspended matter should also be bourne in mind when dealing with speciation studies, particularly where filtration of the water sample is inappropriate due to possible loss of R_4Pb species by adsorption onto suspended particles [112]. In addition, it is also desirable to keep the time between sampling and analysis as short as possible to prevent possible conversion of R_4Pb species into R_3Pb^+ in solution [51]. If dark bottles are used, R_3Pb^+ and R_2Pb^{2+} are fairly stable during transportation, although in all cases it is best to perform the extraction and analysis as soon as possible after sample collection.

Total lead may be determined using one of the techniques discussed in Section 9.2, although the presence of organic materials will require the use of strong oxidising acids such as hot concentrated nitric or perchloric acid to free the lead. The extraction of tetraalkyllead compounds from aqueous samples may be achieved using a range of organic solvents, although benzene and n-hexane have been widely used by many workers. The usual procedure is to add 5–50 ml of solvent to 100–2000 ml of sample in a glass bottle and then shake for about 30 minutes. The two phases are then separated and ready for analysis by GC–AAS. This approach has been used for the direct analysis of tetraalkyllead compounds [75], and has also been used for studies on the interaction between tetraethyllead and seawater [113]. These extractions may be performed in the presence of sodium chloride and sodium diethyldithiocarbamate, and the samples alkylated by addition of Grignard reagent to the extract if separation of R_4Pb, R_3Pb^+, and R_2Pb^{2+} is required. Typical Grignard reagents employed are n-butylmagnesium chloride [114], propylmagnesium chloride [115], and phenylmagnesium chloride [54].

In other cases, however, ionic alkyllead compounds in aqueous samples have been derivatised *in situ* to the R_4Pb form, and then determined by head space analysis. This method reduces potential sample contamination by purging the volatile compounds from solution onto a solid adsorbent, thus minimising sample handling [48]. Detection limits for this technique are comparable to those obtained by solvent extraction, although the approach is not suitable for R_3Pb^+ and R_2Pb^{2+} species.

References on p. 252

9.3.5. Determination of lead in sediments, soils and biological materials

Appreciable quantities of lead may be found in soils and sediments since the very low mobility of lead can give rise to a build up in localised areas. Some lead may be derived from the mineralogy, although anthropogenic inputs often give rise to elevated levels. In orchards for example, the lead content of soil may be high following the application of lead arsenate as an insecticide, and near roads atmospheric deposition (either from dry fallout or wet washout) may be important. Since lead is generally retained in the top few centimeters of soil, and only slowly carried down by leaching due to its interaction with soil colloids it is likely to be quite persistent. The determination of lead in soils and sediments may also be used as an indication of atmospheric or water pollution. In unpolluted soils and sediments a substantial proportion of the lead is firmly bound in the silicate lattices of refractory material and can only be released by use of hydrofluoric acid [116]. Thus hydrofluoric acid–nitric acid mixtures may be required for extraction, although nitric acid or nitric acid–hydrochloric acid is usually adequate to remove most lead from heavily polluted soils. To extract the lead, the soil is digested in the acid which is then evaporated, thus effecting the oxidative decomposition of any organic material present. The digest is finally leached with dilute acid and filtered prior to analysis, usually by AAS. If high levels of organic materials are present, as is the case with many biological materials such as leaves or animal tissue, the use of perchloric acid may be required for complete oxidation following prior digestion with nitric acid. Obviously care must be taken with this procedure to avoid the risk of explosion.

The traditional dissolution techniques such as those discussed above and in Section 9.2.3, tend to be tedious and time consuming. In addition, there is always the possibility of contamination during sample pre-treatment. To avoid these problems attempts have been made to develop direct methods of analysis. One such approach is lead atomisation from soil by slurry introduction electrothermal atomisation AAS [117,118]. Here a modifier is required to delay atomisation until "nearly-isothermal" conditions have been established in the furnace and to remove the interferences due to organic carbon. Various modifiers have been investigated, although the combination of palladium chloride and magnesium nitrate has been found to be particularly well suited to this application [119]. It is, however, important to use the modifier in small amounts to avoid interference from the chloride present within the modifier. The use of graphite capsules which enclose the sample in a graphite cylinder during vapourisation have also been used. In this case the vapours (molecules and/or atoms) diffuse through the porous graphite walls into a flame atomiser [120,121]. However, this approach has been largely confined to research use.

Following density gradient and magnet fractionation, lead compounds in soils may also be identified using x-ray powder diffraction (XRD) [122,123]. The density gradient method may be performed by centrifuging 10 g of sample first in tetrachloro methane and then in mixtures of tetrachloro methane with diiodo methane of increasing density. In such studies almost 80% of the lead has been found in the high-density soil fraction, i.e. > 3.32 g cm^{-3}, and since the major soil components generally have a density lower than this, the technique serves as a preconcentration step. Using XRD several lead species have been reported, e.g. $PbSO_4$, $PbSO_4 \cdot (NH_4)_2SO_4$, Pb_3O_4, $PbO \cdot PbSO_4$ and $2PbCO_3 \cdot Pb(OH)_2$, although it has also been suggested that the majority of lead does not exist in a crystalline form susceptible to XRD analysis [124].

Organolead compounds may be extracted from soils and sediments by one of several procedures. One method involves the vigorous shaking of 5 g of wet sediment with 5 ml of 0.1 M EDTA and 5 ml of hexane in a capped test-tube for 2 hours [124]. The mixture is then centrifuged prior to analysis by GC–AAS. Alternatively, 1 g of sediment may be heated in a glass U-tube at 100°C for 20 minutes to volatilise the tetraalkyl compounds. Here air is passed through the U-tube, and then via a moisture trap, to a U-tube packed with 3% OV-1 Chromosorb W which acts as a sample trap [75]. Again GC–AAS is used for analysis as outlined in Section 9.3.3. For complete separation of R_4Pb, R_3Pb^+ and R_2Pb^{2+}, Chau et al. [125] has reported a method which involves 1–2 g of dried sample (5 g wet) being placed in a capped vial with 3 ml of benzene after the addition of 10 ml of water, 6 g of sodium chloride, 1 g of pottasium iodide, 2 g of sodium benzoate, 3 ml of 0.5 M sodium diethydithiocarbamate and 2 g of coarse glass beads. The vials were shaken for 2 hours and 1-ml aliquots of the benzene extract butylated prior to analysis by GC–ETA–AAS. In this case detection limits of 15 mg g^{-1} Pb were reported. The use of coupled GC–AAS in all of these methods demonstrates the utility of the approach, and indeed this technique has also been used to investigate the methylation of inorganic lead by sediments [126], the data from which gives an important insight into lead transport in esturies.

9.4. CONCLUSIONS

There are many techniques which may be considered suitable for the determination of total lead. However, some of these approaches are not suitable for environmental work due to poor sensitivity or lack of specificity, whilst others are only really suitable for particular applications. In all cases, which ever technique is to be used, care must be taken to ensure representative sampling, and freedom from contamination. There is no doubt that electrothermal atomisation atomic absorption spectroscopy is the most widely used technique for the determination of lead at the present time. This approach offers good detection limits, and most of the interferences encountered with the various matrix types in environmental samples have now been identified and can be eliminated. However, the technique is only single-element, slow, and may need careful optimisation and operator expertise to work well. Thus analysts continue to evaluate other methods for many applications.

X-ray fluorescence is another useful technique, particularly for the determination of airborne lead, since it avoids sample pretreatment of collection filters. The extraordinary detection limits possible using differential pulse anodic stripping voltammetry with glassy carbon/mercury thin film electrodes is also worth noting. Other techniques, however, have proved disappointing. The use of inductively coupled plasma–atomic emission spectrometry for example offers relatively poor detection limits. The choice of which technique to use may be further complicated if speciation data is required. Here, the use of coupled

References on p. 252

gas chromatography–atomic spectroscopy has proved a particularly useful tool since it is both sensitive and selective to lead containing species. The major drawback in this case is the traditional separation of the instrumentation in many commercial laboratories. Clearly, whilst speciation data is not the primary requirement of such laboratories, this approach may not always be available. However, the situation is changing fast as new legislation requires data on specific species and more laboratories seek to establish routine methods for such work.

Looking towards the future, one of the most exciting developments in recent years has been the introduction of ICP–MS instrumentation into many laboratories. The particularly low detection limit for lead and the ability to determine individual isotopes, coupled to the multielement capabilities and sample throughput advantages of ICP–MS, make this a very attractive technique. In addition when compared to traditional techniques such as ETA–AAS better precision may also be obtained. Since ICP–MS may be coupled to both GC and HPLC systems for speciation work, it is certainly a powerful and versatile analytical tool, offering the analyst much potential for the future.

REFERENCES

1 C.C. Patterson, *Arch. Environ. Health*, 11 (1965) 344.
2 N.N. Greenwood and A. Earnshaw, *Chemistry of the Elements*, Pergamon, Oxford, 1986.
3 L. Ebdon, S.J. Hill and R. Ward, *Analyst (London)*, 111 (1986) 1113.
4 L. Ebdon, S.J. Hill and R. Ward, *Analyst (London)*, 112 (1987) 1.
5 R.M. Dagnall and T.S. West, *Talanta*, 11 (1964) 1553.
6 W.J. Price, *Spectrochemical Analysis by Atomic Absorption*, Heydon, London, 1979.
7 H.L. Kahn, G.E. Peterson and J.E. Schallis, *At. Absorption Newslett.*, 7 (1968) 35.
8 H.T. Delves, *Analyst (London)*, 95 (1970) 431.
9 W. Frech and A. Cedergren, *Anal. Chim. Acta*, 88 (1977) 57.
10 A. Cedergren, W. Frech, E. Lundberg and J.A. Persson, *Anal. Chim. Acta*, 128 (1981) 1.
11 J.P. Matousek, *Prog. Anal. At. Spectrosc.*, 4 (1981) 247.
12 B. Welz, *Atomic Absorption Spectrometry*, VCH Publishers, Weinheim, 1985.
13 S.G. Salmon, R.H. Davis and J.A. Holcombe, *Anal. Chem.*, 53 (1981) 324.
14 T.M. Vickrey, G.V. Harrison and G.J. Ramelov, *At. Spectrosc.*, 1 (1980) 116.
15 W. Frech and A. Cedergren, *Anal. Chim. Acta*, 82 (1976) 93.
16 H. Heinrichs, *Z. Anal. Chem.*, 295 (1979) 355.
17 P.N. Vijan and G.R. Wood, *Analyst (London)*, 101 (1979) 966.
18 H.D. Fleming and R.G. Ide, *Anal. Chim. Acta*, 83 (1976) 67.
19 R.F. Sanzolone, *J. Anal. At. Spectrom.*, 1 (1986) 343.
20 R.F. Browner, R.M. Dagnall and T.S. West, *Anal. Chim. Acta*, 50 (1970) 375.
21 U. Jin and M. Taga, *Anal. Chim. Acta*, 143 (1982) 229.
22 J. Kumamaru, F. Nakata, M. Haras and M. Kiboki, *Bunseki Kagaku*, 33 (1984) 624.
23 M. Bonilla, L. Rodriguez and C. Camara, *J. Anal. At. Spectrom.*, 2 (1987) 157.
24 W.M. McCord and J.W. Zemp, *Anal. Chem.*, 27 (1955) 1171.
25 S.L. Tompsett, *Analyst (London)*, 81 (1956) 330.
26 E. Berman, *Amer. J. Clin. Pathol.*, 36 (1961) 549.
27 A.I. Vogel, *A Textbook of Quantitative Inorganic Analysis*, Longman, London, 1978.
28 S. Hancock and A. Slater, *Analyst (London)*, 100 (1975) 422.
29 R. Moss and E.V. Browett, *Analyst (London)*, 91 (1966) 428.
30 D.A. Skoog and D.M. West, *Fundamentals of Analytical Chemistry*, Saunders College Publishing, New York, 4th ed., 1982.

31 *IP Standards for Petroleum and its Products, Part 1, Vol. 1, IP248/71*, Institute of Petroleum, London, 1985.
32 *IP Standards for Petroleum and its Products, Part 1, Vol. 2, IP270/77*, Institute of Petroleum, London, 1986.
33 A.S. Bazhov, *Zh. Anal. Khim.*, 23 (1968) 1640.
34 K. Bamback and J. Cholak, *Ind. Eng. Chem. (Anal. Ed.,)*, 13 (1941) 504.
35 Z. Kublik, *Acta Chim. (Hung.)*, 27 (1961) 79.
36 G.C. Whitnack and C.A. Sasselli, *Anal. Chim. Acta*, 47 (1969) 274.
37 J.L. Guinon and R. Grima, *Analyst (London)*, 113 (1988) 613.
38 L.M. Petrie and R.W. Baier, *Anal. Chem.*, 50 (1978) 351.
39 L. Duic, S. Szechter and J. Srinivasan, *J. Electroanal. Chim.*, 19 (1973) 76.
40 T.R. Copeland and R.K. Skogerbee, *Anal. Chem.*, 46 (1974) 1257A.
41 M.J. Pinchin and J. Newham, *Anal. Chim. Acta*, 90 (1977) 91.
42 A.M.Y. Jaber, G.J. Moody and J.D.R. Thomas, *Analyst (London)*, 113 (1988) 1409.
43 A.M.Y. Jaber, G.J. Moody and J.D.R. Thomas, *Analyst (London)*, 101 (1976) 179.
44 R.M. Harrison, in R. Perry and R.J. Young (Editors), *Handbook of Air Pollution Analysis*, Chapman and Hall, London, 1977.
45 J. de Gyves, M. Baucells, E. Cardellach and J.L. Brianso, *Analyst (London)*, 114 (1989) 559.
46 M. Baucells, G. Lacort, M. Roura and J. de Gyves, *Analyst (London)*, 113 (1988) 1325.
47 N.I. Ward, in A.R. Date and A.L. Gray (Editors), *Applications of Inductively Coupled Plasma Mass Spectrometry*, Blackie and Son Ltd., London, 1989, pp. 189–219.
48 J.R. Dean, L. Ebdon and R. Massey, *J. Anal. At. Spectrom.*, 2 (1987) 369.
49 J.R. Dean, L. Ebdon, H. Crews and R.C. Massey, in C.S. Creaser and A.M.C. Davies (Editors), *Analytical Applications of Spectroscopy*, Royal Society of Chemistry, London, 1988, pp. 305–307.
50 X.R. Wang and R.M. Barnes, *Spectrochim. Acta Part B*, 42 (1987) 139.
51 X.R. Wang and R.M. Barnes, *Spectrochim. Acta. Part B*, 41 (1986) 967.
52 X.R. Wang, M. Viczian, A. Lasztity and R.M. Barnes, *J. Anal. At. Spectrom.*, 3 (1988) 821.
53 M. Radojević, in R.M. Harrison and S. Rapsomanikis (Editors), *Environmental Analysis Using Chromatography Interfaced with Atomic Spectroscopy*, Ellis Horwood, Chichester, 1989, pp. 223–257.
54 A.P. Walton, *PhD. Thesis; Metal Methylation in Estuarine Waters*, 1986, CNAA.
55 C.N. Hewitt and R.M. Harrison, in P.J. Craig (Editor), *Organometallic Compounds in the Environment: Principles and Reactions*, Longman, London, 1986, p. 92.
56 R.M. Harrison, C.N. Hewitt and M. Radojević, *Int. Conf. Chemicals in the Environment, Lisbon, Portugal*, Selper, London, 1986, p. 110.
57 L. Ebdon, R.W. Ward and D.A. Leathard, *Analyst (London), 107 (1982) 129.*
58 S. Rapsomanikis, O.F.X. Donard and J.H. Weber, I*Anal. Chem.*, 54 (1982) 872.
59 D.S. Forsyth and W.D. Marshall, *Anal. Chem.*, 57 (1985) 1299.
60 S.A. Estes, P.C. Uden and R.M. Barnes, *Anal. Chem.*, 53 (1981) 1336.
61 A.W. Bolanowska, *Br. J. Ind. Med.*, 25 (1968) 203.
62 A.W. Bolanowska, *Chem. Anal. (Warsaw)*, 12 (1967) 121.
63 S.A. Estes, P.C. Uden and R.M. Barnes, *Anal. Chem.*, 54 (1982) 2402.
64 D.C. Reamer, W.H. Zoller and T.C. O'Haver, *Anal. Chem.*, 50 (1978) 1449.
65 L. Ebdon and S.J. Hill, in R.M. Harrison and S. Rapsomanikis (Editors), *Environmental Analysis Using Chromatography Interfaced with Atomic Spectroscopy*, Ellis Horwood, Chichester, 1989, pp. 165–188.
66 L. Ebdon, S.J. Hill and P. Jones, *Analyst (London)*, 110 (1985) 515.
67 L. Ebdon, J.I. Garcia Alonso, S.J. Hill and A. Hopkins, *J. Anal. At. Spectrom.*, 3 (1988) 395.
68 L. Ebdon, S.J. Hill and P. Jones, *J. Anal. At. Spectrom.*, 2 (1987) 205.
69 T.M. Vickrey, H.E. Howell, C.V. Harrison and G.J. Ramelow, *Anal. Chem.*, 52 (1980) 1743.
70 H. Koizumi, R.D. McLaughlin and T. Hadeishi, *Anal. Chem.*, 51 (1979) 387.

71 M. Ibrahim, W. Nisamaneepong, D.L. Hans and J.A. Caruso, *Spectrochim. Acta,* 40B (1985) 367.
72 J.W. Robinson and E.D. Boothe, *Spectrosc., Lett.,* 17 (1984) 689.
73 C. Feldman, *Anal. Chem.,* 46 (1974) 1606.
74 W.J. Simmons and J.F. Loneragen, *Anal. Chem.,* 47 (1975) 566.
75 Y.K. Chau, P.T.S. Wong, O. Kramar, G.A. Bengert, R.B. Cruz, J.D. Kinrade, J. Lye and J.C. Van Loon, *Bull. Environ. Contam. Toxicol.,* 24 (1980) 265.
76 R.B. Cruz, C. Lorouso, S. George, Y. Thomassen, J.D. Kinrade, L.R.P. Butler, J. Lye and J.C. Van Loon, *Spectrochim. Acta,* 335B (1980) 775.
77 I.L. Barnes, T.J. Murphy, J.W. Gramlich and W.R. Shields, *Anal. Chem.,* 44 (1972) 2050.
78 P.A. Bertrand, R. Bauer and P.D. Fleischauer, *Anal. Chem.,* 52 (1980) 1279.
79 I.L. Barnes, T.J. Murphy, J.W. Gramilich and W.R. Shields, *Anal. Chem.,* 45 (1973) 1881.
80 C.N. Hewitt and R.M. Harrison, in P.J. Craig (Editor), *Organometallic Compounds in the Environment: Principles and Reactions,* Longman, London, 1986, pp. 160–191.
81 C. Lowig, *Ann. Chem.,* 88 (1853) 318.
82 M.S. Whelen, in *Metal-organic Compounds (Advances in Chemistry Series,* Vol. 23, American Chemical Society, Washington, DC, 1959. p. 82.
83 H.G.J. Overmars and G.M. van der Want, *Chimica,* 19 (1965) 126.
84 *ASTM Standards on Petroleum Products and Lubricants,* American Society for Testing and Materials, Philadelphia, PA, 1967, Method D 526-61, 1961, p. 260.
85 Institute of Petroleum, Method Ip 116/57, modified.
86 J.L. Guinon and R. Grima, *Analyst (London),* 113 (1988) 613.
87 D.T. Coker, *Anal. Chem.,* 47 (1975) 386.
88 Y.K. Chau, P.T.S. Wong and J. Saitoh, *J. Chromatogr. Sci.,* 14 (1976) 162.
89 G. Castello, *Chim. Ind.,* 51 (1969) 700.
90 T.C.S. Ruo, M.L. Selucky and O.P. Strausz, *Anal. Chem.,* 49 (1977) 1761.
91 G. Thilliez, *Anal. Chem.,* 39 (1967) 427.
92 S.J. Hill, L. Ebdon, A. Walton and R. Ward, unpublished data.
93 P. Grandjean and T. Nielsen, *Residue Rev.,* 72 (1979) 97.
94 H.W. Edwards, paper presented at *International Symposium on Health, Paris, 1974.*
95 L.J. Snyder and S.R. Henderson, *Anal. Chem.,* 33 (1961) 1175.
96 L.J. Snyder, *Anal. Chem.,* 39 (1967) 591.
97 D.T. Coker, *Ann. Occup. Hyg.,* 21 (1978) 33.
98 S.E. Birnie and F.G. Noden, *Analyst (London), 105 (1980) 110.*
99 A. Laveskog, *Second International Clean Air Congress,* International Union of Air Pollution Associates, Washington, DC, 1970, p. 549.
100 V. Cantuti and G.P. Cartoni, *J. Chromatogr.,* 32 (1968) 641.
101 H.H. Hill and W.A. Aue, *J. Chromatogr.,* 74 (1972) 311.
102 W.A. Aue anad H.H. Hill, *J. Chromatogr.,* 74 (1972) 319.
103 R.M. Harrison, R. Perry and D.G. Slater, *Atmos. Environ.,* 8 (1974) 1187.
104 B. Kolb, F.H. Kemmner, F.H. Schleser and E. Wiedeking, *Fresenius' Z. Anal. Chem.,* 221 (1966) 166.
105 D.A. Seger, *Anal. Lett.,* 7 (1974) 89.
106 W.R.A. de Jonghe, D. Chakraborti and F.C. Adams, *Anal. Chem.,* 52 (1980) 1974.
107 B. Radziuk, Y. Thomassen, J.C. Van Loon and Y.K. Chau, *Anal. Chim. Acta,* 105 (1979) 255.
108 C.N. Hewitt, R.M. Harrison and M. Radojevic, *Anal. Chim. Acta,* 188 (1986) 229.
109 E. Rohbock, H.W. Georgii and J. Muller, *Atmos. Environ.,* 14 (1980) 89.
110 T. Nielsen, H. Egsgaard, E. Larsen and G. Schroll, *Anal. Chim. Acta,* 124 (1981) 1.
111 R. Breder, H.W. Nurnberg, J. Golimowski and M. Stoeppler, in H.W. Nuonberg (Editor), *Pollutants and their Ecotoxicological Significance,* Wiley, Washington, DC, 1985, p. 205.
112 A.W.P. Jarvie, R.N. Markall and H.R. Potter, *Environ. Res.,* 25 (1981) 241.
113 J.W. Robinson, E.L. Kiesel and I.A.L. Rhodes, *J. Environ. Sci. Health,* A14 (1979) 65.
114 Y.K. Chau, P.T.S. Wong and O. Kramar, *Anal. Chim. Acta,* 146 (1983) 211.

115 M. Radojevic, A. Allen, S. Rapsomanikis and R.M. Harrison, *Anal. Chem.*, 58 (1986) 658.
116 R.M. Harrison and D.P.H. Lexon, *Lead Pollution Causes and Control*, Chapman and Hall, London, 1981.
117 M.W. Hinds and K.W. Jackson, *J. Anal. At. Spectrom.*, 2 (1987) 441.
118 M.W. Hinds and K.W. Jackson, *J. Anal. At. Spectrom.*, 3 (1988) 997.
119 M.W. Hinds, M. Katyal and K.W. Jackson, *J. Anal. At. Spectrom.*, 3 (1988) 83.
120 D.A. Kachov, L.P. Kruglikova and B.V. L'vov, *Zh. Anal. Khim.*, 30 (1975) 238.
121 J. Stupar, *J. Anal. At. Spectrom.*, 1 (1986) 373.
122 K.W. Olson and R.K. Skogerboe, *Environ. Sci. Technol.*, 9 (1975) 227.
123 P.D.E. Biggins and R.M. Harrison, *Environ. Sci. Technol.*, 13 (1980) 336.
124 Y.K. Chau, P.T.S. Wong, G.A. Bengert and O. Kramar, *Anal. Chem.*, 51 (1979) 186.
125 Y.K. Chau, P.T.S. Wong, G.A. Bengert and J.L. Dunn., *Anal. Chem.*, 56 (1984) 271.
126 A.P. Walton, L. Ebdon and G.E. Millward, *Appl. Organomet. Chem.*, 2 (1988) 87.
127 S.A. Estes, P.C. Uden and R.M. Barnes, *J. Chromatogr.*, 239 (1982) 181.
128 P.C. Uden, *Anal. Proc.*, 18 (1981) 181.
129 D. Sommer and K. Ohls, *Fresenius' Z. Anal. Chem.*, 295 (1979) 337.
130 D.S. Treybig and S.R. Ellebracht, *Anal. Chem.*, 52 (1980) 1633.
131 P.T.S. Wong, Y.K. Chau and P.L. Luxon, *Nature (London)*, 253 (1975) 263.
132 J.W. Robinson, E.L. Kiesel, J.P. Goodbread, R. Bliss and R. Marshall, *Anal. Chim. Acta*, 92 (1977) 321.
133 I. Ahmad, Y.K. Chau, P.T.S. Wong, A.J. Carty and L. Taylor, *Nature (London)*, 287 (1980) 716.
134 W. de Jonghe, D. Chakraborti and F. Adams, *Anal. Chim. Acta*, 115 (1980) 89.
135 D. Chakraborti, S.G. Jiange, P. Surkign, W. de Jonghe and F. Adams, *Anal. Proc.*, 18 (1981) 347.
136 L. Chan, *Forensic Sci. Int.*, 18 (1981) 57.
137 W.R.A. de Jonghe, D. Chakraborti and F.C. Adams, in *Proceedings of the 2nd International Conference on Heavy Metals in the Environment, Amsterdam, September 1980*.
138 J.A.J. Thompson, in *Proceedings of the 3rd International Conference on Heavy Metals in the Environment, September 1981, Amsterdam*, CEP Consultants, Edinburgh, 1981, p. 653.
139 T.M. Brueggemeyer and J.A. Caruso, *Anal. Chem.*, 54 (1982) 872.
140 D. Chakraborti, W.R.A. de Jonghe, W.E. van Mol, C. van Cleuvenbergen and F.C. Adams, *Anal. Chem.*, 56 (1984) 2692.
141 C.N. Hewitt and R.M. Harrison, *Anal. Chim. Acta*, 167 (1985) 277.
142 B. Radziuk, Y. Thomassen, L.R.P. Butler, J.C. Van Loon and Y.K. Chau, *Anal. Chim. Acta*, 108 (1979) 31.
143 T.M. Vickrey, H.E. Howell and M.T. Paradise, *Anal. Chem.*, 51 (1979) 1880.

M. Stoeppler (Editor)/*Hazardous Metals in the Environment*
© 1992 Elsevier Science Publishers B.V. All rights reserved

Chapter 10

Mercury

Iver Drabæk

Danish Technological Institute, Department of Environmental Technology, P.O. Box 141, DK-2630 Taastrup (Denmark)

and

Åke Iverfeldt

Swedish Environmental Research Institute, P.O. Box 47086, S-402 58 Gothenburg (Sweden)

CONTENTS

10.1. Introduction 258
    10.1.1. Mercury in general 258
        10.1.1.1. Evolution of methods for mercury determination 258
        10.1.1.2. Evolution in accuracy and confirmed environmental levels . 259
    10.1.2. Toxicity 259
    10.1.3. Environmental levels 261
        10.1.3.1. Biota, soils, and sediments 261
        10.1.3.2. Air and rain 263
        10.1.3.3. Natural waters 263
10.2. Sampling procedures 264
    10.2.1. Mercury interconversion processes of special interest 265
    10.2.2. Air and natural waters 267
        10.2.2.1. Contamination 267
        10.2.2.2. Sampling 269
        10.2.2.3. Preservation 270
        10.2.2.4. Speciation 270
    10.2.3. Stack gases 272
10.3. Determination procedures 273
    10.3.1. Total mercury 273
        10.3.1.1. Methods in general 273
        10.3.1.2. Natural waters 276
        10.3.1.3. Air 277
    10.3.2. Speciation 278
        10.3.2.1. Analyte specific 278
        10.3.2.2. Operational defined 279
    10.3.3. New techniques 280
10.4. Conclusions 281
References 282

10.1. INTRODUCTION

10.1.1. Mercury in general

Mercury is a heavy (density 13.5 g/ml), silver-white liquid at ambient temperatures. It has a high vapour pressure (14 mg/m^3 at 20° and 72 mg/m^3 at 100°C), orders of magnitudes higher than the average permissible concentration of 0.025 mg/m^3 for occupational exposure [1].

Research dealing with mercury and mercury analysis has increased almost exponentially in the past 15 years. The interest in mercury stems primarily from the many serious episodes of mercury pollution, which has created serious hazards for mankind due to the extreme toxicity of methylmercury, e.g. in Minamata (1953–1960), Niigata (1965) and Iraq (1972) (see ref. 2).

It is well known that mercury can exist in a large number of different physicochemical forms having diverse environmental and biological behaviour. The interconversion between these forms controls the environmental mobility of mercury, and determines its biological enrichment and effects. Besides the elemental mercury(0) state, inorganic mercury is found in the environment in the +1 (mercurous) state and in the +2 (mercuric) state. Mercury is capable of making covalent bonds. Organo-mercury compounds (methylmercury and dimethylmercury) are formed in nature by biologically mediated conversion from inorganic forms; the reverse reaction-biological transformation of organo-mercury compounds to inorganic mercury - also takes place. The picture for mercury is further complicated by the high affinity of ionogenic mercury for complexing agents, e.g. chlorides, humic acids and fulvic acids [2]. The interconversion between all these mercury species and complexes - reviewed by several, [2-4] - as well as the high vapour pressure, have significance for the analytical procedures for mercury and its compounds. In order to understand the complicated distribution pattern in nature, to get new insight into the global mercury balances, and to understand the effect of anthropogenic mercury additions to the environment, it is of major importance to be able not only to analyze for mercury at natural environmental levels, but also to discriminate between the different important mercury species. These aspects will therefore be given special attention in this chapter.

10.1.1.1. Evolution of methods for mercury determination

In 1928, Stock and Zimmerman [5] developed a method for determination of mercury concentrations down to 40 mg/kg (deposition of mercury from acid solution onto copper wire, distillation, condensing and measuring of drop size in microscope). Since then various radio-tracer methods - e.g. substoichiometric isotope dilution [6], radiochemical displacement [7] - and colorimetric methods based on dithiozone [8] have evolved, methods which are now all minor importance.

Many of the methods used today were developed in the late 1960's or early 1970's. The fundamental paper on cold vapour atomic absorption spectrophotometry (CVAAS) was published by Poluektov et al. 1964 [9]. Methods based on radiochemical neutron activation analysis (RNAA) [10-12], and the method of Westöö [13,14], which is in general use today for determination of methylmercury, also appeared in this period. The most common methods for determining mercury nowadays are based on CVAAS. Although RNAA methods are still in use, they are less widespread. Other common and sensitive methods are based on atomic fluorescence spectrophotometry (AFS) or plasma induced atomic emission spectrophotometry (AES), e.g. microwave induced plasma-AES (MIP-AES) and direct current plasma-AES (DCP-AES). Details of these analytical methods, as well as methods for determination of different mercury species, will be given later.

10.1.1.2. Evolution in accuracy and confirmed environmental levels

The reliable determination of total mercury in mercury-treated flour at the mg/kg level had been demonstrated by 1966–69 [15]. The same intercalibration also showed that the results were unsatisfactory at μg/kg levels. It was concluded that the bias was probably caused by contamination. Since then the development of more sensitive methods has resulted in a better understanding and awareness of the potential errors in the determination of mercury. As a consequence 'confirmed' levels for mercury in, for example, seawater have shown an almost exponential decline (Fig. 10.1), nearly parallel to the improvement in sensitivity of the CVAAS method, although the distance between the two curves tends to increase. As well as the sensitivity, sample collection, handling and treatment have also been improved. While mercury sampling in air has been under control since 1970, the decline in the concentration levels in rain seems to stop in 1980. A very small increase is indicated after 1980, possibly because better oxidation procedures allow the liberation of mercury from particles and dissolved strong complexes. For remote areas, mercury concentrations around 4 ng/kg have recently been reported [16,17,18].

Since the intercalibration exercises on mercury analysis in 1966–69, it has been demonstrated, by other intercalibration studies on seawater [19] and estuarine water [20], that an increasing number of laboratories are capable of making reliable analyses at the ng/kg level. Problems do however remain - Berman and Boyko [20] reject more than 50% of the submitted analytical results in the statistical treatment of the mercury data.

10.1.2. Toxicity

In general, the toxicity of mercury is subdivided into the following categories in declining order of biological and toxicological activity: alkylmercury salts (methyl, ethyl) > mercury vapour > inorganic mercury-, phenyl-, and methoxyethyl- mercury salts. Arylmercury is largely converted into the inorganic form and handled as such in the body. It is not our intention to focus specifically on toxicological aspects of mercury. For a thorough discussion of toxicity the reader is instead referred to Berlin [21]. However, there are a few topics of special importance, to which we would like to draw attention.

The ability of short-chain alkylmercury compounds to bioaccumulate and their slow metabolism are well known, and in 1972 FAO/WHO recommended a Provisional Tolerable

References on p. 282

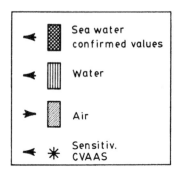

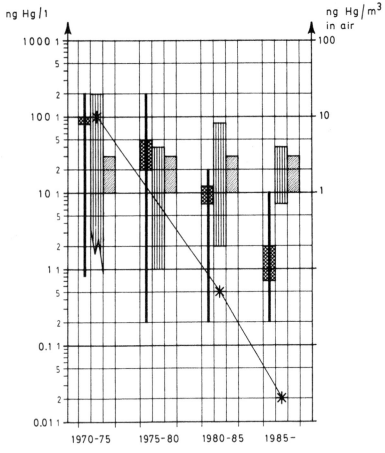

Fig. 10.1. Evolution in sensitivity (CVAAS) and environmental levels for seawater, rain and air since 1970.

Weekly Intake for adults (PTWI) of 300 μg of which only 200 μg should be methylmercury [22]. It is well known that methylmercury is transported from the blood of pregnant women, via the placenta, to the fetus [21]. In New Zealand, new studies on a population of 31 mother-child pairs, in which fish consumption during pregnancy was regular, indicate that the PTWI value should be reduced for pregnant women [23]; see also [25].

About 50% of the children in the examined group showed abnormal or questionable test results in the Denver Developmental Screening Test, compared to only 17% in a reference group. These and later results [24] emphazize the need for the elucidation of the cycling of anthropogenic and natural mercury, and consequently the need for better methods of analysis.

A further factor of both analytical and toxicological importance is the fact that mercury is released from dental amalgams. Most of the elemental mercury vapour entering the lungs is absorbed and passes into the blood stream [21]. A significant correlation has been found between mercury content in the brain and the number of amalgam dental fillings [26]. From measurements of mercury in breath a daily intake of 29 μg mercury, which on a weekly basis approaches the PTWI-value, has been estimated for persons with 12 or more amalgam fillings [27]. The highest mercury concentrations in breath occurred after chewing. Mercury concentrations up to 50 μg/m^3 have been measured in breath [28], i.e. a concentration twice the average permissible concentration for occupational exposure [1]. The contamination risks, especially for acidified samples at ultra-trace levels, are obvious.

10.1.3. Environmental levels

Confirmed values of baseline mercury concentrations in the environment, i.e. levels for remote non-polluted areas, are summarized in Fig. 10.2. Mercury levels found in biota are not included in the figure. In general, agreement is found in the literature concerning mercury concentrations at relatively high levels, as in fish, sediments and soils. In this chapter minor efforts have been directed towards these subjects, and levels have normally been taken from recent reviews.

10.1.3.1. Biota, soils, and sediments

The content of mercury in biota varies from less than 1 μg/kg - in algae [29] - up to approx. 10 mg/kg dry weight (DW), e.g. in sharks [2]. Typical levels in plants are, 0.1–9.5 mg/kg DW; mussels, 0.14–0.75 mg/kg DW; fish (flounders), 50–200 μg/kg DW; mammals, 1–10 mg/kg DW. The percentage of organic mercury varies from approx. 5% (algae) to more than 95% (fish muscle). More specific information on concentrations and percentages of organic mercury can be found in various reviews [2,30-33]; see also refs. [34,35]. It is well known that mercury is bioaccumulated in aquatic food chains. Some investigations indicate that methylmercury and inorganic mercury are bioaccumulated by separate mechanisms [36]. Generally, total mercury concentrations in biota increase with species size or life span. For fish, the concentrations increase with length or weight. However, studies have shown that this relation is valid only for methylmercury, whereas inorganic mercury in fish meat remained constant [37].

For soil, concentrations up to 625 μg/kg have been reported previously [2], [33]. The higher levels have been found in contaminated soil. In top soil containing humus, mercury levels of approx. 200–300 μg/kg, with a maximum at 1000 μg/kg, have been found [38].

References on p. 282

262

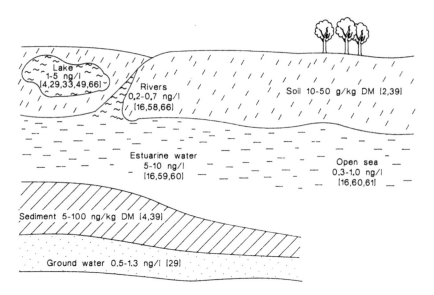

Gas 1-4 ng/m³ [2,3,33,44,47,48,53,76]
Precipitation 5-25 ng/l [16,17,33,46,49,50,52]
Particles 0.001-0.05 ng/m³ [3,33,45,46,47]

Lake 1-5 ng/l [4,29,33,49,66]

Rivers 0.2-0.7 ng/l [16,58,66]

Soil 10-50 g/kg DM [2,39]

Estuarine water 5-10 ng/l [16,59,60]

Open sea 0.3-1.0 ng/l [16,60,61]

Sediment 5-100 ng/kg DM [4,39]

Ground water 0.5-1.3 ng/l [29]

Fig. 10.2. Confirmed mercury concentrations for remote non-polluted areas (DM = dry matter).

Bottom sediments generally constitute the principal sink for mercury. Concentrations as high as 1–10 mg/kg are not rare in polluted areas [2,4,39]. Mercury concentrations are positively correlated with the content of organic matter in the sediment. Methylmercury typically constitutes from less than 1% (unpolluted) to 3% of the total mercury content [2,4,31]. Pore-water mercury concentrations of up to 3.5 μg/kg have been found in contaminated estuarine sediments [40]. Mercury values between 0.005 and 0.037 μg/kg were found in pore-water from sediments in a make-up water reservoir [4].

10.1.3.2. Air and rain

In contrast to the situation for rain, mercury concentrations in air have been agreed upon for almost the last 10 years. Obviously, this is due to the easier sampling procedure for mercury in air and the reduced risk of loss and contamination (compare Section 10.2). Typical levels in air are 1–4 ng/m^3. The predominant form of mercury in air is elemental mercury (probably more that 95% using operationally defined methods [36,41–44]). Measurable amounts of methylmercury have been identified in ambient air [41]. Particulate, bound mercury is normally below 0.1 ng/m^3 and even down to less than 1 pg/m^3 in 'clean' oceanic background areas [33], [45-47]. Concentrations of mercury in air are generally higher in winter (3–4 ng/m^3) than in summer (1.5–2 ng/m^3) [33,44,46]. A concentration of 1 ng/m^3 is representative for the southern hemisphere. For the northern hemisphere, the concentration is typically 20–50% higher [47,62]. A diurnal variation has been found in the upwelling zone, with higher levels in day time than in night time [48].

For rainwater samples, the analytical problems have always been greater than for air samples. The need to stabilize the samples makes them more prone to contamination: lack of stabilization can lead to losses if the samples are not analyzed within a few days, and mercury is largely bound to particles in the rain samples (see Sections 10.2 and 10.3 for discussion). Baseline mercury levels in precipitation vary between 1 and 25 ng/kg (Fig. 10.2). Mean mercury concentrations up to 1140 ng/kg in rain from industrial areas, and as large as 80 ng/kg in samples from relatively unpolluted areas, have been reported [51]. The same authors also report methylmercury levels in rain (5–6% of the total mercury concentration in rural areas; <1% in industrial areas). A great part of mercury in rainwater samples is associated with particles [33,46,52,53]. In order to determine all mercury in such samples heavy oxidation is necessary (see Section 10.3). Ahmed et al. [54] showed that the methylmercury level found in rainwater was unaffected by filtration, i.e. that the methylmercury exists in the form of dissolved complexes.

Brosset [44,55] suggests that an operational defined fraction of mercury in rainwater, defined as the difference between mercury recovered by $NaBH_4$ or $SnCl_2$ treatment, is long-range transported and represents direct anthropogenic emissions. This fraction has also been found to constitute a substantial part of the mercury in stack gases from coal-fired power plants [44,55]. For city waste incineration plants most of mercury in the stack gas is believed to be $HgCl_2$ [56]. As a comparison, the total mercury concentration levels in the stack gases are in the range 3–8 $\mu g/m^3$ for coal-fired power plants [55] and 25–1500 $\mu g/m^3$ for city waste incineration plants [56]. Concentrations of mercury in Greenland ice cores, representing both dry and wet deposition, are believed to be about 10 ng/kg [57].

10.1.3.3. Natural waters

Confirmed baseline levels for natural waters have, in general, decreased during the last decade (see Section 10.1). The reason for this lies mainly in improvements in the analytical technique and in sampling and handling procedures. The present baseline values for freshwater, estuarine water, seawater, riverwater and groundwater are given in Fig. 10.2.

It is important to distinguish between dissolved mercury and mercury bound to suspended matter. With high levels of suspended matter, and in polluted areas, very high concentrations of mercury can be found associated with the particulate fraction. Nelson [59] found a mercury concentration of over 500 ng/kg in the Thames Estuary. Cossa and Noel [60] also stress the importance of the suspended matter fraction; in the Gironde Estuary, a total mercury concentration of 21 ng/kg or more was detected, with peak concentrations of 8 ng/kg, in the reactive mercury fraction. Not all methods used for analysis of open ocean samples, for which the concentration of suspended matter is usually low, are able to determine all mercury associated with particulate matter [60,61]. Determination of total mercury in these samples requires a heavy oxidation step using, for example, BrCl. Also, for waters containing humic or fulvic acids, such as lake and surface run-off water, a procedure including a heavy oxidation step is necessary.

In the open ocean, around 80% of the mercury in water is of the reactive type, i.e. reducible directly with $SnCl_2$ [16,60]. Concentrations of dissolved gaseous mercury (DGM) in seawater of up to 0.04 ng/kg have been found in the equatorial region, and the mercury concentrations can be correlated with the biological productivity [62]. These findings on DGM are in agreement with those of Brosset [44] and Iverfeldt [63,64]. The organic (alkyl) mercury levels in natural waters are generally rather low, at 1–10% of total mercury [35,65,66]. Higher mercury concentrations are found in the Mediterranean than in Atlantic or Pacific seawater [67]. For coastal seawater, lakes, and rivers, a local variation can be found.

10.2. SAMPLING PROCEDURES

Environmental samples containing comparable high mercury concentrations can normally be sampled and stored without the occurrence of severe problems, e.g. contamination or loss of mercury from the samples. The sampling of mercury in fish, sediments and soils is thus only considered to a minor extent in this chapter. See also the discussion on environmental levels in Section 10.1.3. Sediment samples, for example, are collected in the field using ordinary equipment, and stored in a freezer. The optimum preservation is probably by freezing in the field [4]. The technique for preparation of samples from biological material is well established and can normally be performed without serious bias. However, good laboratory practice performed in areas with controlled low mercury levels in the air, is always necessary. On the other hand, an accurate collection and handling procedure for low-level mercury samples from ambient air and various natural waters is most accurately performed if some of the important pathways in mercury's biogeochemical cycle are taken into consideration.

Interconversion processes of mercury are dealt with, in general, in Section 10.2.1: in Section 10.2.2 the aspects of contamination, sampling, preservation and speciation for air and natural waters are treated. In Section 10.2.3 special attention is given to stack gas sampling. Emission of mercury from coal and waste incineration is among the major anthropogenic inputs to the global mercury cycle [3]. In order to be able to estimate the contributions, and to check that emission limits are met, sampling and determination of mercury from stack gases are essential.

10.2.1. Mercury interconversion processes of special interest

The most important natural interconversion processes for mercury are illustrated in Fig. 10.3. Knowledge about these processes is important when determining total mercury concentrations and essential in performing speciated mercury analysis at the ultra-trace level.

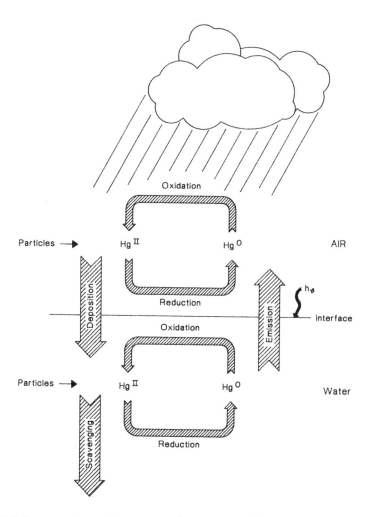

Fig. 10.3. Important neutral interconversion processes for mercury.

References on p. 282

Interconversion of mercury between the elemental (0) and the oxidized (+2) form probably determines the direction of mercury fluxes between the atmosphere and the aquatic-terrestrial compartments of the global mercury cycle, but is to some extent balanced by reduction processes. The transformations between different redox states are also of major importance within the various natural reservoirs. For example, an emanation has been shown of dissolved gaseous mercury from the marine environment to the atmosphere, as a result of the reduction of oxidized mercury forms to elemental mercury in sea water [62,64]. A change in the redox potential and/or a more direct biological action of phytoplankton or bacteria are probably responsible for this reductive reaction. Brosset [44] suggested that mercury is reduced in natural waters, under alkaline conditions (e.g. in sea water), by H_2O_2. There are also several experiments showing that the reduction of oxidized mercury forms is influenced by solar radiation [33,44,63], and that there is a direct atmospheric chemical reduction pathway involving a mercury sulphide complex [33]. When we consider the analytical procedure, the same type of processes as described above may also occur in an unpreserved sample, resulting in loss of mercury from a poorly sealed container. The use of plastic sample containers, e.g. polyethylene or Teflon bottles, may result in loss by diffusion of elemental mercury through the walls. The permeability of Teflon bottles for gaseous mercury has been shown [16,68]. During collection of precipitation samples oxidized mercury forms may be reduced to elemental mercury, with a subsequent emission of a significant part of the mercury. A short collection period is thus preferred. A bias of this type can be avoided if the sample vessel contains for example, an oxidizing reagent and/or stabilising acid, e.g. $HCl + H_2O_2$ [49]. However, this may result in an enhanced dry deposition of mercury in the collected rainwater, giving too high a mercury value in the precipitation sample. Contamination from the acid may also be a problem. A dry deposition of mercury may possibly also occur even if HCl is used as a preservative. Brosset [44] suggests that mercury in ambient air can be oxidised during contact with acidic water. The presence of chloride ions will stabilize the oxidized mercury as an $HgCl_4^{2-}$ complex. Effects of the latter type are described below.

A diffusive flux of mercury species from an aqueous environment to air, or in the opposite direction, depends on the magnitudes of the different partition coefficients (i.e. the Henry's law constants) and the concentration of the respective species in the two phases. In short, there is an important difference between the Henry's law constant for elemental mercury and oxidized mercury forms. The tendency for elemental mercury to be in the gas phase is orders of magnitudes larger than for the other mercury forms [69,70,71]. The only exception is dimethylmercury, with a Henry's law constant similar to that of elemental mercury [72]. More extensive discussions on the magnitude and relevance of the Henry's law constants in describing important parts of the mercury cycle can be found in refs. 3, 44, 45, 62, 63, 73, 74. If we consider elemental mercury in the water phase or, in a broader sense, the dissolved gaseous mercury concentration, if this contains normal mercury levels (about 2 ng/m^3), and a water temperature of 10-20°C, even a concentration of 10 pg of elemental mercury per liter of water results in a supersaturation and an emission to air [16,64]. See also Section 10.1.3.

The second redox reaction of importance in nature as well as in analytical procedures is the oxidation of elemental mercury in the aqueous phase. The occurrence and significance of an atmospheric oxidation have been suggested by Brosset [75,76] and demonstrated by low-level experiments in the laboratory [43,63,77] as well as in field experiments [44,77]. In contrast with the homogeneous gas-phase oxidation, the oxidation in atmospheric water droplets (i.e. cloud, rain and fog) is suggested to be predominant [77]. Basically, oxidizing conditions in the aqueous phase support the diffusive flux of elemental mercury from air to water by keeping the concentration of reduced elemental mercury in water at low or negligible levels. The significance of this for the collection and handling of water samples has been shown [68,78]. Using a Class 100 clean-air bench, Gill and Fitzgerald [16] recorded a steady increase of mercury (2.1 pg $Hg/cm^2/day$) in an HNO_3-acidified seawater sample which was exposed to ambient laboratory air containing high levels of gaseous mercury (145 ng Hg/m^3). A similar type of bias was shown by Brosset [44] during collection of mercury from ambient air. He demonstrated a direct dependence between the amount of oxidized mercury in an aqueous phase containing chloride, after purging with ambient air containing normal oxidant levels, and the retention time of a gas bubble in water. Glass et al. [79] reported a large amount of dry deposited mercury in an oxidizing solution used for comparison with the amounts of wet deposited mercury collected in rain.

Another feature of the mercury cycle, which is important for the design of sampling procedures, is the association between mercury and particles. The particle-associated mercury in rainwater samples collected in Germany amounts to about 60–80% of the total mercury concentration [54]. In central Italy, the values are between 20 and 60% of the total mercury [52]. A similar relation is also found in precipitation samples from the Nordic Countries [33,46]. In addition, the importance of mercury associated with suspended material in turbid coastal waters has been stressed by several authors [50,61,64,67].

Finally, there are a large number of investigations of the occurrence and interconversion of methylmercury compounds in nature. Representative levels of methylmercury in the various environmental reservoirs, and the importance of organic-bound mercury, have been discussed in the previous sections. A number of investigations on suitable preservation methods for a methylmercury sample can be found in the literature, e.g. for rainwater samples [54] and for freshwater samples [4,80].

10.2.2. Air and natural waters

10.2.2.1. Contamination

(i) *Air.* The determination of ambient mercury levels in air implies a preconcentration step. At present, the accumulation of mercury on a gold trap is the most frequently applied method. The use of, for example, acidic oxidative solutions (e.g. $KMnO_4$ in H_2SO_4 solution) normally involves comparable high blank values. The latter method is, however, applicable in measurements of elevated mercury concentration in stack gases, for example (see Section 10.2.3). By careful sampling of mercury in air using the gold trap, contamination problems can be minimized and the blank value can often be neglected. A contribution to the blank value from the trap can be avoided by careful design, cleaning through several heating cycles, and storage of sealed traps in a mercury-free environment. Contamination of the collecting trap during sampling, through back-diffusion from the pump and tubing is best prevented by use of a protective gold trap. Brosset [44] suggests that evasion of mercury from a wet wall can contaminate an air sample: sampling of mercury near walls should thus be avoided.

(ii) *Natural waters.* The contamination problem in determining mercury in natural waters in greater than in the case of air. There are several major contamination sources that must be considered.

The sample collection procedure may result in a serious bias. The importance of an adequate sampling method has been discussed above for the collection of rainwater samples, i.e. the occurrence of a dry deposition of mercury to the sample. For surface and deep waters, only a few studies of possible contamination problems during sampling can be found in the literature: for example, the review by Robertson et al. [4]. Contributions from, for example, the research vessel, the sampler and (as pointed out by Gill and Fitzgerald [81]), the hydrographic cable, must be avoided. The containers and/or funnels used for the collection of surface- and rainwater, or in general for storage of water samples, must be carefully cleaned to prevent contamination [82]. Several cleaning procedures have been suggested [16,17,61,64,81,82].

The reagents used in the analytical procedure are of great importance. It is essential for accurate analysis that reagents with reliable low blank values are used. Frequent replicate determinations of reagent blank values are necessary. Commercial high-purity acids (e.g. Ultrex or Merck-Suprapur brand) can normally be used. A mercury content in HNO_3 of between 20 and 50 ng/kg is normal [61,64], although values from below 20 to more than 4000 ng/kg can be found if acids of different lot numbers are compared [61]. Sub-boiling distillation can be used to prepare high-purity acids [83]: mercury levels in HNO_3 could be reduced to below 20 ng/kg by this process [16].

Laboratory water used in the analytical procedure for rinsing and preparation of solutions may also contribute to the contamination of natural water samples. Gill and Fitzgerald [16] report a three-step purification procedure which gives high-purity water with a mercury concentration lower than 0.6 ng/kg. Nojiri et al. [84] used Milli-Q water with a mercury content of 0.29 ng/kg as high-purity water. The use of a Milli-RO-Milli-Q water system provides an excellent high-purity (= low mercury level) water with a mercury concentration less than 0.1 ng/kg (Iverfeldt, unpublished results). Mercury-free reagent solutions prepared from $SnCl_2$ and $NaBH_4$ can be obtained after extensive purging with nitrogen or helium gas [16,61,64].

The most serious contamination problem arises from the oxidizing reagents used both as preservatives and in the determination of the total mercury concentration [17]. The purification of these solutions is generally very difficult. Bloom and Crecelius [61] describe a self-purifying hot oxidizing reagent, i.e. a modified permanganate-persulfate reagent. Still, they recommend that this technique should be avoided in the analysis of mercury in natural waters. Instead, a cold oxidation reagent, bromine monochloride, is recommended both as a preservative and for total mercury determination in natural waters [61] (see Section 10.3.1).

Finally, the analytical procedure, including handling and purging of the sample, gives an important contribution to the blank value, and procedural blanks must always be determined [16,61,64]. Filtration, a necessary procedural step in the determination of dissolved mercury, may result in a contamination of the sample. Filtration methods and equipment for natural water samples have recently been reported which are probably unbiased [52,54,60,64].

The analytical work should be performed under Class 100 clean-air laboratory conditions, if possible, or in a Class 100 clean-air bench. However, since most of the mercury in laboratory air is in the gas-phase and consequently not removed by the Class 100 filter, the risk of mercury contamination from the laboratory air is high [16]. A typical value in our laboratories is less than 5 ng/m^3. A Class 100 clean-air bench equipped with an activated charcoal filter or coated with a thin layer of colloidal gold reduces the gaseous mercury content in air to less than 1 ng/m^3 [49].

10.2.2.2. Sampling

(i) *Air*. The sampling of total gaseous mercury in ambient air is usually performed by the accumulative gold trap technique [44,46,48,52,62,76,85-88]. The particle-bound mercury is collected on quartz-glass wool in front of a series of three gold traps [44]. Another operational definition of particulate mercury in air is obtained by collection on commercial quartz fiber filters, e.g. Gelman Type A-E [45]. The total precision in the determination of mercury in air - including collection, volume and analytical errors - is best found using parallel sampling trains. Brosset [44] found a reproducibility of 1 to 2% in the concentration range 2–4 ng/m^3. Overall precisions of 14% [87], and less than 10% [45], have been reported for mercury levels of 1–2 and 1 ng/m^3, respectively.

(ii) *Natural waters*. The previous discussion on dry deposition and emanation problems (Section 10.2.1), as well as possible contamination from construction materials, should be kept in mind when sampling rainwater. Various strategies and types of equipment for rainwater sampling have been reported [16,44,51,52,79]. Discrete precipitation events have been collected manually [44,46] or automatically [79], with an oxidizing preservative present [79] or without [44]. Ferrara et al. [52] used an automatic sampling device which was closed with a tight-fitted cover between rainfalls to prevent dry deposition of mercury to the sample. Ahmed et al. [51] collected rainwater over longer periods, with a sampler that consisted of a glass funnel (diameter 19 cm) connected to a 2.5 l brown glass bottle protected against light. As a preservative, they added concentrated HCl to the bottle before collection. In the Scandinavian countries discrete rainfalls have been collected by the use of five large-surface borosilicate (Pyrex) beakers (diameter 19 cm) followed by immediate transfer of the sample to a capped Pyrex bottle [33,44,49]. Gill and Fitzgerald [16] used glass (diameter 38 cm) or Teflon (45 cm) funnels connected via a Teflon block to a 1- or 2-l Teflon bottle.

The collection of sea- or lake-surface waters is usually performed manually. A capped sample bottle of Pyrex or Teflon is opened below the surface on the upwind side of a rubber raft or a fiberglass boat [4,16,61,64,81]. Arm-long plastic gloves should always be used during the collection. Sub-surface water samples are normally collected using a Teflon-lined Go-Flo bottle [4,61,64,84]. A line of polypropylene, or a Kevlar hydrocable is preferred [16]. However, for relatively shallow depths, samples may preferably be pumped directly into the collection bottle using Teflon tubing and a contamination-free pump.

Various filtration strategies have been applied to the determination of dissolved mercury in rain. Ferrara et al. [52] used a funnel with a pretreated filter (Sartorius), while other researchers have filtered the rainwater sample before storage [51]. Ahmed et al. [51] used a Sartorius polycarbonate filter holder and 0.45 μm membrane filters. Iverfeldt [46] used extensively cleaned 0.4 μm Nuclepore polycarbonate membrane filters in combination with a Sartorius polycarbonate filter holder. The latter type of filtration system has also been used for the determination of dissolved mercury in sub-surface coastal seawater by an in-line filtering procedure [64]. Robertson et al. [4] used the in-line filtering approach with a 0.2 μm Nuclepore membrane filter for sub-surface lake-water samples. They also applied different filtration techniques for surface freshwater depending on the sample volume: a Nalgene 0.45 μm sterilization filter unit, with an in-line large-diameter (30 cm) 0.2 μm Nuclepore membrane filter connected to a pump, or a large surface Nuclepore filter cartridge also attached to a pump. Seritti et al. [89] filtered seawater samples through 0.45 μm membrane filters (Sartorius) in a Sartorius filter holder. Cossa and Noel [60] used pretreated (by heating at 400°C for 24 h) glass fiber filters (Whatman GF-F) for filtration of estuarine- and coastal-water samples.

10.2.2.3. Preservation

(i) *Air.* Samples of mercury in air can be stored on gold traps for at least a week [3].

(ii) *Natural waters.* In order to avoid loss of mercury during the sampling period, Ahmed et al. [51] added 25 ml of concentrated HCl to the sampling bottle prior to collection. Bloom and Effler [80] used 10 ml/l of 12 N HCl. After collection, Gill and Fitzgerald [16] added high-purity HNO_3 (to sample pH 1.2) and 15 ppb chloroauric acid as preservatives. Sea- and freshwater samples are often preserved by the addition of purified HNO_3 [16,84]. Other workers have used BrCl as a preservative in the determination of total mercury in sea- as well as in lake-water samples [4,17,61].

10.2.2.4. Speciation

(i) *Air.* Various selective gas-sampling trains for atmospheric mercury speciation have been presented in the literature during recent years. Braman and Johnson [90] initiated these studies by developing a four-step speciation system for mercury in ambient air. The fractionation technique is based on different affinities between the various mercury species present in the air and different collecting materials in sequence. Parts of the results were difficult to reproduce [91]. However, the combination of silver and gold traps has frequently been applied [62,91]. Braman and Johnson's choice of analytical method resulted in an operational definition of the mercury species, and more specific identification techniques have been suggested later [92]. In some applications, it may be possible to use an analyte-specific technique which should then be preferred. Nevertheless, an operational definition of mercury species is often considered most useful.

In the accumulative sampling of mercury species from ambient air, interconversions between mercury forms, and deactivation of absorbing surfaces, are difficult to forsee under the varying chemical conditions that can be found in an air parcel. Dumarey et al. [93] state that the use of silver-coated sand or activated charcoal for collection of atmospheric gaseous mercury should be avoided. They suggested that the collection efficiency depends on the type of mercury species, sampling flow-rate and duration, ageing of the trap, and collection of interfering substances.

Definitions of mercury fractions by application of various speciation techniques for mercury in air are discussed below in greater detail (Section 10.3.2).

(ii) *Natural waters*. Methods for a direct positive identification of methylmercury in natural waters have recently been developed (Section 10.3.2). In general one is recommended to analyze the samples directly within 8 hours or to keep them frozen at –80°C until analysed [17]. In the determination of methylmercury in water, sampling and preservation procedures may be different from those applied for total mercury analysis. Ahmed et al. [51] describe a collection procedure suitable for methylmercury in rainfall, i.e. the use of a darkened glass bottle with HCl added before collection as a preservative (see Section 10.2.2.2 (ii)). The method is based upon the fact that methylmercury decomposes very rapidly when exposed to light [94] and that methylmercury is stable in a rainwater sample for two months if acidified with 5% HCl [54].

Methylmercury in large volume samples (10 l) can be preserved by preconcentration on a chelating resin [4]. Lee [65] describes and analytical procedure where methyl- and ethylmercury in natural water samples of large volume (20 l) are preconcentrated on a sulfhydryl cotton fiber adsorbent. The method was tested for both freshwater lake samples and a snow sample. Special sampling and handling procedures are also necessary in the determination of operationally defined mercury fractions in water (see Section 10.3.2). Preconcentration of samples with acids or oxidizing agents should be avoided. Instead, the various mercury fractions should be transferred to separate gold traps directly after sampling. Robertson et al. [4] showed a rapid interconversion of elemental mercury (0) to mercury (II) upon the addition of acid. In several recent investigations, dissolved gaseous mercury - in most cases probably in the form of elemental mercury - have been determined directly after sample collection [4,62,63,64,80]. Ahmed et al. [54] showed an increase of ionic mercury after acidification of rainwater samples. Consequently, determination of reactive or easily reduced mercury should be performed soon after and acidification [17]. In conclusion, it is recommended that sample preservation and storage are mainly used in total mercury analysis. Also, preservation of samples for methylmercury analysis seems possible, but should be avoided for elemental mercury and dimethylmercury analysis. Preservation of samples for the determination of operational defined fractions could interfere with the speciation and should be avoided.

References on p. 282

10.2.3. Stack gases

Sampling of mercury in stack gases is most often carried out by discrete sampling using some kind of sampling train. Part of the stack gas is drawn through, and mercury collected and determined in the various part of the train. Besides the discrete sampling technique an on-line measurement device has been developed and described [95]. The principle of this method is based on scrubbing of the stack gas stream with water, amalgamation and CVAAS. It is reported that organic substances cause a ghost peak in the determination [95]. Most of the mercury found in stack gases is in the gaseous state. Isokinetic sampling from the stack gas is therefore not necessary, but is recommended. Typically, less than 5% of the mercury is found on particles [96,97]. The amount of mercury found in the particle fraction is, however, dependent on the stack gas and the sampling temperature.

In general, discrete sampling is performed using a sampling train consisting of (1) a heated quartz filter, (2) a condensing impinger which is either empty or contains a washing solution, and (3) either several impingers with a strongly oxidizing agent, or a solid absorbent. The washing solution is often introduced in order to protect an oxidizing solution and/or for speciation purposes. Because a large part of the mercury in stack gases is water soluble, most of the mercury is found in the first impinger [56,98,99]. It is possible to collect all mercury using only an oxidizing solution, but as the lifetime of the solution without any preceding washing of the stack gas is short, only a few applications have appeared in recent literature. Care should be taken not to lose mercury in the first impinger in case of condensation on the impinger walls. Aqua regia is recommended for flushing [56]. As a washing solution, distilled water [98] or 10% sodium carbonate [99] are often used. The non-soluble part of the mercury can be collected in sulfuric acid-potassium permanganate or nitric acid-potassium peroxydisulphate [56,99]. Problems with the oxidizing solutions are not due to the blank values, because high concentrations of mercury are generally encountered in stack gases at the $\mu g/m^3$ level. Instead they are due to instability of the solutions or deterioration of the oxidizing capacity by, e.g. sulphur dioxide in the stack gas [56]. The previously mentioned 10% sodium carbonate washing solution was originally introduced in order to protect the oxidizing capacity of the potassium permanganate solution. The formation of manganese dioxide due to sulphur dioxide can cause a loss of mercury from the potassium permanganate solution [100]. Essentially all of the mercury can, however, be recovered by treatment with acidic $SnCl_2$.

Instead of an oxidizing solution, a solid adsorbent can be used to collect the non-soluble part of the mercury. Gold has been tried by several [98,101], but suffers from limited capacity, which can be further reduced by impurities in the stack gas [56,101]. These impurities are found to be more pronounced for coal-fired power plants than for municipal incineration plants.

A total procedure based on collection in a potassium dichromate solution followed by Tenax and gold has been proposed by Dumarey et al. [96]. The Tenax was used to prevent poisoning of the gold by organic compounds and was found not to adsorb elemental mercury. An approach entirely without the use of solutions was successfully applied by Metzger and Braun [56]. They used Dowex 1x8 followed by iodized charcoal. The Dowex 1x8 was shown to collect $HgCl_2$ quantitatively while elemental mercury passed through to the iodized charcoal. A denuder technique can also be used to distinguish between Hg^0 and $HgCl_2$, for example using KCl coated and Ag coated denuders in series [102].

Today, only operationally defined methods have been reported for speciation of mercury in stack gases. The generally adopted collection methods distinguish between water-soluble and non-water-soluble mercury. In addition to the two techniques mentioned above, the discrimination is most often based on the use of a washing solution in the first impinger, e.g. containing 10% sodium carbonate, followed by two impingers containing potasium permanganate [99]. The water-soluble mercury is thus collected in the first impinger. New investigations comparing this method with the previously mentioned method, based on the Dowex 1x8 adsorber followed by an adsorber with iodized charcoal [56], indicate that the estimation of the water-soluble mercury fraction in the former method might be too low. A possible explanation could be that a disproportionation of mercury salts occurs in the washing solution. The same authors [56] claim that mercury in stack gases from municipal incineration plants is mainly, present as mercury(II) halogenides before stack gas cleaning. Depending on the cleaning process, disproportionation can take place during or after the cleaning process. For coal-fired power plants Brosset [44,55] has performed speciation of the water soluble mercury fraction (selective reduction with $SnCl_2$ and $NaBH_4$), and found 50% of the water-soluble mercury not to be reducible with $SnCl_2$. Thus compounds other than mercury(II) halogenides and elemental mercury seem to present, at least in the stack gases from coal-fired power plants.

10.3. DETERMINATION PROCEDURES

10.3.1. Total mercury

10.3.1.1. Methods in general

A general summary of the applicability of the most frequently used determination methods is shown in Fig. 10.4. As can be seen, rather large concentration ranges are covered by many methods. It should, however, be remembered that special steps and precautions are called for in the lower concentration range, and that a reliable determination of total mercury is closely linked to sampling and sample handling procedures.

In the determination of total mercury levels, mercury compounds are normally converted into mercuric ions. This involves a number of analytical operations, which are associated with methodological and systematic errors. Apart from the sampling and preservation of samples discussed in Section 10.2, decomposition of the sample and separation of the mercury from the matrix are examples of such operations. Contamination, or insufficient control of blank values, can be the result of all these processes and, especially in the ng/kg region, analytical results can be incorrect by orders of magnitudes.

References on p. 282

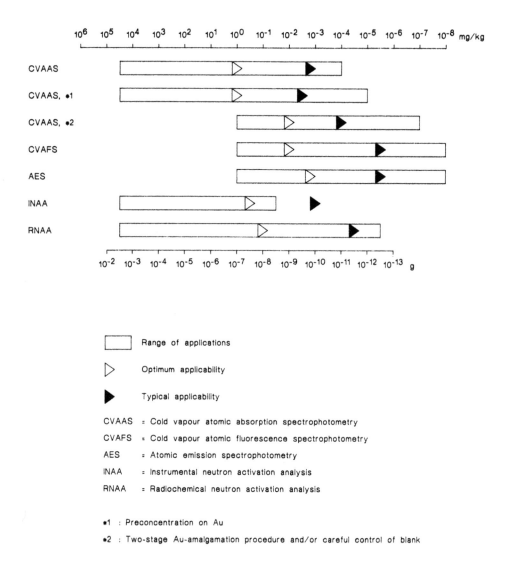

Fig. 10.4. Range of applications of common analytical techniques for determination of mercury.

Most of the reported mercury concentrations over 10 to 20 ng/kg in uncontaminated seawater should thus be viewed with suspicion.

The analysis of solid samples most often involves their decomposition. The preferred method is wet digestion using various combination of acids, e.g. HCl, HClO$_4$, HNO$_3$, H$_2$SO$_4$, K$_2$Cr$_2$O$_7$ [103,104]. Sadiq and Zaidi [105] found that digestion of fish samples (2 g) was most efficiently carried out using a 1:2 mixture of concentrated HNO$_3$:H$_2$SO$_4$ at 80°C for 4–6 hours. A shorter digestion time or lower temperature resulted in reduced mercury values. Louie [106] used concentrated HCl together with HNO_3 and H_2SO_4 for mercury determination in fish. The HCl was found to speed up the digestion process and to allow the temperature to be lowered (70°C). In both the above cases, open systems were used. Open system digestion has been shown to be ineffective for coal samples [107]: reflux and high digestion temperatures were necessary for complete digestion. The optimum procedure involved the use of 20 ml 1:1 $HNO_3/HClO_4$ + 250 mg $(NH_4)_2S_2O_8$ in a Bethge apparatus. The application of an open Kjeldahl flask resulted in 40% loss of mercury. The use of an oxygen combustion bomb with $K_2S_2O_8/H_2SO_4$ achieved complete mineralization of samples with high fat contents, and was followed by cold vapour atomic absorption spectrophotometry (CVAAS) in the determination of mercury in milk products at the ng/g level [108].

In general, the use of wet digestion results in high blank values due to the added reagents. For low mercury concentrations, digestion methods using a minimum of purified reagents are the only methods of choice. An example of such a method is a semi-automated wet digestion ($HClO_3/HNO_3$) followed by CVAAS, with a reported detection limit of 0.5 ng [103]. Another way of minimizing the blank problem is through the use of pyrolysis methods in which mercury is evaporated and subsequently collected on gold [109,110]. A purification step, using a catalytic converter (Ag) and two specific adsorbers for organic substances (silica gel and alumina) was included in the procedure. The pyrolysis train was coupled to a twostage gold-amalgamation CVAAS and the reported detection limit for the system was 5 μg/kg for a 20 mg sample. Larger samples (up to 0.5 g) could be processed using a combustion system which was coupled to a micro-wave-induced atomic emission spectrophotometer, with a reported detection limit of 0.01 ng (0.02 μg/kg) [103].

In neutron activation analysis (NAA), the risk of introducing blanks is excluded provided standards and samples are sealed under controlled conditions. The sensitivity of NAA is high for mercury but, especially for instrumental NAA (INAA), is dependent on sample composition. Reliable INAA-results, down to 10 μg/kg, have been reported for neurological tissue [111]. For radiochemical NAA (RNAA), where matrix interferences are partly eliminated, detection limits down to 2 ng/kg for a 5 g sample (seawater) are feasible [112]. Typical applications of NAA-methods are, however, solid samples at the μg/kg level. The RNAA methods are based on separation of mercury from the matrix, most often after a wet decomposition. Distillation of $HgCl_2$ [10,113] or ion exchange [114] are typical examples. Normal detection limits are in the μg/kg – 0.1 μg/kg range. Other highly sensitive RNAA methods employ a combustion of the sample in an oxygen stream and collection of the evolved mercury in liquid nitrogen [115,116].

The most common determination procedure is CVAAS, for example as in ref. [117]. Mercury in the digested solution is reduced by $SnCl_2$ and the generated mercury vapour swept, for example by a stream of nitrogen, directly through an atomic absorption spectrometer. Atomic absorption is measured for the spin-forbidden resonance line at 253.7 nm, which is caused by the transition $6s^1S_0-6p^3P_1$. Optimization of this technique with respect to the absorption cell, gas flow, and volume of reduction cell, can improve the detection limits to 0.25 ng [118]. The technique is, however, prone to many interferences

References on p. 282

at this low level, from water vapour, halogen ions especially iodide, amino acids or other complexing agents, noble metals, Se and Te which either depress the absorbance signal by reaction with mercury in the solution or enhance it by interferent absorbance [103,119]. The use of $NaBH_4$ increases the risk of interferences in some cases such as Cu, and in others, such as Se and I reduces it [103]. The use of an intermediate amalgamation step on gold decreases the interferences, and ensures a more consistent peak shape during measurements and a better detection limit [120]. Upon leaving the solution, the volatile mercury is collected on a gold trap (see Section 10.2.2), and measured after thermal desorption from the gold. Even better conditions are obtained using the two-stage gold amalgamation procedure. The thermally desorbed mercury is collected on a standardized analytical gold trap, which has been carefully calibrated, and is then desorbed into the spectrometer. Water vapour collected on the first trap can be evaporated, and interferences due to organic compounds, which are oxidized by the first thermal desorption, are avoided by the two-stage amalgamation [121]. Furthermore, errors caused by changes in carrier-gas flow-rate due to differences in trap packing can be eliminated, and more precise measurements can be made without the need for calibration of each gold trap [86]. Especially for the determination of mercury in air, potential interfering substances collected on the first trap can be avoided by using the two-stage amalgamation method [88]. During aeration a white precipitate can be formed from the hydrolysis of Sn^{2+} which absorbs Hg^{2+} [121]. This is dependent on the pH and the type of acid: the precipitate occurs only at pH higher than 1.5 (HCl), 0.8 (H_2SO_4) and 1.1 (HNO_3).

The use of a polypropylene mixing chamber may lead to Sn^{2+} absorption, with subsequent loss of mercury from new samples prior to addition of new reductant [122], and a memory effect in alkaline reductions [123]. The memory effect can be avoided by rinsing with aqua regia.

10.3.1.2. Natural waters

In this paragraph elements of special interest for sample treatment and the determination of mercury in natural water are dealt with.

Since the natural background levels of mercury in natural waters are so low, it is necessary to use some kind of preconcentration in order to determine mercury. Most popular is the amalgamation on gold. The Water Research Centre has described a method for water analysis, based on gold amalgamation and CVAAS using standard equipment (Perkin Elmer), with a detection limit of 2 ng/kg (200 ml sample) [124]. The samples are treated with acid-bromate bromide, and mercury is reduced by $SnCl_2$. For two other commercially available mercury analyzers it is found that the LDC Mercury Monitor is approximately 12 times more sensitive than the Coleman MAS-50 [4]. Several other preconcentration methods based on principles that do not use gold are reported in the literature. Applied to natural waters they often result in too high mercury concentrations as a result of insufficient blank control. They include preconcentration in $KMnO_4$ [125], flotation of dithizone-mercury complexes [126], adsorption on XAD-2 resin [127], and extraction with APDC-chloroform [128]. Detection limits in water analysis can be reduced by carefully controlling the blank, often in combination with two-stage amalgamation. Detection limits between 0.1 and 0.01 ng (0.1–0.05 ng/kg) are reported [16,60,61, 129].

CVAAS using the vacuum-ultraviolet absorption at 185.0 nm is theoretically more sensitive than absorption spectrophotometry at 253.7 nm, but is experimentally more difficult due to absorption interference by molecular oxygen and background absorption by molecules produced in the atomization process. Detection limits down to 0.01 ng (0.1 ng/kg) have been obtained [130].

Some of the most sensitive determination methods are based on various forms of plasma atomic emission spectrophotometry (AES). Detection limits down to 1 pg (0.01 ng/kg) can be obtained with a modification of the original direct current plasma AES of Braman and Johnson [90]: see, for example, Iverfeldt and Lindqvist [70] and Iverfeldt [64]. For a microwave induced AES, an instrumental detection limit of 0.5 pg is reported [84]. The most promising technique, which is also inexpensive and easy to handle, is atomic fluorescence spectrophotometry (AFS). The AFS at 253.7 nm has been used as the determination procedure, in conjunction with the cold vapour-gold principle, for analysis of natural waters [41,59,131]. Detection limits down to 0.5 pg have been reported [41]. The applicability of the method is currently determined by the standard deviation of the procedural blank. The use of inductively-coupled plasma mass spectrometry (ICP-MS) for mercury determination has also been documented [132]. A detection limit of 8 pg for mercury in water has been found (determined as 3 times the standard deviation of the blank).

For samples having a high particle content or a high humus content, oxidation by HNO_3 alone [60], after ageing or after treatment with UV [89], followed by $SnCl_2$ or $NaBH_4$ reduction, is not sufficient to liberate all of the mercury. Oxidation using BrCl in HCl or $HClO_4$ should be used [4,54,61,64]. High particle contents can be found in rainfall, estuarine- and freshwater samples; humic and fulvic acids can also be found in the latter. It is important to be aware of the different reactivities of mercury compounds when analyses are performed. These differences can be, and are, often used operationally to distinguish between various mercury forms; cf. Section 10.3.2.

Interference-free determination of mercury in water, using a gold-plated piezoelectric crystal detector, is described by Ho and Guilbault [133]. The detection limits are, however, a little too high for general applicability (5 ng). An interesting approach employing direct amalgamation on gold in contact with the water sample has been applied by Neske et al. [134]. A detection limit of 0.5 ng/kg was found for a 1 l sample.

10.3.1.3. Air

Most of the analytical determination procedures for natural waters mentioned in Section 10.3.1.2 are also used to determine mercury in air [44,46,52,76,77,86]. The methods are applied after collection of the mercury from the air into a series of traps. For a more extensive discussion, see Section 10.2.2.

References on p. 282

Other analytical procedures used to determine mercury in air are the Barringer atmospheric mercury monitor, with speciation capability (see Section 10.3.2), which is based on AFS [135], and the Jerome gold film mercury monitor, which is based on the measurement of the increase in electrical resistance as a result of amalgamation [136]. The Jerome apparatus has also been used to determine mercury in sediments after digestion [137]. A detection limit of 0.05 mg/g, for 1 g of sediments, is reported.

10.3.2. Speciation

10.3.2.1. Analyte specific

A positive identification of the specific compound is always the best approach in determining mercury fractions. It should, however, be kept in mind that even a preconcentration of mercury in air sampling, or the filtration of water samples, are operational defined steps. In the present chapter we consider this to be a part of the sampling procedure. The discussion on analyte-specific methods is focused on methylmercury because of the great environmental importance of this species (see Section 10.1).

(i) *In general and biological material.* Separation of specific mercury compounds requires some sort of chromatographic step, of which gas chromatography (GC) is the preferred one. Westöö developed the first and still widely employed method for analysis of methylmercury in biological materials [13,14]. The method is based on benzene-cysteine extraction and packed-column GC. From this first approach various GC-based techniques have been elaborated [138,139,141]. Some general problems with the packed-column GC method are the tricky conditioning of the column, and the difficulties in obtaining a stable column. The use of capillary columns is still not widespread due to the lack of suitable commercially available columns. Proper optimization of, for example, a CP-SIL 8 capillary column seems to be promising, however [141].

The original detector used in the Westöö procedure was an electron-capture detector (EC). Later, more specific and sensitive detectors were combined with the GC-technique, e.g. a mass spectrometer (MS) [142], an atomic absorption spectrophotometer (AAS) [143,144], various atomic emission spectrometers (AES) such as microwave-induced plasma emission (MIP) [4,72,145] and direct-current plasma (DCP) [146]. The specificity of these detection principles allows the preceding extraction and cleaning steps to be simplified. An example of a very simple method for determination of methylmercury in biological material is the direct use of head-space capillary GC-MIP on homogenates treated with sulfuric acid and a halide [147]. Examples of other chromatographic techniques are thin-layer chromatography of dithizone extracts combined with AAS [148], high-performance liquid chromatography (HPLC) in combination with inductively coupled plasma AES [149] or MIP-AES [150], and HPLC of dithizonate extracts combined with spectrophotometry [151]. In general, HPLC in combination with an AES method is not as sensitive as the GC analogue. In another elaboration of the Westöö approach total organic mercury is determined in the extract by, for example, neutron activation analysis [152,153]. CVAAS or ICP-MS have also been applied on such extracts [154].

(ii) *Air.* Ballantine and Zoller [92] report a method for the collection and positive identification of methylmercury (MM) and dimethylmercury (DMM) in the atmosphere, with a detection limit of 0.1 ng/m^3. They used Chromosorb 101 for preconcentration of MM, a cryogenic trap ($-80°C$) for DMM, and detection by GC-MIP. The method has been tested in the field in the near vicinity of a power plant. The detection limit of the method is probably insufficient for mercury sampling in background air. In general, the separation of mercury species in air is very difficult to perform [3] and the widespread existence and the amount of gaseous alkylmercury compounds in background air may still be disputed, even though the existence of MM has been shown [41].

(iii) *Natural waters.* Lee [65] reported a promising method for methylmercury determination at sub-nanogram per liter levels, based on preconcentration on a sulfhydryl cotton fiber (SCF) prior to benzene extraction and detection by GC–ECD. The detection limit, 0.04 ng/kg, was derived for a 20 l sample. Later, both the preconcentration and the analytical step of this method have been developed, e.g. by use of a capillary column [155]. Chiba et al. [145] report rather high methylmercury values in seawater determined by benzene-cysteine extraction and GC-MIP analysis. From a combination of preconcentration of mercury bound to macromolecular organic material on an XAD-2 resin, direct extraction of free alkylmercury species in seawater, and the GC-ECD method, Suzuki and Sugimura [127] suggest that the methylmercury found is a secondary product derived from the analytical procedure. May et al. [156] successfully used an acid extraction followed by ion exchange, UV irradiation and CVAAS. Robertson et al. [4] describe a preconcentration method for methylmercury which uses a chelating resin and a GC-MIP detector. In the field, however, they found 20–30 times lower methylmercury values in some filter freshwater samples by this method than by a direct extraction-GC-MIP determination. Robertson et al. [4] suggest that methylmercury is either lost from the resin or converted to an undetectable mercury form during storage.

Yamamoto et al. [66] report a detection limit of 0.005 ng/kg for methylmercury. They used the original Westöö extraction, starting with a 3.6 l sample. After the last cysteine extraction methylmercury was transferred into dithizone/chloroform and the methylmercury content determined by AAS. Another approach is described by Bloom [157]. He uses sodium tetraethylborate for aqueous phase ethylation of methylmercury, collection on a graphitic carbon column, thermal desorption, and cryogenic GC with CVAFS detection: the detection limit was 0.6 pg (0.003 ng/kg) for a 200 ml sample.

10.3.2.2. Operational defined

In contrast to the analyte-specific methods, operational defined methods do not positively separate and identify an analyte. Instead speciation is carried out by operations, i.e. groups that behave alike, chemically or physically are speciated.

(i) *Biological material.* It should be noted that the determination of total organic mercury by, for example, an extraction followed by CVAAS is strictly speaking an operational defined method. Other operational defined methods for mercury compounds are not often applied to biological materials. Oda and Ingle [158] describe a CVAAS method where selective reduction is performed on a KOH digest using $SnCl_2$ and $NaBH_4$. Inorganic mercury is determined by the $SnCl_2$ reduction, and organomercury compounds in the same solution as the mercury are recovered afterwards by $NaBH_4$ treatment. The reported detection limits are in the ppb range. Konoshi and Takahashi [159] describe a method to determine inorganic mercury selectively in the presence of organomercury compounds by H_2O_2 reduction in strong alkaline solution. Yeast is used to accelerate the oxidative liberation of the inorganic mercury from organic substances.

(ii) *Air*. Brosset [44] describes a method for the determination of operationally defined mercury fractions in air, based on the pronounced difference in the Henry's Law constants of various mercury compounds. The apparatus consists of a wash bottle containing pure water in front of an accumulative gold trap. A group of mercury species (Type I) with rather high Henry's Law constants (about 0.3) and including elemental and dimethylmercury, pass through the water and are preconcentrated on the gold trap. Another group (Type II) with Henry's Law constants orders of magnitude lower (less than 10^{-4}) including inorganic mercury(II) and methylmercury(II) compounds, are collected in the water. The latter group is subdivided into Hg-II, i.e. mercury compounds which can be recovered by $NaBH_4$ and collected on a gold trap after subsequent purging of the water, and Hg-IIa compounds which are reduced by $SnCl_2$ and recovered on a gold trap.

Other methods for operational defined speciation in air are based on the various collecting capacities of commercial polymers, and silver and gold surfaces. Schroeder and Jackson [91] designed a sequential preconcentration method using Chromosorb treated with HCl vapour, Tenax GC, Carbosieve B, and a gold wire. Kim and Fitzgerald [62] used a silver-gold collecting train to differentiate between total gaseous mercury and organic gaseous mercury in air. The results of these methods could, however, be disputed.

(iii) *Natural waters*. The operational definition of mercury species in natural waters is based on the type of pretreatment necessary to volatilize and collect mercury on a gold trap. Dissolved gaseous mercury (DGM), also named 'volatile' mercury or Hg-I, is directly transferred to a gold trap by a purging of the water sample in the field [62,63,64]. Robertson et al. [4] defined 'elemental' mercury by direct purging and collection using the combination of silver-gold taps. Kim and Fitzgerald [62] also used the silver-gold approach in separating dissolved organic mercury from dissolved gaseous mercury. The relevance of this separation is unclear. 'Reactive' mercury (alternatively 'easily reduced' or Hg-IIa) is reduced to elemental mercury by $SnCl_2$ before purging [16,44,60,61,64,89]. Sodium borohydride, $NaBH_4$ is used for the recovery of 'reactive + non-reactive' mercury or Hg-II [44,64]. Methods for total mercury determination in natural waters are described above.

10.3.3. New techniques

It is our experience that, in order to achieve accurate and precise results for mercury, especially at the ultra-trace level, dedicated apparatus is the optimum solution. HPLC in combination with inductively coupled plasma mass spectrometry (HPLC-ICP-MS) has been tried for organomercury determination. Although the instrumentation is rather expensive the approach looks promising. Promising experiments in measuring mercury emissions from waste disposal sites in Sweden have been carried out using differential optical absorption spectrometry - DOAS: see Platt and Perner [160] for a description of the basic method and Edner et al. [161] for the applicability in mercury analysis. The DOAS method uses differential optical absorption over a long path through the atmosphere. The detection limit is approximately 5 ng/m^3; the method can therefore only be applied for elevated mercury concentrations in air. Results from field measurements of mercury by this technique have not yet been published in the open literature.

10.4. CONCLUSIONS

For almost any element special care and precautions are prescribed in sampling and analysis. For mercury the physical characteristics make this even more necessary than for most other elements. Mercury has a high vapour pressure, a fact that is utilized in most of the determination procedures (cold vapour techniques). Mercury forms amalgams, a fact that is utilized in sampling as well as in the analytical techniques (the use of gold traps). Mercury forms covalent bonds, and organo-mercury compounds are formed by biologically-mediated processes in nature. The great and early interest in mercury stems from the extreme toxicity, especially of methylmercury, a compound which newer investigations suggest may be even more toxic than first believed. The catastrophes in the 1950's meant tremendous efforts in method development, and mercury was one of the first, if not the first, element for which speciation became common.

The special characteristics of mercury cause its anthropogenic pollution to be of more than just national concern. Mercury is spread through atmospheric transport, and during the last decade many efforts have been directed towards an understanding of its environmental cycling, which is very much governed by subtle redox processes between water-soluble mercury(II) compounds and mercury(0). A better understanding of these processes improves both the sampling and sample handling for mercury, because the same processes govern its loss from samples as well as contamination at ultra-trace levels.

The general state of mercury analysis in 1988–1990 that, although mercury analysis is still difficult, many laboratories are capable of doing analysis at ultra-trace levels. The detection limits for state-of-the-art procedures are determined by precision and method blank values. Sensitivities for various cold vapour techniques, e.g. CVAAS, CVAFS, and MIP-AES, and the control of blank values, are generally so good that we believe the baseline values summarized in this Chapter to be correct. The decline in 'confirmed' values for mercury in seawater has stopped.

Many aspects of the mercury cycling in the environment still have to be clarified. The roles of particles and organic complexes are being recognized as important and research on these is continuing. We believe that the important steps forward will be taken using dedicated instruments. Experience has shown that research on mercury, especially at the ultra-trace level, poses special demands on the instruments that are difficult to fulfill with general-purpose instruments. Cold-vapour techniques using gold traps and atomic absorption/emission, and especially atomic fluorescence spectrometry, are now the dominant determination procedures and will still be in the near future.

A full understanding of the mercury cycle requires speciation capabilities. The best approaches are based on the analyte-specific methods, using some kind of chromatographic techniques (GC) in combination with, for example, AES or AFS. At the ultra-trace level, the lack of sensitivity and difficulties in controlling the method blank mean that much important research will be carried out on the basis of operational defined methods. In general, research on speciation, and analyses performed at ultra-trace level, are hampered by the scarcity of suitable certified reference materials.

REFERENCES

1 WHO, *Recommended Health - based Limits in Occupational Exposure to Heaavy Metals, Technical Report Series*, World Health Organization, Geneva, 1980, No. 647.
2 J.O. Nriagu (Editor), *The Biogeochemistry of Mercury in the Environment*, Elsevier-North-Holland Biomedical Press, 1979.
3 O. Lindqvist and H. Rodhe, *Tellus*, 37B (1985) 136-159.
4 D.E. Robertson, D.S. Sklarew, K.B. Olsen, N.S. Bloom, E.A. Crecelius and C.W. Apts, *Measurement of bioavailable mercury species in fresh water and sediments*, Electric Power Research Institute, Palo Alto, CA, 1987, Report EPRI EA-5197.
5 A. Stock and W. Zimmermann, *Z. Angew. Chem.*, 41 (1928) 1336-1337.
6 J. Ruzicka and C.G. Lamm, *Talanta*, 16 (1969) 157-168.
7 B.V. Acharya and B. Rangamannar, *J. Radioanal. Nucl. Chem., Letters*, 95 (1985) 127-130.
8 Analytical Methods Committee, *Analyst (London)*, 90 (1965) 515-530.
9 N.S. Poluektov, R.A. Vitkun and Y.V. Zelyukova, *Z. Anal. Khim.*, 19 (1964) 937-942.
10 B. Sjostrand, *Anal. Chem.*, 36 (1964) 814-819.
11 V. Weiss and T.E. Crozier, *Anal. Chim. Acta*, 58 (1972) 231-233.
12 H.A. Van Der Sloot and H.A. Das, *Anal. Chim. Acta*, 73 (1974) 244-325.
13 G. Westöö, *Acta Chem. Scand.*, 20 (1967) 1790-1800.
14 G. Westöö, *Acta Chem. Scand.*, 22 (1968) 2277-2280.
15 A. Tugsavul, D. Merten and O. Suschny, *The Reliability of Low-level Radiochemical Analysis. Results of Intercomparisons Organized by the Agency during the Period 1966- 1969*, IAEA Report, 1970.
16 G.A. Gill and W.F. Fitzgerald, *Mar. Chem.*, 20 (1987) 227-243.
17 N.S. Bloom and C.J. Watras, *Sci. Total Environ.*, 87/88 (1989) 199-207.
18 E. Uchino, T. Kosuga, S. Kanishi and M. Nishimura, *Envir. Sci. Technol.*, 21 (1987) 920-922.
19 J. Olafsson, *Mar. Chem.*, 11 (1982) 129-142.
20 S.S. Berman and V.J. Boyko, *ICS Sixth Intercalibration for Trace Metals in Estuarine Water (JMG 6-TM-SW)*, ICES Cooperative Research Report No. 152, International Council for the Exploration on the Sea, Copenhagen, 1988.
21 M. Berlin, in L. Friberg, G.F. Nordberg and V.B. Vouk (Editors), *Handbook on the Toxicology of Metals, Vol. II; Specific Metals*, Elsevier, Amsterdam, 2nd edn., 1986, Ch. p. 387.
22 FAO-WHO, *Technical Report Series*, No. 505, 1972. World Health Organization.
23 T. Kjellström, P. Kennedy, S. Wallis and C. Mantell, *Nat. Swedish Environ. Protec. Board*, 1986, SNV PM 3080.
24 T. Kjellström, P. Kennedy, S. Wallis and C. Mantell, *Nat. Swedish Environ. Protec. Board*, 1989, PM 3642.

25 WHO, *Technical Report Series,* No. 776, 1989, World Health Organisation.
26 L. Friberg, L. Kellman, B. Lind and M. Nylander, *Lakartidningen,* 7 (1986) 519-522.
27 M.J. Vimy and F.L. Lorscheider, *J. Dental Res.,* 64 (1985) 1070-1075.
28 J.E. Patterson, B.G. Weissberg and P.J. Dennison, *Bull. Environ. Contam. Toxicol.,* 34 (1985) 459-468.
29 K. May and M. Stoeppler, in *Proc. Int. Conf. Heavy Metalas in the Environment, Heidelberg, September 1983,* Vol. 1, CEP Consultants Ltd., Edinburgh, 1983, pp. 241-244.
30 T. Andersson, A. Nilsson, L. Hakanson and I. Brydsten, *Nat. Swedish Environ. Protec. Board,* 1987, Rep. 3291.
31 P.J. Craig, in P.J. Craig (Editor), *Organometallic Compounds in the Environment - Principles and Reactions,* Longman Group Ltd., Harlow, U.K., 1986, Ch. 2, p. 65.
32 M. Bernhard, *Mercury in the Mediterranean,* UNEP Regional Seas Reports and Studies, No. 98, 1988.
33 O. Lindquist, K. Johansson, M. Aastrup, A. Andersson, L. Bringmark, G. Hovsenius, L. Håkanson, Å. Iverfeldt, M. Meili and B. Timm, *Water, Air Soil Pollut.,* 55 (1991) xi-xiii.
34 K. May, K. Reisinger, B. Torres and M. Stoepler, *Fresenius' Z. Anal. Chem.,* 320 (1985) 646.
35 K. May, R. Ahmed, K. Reisinger, B. Torres and M. Stoeppler, in T.D. Lekas (Editor), *Proc. Int. Conf. Heavy Metals in the Environment, Athens - September, 1985,* CEP Consultants Ltd., Edinburgh, 1985, Vol. 1, pp. 513-515.
36 H.U. Riisgard, T. Kiørboe, F. Møhlenberg, I. Drabaek and P. Pheiffer Madsen, *Mar. Biol.,* 86 (1985) 55-62.
37 R. Capelli and V. Minganti, *Sci. Total Environ.,* 63 (1987) 83-99.
38 A. Andersson, *Nat. Swedish Environ. Protec. Board,* 1985, Rep. 3042.
39 O. Lindqvist, A. Jernelov, K. Johansson and H. Rohde, *Nat. Swedish Environ. Protec. Board,* 1984, PM. 1816.
40 M.H. Bothner, R.A. Jahnke, M.L. Peterson and R. Carpenter, *Geochim. Cosmochim. Acta,* 44 (1980) 273-285.
41 N.S. Bloom and W.F. Fitzgerald, *Anal. Chim. Acta,* 208 (1988) 151-161.
42 C. Brosset and Å. Iverfeldt, *Water, Air Soil Pollut.,* 43 (1989) 147-168.
43 C. Brosset and E. Lord, *Water, Air Soil Pollut.,* 56 (1991) in press.
44 C. Brosset, *Water, Air Soil Pollut.,* 34 (1987) 145-166.
45 W.F. Fitzgerald, in P. Baut-Menard (Editor), *The Role of Air-Sea Exchange in Geochemical Cycling,* Reidel, Dordrecht, 1986, pp. 363-408.
46 Å. Iverfeldt, *Water, Air Soil Pollut.,* 56 (1991) in press.
47 F. Slemr, G. Schuster and W. Seiler, *J. Atmos. Chem.,* 3 (1985) 407-434.
48 W.F. Fitzgerald, G.A. Gill and J.P. Kim, *Science,* 244 (1984) 597-599.
49 G. Mierle, *Environ. Toxicol. and Chem.,* 9 (1990) 843-851.
50 G.E. Glass, J.A. Sorensen, K.W. Schmidt and G.R. Rapp, Jr., *Environ. Science Technol.,* 24 (1990) 1059-1069.
51 R. Ahmed, K. May and M. Stoeppler, *Sci. Total Environ.,* 60 (1987) 249-261.
52 R. Ferrara, B. Maserti, A. Petrosino and R. Bargagli, *Atmos. Environ.,* 20 (1986) 125-128.
53 R. Ahmed and M. Stoeppler, *Anal. Chim. Acta,* (1987) 109-113.
54 R. Ahmed, K. May and M. Stoeppler, *Fresenius' Z. Anal. Chem.,* 326 (1987) 510-516.
55 C. Brosset, in *Proc. Int. Conf. Coal Fired Power Plants,* Copenhagen, August 16-18, 1982, F. Rasmussen, Copenhagen, 1982, pp. 460-467.
56 M. Metzger and H. Braun, *Chemosphere,* 16 (1987) 821-832.
57 H. Appelquist, K. Ottar Jensen, T. Sevel and C. Hammer, *Nature,* 273 (1978) 657-659.
58 Å. Iverfeldt and K. Johansson, in: B. Moldan and T. Paces (Editors), *Proc. Int. Workshop on Geochem. and Mon. in Repr. Basins in Prague 1987* - Geol. Survey, Prague, Czechoslovakia.
59 L.A. Nelson, *Environ. Technol. Letters,* 2 (1981) 225-232.
60 D. Cossa and J. Noel, *Mar. Chem.,* 20 (1987) 389-396.

61 N.S. Bloom and E.A. Crecelius, *Mar. Chem.*, 14 (1983) 49-59.

62 J.P. Kim and W.F. Fitzgerald, *Science*, 231 (1986) 1131-1133.

63 Å. Iverfeldt, *Structural, thermodynamic and kinetic studies of mercury compounds; applications within the environmental mercury cycle.* Ph.D. Thesis, Chalmers University of Technology and University of Göteborg, Göteborg, Sweden, 1984, 48 pp.

64 Å. Iverfeldt, *Mar. Chem.*, 23 (1988) 441-456.

65 Y.H. Lee, *Int. J. Environ. Anal. Chem.*, 29 (1987) 263-276.

66 J. Yamamoto, Y. Kaneda and Y. Hikasa, *Int. J. Environ. Anal. Chem.*, 16 (1983) 1-16.

67 R. Ferrara, A. Seritti, C. Barghigiani and A. Petrosino, *Mar. Chem.*, 18 (1986) 227-232.

68 J. Cragin, *Anal. Chim. Acta*, 110 (1979) 313-319.

69 I. Sanemasa, *Bull. Chem. Soc. Jpn.*, 48 (1975) 1795-1798.

70 Å. Iverfeldt and O. Lindqvist, *Atmos. Environ.*, 16 (1982) 2917-2925.

71 Å. Iverfeldt and I. Persson, *Inorg. Chim. Acta*, 103 (1985) 113-119.

72 Y. Talmi and R.E. Mesmer, *Water. Res.*, 9 (1975) 547-552.

73 Å. Iverfeldt and O. Lindqvist, in W. Brutsaert and G.H. Jirka (Editors), *Proc. of the International Symposium on Gas Transfer at Water Surfaces*, Reidel, Dordrecht, 1984, pp. 533-538.

74 D. Mackay and W.Y. Shiu, in W. Brutsaert and G.H. Jirka (Editors), *Proc. of the International Symposium on Gas Transfer at Water Surfaces*, Reidel, Dordrecht, 1984, pp. 3-16.

75 C. Brosset, *Water, Air Soil Pollut.*, 16 (1981) 253-255.

76 C. Brosset, *Water, Air Soil Pollut.*, 17 (1982) 37-50.

77 Å. Iverfeldt and O. Lindqvist, *Atmos. Environ.*, 20 (1986) 1567-1573.

78 H. Morita, T. Mitsuhashi, H. Sakurai and S. Shimomura, *Anal. Chim. Acta*, 153 (1983) 351-355.

79 G.E. Glass, E.N. Leonard, W.H. Chan and D.B. Orr, *J. Great Lakes Res.*, 12 (1986) 37-51.

80 N. Bloom and S.W. Effler, *Water, Air and Soil Pollut.*, 53 (1990) 251-265.

81 G.A. Gill and W.F. Fitzgerald, *Deep-Sea Res.*, 32 (1985) 287-297.

82 K. Matsunaga, S. Konishi and M. Nishimura, *Environ. Sci. Technol.*, 13 (1979) 63-65.

83 J.M. Mattinson, *Anal. Chem.*, 44 (1972) 1715-1716.

84 Y. Nojiri, A. Otsuki and K. Fuwa, *Anal. Chem.*, 58 (1986) 544-547.

85 F. Slemr, W. Seiler, C. Eberling and P. Roggendorf, *Anal. Chim. Acta*, 110 (1979) 35-47.

86 W.F. Fitzgerald and G.A. Gill, *Anal. Chem.*, 51 (1979) 1714-1720.

87 F. Slemr, W. Seiler and G. Schuster, *J. Geophys. Res.*, 86 (1981) 1159-1166.

88 W.H. Schroeder, M.C. Hamilton and S.R. Stobart, *Revs. Anal. Chem.*, 8 (1985) 179-209.

89 A. Seritti, A. Petrosino, R. Ferrara and C. Barghigiani, *Environ. Technol. Lett.*, 1 (1980) 50-57.

90 R.S. Braman and D.L. Johnson, *Environ. Sci. Technol.*, 8 (1974) 996-1003.

91 W.H. Schroeder and R.A. Jackson, *Int. J. Environ. Anal. Chem.*, 22 (1985) 1-18.

92 D.S. Ballantine and W.H. Zoller, *Anal. Chem.*, 56 (1984) 1288-1293.

93 R. Dumarey, R. Dams and J. Hoste, *Anal. Chem.*, 57 (1985) 2638-2643.

94 R. Ahmed and M. Stoeppler, *Analyst (London)*, 111 (1986) 1371-1374.

95 M. Sadakata, *Bunseki Kagaku*, 34 (1985) 276-282.

96 R. Dumarey, R. Heindryckx and R. Dams, *Environ. Sci. Techn.*, 15 (1981) 206-209.

97 H. Vogg, H. Braun, M. Metzger and J. Schneider, *Waste Management Res.*, 4 (1986) 65-74.

98 C. Brosset, *Swedish State Power Board*, KHM Rep. 76, 1983.

99 J.G.T. Bergstrom, *Waste Management Res.*, 4 (1986) 57-64.

100 C. Feldman, *Anal. Chem.*, 46 (1974) 99-102.

101 C. Baldeck, G.W. Kalb, *Amer. Environ. Protection Agency*, 1973, PB-220, p. 323.

102 D. Klockow, V. Siemens and K. Larjava, *VDI Berichte*, No. 838, 389-399, VDI Verlag, 1990.

103 G. Kaiser, D. Gotz, G. Tölg, G. Knapp, B. Maichin and H. Spitzy, *Fresenius' Z. Anal. Chem.*, 291 (1978) 278-291.
104 G. Kaiser and G. Tölg, in O. Hutzinger (Editor), *The Handbook of Environmental Chemistry, Volume 3-Part A, Mercury,* Springer Verlag, Berlin, Heidelberg, 1980, pp. 1-58.
105 M. Sadiq and T.H. Zaidi, *Int. J. Environ. Anal. Chem.*, 16 (1983) 57-66.
106 H.W. Louie, *Analyst (London),* 108 (1983) 1313-1317.
107 R. Dumarey, P. Verbiest and R. Dams, *Bull. Soc. Chim. Belg.*, 94 (1985) 351-357.
108 H. Narasaki, *Anal. Chim. Acta,* 125 (1981) 187-191.
109 R. Dumarey and R. Dams, *Mikrochim. Acta,* (1984) 191-198.
110 B. Rozanska and E. Lachowicz, *Anal. Chim. Acta,* 175 (1985) 211-217.
111 W.D. Ehmann, W.R. Markesbery, T.I.M. Hossain, M. Alauddin and D.T. Goodin, *J. Radioanal. Chem.*, 70 (1982) 57-65.
112 K.O. Jensen and V. Carlsen, *J. Radioanal. Chem.*, 47 (1978) 121-134.
113 I. Drabaek, v. Carlsen and L. Just, *J. Radioanal. Nucl. Letters,* 103 (1986) 249-260.
114 H.F. Haas and V. Krivan, *Fresenius' Z. Anal. Chem.*, 324 (1986) 13-18.
115 E. Orvini and M. Gallorini, *N.B.S. Special Publication,* 422 (1976) 1233-1240.
116 L. Kosta and A.R. Byrne, *Talanta,* 16 (1969) 1297-1303.
117 W.R. Hatch and W.L. Ott, *Anal. Chem.*, 40 (1968) 2058-2087.
118 O. Donard and P. Pedemay, *Anal. Chim. Acta,* 153 (1983) 301-305.
119 R.F. Suddendorf, *Anal. Chem.*, 53 (1981) 2234-2236.
120 E. Temmerman, R. Dumarey and R. Dams, *Stud. Environ. Sci.*, 29 (1986) 745-748.
121 E. Temmerman, R. Dumarey and R. Dams, *Anal. Letters,* 18 (1985) 203-216.
122 A. Kuldvere, *Analyst (London),* 107 (1982) 179-184.
123 H.B. MacPherson and S.S. Berman, *Analyst (London),* 108 (1983) 639-641.
124 S. Blake, Water Research Centre, 1985, Environment, TR 229.
125 M.P. Bertenshaw and K. Wagstaff, *Analyst (London),* 107 (1982) 664-672.
126 X. Feng and D.E. Ryan, *Intern. J. Environ. Anal. Chem.*, 19 (1985) 273-280.
127 Y. Suzuki and Y. Sugimura, in A.C. Sigleo and A. Haatton (Editors), *Marine and Estuarine Geochemistry,* Lewis Publishers, 1985, pp. 259-273.
128 M. Filippelli, *Analyst (London),* 109 (1984) 515-517.
129 R. Ahmed and M. Stoeppler, in M. Stoeppler and M. Durbeck (Editors), *Contributions to Environmental Specimen Banking,* Jul-Spez-349, 1986, KFA Julich.
130 K. Tanabe, J. Takahashi, H. Haraguchi and K. Fuwa, *Anal. Chem.*, 52 (1980) 453-457.
131 R. Ferrara, A. Seritti, C. Barghigiani and A. Petrosino, *Anal. Chim. Acta,* 117 (1980) 391-395.
132 C. Haraldsson, S. Westerlund and P. Öhman, *Anal. Chim. Acta,* 221 (1989) 77-84.
133 M.H. Ho and G.G. Guilbault, *Anal. Chim. Acta,* 130 (1981) 141-147.
134 P. Neske, A. Hellwig, L. Dornheim and B. Thriene, *Fresenius' Z. Anal. Chem.*, 318 (1984) 498-501.
135 W.H. Schroeder and R. Jackson, *Chemosphere,* 13 (1984) 1041-1051.
136 W.H. Schroeder and R.A. Jackson, *Chemosphere,* 16 (1987) 183-199.
137 A. Mudroch and E. Kokotich, *Analyst (London),* 112 (1987) 709-710.
138 J.A. Rodriguez-Vazquez, *Talanta,* 25 (1978) 299-310.
139 C.J. Cappon, *LC-GC,* 5 (1987) 400-418.
140 M. Horvat, A.R. Byrne and K. May, *Talanta,* 37 (1990) 207-212.
141 *Community Bureau of Reference,* EEC, unpublished data.
142 S. Ohkoshi, T. Takahashi and T. Sato, *Bunseki Kagaku,* 22 (1973) 593-595.
143 R. Bye and P.E. Paus, *Anal. Chim. Acta,* 107 (1979) 169-175.
144 R. Dumarey, R. Dams and P. Sandra, *J. High Resolut. Chromatogr. Chromatogr. Commun.*, 5 (1982) 687-689.
145 K. Chiba, K. Yoshida, K. Tanabe, H. Haraguchi and K. Fuwa, *Anal. Chem.*, 55 (1983) 450-453.
146 K.W. Panaro, D. Erickson and I.S. Krull, *Analyst (London),* 112 (1987) 1097-1105.
147 G. Decadt, W. Baeyens, D. Bradley and L. Goeyens, *Anal. Chem.*, 57 (1985) 2788-2791.

148 A. Kudo, H. Nagase and Y. Ose, *Water Res.*, 16 (1982) 1011-1015.
149 I.S. Krull, D.S. Bushee, R.G. Schleicher and S.b. Smith, Jr., *Analyst (London)*, 111 (1986) 345-349.
150 D. Kollotzek, D. Oechsle, G. Kaiser, P. Tschopel and G. Tölg, *Fresenius' Z. Anal. Chem.*, 318 (1984) 485-489.
151 W. Langseth, *Anal. Chim. Acta*, 185 (1986) 249-258.
152 E. Orvini and M. Gallorini, *J. Radioanal. Chem.*, 71 (1982) 75-95.
153 I. Drabaek and V. Carlsen, *Int. J. Anal. Chem.*, 17 (1984) 231-239.
154 S.S. Berman, K.W.M. Siu, P.S. Maxwell, D. Beauchemin and V.P. Clancy, *Fresenius' Z. Anal. Chem.*, 333 (1989) 641-644.
155 Y.H. Lee and J. Mowrer, *Anal. Chim. Acta*, 221 (1989) 259-268.
156 K. May, M. Stoeppler and K. Reisinger, *Toxicol. Environ. Chem.*, 13 (1987) 153-159.
157 N. Bloom, *Can. J. Fish. Aquat. Sci.*, 46 (1989) 1131-1140.
158 C.E. Oda and J.D. Ingle, Jr., *Anal. Chem.*, 53 (1981) 2305-2309.
159 T. Konishi and H. Takahashi, *Analyst (London)*, 108 (1983) 827-834.
160 U. Platt, D. Perner and H.W. Patz, *J. Geophys. res.*, 84 (1979) 6329-6335.
161 H. Edner, A. Sunesson, S. Svanberg, L. Uneus and S. Wallin, *Appl. Opt.*, 25 (1986) 403-409.

M. Stoeppler (Editor)/*Hazardous Metals in the Environment*
© 1992 Elsevier Science Publishers B.V. All rights reserved
287

Chapter 11

Arsenic

Kurt J. Irgolic

Karl-Franzens-Universität Graz, Institut für Analytische Chemie, Universitätsplatz 1, A-8010 Graz (Austria)

CONTENTS

11.1. Introduction 288
11.2. Total arsenic versus arsenic compounds 290
11.3. Arsenic compounds in the environment 292
11.4. Total arsenic determinations 294
    11.4.1. Digestion 295
    11.4.2. Nuclear methods for the determination of total arsenic 296
        11.4.2.1. Instrumental neutron activation analysis 297
        11.4.2.2. Chemical separation before or after neutron activation 297
    11.4.3. Atomic methods for the determination of total arsenic 300
        11.4.3.1. X-ray fluorescence with X-ray excitation 301
        11.4.3.2. Proton-induced X-ray emission 302
        11.4.3.3. Atomic fluorescence spectrometry 303
        11.4.3.4. Flame atomic absorption and emission spectrometry 303
        11.4.3.5. Electrothermal atomic absorption spectrometry 304
            11.4.3.5.1. Interferences 304
            11.4.3.5.2. Applications of GFAAS for the direct determination of total arsenic 308
        11.4.3.6. Plasma emission spectrometries 309
    11.4.4. Molecular methods for the determination of total arsenic 310
        11.4.4.1. Electrochemical methods 310
            11.4.4.1.1. Amperometry and coulometry 311
            11.4.4.1.2. Polarography 311
            11.4.4.1.3. Stripping voltammetry 311
        11.4.4.2. Colorimetric methods 312
            11.4.4.2.1. Molybdenum-blue method 312
            11.4.4.2.2. Silver diethyldithiocarbamate method 313
        11.4.4.3. Hydride generation method (HG) 314
            11.4.4.3.1. HG-gas chromatography 315
            11.4.4.3.2. HG-colorimetry 315
            11.4.4.3.3. HG-atomic fluorescence spectrometry 315
            11.4.4.3.4. HG-flame atomic absorption spectrometry 316
            11.4.4.3.5. HG-quartz tube atomizer-atomic absorption spectrometry 316

288

11.4.4.3.6. HG-electrothermal atomic absorption spectrometry 317
11.4.4.3.7. HG-microwave-induced plasma atomic emission spectrometry 318
11.4.4.3.8. HG-molecular emission spectrometry 318
11.4.4.3.9. HG-DC-plasma atomic emission spectrometry . . . 319
11.4.4.3.10. HG-helium glow discharge atomic emission
spectrometry 319
11.4.4.3.11. HG-inductively coupled argon plasma emission spectrometry 319
11.4.4.3.12. Hydride generation method: interferences 320
11.5. Determination of arsenic compounds 321
11.5.1. Determination of arsenite and arsenate 322
11.5.1.1. Electrochemical methods for the determination of arsenite
and arsenate 322
11.5.1.2. Separation of arsenite and arsenate by extraction 323
11.5.1.3. Separation of arsenite and arsenate by precipitation 324
11.5.1.4. Determination of arsenite and arsenate by colorimetry . . . 324
11.5.1.5. Determination of arsenite and arsenate by the hydride
generation method 325
11.5.1.6. Chromatographic separation of arsenite and arsenate 325
11.5.2. Determination of methylarsenic compounds 326
11.5.2.1. Separation and determination of arsenite, arsenate, and
methylarsenic compounds by ion-exchange chromatography 327
11.5.2.2. Separation and determination of arsenite, arsenate, and
methylarsenic compounds by hydride generation 327
11.5.2.3. Gas chromatographic separation of volatile methylarsenic
derivatives other than arsines 332
11.5.2.4. Extraction procedures for the separation of inorganic and
organic arsenic compounds 334
11.5.2.5. Determination of alkylarsines and triphenylarsine 334
11.5.2.6. Identification of tetramethylarsonium cation 335
11.5.3. Determination of arsenobetaine and arsenocholine 335
11.5.3.1. Separation, purification and isolation of arsenobetaine 335
11.5.3.2. Identification of arsenobetaine 336
11.5.3.3. Determination of arsenocholine 338
11.5.4. Determination of dimethyl(ribosyl)arsine oxides 338
11.6. Standards for the determination of arsenic 340
References 340

11.1. INTRODUCTION

Arsenic compounds, such as diarsenic trisulfide (orpiment), tetraarsenic tetrasulfide (realgar), and arsenic trioxide have been known – at least in impure form – for several millenia [1,2]. The toxic effects of arsenic-containing ores were recognized by observation of the workers pursuing rich veins in the bowels of the earth. The life span of early miners was very short. The ancient Greeks preferred to have slaves work the arsenic-sulfide bearing formations. The peculiar properties of arsenic trioxide – life threatening at oral doses of approximately 100 mg [3], but beneficial to the appearance of horses and men under well-controlled conditions [4] – were probably discovered by people operating crude roasting facilities for arsenic-containing sulfide ores and people working and living near such facilities. Knowledge about these arsenic compounds and their properties reached Europe with the advancing armies of the Islamic conquerors.

The mountainous regions in Europe are rich in sulfidic ore deposits. The arsenic concentrations in these ores may range from traces to several percent [5]. Veins consisting entirely of arsenopyrite, FeAsS, and mined for their arsenic content are not rare. The first step in the beneficiation of sulfidic ores is roasting. The ore, broken into small pieces, is heated in air to temperatures at which the sulfide-sulfur is oxidized to sulfur dioxide and the metal components converted into their oxides. Under these conditions the arsenic in the ores forms arsenic trioxide, a substance that sublimes at 180°C at atmospheric pressure and has a considerable vapor pressure even at lower temperatures [6]. The arsenic trioxide condenses onto cool surfaces, for instance in the smoke stack or in flues attached to the roasting furnace, or escapes through the stack and accumulates as a white powder on the vegetation and the soil surrounding the roasting facility. In this way arsenic trioxide became easily accessible to a large number of people. The colloquial term for arsenic trioxide (also called "white arsenic") coined by the peasants in the Austrian and German mountains was "Hüttrach" (smelter house smoke). As mining activities increased and with them the availability of "Hüttrach", the use of this mineral poison for criminal activities began to spread. Whereas murder by premeditated exposure to arsenic trioxide via oral ingestion was rare in the thirteenth century, it must have been a rather common occurrence during the fourteenth and fifteenth centuries, because governmental authorities passed specific laws to restrain these illegal activities. "Hüttrach" was used to settle political differences, speed up transfer of wealth through inheritances, and solve domestic problems in an irrevocable manner [7]. Arsenic trioxide was almost ideal for these purposes: as a white powder it looks similar to other materials used in the preparation of food; mixed with prepared foods it is without taste; it will dissolve in drinks to produce a lethal dose; and no sufficiently selective and sensitive method existed to prove its presence in foods or the tissues of the victims. After Marsh had perfected in 1836 his test for inorganic arsenic based on the reduction of arsenite to arsine and the thermal decomposition of arsine to elemental arsenic, murders by arsenic trioxide declined precipitously. With the closing of many mines and roasting facilities, access to arsenic trioxide became more difficult.

Arsenic compounds can have beneficial influences on human and animal life. Beginning with Paracelsus, doctors prescribed a variety of formulations containing inorganic arsenic compounds to reduce fever, prevent black death, heal boils, cure ailments of the lungs, fight cancer [8], treat chronic myelocytic leukemia, psoriasis, bronchial asthma and dermatoses, and improve general appearance and well-being [9]. "Arsenic eaters", reported to have consumed as much as one-half gram of arsenic trioxide at a time on a regular basis, are said to have acquired the habit to protect themselves from illnesses, to relieve asthmatic pains, to increase their sexual potency, to prevent conception (taken by women), to raise their feeling of general well-being, and to better endure fatigue [10]. Although these claims have never been scientifically substantiated, the beneficial effects of

References on p. 340

aromatic arsonic acids as feed additives [9] and the curative powers of bis(4-hydroxy-3-aminophenyl)diarsene and 4-hydroxy-3-aminophenyl[4-hydroxy-(3-sodium sulfoxylato-methylamino)phenyl]diarsene (salvarsane, neosalvarsane) as antisyphylitic agents [11] are well documented.

The lingering memories of the criminal use of arsenic trioxide in the distant past causes the public mind to equate the term "arsenic" with the term "poison" adding for good – but not justified – measure adjectives such as potent, horrible, and worst. This attitude, bordering on arsenophobia, overshadows the beneficial uses of arsenic and its compounds and prevents reasoned evaluations of its impact when it is detected in environmental samples. Often, the search for the culprit in a poisoning episode ends abruptly when arsenic is found. The British beer-poisoning episode that caused 1000 deaths in 1900 may serve as an example [12]. When arsenic was discovered in the beer, the case was closed with the setting of limits for arsenic in brewery feedstocks. Later, the symptoms were identified as typical of selenium poisoning [9].

The bad reputation of arsenic that arose from the misuse of easily available arsenic trioxide may be slowly waning and may be replaced by a more reasonable attitude. Since 1975 evidence has been accumulating that points toward an essential role for arsenic in life-processes. The experiments were carried out with chicks, rats, hamsters, goats, and minipigs [13]. The metabolic role of arsenic is unknown. An arsenic requirement for man cannot be estimated with certainty. For animals, 50 ng of arsenic per gram of diet may be just adequate to prevent deficiencies [14]. W. Mertz states [15]: "If such findings could be extrapolated to man (50 ppb in 500–600 g of dry diet) a minimum daily requirement would be estimated at 25 to 30 μg, an amount not furnished by the typical diet, of which the Market Basket Survey [in the U.S.A.] is representative. Although the initial immediate consequences of deficiency of most trace elements are mild, the fact that severe disease, sudden heart-death, appeared in the third generation of arsenic-deficient goats, should serve as a strong motivation for intense future research". These developments should encourage investigations that attempt to uncover not only hazards associated with exposure to arsenic but also to define its essential role in life-processes.

11.2. TOTAL ARSENIC VERSUS ARSENIC COMPOUNDS

Historical events make it understandable that arsenic is perceived as a life-threatening poison. Events in the recent past, such as the Morinaga milk incident (Japan, 1955; in which infants received milk containing arsenate) and the soy sauce incident (Japan, 1956) [16] strengthened this perception. Certain arsenic compounds are toxic and can cause acute and chronic poisonings [17]. Until very recently, the effects of arsenic were discussed in terms of "total arsenic". Such blanket generalizations cannot be defended. Arsenic is a metalloid with a rich chemistry and forms a large number of inorganic and organic compounds. Each one of these compounds has its own properties and affects biological systems in its own way. To state that arsenic is toxic is equivalent to the proclamation that carbon is a poison. Just as certain carbon compounds (such as carbon monoxide, oxalic acid, alkaloids, and snake venoms) are toxic and others are life-support-ing (such as carbohydrates, lipids, amino acids, and vitamins), so are arsenic com-pounds.

Most of the arsenic-oriented analytical work is still directed toward the determination of total arsenic concentrations in environmental samples. This situation probably came about because the instrumental analytical techniques that were developed during the past forty years were element-oriented. Early qualitative and quantitative analytical procedures were compound-specific. Qualitative tests [18] for arsenite and arsenate and quantitative meth-ods [19] allowed trivalent and pentavalent inorganic arsenic to be differentiated. The detection limits of these methods are too high to be useful for the determination of arsenic compounds in biological materials. Instrumental methods, such as flame emission and absorption spectrometry, graphite furnace atomic absorption spectrometry, plasma atomic emission spectrometry, and neutron activation analysis, lowered the detection limits for arsenic to nanogram and subnanogram levels. However, they provided results in terms of "total arsenic", because the arsenic-containing molecules are atomized during the analysis (in absorption and emission spectrometries) or the methods are insensitive to the chemical environment of arsenic (in nuclear techniques). These methods dominated ana-lytical investigations directed toward trace elements, fostered an attitude that neglected consideration of trace element compounds, and led to the establishment of exposure limits in terms of "total trace element" concentrations [20,21]. For instance, the maximum allowable arsenic concentration in U.S. drinking waters is 50 μg total arsenic per liter. In aerobic water the predominant arsenic compound is arsenate. If the drinking water limit were applied to seafood that often contains arsenic at milligram per kilogram levels, consumption of seafood would have to cease. Setting limits in terms of arsenic com-pounds, with proper consideration of the possibility that they may be changed by metabolic processes, is the rational approach as soon as sufficient and reliable analytical and toxicological data are available. Such data are now available, for arsenic compounds, whose toxicities vary widely. Among arsenic compounds common in the environment, arsenite is the most toxic (LD_{50} 10 mg/kg, rats) [22]. Arsenate (LD_{50} 100 mg/kg, rats), methylarsonic acid (LD_{50} 700 mg/kg, mice), dimethylarsinic acid, and aromatic arsonic acids, are much less toxic than arsenite. Arsenobetaine, the arsenic compound ubiquitous in marine animals, was without effect when given to mice orally at a dose of 400 mg/kg body weight [23]. Rat embryos were also unaffected by arsenobetaine [24].

Environmental scientists who investigate the detrimental and beneficial effects of "ar-senic" must focus their attention on arsenic compounds and consider the determination of total arsenic only as the first step of an analysis. A large percentage of the approximately 1500 papers on arsenic and the biological environment published during the past twenty years in 650 different journals consider only "total arsenic". With more attention to arsenic compounds the understanding of their interactions with biological systems will deepen. This should allow regulatory decisions to be based on well-founded scientific knowledge, lead to better methods for treatment of arsenic toxicoses, and provide welcome insights into arsenic's life-supporting functions.

References on p. 340

The analytical methodologies for accurate and precise determinations of total arsenic and a variety of arsenic compounds are now available. The following sections will discuss these methodologies and point out their applicabilities, drawbacks, and advantages. The coverage cannot be encyclopaedic; the abundance of publications and the limitation of space prevent such a desirable undertaking. However, the methods that are widely used and those that promise to lead to new insights will be discussed.

11.3. ARSENIC COMPOUNDS IN THE ENVIRONMENT

Knowledge of the chemical nature of arsenic compounds present in biological systems, and the environment with which they are in contact, is necessary when they are to be identified and determined, and when the total arsenic concentrations are to be measured. The results of total arsenic determinations are often influenced by the type of arsenic compounds present.

Figure 11.1 contains the structural formulae and names of the arsenic compounds that are known to be associated with organisms and which may be encountered in their local environment. In the soil and sedimentary environments the inorganic and organic arsenic acids may be present as salts of available cations in the form of precipitates or adsorbates on hydrous oxides, clays, and other surface active materials. Otherwise these arsenic compounds are largely found in aqueous media.

Arsenous acid, H_3AsO_3, which contains trivalent arsenic, is a weak acid with a first dissociation constant of 8.5×10^{-10} [25]. In the aqueous environment, in intracellular fluids, and in the extracellular aqueous medium, characterized by pH values not far from neutral, the undissociated acid, H_3AsO_3, will be the predominant species, with the dihydrogen arsenite ion, $H_2AsO_3^-$, increasing in relative concentration at pH values above 7.

Arsenic acid, H_3AsO_4, which contains pentavalent arsenic, is a tribasic acid ($K_1 = 5.7 \times 10^{-3}$, $K_2 = 1.1 \times 10^{-7}$, $K_3 = 3.2 \times 10^{-12}$) [26] comparable in strength to phosphoric acid. Hydrogen arsenate ($HAsO_4^{2-}$) and dihydrogen arsenate ($H_2AsO_4^{2-}$) are the predominant species in an aqueous solution of arsenic acid around pH 7. The terms "arsenite" [arsenate(III)] and "arsenate" [arsenate(V)] are commonly used as names for the species formed from arsenous acid and arsenic acid. This time-honored and convenient practice does not accurately identify the arsenic compounds present in an aqueous solution. The species-distribution is defined by the pH of the medium and the presence of cations that might form insoluble arsenites and arsenates.

Arsenite and arsenate are the most common arsenic compounds in the environment. Arsenite is formed as the weathering product of arsenic-containing sulfidic ores. In aerobic waters arsenate is the thermodynamically stable form of inorganic arsenic. However, for example, the biological activity in the aerobic surface layer of the ocean is known to reduce arsenate to arsenite [27]. Without intervention of organisms, arsenite and arsenate are expected to be the only arsenic compounds in the environment.

Methylarsonic acid ($K_1 = 1.1 \times 10^{-4}$, $K_2 = 2.8 \times 10^{-9}$) [28,29] and dimethylarsinic acid ($K = 5.4 \times 10^{-7}$) [29] are widely distributed in the environment. Although the use of

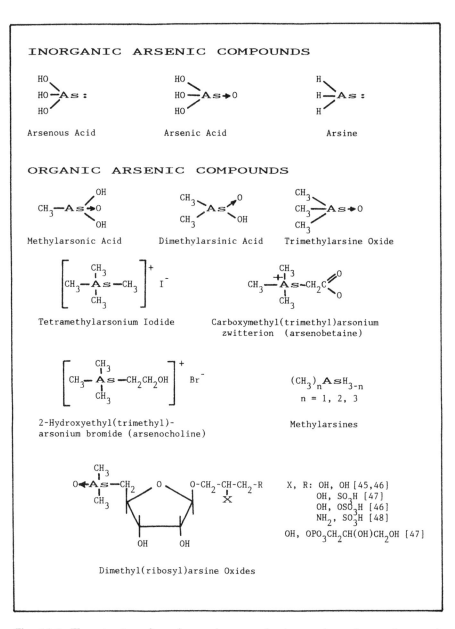

Fig. 11.1. The structure formulae and names for inorganic and organic arsenic compounds identified in nature.

these organic arsenic compounds as herbicides is responsible for a part of the environmental load, on a global scale most of these methylated, pentavalent arsenic compounds are of biological origin [30], formed from arsenite by successive methylations. The undissociated acids and their mono-anions are the predominant species in neutral aqueous solutions.

Trimethylarsine oxide, $(CH_3)_3AsO$, a white, crystalline, very hygroscopic compound, that may exist in solution as trimethylarsine dihydroxide or hydroxy(trimethyl)arsonium hydroxide, was identified in estuarine catfish [31], Baltic seafish [32], and as a microbial degradation product of arsenobetaine [33].

Methylarsines, $(CH_3)_nAsH_{3-n}$ (n = 1, 2, 3), the reduction products of methylarsonic acid, dimethylarsinic acid, and trimethylarsine oxide, were detected over soils treated with methylarsonate [34] and over sediments containing methylated arsenic compounds [35]. Arsine, the end-product of the reduction of arsenite and arsenate, was detected in the headspace of culture flasks containing bacteria (*Aeromonas* sp., *E. coli*, *Flavobacterium* sp.) and inorganic arsenic [35]. In contrast to the extreme air-sensitivity of methylarsines in bulk, trimethylarsine at trace concentration reacts slowly with oxygen and may travel considerable distances in aerobic environments without undergoing chemical changes [36].

The tetramethylarsonium cation was detected for the first time in 1987 in the clam *Meretrix lusoria* [37], the sea hare, *Aplysia kuriodai* [38], and the sea anemone, *Parasicyonis actinostoloides* [38]. This compound is acutely toxic to mice, with LD_{50} values (for tetramethylarsonium iodide) of 890 (p.o.), 175 (i.p.), and 82 (i.v.) [39].

Arsenobetaine is ubiquitous in marine animals and is present at mg/kg levels [33,40], often as the major arsenic compound. Arsenobetaine appears to be non-toxic [23,24]. Arsenocholine has been found only infrequently [41,42,43] in marine organisms. It is likely that arsenocholine is rapidly oxidized in the body to arsenobetaine by an enzyme-catalyzed reaction [44].

Dimethyl(ribosyl)arsine oxides were isolated from brown kelp, *Ecklonia radiata*, [45,47] the giant clam, *Tridacna maxima*, [45,47] the brown seaweed, *Laminaria japonica*, [49] and the green seaweed, *Codium fragile* [50]. These ribosylarsine oxides may be the precursors of arsenobetaine [51,52].

In addition to the arsenic compounds in Fig. 11.1, arsenic- containing lipids are probably common in marine algae [53,54]. Firm experimental proof for the constitution of such lipids is lacking, although attachment of an arsenocholine [53] or a dimethyl(ribosyl)arsine oxide group [Fig. 11.1, X = OH, R = $HOCH_2CH(OH)CH_2OPO_3$] [54] to a phosphatidic acid appears to be reasonable.

As the search for arsenic compounds in biological systems widens, additional compounds that are part of the natural arsenic cycle will be identified. Thus far, most of the investigations have been carried out with marine organisms because of their importance to the human diet. Whether arsenic compounds structurally more complex than simple methylarsenic derivatives exist in terrestrial and fresh water organisms is not known.

11.4. "TOTAL ARSENIC" DETERMINATIONS

Determination of "total arsenic" in biological and environmental samples is an important first step in the analytical work leading to the identification and quantification of arsenic compounds. The more involved an analytical procedure, the greater is the need for a good mass balance with respect to total arsenic. Although such determinations have been carried out on a large number of samples of varying complexity for a long time, they cannot be considered to be completely routine when inorganic and organic arsenic compounds are expected to be present in a sample. Whereas arsenic determinations in matrices approaching distilled water generally do not cause any problems, such determinations are not easy in the presence of organic materials or an excess of various anions and cations. For instance, a round-robin experiment with synthetic human urine containing arsenate at 642 or 200 micrograms of arsenic per liter, using atomic absorption spectrometry, showed that seventeen of the forty-six specially qualified laboratories obtained results more than three standard deviations from the true values [55]. Synthetic urine is not a very complex matrix. Most of the methods for the determination of total arsenic are influenced by the molecular forms in which it occurs. The type and concentrations of the arsenic species are not known for most samples. Standard addition using, for example, sodium arsenite will not improve the situation unless arsenite is the only arsenic compound in the sample. Therefore, it is advisable to digest samples with oxidizing acids in order to convert all arsenic compounds into arsenate. Such a digestion is especially important for samples of biological materials, in which several organic arsenic compounds may be present.

The methods for the determination of total arsenic concentrations will be classified according to the properties upon which the analytical procedures are based: i.e., those using nuclear methods, atomic methods, and molecular methods.

11.4.1. Digestion

The digestion of arsenic-containing samples must be conducted under conditions that convert all organic arsenic compounds to inorganic arsenic and prevent its loss. In a medium of high chloride ion concentration arsenic may be volatilized as arsenic trichloride. Similar losses may occur in media containing fluoride, bromide, or iodide. At the elevated temperatures encountered, for instance, during dry ashing or upon evaporation of a digest to dryness, arsenic oxides may sublime. Arsenic pentoxide will decompose above 300°C to arsenic trioxide, with release of oxygen. Arsenic trioxide may be lost even below 180°C.

The digestion of biological materials is performed in an oxidizing environment by heating the sample with a mixture of magnesium oxide and magnesium nitrate [56–58], or with concentrated nitric acid, mixtures of nitric and perchloric acid, of nitric and sulfuric acid, or of nitric, sulfuric and perchloric acids. A range of temperatures and times can be used, and vessels such as open beakers and Teflon bombs [59,60] can be employed. Alternatively, the sample can be heated with oxygen in a combustion bomb [61]. The literature is replete with reports claiming success as well as failures in attempts to mineralize biological samples in preparation for the determination of total arsenic [59]. Most of the digestions have been carried out without knowledge of the nature of the arsenic compounds present in the samples. Some organic arsenic compounds are now known not to be easily converted to arsenate. For example, methylarsonic acid and dimethyl-arsinic acid were not completely decomposed by a mixture of concentrated nitric and perchloric acids at 100°C during 30 minutes [62]. Phenylarsonic acid survived three hours at 160°C in concentrated nitric acid [63]. Digestions of sole, halibut, and shrimp samples containing arsenobetaine converted only 35 to 60% of the arsenic material to arsenate [57]. A comparison of ten digestion procedures for soils, in which total arsenic had to be determined by hydride generation-atomic absorption spectrometry, showed eight methods to be inadequate. Best results were obtained with sulfuric acid/hydrogen peroxide and a nitric/sulfuric acid leaching procedure [64]. In a procedure for the digestion of chemically very resistant arsenic compounds marine algae were treated with concentrated nitric acid and then with perchloric and sulfuric acids with programmed increases of temperature from 20 to 310°C. The completeness of digestion was checked by comparisons with arsenic concentrations determined by neutron activation analysis of the undigested material [65].

Much of the confusion about the reliability and repeatability of digestion procedures which exists in the literature derives from the use of inappropriate methods. The analytical methods used for the determination of total arsenic in the original samples and in the digests often respond only to inorganic arsenic. A comparison of arsenic concentrations in a sample and in its digest is meaningful only when the concentration in the sample was determined by a species-independent method, such as neutron activation analysis, and that in the digest by a method applicable only to arsenate. Recovery studies, which are often performed with standard reference materials, or by the addition of an arsenic compound that in most cases is arsenate, do not provide reliable information. Experiments of this type can be successful only when the matrices and the arsenic compounds in them are the same. Thus, complete recovery of arsenic, for instance from NBS SRM Oyster Tissue, by a particular digestion method does not imply that under the same conditions complete recovery is achievable, for instance, from a seaweed sample.

Recently, a closed-vessel microwave digestion system was described that uses microwave energy as the heat source. This system has the ability to reduce digestion time, increase efficiency, and prevent loss of analytes [66]. Such a system was employed to digest tissues from marine animals and plants with mixtures of nitric and sulfuric acids in the presence of nickel(II) nitrate, in preparation for the determination of total arsenic [67]. More experience must be accumulated before the microwave digestion system can be claimed to solve the problems encountered with arsenic compounds in biological materials.

11.4.2. Nuclear methods for the determination of total arsenic

In principle, nuclear properties are ideal for the determination of total arsenic. These properties do not depend on the chemical environment of the arsenic atom or on any other conditions that influence atomic or molecular properties. Therefore, nuclear methods truly measure total arsenic irrespective of how many different arsenic compounds are present in the sample. Digestion, which often introduces systematic errors, is unnecessary. Nuclear activation analysis exploits the n, γ-reaction (Eqn. 11.1) of arsenic that is present in nature only as As-75.

$$As^{75}_{33} + n^1_0 \rightarrow As^{76}_{33} \rightarrow Se^{75}_{33} + e^0_{-1} + \gamma \qquad (11.1)$$

The arsenic isotope As-76, formed by irradiation of the sample with a flux of 10^{12} neutrons $cm^{-2}s^{-1}$ for several hours, is a β-emitter and transmutes to the stable selenium isotope Se-76. Associated with this transmutation is the emission of γ-photons. The most intense emission at 559 keV is used for analytical purposes. The detection limits achievable by this method depend on the composition of the matrix but are of the order of one nanogram of arsenic. This method requires a neutron source, generally a nuclear reactor, and counting equipment [using a Ge(Li) detector].

11.4.2.1. Instrumental neutron activation analysis

The direct, non-destructive determination of arsenic by neutron activation analysis, without pretreatment or digestion of the sample, suffers from several interferences. Biological specimens generally have a high sodium content. A high γ-activity by Na-24 (half-life = 15 hr) must be allowed to decay for several days before the arsenic activity can be counted. The decay of As-76 during this period of approximately five half-lives degrades the detection limit of the method to ppm levels. The As-76 γ-peak (55.9 keV) appears between the Br-82 (554, 619 keV, 35.3 h) and Sb-122 (564 keV, 67.2 h) peaks. Because of the similar half-lives of these three isotopes, the interferences by bromine and antimony – if present – cannot be avoided by allowing the Br- and Sb-activities to decay. A high-resolution detector, such as a Ge(Li) detector must be used. When the selenium or bromine concentration of a sample is high, the potential interference of the Se-76 (n, p) and Br-79 (n, α) reactions producing As-76 should be considered [68,69].

Because of the interferences likely to be encountered in the non-destructive NAA determination of arsenic in biological materials, the method has not been used very often [60]. However, arsenic was successfully determined in soils [70], water [71], plant materials [72], mushrooms [73], and U.S. National Bureau of Standards Standard Reference Materials [68,74].

11.4.2.2. Chemical separation before or after neutron activation

To avoid the interferences that plague the non-destructive neutron activation analyses, arsenic can be separated from the sample before or after activation. The separation techniques are chemical in nature and the reactions are compound-specific. For complete recovery, arsenic must be in the chemical form appropriate to the separation process chosen. Therefore, radiochemical analysis procedures often include the oxidative digestion of biological materials and the concomitant conversion of all arsenic compounds to

TABLE 11.1.

RADIOCHEMICAL PROCEDURES FOR THE SEPARATION AND DETERMINATION OF TOTAL ARSENIC AFTER IRRADIATION OF AQUEOUS, BIOLOGICAL, AND SOIL SAMPLES WITH NEUTRONS

Sample	Digestion	Separation	Material Counted	DL	Other Elements Determined by NAA	Ref.
Glacial ice	conc. H_2SO_4	as As_2S_3 by addition of H_2S, separated from Cd,Cu,Hg,Mn	As_2S_3	0.4 ng	Cd,Cu,Hg,Mn	79
Freshwater	none; acidification with HNO_3	ion exchange on Al_2O_3; Na-24 washed from column with 1 M NaCl in 0.1 M HNO_3	Arsenic on Al_2O_3	0.05 mg/l	Sb,W	80
		Anion exchange (arsenate); As_2S_3 precipitation with thioacetamide; dissolution in HNO_3	arsenic in HNO_3 solution	0.001 μg/l	Cd,Co,Hg,Mo, Zn	81
Freshwater	$K_2S_2O_8$/HCl	reduction to AsH_3 by $NaBH_4$; AsH_3 collected on activated carbon	arsenic on activated carbon; γ-counting or Cerenkov counting	1 ng/l 0.1 ng/l	Sb	82
Aerosol (membrane filter)	conc. H_2SO_4/HNO_3/aq. KMnO₄, Br₂, As-74 spike	$NH_2OH\cdot HCl$/HCl/CuCl; extract $AsCl_3$ into benzene; strip $AsCl_3$ into water; precipitate As as As_2S_3 (thioacetamide)	As_2S_3	1 ng	–	83
Tobacco leaves, clam tissue, NBS Orchard Leaves	conc. HNO_3/$HClO_4$/ H_2O_2/carrier	distillation of $AsCl_3$; then precipitation as As_2S_3	$As2S_3$	–	Hg	84

Sample	Decomposition	Treatment	Form collected	Detection limit	Interferences	Ref.
Biological materials, NBS Bovine Liver, spinach, pine needles, animal bone, muscle	dry ashing with $Mg(NO_3)_2$	reduction with KI/ascorbic acid/$NaBH_4$ to arsine; AsH_3 collected on carbon absorber	arsenic on carbon absorber	–	Sb	85
Land snails, Periwinkles	conc. H_2SO_4/HBr/H_2O_2	distillation of $AsBr_3$	$AsBr_3$ in aqueous HBr	–	Au,Hg,Sb	86
NBS Wheat Flower, NBS Bovine Liver	H_2SO_4/HNO_3/$HClO_4$, carrier	extraction of AsI_3 from aqueous phase into toluene	AsI_3 in toluene	–	V,Mo	87
NBS Orchard Leaves, NBS Bovine Liver	HNO_3/$HClO_4$/HF	passage of 1 M HNO_3 solution through hydrated MnO_2 column	arsenic on MnO_2	–	Sb,Se,Cr,Cd,Cu	88
NBS Orchard Leaves	H_2SO_4/HNO_3	addition of ascorbic acid to remove Se; removal of heavy metals by cupferron; precipitation of As as As_2S_3 in filtrate (thioacetamide); dissolution of As_2S_3 by ammonium sulfide	arsenic in ammonium sulfide solution	1 mg/kg	Mn,Se	89
NBS River Sediment, NBS Oyster Tissue, NBS Coal	H_2SO_4/HNO_3	removal of $HgCl_2$ by distillation from $HClO_4$ solution; distillation of $AsBr_3$ from HCl/HBr solution; reduction to As by $NH_4H_2PO_2$	As on filter	2 ng/g	Hg,Se	90
Soil	$NaOH$/Na_2O_2 fusion/carrier	precipitation of arsenic as As_2S_3 by thioacetamide from HCl solution	As_2S_3 on filter	–	Se	91

References on p. 340

arsenate. Selective extraction or ion exchange chromatography can be used to collect the arsenic and separate it from interfering elements such as sodium and antimony.

Arsenic was determined in freshwater and saltwater by neutron activation analysis after its separation from the matrix by co-crystallization with thionalide. This procedure requires arsenic to be in the trivalent state. Therefore, the sample was heated with ascorbic acid which reduces arsenate to arsenite. The precipitate was irradiated and then counted [75]. Alternatively, the precipitate was mixed with appropriate carriers, dissolved in nitric acid, then copper and antimony were removed as oxinates, and arsenic in the filtrate co-precipitated with iron(III) hydroxide. The iron-hydroxide precipitate was counted [76]. These procedures have a detection limit of 0.01 μg of arsenic per liter ($\sim$1 nmol/l).

Arsenic was concentrated from seawater by co-precipitation with iron(III) hydroxide, and flotation of the precipitate with the aid of sodium dodecyl sulfate. The collected precipitate was irradiated for 16 hours. The detection limit was 0.06 μg/l [77].

Arsenic in milk powder was determined by first ashing the powder in a low temperature oxygen plasma, then irradiation of the ash, followed by separation of arsenic from copper and zinc by ion exchange and extraction procedures [78].

Post-irradiation treatment of samples such as water and biological materials is frequently used to eliminate the interference by Na-24 and Sb-122 and to separate arsenic from other nuclides. These separations facilitate the simultaneous determination of elements and improve the detection limits for arsenic. Examples of such procedures are listed in Table 11.1. Anion exchange separations, extraction of arsenic trihalides into organic solvents, removal of arsenic from the sample by distillation of arsenic trihalides, precipitation of arsenic trisulfide, and formation of volatile arsine (AsH_3) have all been employed. Owing to the separation of arsenic from other, interfering constituents in a sample by the radiochemical techniques, high-resolution γ-detectors are not required and NaI (Tl)-based systems can be used [81].

Although the radiochemical methods achieve detection limits as low as 1 ng/l, they have not been widely used in routine determinations of total arsenic in environmental samples. The requirement of access to a nuclear reactor and a laboratory equipped for radiochemical work prevents a wider use of these techniques. Many of the papers report the results of efforts to develop methods using standard reference materials as samples, but few publications describe the application of the methods to environmental samples (Table 11.1).

11.4.3. Atomic methods for the determination of total arsenic

Atomic methods for the determination of total arsenic use spectrophotometric techniques based on electronic transitions within the arsenic atom. The arsenic atom has 28 core electrons residing in the completely filled K, L, and M shells and five valence electrons in the N shell. Electronic transitions associated with valence shell electrons occur in the visible and ultraviolet region and are influenced by other atoms bonded to arsenic. Therefore, methods based on such transitions (i.e., flame absorption and emission, graphite furnace absorption, atomic fluorescence, and plasma emission) require the presence of uncombined atoms. The atoms are formed from molecular species at the high temperatures to which the analytes are exposed. The transitions which become possible after one of the core electrons is dislodged by irradiation of the sample with X-rays or protons are influenced very little – if at all – by the chemical environment of the arsenic atom. Therefore, X-ray fluorescence and particle-induced X-ray emission spectroscopies do not require that arsenic be present in atomic form.

The atomic methods for the determination of total arsenic, especially graphite furnace atomic absorption spectrometry, are much used in the analysis of environmental samples. These analytical techniques are not interference-free and each technique has its specific problems that must be considered when arsenic present at very low concentrations has to be determined and when accuracy and precision are of importance.

11.4.3.1. X-ray fluorescence with X-ray excitation

The irradiation of a sample with X-rays, often produced using a molybdenum X-ray tube, removes a core electron from an atom. The vacancy is filled by an electron from a higher shell, resulting in the emission of an X-ray photon. The wavelength or the energy and the intensity of such an emission are measured by wavelength- or energy-dispersive detectors.

X-ray fluorescence spectroscopy, a non-destructive and compound-independent technique, unfortunately has high detection limits of approximately one milligram per liter for unprocessed samples and is not suitable for the determination of traces of arsenic. To overcome this disadvantage, separation and preconcentration techniques have been developed for metal ions [92] and for arsenite and arsenate. These techniques collect the desired precipitate in a thin layer on a support and thus provide optimal conditions for the excitation with concomitant reduction of the self-absorption of the emitted X-rays. Under these conditions the attenuation of the emitted X-rays by the sample might become negligible.

Biological matrices usually require digestion and conversion of all arsenic compounds to inorganic arsenic. Reagents employed for the pre-concentration of arsenic are ammonium pyrrolidinedithiocarbamate [93,94], diethyldithiocarbamate [95,96], dibenzyldithiocarbamate [97], thioacetamide [98], thionalide [99], 1-(2-pyridylazo)-2-naphthol [95], and sodium molybdate [100]. The precipitates can either be extracted into an organic solvent that is subsequently evaporated [95] or collected on a support such as a filter. The precipitation can also be carried out on quartz glass as the support, with small volumes of the solution to be analyzed [94]. This procedure avoids the filtration step. The arsenic must be present as arsenite when dithiocarbamates are used as the precipitating agents. Arsenate can be reduced to arsenite by ascorbic acid [96]. The extraction of arsenic as arsenomolybdate into a polyurethane foam [100] will be possible only when all of the arsenic is in the form of arsenate. The precipitation of arsenic as arsenic sulfide with thioacetamide [98] as reagent is effective for arsenite and arsenate. The dithiocarbamate precipitates form in solutions of pH 5. Cations that form insoluble hydroxides under these

References on p. 340

conditions might interfere – when present in excess – with the formation of dithiocarbamates.

The preconcentration of arsenic improves the detection limits of the X-ray fluorescence method into the sub-ppm range. Further improvements were achieved by using an intense beam of monochromatic X-rays [99]. When the dithiocarbamate precipitate was deposited on an optically flat silanized quartz surface and the geometry was set to achieve total reflection without scattering of the exciting X-ray beam by the sample support, a detection limit of 20 pg of arsenic was achieved using 100 μl of the original sample. This detection limit of 0.2 mg/l is valid only for specimens of low matrix content. With increasing matrix the detection limits deteriorate owing to an increasing background caused by radiation scattered by the matrix [94]. The detection limits could probably be further improved by using intense monochromatic X-rays. A detection limit of 0.05 mg/l was claimed for the arsenomolybdate/polyurethane foam method [100].

X-ray fluorescence was used to determine arsenic, mainly in water samples [93,94,98,99]. A few applications of this technique to marine biological materials [101], vegetables [98], meat [96,102], soil [98,103], and air [98] were reported.

For the quantitative determination of arsenic the As K_α line (1.177 Å) is most often used. Lead (Pb L_α 1.175 Å) and (Bi L_α 1.114 Å) may interfere with the arsenic determination [96,98]. In these cases the As K_β line may be used [96].

Despite the preconcentration of arsenic and spectrometric improvements, X-ray fluorescence does not have the low- and sub-nanogram (μg/l) detection limits that other methods, employing less expensive instrumentation, offer. These facts, the high cost of X-ray fluorescence spectrometers, and matrix interferences have prevented a wider use of X-ray fluorescence for the determination of total arsenic in biological samples.

11.4.3.2. Proton-induced X-ray emission

Vacancies in core shells of an atom can be created by bombarding a sample with particles such as protons. The electrons from higher shells then fill these vacancies with the emission of X-rays, whose intensity and wavelength or energy is measured in the same way as in X-ray fluorescence spectrometry. The advantage of the proton-induced-X-ray emission (PIXE) method derives from the large cross section for characteristic X-ray production [104]. The detection limit of this method is set by the broad bremsstrahlung continuum that has its maximal intensity at the low-energy end of the X-ray spectrum. The bremsstrahlung is generated by the deceleration of electrons that are removed from the atoms within the sample by the incident protons. For successful PIXE analyses, the samples must be digested to simplify the matrix and treated appropriately to produce very thin targets for irradiation by protons. The required proton beam can be generated by a Van de Graaf accelerator. A proton energy of approximately 3 MeV was found to be optimal [105]. Biological materials were wet-digested with nitric acid, and two drops of the digest placed on a carbon foil to evaporate. This procedure gave detection limits of 0.01 mg/l in liquid and 0.1 mg/kg wet weigh for biological specimens [105]. Because PIXE provides total arsenic concentrations irrespective of the chemical form of arsenic, the digestion procedure does not have to convert all arsenic compounds to arsenite or arsenate, unless preconcentration techniques are used.

The PIXE method was used for the determination of total arsenic in urine, in horse kidney, horse liver, chicken meat, NBS Orchard Leaves [105], wood [106], and fish and crabs [107].

11.4.3.3. Atomic fluorescence spectroscopy

In atomic fluorescence spectroscopy, arsenic compounds are thermally decomposed into their constituent atoms. Irradiation of the assemblage of gaseous atoms with light emitted, for instance, by a hollow-cathode or an electrode-less discharge arsenic lamp excites arsenic atoms selectively from their ground state to excited states. Return to the ground states causes light characteristic of arsenic to be emitted (189.0, 193.7, 197.2, 235.0 nm and other [108]). Valence electrons are involved in these transitions. The most intense emission line is the 189.0 nm line. Because only arsenic atoms are excited and only they emit light, monochromators are not needed for atomic fluorescence spectroscopy. Non-dispersive atomic fluorescence spectrometers measure the intensity of the emitted light at 90° to the excitation beam. These systems are severely affected by light scattered by any solid particles in the gas which contains the arsenic atoms. Therefore, samples with arsenic in complex matrices cannot be directly introduced into the atomizer. The sample must be digested or the arsenic removed from the matrix. The most common method for the separation of arsenic from the sample is hydride generation (see section 11.4.4.3). The gaseous arsines are flushed from the sample into a flame [109–113] or a heated quartz tube [114]. Detection limits of 0.1 milligram of arsenic per liter [108], two nanograms [109], and 0.015 nanograms [114] are reported. This method was used to determine arsenic in seawater, river water, and tap water [113,114], waste-waters [111], soil digests [112], and plants [114] after separation of the arsenic as arsine.

11.4.3.4. Flame atomic absorption and emission spectrometry

The lines in the atomic spectrum of arsenic are in the far ultraviolet region [189.0, 193.7, 197.2 nm), where the flame gases also absorb appreciably. At 193.7 nm air-acetylene flames have a transmittance of 28%, nitrogen-hydrogen flames have 42%, argon-hydrogen flames 82% [115], and a nitrogen-shielded acetylene-nitrous oxide flame has 100% [116]. Extensive chemical interferences generally plague flame atomic absorption spectrometric determinations of arsenic. The detection limits are rarely better than 1 mg/l but often worse [117]. For these reasons, direct introduction of an aqueous sample into the flame has hardly been used at all for the determination of arsenic during the last thirty years. A flame Zeeman atomic absorption spectrometric approach, in conjunction with thermal evolution of arsenic, has been used successfully for the determination of arsenic in the NBS Oyster Tissue Standard Reference Material [118], and deserves further investigation.

Pre-concentration of arsenic can bring its concentration into the range suitable for flame atomic absorption spectrometry. For instance, extraction of the arsenic by trioctyl-amine, trioctylphosphine oxide, dithiocarbamates [119], or dioctyl dinitrates [120] into an organic solvent, back-extraction of arsenic into an aqueous phase, and aspiration of the aqueous solution into an air-acetylene flame made arsenic detectable at a concentration of 20 micrograms per liter of original sample [119]. These methods were not applied to environmental samples. A commonly used pre-concentration method, involving reduction to arsines which are collected in traps prior to introduction into the flame, is discussed in the section on hydride generation.

Flames are not hot enough to excite arsenic atoms in sufficient numbers to make flame atomic emission spectrometry a viable method for the determination of arsenic. Emission spectrometry becomes possible with the much hotter plasmas (microwave, DC, or inductively coupled argon plasmas).

11.4.3.5. Electrothermal atomic absorption spectrometry

Electrothermal or graphite furnace atomic absorption spectrometry (GFAAS) has found wide application for the determination of trace elements in water, soil, and biological samples. The detection limits are excellent and, under favorable conditions, may reach picogram levels, corresponding to ppb concentrations. The samples are generally injected into the graphite cuvette as solutions having volumes from 10 to 100 μl. The ashing of the injected sample at elevated temperatures simplifies the matrix, but may also cause losses of analyte. The atomization is carried out at temperatures between 2000 and 3000°C. These temperatures produce a collection of gaseous atoms that are exposed to the radiation of hollow cathode or electrode-less discharge lamps. If the ashing step was performed under the proper conditions, little material should be left that might vaporize during atomization and interfere with the analysis. However, "dirty samples" with matrices far removed in composition from distilled water, such as organic-rich extracts from biological specimens, increase the background and the detection limits. Background correction devices available for atomic absorption spectrometers will overcome many of these problems. The background correction based on the Zeeman effect has been proved to be very powerful.

Graphite furnace atomic absorption spectroscopy is not interference-free and can be influenced by the molecular form in which arsenic is present int he sample. A survey of the literature indicates that only samples of benign composition, such as rainwater and fresh-water with low mineral content, are injected directly without pre-treatment. Almost all other samples are either digested to remove organic material or extracted to separate the arsenic from the matrix. The determination of arsenic in a sample with matrix is difficult and often unreliable.

11.4.3.5.1. Interferences

The literature on interferences [121,122] in GFAAS is almost as voluminous as the literature on the application of this technique to the determination of elements. This fact suggests that great care must be exercised in determining arsenic. This care must increase as the sample matrices become more complex.

After a sample has been injected into the graphite furnace, the solvent is removed during the drying cycle at temperatures generally not much higher than 100°C. Arsenic is not likely to be lost during this cycle unless arsenic trichloride is present. The formation of arsenic trichloride is favored by a high-chloride medium and high acidity, particularly in hydrochloric acid solution.

The ashing cycle, during which organic materials should be completely removed by thermal decomposition to gaseous substances, and unwanted inorganic materials volatilized, has been carried out at temperatures ranging from 300 to 1500°C. The temperature must be chosen to avoid losses of arsenic. The maximal permissible temperature is greatly influenced by the properties of the arsenic compounds present in the sample, by the types and concentrations of other cations and anions, by organic matter, and by the properties of the graphite container. For instance, standard solutions prepared from arsenic pentoxide could not be analyzed reliably using any of the conditions tried for ashing and atomization, whereas standard solutions of disodium hydrogen arsenate and arsenic trioxide gave reliable results [123]. At elevated ashing temperatures the matrix may react with the arsenic compounds and form arsenic derivatives that may have higher or lower volatility than the original compounds. A medium containing high concentrations of sulphate may produce sulfide in the reducing environment of the graphite furnace and cause the formation of volatile arsenic sulfides that may be lost. A reduction of the signal for arsenic was observed in the presence of sulfate [124]. Interactions of arsenic with the matrix or with graphite [125] may form rather involatile arsenic compounds and prevent losses of arsenic during ashing. However, these involatile compounds must be volatile enough to reach the gas phase during the atomization.

The influence of a number of cations and anions on the direct determination of arsenic was investigated [124,126–131]. These matrix ions may enhance or reduce the arsenic signal. It is not always clear whether the interferences arise during the ashing cycle, during the atomization cycle, or both. Knowledge about the interference by one particular salt in the absence of other matrix components cannot be confidently used to evaluate interferences in the presence of other salts. For instance, sodium sulfate and phosphoric acid did not influence the arsenic signal when present singly. However, when several additional salts were added to the sample, the arsenic signal was severely depressed [124]. A detailed investigation of the influence of phosphate on the As(arsenite) and As(arsenate) signals showed that the signal was not only dependent on the phosphate concentration, but also on the phosphate/arsenic ratio [124] (Fig. 11.2).

An additional complication arises when a sample contains more than one arsenic compound and is analyzed without digestion. Arsenic compounds do not all have the same volatility at a particular temperature. Rather volatile arsenic compounds may be lost during ashing and rather involatile compounds may not be efficiently decomposed and atomized during the atomization cycle. When distilled water solutions of various arsenic compounds were analyzed using an ashing temperature of 200°C, and atomized at temperatures ranging from 2000 to 2800°C, large differences in signal intensities were

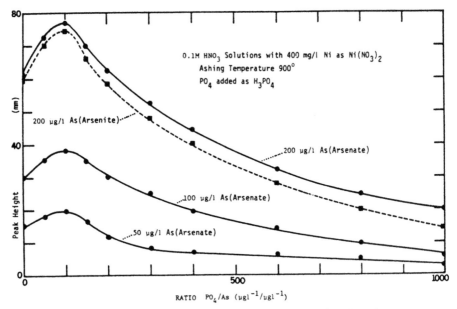

Fig. 11.2. The dependence of the GFAAS arsenic signal on the phosphate concentration and the phosphate/arsenic ratio at an ashing temperature of 900°C (reproduced with permission, from ref. 124).

observed although the same amount of arsenic (10 ng) was introduced into the graphite furnace (Fig. 11.3). To overcome these difficulties in the direct determination of total arsenic by GFAAS, matrix modifiers are added [121,122] to prevent volatilization of arsenic during ashing and ensure uniform decomposition and atomization. Nickel(II) salts, commonly nickel(II) nitrate [124,125,132,133], silver nitrate [134], palladium [135–137], lanthanum [125], and platinum [130] were used with varying success. The most widely used matrix-modifiers are nickel salts. Little is known about the effects of matrix-modifiers on organic arsenic compounds [124].

To correct for non-analyte signals, instruments often have a background correction device such as a deuterium, Zeeman, or Smith-Hietfje background corrector. Many – but not all – spectral interferences can be avoided in this manner [122,123,138–140].

The standard addition technique is frequently used to compensate for interferences and the systematic errors they cause [141]. When GFAAS is employed for the determination of total arsenic, the standard addition technique will be beneficial only when the arsenic compound added is identical to that in the sample. If more than one arsenic compound is present in the sample, the standard addition solution should ideally contain the same arsenic compounds in the concentration ratios characteristic of the sample. Preparation of such a solution requires knowledge of the composition of the sample.

Calibration curves are needed for the quantitative determination of arsenic. Because the GFAAS signal depends on the chemical nature of arsenic compounds (Fig. 11.3), the calibration curves must be established with standard substances identical to the arsenic

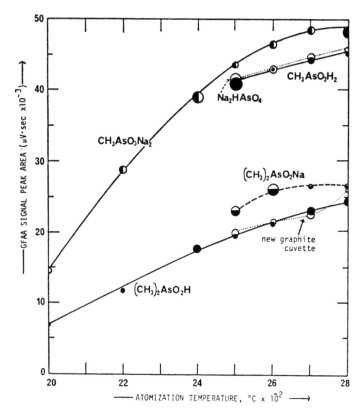

Fig. 11.3. The dependence of GFAAS peak areas (Perkin-Elmer GFAAS) on the chemical nature of the arsenic compound and the atomization temperature (10 ng As, distilled water matrix, ashing 200°/5 sec) (reproduced with permission, from ref. 384).

compounds in the sample. If several different compounds are present, quantitative determinations become problematic. Calibration curves must checked frequently, because the properties of the graphite cups, tubes, rods, and platforms change with use, although deterioration and interaction of the graphite with the analyte can be ameliorated by coatings such as glassy carbon, tantalum, niobium, or molybdenum [121,122,125,142,143].

Graphite furnace atomic absorption spectrometry, although a sensitive method for the determination of arsenic, must be used with caution even when samples representing matrices as benign as freshwater are analyzed. Losses of arsenic may occur during the ashing cycle, even in the presence of nickel salts, at ashing temperatures above 900°C. The maximum, usable ashing temperature is determined by the composition of the matrix. In order to obtain reliable results, an ashing curve should first be obtained. The calibration of the instrument and the analysis of the samples after mineralization should then be carried out at an ashing temperature corresponding to the high-temperature part of the plateau in the ashing curve [127].

11.4.3.5.2. Applications of GFAAS for the direct determination of arsenic

The many potential and often real obstacles to the direct determination of total arsenic with GFAAS make it understandable that this method has been little used for the direct determination of arsenic. Examples of samples which allowed the successful determination of arsenic without digestion are: extracts from animal feeds [144], soft drinks [145], aqueous extracts from winter flounder [146], wastewater [147], and freshwater [124].

The organic matrix of a sample can be removed by an appropriate digestion procedure. When the digestion is carried out by heating with nitric, sulfuric, or/and perchloric acid, anions derived from volatile acids such as chloride and bromide will also be removed from the sample. However, cations will remain as nitrates, sulfates, or perchlorates. Therefore, digestion will not prevent interferences caused by the cationic constituents of the sample. Total arsenic was determined by GFAAS in the digests obtained from soils ($H_2SO_4/HClO_4/Na_2MoO_4$ [135], or aqua regia HF [131]), sediments [131,136], blood and urine from rats (dry ashing) [130], animal tissues (dry ashing) [148], animal tissue (HNO_3, bomb) [139,149], plant materials ($H_2SO_4/HClO_4/Na_2MoO_4$) [135], extracts from fish [146], and seafood (HNO_3/H_2SO_4, microwave) [67].

Matrix interferences can also be avoided by the separation of arsenic from the original sample via extraction or distillation. To obtain values for total arsenic concentrations, all the arsenic in the sample must be in the chemical form suitable for the separation procedure. For instance, total arsenic in a lake water was determined after ultraviolet photooxidation to decompose organic arsenic compounds, reduction of the arsenate to arsenite by sodium hydrogen sulfite, extraction of the arsenite with diethylammonium diethyldithiocarbamate into carbon tetrachloride, and the placing of aliquots of the extract into a carbon rod atomizer [150]. Arsenic was similarly determined in wet-ashed animal tissue [151], dry-ashed food materials [57], and water (no digestion) [142] using dithiocarbamates as the extracting agents. Arsenic was separated by distillation as arsenic trichloride from water samples and digested biological materials, then the distillates were treated with nitric acid, and the resulting solutions analyzed by GFAAS [152]. The sorption of arsenic from solution by a hydrophilic glycolmethacrylate gel with bound thiol groups was investigated. The centrifuged resins were suspended in water and aliquots analyzed by GFAAS. The method was used to determine arsenic in river water [153].

Recently, total arsenic was determined successfully in standard reference materials and marine organisms by placing the powdered samples (0.1 to 1.0 mg) into a miniature cup which was put into the furnace. A matrix-modifier solution (10 μl, 60 μg Ni^{2+}, 4M HNO_3, 3M H_2SO_4) was dropped into the cup. The powder was then dried, ashed, and atomized. The method has a detection limit of approximately one nanogram arsenic [154] and deserves to be investigated in more detail. It would make a separate digestion step unnecessary. However, the small sample size of one milligram might cause problems with respect to the homogeneity of the material to be analyzed.

11.4.3.6. Plasma emission spectrometries

The most intense emissions of atomic arsenic are in the far ultraviolet spectral region at 189.0, 193.7, and 197.2 nm. Temperatures in excess of those achievable in flames and in graphite furnaces are needed for efficient excitation of arsenic atoms to give analytically useful intensities of emission lines. Such temperatures in excess of 5000 K are reached in plasmas. Direct current plasmas (DCP) and inductively coupled plasmas (ICP) have become convenient sources for emission spectrometry. The possibility of simultaneous or sequential determination of many elements, and detection limits in the low microgram per liter range, have made DCP and ICP spectrometry valuable tools for the determination of trace elements [155]. A variety of commercial instruments, that are rather costly, is on the market. The sample is aspirated as an aqueous solution or the analyte is injected into the plasma as a gas, for example as arsine. The detection limits of these systems for arsenic are in the range of 10 to 50 μg/l with approximately 10 ml as the minimum volume required for aspiration. At the high temperatures in the plasma all arsenic compounds are efficiently atomized and the instrument response is largely independent of the chemical nature of the arsenic compound present in the sample. Digestion of the samples to convert arsenic compounds into arsenate is not always required.

Publications reporting the use of plasma spectroscopy for the determination of total arsenic are not numerous. The reasons may be the high expense of the instruments and the reliability of the results obtained. Routine analyses without problems are not prime material for publication. Examples of materials in which total arsenic was determined by inductively coupled plasma spectroscopy are digests from bones, feathers, and livers of marsh birds (ICP) [156]; digests of marine sediments and animal tissues [157,158], and freshwaters [124].

Whether DC plasma emission spectrometers or ICP spectrometers perform better is still being debated. Both instruments readily accept gaseous and liquid samples. Slurries were successfully nebulized and even solid samples have been introduced into the plasma. A variety of nebulizing systems is available which can be used for special matrices or which increase the efficiency of aerosol formation. Plasma emission spectrometry is much less prone to interferences than electrothermal atomic absorption spectrometry. However, plasma emission methods are not interference-free. High concentrations of aluminum, calcium, iron, and other elements cause spectral interferences, some of which can be corrected by software-controlled background correction. Carefully performed plasma emission determination of total arsenic in digested samples will produce reliable results. The multi-element capabilities of the instruments provide information about the components of the matrix and potential interferences. The ICP–MS systems that have been available for a few years generally have 100-fold better detection limits for most elements than optical ICP instruments. However, the ICP-MS detection limit for arsenic in a chloride-containing matrix is probably not much lower than the optical ICP detection limit because of Ar-Cl species with the same m/z value as arsenic.

Microwave-induced plasmas (MIP) cannot accomodate liquid samples easily. Therefore, MIP emission spectrometers are generally used only for gaseous samples. MIP instruments as well as ICP instruments have frequently been used in combination with hydride generation systems that convert arsenic compounds to volatile arsines. These methods for the determination of total arsenic are discussed in the section on hydride generation.

11.4.4. Molecular methods for the determination of total arsenic

Molecular methods for the determination of total arsenic are based on the properties of arsenic-containing molecules. An arsenic-containing molecule is defined as a group of atoms, among which there is at least one arsenic atom, that are linked by bonds of sufficient strength to make this assembly identifiable and determinable by the analytical technique chosen. The hydrogen arsenate anion, $HAsO_4^{2-}$, the undissociated arsenic acid, H_3AsO_4; arsenous acid, H_3AsO_3; the dihydrogen arsenite anion, $H_2AsO_3^-$; the arsenocholine cation $[(CH_3)_3AsCH_2CH_2OH]^+$; and the arsenobetaine zwitterion, $(CH_3)_3As^+CH_2COO^-$, are examples of arsenic-containing molecules. According to this definition, molecules do not have to be uncharged.

A method for the determination of total arsenic must be applicable to all of the different arsenic-containing molecules in the sample. If the method responds only to one particular molecule, all of the arsenic in the sample must be present in this form or must be converted to this form. A molecule-based analytical method with universal response to all the arsenic compounds known to exist in biological systems has not yet been developed, although such methods do exist for closely related arsenic compounds such as the series of arsines, $(CH_3)_n AsH_{3-n}(n = 0 - 3)$. Because the types of arsenic compounds present in a sample are generally not known, the complete conversion of all arsenic compounds to arsenate by digestion of the sample is the recommended and safe procedure to follow. Should arsenite be required, arsenate can be easily reduced.

The molecule-based methods for the determination of total arsenic include electro-chemical, colorimetric and chromatographic techniques, and the many variants of the hydride generation technique. The last is the most widely used method.

11.4.4.1. Electrochemical methods

A variety of electrochemical techniques is available for the determination of arsenic. These techniques depend on the unimpeded functioning of electrodes. To avoid contamination of the electrodes by organic substances, the samples are generally mineralized and, if necessary, the arsenic is separated from the matrix by extraction. The electrochemical methods for the determination of inorganic arsenic may require arsenite or arsenate. Therefore, the sample must be properly treated to convert the arsenic compounds into the required form. For example, arsenate is reduced to arsenite with sulfur dioxide, sulfites, hydrazine, or acidified potassium iodide.

11.4.4.1.1. Amperometry and coulometry

Amperometric and coulometric titration techniques [159] are not generally useful for the determination of traces of arsenic because of high detection limits, although arsenic recovered from organic matter has been determined in this manner [160]. The coulometric titration of arsenite with iodine in soil digests, with biamperometric endpoint indication, was found to be as accurate as, but more precise than, the colorimetric arsenomolybdate method for amounts of arsenic of approximately 20 μg [161].

11.4.4.1.2. Polarography

Although arsenate and arsenite can be reduced polarographically [162], almost all analytical work has been carried out with arsenite, because polarographic waves for arsenate are obtained only in concentrated acids. Arsenate is polarographically inactive under conditions (1.5 M HCl) which are suitable for the determination of arsenite. For the determination of total arsenic in a sample, all organic arsenic compounds must be converted to inorganic arsenic compounds and arsenate must be reduced to arsenite. Rapid linear sweep polarography with oscilloscopic readout was used to determine arsenite in raw well water, and spring water, acidified with sulfuric acid. The detection limit using the peak at –0.97 V *versus* the mercury pool was 5 micrograms of arsenic per liter [163].

Differential pulse polarography of arsenite in 1 M hydrochloric acid was found to have a detection limit of 0.3 μg/l [164], which is much better than that achievable with DC-polarography. The method was used to determine arsenite using the reduction peak at approximately –0.45 V in tap water, well water and river water [165–167], in water reference standards [168], and in extracts from fish and shellfish [169]. Interferences from Pb(II), Sn(II), Sn(IV), Tl(I), Tl(III) [164,167], Cu(II) [166], and methylarsonic acid [166,168] were observed. These interferences can be avoided by removing the troublesome cation with a Chelex-100 ion-exchange column [166], extraction of arsenic as arsenic trichloride [169], or by recording the baseline after oxidation of arsenite by cerium(IV) to arsenate and subtracting the background from the polarographic peak [164,168].

An indirect determination of arsenate is based on the formation of arsenomolybdate and its selective extraction into an organic solvent, then stripping of the arsenomolybdate into an aqueous basic medium, acidification, and DC-polarographic determination of molybdenum. This method achieves an enhancement factor of 12 and a detection limit of 5 μg/l of arsenic. Sequential solvent extraction allows the determination of arsenate, phosphate, and silicate [170].

11.4.4.1.3. Stripping voltammetry

Reduction of arsenite on an electrode of sufficiently negative potential (–0.5 V) can be used to strip arsenic from the solution and to collect it on the electrode. The amount of arsenic deposited increases with time. In most applications the electrolysis is carried out for a few minutes. The hanging-drop mercury electrode was found to be unsatisfactory [171]. Solid gold, or gold-plated graphite, are the preferred collecting electrodes. The elemental arsenic accumulated on the gold, the Cu$_3$As deposited on the surface of the mercury drop when copper(II) ions [172,173] are present in solution, or the As$_2$Se$_3$

similarly formed in the presence of selenite [174], can then be stripped at a more negative potential (cathodic stripping) or at a more positive potential (anodic stripping) than used for the deposition, with a large gain in signal intensity. Detection limits as low as 20 ng/l of arsenic were reported for differential pulse cathodic voltammetry [175]. The samples to be analyzed by these methods must be digested unless they are free of organics and low in mineral matter.

Cathodic stripping voltammetry was used to determine arsenic in NBS SRM Orchard Leaves (using dry ashing with magnesium nitrate, removal of copper and other metal ions by extraction with dithizone/CCl_4, and reduction with hydrazine) [174], and in NBS SRM Oyster Tissue (with separation of arsenic with a tetraalkylammonium hydroxide-based extractant and arsenate reduction with NH_2OH HCl) [173].

Anodic stripping voltammetry was found useful for the determination of arsenic as arsenite in seawater, tap water, and polluted estuarine water (with no digestion only acidification) [176], clean seawater [177], NBS Standard water sample, and polluted water samples (with acid digestion and separation of arsenic as $AsCl_3$) [178], NBS SRM Orchard Leaves, NBS SRM Bovine Liver, and ketchup (with acid digestion, and separation of arsenic as $AsCl_3$) [179]. A computerized potentiometric stripping method that uses gold(III) to chemically oxidize deposited arsenic has a detection limit of 0.3 μg/l [180]. Anodic stripping voltammetry was adapted to operate in flow systems with a stationary gold electrode. A detection limit of 20 ng/l was achieved [173].

The anionic components of a sample can be controlled by the digestion procedure and can be chosen to optimize the procedure. Cationic components may cause serious interferences. In such cases, the arsenic must be separated from the matrix, for instance by extraction or by ion chromatography.

11.4.4.2. Colorimetric methods

The reactions of arsenic compounds with color-producing reagents can be used to determine arsenic spectrophotometrically. The simplicity of these methods and the low cost of the required instruments make them very popular, in spite of the high detection limit of approximately one microgram of arsenic. A number of color-producing reagents has been discovered for arsenic. However, only the molybdenum-blue method and the silver diethyldithiocarbamate method have found wide application.

11.4.4.2.1. Molybdenum-blue method

The molybdenum-blue method requires arsenic to be present as arsenate. Samples digested with oxidizing acids will have all arsenic converted into arsenate. Arsenite can be oxidized to arsenate, for example with iodine. It is often desirable to separate the arsenic from the matrix by distillation of arsenic trichloride or arsenic tribromide [181], reduction to volatile arsine [161,182,183,184], extraction [185], or precipitation as arsenic trisulfide [186], and reconversion of the isolated arsenic compound to arsenate. The solution containing arsenic as arsenate is acidified with sulfuric or hydrochloric acid to hydrogen ion concentrations in the range of 0.05 M to 2.0 M. Addition of sodium or ammonium molybdate causes the formation of an arsenate-containing molybdenum heteropoly acid that is reduced to molybdenum-blue with agents such as tin(II) chloride, sodium disulfite [186], ascorbic acid, or hydrazine. Hydrazine sulfate appears to be the best reducing agent. The intense blue solution has an absorption maximum at 840 nm (ε = 25,000). Other wavelenghts have also been used for quantitative determinations. The detection limits are approximately one microgram of arsenic [187].

The molybdenum-blue method with spectrophotometric detection was used to determine total arsenic in soil digests [161], in freshwater [182,186], seawater [182], effluents [182], animal tissues [184,188], hair [189], marine organisms [183], and sediments [183,186]. Instead of a spectrophotometer, a flame [190,191] or graphite furnace [192,193] atomic absorption spectrometer can be used as the detector for the molybdenum associated wit the arsenic. The arsenomolybdate is extracted into an organic solvent, the extract freed from all other molybdenum species, the heteropoly acid destroyed by shaking with an aqueous basic solution and the molybdate that is now in the aqueous phase determined by atomic absorption spectrometry. A detection limit of 25 pg of arsenic in 0.4 ml of ultrapure water (6 ng/l) was reported [193].

Unless arsenic is cleanly separated from the sample matrix, the molybdenum-blue method may give erroneous results because of interferences from cations and anions. Thirty five elements form heteropoly acids [192]. Among these, phosphorus as phosphate and silicon as silicate are most likely to be present in environmental samples and interfere with the determination of arsenate. Methods based on solvent extraction [194], suppression of the formation of other heteropoly acids [195], reduction of arsenate to arsenite [196], or separation of the heteropoly acids by liquid chromatography [197] were suggested for avoiding these interferences.

Radiometric methods based on the co-extraction of W-185 tracers with the arsenomolybdate [198] or the extraction of Mo-99 tagged arsenomolybdate [199] into an organic solvent allowed arsenic to be determined with detection limits of approximately 1 μg.

11.4.4.2.2. Silver diethyldithiocarbamate method

Arsine, AsH$_3$, generated through reduction of arsenite and arsenate by zinc [200] or sodium borohydride [201] in an aqueous medium of pH 1 is flushed by the evolving hydrogen or an inert gas into a solution of silver diethyldithiocarbamate in pyridine (or a similar base) or pyridine/chloroform [202–204]. The intensity of the red color formed is measured at 520 nm. The detection limit is approximately one microgram of arsenic. This method has been used mainly for the determination of arsenic in water samples [200]. Hydrochloric acid-releasable inorganic arsenic in soils [205], arsenic in dental cements [206], and total arsenic in marine organisms (with acid digestion) [207] and bovine liver (with dry ashing) [208] were also determined by this spectrophotometric technique. For this method to provide reliable total arsenic concentrations, all arsenic must be in an inorganic form (as arsenite or arsenate). Samples suspected to contain organic arsenic compounds must be digested. If methylarsonic acid, dimethylarsinic acid, or trimethylarsine oxide are present, methylarsines will be formed that react with the dithiocarbamate solution to produce colored products with absorptions different from the product arising

References on p. 340

from arsine [209]. The total arsenic concentrations will then be incorrect. If non-reducible organic arsenic compounds are in the sample, the total arsenic concentrations found by this method will be low.

11.4.4.3. Hydride generation

Hydride generation is the most frequently used method for the determination of total arsenic in environmental or biological samples. Several attractive features are combined in this method. After digestion of a sample to convert all arsenic compounds to arsenate, the arsenic is separated from the digest by reduction to volatile arsine, AsH_3. This is flushed by an inert gas into a cooled trap, where it is collected. This step allows arsenic to be concentrated before detection. If other hydrides such as stannane, stibine, or germane are also present, the arsine can be separated by gas chromatography. Arsine is then volatilized by warming the trap and passed to a detector. Great flexibility exists with respect to the choice of a detector. The selection of a detector is guided by the detection limits that are demanded by the analytical task and the instruments available in the laboratory. An arsenic-specific detector is desirable. Absolute detection limits of 1 ng of arsenic are easily achievable. Even lower detection limits are possible by careful design and operation of a hydride generation system. The major components of a hydride generation system are shown schematically in Fig. 11.4. Not all of the components are needed for all tasks. If the sample does not contain organic compounds that might interfere with the hydride generation, and is free of organic arsenic compounds, the digestion step can be omitted. The hydride collector is not necessary when sufficient arsenic is present in the sample and the reduction rapidly sends all the arsine to the detector during a short period. If no other hydride-forming elements are present in the sample, the hydride separator is not needed.

Fig. 11.4. The components of a hydride generation system.

Because the composition of arsenic compounds in a sample is generally not known, the sample should be digested in a manner that ensures the complete conversion of all arsenic compounds into arsenate. In an aqueous acidic medium of pH 1 or lower arsenate and arsenite are reduced to arsine. Zinc/hydrochloric acid and aqueous solutions of sodium borohydride are the most common reducing agents. When zinc/HCl is used, the arsenate is pre-reduced to arsenite, for example with sulfur dioxide, sulfite, or hydrazine sulfate (eqn. 11.2). During the reduction interferences may occur that inhibit the formation of

$$H_3AsO_4 \xrightarrow{\text{red.}} H_3AsO_3 \xrightarrow{\text{red.}} AsH_3 \tag{11.2}$$

$$\underline{\hspace{2cm} \text{red.} \hspace{2cm}}\uparrow$$

arsine or prevent it from reaching the detector by forming insoluble arsenides. At pH 1 methylarsonic acid, dimethylarsinic acid, and trimethylarsine oxide are also reduced to their respective arsines. If the detector response is independent of the molecular form in which arsenic occurs, total arsenic can be determined by hydride generation without conversion of the methylarsenic compounds to arsenate. If the detector response is compound-dependent, incorrect results for total arsenic will be obtained [210].

The many variants of the hydride generation technique are best classified according to the detector used. This classification is employed in the following discussion of the hydride generation technique for the determination of total arsenic. The identification and determination of arsenic compounds by hydride generation is presented in Sections 11.5.1.5, 11.5.2.2, 11.5.3.2 and 11.5.4.

11.4.4.3.1. HG-gas chromatography

Gas chromatographs have not often been used as detectors. Flame ionization detectors do not respond to small amounts of arsine [211]. With a thermal conductivity detector the detection limit was 10 μg/l [211] and with a gold gas-porous electrode detector 40 μg/l [212].

Arsine, stannane, and stibine were generated by reduction with sodium borohydride, swept from the aqueous sample onto a Porapak Q column and separated [212]. Arsine, stibine, and germane were formed by addition of sodium borohydride tablets to natural water samples acidified with sulfuric acid to pH 2, collected in a silica-gel-filled Dry ice/acetone cooled trap, the trapped compounds volatilized onto the silica gel GC column by heating the trap to 140°, and the hydrides separated using helium as the carrier gas. An HG-GC method was successfully used to determine arsenic in food-grade phosphoric acid with a detection limit of 0.5 ng of arsenic [213].

11.4.4.3.2. HG-colorimetry

Colorimetric detection was combined with hydride generation. After reduction of inorganic arsenic to arsine, the arsine was passed into an aqueous solution of KI/I_2 to oxidize it to arsenate. This was to be determined by the molybdenum-blue method (Section 11.4.5.2.1). This method was used to determine arsenic in seawater, potable waters and effluents [182], in soil digests [161], and in biological materials [183,188,189].

The hydride generation-silver diethyldithiocarbamate method is discussed in Section 11.4.4.2.2.

11.4.4.3.3. HG-atomic fluorescence spectrometry

With an atomic fluorescence spectrometer as the arsenic-specific detector, the generated arsine is directly swept into the atomizer, which in most cases is an argon-hydrogen/entrained air flame [109–113,214–216]. However, an electrically heated quartz-tube atomizer has also been used [114]. The HG-atomic fluorescence method was successfully applied to fresh water [113,114], wastewater [111], soil digests [112], plant materials [114], and coal [214]. Although sodium borohydride is the most commonly used reducing agent, zinc/tin(II) chloride/potassium iodide [111] and electrochemical reduction [114] have also been employed. The direct determination of arsenic by atomic fluorescence spectrometry is discussed in Section 11.4.3.3.

11.4.4.3.4. HG-flame atomic absorption spectrometry

Although flame atomic absorption spectrometry is not well suited for the determination of total arsenic by direct aspiration of a liquid sample into the flame (Section 11.4.3.4), this method does have detection limits of a few nanograms of arsenic when coupled with a hydride generator. Most applications of the HG-flame atomic absorption method do not use a trap for the arsine. Instead, it is swept by an inert gas directly into the hydrogen air entrained flame. However, traps can be used to collect arsines: packed and unpacked U-tubes cooled with liquid nitrogen [217,218], and balloons [219–221] have been used. Arsenic hollow cathode lamps or ELD lamps supplied the radiation. The most sensitive resonance line at 193.7 nm was predominantly chosen for the determinations, although the 192.7 nm line was also used [218].

Total arsenic was determined by the HG-flame atomic absorption method in the following samples: standard water samples [222], water [223], urine and whole blood of rats [223], sediments [224,225], NBS SRM Orchard Leaves [226], plant materials [227], molasses [225], sugar beet pulp [221], tobacco [217], food items [226,228], algae [229], fish [226,228], eggs [226], NBS SRM Bovine Liver and tuna [226].

11.4.4.3.5. HG-quartz tube atomizer-atomic absorption spectrometry

The sample in which total arsenic is to be determined must contain all the arsenic in the form of arsenite or arsenate. With the exception of freshwater, all samples must be digested. Arsine is then formed by reduction of inorganic arsenic in 1 M HCl with sodium borohydride. Because arsenate appears to be reduced more slowly than arsenite [230–232], a pre-reduction of arsenate to arsenite by KI/SnCl$_2$, KI/ascorbic acid, or KI/HCl is recommended [230,233]. However, arsenite and arsenate were reported to give equivalent signals in 0.5 to 1.5 M hydrochloric acid media [234]. The reduction rates are certainly dependent on the nature of the mineral acid, the acid concentration, and the vessel in which arsine is generated. When a trap is used to collect the arsine, the rate differences are of no consequence [235]. Most systems, however, do not use a trap but inject the arsine directly into a T-shaped quartz tube that serves as atomizer. The tubes may be of several types [236] but are generally 10–20 cm long with inner diameters between 7 and 15 mm. The quartz tube is heated electrically [232,237,238] or with a flame (acetylene-air [236], argon-hydrogen-entrained air [239–241], hydrogen-oxygen [233], dinitrogen oxide-acetylene [242]) to temperatures between 800 and 1100°C. A small amount of air mixed into the arsine-containing gas, upstream of the arsine generator, was found to increase the signal intensity [232]. Glass surfaces should be silanized and the system conditioned by performing several analysis cycles [243]. Instead of the quartz tube, a tube fashioned from graphite paper was successfully used [244]. Because the vessels in which the arsine is generated can hold up to 50 ml of sample, the concentration corresponding to the absolute detection limit of 0.1 ng is 2 ng/l.

Automated systems based on flow injection analysis were developed for the determination of total arsenic [238,241,245–149]. Some of these will handle 120 samples per hour [246].

The manual and automated systems have been used to determine total arsenic in the following samples: reference water samples [250], riverwater [239,251], lake water [231], run-off [232], seawater [252], soils [232,251,253], urban particulates [235], sludges [254], river sediments [250], marine sediments [255], apples [256], single cell protein [237], marine plankton feeders, marine herbivores and marine carnivores [257], fish [247,258], fish solubles [259], fish meal [260], salmon [247], lobster, scallops, plaice [261], brown algae [252], beverages [233], wine, fruit juices [233], food items [233,260], fat materials [61], raisins [233], vegetation [253], and in the NBS Standard Reference Materials Orchard Leaves [235,239,240,246,247,252,254,255,259], Tomato Leaves [259], Wheat Flour [239,246], Rice Flour [246,259], Oyster [255], Tuna [255,260], and Bovine Liver [240,247].

Instead of the commonly used reduction of arsenite and arsenate by sodium borohydride in an arsine generation vessel the sample can be passed through a column filled with zinc [257,262]. Another promising approach uses a chromatographic column [10% OV-101 on 80–100 mesh Chromosorb W) pretreated with sodium borohydride solution. Injection of the arsenic-containing solution leads to the formation of arsine that is flushed into the quartz tube furnace. The detection limit is claimed to be at least one nanogram of arsenic [263].

11.4.4.3.6. HG-electrothermal atomic absorption spectrometry

Arsine generated from appropriately digested samples by reduction of arsenite or arsenate with zinc/hydrochloric acid in a zinc reduction column [262,264], or more commonly by sodium borohydride in an aqueous medium of low pH, can be passed into a graphite tube kept at 370°C [265] or 600°C [266]. At these temperatures arsine decomposes to hydrogen and arsenic. The arsenic deposited on the walls of the tube is then atomized at temperatures of 2400 to 2600°C. The limits of detection (3 σ) were matrix-dependent and were 0.006 ng/g for seawater, 220 ng/g for sediments, and 150 ng/g for biological materials [266]. Alternatively, the arsine can be passed into the graphite tube already at the atomization temperature. This approach was used for the determination of arsenite in standard solutions [264] and of total arsenic in tea leaves and orchard leaves [267], in the NBS Standard Reference Materials Bovine Liver, Pine Needles, Oyster Tissue, and Orchard Leaves [268], and in freshwater [269]. If desired the arsine can first be collected in a trap cooled in liquid nitrogen and then volatilized into the hot graphite tube [269]. The absolute detection limits achieved with these systems are approximately 0.2 ng of arsenic.

The generated arsine can also be passed into trapping solutions that retain the arsenic for subsequent injection of an aliquot into the graphite furnace. Solutions of silver diethyldithiocarbamate and ephedrine in chloroform [270] and iodine/potassium iodide in water

References on p. 340

[271] were used as trapping agents. Detection limits of 10 ng of arsenic (0.2 ppb, 50 ml sample) were achieved with freshwater samples [270].

11.4.4.3.7. HG-microwave-induced plasma atomic emission spectrometry

Arsine introduced into a microwave-induced plasma (helium, argon) is decomposed. The gaseous arsenic atoms are excited and can be determined by the light emitted upon their return to lower energy levels. The analytical systems must minimize the transfer of water and gases other than the plasma and analyte gases into the microwave cavity. High gas loads from water vapor or hydrogen destabilize the plasma and might extinguish it.

Inorganic arsenic is reduced to arsine by sodium borohydride at low pH [272–276], or by zinc/hydrochloric acid after conversion of arsenate to arsenite by tin(II) chloride [277]. The gases pass through a drying tube to remove water, and into a trap cooled by liquid nitrogen to collect the analyte gases and separate them from hydrogen. Upon heating of the trap the analyte gases are passed through a chromatographic column (Chromosorb 102) to separate them from hydrogen and from each other on their way into the plasma [272,273]. The microwave cavities (Beenekker, Evenson quarter wave) are operated at 100 watts. With a direct reading polychromator several hydride-forming elements such as As, Sb, Ge, Sn [272], and Se [275] were determined simultaneously. The absolute detection limits of these systems for arsenic are in the range of 4 to 40 ng (sample size 20 ml).

These systems were used to determine total arsenic in water, plant leaves [277], and NBS SRM Enriched Flour and Orchard Leaves [273,275]. Problems with the stability of the microwave plasmas are probably responsible for the fact that these systems have not been used more often.

A hydride generation-capacitively coupled microwave plasma atomic emission spectrometry system (detection limit 25 ng As) was used for the determination of arsenic in digested sewage slude [278].

11.4.4.3.8. HG-molecular emission spectrometry

Arsine generated by sodium borohydride reduction of inorganic arsenic was transported by an oxygen-nitrogen mixture into a molecular emission cavity heated by a hydrogen-nitrogen flame. The molecular emissions in the range 330–350 nm, believed to arise from AsO, were measured [279,280]. The detection limit is 0.2 microgram of arsenic [279,280]. Interferences from cations can be avoided by adding EDTA to the samples [281]. With a gas chromatograph between the hydride generator and the emission cavity the hydrides of arsenic, antimony, tin, and germanium can be separated and the elements subsequently determined by molecular emission spectrometry at a single wavelength (490 nm) [280,282]. A hydrogen-oxygen flame burning within the cavity gave better detection limits than did external heating [282]. A flow injection system was also developed and tested with NBS Orchard Leaves [283].

11.4.4.3.9. HG-DC plasma atomic emission spectrometry

Analytical systems combining hydride generation with DC-plasma atomic emission spectrometers have only been used infrequently for the determination of total arsenic. Seawater and wastewater samples have been treated with zinc and hydrochloric acid and the evolved arsine (and stibine) collected in a trap cooled by liquid nitrogen and thus separated from the gaseous hydrogen. The detection limit for arsenic was 8 ng [282]. Continuous hydride generation systems were coupled to DC-plasma emission spectrometers for the determination of total arsenic in water, digested canned tuna fish [285], soil extracts, and digested reference materials (NBS Orchard Leaves, NIES, Japan, No. 6 Mussel) [286] with detection limits in the 4 to 10 ppb range. These systems can be used for the simultaneous determination of all hydride-forming elements.

11.4.4.3.10. HG-helium glow discharge atomic emission spectrometry

The helium glow discharge is a convenient and easily constructed light source for the determination of arsenic by atomic emission [287]. The arsine generated by sodium hydride reduction is freed from aerosols in an impinger, passed through a tube filled with solid sodium hydroxide, and collected in a trap cooled by liquid nitrogen. The condensed arsine is then volatilized in a stream of helium into a helium plasma established between two tungsten/2% thorium dioxide helium-arc welding electrodes (1100 V DC breakdown, 400 V operating voltage, 40 mA). The emitted light (228.8 nm) is passed into a monochromator and measured by a photomultiplier tube. The detection limit is 0.1 ng of arsenic [288,289]. This method was used to determine total arsenic (after appropriate digestion of the sample when necessary) in seawater, freshwater, marine sludges [287], Coal, Coal Fly Ash, Orchard Leaves, Bovine Liver (NBS Standard Reference Materials) [288], and lignite [289].

11.4.4.3.11. HG-inductively coupled argon plasma atomic emission spectrometry (ICP)

The robustness of inductively coupled argon plasmas and the multi-element capabilities of the plasma emission spectrometers are attractive characteristics for detectors to be coupled with hydride generation systems. The availability of continuous hydride generators led to the development of flow injection systems that use simultaneous [290,292–297,298] or sequential [297,299] inductively coupled argon plasma emission spectrometers as detectors for arsenic and other hydride forming elements. Optimization of the entire system [297,300–302] led to detection limits as low as 0.02 ng/ml of arsenic [297]. Digestion with persulfate in acid solution followed by heating at 95°C in 6 M hydrochloric acid was found to be the best mineralization procedure [303]. The automated analysis systems can perform 200 determinations per hour [299]. The HG-ICP system can also be operated in the batch mode using, for example, the syringe hydride technique [294].

The HG-ICP technique was used to determine total arsenic in freshwater [295,298, 304], seawater [295], wastewater [291,296,303], soils [292,304], sediments [292,299, 303,304], cocoa beans and nibs [305], NBS Orchard Leaves [293,294,297,299,306], Citrus Leaves [305], Tomato Leaves [294,305], cabbage [294], rice flour, wheat flour, spinach [297], pasture herbage [293], Bowen's kale [293,294], fish tissue [304], oyster tissue, scallops, plaice, and lobster [306]. The multi-element capabilities of ICP spectrometers allowed arsenic and the following hydride-forming elements to be determined simultaneously: Sb [290,292,297,303,304,307], Bi [290,292,293,297,304,307], Se [290, 297,298,303,304,307], Te [290,307], Ge [297], and Sn [297,304,307].

The combination of a hydride generation system with an inductively coupled argon plasma-mass spectrometer system was shown to have a detection limit for arsenic of 0.005 ng/ml that is much better than the detection limit of 0.8 ng/ml achieved with the hydride generation-ICP atomic emission spectrometer system [308].

11.4.4.3.12. Hydride generation method: interferences

Hydride systems for the determination of total arsenic consist of a hydride generation vessel, drying tubes and traps, and the detector. Interferences with the arsenic determination may occur during the formation of arsine in the hydride generation vessel, during transport to the detector, or in the detector. Many reports in the literature address the influence of the mineral acids on the reduction to arsine [225,309–312], and the effects of the volume of the sample [310], the nature of the reducing agent [310,313], and the presence of anions such as dichromate, chromate, permanganate, molybdate, vanadate, and peroxydisulfate [227,314,315], and cations [224,225,227,229,264,267,287,296,299, 309,310,313,314,315–320]. The degree of interference with the determination of arsenic is determined by the composition of the medium in which the reduction occurs, and by the concentration of the interfering cations. Although literature reports are not always in agreement concerning interference by cations, the cations of the following elements were found to depress arsenic signals: Ag, Au, Cd, Co, Cr, Cu, Fe, Hg, Ni, Pb, Pd, and Pt [299]. Some of these interferences can be avoided by adding complexing agents to the arsenic-containing solutions. Such complexing agents are thiosemicarbazide and 1,10-phenanthroline for Cu, Ni, Pt, Pd [267,316], EDTA [317], thiourea for elements of the Co, Ni, and Cu groups [318], and potassium iodide [296,299,319]. Interfering cations were separated from arsenic on strongly acidic ion exchangers [317] or by co-precipitation of arsenic with lanthanum hydroxide [319]. Combinations of metal ions may act in quite a different way from that expected from the sum of their individual effects. For instance, the addition of Cu(II) removes interferences due to selenium [321], and the addition of Fe(III) prevents interferences from lead [225]. The interferences caused during the formation of arsine were explained by competition for the reducing agent, for example by chromate, dichromate, permanganate, and reducible cations, or by the formation of complexes of cations with arsine, by the precipitation of metal arsenides, or by the decomposition of arsine on finely divided metals [299].

The hydride-forming elements in the main groups IV, V, and VI may interfere with the arsenic determination during the reduction phase or in the detector [320]. The atomization of arsine in the quartz tube furnace is caused not by thermal decomposition but by collision with hydrogen radicals [322]. Substances that scavenge hydrogen radicals or catalyze recombination will interfere with this atomization and depress the arsenic signal. Interferences by antimony, selenium, and tin were found to occur in the quartz tube furnace [320]. Tin tended to deposit on the walls of the quartz tube and make the determination of arsenic impossible [320].

Most of the investigations of interferences were performed in such a manner as to make difficult the exact identification of the place where the interferences occur. More investigations should be carried out with the powerful radiotracer technique [311,320]. Determinations of total arsenic by the hydride generation technique must be approached with appropriate consideration of potential interferences from the matrix.

11.5. DETERMINATION OF ARSENIC COMPOUNDS

Most of the methods available for the determination of total arsenic require that all the inorganic and organic arsenic compounds present in a sample be converted to inorganic arsenite or arsenate. Even methods such as atomic emission and absorption spectro-metry which exploit atomic properties give more precise and accurate results when only one arsenic compound is present. The conversion of arsenic compounds to inorganic arsenic is accomplished by digesting the samples in an oxidizing medium (Section 11.4.1). Arsenate formed during the digestions can be reduced to arsenite if required.

When arsenic compounds are to be identified and determined, destructive digestion must not be used. Samples in which arsenic compounds will be determined must have their total arsenic and each inorganic and organic arsenic compound preserved.

Aqueous samples collected for the determination of trace elements are preserved by acidification, although the literature contains contradictory statements [323]. Acetic acid was suggested for the preservation of arsenic-containing solutions [224]. Losses of ar-senic up to 70% within one week were observed from seawater acidified with 9 ml of concentrated hydrochloric acid per liter [75]. Arsenite was not lost from well water sam-ples kept for one week in polystyrene vials with snap caps of polyethylene [163]. Dilute solutions of arsenite or arsenate (20 μg/ml) in distilled water remained unchanged for at least 50 days when stored in the light or in the dark in borosilicate glass, soda glass, or polyethylene containers [324]. However, filtered seawater containing 2 μg/l of arsenic lost this during storage in soda glass, Pyrex, or polyethylene bottles [325]. Frozen solutions of 1 mg/l arsenic did not lose arsenic on freeze-drying [326]. Natural water samples stored below $-15°C$ or under dry ice did not lose arsenite [327]. Seawater samples kept in a refrigerator at 4°C showed no change in the arsenite/arsenate ratio within ten days [328]. For longer storage the samples had to be frozen quickly with crushed dry ice within 15 minutes of collection: they could then be stored frozen. In samples frozen in a freezer spurious losses of arsenite were observed [329]. Solutions of arsenite and arsenate in distilled demineralized water were observed to have a constant arsenite/arsenate ratio for at least six weeks even in the presence of redox reagents [oxygen, hydrogen sulfide, iron(III)] which are common in natural waters [330,331]. However, phosphate-buffered arsenite solutions at concentrations below 100 μg/l of arsenic were reported to be spontaneously oxidized to arsenate within a week [288]. Addition of ascorbic acid (0.1% by weight) to aqueous arsenite solutions appears to prevent oxidation to arsenate

[288,323]. Methylarsine, dimethylarsine, and trimethylarsine were stable in aqueous solutions for a few days when stored in air-tight containers. Methylarsonic acid and dimethylarsinic acid were lost measurably from untreated water samples after three days. These losses were prevented by acidification of the samples with hydrochloric acid to a concentration of 0.05 M [327]. At this time a reliable method for the preservation of arsenic compounds in water samples is not available. The reactivity of arsenite, of arsenate, and perhaps of organic arsenic compounds might be influenced by the presence of metal ions, anions, and other matrix components.

When samples such as soils, sediments, and biological tissues are to be analyzed, the arsenic compounds must first be extracted. The extraction should remove arsenic compounds from the matrix quantitatively and unchanged. Whether the extraction procedures fulfill these requirements has not yet been investigated in detail. For the extraction of arsenite and arsenate from plant materials, the material (2 g) was heated for five minutes with 10 ml nitric acid (water: conc. nitric acid, 1:2) and then for 15 minutes with 10 ml sulfuric acid (water: conc. sulfuric acid, 1:2) [227]. This procedure was claimed to cause oxidation of arsenite to arsenate [332,333] and reduction of arsenate to arsenite in experiments with egg yolk [333]. Arsenic(III) (1–4 μg/ml) was found to be stable for up to two weeks in interstitial waters separated from lake sediments after acidification of the water with concentrated hydrochloric acid to pH 2 and then deoxygenation with nitrogen. These samples were kept in well sealed glass containers at room temperature. When the acidified samples were stored at 0°C deoxygenation was not necessary [334].

11.5.1. Determination of arsenite and arsenate

Several of the methods available for the determination of total arsenic are based on reactions specific for arsenite or arsenate (Section 11.4.4). Electrochemical methods (Section 11.4.4.1), specifically polarography, respond to arsenite only [164]. In the presence of a polyhydroxy compound, such as mannitol, arsenate also becomes electroactive [165]. The molybdenum-blue method (Section 11.4.4.2.1) is specific for arsenate. The hydride generation method (Section 11.4.4.3), with sodium borohydride as the reducing agent, can be used to determine arsenite and arsenate. Additionally, chromatography, extraction, and precipitation methods are available for the separation of arsenite and arsenate. Because arsenite can be oxidized to arsenate and arsenate reduced to arsenite, methods specific for one of these two arsenic compounds can be used for the other after an appropriate redox conversion. Whereas the determination of arsenite and arsenate in water samples can be carried out quite easily, their determination in biological samples is troubled by lack of reliable procedures that extract arsenite and arsenate unchanged from tissues.

11.5.1.1. Electrochemical methods for the determination of arsenite and arsenate

Arsenite at concentrations as low as 5 μg/l of arsenic was determined in raw well waters, spring waters, and other drinking waters by single-sweep polarography at –0.97

V. Acidification of the samples with sulfuric acid was the only pretreatment required [163]. Differential pulse anodic stripping voltammetry with a detection limit of 0.2 μg/l of arsenic allowed arsenite and arsenate to be measured in seawater, estuarine water, and tap water. Arsenic from arsenite only is deposited from the water sample acidified with sulfuric acid on a rotating gold electrode, at −0.3 V versus the normal calomel electrode. The arsenic is determined by anodic stripping. The electro-inactive arsenate is reduced to arsenite by sulfur dioxide at an elevated temperature, and the total inorganic arsenic found by a second deposition-stripping cycle. Arsenate concentrations are obtained as the difference between total inorganic arsenic and arsenite [176]. When a sample contains interfering amounts of organic matter, only total inorganic arsenic can be determined, because treatment with hydrogen peroxide followed by UV-irradiation to mineralize organic matter will oxidize arsenite to arsenate [176].

Differential pulse polarography can be used similarly to determine arsenite (−0.3 to −0.5 V depending on the supporting electrolyte) and total inorganic arsenic, after the reduction of arsenate by sulfite [166,168,335], with detection limits of a few micrograms of arsenic per liter. Total inorganic arsenic can be determined as arsenate (−0.55 V) in 2.0 M aqueous perchloric acid in the presence of mannitol [165], after the oxidation of arsenite by chlorine water. The differential pulse polarographic methods were successfully used for the determination of arsenite, arsenate, and total inorganic arsenic in freshwater [165,166,168].

11.5.1.2. Separation of arsenite and arsenate by extraction

Dithiocarbamates, such as sodium diethyldithiocarbamate [336], ammonium pentamethylenedithiocarbamate [337,338], and ammonium sec-butyl dithiophosphate [339–341] react selectively with arsenite in buffered acid solution. The products of these reactions can be extracted with methyl isobutyl ketone, nitrobenzene [327], chloroform [338,342], hexane [339,340], or carbon tetrachloride [326,341]. The arsenic in the extracts is then determined by injection of an aliquot of the organic phase into a graphite furnace atomic absorption spectrometer [337,338,340] or by mineralizing the organic phase [339,342] and using aliquots of the resulting solutions for graphite furnace atomic absorption spectrometry [339] or hydride generation [341,342]. For the radioanalytical determination of arsenite, arsenic in carbon tetrachloride(diethyldithiocarbamate) was stripped into an aqueous solution of nitric acid/sodium nitrite, the arsenite oxidized to arsenate, and the W-185 tungstomolybdoarsenate heteropoly acid formed by the addition of sodium tungstate (W-185) and ammonium molybdate. Only the W-Mo-As heteropoly acid was extracted into 1,2-dichloroethane containing tetraphenylarsonium chloride. The W-185 activity of the evaporated organic phase was counted [336]. After the removal of arsenite, arsenate can be reduced to arsenite using titanium trichloride [338] or potassium iodide [342], and the arsenite that corresponds to the original arsenate determined. These methods were employed to determine arsenite, and in a few instances arsenate, in riverwater [336,337,339,340], tap water [336,339], well water [340] wastewaters [337], lake water, rainwater [339], and seawater [337–339]. Antimony(III) and (V), selenium (IV) and

(VI), and tellurium(IV) and (VI), as well as arsenic(III) and (V), were determined in seawater by dithiocarbamate extraction and graphite furnace atomic absorption spectrometry [338].

Arsenite was extracted into benzene, as arsenic trichloride, from homogenized fish and shellfish tissues strongly acidified with concentrated hydrochloric acid. The benzene layer was extracted with water and the arsenite in the aqueous phase quantified by differential pulse polarography. Arsenate left in the homogenates was reduced to arsenite with copper(I) chloride and the procedure repeated [167].

The detection limits reported for these methods fall in the range 0.2 [336] to 60 nanograms of arsenic per liter [338,339,340,342].

11.5.1.3. Separation of arsenite and arsenate by precipitation

Thionalide, N-(2-naphthyl)mercaptoacetamide, is known to precipitate arsenite preferentially. Arsenite solutions (obtained by digestions of orchard leaves and bovine liver, and reduction of arsenate to arsenite by ascorbic acid) were treated with thionalide and the precipitate reacted with phenylmagnesium bromide to convert all arsenic to triphenylarsine. Antimonite was similarly transformed into triphenylstibine. The triphenyl compounds were extracted into diethyl ether, separated by gas chromatography and detected by light emission from a microwave-induced plasma [343]. This technique allowed the detection of twenty picograms of arsenic and was used only for the determination of total arsenic, after digestion of biological materials, and of total inorganic arsenic in water samples after the reduction of arsenite.

Arsenite was separated from seawater by passing the sample (pH 7) through a silica gel column impregnated with thionalide: only the arsenite was retained, arsenate passing through the column. Arsenate was reduced to arsenite by sodium sulfite/potassium iodide. The arsenite thus obtained was collected on another column. It was eluted from the column by a sodium borate (0.01 M)/sodium hydroxide (0.01 M)/iodine (10 mg/l) solution and the arsenic in the effluent determined by the silver diethyldithiocarbamate method [344].

11.5.1.4. Determination of arsenite and arsenate by colorimetry

Arsenite reacts quantitatively with potassium iodate in sulfuric acid solution with liberation of iodine. Extraction of the iodine into chloroform and measurement of the absorbance of the resulting solution at 520 nm allowed arsenite to be determined in drinking water, lake water and river water with a detection limit of 2 μg/l of arsenic [345].

A spectrophotometric method based on the formation of molybdenum-blue by reduction of molybdoarsenate heteropoly acid (Section 11.4.4.2.1) was developed for the determination of arsenate, arsenite, and phosphate in seawater and freshwater [346]. Molybdenum-blue formed in the untreated water (absorbance measurement at 865 nm) corresponds to arsenate and phosphate. After oxidation of arsenite to arsenate by potassium iodate, the total inorganic arsenic and phosphate are determined. After reduction of arsenate to arsenite by an acidic solution of sodium disulfite and sodium thiosulfate [347], phosphate is determined. A flow injection version of this method was developed [348].

The determination of arsenite and arsenate by the silver diethyldithiocarbamate method is discussed in Section 11.5.1.5.

11.5.1.5. Determination of arsenite and arsenate by the hydride generation method

Arsenite and arsenate can be reduced by sodium borohydride to arsine, AsH_3. Arsine boils at $-55°C$ and can be easily flushed from the reaction mixture and passed to an appropriate detector. Arsenite is separated from arsenate by pH-controlled reduction with sodium borohydride. In buffered aqueous solutions (using acetate or citrate buffers) of pH 4–6, only arsenite is reduced to arsine, whereas arsenite and arsenate are converted to arsine in solutions of pH 1 or less. In this manner arsenite and arsenate can be determined sequentially or arsenite can be determined in one aliquot of a sample and the sum of arsenite and arsenate in another aliquot [239,334,349–353]. Zinc in 1.3 M hydrochloric acid will also preferentially reduce arsenite [354]. The concentration of arsenate or of the total inorganic arsenic can be obtained after reduction of the arsenate to arsenite. The detection limits of these systems depend on the detector and lie in the range of nanograms to micrograms of arsenic.

The methods outlined above were used to determine arsenic and arsenate in fresh waters [227,354,349,352] and in seawater [350]. The arsine was passed into a solution of silver diethyldithiocarbamate for the colorimetric determination of arsenic (Section 11.4.4.2.2), or into the flame of a flame atomic absorption spectrometer [227,350] (Section 11.4.4.3.4), into a heated quartz tube mounted on an atomic absorption spectrometer [352,355] (Section 11.4.4.3.5), or onto a filter paper impregnated with silver nitrate for subsequent neutron activation analysis [349].

Orchard leaves [227], rat tissues, and rat blood [234] were extracted with a mixture of nitric/sulfuric acid. The arsenite and arsenate extracted were determined by hydride generation with sodium borohydride using a quartz tube [334] or a flame [227] atomic absorption spectrometer as detector. The extraction procedure may have changed the arsenite/arsenate ratios in these biological samples [334].

11.5.1.6. Chromatographic separation of arsenite and arsenate

Ion chromatography with conductivity detection can be used for the separation and determination of arsenate in the presence of other anions [356–359]. Arsenite, because of its weakly acidic character, is silent in the conductivity detector. However, arsenite can be oxidized to arsenate with hydrogen peroxide, and the total inorganic arsenic determined as arsenate [358]. Aqueous solutions of sodium carbonate [357], carbonate/bicarbonate [356], and carbonate/sodium hydroxide were recommended as mobile phases. With a pre-concentration column [357] or recycling [356] detection limits of 20 ng of arsenic were achieved. Arsenate [357,358] and arsenite [358] were determined in this manner in fresh water samples.

Arsenite and arsenate in aqueous samples were separated on an IRA-458 ion exchange resin in the chloride form (arsenate retained, eluted with 0.5 M HCl, fractions
analyzed by electrothermal atomic absorption) [360], on a strong anion-exchange resin
(Dowex 1x8, acetate form, arsenate retained, eluted with 0.12 M HCl, fractions analyzed
by electrothermal atomic absorption) [361], on a "thiol-cotton" column (arsenite retained,
desorbed with concentrated HCl, analyzed by the hydride generation-quartz tube atomizer
method; arsenate reduced by KI/ascorbic acid) [362], on an ion-pair C_2 reversed-phase
column (HPLC, water–methanol 95:5, tetraheptylammonium bromide, inductively coupled
plasma atomic emission detection) [363], or on a strong anion-exchange column (Nu-
cleosil-NHMe$_2$ on silica support, HPLC, ammonium acetate and ammonium dihydrogen
phosphate solutions as mobile phases, inductively coupled plasma atomic emission detection) [364].

Orchard leaves were extracted with 0.5 M perchloric acid. The extract with its pH
adjusted to 5 was passed through two ion exchange columns [cation AG 50Wx4(H^+)
connected to an anion AG3x4 (Cl$^-$) column]. Arsenite was eluted by water and arsenate
by an aqueous 2 M sodium chloride solution. Arsenic was determined in the fractions by
hydride generation-quartz tube atomization [365].

11.5.2. Determination of methylarsenic compounds

Methylarsenic compounds considered in this section are methylarsonic acid, dimethylarsinic acid, trimethylarsine oxide, tetramethylarsonium salts, methylarsine,
dimethylarsine, and trimethylarsine (Fig. 11.1). All of these compounds occur naturally
and methylarsonic acid and dimethylarsinic acid were used as herbicides. Methods are
available to separate these organic arsenic compounds from each other and from the
inorganic arsenite and arsenate. The most frequently used methods include ion-exchange
chromatography, pH-controlled hydride generation coupled with a separation method for
the arsines, extraction methods selective for inorganic and organic arsenic compounds,
and gas chromatography of volatile derivatives of the arsenic compounds. "Benign" matrices such as freshwater do not pose any problems for the analyst facing the task of
identifying and quantifying arsenic compounds. With matrices such as sediments and
biological tissues from plants and animals, the extraction of the arsenic compounds into
an aqueous or organic solvent is the weak link in the analysis sequence. Proof that the
arsenic compounds were not changed during the extraction is often lacking. Any extraction or determination procedure that includes a reduction or oxidation step will very
probably provide correct information about the number of methyl groups bonded to the
arsenic atom, but cannot show whether the compound contained trivalent or pentavalent
arsenic. For example, a solution that produces dimethylarsine upon reduction with sodium
borohydride may have contained dimethylarsinic acid, $(CH_3)_2AsOOH$, or dimethyl(hydroxy)arsine, $(CH_3)_2AsOH$.

11.5.2.1. Separation and determination of arsenite, arsenate, and methylarsenic compounds by ion-exchange chromatography

Cation exchange resins, anion-exchange resins, and reverse-phase materials have been used as stationary phases to separate arsenite, arsenate, methylarsonic acid, dimethylarsinic acid, and trimethylarsine oxide. As mobile phases that control the ionization of the analytes and influence their retention, water, acetic acid, acetate buffers, trichloroacetic acid, hydrochloric acid, aqueous solutions of ammonium carbonate, ammonium hydrogen carbonate, ammonia, sodium hydroxide, and sodium dihydrogen phosphate were used sequentially or in a gradient operation. With reverse phase columns aqueous dimethylformamide, acetonitrile, methanol, or acetic acid were employed with sodium heptanesulfonate, tetrahepthylammonium nitrate, or hexadecyltrimethylammonium bromide as ion-pairing reagents. The matrices in which these arsenic compounds were determined include water, urine, and extracts from plant tissues, animal tissues, fresh water sediments, and estuarine sediments. The absolute detection limits of these analysis systems were reported to be as low as two nanograms of arsenic [370,371]. All five arsenic compounds (arsenite, arsenate, methylarsonic acid, dimethylarsinic acid, trimethylarsine oxide) were separated on an Aminex A-27 (Bio-Rad) anion exchange resin with a gradient of water to 0.2 M aqueous ammonium carbonate [379,380]. A summary of the separation of inorganic and methylarsenic compounds achieved by ion exchange chromatography is given in Table 11.2.

11.5.2.2. Separation and determination of arsenite, arsenate, and methylarsenic compounds by hydride generation

The pH-controlled reduction of arsenite (pH 4–6) and arsenate (pH $\leq$ 1) by sodium borohydride is described in Section 11.5.1.5. At the higher pH, at which arsenite but not arsenate is reduced, methylarsonic acid, dimethylarsinic acid, and trimethylarsine oxide are also not affected by sodium borohydride. These methylated arsenic compounds are reduced to the corresponding arsines at pH 1 and even higher acid concentrations. When the pH-controlled reduction is coupled with a separation of the generated arsines all five arsenic compounds can be determined. Systems for such analyses consist of a hydride generator and, in many arrangements, a trap (commonly cooled by liquid nitrogen), a gas chromatograph, and a detector. In order to achieve the low pH required for the reduction of methylarsenic compounds, oxalic acid, hydrochloric acid (1 to 6 M), nitric acid, and sulfuric acid were investigated [320]. The signal intensity for the methylarsines depends on the nature of the acid and its concentration. Hydrochloric acid is the most commonly used agent for acidification.

This method, introduced by Braman and his coworkers for the determination of arsenic compounds in environmental samples [388–390], uses sodium borohydride (1 to 10% w/v) in 0.1 M aqueous sodium hydroxide solution as the reducing agent. Zinc/hydrochloric acid in a reduction column does not quantitatively reduce the methylarsenic acids to the corresponding methylarsines unless potassium iodide and ascorbic acid are pre-

TABLE 11.2.

SEPARATION AND DETERMINATION OF ARSENITE, ARSENATE, METHYLARSONIC ACID, DIMETHYLARSINIC ACID AND TRIMETHYLARSINE OXIDE BY ION-EXCHANGE CHROMATOGRAPHY

Abbreviations: AB = arsenobetaine; AC = arsenocholine; III = arsenite; V = arsenate; M = methylarsonic acid; M2 = dimethylarsinic acid; M3 = trimethylarsine oxide; HTAB = hexadecyl(trimethyl)ammonium bromide; AX = anion exchange, CX = cation exchange; RP = reverse phase; THAN = tetraheptylammonium nitrate; DDP = differential pulse polarography; GFAAS = graphite furnace atomic absorption spectrometer; HPLC = high-performance liquid chromatograph; HPLC–GFAAS-I = automated HPLC–GFAAS system based on autosampler interface [386]; HPLC–GFAAS-II = automated HPLC–GFAAS system based on electronic interface [387]; HG-AgDDC = hydride generation followed by determination of arsenic by colorimetry; HG-QT = hydride generation followed by atomic absorption determination in a heated quartz tube; XRF = X-ray fluorescence.

Column material	Mobile phase	Arsenic compounds separated[a]/matrix	Detector (detection limit)[b]	Remarks	Ref.
CX AG50Wx8/column	1. 0.2 M CCl₃COOH (30 ml) 2. 1.0 M ammonium acetate (70 ml)	(III, V), M, M2 River water, sediment extracts	GFAAS (2 μg/l) 5 ml-fractions	V, M2 had same calibration curves; column pH < 1.5 at beginning	366
CX Dowex 50Wx8/column	1. 0.2 M CCl₃COOH pH 1.7 V 2. 1.8 M ammonium acetate M pH 6.5 3. 1.0 M ammonium acetate M2 pH 1.8	Water, extracts from pond sediments	HG-AgDDC	After digestion of the fractions and reductions V → III	367
CX Dowex 50Wx8/column	1. 0.2 M NaCl, HCl to pH 1.5 V 2. 1.8 M NaCl, pH 6.5 M 3. 1.0 M NaOH M2	Run-off water, sediment extracts	HG-QT	Acetate matrices interfere with As determination	368

Column	Eluent	Species	Sample	Detection	Comments	Ref.
CX Dowex 50Wx8, H$^+$/column	1. 0.5 M HCl 2. H$_2$O 3. 1.5 M ammonia	(III, V) M M2	Extracts from algae, molluscs, estuarine sediments	HG-GFAAS Zn-reductor	method uses lengthy extraction determination	369
CX Dowex 50Wx8, H$^+$/column	1. 0.02 M AcOH 2. 1.0 M NaOH	III, V, M M2	Fly-ash pond water	DPP (8 ppb)	after mineralization of fractions and reductions V → III	335
AX AG1x8, acetate/column	1. 0.01 M acetate buffer pH 4.7 2. 0.1 M acetate buffer pH 4.7	III, M2 M, V	Reference water			335
AX Zipax connected to CX BAX-10, 4$^+$	1. 0.0004% sulfuric acid 2. 0.1 M ammonium carbonate	III, M2, M, V	Soil water, bottled water	continuous flow HG-QT (2 ng)	As compounds give different calibration curves	370 371
AX Dowex 1x4, acetate/column	1. 0.1% AcOH 2. 5.0% AcOH 3. 1.0 M HCl	M2 M (III, V)	Extracts from lake sediments, surface soils	GFAAS of 130-drop fractions	method uses lengthy extraction procedure	372 373
CX Dowex 50Wx2/column	0.1 M pyridine	AB, M3	Water	GFAAS		374
CX AG 50Wx8 connected to AX AG 1x8/column	1. 0.006 M CCl$_3$COOH 2. 0.2 M CCl$_3$COOH 3. 1.5 M ammonia 4. 0.2 M CCl$_3$COOH	III, M V M2	Lake water, interstitial water from river sediments	GFAAS of fractions (10 ppb)		375
CX AG 50Wx8/column	1. 0.5 M HCl 2. H$_2$O 3. 5% ammonia 4. 20% ammonia	III, V M M2	Plasma and urine of dogs, red blood cells	γ-counter, XRF	dogs were given As-74 arsenate i.v.	376

(continued on p.)

TABLE 11.2. (continued)

Column material	Mobile phase	Arsenic compounds separated[a]	/matrix	Detector (detection limit)[b]	Remarks	Ref.
CX AG 50Wx8 H$^+$ (Bio-Rad)	1. 0.5 M HCl 2. H$_2$O 3. 1.5 M HClO$_4$	(III, V) M M2	Urine, faeces from mice	γ-counter	mice were given As-74 M2 orally; TLC, electrophoretic separation of AB, AC, III, V, M3, M2, M	378
AX Aminex A-27 (Bio-Rad) HPLC	gradient H$_2$O $\rightarrow$ 0.2 M ammonium carbonate	M3, III, M2, M, V	Extracts from vegetables	HPLC-GFAAS-I	As compounds separated from extract by conversion to hydrides, then oxidation	
AX Aminex A-27/HPLC	gradient H$_2$O $\rightarrow$ 0.2 M ammonium carbonate	M3, III, M2, M, V	Air, water, soil extracts	HPLC-GFAAS-I		
Dionex AX low capacity/HPLC	gradient H$_2$O/MeOH (80:20) $\rightarrow$ 0.02 M ammonium carbonate/CH$_3$OH (85:15)	III, M2, M, V	Standards	HPLC-GFAAS-I (0.25 mg/l)	arsenate peak influenced by solvent	381
Dionex AX	gradient 0.025 M Na$_2$B$_4$O$_7$ $\rightarrow$ 0.003 M NaHCO$_3$/0.0024 M Na$_2$CO$_3$	M2, III, M, 4-H$_2$NC$_6$H$_4$AsO$_3$H$_2$, V		HG-QT automated system (10 μg/l)	other mobile phases do not resolve all five compounds	382
AX SRA 70 (BDH Chem. Ltd.)/column	1. Na$_2$CO$_3$/CO$_2$ pH 5.5 2. Na$_2$CO$_3$ satd. with CO$_2$, 10g/l NH$_4$Cl	III, M2 V, M	Extracts from sediments, lake weeds	HG-flame AA	Other resins, mobile phases investigated; pretreatment of resin with 1 M HNO$_3$, 0.1 M EDTA pH 5 prevents III $\rightarrow$ V conversion	383

System	Mobile phase	Compounds[a]	Sample	Technique	Comments	Ref.[b]
AX LiChrosorb/HPLC	0.05 M NaH$_2$PO$_4$	M2, M, 4-H$_2$NC$_6$H$_4$AsO$_3$H$_2$		HPLC–GFAAS-I	Arsenate too strongly retained	384
	0.03 M NH$_4$OAc/0.045 M AcOH pH 4.4	III, M2, M	Soil extracts, water	HPLC–GFAAS-I		
RP-C18 Lichrosorb (Altex)	1. H$_2$O/MeOH 75:25 satd. with THAN 2. MeOH	III, M2, M / V	Standards	HPLC–GFAAS-II	Phosphate depressed signal	384
RP Hamilton resin-based PRP-1/HPLC	1. H$_2$O pH 9 (NaOH), 0.002 HTAB 2. H$_2$O/AcOH 99:1 3. H$_2$O/DMF 90:10	III, M2 / M, V, selenite, phosphate C$_6$H$_5$AsO$_3$H$_2$		HPLC–ICP-AES (130 μg/l)	Detector is multielement specific; response independent of As-compounds	385
RP C18 TSK gel ODS-120T/HPLC	W/MeCN/AcOH 95:5:6 0.005 M Na heptanesulfonate	AB, M3	Water inocculated with coastal ocean sediments and AB added	GFAAS		422

[a] The compounds are listed in the order they elute. Compounds in parentheses, e.g., (III, V) elute together.
[b] Detection limits for the systems are given in terms of elemental arsenic.

References on p. 340

sent [391]. The performance of the hydride generation system was improved by placing a trap cooled by ice/salt/water [392] or a trap filled with magnesium perchlorate [393] to collect water, a lead-acetate trap to remove hydrogen sulfide, and a sodium hydroxide-filled trap to absorb carbon dioxide in front of the liquid nitrogen trap. These gases interfere with the detection devices. The liquid nitrogen trap, in which all the arsines condense, serves to preconcentrate the arsines and makes the methods independent of the kinetics of the reduction reaction. After all of the arsenic compounds are reduced and the arsines collected, the liquid nitrogen Dewar is removed and the trap electrically heated or warmed in a water bath to volatilize the arsines into the detector in a stream of helium [269,327,388–390,394,395]. The arsines reach the detector in the order of increasing boiling point and molecular mass. If a better and more reliable separation of the arsines is desired, a gas chromatograph can be placed between the liquid nitrogen trap and the detector [327,393,395]. Chromosorb 101 [396] with 16.5% silicone oil DC-550 [393], or Chromosorb W AW DMCS with 5% PEG-20M [327], served as stationary phases. The gaseous arsines were passed into helium-glow discharges [388–390,392,394], microwave induced plasmas [396], or inductively coupled argon plasmas [312] for detection by atomic emission, or into a flame [210,397], a quartz-tube atomizer [312,327,395,398], or a graphite furnace [269] for detection by atomic absorption, or into a mass spectrometer [393], a flame ionization detector (FID) [327], or an electron capture detector (ECD). The FID detector is less sensitive but has a larger dynamic range than the ECD [327]. The generated arsines were also introduced directly (without collection in a cooled trap) into the detectors [269,320,397]. Detection limits of 30 pg of arsenic were reported for the system with the mass spectrometric detector [393] and 200 pg for the other systems [327,398]. The arsines can also be collected in cold heptane [393] or cold toluene [396] and these solutions injected into a gas chromatograph.

The foregoing systems were used to identify and determine inorganic and methylarsenic compounds in freshwaters [269,327,328,388,393,398], geothermal water [392], seawater [327,328], urine [269,388,392,399], air particulates [389], algae [328,398] egg shells, sea shells, limestone [388], molluscs [398], animal tissues [392], animal faeces [399], soils [393,394], plants [393], river sediments [400], estuarine sediments [395], oil shale, bitumen, kerogen, retort water [401], and herbicide formulations [401].

11.5.2.3. Gas chromatographic separation of volatile methylarsenic derivatives other than arsines

Methylarsonic acid, dimethylarsinic acid, arsenate, and arsenite can be converted to derivatives that are sufficiently volatile for gas chromatographic separation by reacting them with a trimethylsilylating agent [402,403] (eqn. 11.3), methyl thioglycolate [404] (eqn. 11.4), 2,3-dimercaptopropanol [405] (eqn. 11.5), or diethylammonium diethyldithiocarbamate [406] (eqn. 11.6). Gas chromatographic separation of the silylated arsenic compounds (eqn. 11.3) and detection by flame ionization gave detection limits of 0.1 ng of arsenic. The working curves were linear over three to five orders of magnitude [402].

$$(CH_3)_2As{\small\begin{array}{c}O\\\\OH\end{array}} \quad\longrightarrow$$

$$As(OH)_3 \quad\longrightarrow$$

$$O=As{\small\begin{array}{c}OH\\OH\\OH\end{array}} \quad\longrightarrow$$

$$CF_3-C{\small\begin{array}{c}O-Si(CH_3)_3\\\\N-Si(CH_3)_3\end{array}}$$

$$(CH_3)_2As{\small\begin{array}{c}O\\\\O-Si(CH_3)_3\end{array}}$$

$$As[O-Si(CH_3)_3]_3 \qquad (11.3)$$

$$O=As[O-Si(CH_3)_3]_3$$

Arsenic compounds in water, animal urine, and whole blood were reacted with methyl thioglycolate (eqn. 11.4). The arsenic derivatives were extracted into cyclohexane and separated by gas chromatography with a flame ionization or a thermionic specific detector. The detection limits were 0.2 ng of arsenic [404].

$$(CH_3)_n As{\small\begin{array}{c}O\\\\(OH)_{3-n}\end{array}} + (5-n)\ HS-CH_2COOCH_3 \longrightarrow (CH_3)_n As(SCH_2COOCH_3)_{3-n} +$$

$$+ (H_3COOCCH_2S-)_2 \qquad (11.4)$$

For the reaction of arsenic compounds with 2,3-dimercaptopropanol to give arsadithiocyclopentanes (eqn. 11.5), the pentavalent arsenic compounds were reduced first to trivalent derivatives. The heterocyclic products were extracted into benzene and the extract introduced into a gas chromatograph with a flame photometric detector. The detection limit was 0.04 ng of arsenic. This method was used to determine arsenic compounds in urine [405].

$$CH_3AsO_3H_2 \xrightarrow{SnCl_2/KI} (CH_3)As(OH)_2 \xrightarrow[benzene]{\overset{SH}{HOCH_2CHCH_2SH}} CH_3-As{\small\begin{array}{c}S-CH_2\\\ |\\S-CH\\|\\CH_2OH\end{array}} \quad (11.5)$$

The dithiocarbamates prepared from trivalent arsenic compounds (eqn. 11.6) were extracted into benzene or hexane and chromatographed on silanized 5% OV-17 on Anakrom AS, with electron capture detection. This method was used to determine arsenic compounds in water and urine samples with detection limits in the range of 15 to 73 nanograms of arsenic per milliliter [406].

$$(CH_3)_n As{\small\begin{array}{c}O\\\\(OH)_{3-n}\end{array}} \xrightarrow[H_2SO_4]{Na_2S_2)_5/KI} (CH_3)_n As(OH)_{3-n}$$

$$(CH_3)_n As[S_2CN(C_2H_5)_2]_{3-n} \xleftarrow[benzene]{(C_2H_5)_2NH_2^+ [(C_2H_5)_2NCS_2]^-} \qquad (11.6)$$

References on p. 340

The determination of dimethylarsinic acid by gas chromatography, after reduction to dimethyliodoarsine, was also investigated and applied to water and soils [403].

11.5.2.4. Extraction procedures for the separation of inorganic and organic arsenic compounds

Treatment of samples containing arsenite, arsenate, or methylarsenic compounds with concentrated hydrochloric acid to produce a 9 M HCl solution converts arsenite to arsenic trichloride that can be selectively extracted into dichloromethane [407] or toluene [408, 409], or distilled from the sample [410]. The arsenic in the organic phase is back-extracted into water and determined by atomic absorption spectrometry [407–409] or the silver diethyldithiocarbamate method [410]. To determine arsenate, the arsenite-free sample is treated with potassium iodide to reduce arsenate to arsenite and the extraction repeated [408,409]. The aqueous hydrochloric acid phase is either digested for the determination of total non-inorganic arsenic [408,409], or treated with sodium borohydride for the identification of methylarsines [407].

When samples 8 M with respect to hydrochloric acid were heated to 60–70°C, then cooled, treated with potassium iodide, and extracted with chloroform, inorganic arsenic and dimethylarsenic compounds were transferred as iodides into the organic phase. Re-extraction of the organic phase with distilled water affected only arsenic triiodide. When aqueous dichromate was used as an extractant, organic and inorganic arsenic were found in the aqueous phase for determination by graphite furnace atomic absorption spectrometry [411].

The preceding extraction methods were used to determine inorganic and "organic" arsenic in water, urine [411], marine algae [408,410], NBS Orchard Leaves [407,408], marine animals [407–409], and pig liver and kidney [407].

11.5.2.5. Determination of alkylarsines and triphenylarsine

Methylarsines are formed by the action of microorganisms on arsenic compounds. These methylarsines can be stripped from solution by an inert gas and condensed in a liquid nitrogen trap [327], collected in a polyethylene bottle attached to a conical-bottomed gas sampler [412], or flushed into an aqueous absorbing solution containing potassium iodide and iodine [413]. The three methylarsines are then separated by gas chromatography [327,413], and detected by a quartz-tube atomizer/atomic absorption system [414]. Dimethylarsine and trimethylarsine were identified as emanating from soils treated with arsenic compounds [413].

Trimethyl-, triethyl-, and triphenylarsine were separated by capillary gas chromatography and detected using a microwave plasma system [415]. Systems using a graphite furnace atomic absorption spectrometer as an arsenic-specific detector for a gas chromatograph and a liquid chromatograph were shown to respond to trimethylarsine [416] and triphenylarsine [386], respectively. These analytical systems have not yet been used for the determination of arsines in environmental samples.

11.5.2.6. Identification of tetramethylarsonium cation

A methanol extract from the gills of the clam *Meretrix lusoria* was purified by chromatography on Dowex 50, by gel filtration, and chromatography on a weakly acidic cation exchange resin (Bio-Rex 70). the purified compound was chromatographed on silica gel thin layers and then identified as tetramethylarsonium cation by proton NMR spectroscopy, FAB mass spectrometry, and liquid chromatography [417]. Tetramethylarsonium cation was identified by high pressure liquid chromatography (using Nucleosil 10 SA with 0.1 M pyridine-formic acid of pH 3.1) in urine from mice given tetramethylarsonium iodide orally. An inductively coupled argon plasma atomic emission spectrometer served as the arsenic-specific detector [418].

11.5.3. Determination of arsenobetaine and arsenocholine

Arsenobetaine can exist in one of two forms depending on the pH of the medium. In acidic solution the predominant form will be the positively charged arsonium cation, whereas in basic solution the zwitterionic form will be present (eqn. 11.7). The purification of arsenobetaine by ion exchange column chromatography takes advantage of the properties of these two forms.

$$[(CH_3)_3\overset{+}{As}-CH_2COOH]^+X^- \underset{H^+}{\overset{OH^-}{\rightleftharpoons}} [(CH_3)_3\overset{+}{As}-CH_2COO^-]^0 \quad (11.7)$$

$$\text{arsonium form} \qquad\qquad\qquad \text{zwitterion}$$

11.5.3.1. Separation, purification and isolation of arsenobetaine

Samples of tissues from marine animals (most often muscle, or liver) or plants, ranging in mass from grams to several kilograms, were extracted with chloroform/methanol (2:1) [419–422], methanol [417,423–431], or acetonitrile [432]. The chloroform/methanol extracts were mixed with water to cause phase separation. The aqueous methanol, the methanol, and the acetonitrile solutions that contained the arsenobetaine were then evaporated. The residues were dissolved in water and the resulting solutions defatted by extraction with diethyl ether. The defatted aqueous phases were shaken with liquefied phenol to transfer arsenobetaine into the organic phase and separate it from inorganic salts [427]. The aqueous layers were discarded. After the addition of diethyl ether to the phenol, the phenol/ether phases were shaken with water. The aqueous layers now containing the arsenobetaine were separated, washed with diethyl ether, and then evaporated to dryness. The residues were dissolved in a small amount of water and these solutions chromatographed. Passage through anion exchange columns (Dowex 2x8, AG 1x2, OH-form; water as mobile phase) removes anions from the solution: the arsenobetaine is not retained. Cations are then separated using cation exchange columns, (Dowex 50Wx8, Zeokarb 225 [423], Dowex 50x2, Amberlite CG-50, or Amberlite IRC-50 in the H^+-form). When aqueous solutions of arsenobetaine are placed on these columns, the cationic arsenobetaine is retained. After water has been passed through the columns, the arseno-betaine is eluted with aqueous ammonia (0.3 to 3.0 M). For further purification the arsenic-containing fractions are rechromatographed on activated carbon [419,420,433], or Dowex 50Wx8 (0.1 M pyridine/formic acid buffer, pH 3.1 as mobile phase) [419,421,429,431,433], on Sephadex LH-20 [427,434], Sephadex G25F [419,420,433], or BioGel P-2 [421,426,429,431]. If necessary, the chromatographic steps are repeated. As a final purification step thin layer chromatography on cellulose or silica gel is used [423–425, 430–435]. Not all of these purification steps need be applied and investigators frequently omitted several of them. The fractions containing arsenic are generally identified through analysis of appropriate aliquots by graphite furnace atomic absorption spectrometry. These extraction and purification procedures produce analytically pure arsenobetaine in milligram quantities as white crystals that can be recrystallized from acetone [423,425].

11.5.3.2. Identification of arsenobetaine

Pure samples of arsenobetaine isolated from marine organisms were identified as the zwitterion of carboxymethyl(trimethyl)arsonium monohydrate by single-crystal X-ray crys-tallography [423,425] and by comparison of their infrared [419,420,424,433,436] and nuclear magnetic resonance [424,426,429,430,434] spectra with those of synthetic arseno-betaine [423,425]. The R_f-values from thin-layer chromatography with various mobile phases [419,424,431–433,436], fast atom bombardment mass spectrometry [417,419–421,430,433,436,437], and field desorption mass spectrometry [421,429,432,438] were also used.

A direct identification of arsenobetaine in extracts is desirable to avoid the time-con-suming purification steps. Experiments with synthetic arsenic compounds proved that arsenobetaine can be separated by liquid chromatography from other arsenic compounds (Table 11.3). Reversed-phase columns, anion-exchange columns, and cation-exchange columns served as the stationary phases. Chromatography of extracts from crab meat at various stages of purification on a C_{18} reversed-phase column with heptanesulfonate as ion-pairing reagent showed, that impure extracts contain materials that compete with the heptanesulfonate and produce several arsenic-containing bands with retention times which do not match synthetic arsenobetaine. The competition among ion-pairing reagents can be influenced in favor of heptanesulfonate by further purification of the extracts [428], by increasing the concentration of the heptanesulfonate, or by decreasing the volume of extract placed on the column [427]. Extracts from a variety of marine carnivores, herbi-vores, and plankton feeders were purified by passage through a cation-exchange column (Dowex 50Wx8) and an anion exchange column (Dowex 1x8). Arsenobetaine in the puri-fied extracts was successfully identified by high-performance liquid chromatography on a reversed-phase column [422]. Cation- and anion-exchange columns appear not to be as sensitive to impurities in the extracts as are the reversed-phase columns. Purification of

TABLE 11.3.

SEPARATION AND DETERMINATION OF ARSENOBETAINE AND ARSENOCHOLINE BY LIQUID CHROMATOGRAPHY

Abbreviations: AX = anion exchange; CX = cation exchange; III = arsenite; V = arsenate; AB = arsenobetaine; AC = arsenocholine; M = methylarsonic acid; M2 = dimethylarsinic acid; M3 = trimethylarsine oxide; GFAAS-A = graphite furnace atom absorption spectrometer coupled to HPLC; GFAAS = graphite furnace atomic absorption spectrometer with manual sample introduction; ICP-A = inductively coupled argon plasma emission spectrometry interfaced with HPLC.

Column material	Mobile phase	Arsenic compounds separated	Matrix	Detector (detection limit)	Ref.
C$_{18}$ reversed-phase	0.005 M heptanesulfonate in H$_2$O/CH$_3$CN/AcOH 95/5/6 (v/v/v)	(III, V) AB AC	Standards	GFAAS-A	439
Hamilton PRP-1 reversed-phase	0.05 M heptanesulfonate in 2.5% aqueous AcOH	M M2 AB AC	standards crab extracts	ICP-A (75 ng, 750 ng/ml)	427
Nucleosil-N(CH$_3$)$_3$-10 AX (Nagel)	0.05 M phosphate buffer (Na$_2$HPO$_4$/NaH$_2$PO$_4$ 1:1)	AB III M M2 V	Standards, extract from *Hizikia fusiforme*	ICP-A (30 ng)	440
Nucleosil-SO$_3$H-10 CX (Nagel)	0.05 M phosphate buffer	(III, V) M M2 AB	Standards (30 ng)	ICP-A	440
Nucleosil-10SB AX	0.025 M phosphatea buffer, pH 7.4	AB IV M M2 V	Standards, extracts from shark muscle, shark liver	ICP-A	441
Nucleosil-10SA CX	0.025 M phosphate buffer, pH 7.4	(V, M) M2 III AB		ICP-A	441
C-18 reversed-phase Shodes ODS	0.002 M aqueous heptanefulfonate with 0.5% AcOH, 5% CH$_3$CN	III AB AC	Standards, crab extract	GFAAS-A	428
Nucleonic-10SA CX	0.1 M pyridine/formic acid	M2 AB AC	Standards, purified extracts from clam *Meretrix lusoria*	ICP-A	417
TSK Gel ODS-120T	0.0056 M heptanesulfonate in H$_2$O/CH$_3$CN/AcOH 95:5:6 (v/v/v)	III, V, M) M2 AB M3	Standards, extracts from marine organisms	GFAAS	422

References on p. 340

the extracts from tissues of the clam *Meretrix lusoria* by passage through a cation-exchange column (Dowex 50x2) was sufficient for the identification of arsenobetaine by HPLC on a Nucleosil-10SA cation exchange column [417]. The aqueous phases obtained by addition of water to the chloroform/methanol extracts from the muscle and the liver of the shark *Prionace glauca* were chromatographed without difficulties on Nucleosil ion exchange columns [441]. Similarly, the aqueous phase from the extraction of powdered seaweed (*Hizikia fusiforme*) with phosphate buffer was chromatographed on a Nucleosil anion-exchange column. Arsenobetaine, arsenite, and arsenate were identified [440].

A very sensitive automated system based on hydride generation was developed for the determination of arsenite/arsenate, methylarsonic acid, dimethylarsinic acid, trimethylarsine oxide, and arsenobetaine. Aliquots of the methanolic extracts from marine organisms are introduced into the hydride generator. The arsenic compounds are reduced by sodium borohydride in hydrochloric acid. The generated arsines are collected in a liquid-nitrogen trap and then volatilized into a gas chromatograph for separation. The effluents from the gas chromatograph are routed to a mass spectrometer operating in the select-ion-monitoring mode [76 for AsH_3, 90 for CH_3AsH_2 and $(CH_3)_2AsH$, and 103 for $(CH_3)_3As$]. Under these conditions arsenobetaine is not converted to a volatile arsine. Heating the extract with 2 M aqueous sodium hydroxide at 85°C for three hours converts arsenobetaine to trimethylarsine oxide. The digested extract is analyzed as described for the undigested extract. The difference between the trimethylarsine concentrations before and after digestion gives the concentration of arsenobetaine. This method was used to identify arsenobetaine in shell fish, fish, crustaceans and sea weeds [442]. The detection limit is 0.5 ng of arsenic.

11.5.3.3. Determination of arsenocholine

Arsenocholine does not appear to be as ubiquitous as arsenobetaine. Arsenocholine was detected in shrimp. Extraction of the shrimp with acetonitrile [432] or methanol [430] and purification of the extracts by column chromatography produced samples in which arsenocholine was identified by cation-exchange chromatography [430,432], thin-layer chromatography, and pyrolysis-mass spectrometry [430]. Chromatographic experiments with synthetic compounds showed that arsenocholine can be separated from arsenobetaine and other arsenic compounds on reversed-phase columns [427,428,439] and cation-exchange columns [417] (Table 11.3). Arsenocholine is not converted to trimethylarsine oxide by hot 2 M aqueous sodium hydroxide.

11.5.4. Determination of dimethyl(ribosyl)arsine oxide

Dimethyl(ribosyl)arsine oxides (Fig. 11.1, Section 11.3) were identified in brown kelp, *Ecklonia radiata* [45,47], in the kidney of the giant clam, *Tridacna maxima* [46], in the seaweeds, *Hizikia fusiforme* [48,442], *Laminaria japonica* [442,443], and *Codium fragile* [444]. The arsenic compounds were extracted from the organisms with methanol. The residues obtained upon evaporation of the extracts were dissolved in water and the aqueous solutions defatted by extraction with diethyl ether. The solutions were chromatographed on Sephadex LH-20 columns with water as the mobile phase. The arsenic-containing fractions were further chromatographed on Sephadex DEAE A25 (0.05 M Tris-buffer, pH 7.6). Additional chromatography on Sephadex DEAE, Sephadex G-15, and the anion exchange column Unisil Q-NH2, as appropriate, led to pure compounds. The dimethyl(ribosyl)arsine oxides were identified by X-ray crystallography [46], by NMR spectroscopy [45–48,443,444], by infrared spectroscopy [45], and thin layer chromatography [46]. Chromatography of the crude methanolic extract from *Laminaria japonica* on Nucleosil 5SB(AX) and Asahipack GS 220 (GP), with 0.05 M phosphate buffer of pH 6.8 as the mobile phase, indicated that the dimethyl(ribosyl)arsine oxides can be identified by high pressure liquid chromatography with an inductively coupled plasma atomic emission spectrometer as the arsenic-specific detector [444].

The conversion of dimethyl(ribosyl)arsine oxides to dimethylarsinic acid by hot 2 M aqueous sodium hydroxide solution can be used to detect and determine these compounds by the hydride generation technique, as described for arsenobetaine in Section 11.5.3.2.

TABLE 11.4.

ENVIRONMENTAL AND BIOLOGICAL STANDARD REFERENCE MATERIALS CERTIFIED FOR TOTAL ARSENIC [445]

SRM No.	Name	Total arsenic (μg/g)
1643b	Trace Element in Water	0.049[a]
1648	Urban Particulate Matter	115 ± 10
1632b	Trace Elements in Bituminous Coal	3.72
1634b	Trace Elelments in Fuel Oil	0.12
1635	Trace Elements in Subbituminous Coal	0.42 ± 0.15
1633a	Trace Elelments in Coal Fly Ash	145 ± 15
1646	Estuarine Sediment	11.6
1566a	Oyster Tissue	14.0 ± 1.2
1567a	Wheat Flour	0.006[a]
1568	Rice Flour	0.41 ± 0.05
1573	Tomato Leaves	0.27 ± 0.05
1575	Pine Needles	0.21 ± 0.04
1572	Citrus Leaves	3.1 ± 0.3
1577a	Bovine Liver	0.047
RM50	Albacore Tuna	3.3[a]
RM8431	Mixed Diet	0.924[a]
1549	Nonfat Milk Powder	0.0019[a]
2670	Freeze Dried Urine	0.015 μg/ml; 0.48 μg/ml

[a] Value not certified.

References on p. 340

340

11.6. STANDARDS FOR THE DETERMINATION OF ARSENIC

Certified standards are of great help to analysts in establishing the accuracy of analytical methods. The U.S. National Bureau of Standards issued environmental and biological Standard Reference Materials, for which total arsenic concentrations are certified (Table 11.4). Examples of other standards are Scallops (5.6 μg/g), Swordfish (3.5 μg/g) (National Research Council of Canada), Animal Muscle (0.007 μg/g) (International Atomic Energy Agency), and Standard Reference Soil SO-4 (7.1 μg/g) (Canada Centre for Mineral and Energy Technology). Standards for individual arsenic compounds are not available.

REFERENCES

1 M.E. Weeks, *Discovery of the Elements*, Journal of Chemical Education, Easton, Pennsylvania, 6th edn., 1960, p. 92.
2 R.M. Allesch, *Arsenik*, Ferd. Kleinmayr, Klagenfurt, Austria, 1959, p. 13.
3 W.D. Buchanan, *Toxicity of Arsenic Compounds*, Elsevier, Amsterdam, 1962, p. 16.
4 R.M. Allesch, *Arsenik*, Ferd. Kleinmayr, Klagenfurt, Austria, 1959, p. 213.
5 A.M. Lansche, *Arsenic, in Mineral Facts and Problems*, U.S. Bureau of Mines Bull., 630 (1964).
6 *Gmelin's Handbuch der Anorganischen Chemie, Arsen*, System No. 17, 8th edn., 1952, p. 264.
7 R.M. Allesch, *Arsenik*, Ferd. Kleinmayr, Klagenfurt, Austria, 1959, pp. 256-278.
8 R.M. Allesch, *Arsenik*, Ferd. Kleinmayr, Klagenfurt, Austria, 1959, pp. 239-250.
9 D.V. Frost, *Fed. Proc., Fed. Am. Soc. Expl. Biol.*, 26 (1967) 194.
10 K.-H. Most, *Arsen als Gift und Zaubermittel in der deutschen Volksmedizin mit besonderer Berücksichtigung der Steiermark*, Dissertation, Philsophische Fakultät, Universität Grraz, Graz, 1939, Doktorats-Akten Z 2430.
11 E. Baümler, *Paul Ehrlich: Forschen für das Leben*, Societätsverlag, Frankfurt, West Germany, 1979; and references therein.
12 H.S. Sutterlee, *New Engl. J. Med.*, 263 (1960) 676.
13 E.O. Uthus and F.H. Nielsen, *Nutr. Res.*, 7 (1987) 1061, and references therein.
14 M. Anke, M. Grün, M. Partschefeld, B. Groppel and A. Hennig, in M. Kirchgessner (Editor), *Trace Element Metabolism in Man and Animals-3*, Freising-Weihenstephan, 1978, p. 248.
15 W. Mertz, *Implications of the New Trace Elements for Human Health*, in M. Anke, H.-J. Schneider and C. Brückner (Editors), *Arsen 3. Spurenelement Symposium*, Schiller Universität Jena, East Germany, 1980, p. 14.
16 G. Pershagen, in B.A. Fowler (Editor), *The Epidemiology of Human Arsenic Exposure*, in *Biological and Environmental Effects of Arsenic*, Elsevier, Amsterdam, 1983, p. 199; and references therein.
17 N. Ishinishi, K. Tsuchiya, M. Vahter and B.A. Fowler, in L. Friberg, G.F. Nordberg and V. Vouk (Editors), *Handbook on the Toxicology of Metals*, Elsevier, Amsterdam, 2nd edn., 1986, p. 43.
18 F.P. Treadwell, *Kurzes Lehrbuch der Analytischen Chemie, Vol. 1, Qualitative Analyse*, Franz Deuticke, Wien, Austria, 17th edn., 1940, p. 223.
19 F.P. Treadwell, *Kurzes Lehrbuch der Analytischen Chemie, Vol. 2, Quantitative Analyse*, Franz Deuticke, Wien, Austria, 12th edn., 1949.
20 E. Merian (Editor), *Metalle in der Umwelt*, Verlag Chemie, Weinheim, 1984, p. 671.
21 U. Ewers, in E. Merian (Editor), *Metalle in der Umwelt*, Verlag Chemie, Weinheim, 1984, p. 263.
22 K.S. Squibb and B.A. Fowler, in B.A. Fowler (Editor), *Biological and Environmental Effects of Arsenic*, Alsevier, Amsterdam, 1983, p. 233.

23 M. Vahter, E. Marafante and L. Dencker, *Sci. Total Environ.*, 30 (1983) 197.

24 T.R. Irvin and K.J. Irgolic, *J. Appl. Organomet. Chem.*, 2 (1988) 509.

25 A.E. Martell and R.M. Smith, *Critical Stability Constants, Vol. 5, First Supplement,* Plenum Press, New York, 1982, p. 409.

26 A.E. Martell and R.M. Smith, *Critical Stability Constants, Vol. 5, First Supplement,* Plenum Press, New York, 1982, p. 410.

27 D.L. Johnson, *Nature,* 240 (1972) 44.

28 E.A. Lewis, L.D. Hansen, E.J. Baca and D.J. Temer, *J. Chem. Soc., Perkin Trans. 2,* (1976) 125.

25 A.E. Martell and R.M. Smith, *Critical Stability Constants, Vol. 5, First Supplement,* Plenum Press, New York, 1982, p. 445.

30 R.S. Braman, in B.A. Fowler (Editor), *Biological and Environmental Effects of Arsenic,* Elsevier, Amsterdam, 1983, p. 141.

31 J.S. Edmonds and K.A. Francesconi, *Sci. Total Environ.*, 64 (1987) 317.

32 H. Norin, A. Christakopoulos and M. Sandstroem, *Chemosphere,* 14 (1985) 313.

33 K. Hanaoka, H. Yamamoto, K. Kawashima, S. Tagawa and T. Kaise, *J. Appl. Organomet. Chem.*, 2 (1988) 371.

34 R.S. Braman, in E.A. Woolson (Editor), *Arsenical Herbicides (ACS Symp. Ser.,),* Vol. 7, 1975, p. 108.

35 Y.K. Chau and P.T.S. Wong, in F.E. Brinckman and J.M. Bellama (Editors), *Organometals and Organometalloids (ACS Symp. Ser.,* Vol. 82), 1978, p. 39.

36 G.E. Parris and F.E. Brinckman, *Environ. Sci. Technol.*, 10 (1976) 1128.

37 K. Shiomi, Y. Kakehashi, H. Yamanaka and T. Kikuchi, *J. Appl. Organomet. Chem.*, 1 (1987) 177.

38 K. Shiomi, M. Aoyama, H. Yamanaka and T. Kikuchi, *Comp. Biochem. Physiol.*, 90C (1988) 361.

39 K. Shiomi, Y. Horiguchi and T. Kaise, *J. Appl. Organomet. Chem.*, 2 (1988) 385.

40 K.J. Irgolic, in A.V. Xavier (Editor), *Frontiers in Bioinorganic Chemistry,* VCH Publishers, Weinheim, 1986, p. 339; and references therein.

41 H. Norin and A. Christakopoulos, *Chemosphere,* 11 (1982) 287.

42 H. Norin, R. Ryhage, A. Christakopoulos and M. Sandstroem, *Chemosphere,* 12 (1983) 299.

43 K. Shiomi, M. Orii, H. Yamanaka and T. Kikuchi, *Nippon Suisan Gakkaishi,* 53 (1987) 103.

44 E. Marafante, M. Vahter and L. Dencker, *Sci. Total Environ.*, 34 (1984) 223.

45 J.S. Edmonds and K.A. Francescani, *Nature,* 289 (1981) 602.

46 J.S. Edmonds, K.A. Francesconi, P.C. Healy and A.H. White, *J. Chem. Soc., Perkin Trans.1,* (1982) 2989.

47 J.S. Edmonds and K.A. Francesconi, *J. Chem. Soc., Perkin Trans. 1,* (1983) 2375.

48 J.S. Edmonds, M. Morita and Y. Shibata, *J. Chem. Soc., Perkin Trans. 1,* (1987) 577.

49 Y. Shibata, M. Morita and J.S. Edmonds, *Agric. Biol. Chem.*, 51 (1987) 391.

50 K. Jin, T. Hayashi, Y. Shibata and M. Morita, *J. Appl. Organomet. Chem.*, 2 (1988) 365.

51 J.S. Edmonds, K.A. Francesconi and J.A. Hansen, *Experientia,* 38 (1982) 643.

52 J.S. Edmonds and K.A. Francesconi, *J. Appl. Organomet. Chem.*, 2 (1988) 297.

53 N.R. Bottino, E.R. Cox, K.J. Irgolic, S. Maeda, W.J. McShane, R.A. Stockton and R.A. Zingaro, in F.E. Brinckman and J.M. Bellama (Editors), *Organometals and Organometalloids (ACS Symp. Ser.,* Vol. 82), 1978, p. 116.

54 A.A. Benson and P. Nissen, *Dev. Plant Biol.*, 8 (1982) *(Biochem. Metab. Plant Lipids)* 121.

55 P.W. Enders, M. Geldmacher-von Mallinckrodt and D. Stamm, *J. Forensic Med. (Istanbul),* 1 (1985) 133.

56 G.K.H. Tam and H.B.S. Conacher, *J. Environ. Sci. Health, Part B,* B12 (1977) 213.

57 R.W. Dabeka and G.M.A. Lacroix, *Can J. Spectrosc.*, 30 (1985) 154.

58 B.T. Hunter, *J. Assoc. Off. Anal. Chem.*, 60 (1986) 493.

59 W.R. Penrose, *CRC Crit. Rev. Environ. Control,* 4 (1974) 465.

60 Y. Talmi and C. Feldman, in E.A. Woolson (Editor), *Arsenical Pesticides (ACS Symp. Ser.,* Vol. 7), 1975, p. 13.
61 H. Narasaki, *Anal. Chem.,* 57 (1985) 2481.
62 J. Jaganathan, M.S. Mohan, R.A. Zingaro and K.J. Irgolic, *J. Trace Microprobe Techn.,* 3 (1985) 345.
63 O. Oster and W. Prellwitz, *Fortschr. Atomspektrom. Spurenanal.,* 1 (1984) 275.
64 N.G. Van der Veen, H.J. Keukens and G. Vos, *Anal. Chim. Acta,* 171 (1985) 285.
65 M. Stoeppler, M. Burow, F. Backhaus, W. Schramm and H.W. Nürnberg, *Mar. Chem.,* 18 (1986) 321.
66 H.M. Kingston and L.B. Jassie, *Anal. Chem.,* 58 (1986) 2534.
67 S. Tsukada, R. Demura and I. Yamamoto, *Eisei Kagaku,* 31 (1985) 37.
68 R.A. Nadkarni, *Radiochem. Radioanal. Lett.,* 19 (1974) 127.
69 G. Kennedy, J.-L. Galimer and L. Zikovsky, *Can. J. Chem.,* 64 (1986) 790.
70 J.R. Greisman, W.E. Carey, W.A. Gould and E.K. Alban, *J. Food Sci.,* 34 (1969) 295.
71 J. Schneider and R. Geisler, *Z. Anal. Chem.,* 267 (1973) 270.
72 M.J. Minski, C.A. Girling and P.J. Peterson, *Radiochem. Radioanal. Lett.,* 30 (1977) 179.
73 A.R. Byrne and M. Tusek-Znidaric, *Chemosphere,* 12 (1983) 1113.
74 L.E. Wangen and E.S. Gladney, *Anal. Chim. Acta,* 96 (1978) 271.
75 B.J. Ray and D.L. Johnson, *Anal. Chim. Acta,* 62 (1972) 196.
76 S. Gohda, *Bull. Chem. Soc. Japan,* 45 (1972) 1704.
77 R.S. Shreedhara Murthy and D.E. Ryan, *Anal. Chem.,* 55 (1983) 682.
78 J.P. Lacroix, J. Steinier and P. Dreze, *Anal. Lett.,* 6 (1973) 565.
79 H.V. Weiss and K.K. Bertine, *Anal. Chim. Acta,* 65 (1973) 253.
80 E.S. Gladney and J.W. Owens, *Anal. Chem.,* 48 (1976) 2220.
81 K. Lenvik, E. Steinnes and A.C. Pappas, *Anal. Chim. Acta,* 97 (1978) 295.
82 L. Maggi, S. Meloni, G. Queirazza and N. Genova, *J. Trace Microprobe Techn.,* 1 (1983) 369.
83 K. Iwashima, M. Fujita and N. Yamagata, *Radioisotopes,* 25 (1976) 443.
84 Y. Maruyuma and Y. Nagaoka, *Radioisotopes,* 28 (1979) 306.
85 D. Hoede and H.A. van der Sloot, *Anal. Chim. Acta,* 111 (1979) 321.
86 C.L. Ndiokwere, *Radioisotopes,* 32 (1983) 117.
87 A.R. Byrne, *Radiochem. Radioanal. Lett.,* 52 (1982) 99.
88 M. Gallorini, R.R. Greenberg and T.E. Gills, *Anal. Chem.,* 50 (1978) 1479.
89 K. Heydorn and E. Damsgaard, *Talanta,* 20 (1973) 1.
90 I. Drabaek, V. Carlson and L. Just, *J. Radioanal. Nucl. Chem. Lett.,* 103 (1986) 249.
91 O.J. Kronberg and E. Steinnes, *Analyst,* 100 (1975) 835.
92 A. Disam, P. Tschöpel and G. Tolg, *Z. Anal. Chem.,* 295 (1979) 97.
93 F.J. Marcie, *Environ. Sci. Technol.,* 1 (1967) 164.
94 J. Knoth and H.Schwenke, *Fresenius' Z. Anal. Chem.,* 294 (1979) 273.
95 H. Taylor and F.E. Beanish, *Talanta,* 15 (1968) 1497.
96 E. Scheubeck, C. Jörrens and H. Hoffmann, *Fresenius' Z. Anal. Chem.,* 303 (1980) 257.
97 H.R. Lindner, H.D. Seltner and B. Schreiber, *Anal Chem.,* 50 (1978) 896.
98 T.M. Reymont and R.J. Dubois, *Anal. Chim. Acta,* 56 (1971) 1.
99 Y. Talmi and D.T. Bostick, *J. Chromatogr. Sci.,* 13 (1975) 233.
100 A.S. Khan and A. Chow, *Talanta,* 31 (1984) 304.
101 G. Lunde, *J. Sci. Food Agr.,* 24 (1973) 1021.
102 E. Forschner, *Fleischwirtschaft,* 54 (1974) 529.
103 G.H. Smith, O.L. Lloyd and F.H. Hubbard, *Arch. Environ. Health,* 41 (1986) 120.
104 R.L. Walter, R.D. Willis, W.F. Gutknecht and J.M. Joyce, *Anal. Chem.,* 46 (1974) 843.
105 J.L. Campbell, B.H. Orr, A.W. Herman, L.A. McNelles, J.A. Thomson and W.B. Cook, *Anal. Chem.,* 47 (1975) 1542.
106 N. Kojima, Y. Katayama, Y. Ishimaru, K. Minato, N. Okada and A. Aoki, *Proc. 5th Symp. Ion Beam Technol.,* Dec. 5-6, 1986, *Hodsei Univ. Suppl.,* 6 (1987) 109.
107 B.A. Fowler, R.C. Fay, R.L. Walter, R.D. Willis and W.F. Gutknecht, *Environ. Health Perspect.,* 12 (1975) 71.

108 R.M. Dagnal, K.C. Thompson and T.S. West, *Talanta*, 15 (1968) 677.
109 K. Tsuji and K. Kuga, *Anal. Chim. Acta*, 72 (1974) 85.
110 K. Tsuji and K. Kuga, *Anal. Chim. Acta*, 97 (1978) 51.
111 T. Nakahara, S. Kobayashi and S. Musha, *Anal. Chim. Acta*, 104 (1979) 173.
112 J. Azad, G.F. Kirkbright and R.D. Snook, *Analyst*, 105 (1980) 79.
113 A. D'Ulivo, R. Fuoco and P. Papoff, *Talanta*, 32 (1985) 103.
114 V.I. Rigin, *Zh. Anal. Khim.*, 33 (1978) 1996; engl.: 1510.
115 R.G. Lewis, *Residue Rev.*, 68 (1977) 123; and references therein.
116 G.F. Kirkbright and L. Ranson, *Anal. Chem.*, 43 (1971) 1238.
117 N.W. Bower and J.D. Ingle, Jr., *Anal. Chem.*, 49 (1977) 574.
118 T. Sakai, S. Hanamura and J.D. Winefordner, *Anal. Chim. Acta*, 170 (1985) 237.
119 J.C. Chambers and B.E. McClellan, *Anal. Chem.*, 48 (1976) 2061.
120 V.M. Shkinev, I. Khavezov, B.Ya. Spivakov, S. Mareva, E. Ruseva, Yu.A. Zolotov and N. Iordanov, *Zh. Anal. Chem.*, 36 (1981) 896; engl.: 606.
121 W. Slavin and D.C. Manning, *Prog. Anal. Atom. Spectrosc.*, 5 (1982) 243.
122 J.P. Matousek, *Progr. Anal. Atom. Spectrosc.*, 4 (1981) 247.
123 D. Chakraborti, W. De Jonghe and F. Adams, *Anal. Chim. Acta*, 119 (1980) 331.
124 D. Chakraborti, K.J. Irgolic and F. Adams, *Intern. J. Environ. Anal. Chem.*, 17 (1984) 241.
125 J. Koreckova, W. Frech, E. Lundberg, J.A. Persson and A. Cedergren, *Anal. Chim. Acta*, 130 (1981) 267.
126 P.R. Walsh, J.L. Fasching and R.A. Duce, *Anal. Chem.*, 48 (1976) 1014.
127 F.D. Pierce and H.R. Brown, *Anal. Chem.*, 49 (1977) 1417.
128 K. Saeed and Y. Thomassen, *Anal. Chim. Acta*, 130 (1981) 281.
129 K.W. Riley, *Atom. Spectrosc.*, 3 (1982) 120.
130 J. Bauslaugh, B. Radziuk, K. Saeed and Y. Thomassen, *Anal. Chim. Acta*, 165 (1984) 149.
131 M.A. Lovell and J.G. Farmer, *Intern. J. Environ. Anal. Chem.*, 14 (1983) 181.
132 R.D. Ediger, *Atom. Absorpt. Newslett.*, 14 (1975) 127.
133 F. Puttemans and D.L. Massart, *Mikrochim. Acta [Wien]*, (1984) I, 261.
134 R.F. Sanzolone and T.T. Chao, *Anal. Chim. Acta*, 128 (1981) 225.
135 Shan Xiao-Quan, Ni Zhe-Ming and Zhang Li, *Anal. Chim. Acta*, 151 (1983) 179.
136 Shan Xiao-Quan, Ni Zhe-Ming and Zhang Li, *Atom. Spectrosc.*, 5 (1984) 1.
137 G. Schlemmer and B. Welz, *Spectrochim. Acta, Part B*, 41B (1986) 1157.
138 L. Ebdon and H.G.W. Parry, *J. Anal. Atom. Spectrom.*, 2 (1987) 131.
139 F.J. Fernandez, S.A. Meyers and W. Slavin, *Anal. Chem.*, 52 (1980) 741.
140 B. Welz and G. Schlemmer, *J. Anal. Atom. Spectrosc.*, 1 (1986) 119.
141 B. Welz, *Fresenius' Z. Anal. Chem.*, 325 (1986) 95.
142 A. Griaud and C. Fouillac, *Anal. Chim. Acta*, 167 (1985) 257.
143 P. Hocquellet, *Analusis*, 6 (1978) 426.
144 G.M. George, L.J. Frahn and J.P. McDonnell, *J. Assoc. Off. Anal. Chem.*, 65 (1982) 711.
145 J.S. Chia and S.T. Chan, *MARDI Res. Bull.*, 13 (1985) 194.
146 H. Freeman, J.F. Uthe and B. Fleming, *At. Absorpt. Newslett.*, 15 (1976) 49.
147 J.T. Kinnard and M. Gales, Jr., *J. Environ. Sci. Health, Part A*, 16A (1981) 27.
148 D.B. Lo and R.L. Coleman, *At. Absorpt. Newslett.*, 18 (1979) 10.
149 M. Hoenig and P. van Hoeyweghen, *Int. J. Environ. Anal. Chem.*, 24 (1986) 193.
150 K.C. Tam, *Environ. Sci. Technol.*, 8 (1974) 734.
151 F. Puttemans, P. van den Winkel and D.L. Massart, *Anal. Chim. Acta*, 149 (1983) 123.
152 C.L. Ho, S. Tweedy and C. Mahan, *J. Test. Eval.*, 12 (1984) 107.
153 Z. Slovak and B. Docekal, *Anal. Chim. Acta*, 117 (1980) 293.
154 I. Atsuya, K. Itoh, K. Atsuka and K. Jin, *Fresenius' Z. Anal. Chem.*, 326 (1987) 53.
155 R.M. Barnes, *CRC Crit. Rev. Anal. Chem.*, 7 (1978) 203.
156 S.L. Hall and F.M. Fisher, Jr., *Bull Environ. Contam. Toxicol.*, 35 (1985) 1.
157 E. de Oliveira, J.W. McLaren and J.J. Berman, *Anal. Chem.*, 55 (1983) 2047.
158 A. Brzezinska-Paudyn, J. van Loon and R. Hancock, *At. Spectrosc.*, 7 (1986) 72.

159 R.F. Skanieczny, *Arsenic*, in I.M. Kolthoff and P.J. Elving (Eds.), *Treatise of Analytical Chemistry*, John Wiley & Sons, New York, 1978, p. 205.
160 S.J. Sandhu, S.S. Pahil and K.D. Sharma, *Talanta*, 20 (1973) 329.
161 E.A. Woolson, J.H. Axley and P.C. Kearney, *Soil Sci.*, 111 (1971) 158.
162 J.P. Arnold and R.M. Johnson, *Talanta*, 16 (1969) 1191.
163 G.C. Whitnack and R.G. Brophy, *Anal. Chim. Acta*, 48 (1969) 123.
164 D.J. Meyers and J. Osteryoung, *Anal. Chem.*, 45 (1973) 267.
165 D. Chakraborti, R.L. Nichols and K. Irgolic, *Fresenius' Z. Anal. Chem.*, 319 (1984) 248.
166 P.L. Buldini, D. Ferri and Q. Zini, *Mikrochim. Acta [Wien]*, I (1980) 71.
167 M.A. Reed and R.J. Stolzberg, *Anal. Chem.*, 59 (1987) 393.
168 F.T. Henry, T.O. Kirch and T.M. Thorpe, *Anal. Chem.*, 51 (1979) 215.
169 J. Reinke, J.F. Uthe, H.C. Freeman and J.R. Johnston, *Environ. Lett.*, 8 (1975) 371.
170 R. Kannan, T.V. Ramakrishna and S.R. Rajaagopalan, *Talanta*, 32 (1985) 419.
171 G. Forsberg, J.W. O'Laughlin, F.G. Megargle and S.R. Kortyohann, *Anal. Chem.*, 47 (1975) 1586.
172 G. Henze, A.P. Joshi and R. Neeb, *Fresenius' Z. Anal. Chem.*, 300 (1980) 267.
173 M. Rozali Bin Othman, J.O. Hill and R.J. Magee, *J. Eletroanal. Chem.*, 168 (1984) 219.
174 W. Holak, *Anal. Chem.*, 52 (1980) 2189.
175 J. Wang and B. Green, *J. Electroanal. Chem.*, 154 (1983) 261.
176 F.A. Bodewig, P. Valenta and H.W. Nürnberg, *Fresenius' Z. Anal. Chem.*, 311 (1982) 187.
177 C.-W. Whang, J.A. Page and G. van Loon, *Anal. Chem.*, 56 (1984) 539.
178 P.C. Leung, K.S. Subramanian and J.C. Meranger, *Talanta*, 29 (1982) 515.
179 P.H. Davis, G.R. Dulude, R.M. Griffin, W.R. Matson and E.W. Zink, *Anal. Chem.*, 50 (1978) 137.
180 D. Jagner, M. Joserson and S. Westerlund, *Anal. Chem.*, 53 (1981) 2144.
181 Analytical Methods Committee, *Analyst*, 100 (1975) 54.
182 M.G. Haywood and J.P. Riley, *Anal. Chim. Acta*, 85 (1976) 219.
183 W.A. Maher, *Analyst*, 108 (1983) 939.
184 W.H. Buttrill, *J. Assoc. Off. Anal. Chem.*, 56 (1973) 1144.
185 P. Pakalns, *Anal. Chim. Acta*, 47 (1969) 225.
186 P.F. Reay, *Anal. Chim. Acta*, 72 (1974) 145.
187 V.I. Bogdanova, *Mikrochim. Acta [Wien]*, II (1984) 317.
188 R.J. Evans and S.L. Bandemer, *Anal. Chem.*, 26 (1954) 595.
189 G. Eichhorn, W. Wolf and S.A. Berger, *Mikrochim. Acta [Wien]*, I (1976) 135.
190 R.S. Danchik and D.F. Boltz, *Anal. Lett.*, 1 (1968) 901.
191 Y. Yamamoto, T. Kumamari, Y. Hayashi, M. Kanke and A. Matsui, *Talanta*, 19 (1972) 1633.
192 J.F. Tyson and G.W. Stewart, *Anal. Proc. (London)*, 18 (1981) 184.
193 V. Rozenblum, *Anal. Lett.*, 8 (1975) 549.
194 M.J. Pedrosa and J. Paul, *Microchem. J.*, 19 (1974) 314.
195 R.E. Stauffer, *Anal. Chem.*, 55 (1983) 1205.
196 G. Ackerman and J. Köthe, *Fresenius' Z. Anal. Chem.*, 323 (1986) 271.
197 M. Braungart and H. Rüssel, *Chromatographia*, 19 (1984) 185.
198 A. Zeman and J. Prasilova, *Radiochem. Radioanal. Lett.*, 43 (1980) 329.
199 N.D. Thaker and D.N. Patkar, *Radiochem. Radioanal. Lett.*, 48 (1981) 185.
200 *Standard Methods for the Examination of Water and Wastewater*, American Public Health Association, Washington, D.C., 15 edn., 1981, p. 174.
201 M. Basadre-Pampin, A. Alvarez-Devesa, F. Bermejo-Martinez, M.H. Bollain-Rodriquez, *Microchem. J.*, 23 (1978) 360.
202 P.K. Gupta and P.K. Gupta, *Microchem. J.*, 33 (1986) 243.
203 E. Gastinger, *Mikrochim. Acta [Wien]*, (1972) 526.
204 J.F. Kopp, *Anal. Chem.*, 45 (1973) 1786.
205 S.S. Sandhu, *Analyst*, 106 (1981) 311.
206 D. Brune and H. Beltesbrekke, *Swed. Dent. J.*, 4 (1980) 125.

207 K. Ikata, *Nippon Suisan Gakkaishi*, 53 (1987) 1883.
208 R.T. Hunter, *J. Assoc. Off. Anal. Chem.*, 69 (1986) 493.
209 S.A. Peoples, J. Lakso and T. Lais, *Proc. West. Pharmacol. Soc.*, 14 (1971) 178.
210 T.A. Hinners, *Analyst*, 105 (1980) 751.
211 R.K. Skogerboe and A.B. Bejmuk, *Anal. Chim. Acta*, 94 (1977) 297.
212 P.R. Gifford and S. Bruckenstein, *Anal. Chem.*, 52 (1980) 1028.
213 T.H. Ding, Z.Y. Wu, Z.M. Dong, S.P. Li and J.B. Wei, *Sepu*, 5 (1987) 197.
214 K.C. Thompson, *Analyst*, 100 (1975) 307.
215 L. Ebdon and J.R. Wilkinson, *Anal. Chim. Acta*, 194 (1987) 177.
216 L. Ebdon, J.R. Wilkinson and K.W. Jackson, *Anal. Chim. Acta*, 136 (1982) 191.
217 H.R. Griffin, M.B. Hocking and D.G. Lowery, *Anal. Chem.*, 47 (1975) 229.
218 J.R. Liddle, R.R. Brooks and R.D. Reeves, *J. Assoc. Off. Anal. Chem.*, 63 (1980) 1175.
219 F.J. Fernandez and D.C. Manning, *At. Absorpt. Newsl.*, 10 (1971) 86.
220 D.C. Manning, *At. Absorpt. Newslett.*, 10 (1971) 123.
221 A.W.M. Huijbregts, D. Hibbert, R.T. Phillipson, H. Schiwek and G. Steinle, *Zuckerindustrie (Berlin)*, 110 (1985) 797.
222 D.R. Corbin and W.M. Barnard, *At. Absorpt. Newsl.*, 15 (1976) 116.
223 D.R. Webb and D.E. Carter, *J. Anal. Toxicol.*, 8 (1984) 118.
224 J. Guimont, M. Pichette and N. Rheaume, *At. Absorpt. Newsl.*, 16 (1977) 53.
225 S. Terashima, *Anal. Chim. Acta*, 86 (1976) 43.
226 J.A. Fiorino, J.W. Jones and S.G. Capar, *Anal. Chem.*, 48 (1976) 120.
227 J. Aggett and A.C. Aspell, *Analyst*, 101 (1976) 341.
228 H. Weidich and W. Pfannhauser, *Fresenius' Z. Anal. Chem.*, 276 (1975) 61.
229 S. Peats, *At. Absorpt. Newslett.*, 18 (1979) 118.
230 B. Welz and M. Melcher, *Analyst*, 109 (1984) 573.
231 H.W. Sinemus, M. Melcher and B. Welz, *At. Spectrosc.*, 2 (1981) 81.
232 R.D. Wauchope, *At. Absorpt. Newslett.*, 15 (1976) 64.
233 D.D. Siemer, R.K. Vitek, R. Kateel and W.C. Houser, *Anal. Lett.*, 10 (1977) 357.
234 H. Agemian and V. Cheam, *Anal. Chim. Acta*, 101 (1978) 193.
235 M. Yamamoto, Y. Yamamoto and T. Yamashige, *Analyst*, 109 (1984) 1461.
236 B.T. Sturman, *Appl. Spectrosc.*, 39 (1985) 48.
237 J.F. Chapman and L.S. Dale, *Anal. Chim. Acta*, 111 (1979) 137.
238 H. Narasaki, *Fresenius' Z. Anal. Chem.*, 321 (1985) 464.
239 S. McCabe and J.M. Ottaway, *Anal. Proc. (London)*, 23 (1986) 16.
240 J.C. Van Loon and E.J. Brooker, *Anal. Lett.*, 7 (1974) 505.
241 F.J. Schmidt, J.L. Royer and S.M. Muir, *Anal. Lett.*, 8 (1975) 123.
242 D.E. Fleming and G.A. Taylor, *Analyst*, 103 (1978) 101.
243 N.E. Parisis and A. Heyndrickx, *Analyst*, 111 (1986) 281.
244 K. Diettrich and R. Mandry, *Analyst*, 111 (1986) 269.
245 R.H. Ward and P.B. Stockwell, *J. Autom. Chem.*, 5 (1983) 193.
246 M. Yamamoto, M. Yasuda and Y. Yamamoto, *Anal. Chem.*, 57 (1985) 1382.
247 H. Agemian and R. Thomson, *Analyst*, 105 (1980) 902.
248 M.H. Arbab-Zavar and A.G. Howard, *Analyst*, 105 (1980) 744.
249 H. Narasaki and M. Ikeda, *Anal. Chem.*, 56 (1984) 2059.
250 M. Fishman and R. Spencer, *Anal. Chem.*, 49 (1977) 1599.
251 B.J.A. Haring, W. van Delft and C.M. Bom, *Fresenius' Z. Anal. Chem.*, 310 (1982) 217.
252 M. Stoeppler, M. Burow, F. Backhaus, W. Schramm and H.W. Nürnberg, *Marine Chem.*, 18 (1986) 321.
253 P.N. Vijan, A.C. Rayner, D. Sturgis and G.R. Wood, *Anal. Chim. Acta*, 82 (1976) 329.
254 R.G. Smith, J.C. Van Loon, J.R. Knechtel, J.L. Fraser, A.E. Pitts and A.E. Hodges, *Anal. Chim. Acta*, 93 (1977) 61.
255 H.B. MacPherson and S.S. Berman, *Can J. Spectrosc.*, 28 (1983) 131.
256 J. Aggett and A.C. Aspell, *Environ. Pollution, Ser. A*, 22 (1980) 39.
257 W.A. Maher, *Comp. Biochem. Physiol.*, 82C (1985) 433.
258 P.J. Brooke and W.H. Evans, *Analyst*, 106 (1981) 514.

259 S.E. Raptis, W. Wegscheider and G. Knapp, *Mikrochim. Acta [Wien]*, I (1981) 93.
260 W.H. Evans, F.J. Jackson and D. Dellar, *Analyst*, 104 (1979) 16.
261 B. Welz and M. Melcher, *Anal. Chem.*, 57 (1987) 427.
262 W.A. Maher, *Talanta*, 29 (1982) 532.
263 S. Clark, J. Ashby and P.J. Craig, *Analyst*, 112 (1987) 1781.
264 W.A. Maheer, *Comp. Biochem. Physiol.*, 80C (1985) 199.
265 G. Drasch, L.V. Meyer and G. Kouert, *Fresenius' Z. Anal. Chem.*, 304 (1980) 141.
266 R.E. Sturgeon, S.N. Willie and S.S. Berman, *J. Anal. At. Spectrom.*, 1 (1986) 115.
267 T. Inui, S. Terada and H. Tamura, *Fresenius' Z. Anal. Chem.*, 305 (1981) 189.
268 E.O. Uthus, M.E. Collings, W.E. Cornatzer and F.H. Nielsen, *Anal. Chem.*, 53 (1981) 2221.
269 A.U. Shaikh and D.E. Tallman, *Anal. Chim. Acta*, 98 (1978) 251.
270 A.U. Shaikh and D.E. Tallman, *Anal. Chem.*, 49 (1977) 1093.
271 J.A. Hagen and R.J. Lovett, *At. Spectrosc.*, 7 (1986) 69.
272 K.J. Mulligan, M.H. Hahn and J.A. Caruso, *Anal. Chem.*, 5 (1979) 1935.
273 W.B. Robbins, J.A. Caruso and F.L. Fricke, *Analyst*, 104 (1979) 35.
274 F.L. Fricke, W.B. Robbins and J.A. Caruso, *J. Assoc. Off. Anal. Chem.*, 60 (1978) 1118.
275 W.B. Robbins and J.A. Caruso, *J. Chromatogr. Sci.*, 17 (1979) 360.
276 N.W. Barnett, L.S. Chen and G.F. Kirkbright, *Spectrochim. Acta*, 39B (1984) 1141.
277 F.E. Lichte and R.K. Skogerboe, *Anal. Chem.*, 44 (1972) 1480.
278 I. Atsuya and K. Akatsuka, *Spectrochim. Acta*, 36B (1981) 747.
279 R. Belcher, S.L. Bogdanski, S.A. Ghonaim and A. Townshend, *Anal. Chim. Acta*, 72 (1974) 183.
280 R. Belcher, S.L. Bogdanski, E. Henden and A. Townshend, *Anal. Chim. Acta*, 92 (1977) 33.
281 R. Belcher, S.L. Bogdanski, E. Henden and A. Townshend, *Analyst*, 100 (1975) 522.
282 E. Henden, *Anal. Chim. Acta*, 173 (1985) 89.
283 M. Burguera and J.L. Burguera, *Analyst*, 111 (1986) 171.
284 A. Miyazaki, A. Kimura and Y. Umezaki, *Anal. Chim. Acta*, 90 (1977) 119.
285 K.W. Panaro and I.S. Krull, *Anal. Lett.*, 17 (1984) 157.
286 L. Ebdon and S.T. Sparkes, *Microchem. J.*, 36 (1987) 198.
287 R.S. Braman, L.J. Justen and C.C. Foreback, *Anal. Chem.*, 44 (1972) 2195.
288 C. Feldman, *Anal. Chem.*, 51 (1979) 664.
289 P.J. Clark, R.A. Zingaro, K.J. Irgolic and A.N. McGinley, *Intern. J. Environ. Anal. Chem.*, 7 (1980) 295.
290 M. Thompson, B. Pahlavanpour, S.J. Walton and G.F. Kirkbright, *Analyst*, 103 (1978) 568.
291 J.A.C. Broekaert and F. Leis, *Fresenius' Z. Anal. Chem.*, 300 (1980) 22.
292 B. Paahlavanpour, M. Thompson and L. Thorne, *Analyst*, 105 (1980) 756.
293 B. Pahlavanpour, M. Thompson and L. Thorne, *Analyst*, 106 (1981) 467.
294 C.J. Pickford, *Analyst*, 106 (1981) 464.
295 M. Thompson, B. Pahlavanpour and L.T. Thorne, *Water Res.*, 15 (1981) 407.
296 T. Nakahara, *Anal. Chim. Acta*, 131 (1981) 73.
297 M.H. Hahn, K.A. Wolnik, F.L. Fricke and J.A. Caruso, *Anal. Chem.*, 54 (1982) 1048.
298 R.R. Liversage, J.C. Van Loon and J.C. De Andrade, *Anal. Chim. Acta*, 161 (1984) 275.
299 E. Pruszkowska, P. Barrett, R. Ediger and G. Wallace, *At. Spectrosc.*, 4 (1983) 94.
300 L.R. Parker, Jr., N.H. Tioh and R.M. Barnes, *Appl. Spectrosc.*, 39 (1985) 45.
301 G.S. Pyen, S. Long and R.F. Browner, *Appl. Spectrosc.*, 40 (1986) 246.
302 R.C. Fry, M.B. Denton, D.L. Windsor and S.J. Northway, *Appl. Spectrosc.*, 33 (1979) 399.
303 D.D. Nygaard and J.H. Lowry, *Anal. Chem.*, 54 (1982) 803.
304 P.D. Goulden, D.H.J. Anthony and K.D. Austen, *Anal. Chem.*, 53 (1981) 2027.
305 C.K. Lee and K.S. Low, *Pertanika*, 10 (1987) 69.
306 A. Brzezinska-Paudyn, J. Van Loon and R. Hancock, *At. Spectrosc.*, 7 (1986) 72.
307 R.C. Hutton and B. Preston, *Analyst*, 108 (1983) 1409.

308 M.J. Powell, D.W. Boomer and R.J. McVicars, *Anal. Chem.*, 58 (1986) 2867.
309 H.K. Kang and J.L. Valentine, *Anal. Chem.*, 49 (1977) 1829.
310 T. Maruta and G. Sudoh, *Anal. Chim. Acta*, 77 (1975) 37.
311 K. Petrick and V. Krivan, *Anal. Chem.*, 59 (1987) 2476.
312 R.K. Anderson, M. Thompson and E. Culbard, *Analyst*, 111 (1986) 1143.
313 R.M. Brown, Jr., R.C. Fry, J.L. Moyers, S.J. Northway, M.B. Denton and G.S. Wilson, *Anal. Chem.*, 53 (1981) 1560.
314 F.D. Pierce and H.R. Brown, *Anal. Chem.*, 48 (1976) 693.
315 S.S. Sandhu and P. Nelson, *Anal. Chem.*, 50 (1978) 322.
316 G.F. Kirkbright and M. Taddia, *Anal. Chim. Acta*, 100 (1978) 145.
317 M. Mausbach, *GIT Fachz. Lab.*, 23 (1979) 898.
318 C.J. Peacock and S.C. Singh, *Analyst*, 106 (1981) 931.
319 M. Thompson, B. Pahlavanpour, S.J. Walton and G.F. Kirkbright, *Analyst*, 103 (1978) 705.
320 K. Patrick and V. Krivan, *Fresenius' Z. Anal. Chem.*, 327 (1987) 338.
321 B. Welz and M. Melcher, *Anal. Chim. Acta*, 131 (1981) 17.
322 B. Welz and M. Melcher, *Analyst*, 108 (1983) 213.
323 K.J. Irgolic, *Speciation of Arsenic Compounds in Water Supplies*, Report 1982, EPA/600/1-82/010, Order No. PB82-257817, NTIS, 125 pp., p. 7, *Chem. Abstr.*, 98 (1983) 113378.
324 A.A. Al-Sibaai and A.G. Fogg, *Analyst*, 98 (1973) 732.
325 J.E. Portmann and J.P. Riley, *Anal. Chim. Acta*, 31 (1964) 509.
326 S.H. Harrison, P.D. LaFleur and W.H. Zoller, *Anal. Chem.*, 47 (1975) 1685.
327 M.O. Andreae, *Anal. Chem.*, 49 (1977) 820.
328 M.O. Andreae, *Deep-Sea Res.*, 25 (1978) 391.
329 M.O. Andreae, *Limnol. Oceanogr.*, 24 (1979) 440.
330 J.A. Cherry, A.U. Shaikh, D.E. Tallman and R.V. Nicholson, *J. Hydrol.*, 43 (1979) 373.
331 D.E. Tallman and A.U. Shaikh, *Anal. Chem.*, 52 (1980) 199.
332 B.I. Diamondstone and R.W. Burke, *Analyst*, 102 (1977) 613.
333 P. Weigert and A. Sappl, *Fresenius' Z. Anal. Chem.*, 316 (1983) 306.
334 J. Aggett and M.R. Kriegman, *Analyst*, 112 (1987) 153.
335 F.T. Henry and T.M. Thorpe, *Anal. Chem.*, 52 (1980) 80.
336 J. Stary, A. Zeman, K. Kratzer and J. Prasilova, *Intern. J. Environ. Anal. Chem.*, 8 (1980) 49.
337 T. Kamada, *Talanta*, 23 (1976) 835.
338 C.H. Chung, E. Iwamoto, M. Yamamoto and Y. Yamamoto, *Spectrochim. Acta*, 39B (1984) 459.
339 D. Chakraborti, W. deJonghe and F. Adams, *Anala. Chim. Acta*, 120 (1980) 121.
340 D. Chakraborti, K.J. Irgolic and F. Adams, *J. Assoc. Off. Anal. Chem.*, 67 (1984) 277.
341 D. Chakraborti, F. Adams and K.J. Irgolic, *Fresenius' Z. Anal. Chem.*, 323 (1985) 340.
342 S.A. Amankwah and J.L. Fasching, *Talanta*, 32 (1985) 111.
343 Y. Talmi and V.E. Norvell, *Anal. Chem.*, 47 (1975) 1510.
344 K. Terada, K. Matsumoto and T. Inaba, *Anal. Chim. Acta*, 158 (1984) 207.
345 S.S. Sandhu, *Analyst*, 101 (1976) 856.
346 D.L. Johnson and M.E.Q. Pilson, *Anal. Chim. Acta*, 58 (1972) 289.
347 D.L. Johnson, *Environ. Sci. Technol.*, 5 (1971) 411.
348 P. Linares, M.D. Luque de Castro and M. Valcarcel, *Anal. Chem.*, 58 (1986) 120.
349 E. Orvini, M. Gallorini and M. Spezali, *NATO Conf. Ser.*, [Ser.] 1, 1983, 6 *(Trace Element Speciation Surf. Waters Its Ecol. Implic.)* 241.
350 M. Yamamoto, K. Urata, K. Murashige and Y. Yamamoto, *Spectrochim. Acta*, 36B (1981) 671.
351 A.G. Howard and M.H. Arbab-Zavar, *Analyst*, 105 (1980) 338.
352 S. Nakashima, *Analyst*, 104 (1979) 172.
353 W. Vycudilik, *Arch. Toxicol.*, 36 (1976) 177.
354 W.H. Clement and S.D. Faust, *Environ. Lett.*, 5 (1973) 155.
355 S. Nakashima, *Analyst*, 103 (1978) 1031.

348

356 T.B. Hoover and G.D. Yager, *Anal. Chem.*, 56 (1984) 221.
357 Yu.Z. Zolotov, O.A. Shpigun and L.A. Bubchikova, *Fresenius' Z. Anal. Chem.*, 316 (1983) 8.
358 L.D. Hansen, B.E. Richter, D.K. Rollins, J.D. Lamb and D.J. Eatough, *Anal. Chem.*, 51 (1979) 633.
359 R. Steiber and R. Merrill, *Anal. Lett.*, 12 (1979) 273.
360 D. Clifford, L. Ceber and S. Chow, *Proc.-AWWA Water Qual. Technol. Conf.*, 1984, 11th 223; *Chem. Abstr.*, 102 (1985) 100494.
361 W.H. Ficklin, *Talanta*, 30 (1983) 371.
362 M.Q. Yu, G.Q. Liu and Q. Jin, *Talanta*, 30 (1983) 265.
363 W. Nisamaneepong, M. Ibrahim, T.W. Gilbert and J.A. Caruso, *J. Chromatogr. Sci.*, 22 (1984) 473.
364 J.P. McCarthy, J.A. Caruso and F.L. Fricke, *J. Chromatogr. Sci.*, 21 (1983) 389.
365 F.A. Austenfeld and R.L. Berghoff, *Plant Soil*, 64 (1982) 267.
366 D.G. Iverson, M.A. Anderson, T.R. Holm and R.R. Stanforth, *Environ. Sci. Technol.*, 13 (1979) 1491.
367 M. Yamamoto, *Soil Sci. Soc. Am. Proc.*, 39 (1975) 859.
368 R.D. Wauchope and M.Yamamoto, *J. Environ. Qual.*, 9 (1980) 597.
369 W.A. Maher, *Anal. Chim. Acta*, 126 (1981) 157.
370 C.T. Tye, S.J. Haswell, P. O'Neill and K.C. Bancroft, *Anal. Chim. Acta*, 169 (1985) 195.
371 S.J. Haswell, P. O'Neill and K.C. Bancroft, *Talanta*, 32 (1985) 69.
372 T. Takamatsu, R. Nakata, T. Yoshida and M. Kawashima, *Jpn. J. Limnol.*, 46 (1985) 93.
373 T. Takamatsu, H. Aoki and T. Yoshida, *Soil Sci.*, 133 (1982) 239.
374 T. Kaise, K. Hanaoka and S. Tagawa, *Chemosphere*, 116 (1987) 2551.
375 A.A. Grabinski, *Anal. Chem.*, 53 (1981) 966.
376 K.H. Tam, S.M. Charbonneau, F. Bryce and G. Lacroix, *Anal. Biochem.*, 86 (1978) 505.
377 G.K.H. Tam, S.M. Charbonneau, G. Lacroix and F. Bryce, *Bull. Environ. Contam. Toxicol.*, 22 (1979) 69.
378 E. Marafante, M. Vahter, H. Norin, J. Evall, M. Sandström, A. Christakopoulos and R. Ryhage, *J. Appl. Toxicol.*, 7 (1987) 111.
379 R.A. Pyles and E.A. Woolson, *J. Agr. Food Chem.*, 30 (1982) 866.
380 E.A. Woolson, Pestic Chem.: Human Welfare Environ., *Proc. Int. Congr., Pestic Chem.*, 5th 1982 (Pub. 1983), Pergamon Press, 4 (1983) 79.
381 E.A. Woolson and N. Aharonson, *J. Assoc. Off. Anal. Chem.*, 63 (1980) 523.
382 G.R. Ricci, L.S. Sheppard, G. Colovos and N.E. Hester, *Anal. Chem.*, 53 (1981) 610.
383 J. Aggett and R. Kadwani, *Analyst*, 108 (1983) 1495.
384 F.E. Brinckman, K.L. Jewett, W.P. Iverson, K.J. Irgolic, K.C. Ehrhardt and R.A. Stockton, *J. Chromatogr.*, 191 (1980) 31.
385 K.J. Irgolic, R.A. Stockton, D. Chakraborti and W. Beyer, *Spectrochim. Acta*, 38B (1983) 437.
386 F.E. Brinckman, W.R. Blair, K.L. Jewett and W.R. Iverson, *J. Chromatogr. Sci.*, 15 (1977) 493.
387 S.J. Haswell, R.A. Stockton, K.C.C. Bancroft, P. O'Neill, A. Rahman and K.J. Irgolic, *J. Autom. Chem.*, 9 (1987) 6.
388 R.S. Braman and C.C. Foreback, *Science*, 182 (1973) 1247.
389 D.L. Johnson and R.S. Braman, *Chemosphere*, 6 (1975) 333.
390 R.S. Braman, D.L. Johnson, C.C. Foreback, J.M. Ammon and J.L. Bricker, *Anal. Chem.*, 49 (1977) 621.
391 W.A. Maher, *Spectrosc. Lett.*, 16 (1983) 865.
392 E.A. Crecelius, *Anal. Chem.*, 50 (1978) 826.
393 Y. Odanaka, N. Tsuchiya, O. Matano and S. Goto, *Anal. Chem.*, 55 (1983) 929.
394 M.S. Mohan, R.A. Zingaro, P. Micks and P.J. Clark, *Intern. J. Environ. Anal. Chem.*, 11 (1982) 175.

395 L. Ebdon, A.P. Walton, G.E. Millward and M. Whitfield, *J. Appl. Organomet. Chem.*, 1 (1987) 427.
396 Y. Talmi and D.T. Bostick, *Anal. Chem.*, 47 (1975) 2145.
397 J.S. Edmonds and K.A. Francesconi, *Anal. Chem.*, 48 (1976) 2019.
398 A.G. Howard and M.H. Arbab-Zavar, *Analyst*, 106 (1981) 213.
399 H. Yamauchi and Y. Yamamura, *Toxicol. Appl. Pharmacol.*, 74 (1984) 134.
400 J.M. Brannon and W.H. Patrick, Jr., *Environ. Sci. Technol.*, 21 (1987) 450.
401 J. Jaganathan, M.S. Mohan, R.A. Zingaro and K.J. Irgolic, *J. Trace Microprobe Techn.*, 3 (1985) 345.
402 F.T. Henry and T.M. Thorpe, *J. Chromatogr.*, 166 (1978) 577.
403 S.J. Sederquist, D.G. Crosby and J.B. Bowers, *Anal. Chem.*, 46 (1974) 155.
404 B. Beckermann, *Anal. Chim. Acta*, 135 (1982) 77.
405 S. Fukui, T. Hirayama, M. Nohara and Y. Sakagami, *Talanta*, 30 (1983) 89.
406 E.H. Daugherty, Jr., A.W. Fitchett and P. Mushak, *Anal. Chim. Acta*, 79 (1975) 199.
407 H. Münz and W. Lorenzen, *Fresenius' Z. Anal. Chem.*, 319 (1984) 395.
408 A. Yasui, C. Tsutsumi and S. Toda, *Agric. Biol. Chem.*, 42 (1978) 2139.
409 A. Shinagawa, K. Shiomi, H. Yamanaka and T. Kikuchi, *Bull. Jpn. Soc. Sci. Fish.*, 49 (1983) 75.
410 J.N.C. Whyte and J.R. Englar, *Bot. Marina*, 26 (1983) 159.
411 A.W. Fitchett, E.H. Daughtrey, Jr. and P. Mushak, *Anal. Chim. Acta*, 79 (1975) 93.
412 Y.K. Chau, W.J. Snodgrass and P.T.S. Wong, *Water Res.*, 11 (1977) 807.
413 E.A. Woolson, *Weed Sci.*, 25 (1977) 412.
414 Y.K. Chau and P.T.S. Wong, in *Environmental Analysis*, Academic Press, New York, 1977, p. 215.
415 K.B. Olsen, D.S. Sklarew and J.C. Evans, *Spectrochim. Acta*, 40B (1985) 357.
416 G.E. Parris, W.R. Blair and F.E. Brinckman, *Anal. Chem.*, 49 (1977) 378.
417 K. Shiomi, Y. Kakehaashi, H. Yamanaka and T. Kikuchi, *J. Appl. Organomet. Chem.*, 1 (1987) 177.
418 K. Shiomi, Y. Horiguchi and T. Kaise, *J. Appl. Organomet. Chem.*, 2 (1988) 385.
419 T. Kaise, S. Watanabe, K. Ito, K. Hanoaka, S. Tagawa, T. Hirayama and S. Fukui, *Chemosphere*, 16 (1987) 91.
420 K. Hanaoka, H. Kobayashi, S. Tagawa and T. Kaise, *Comp. Biochem. Physiol.*, 88C (1987) 189.
421 J.B. Luten, G. Riekwel-Booy, J.V.d. Greef and M.C. ten Noever de Brauw, *Chemosphere*, 12 (1983) 131.
422 K. Hanaoka, H. Yamamoato, K. Kawashima, S. Tagawa and T. Kaise, *J. Appl. Organomet. Chem.*, 2 (1988) 371.
423 J.S. Edmonds, K. Francesconi, J.R. Cannon, C.L. Raston, B.W. Shelton and A.H. White, *Tetrahedron Lett.*, (1977) 1543.
424 K. Shiomi, A. Shinagawa, H. Yamanaka and T. Kikuchi, *Bull. Jpn. Soc. Sci. Fish.*, 49 (1983) 79.
425 J.R. Cannon, J.S. Edmonds, K.A. Francesconi, C.L. Raston, J.B. Saunders, B.W. Skelton and A.H. White, *Aust. J. Chem.*, 34 (1981) 787.
426 K. Shiomi, A. Shinagawa, T. Igarashi, H. Yamanaka and T. Kikuchi, *Experientia*, 40 (1984) 1247.
427 K.A. Francesconi, P. Micks, R.A. Stockton and K.J. Irgolic, *Chemosphere*, 14 (1985) 1443.
428 S. Matsuto, R.A. Stockton and K.J. Irgolic, *Sci. Total Environ.*, 48 (1986) 133.
429 K. Shiomi, A Shinagawa, K. Hirata, H. Yamanaka and T. Kikuchi, *Agric. Biol. Chem.*, 48 (1984) 2863.
430 H. Norin, R. Ryhage, A. Christakopoulos and M. Sandström, *Chemosphere*, 12 (1983) 299.
431 K. Shiomi, A. Shinagawa, M. Azuma, H. Yamanaka and T. Kikuchi, *Comp. Biochem. Physiol.*, 74C (1983) 393.
432 H. Norin and A. Christakopoulos, *Chemosphere*, 11 (1982) 287.
433 K. Hanaoka and S. Tagawa, *Bull. Jpn. Soc. Sci. Fish.*, 51 (1985) 681.
434 J.S. Edmonds and K.A. Francesconi, *Chemosphere*, 10 (1981) 1041.

435 J.R. Cannon, J.B. Saunders and R.F. Toia, *Sci. Total Environ.*, 31 (1983) 181.
436 K. Hanaoka, T. Fujita, M. Matsuura, S. Tagawa and T. Kaise, *Comp. Biochem. Physiol.*, 86B (1987) 681.
437 K. Hanaoka, H. Matsuda, T. Kaise and S. Togawa, *J. Shimonoseki Univ. Fish.*, 35 (1986) 37.
438 J.B. Luten, G. Riekwel-Booy and A. Rauchbaar, *Environ. Healtah Perspect.*, 45 (1982) 165.
439 R.A. Stockton and K.J. Irgolic, *Intern. J. Environ. Anal. Chem.*, 6 (1979) 313.
440 M. Morita, T. Uehiro and K. Fuwa, *Anal. Chem.*, 53 (1981) 1806.
441 S. Kurosowa, K. Yasuda, M. Taguchi, S. Yamazaki, S. Toda, M. Morita, T. Uehiro and K. Fuwa, *Agric. Biol. Chem.*, 44 (1980) 1993.
442 T. Kaise, H. Yamauchi, T. Hirayama and S. Fukui, *J. Appl. Organomet. Chem.*, 2 (1988) in press.
443 Y. Shibata, M. Morita and J.S. Edmonds, *Agric. Biol. Chem.*, 51 (1987) 391.
444 K. Jin, T. Hayashi, Y. Shibata and M. Morita, *J. Appl. Organomet. Chem.*, 2 (1988) 365.
445 *NBS Standard Reference Materials, Catalog 1990-91, NBS Publication 260*, U.S. Department of Commerce, National Bureau of Standards; Gaithersburg, MD.

M. Stoeppler (Editor)/*Hazardous Metals in the Environment*
© 1992 Elsevier Science Publishers B.V. All rights reserved

Chapter 12

Thallium

Manfred Sager

Geotechnical Institute, BVFA-Arsenal, A-1030 Wien (Austria)

CONTENTS

12.1. Introduction 352
12.2. Occurrence of thallium 352
    12.2.1. Minerals and rocks 352
    12.2.2. Biological samples 353
12.3. Sample decomposition methods 354
    12.3.1. General dissolution procedures 355
    12.3.2. Special combined decomposition/separation techniques 356
        12.3.2.1. Volatilization in a stream of carrier gas 356
        12.3.2.2. Volatilization without carrier gas stream 356
    12.3.3. Selective dissolution 357
12.4. Separation methods 357
    12.4.1. Co-precipitation 357
    12.4.2. Solvent extraction 358
        12.4.2.1. Solvent extraction of monovalent thallium 358
        12.4.2.2. Thallium(III) halides and reagent solvents 359
        12.4.2.3. Thallium(III) halides extracted with reagents into inert solvents 359
        12.4.2.4. Solvent extraction of thallium(III) without halides 359
    12.4.3. Non-chromatographic enrichment on solids packed in columns . . . 359
        12.4.3.1. Inorganic sorbents 360
        12.4.3.2. Active carbon 360
        12.4.3.3. Cation-exchange resins 360
        12.4.3.4. Anion-exchange resins 360
        12.4.3.5. Liquid–liquid column methods 361
    12.4.4. Ion chromatography 361
12.5. Final determination methods 361
    12.5.1. Atomic emission spectrometry (arc and spark methods) 361
    12.5.2. Inductively coupled plasma–atomic emission spectrometry (ICP–AES) 362
    12.5.3. Microwave-induced plasma 363
    12.5.4. Mass spectrometry 363
    12.5.5. ICP with mass spectrometric detection (ICP–MS) 364
    12.5.6. Atomic absorption spectrometry (AAS) 364
        12.5.6.1. Flame AAS 364
        12.5.6.2. Graphite furnace AAS 365
    12.5.7. Electroanalytical determinations 366
        12.5.7.1. Hanging mercury drop electrode 367
        12.5.7.2. Mercury thin film electrode 368
    12.5.8. Spectrophotometric and fluorometric methods 368
References 369

12.1. INTRODUCTION

For the practical use of inorganic and environmental analytical chemistry by non-specialists it is necessary to have a rapid overview of the 'state of the art' of currently available methods. In this essay, rather common methods are treated, which are applicable to the analysis of thallium. Much is also known about the toxicology of this element, something about its occurrence, and much less about its ecology. The technological importance of thallium remains low. The literature up to 1983, including physical constants, organic reagent formulas, and technical uses, has been extensively reviewed in a booklet [1].

Less common analytical methods, or those which are not suitable for the economical analysis of thallium, are omitted. These include atomic fluorescence, solid-sampling atomic absorption, X-ray fluorescence, activation analysis, gas chromatography, electrolytic deposition, and ion-sensitive electrodes. On the other hand, the rapid development of analytical devices for atomic emission and mass spectrometry provoked many interesting papers, giving both methods and analytical figures from the environment, which justify a new survey.

For users who are not familiar with every analytical technique, short introductory explanations and recommendations are given. A table of certified standard reference materials is added to prove the own results (Table 12.2).

12.2. OCCURRENCE OF THALLIUM

12.2.1. Minerals and rocks

In silicates, isomorphism of anhydrous monovalent Tl in the lattice with anhydrous Rb is of great importance. In primary silicates, the ratio Rb:Tl is about 300:1. This also leads to geochemical affinities to sites of K. The Tl content is clearly higher in acid rocks than in basic ones. It is also low in soils, sediments, and materials of sedimentary origin, especially in carbonates, but also in clay minerals if they are not contaminated from ore processing or flue dusts (Tables 12.1 and 12.2).

In sulfides, Tl occurs at levels above its average abundance in the earth's crust. It often accompanies Ag and Pb, but is also found in deposits of Zn, Cd, As, Sb and Hg. The major part of Tl in the earth's crust is bound to pyrite. True Tl minerals are rarely found.

Because Tl is volatile, both under reducing and oxidizing conditions, in the elementary form or as the chloride, nitrate, or oxide, it is enriched in the flue dust of smelters, cement plants, brickworks etc. Similarly, Tl from coal is probably volatilized during combustion; crude oil contains lower amounts of Tl. This can lead to the contamination of top-soil layers due to atmospheric deposition, and thus to elevated levels in plants growing on it.

TABLE 12.1.

RANGE OF OCCURRENCE OF THALLIUM (NON-ANTHROPOGENICALLY POLLUTED AREAS

Material	Location		Range of Tl (μg/g)	Ref.
Granite	South West Africa	Precambrian	0.35 – 0.74	2
		Cambrian	0.65 – 1.61	2
		Mesozoic	0.60 – 3.60	2
	Cornwall	Biotitic	2.7 – 4.0	3
	Baikal massive	Batholitic	1.1 – 1.4	4
Basalt	West Germany		0.021 – 0.058	5
Limestone	West Germany		<0.005 – 0.19	6
Gypsum	West Germany		0.04 – 0.13	6
Iron ores			0.006 – 1.9	6
Surface sediment of the Pacific Ocean:				7
	volcanic and biogenic terrigenous		0.27 – 0.7	
	pelagic red clays		0.95 – 1.25	
Deep sea nodules	Atlantic ocean		87 – 118	8
	Pacific ocean		23 – 226	9
Coals used in Austria		pit coals	>0.1 – 1.5	10
		brown coals	>0.1 – 0.5	
		coal fly ash	>0.1 – 0.7	
		coal slags	>0.1 – 3.0	
Fuel oil			0.02 – 0.10	6
Soils	West Germany		0.17 – 0.53	11
Fertilizers	West Germany		0.02 – 0.57	11
Plants for animal feeding			0.02 – 0.08	12
Vegetables (edible parts)	West Germany		0.03 – 0.20	12
Red wines	West Germany		0.09 – 0.48	11
White wines	West Germany		0.04 – 0.3	11
Milk	West Germany		0.010 – 0.033	12

12.2.2. Biological samples

In living biota, Tl is often transported and treated like K. It interferes with Na/K-regulated functions. Analytical procedures have now reached the low levels at which Tl occurs in biological materials, and many more data can be expected (Tables 12.1 and 12.2).

References on p. 369

TABLE 12.2.

STANDARD REFERENCE MATERIALS [13] (Recommended values only)

Tl (μg/g)	Material	Tl (μg/g)	Material	
5.7	Fly ash NBS-1633a	1.0	Granodiorite JG-1	
4.0	Fly ash NBS-1633	0.94	Soil GSS-4	
2.9	Sediment GSD-11	0.93	Stream sediment GSD-7	
2.4	Soil GSS-6	0.78	Stream sediment GSD-8	
1.93	Granite GSR-1	0.71	Shale GSR-5	
1.9	Stream sediment GSD-2	0.62	Soil GSS-2	
1.8	Sediment GSD-12	0.61	Stream sediment GSD-1	
1.6	Soil GSS-5	0.59	Soil GSS-8	
1.44	River sediment NBS-1645	0.58	Stream sediment GSD-3	
1.43	Granodiorite GSP-1	0.54	Obsidian NBS-278	
1.23	Granite G-1	0.5	Estuarine Sediment NBS-1646	
1.2	Stream sediment GSD-4	0.49	Sediment GSD-9	
1.2	River Sediment NBS-2704	0.48	Soil GSS-3	
1.16	Stream sediment GSD-5	0.36	Sandstone GSR-4	
1.08	Stream sediment GSD-6	0.3	Basalt BCR-1	
1.0	Soil GSS-1	0.21	Sediment GSD-10	
		0.114	Diabase W-1	
		0.248	Mussels II	
		0.0272	Alga II	Environ-
		0.0203	Earthworm	mental
		0.0143	Alga I	Specimen
		0.0076	Poplar leaves	Bank [14]
		0.0029	Pig liver	

12.3. SAMPLE DECOMPOSITION METHODS

For the trace analysis of thallium at non-elevated levels there are hardly any methods which enable accurate determinations from the original solids or liquids, without further treatment.

Among the methods applied directly to solids, only arc-excitation for atomic emission spectrography is sensitive enough to yield analytical values from natural samples such as granitic rocks. Direct neutron activation is not possible because [204]Tl, a pure β-emitter, is formed. Spark-source mass spectrometry, and solid-sampling furnace AAS, require very low weights of sample and so require high sample homogeneity during sieving, grinding and milling: this is a lot of effort for series of routine analyses. Finally, X-ray fluorescence studies of the original samples is either too insensitive or, in the case of ore analysis is uncertain, because of the lack of standard materials containing high amounts of Tl.

Most methods for separation and determination start from a completely mineralized sample solution. Problems of sample homogeneity are avoided if relatively large amounts, e.g. half a gram, are dissolved.

12.3.1. General dissolution procedures

The general aim of a sample decomposition method is the destruction of organic compounds and the dissolution of inorganic solids, without volatilization losses. The resultant solution should be suitable for multi-element analysis. When heating the samples, the volatility of Tl compounds has to be considered. Many compounds of trivalent Tl decompose at 400–500°C (Table 12.3). Residual organic compounds interfere, especially in photometric and electroanalytical methods (see below).

TABLE 12.3.

BOILING POINTS (°C) OF INORGANIC THALLIUM COMPOUNDS

Tl-metal	1457	
TlNO$_3$	430	
TlF	655	
TlCl	720	
TlBr	815	
Tl$_2$O	1080	
Tl$_2$O$_3$	875	(in oxygen; see ref. 15)

The usual methods of wet decomposition with acid mixtures can be applied using, for example, HNO$_3$/HF for rock and soil samples [16–19], aqua regia, and also HNO$_3$/H$_2$O$_2$ for biological matrices [20]. The procedure can be accelerated with microwaves. Sulfide concentrates are easily decomposed with Br$_2$/HCl [21]. However, fuming of cement samples with HF in open vessels leads to volatilization losses above 140°C, and with HBr above 160° [22].

To ensure complete destruction of organics from sediments and soils, decomposition in a pressure bomb with HNO$_3$ or HNO$_3$/HF is preferred. Coal samples are very resistant and need at least 14h at 140°C in the pressure bomb [23]; higher temperatures cannot be applied because of too much pressure inside the bomb. As an alternative to wet oxidation of organics, low-temperature ashing in an oxygen plasma is possible: the ash can be dissolved in nitric acid [14].

Alkaline fluxes may suffer from problems with volatilization and also with Tl blanks from the K_2CO_3, and are thus rarely used.

For subsequent determinations by AAS, enzymatic solubilisation of tissues in alkaline solutions is sufficient — for example, by the proteolytic enzyme subtilisin in the presence of trizma base [2-amino-2-hydroxymethylpropane(1,3)diol] at 55°C [24].

12.3.2. Special combined decomposition/separation techniques

The volatility of the elements or of their oxides or chlorides can be used to allow the evaporation of Tl and some other trace elements selectively from solid samples. Starting with up to 1 g of solid, the condensed traces can be dissolved in 1 ml or less. As well as the separation from many matrix elements, this means an appreciable enrichment compared to conventional general decomposition. Much more time per sample is needed, however, and the procedure can only be used for the determination of a limited number of trace elements.

12.3.2.1. Volatilization in a stream of carrier gas

Either oxygen or hydrogen, or hydrogen diluted with nitrogen, can be used as the carrier gas. The sample is weighed into a boat, which is inserted into a tube and heated in an oven at 1000°C, gradually increasing to 1200 °C. A stream of carrier gas reacts with the sample and transports the analytes to a condensation area, from which they are eluted with strong acid afterwards. The method is suitable for silicates, oxides, carbonates, and metals; melting of the sample can prevent complete volatilization [6,25,26]. Alternatively, Tl can be volatilized from rock samples after admixture with cellulose/$MgCl_2$, and from dried biological samples after admixture with silicic acid, by combustion in oxygen in a special quartz apparatus (Trace-0-Mat[®]). Volatile traces are condensed on a cold-finger and dissolved in nitric acid. The dead volume and blanks and the Trace-0-Mat are much lower than with conventional tube furnaces [27].

12.3.2.2. Volatilization without a carrier gas stream

Upon admixture with desiccated $MgCl_2$, Tl is volatilized by fusion up to 1250°C in a quartz tube extended to a capillary until it is outside the furnace. Finally the capillary is cut off, and eluted with a small volume of acid. The Tl is accompanied by, for example, Cd, Sn, Pb, and Bi [27,28].

12.3.3. Selective dissolution

The environmental mobility of the constituents of sediments, rocks, soils, etc., can be estimated by selective dissolution in suitable reagents, thus imitating a change in pH and redox conditions. The mobility is due to the speciation in the solid phase.

In the stream sediment of the Lenne, near a smelter, Tl was found to be very immobile, at a total of 2 μg/g. Only 15% was dissolved by 1 M HNO$_3$, and less than 1% at pH > 2 [29]. In the leaching sequence, after Tessier/Förstner, Tl from uncontaminated soils was found mainly in the residual organic and silicatic fractions. In contaminated samples, however, exchangeable amounts were also present [30]. In stream sediments from Lower Austria [31], at a level below 1 μg/g, Tl has been largely found very immobile, partially bound to Fe/Mn hydroxides, and partially only releasable with hot HNO$_3$. In parallel investigations, it has been found to be evenly distributed among alkali- and acid-mobile fractions. Soluble thallium was completely absorbed by the sediments from tap water overnight, it was highly exchangeable against neutral salts [31]. Figures for the selective detection of dimethylthallium in environmental samples are not yet available.

12.4. SEPARATION METHODS

If concomitant elements interfere with the final determination, or the detection limits are too low for the samples to be analyzed, it becomes necessary to employ separation methods, increasing the time of analysis and thus the costs. For thallium, where low amounts are to be expected, and many determination methods suffer from interferences, separation steps often have to be considered. According to the author's experience, losses owing to incomplete recovery of thallium are more probable than the introduction of blanks. Within this context, special problems of the separation of various Tl species which may occur in the environment have to be dealt with. Chromatographic methods, which combine the separation and determination of trace elements, with the additional capabilities of speciation studies, have recently attracted interest.

12.4.1. Co-precipitation

Trace constituents can rarely be precipitated as their own compounds, because their amounts are usually too low to exceed the solubility product of the respective solids. They can be sorbed, however, on a solid collector which is precipitated from the sample solution, and leads to a uniform solid matrix. Co-precipitation can be performed simultaneously for a large number of samples, with subsequent separation of the collector and the desired trace constituents by centrifugation. Co-precipitation reactions need not be at equilibrium. They depend on the collector employed, on a accompanying ions, and the duration and temperature of the precipitation which influence the surface of the newly formed precipitate. Thus, co-precipitation behaviour can change significantly in the presence of strong buffer solutions, fatty acids, or surfactants. Complexation of the collec-

References on p. 369

tor-ions or adsorption of organics on the precipitate can lead to incomplete recovery of the desired traces, and even prevent the desired precipitation.

Co-precipitation methods have primarily been designed for enrichments from large environmental water samples, and for the determination of impurities in metals, and heavy metals in salts and biological matrices. In the analysis of water and biological samples, co-precipitation is employed to separate Tl and other trace elements from large amounts of water, alkali- and alkaline earth metals, and halides. As solid collectors, $MnO_2.xH_2O$, $Fe(OH)_3$, $Al(OH)_3$, $Zr(OH)_4$, and $Mg(OH)_2$ have been tested in detail: they can easily be dissolved in nitric acid [25,31-35]. From NH_4Cl and HCl, Tl is not co-precipitated with $MnO_2 \cdot xH_2O$. Hydroxylamine and oxalate have to be removed first [31]. If iron hydroxide is precipitated with ammonia instead of NaOH (as usual for co-precipitation), Ga, In and Tl are co-precipitated, but Cu, Zn, Cd, Ag, Hg and Cr(VI) are largely left in solution: this is important for the enrichment from sulfide ores [36]. Similarly, thallium is quantitatively sorbed on ZnS at pH 5.6, on Bi_2S_3 from dilute sulfuric acid, and on CuS, whereas CdS is only partially effective [37,38].

Among common organic derivatives, dithiocarbamates precipitated at pH 4–7 have been used for multi-element collection of heavy metals including thallium [39,40]. A reductive precipitation method, using $NaBH_4$ to collect 16 elements from environmental water samples, with Fe and Pd as carrier, has been also described. The added Pd is useful as a matrix modifier for subsequent furnace AAS [41]. Precipitates of AgCl, AgI, and Hg_2Cl_2 are quite selective collectors for the enrichment of thallium traces from environmental waters, especially in the presence of EDTA [42–46]. Selective sorption of Tl(I) from neutral solution has been achieved on mixed K-Co/Ni-ferrocyanides on a silica gel support [47].

12.4.2. Solvent extraction

Dissolved species are distributed between two immiscible liquid phases, usually an aqueous and an organic one. It is a rapid technique for a small series (up to 20 samples), with low-cost equipment. Enrichment factors of 5 are easily achieved. It is well documented for thallium, because most of the publications dealing with separation methods employ this technique. For large numbers of samples, however, environmental regulations about the waste organic solvents, such as halogenated hydrocarbons or benzene, have to be considered.

12.4.2.1. Solvent extraction of monovalent thallium

Complexes of monovalent thallium are usually weak. The selectivity can be increased by masking accompanying cations with cyanide, EDTA, or citrate. For the isolation of thallium together with other heavy metals, dithiocarbamate, dithizone, thiooxine or diantipyrylthiourea in weak alkaline media can be used [48–53]. Solvent extraction of Tl(I) with 2-mercaptobenzothiazole into chloroform can be made quite selective in the presence of tartrate and cyanide at pH 6–9 [54]. Selective separations can be achieved with crown ethers, like dicyclohexano-18-crown-6 in toluene [55], and dibenzo-18-crown-5 picrate in chloroform or benzene [56,57]; thallium is accompanied only by K, Ag and Pb.

12.4.2.2. Thallium(III) halides and reagent solvents

It is not necessary to use special organic reagents for the separation of thallium. Anionic Tl(III) complexes with halides or thiocyanate react with Lewis bases to give solvent-extractable ternary complexes. As the Lewis bases, one can use solvents like ethers (diisopropyl ether, dibutyl ether), esters (amyl acetate), ketones (methyl isobutyl ketone, cyclohexanone), or amines (trioctylamine, diphenylamine, benzylaniline). Tetrachlorothallate(III) is quantitatively extracted into diethyl-, diisopropyl-, and dibutyl ethers within the broad range of 0.8–10 M HCl [58,59]. With regard to volatility and toxicity, diisopropyl ether is the solvent of choice, but methyl isobutyl ketone yields the most sensitive signal in subsequent flame AAS of the extract [60]. Extraction of tetrabromothallate is even more selective. From 0.5 M HBr, only Au and some Sb, Sn, and Hg are extracted together with Tl, but they can be stripped with hydroxylamine/0.5 M HBr [61]. Back-extraction into an aqueous phase can be achieved with 5% ascorbic acid or ammoniacal sulphite [62–64].

12.4.2.3. Thallium(III) halides extracted with reagents into inert solvents

Ternary complexes of Tl(III) are extracted together with halides and reagent bases, from acid solution into relatively inert solvents. Examples are the extraction from 0.2 M HBr with tri-n-octylamine into benzene [65], from 0.001 M HCl with PAN into chloroform [66], with rhodamine-B into benzene (see photometric determinations), and from 3–8 M HCl with dioctyl sulfoxide into benzene [68,69]. As in the case of reagent solvents, Au, Sb and Hg are difficult or impossible to separate from Tl(III). Dimethylthallium is extracted from ammoniacal EDTA solution into CHCl$_3$ by PAN [63].

12.4.2.4. Solvent extractions of thallium(III) without halides

Thiocyanate complexes of Tl(III) are formed as with chloride and bromide. They cannot be used in strong acid solution because of decomposition. The Tl is extracted from 0.04 M KSCN into pyridine, together with Al, V, Fe, Cu, Mo, Ag, In and U [70], but is not extracted from 1 M NH$_4$SCN with dipentyl sulfoxide into benzene [68]. Similarly, Tl(III) is extracted from 0.1 M malonate at pH 4 with the liquid ion-exchanger Amberlite LA-2 into xylene [67], and is separated from Zn, Cd, Pb, Fe, Ga, and In. Also, Tl is rather selectively extracted into butanol from 0.1 M acetate buffer pH 5.5, using the liquid cation exchanger Versatic 10: only Fe, Cu and Zn go with the Tl [71].

12.4.3. Non-chromatographic enrichment on solids packed in columns

Separation on solids packed in columns is a rather old technique. In to contrast precipitation and liquid–liquid extraction, the sample is divided not just into two parts, but

References on p. 369

into many fractions collected from the column: this makes it possible to use non-selective determination procedures. There are current attempts to adapt the column methods to chromatographic systems (see Section 12.4.4). The observed processes can act as models for the penetration of contaminants through sediments or soils.

12.4.3.1. Inorganic sorbents

Inorganic sorbents can be easily prepared in the laboratory. The same substance can act as a cation-exchanger in acid and as an anion-exchanger in alkaline solution. They are not damaged by radioactive irradiation. Hydrated zirconium oxide and zirconium phosphate selectively sorb Tl(III) from dilute HCl solution [72,73]. Univalent thallium can be reversibly enriched on zirconium molybdatophosphate and zirconium antiamonate [74–77]. Also, PbO_2 sorbs univalent thallium at pH 1–12, releasing it with EDTA in alkali [78].

12.4.3.2. Active carbon

About 90% of Tl(III) is sorbed onto active carbon from 3–5 M HCl: it is accompanied by some Cu, Cd, and Sb. Similarly, Fe and Ga are fixed from more concentrated HCl. Desorption from the active carbon could only be achieved by fuming twice with HNO_3 [18,79]. For the multi-element pre-concentration of Tl^+ together with other heavy metals on active carbon, diethyldithiocarbamate or ethyl xanthate complexes, formed in weakly alkaline solution, are suitable. To increase the selectivity for thallium, other ions can be masked with citrate, tartrate, or fluoride [80,81].

12.4.3.3. Cation-exchange resins

In the presence of complexing agents (tartrate, citrate, EDTA), many cations are masked towards uptake by cation-exchange resins, but monovalent thallium is selectively retained on the column [82]. Trivalent thallium in HCl and HBr solutions is anionic, and therefore usually passes through the column, but traces can be adsorbed on the matrix of the resin [83–86]. For the enrichment of thallium(III) solutions obtained from silicate decomposition in 1 M HCl, a special crown ether polymer, poly(dibenzo-18-crown-6), has been synthesized [87].

12.4.3.4. Anion-exchange resins

The chloro- and bromo complexes of trivalent thallium are retained on strongly basic anion-exchange resins, which can be utilized for their enrichment from sea water. Most other cations are released from the column with dilute nitric acid, but thallium is regained upon reduction, e.g., with aqueous sulfur dioxide [17,86,88–90]. The addition of acetone stabilizes the chloro-complexes and thus increases the retention time if a chromatographic design is used [91]. Only Tl and Au are sorbed from 1–3 M HCl onto the polystyrene-divinylbenzene resin, Amberlite XAD-4. Polyacrylate Amberlite XAD-7 selectively takes Tl, Au and Sb from 1–2 M HCl [79].

12.4.3.5. Liquid–liquid column methods

A reactive liquid reagent is made stationary by sorption on an inert carrier, like the high-molecular-mass amine Aliquat 336 on a silica gel column [92]. Thallium(III) can thus be selectively separated from citrate media. By the adsorption of organic solvents upon solid granules packed within a column, liquid–liquid extraction methods can be adapted to column separations. This may gain importance in the development of future chromatographic methods. The enrichment of Tl(III) from 1 M HBr in methyl isobutyl ketone, from HCl in TBP, or of Tl-diethyldithiocarbamate in CCl_4, sorbed on inert granules, have been reported [93,94].

12.4.4. Ion-chromatography

The strongly acidic cation-exchanger, Nucleosil 10-SA, has been used for the ion-chromatographic separation of various cations prior to on-line colour-reaction with PAR at pH 11. Under optimized conditions, i.e., using 0.12% NaCl at pH 2.25, the separation and determination of Tl, Fe, Ga, Cu, In, Pb, and Zn have been achieved in the analysis of ores [36].

12.5. FINAL DETERMINATION METHODS

12.5.1. Atomic emission spectrometry (arc and spark methods)

The finely ground solid sample is applied to a graphite or carbon electrode, and excited with either direct current (4–13 A) or alternating current (20–22 A). The emitted spectrum is photographed on a plate or film. The main advantage of arc- and spark-emission methods for the analysis of solids is that they start with the solids themselves. The photo-plate is like a fingerprint, yielding a complete qualitative analysis for any sample: this is valuable in ore prospecting or for environmental material. No dissolution or separation methods are needed. However, the combination with dissolution and enrichment methods, or co-precipitation with Fe or Mn oxides, can greatly improve detection limits [32]. For thallium, the most sensitive spectral lines are at 535.05 and 377.57 nm. To correct for matrix effects, a known amount of another element is added: this should show similar spectroscopic behaviour, with a nearby line of similar intensity, and volatilization during excitation. Non-conducting samples, e.g., rocks, are mixed with graphite. To avoid molecular bands, organic substances have to be dry-ashed before analysis [95,96]. The addition of spectrochemical buffers favours constant atomization and quenches matrix interferences. Cyanide emission, which may interfere with the line at 377.57 nm is effectively quenched by arcing in the presence of an alkali-metal-rich vapour. The natural Na and K contents in granite and related rocks are usually sufficient. Dilution of the sample with 25% K_2SO_4 has been found to cope best with this background [97]. At lower arc temperatures, emission of non-volatile Ni at 377.557 nm is suppressed [25,97]. Addition of antimony sulfide enhances the sensitivity of the emission of Tl and many other elements. Matrix modification with a mixture of Al_2O_3–Sb_2S_3–NaCl = 8:1:1 allows the determination down to 0.3 $\mu g/g$ Tl in rocks, soils and even sulfide minerals (sphalerite, galenite) with a single calibration [98,99] and Bi as internal standard. At the less sensitive line at 276.8 nm, maximum emission intensity could be achieved by the addition of a mixture containing NaCl–graphite (10:90) [100]. For the determination of down to 1 $\mu g/g$ in plant ash at 276.8 nm, a spectral buffer of Al_2O_3–$CaCO_3$–K_2CO_3 (7:3:1) has been used [95].

12.5.2. Inductively coupled plasma–atomic emission spectrometry (ICP–AES)

In this technique the sample solution is aspirated into a toroidal plasma, where atomization and excitation occur. The emitted light is either measured simultaneously at previously selected atomic lines, or the lines are sequentially scanned. The sensitivity depends on many parameters of the apparatus used, such as the type of nebulizer, optics, power, gas-flows, and observation height. For thallium, there is some choice between different lines, but sensitivity is rather poor in any case, and the direct application to decomposition solutions of rocks, soils, and biological samples is limited. At an excitation frequency of 50 MHz, the detection limits are better than at 100 MHz [101]. Thallium is easily excitable, so optimum ratios of signal to background are obtained at low powers [102].

The most sensitive emission line in the ICP is probably at 190.864 nm [101,103], but vacuum conditions are needed. The line at 276.787 nm, which is prominent for atomic absorption, is frequently used for emission also. For solutions from the decomposition of soil or sediments, a detection limit of 0.5 mg/l has been reported: this has to be corrected for Fe, Mn and Cr [104]. In pure 0.1% HNO_3, 0.12 mg/l has been achieved as the detection limit for this line, which is the worst among 19 elements studied [105]. Sensitivity can be improved to 0.05 mg/l, by thermospray vaporization, but the effects of salts and buffer substances on the thermospray design were not investigated. Spectral interference occurs from a greater than 100-fold excess of Fe (at 276.75 nm). If Tl at 276.787 nm is measured in the second order, there might be an overlap with Ba at 553.548 nm [106]. Spectral interferences at 351.924 nm have not been reported. Whereas the sensitivity for other Tl-lines has been found to be relatively low [103], other authors have found this line to be the second most sensitive one [107]. The only application reported for this line has been for glass, after fusion with NaOH, and using ultrasonic nebulization [108]. The line at 377.572 nm, which is often used in DC arc excitation methods, may not be resolved from Ca (377.59 nm), Ni (377.56 nm), Ti (377.61 nm), and V (377.62 nm, 377.57 nm, 377.52 nm) [109], and especially Ar at 377.545 nm [105]. Therefore, no practical application of this line has been found in ICP–AES.

At 535.046 nm there are hardly any spectral interferences. For technical zinc, a detection limit of 2 $\mu g/ml$ in the sample solution could be achieved [110]. Matrix zinc depresses the analytical signal. The transparency of the optics may not be ideal in this spectral region for any device. For routine analysis of rocks and soils, without preconcentration, the detection using this line is insufficient, in the author's experience. Matrix Na up to 0.5% depresses the signal by up to 10%, with considerable variations from day to day. At the cost of losing sensitivity, a zone of minimum interference could be located in the plasma. The overall optimization of parameters was not affected by a change from Meinhard to cross-flow nebulizer. Unfortunately, the wavelength used in this work is not given [111].

Detection limits can be significantly improved by separation and preconcentration techniques (preferably multi-element, to use the multi-element capabilities of the method). The plasma is, however, easily distorted and extinguished by organic solvents.

12.5.3. Microwave-induced plasma

In a toroidal-shaped microwave-induced plasma (MIP), the sensitivity of Tl at 377.57 nm is fairly good, and the detection limit of 15 ng/ml with pneumatic Meinhard-nebulization is superior to the ICP. As the power of an MIP is lower than that of an ICP, matrix influence can be severe. Unlike the intensities of atomic and ionic lines of most elements, the Tl-sensitivity does not change with power within the range 70–100 W. The stability of the plasma is not affected by HCl and HNO_3 but the maximum tolerable amount of alkali salts is 0.7–1 mg/l, and micrograms of organic solvents can extinguish the plasma [112].

12.5.4. Mass spectrometry

Mass spectrometry is one of the most powerful multi-element determination methods. Besides emission spectrometry (arc method) and PIXE (proton induced X-ray emission), it is the only technique which allows the estimation of non-pollutional levels of thallium in a multi-element run. The high costs of the instrumentation and the small sample size have, however, limited its application.

If a spark ion source is used for ion generation, biological samples have to ashed, preferably with high-frequency oxygen, because organic degradation products interfere with the interpretation of the spectra. The resulting ash is formed into an electrode, together with graphite or carbon. The accuracy can be improved by spiking with ^{203}Tl, taking into account the natural isotope ratio ^{203}Tl/^{105}Tl = 3:7 [113]. The low ionization energy (6.07 eV) makes Tl an element which is suitable for easy thermal ionization from, for example, a Re filament (ReO^+ is isobaric with Tl, but obviously this does not matter); ionization has been achieved by 700°C, whereas neighbouring Pb needs higher temperatures [114,115]. Biological material and sewage sludge have been analyzed in this way, after wet ashing with HNO_3/HF or low-temperature ashing in O_2. The method was so sensitive that blanks from the procedure, carried out in a clean room with ultrapure chemicals, could be determined as 0.05–0.3 ng Tl [14,116]. For analysis of water samples, cathodic electrodeposition was used from ammonium tartrate buffer onto the Re filament [114].

Using ionization with Ar^+ (secondary ion mass spectrometry), it is possible to determine dimethylthallium beneath inorganic thallium, preferably after extraction with PAN from ammoniacal EDTA into $CHCl_3$. Identification is from the molecular mass and the isotope ratio $^{203}Tl/^{205}Tl = 3:7$ [63].

For the analysis of electrically conducting surfaces, secondary ion mass spectrometry with O_2^+ or Ar^+ allows the detection of about 0.1 $\mu g/g$ in 1 μm^2; oxygen bombardment seems to yield less matrix-dependent signals [117].

12.5.5. ICP with mass spectrometric detection (ICP–MS)

This new technique, which has gained application for some routine work during recent years, offers the advantage of lower detection limits than ICP–AES, simple spectra, and linear calibration over 5 orders of magnitude [118]. The ion/atom ratio produced in the plasma greatly influences the height of the signal. To cope with ionization matrix effects, isotope dilution methods have been applied. During measurement, a high vacuum has to be maintained in the detection system. Thallium is measured as ^{205}Tl; the ^{203}Tl isotope can be used as a spike for isotope dilution [119]. Easily ionizable elements reduce the signal, but Tl is less affected than Y and Li [120]. In the routine analysis of lake waters, a detection limit of 0.1 $\mu g/l$ has been achieved, which is due to the instrument and not to the blanks. In lake-waters from the Eastern Lake Survey (USA), Tl was detected (without preconcentration) in only 8% of the samples [121,122]; Bi was used as an internal standard. Other authors achieved a sensitivity of 0.01 $\mu g/g$ in natural waters from a granitic region (Fichtelgebirge; Germany), with ^{103}Rh as internal standard [118]. Sample introduction into the ICP by electrothermal vaporization leads to a dramatic improvement in analyte transport efficiency. Losses during the ashing stage are overcome by isotope dilution with ^{203}Tl. Chemical equilibrium between native and spike material has been achieved by heating samples of geological materials, with concentrated HCl prior to decomposition [119]. Similarly, a direct current plasma has been proposed as an ion-source for mass spectrometry, but no application has been given [123].

12.5.6. Atomic absorption spectrometry

12.5.6.1. Flame AAS

Flame AAS is a very common single-element technique. For thallium, determinations at the 276.78 nm line, with an acetylene/air flame are used throughout, whereas 535.05 nm has been used in flame emission. In flame AAS, matrix problems are very low, but the sensitivity with regard to the low level of occurrence is poor. The sensitivity can be increased by mounting a slotted quartz tube on the burner head (STAT = "slotted tube atom trap") [124]: this leads to a detection limit of about 20 $\mu g/g$ in the solid sample, after acid decomposition and direct aspiration. As well as in the screening of ore samples, such as sphalerites and pyrites, enrichment methods are essential for the use of flame AAS. The sensitivity for thallium in flame-AAS is up to 7 times greater using methyl isobutyl ketone extracts than aqueous solutions. One can use this solvent extraction of thallium from 0.1 M HBr [19], or the xanthate at pH 8 [125], or the iodide with tri-n-octylphosphine oxide [126,127] into methyl isobutyl ketone (MIBK), and direct aspiration of the organic phase into the flame. Thus, with large sample weights and small volumes of organic solvent, detection limits can be achieved of 0.1 μg/g in steel, or 0.2 μg/g in geological samples. Using diethyldithiocarbamate, both dimethylthallium and inorganic thallium are extracted into MIBK from aqueous solutions of pH 7–8, and are determined with equal sensitivity [128]. In the latter case, Cd at more than 60 μg/ml enhances the Tl signal.

The absolute amount of thallium determined by flame AAS can be reduced by injection of micro-volumes [129,130], or the vaporisation from a Pt loop which is rapidly inserted into the flame [39]. For determinations near the detection limit, in the presence of high salt concentrations, care has to be taken for the stability of the flame during the sudden injection.

12.5.6.2. Graphite furnace AAS

Atomic absorption determination in the graphite furnace is a rather slow single-element method, but it can easily be automated. For thallium, the detection limits achieved in poor solutions at 276.8 nm, are fairly good (0.5 μg/l), but severe interferences from chloride and fluoride often necessitate separations from the original matrix. Because the sensitivity decreases drastically in the presence of excess chloride, fluoride, or perchlorate [131], mere standard addition is inadequate. Other cations, especially Fe, in large excess also depress the signal [132]. Suitable separation methods have to be applied, e.g. solvent extraction, co-precipitation, or volatilization directly from an inorganic solid sample [26,28]. Because no severe spectral overlaps are known, background correction with a deuterium lamp is usually sufficient, especially after use of the separation methods.

In pure solutions, charring can be done up to about 500°C, and atomization at about 2300°C, but the latter figure may depend on the instrumentation used. In the presence of halides the permissible charring temperature is reduced: matrix modifiers, on the other hand, permit charring at higher temperatures (see later). The chloride interferences on Tl are partly caused by volatilization of TlCl in the pyrolysis stage, and partly by the formation of TlCl in the gas phase during atomization [133]. Low levels of halides can be volatilized from dilute sulfuric acid prior to charring of the thallium [134–137]. In 0.02 M EDTA–NH$_3$ at pH 10, the tolerance towards chloride is 400 times greater than in 0.1 M HNO$_3$ [138]. For solutions from the decomposition of human hair with HNO$_3$–H$_2$O$_2$, non-linear calibration has been reported [20].

It has been found that Tl does not form carbides during the determination cycle. Coated tubes usually yield lower signals, depending on the matrix [139]. The performance of coated tubes is very bad if organic solvent extracts are used, such as the extract of diethyldithiocarbamate in toluene [140,141]. In the latter case, nearly double the sensitivity was obtained by inserting a tantalum foil into the tube, to avoid soaking the graphite with toluene. For the dithiocarbamate, atomization at only 1300°C was sufficient [140]. Because the volatility of organic chelates is low, and the volatility of halides is the same, organic phases obtained from solvent extractions have been put directly into the furnace. In this case, however, carry-over in the auto-sampler has to be prevented by using an

References on p. 369

organic solvent, or at least an isopropanol–water mixture in the wash bottle. Back-extraction may be preferable. Coating of the graphite tube with $ZrOCl_2$ resulted in some improvement with respect to interferences from other cations, but this was insufficient for the analysis of steel samples [142]. Replacing the graphite tube with molybdenum resulted in increased heating rates, and allowed matrix modification with thiourea with a continuum background correction: this is not possible with a graphite tube [143], and allowed improvements in the presence of matrix Cd, Zn, and Cu. The addition of some hydrogen to the shielding gas enhanced the signal without the use of thiourea, possibly by preventing oxidation of the molybdenum tube.

As an alternative to separation methods, the addition of various matrix modifiers has been tried, especially to volatilize the halides. After the decomposition of plants with $H_2SO_4/HNO_3/H_2O_2$, the slope of Tl in the resulting solution was only 3–25% based on height, and 17–36% based on area. The performance was significantly improved by adding ascorbic acid and using a L'vov platform [144]. The addition of Li nitrate, ammonium salts, or phosphate was not successful [145–147]. When matrix modification was effected with 500 $\mu g/ml$ Ni (as the nitrate) [148], 1500 $\mu g/ml$ Pd and 1000 $\mu g/ml$ Mg (as the nitrates) [133], or just Mg nitrate [149], Zeeman-background compensation was used throughout. The Pd/Mg matrix modifier permits charring up to 1000°C, and 2000 $\mu g/ml$ of Pd even permits up to 1400°C, but Ni nitrate has little effect (charring < 600°C) [148]. Thus, seawater and urine samples have been analyzed without prior separation. Similarly, $LiBO_2$ used as a matrix after the fusion of siliceous material and fly-ash allows charring up to 1000°C in coated tubes with a platform. The addition of 1% sulfuric acid copes with halides; Zeeman-compensation and peak-area measurements are necessary [150]. The Tl line at 276.8 nm exhibits an anomalous Zeeman pattern. The relative absorbance passes a maximum at 9 kG field strength, and declines at higher field strengths [151].

12.5.7. Electroanalytical determinations

In the most sensitive electroanalytical methods, exclusively treated within this context, the analyte ion is electrodeposited on an electrode from an electrically conducting sample solution. The current, and potential for subsequent re-dissolution depend on the concentration and the kind of ion to be determined. For thallium, the reversible redox-couple Tl^+/Tl^0 at about –0.5 V versus a saturated calomel electrode, is used [152]. Infinite tolerance towards alkali, alkaline earths and halides are great merits for the analysis of seawater, brines, and biological materials. Because of the preconcentration step included, the thallium determination is more sensitive than atoms-spectrometric methods. For thallium, the multi-element capabilities of the method can hardly be used, because lead and frequently cadmium have to be masked with an excess of complexants, which leaves only Tl in the potential range of determination. As thallium complexes are rather weak, the electroanalytical thallium signal is nearly unaffected by complexants in solution. Complexes of metals with a more positive standard electrode potential than Tl, however, can be shifted into the range of Tl deposition by complexation [153–155]. If the potential of a reversible redox-couple present in the solution is passed in the stripping, the background current is much enhanced [27,156]. Surfactants provoke shifts of the signal, which can be used for electrochemical masking (see below). Residual organics from incomplete digestion of, for example, coal and rubber can, however, yield ghost-peaks [6]. Combustion techniques avoid this problem.

The equipment is quite cheap, but the method can only be partly automated. Additions of standards have to be used throughout, and the through-put of analyses is low.

12.5.7.1. Hanging mercury drop electrode

With Hg, Tl forms an amalgam which is stable within a broad concentration range. Because Tl^+ is deposited about 1.6 times faster than Tl^{3+}, it is preferable to start with monovalent Tl and to add standards. Aliphatic and aromatic Tl compounds can also be reduced at the hanging mercury drop electrode. The peak for dimethylthallium appears separately from inorganic thallium, but phenylthallium coincides [157–160]. In dilute HCl solution there are overlaps of Tl with Pb, and partially with Cd and Sn. To avoid co-deposition of Sn, Cd, and partially Pb, the deposition potential has to be kept as low as possible, at least –0.8V (versus saturated calomel) in dilute HCl [161]. Any Fe(III) has to be reduced with hydroxylamine [162], and EDTA, citrate, or tartrate are conducting electrolytes well suited for complexing interferents. For environmental water samples, the determination of thallium in acetate buffer at pH 4.5, in the presence of EDTA, has adequate tolerance towards accompanying Pb, Cd and other metals [163,164]. The tolerance towards Pb is greatly enhanced by the addition of neutral (e.g., polyethyleneglycol derivatives) or cationic surfactants (high molecular amines or phosphonium salts), and by shifting to higher pH values. In citrate–EDTA at pH 7, a 10 000-fold excess of Pb does not interfere with the Tl peak, but the sensitivity is slightly reduced. The direct use of solutions from the decomposition of rock samples, however, is hampered by larger excesses of Cu, Fe, or Ti, which make the Tl peak disappear in the background current; these have to be separated beforehand [27].

The influence of surfactants on the thallium signal is low, whereas other peaks, which are close together in dilute mineral acid, are moved apart. For matrix Bi and Pb, the addition of tetrabutylammonium chloride to 0.2 M EDTA at pH 4.5 is suitable [165–167]. For determination of Tl in matrix Cu, the addition of surfactants containing an ethylene oxide chain completely suppresses the peak of Cu in 0.2 M EDTA at pH 4.5 [168]. The addition of polyethylene glycol to EDTA / tartrate enables the determination of Tl in matrix Cd [169]. Excess Fe is electrochemically masked by Rokafenol N-3 in oxalate as the supporting electrolyte [170]. Anodic stripping voltammetry has been performed on the organic layer obtained by extracting the dithiocarbamates into benzene from neutral solution, with the addition of methanol and $NaClO_4$ as supporting electrolyte. There is nearly no difference between peaks from Tl dithiocarbamate and inorganic Tl, and 1000-fold excesses of In and Cd did not interfere [171].

References on p. 369

12.5.7.2. Mercury thin film electrode

By the electrolytic deposition of Hg at pH < 4 onto an inert electrode, a layer of metallic mercury is generated, which serves as a new electrode surface for subsequent use. In contrast with the hanging mercury drop electrode, Tl and Pb are resolved in the absence of complexants and surfactants, but Cd coincides [154]. A determination of Tl in technical Ni has been made in 1 *M* tartaric acid [172]. Using alkaline EDTA–citrate solution, Tl could be determined in ores and natural waters [173].

A change of the electrolyte between the in-situ generation of the thin-film electrode with electrodeposition of the analyte, and the subsequent potentiometric stripping, allows the separate optimization of both procedures. This is preferably done in a flow-injection system, which permits automation of the method. Among 40 different electrolytes investigated, the best separation of Sn, Cd, Pb, and Tl peaks was achieved by stripping in 5 *M* NH_3/0.5 *M* KOH, where the deposition is difficult [174].

12.5.8. Spectrophotometric and fluorimetric methods

Simple colour reactions of thallium with PAR, PAN or similar reagents are neither sensitive nor selective. They can be applied only as general post-column colour reactions for ion chromatography [36]. Determinations using the ternary complexes between Tl(III) and Cl-, Br-, or SCN- and a cationic dye, can be made quite selective, because of the preceding separation. They can also be sensitive, because of the possibility of choosing a very strong absorbing dye. The ternary complex has nearly the same spectrum as the dye, but it is separable from the excess dye by extraction into an organic solvent from an acid solution. Cross-interferences are usually similar, whether the dye is measured by photometry or fluorimetry. If a preceding enrichment is done by volatilization, or at least by solvent extraction, environmental background levels can be approached using simple equipment. Prior to the determination, Tl has to be oxidized. Chlorine, bromine, Ce(IV), or peroxydisulfate are suitable in the cold; The excess oxidant has to be destroyed or separated before contact with the dye. Severe interferences are caused by incompletely destroyed organics (from humic substances, chlorophyll, etc.) which carry carboxylic groups, because these form extractable ion-associates with the cationic dye. Thus, for decomposition of soil samples, etc., a combustion, volatilization or, at least a pressure decomposition method, has to be used, prior to a photometric analysis. Among inorganics, the main positive interferences derive from Au, Hg, Sb, and Ga; the co-extraction of tungstate, chromate, borate, perchlorate, thiocyanate and iodide are less frequently reported. A large excess of fluoride depresses the signal. Among the dyes, brilliant green, methyl violet, crystal violet, methyl green, rhodamine B, safranine T, etc., have been used (see, e.g., [175–182]). A complete survey of selectivities and interferences would go beyond the scope of this work. High selectivity could, for example, be achieved by oxidation of the sample with Ce(IV), extraction of Tl from 0.5 *M* HBr into diisopropyl ether, cleaning the extract with hydroxylamine–HBr from co-extracted Hg, Sb, and Au, and finally shaking with acidified aqueous rhodamine B dye solution. Only an amount of dye equivalent to TlBr$_4$⁻ moves to the organic layer, and can be measured at 555 nm [61]. Adaptation of ternary-complex methods to flow-injection or continuous flow techniques has not yet been reported.

REFERENCES

1 M. Sager, *Spurenanalytik des Thalliums*, G. Thieme, Stuttgart, 1984.
2 G. Siedner, *Geochim. Cosmochim. Acta*, 32 (1968) 1303-1315.
3 J.R. Butler, *Geokhimiya*, 6 (1962) 514-523.
4 C. Dupuy, M. Fratta and D.M. Shaw, *Earth Plan. Sci. Lett.*, 19 (1973) 209-212.
5 H. Heinrichs, *Dissertation*, Göttingen 1975.
6 H. Gorbauch, H.H. Rump, G. Alter and C.H. Schmitt- Henco, *Fresenius' Z. Anal. Chem.*, 317 (1984) 236-240.
7 O.A. Dvoretskaja and T.F. Bojko, *Litol. Pol. Iskop.*, 3 (1979) 95-104.
8 S.E. Kirchner, H. Oona, S.J. Perron, Q. Fernando, Jong-Hae Lee and H. Zeitlin, *Anal. Chem.*, 52 (1980) 2195-2201.
9 J.M. Iskowitz, J.J.H. Lee, H. Zeitlin and Qu. Fernando, *Mar. Min.*, 3 (1982) 285-295.
10 K. Augustin-Gyurits and E. Schroll, *Bericht der BVFA-Arsenal* 1989.
11 H. Eschnauer, V. Gemmer-Colos and R. Neeb, *Z. Lebensm. Unters. Forsch.*, 178 (1984) 453-460.
12 W. Scholl, *Landw. Forsch., Kongressband*, 1980, 275-285.
13 K. Govindaraju, *Geostandards Newslett.*, 13, July 1989.
14 E. Waidmann, K. Hilpert, J.D. Schladot and M. Stoeppler, *Fresenius' Z. Anal. Chem.*, 317 (1984) 273-277.
15 D. Cubicciotti and F.J. Keneshea, *J. Phys. Chem.*, 7 (1967) 808-814.
16 R. Bock, *Handbook of Decomposition Methods in Analytical Chemistry*, Int. Textbook Comp. Ltd., London, 1st ed., 1979.
17 A.D. Matthews and J.P. Riley, *Anal. Chim. Acta*, 48 (1969) 25-34.
18 H. Koshima and H. Onishi, *Anal. Sci.*, 1 (1985) 237-240.
19 A.E. Hubert and T.T. Chao, *Talanta*, 32 (1985) 568-570.
20 J.F. Chapman and B.E. Leadbeatter, *Anal. Lett.*, 13 (1980) 439-446.
21 E.A. Jones and A.F. Lee, *Rep. Natl. Inst. Metall.*, 2022 (1979) pp. 38.
22 W. Rechenberg, *Zement-Kalk-Gips.*, 35 (1982) 90-95.
23 M. Sager, in preparation.
24 R.C. Carpenter, *Anal. Chim. Acta*, 125 (1981) 209-213.
25 W. Geilmann and K.H. Neeb, *Fresenius' Z. Anal. Chem.*, 165 (1959) 251-268.
26 H. Heinrichs, *Fresenius' Z. Anal. Chem.*, 294 (1979) 345-351.
27 I. Liem, G. Kaiser, M. Sager and G. Tölg, *Anal. Chim. Acta*, 158 (1984) 179-197.
28 M. Sager, *Mikrochim. Acta*, 1984 I, 461-466.
29 K. Günther, W. Henze and F. Umland, *Fresenius' Z. Anal. Chem.*, 327 (1987) 301-303.
30 H. Lehn and J. Schoer, *Proc. Int. Conf. Heavy Met. Environm.*, Athens, 1985, pp. 286-288.
31 M. Sager, *Mikrochimica Acta*, 1991, in press.
32 A.P. Kreshkov, E.A. Kuchkarev and T.Ya. Davydova, *Tr. Mosk. Khim. Tekhnol. Inst.*, 85 (1975) 185-187.
33 L.S. Veselago and A.I. Novikov, *Radiokhimiya*, 3 (1977) 285-288.
34 G.F. Efremov and K.P. Stolyarov, *Uch. Zap. Lgu., Ser. khim. nauk.*, 18 (1959) 99-105.
35 G.V. Efremov and I.P. Alexeeva, *Uch. Zap. Lgu., Ser. Khim. Nauk.*, 15 (1957) 87-91.
36 D. Yan, J. Zhang and G. Schwedt, *Fresenius' Z. Anal. Chem.*, 331 (1988) 601-606.
37 N.A. Rudnev and A.A. Mazur, *Zh. Anal. Khim.*, 12 (1957) 433-442.
38 N.A. Rudnev, G.I. Malofeeva and V.S. Rasskazova, *Zav. Lab.*, 27 (1961) 20-21.
39 H. Berndt and J. Messerschmidt, *Fresenius' Z. Anal. Chem.*, 299 (1979) 28-32.
40 S. Brüggerhoff, E. Jackwerth, B. Raith, A. Stratmann and B. Gonsior, *Fresenius' Z. Anal. Chem.*, 311 (1982) 252

41 S. Nakashima, R.E. Sturgeon, S.N. Willie and S.S. Berman, *Anal. Chim. Acta*, 207 (1988) 291-299.
42 Yu.V. Morachevskii and G.V. Efremov, *Nauchn. Dokl. Vyssh. Shkol., Ser. Khim. Khim. Tekhnol.*, 4 (1958) 706-709.
43 E. Jackwerth and G. Graffmann, *Fresenius' Z. Anal. Chem.*, 241 (1968) 96-110.
44 V.P. Borovitskii, A.D. Miller and V.N. Shemyakin, *Geokhimiya*, 4 (1966) 483-488.
45 H. Berndt and E. Jackwerth, *Fresenius' Z. Anal. Chem.*, 282 (1976) 427-433.
46 H. Berndt and E. Jackwerth, *Fresenius' Z. Anal. Chem.*, 283 (1977) 15-21.
47 V.V. Alikin, V.V. Vol'khin and S.S. Kolesova, *Izv. Akad. Nauk. Turkm. SSR, Ser. Fiz.-Tekh., Khim. i Geol. nauk.*, 1984, 59-63.
48 S.V. Freger, M.I. Ovrutskii and G.S. Lisetskaya, *Zh. Anal. Khim.*, 29 (1974) 19-23.
49 R. Keil, *Fresenius' Z. Anal. Chem.*, 309 (1981) 181-185.
50 G. Ackermann and W. Angermann, *Fresenius' Z. Anal. Chem.*, 250 (1970) 353-357.
51 A. Dyfvermann, *Anal. Chim. Acta*, 21 (1959) 357-365.
52 V.V. Bagreev and Yu.A. Zolotov, *Zh. Anal. Khim.*, 17 (1962) 852-857.
53 M.I. Degtev, L.I. Toropov and Yu.A. Makhnev, *Zh. prikl. Spektr.*, 48 (1988) 278-282.
54 R.K. Itawi and Z.B. Turel, *J. Radioanal. Nucl. Chem., Lett.*, 87 (1984) 151-160.
55 G.A. Clark, R.M. Izatt and J.J. Christensen, *Sep. Sci. Technol.*, 18 (1983) 1473-1482.
56 T. Sekine, H. Wakabayashi and Y. Hasegawa, *Bull. Chem. Soc. Jpn.*, 51 (1978) 645-646.
57 T. Maeda, K. Kimura and T. Shono, *Fresenius' Z. Anal. Chem.*, 298 (1979) 363-366.
58 H.M. Irving and F.J.C. Rossotti, *Analyst (London)*, 77 (1953) 801-812.
59 Yu.A. Zolotov, I.A. Alimarin and A.I. Sukhanovskaya, *Zh. Anal. Khim.*, 20 (1965) 165-171.
60 C.M. Elson and C.A.R. Albuquerque, *Anal. Chim. Acta*, 134 (1983) 393-396.
61 M. Sager and G. Tölg, *Mikrochim. Acta*, 1982 II, 231-245.
62 G. Cammann and J.T. Andersson, *Fresenius' Z. Anal. Chem.*, 310 (1982) 45-50.
63 K. Günther and F. Umland, *Fresenius' Z. Anal. Chem.*, 333 (1989) 6-8.
64 M. Sager, *Enquete*, Arbeitsgemeinschaft Landwirtschaftlicher Bundesanstalten in Österreich, Fachgruppe Pflanzenanalyse, Eisenstadt 1987.
65 I. Tsukahara, M. Sakakibara and T. Yamamoto, *Anal. Chim. Acta*, 83 (1976) 251-258.
66 K.A. Rao, B.V. Acharya, N. Muralidhar and B. Rangamannar, *Radioanal. Nucl. Chem.*, 94 (1985) 351-355.
67 R.R. Rao and S.M. Khopkar, *Bunseki Kagaku*, 33 (1984) E473-E479.
68 A.S. Reddy and M.L.P. Reddy, *J. Radioanal. Nucl. Chem.*, 87 (1984) 397-403.
69 A.S. Reddy and M.L.P. Reddy, *J. Radioanal. Nucl. Chem.*, 94 (1985) 259-264.
70 R.S. Ramakrishna and M.E. Fernandopulle, *Anal. Chim. Acta*, 60 (1972) 87-92.
71 S.B. Kar and U.S. Ray, *J. Indian Chem. Soc.*, 62 (1985) 701-703.
72 G. Alberti, M.G. Bernasconi, U. Costantino and J.S. Gill., *J. Chromatogr.*, 132 (1977) 477-484.
73 N.J. Singh and S.N. Tandon, *Anal. Lett.*, 10 (1977) 701-708.
74 N.J. Singh and S.N. Tandon, *Ind. J. Chem.*, 19A (1980) 502-504.
76 A.K. Jain, S.K. Srivastava, R.P. Singh and S. Agrawal, *J. Appl. Chem. Biotechnol.*, 28 (1978) 626-632.
77 A.K. Jain, S. Agrawal and R.P. Singh, *J. Radioanal. Chem.*, 60 (1980) 111-119.
78 M. Miyazaki, S. Itoh and S. Nagai, *Bunseki Kagaku*, 37 (1988) 544-548.
79 X.G. Yang and E. Jackwerth, *Fresenius' Z. Anal. Chem.*, 331 (1988) 588-593.
80 E. Jackwerth and H. Berndt, *Anal. Chim. Acta*, 74 (1975) 299-307.
81 M. Kimura, *Talanta*, 24 (1977) 194-196.
82 L.B. Ginzburg and E.P. Shkrobot, *Zav. Lab.*, 11 (1955) 1289-1294.
83 F.W.E. Strelow and A.H. Victor, *Talanta*, 19 (1972) 1019-1023.
84 R.G. Boehmer and P. Pille, *Talanta*, 24 (1977) 521-523.
85 F.W.E. Strelow, *Talanta*, 27 (1980) 231-236.
86 G. Pfrepper, *J. Chromatogr.*, 110 (1975) 133-140.
87 H. Koshima and H. Onishi, *Analyst (London)*, 114 (1989) 615-617.
88 R.R. Brooks, *Analyst (London)*, 85 (1960) 745-751.
89 G.V. Efremov, M.N. Zvereva and Ts. Tsedevsuren, *Zav. Lab.*, 28 (1962) 155-161.

90 G.E. Batley and T.M. Florence, *Electroanal. Chem. Interf. Electrochem.,* 61 (1975) 205-211.
91 E.A. Jones and A.F. Lee, *Rep. Natl. Inst. Metall. 2022,* (1979) pp. 38.
92 S.K. Sahoo and S.M. Khopkar, *Analyst (London),* 113 (1988) 1781-1783.
93 R.A. Kuznetsov, *Zh. Anal. Khim.,* 33 (1978) 2062-2064.
94 J.S. Fritz, R.T. Frazee and G.L. Latwesen, *Talanta,* 17 (1970) 857-864.
95 J. Parle and G.A. Fleming, *Ir. J. Agric. Res.,* 16 (1977) 49-55.
96 D.R. Scott, W.A. Loseke, L.E. Holboke and R.J. Thompson, *Appl. Spectr.,* 30 (1976) 392-405.
97 J.P. Willis, M. Kaye and L.H. Ahrens, *Appl. Spectr.,* 18 (1964) 84-87.
98 A.I. Kuznetsova and N.L. Chumakova, *Zh. Anal. Khim.,* 43 (1988) 2183-2190.
99 A.I. Kuznetsova and N.L. Chumakova, *Geostand. Newsl.,* 13 (1989) 269-272.
100 L.I. Serdobova, T.A. Chemleva and N.A. Bol'shakova, *Zh. Anal. Khim.,* 34 (1979) 1258-1265.
101 P.W.J.M. Boumans, *Spectrochim. Acta,* 12 (1989) 1285-1296.
102 A. Montaser, V.A. Fassel and J. Zalewsky, *Appl. Spectr.,* 35 (1981) 292-302.
103 R. Barnes, *ICP Inf. Newsl.,* 5 (1980) 435.
104 W. Jäger and I. Horn, *Z. Wasser-Abwasser- Forsch.,* 20 (1987) 138-141.
105 R. Peng, J.J. Tiggelman and M.T.C. De Loos- Vollebregt, *Spectrochim. Acta,* 45B (1990) 189-199.
106 T.A. Anderson and M.L. Parsons, *Appl. Spectrosc.,* 38 (1984) 625-634.
107 M.T. Dorado Lopez, A. Gómez Coedo and R. Gallego Andreu, *Rev. Metal. Madrid,* 18 (1982) 69-80.
108 M. Myunks, *Zav. Lab.,* 34 (1968) 165-169.
109 R.I. Botto, *ICP Inf. Newsl.,* 6 (1981) 527-543.
110 A. Gomez Coedo, M.T. Dorado Lopez and J.M. Sistiaga, *Rev. Met. Madrid,* 15 (1979) 379-385.
111 L. Ebdon and R. Carpenter, *Anal. Chim. Acta,* 209 (1988) 135-145.
112 D. Kollotzek, P. Tschöpel and G. Tölg, *Spectrochim. Acta,* 39B (1984) 625-636.
113 E.I. Hamilton, M.J. Minsky and J.J. Cleary, *Sci. Total Environm.,* 1 (1972/73) 341-374.
114 J. Trettenbach and K.G. Heumann, *Fresenius' Z. Anal. Chem.,* 322 (1985) 306-310.
115 K.G. Heumann, P. Kastenmayer and H. Zeininger, *Fresenius' Z. Anal. Chem.,* 306 (1981) 173-177.
116 E. Weinig and P. Zink, *Arch. Toxikol.,* 22 (1967) 255-274.
117 G. Stingeder, personal communication.
118 B. Sansoni, W. Brunner, G. Wolff, H. Ruppert and R. Dittrich, *Fresenius' Z. Anal. Chem.,* 331 (1988) 154-169.
119 Ch.J. Park and G.E.M. Hall, *J. Anal. At. Spectrom.,* 3 (1988) 355-361.
120 D.C. Gregoire, *Spectrochim. Acta,* 42B (1987) 895-907.
121 J.M. Henshaw, E.M. Heithmar and T.A. Hinners, *Anal. Chem.,* 61 (1989) 335-342.
122 D.J. Douglas and J.B. French, *Anal. Chem.,* 53 (1981) 37-41.
123 A.L. Gray, *Analyst (London),* 100 (1975) 289-299.
124 B. Milner, *Int. Lab.,* Nov/Dec. (1983) 30-32.
125 M. Aihara and M. Kiboku, *Bunseki Kagaku,* 29 (1980) 243-247.
126 K.E. Burke, *Appl. Spectr.,* 28 (1974) 234-237.
127 G. Staats and E. Weichert, personal communication, A.G. Dillinger Hüttenwerke.
128 J.M. Morgan, J.R. McHenry and L.W. Masten, *Bull. Environm. Contam. Toxicol.,* 24 (1980) 333-337.
129 H. Berndt and E. Jackwerth, *Spectrochim. Acta,* 30B (1975) 169-177.
130 H. Berndt and W. Slavin, *At. Abs. Newsl.,* 17 (1978) 109-112.
131 K.H. Sauer and S. Eckhard, *Mikrochim. Acta,* 1981/Suppl. 9, 87-98.
132 G. Machata and R. Binder, *Z. Rechtsmed.,* 73 (1973) 29-34.
133 B. Welz, G. Schlemmer and J.R. Mudakavi, *Anal. Chem.,* 60 (1988) 2567-2572.
134 C.W. Fuller, *Anal. Chim. Acta,* 81 (1976) 19-202.
135 O. Kujirai, T. Kobayashi and E. Sudo, *Fresenius' Z. Anal. Chem.,* 297 (1979) 398-403.
136 G.O. Welcher, O.H. Kriege and J.Y. Marks, *Anal. Chem.,* 46 (1974) 1227-1231.
137 W. Slavin, G.R. Carnrick and D.C. Manning, *Anal. Chim. Acta,* 138 (1982) 103-110.

372

138 K. Matsusaki and T. Yoshino, *Bunseki Kagaku*, 35 (1986) 931-934.
139 M. Beaty, W. Barnett and Z. Grobenski, *At. Spectr.*, 1 (1980) 72-77.
140 H.A. Chandler and M. Scott, *At. Spectr.*, 5 (1984) 230-233.
141 N.P. Kubasik and M.T. Volosin, *Clin. Chem.*, 19 (1973) 954-958.
142 W. Schmidt and F. Dietl, *Fresenius' Z. Anal. Chem.*, 315 (1983) 687-690.
143 M. Suzuki and K. Ohta, *Fresenius' Z. Anal. Chem.*, 322 (1985) 480-485.
144 M. Hoenig, P.O. Scokart and P. Van Hoeyweghen, *Anal. Lett.*, 17 (1984) 1947-1962.
145 B.V. L'vov, L.A. Pelieva and A.I. Sharnopol'skii, *Zh. prikl. Khim.*, 28 (1978) 19-22.
146 T.R. Dulski and R.R. Bixler, *Anal. Chim. Acta*, 91 (1977) 199-209.
147 S. Callio, *At. Spectr.*, 1 (1980) 80-81.
148 D.A. Bass, *Am. Lab.*, December 1989.
149 D.C. Paschal and G. Bailey, *J. Anal. Toxicol.*, 10 252-254.
150 M. Grognard and M. Piolon, *At. Spectr.*, 6 (1985) 142-143.
151 F.J. Fernandez, S.A. Myers and W. Slavin, *Anal. Chem.*, 52 (1980) 741-746.
152 A.J. Bard, *Encyclopedia of Electrochemistry of the Elements*, Vol. 4, Marcel Dekker, New York, 1975.
153 R. Pribil, *Analytical Applications of EDTA and Related Compounds*, Pergamon Press, Oxford, 1972.
154 J.P. Roux, O. Vittori and M. Porthault, *Analusis*, 3 (1975) 411-416.
155 A. Rigo, M. Cherido, E. Argese, P. Viglino and C. Dejak, *Analyst (London)*, 106 (1981) 474-478.
156 L. Kryger, *Anal. Chim. Acta*, 120 (1980) 19-30.
157 J.S. Di Gregorio and M.D. Morris, *Anal. Chem.*, 40 (1968) 1286-1291.
158 R.G. Dhaneshwar and L.R. Zarapkar, *Analyst (London)*, 105 (1980) 386-390.
159 L.K. Hoeflich, R.J. Gale and M.L. Good, *Anal. Chem.*, 55 (1983) 1591-1595.
160 M. Hoenig, P.O. Scokart and P. Van Hoeyweghen, *Anal. Lett.*, 17 (1984) 1947-1962.
161 R. Neeb and I. Kiehnast, *Fresenius' Z. Anal. Chem.*, 241 (1968) 142-155.
162 V. Gemmer-Colos, I. Kiehnast, J. Trenner and R. Neeb, *Fresenius' Z. Anal. Chem.*, 306 (1981) 144-149.
163 J.E. Bonelli, H.E. Taylor and R.K. Skogerboe, *Anal. Chim. Acta*, 118 (1980) 243-256.
164 N. You and R. Neeb, *Fresenius' Z. Anal. Chem.*, 314 (1983) 394-397.
165 A. Ciszewski, *Talanta*, 32 (1985) 1051-1054.
166 A. Ciszewski and Z. Lukaszewski, *Talanta*, 30 (1983) 873-875.
167 A. Ciszewski and Z. Lukaszewski, *Anal. Chim. Acta*, 146 (1983) 51-59.
168 A. Ciszewski and Z. Lukaszewski, *Talanta*, 32 (1985) 1101-1104.
169 Z. Lukaszewski, M.K. Pawlak and A. Ciszewski, *Talanta*, 27 (1980) 181-185.
170 Z. Lukaszewski, A. Ciszewski and A. Szymanski, *Chem. Anal.*, 32 (1987) 903-912.
171 J. Labuda and M. Vaničkova, *Anal. Chim. Acta*, 208 (1988) 219-230.
172 L.G. Petrova, V.I. Ignatov and E.Ya. Neiman, *Zh. Anal. Khim.*, 37 (1982) 22-26.
173 Sh. Wang and Sh. Li, *Yankuang Ceshi*, 7 (1988) 31-35; from *Chem. Abs.*, 110 (1988) 127729q.
174 G. Schulze, W. Bönigk and W. Frenzel, *Fresenius' Z. Anal. Chem.*, 322 (1985) 255-260.
175 Z. Marczenko, H. Kalowska and M. Mojski, *Talanta*, 21 (1974) 93-97.
176 K. Vadasdi, P. Buxbaum and A. Salamon, *Anal. Chem.*, 43 (1971) 318-322.
177 E.L. Kothny, *Analyst (London)*, 94 (1969) 198.
178 J.F. Woolley, *Analyst (London)*, 83 (1958) 477-479.
179 R.E. Van Aman and J.H. Kanzelmeyer, *Anal. Chem.*, 33 (1961) 1128-1129.
180 V. Miketukova and J. Kohliček, *Fresenius' Z. Anal. Chem.*, 208 (1965) 7-15.
181 Z. Gregorowicz, J. Ciba and B. Kowalczyk, *Talanta*, 28 (1981) 805-808.
182 J.A. Mikaelyan, Zh.V. Artsruni and A.G. Khachatryan, *Zh. Anal. Khim.*, 44 (1989) 749-751.

M. Stoeppler (Editor)/*Hazardous Metals in the Environment*
© 1992 Elsevier Science Publishers B.V. All rights reserved

Chapter 13

Chromium

Nancy J. Miller-Ihli

USDA, Nutrient Composition Laboratory, Beltsville, MD 20705 (U.S.A.)

CONTENTS

13.1. Introduction 374
    13.1.1. Natural occurrence 374
    13.1.2. Production, consumption, and uses of chromium 375
    13.1.3. Anthropogenic sources of chromium 376
        13.1.3.1. Atmospheric emissions 376
        13.1.3.2. Chromium in water 377
    13.1.4. Essentiality and toxicity 377
13.2. Sampling procedures 379
    13.2.1. Sample collection 380
        13.2.1.1. Tools and containers for sampling 380
        13.2.1.2. Cleaning of containers 382
    13.2.2. Sample handling and storage 383
        13.2.2.1. Contamination control 383
        13.2.2.2. Sample homogenization 384
    13.2.3. Sample digestion 385
        13.2.3.1. Wet ashing 385
        13.2.3.2. Dry ashing 386
        13.2.3.3. Other techniques 387
    13.2.4. Preparation and storage of standard solutions 387
13.3. Determination procedures 387
    13.3.1. Atomic absorption spectrometry 388
        13.3.1.1. Flame atomic absorption spectrometry 389
        13.3.1.2. Graphite furnace atomic absorption spectrometry 389
    13.3.2. Inductively coupled plasma spectrometry 392
        13.3.2.1. Inductively coupled plasma atomic emission spectrometry . 392
        13.3.2.2. Inductively coupled plasma mass spectrometry 393
    13.3.3. Other techniques 394
        13.3.3.1. Neutron activation analysis 394
        13.3.3.2. Isotope dilution mass spectrometry 394
        13.3.3.3. Colorimetric methods 395
        13.3.3.4. Chromatographic methods 395
        13.3.3.5. Miscellaneous methods 396
    13.3.4. Speciation 396
    13.3.5. Accuracy and reference materials 397
References 400

13.1. INTRODUCTION

Chromium is an element which has several roles in our daily lives. At low concentrations it is an essential micronutrient in human and animal nutrition and at high concentrations it is a known carcinogen when present as chromate [1]. The accurate determination of chromium at very low concentrations in "unexposed" biological materials such as tissues, blood, urine, food, water, and air, is extremely challenging and methods are under continual development [2]. Only in the last few years have analysts been able to develop consensus "normal" chromium concentrations for urine and serum because of the low levels of chromium found in these biological fluids (0.1-0.5 ng/ml) [3]. At these levels methods with good sensitivity are required and care must be taken to avoid contamination during sample collection, preparation, and analysis.

Chromium occurs naturally in the environment and is an abundant element in the earth's crust. Chromium occurs in oxidation states ranging from Cr^{2+} to Cr^{6+}. Elemental chromium is rarely found in nature. Chromite, $FeOCr_2O_3$, is the only major commercial chromium mineral and its deposits are mainly found in Rhodesia, the U.S.S.R., the Republic of South Africa, and the Philippines. Widespread use of chromium in products, as well as the toxicity of the hexavalent and trivalent forms have brought about growing concern about chromium emissions into the atmosphere [1].

13.1.1. Natural occurrence

Chromium occurs ubiquitously in nature with background levels in air typically being less than 1 ng/m^3 [1,2]. The chemical forms of chromium in air are not known. The principle natural sources of air chromium include windblown dust and volcanoes. Other natural sources such as airborne sea salt and smoke from forest fires, as well as biogenic emissions from vegetation do not seem to be such significant sources [1]. The variable chromium concentrations in different soils and rocks influence the natural atmospheric emissions of the element. Balsberg-Pahlsson et al. reviewed bulk chromium concentration ranges in various types of rocks and soil as did Fleischer [1,4]. They concluded that average chromium levels were as follows: basaltic igneous 40–600 $\mu g/g$, granitic igneous 2–100 $\mu g/g$, quartz 3–200 $\mu g/g$, limestone 10–60 $\mu g/g$, shale 20–600 $\mu g/g$, sandstone 35–90 $\mu g/g$, phosphorite 300–3000 $\mu g/g$, sedimentary iron ore 150–800 $\mu g/g$, soil 80–200 $\mu g/g$. Although an average soil concentration is considered to be 100 $\mu g/g$, the metal contents of some soils may exceed 1000 $\mu g/g$ and these soils typically are derived from basalt and serpentine rocks [1,2,5]. Chromium is also found in coal at the 5–10 $\mu g/g$ level [6].

Naturally occuring chromium concentrations in water arise from mineral weathering processes, soluble organic chromium species, precipitation, and sediment load. Chromium from industrial polution is also present and typically inseparable [7]. Typical concentrations range from a fraction of 1 ng/ml to a few ng/ml. Data from rivers around the world indicated levels of less than 1 ng/ml of chromium [8-12]. In seawater, the mean chromium concentration reported by Bowen was 0.3 ng/ml [13]. Australian saline waters analyzed suggest that 62–87% of the labile chromium present was Cr(VI) and concentrations determined were all less than 1 ng/ml [14]. The exact chemical forms in which chromium is present in water are not known. Theoretically, chromium can persist in the hexavalent state in water with a low content of organic matter. In the trivalent form, chromium forms insoluble compounds at the natural pH of water, unless protected by complex formation. The distribution between the trivalent and hexavalent state is not known [2].

No data indicate any significant contamination of the environment from natural sources. Large volcanoes or forest fires could certainly contribute to the concentration of chromium in the air and water supplies and areas with chromium deposits may have elevated chromium levels. These natural sources, however, do not contribute enough chromium to pose a hazard to human or animal health [2].

13.1.2. Production, consumption, and uses of chromium

Chromium is the 21st most abundant element in the earth's crust with an average concentration of 100 μg/g. Although approximately 40 chromium-containing minerals are known, the only one of economic importance is chromite ($FeOCr_2O_3$). Chromium ores are often classified based on their elemental composition and industrial use into: high-chromium (46–55% Cr_2O_3, Cr:Fe > 2.1) used for metallurgical processes, high-iron (40–46% Cr_2O_3, Cr:Fe 1.5–2.1) used for chemical and metallurgical purposes, and high-aluminum (33–38% Cr_2O_3, 22–34% Al_2O_3) used widely as refractory material [1]. More than 1 billion tons (430 million tons of chromium metal) of the world's chromite reserves are located in South Africa. The world mine production of chromium in 1983 totaled 8.1 million tons [15]. Because more than 90% of the chromium reserves are in developing countries, and East-Europe and the U.S.S.R. countries and the principal consumer nations produce very little, chromium is a classic example of a strategic mineral as evidenced by a review of the historical trend of chromite production. Chromium is stockpiled by industrialized countries with about 5 million tons being stored in the U.S.A. [16].

The demand for chromium is dictated by the economic conditions of the steel industry which dominates the use of this metal. The three principal industrial uses for chromium each year are 76% for metallurgical, 13% for refractory, and 11% for chemical applications [17]. The metallurgical industry uses 44% of the total chromium it utilizes for the production of stainless and heat resistant steels [18]. The use of chromium for the production of refractories is the result of the high melting point (2040°C) and the chemical inertness of chromite. Chromite is, therefore, useful in the production of refractory bricks, and mortars used in high-temperature furnaces [2]. The chemical industry produces more than 70 different chromium compounds for commercial use. Those produced in the largest quantities include: sodium chromate, potassium chromate, potassium dichromate, ammonium dichromate, chromic acid, and basic chromic sulfate used primarily for tanning leather [1,19,20]. Chromium compounds are used to prevent corrosion, improve product durability, and to provide high quality paints. Pigments for paints are often made from chromate and are used to provide color (lead chromates) or for corrosion inhibition (zinc chromates) [15,21]. Chromates are also used as rust and corrosion inhibitors in engines and chromium compounds are added to antifreeze to inhibit corrosion and stop the growth of algae [1]. Chromic acid is used primarily for plating and is also used for photoengraving and offset printing. Many chromium compounds play a significant role in research serving as catalysts for chemical reactions and are commonly found in research laboratories.

13.1.3. Anthropogenic sources of chromium

Industrial activities result in significant amounts of chromium being released into the environment and may pose a potential health hazard to workers exposed to some chromium compounds.

13.1.3.1. Atmospheric emissions

The principal industrial use of chromium in the form of chromite ore, is the metallurgical industry. As such, this industry is the dominant source of anthropogenic chromium emission into the atmosphere. Chromium emissions from the steel industry are most often in the form of particles. Steelmaking utilizes a variety of furnace types with the most common being the electric-arc furnace. Furnaces are equipped with emission control devices including electrostatic precipitators, venturi scrubbers, and bag filters. Of these, the electrostatic precipitator is the most efficient with a chromium emission factor of 14-36 g chromium/ton of steel produced [21]. The particle size of electric-arc furnace emissions containing chromium are usually 10 μm or larger. Metallic chromium produced commercially utilizes processes which do not emit significant quantities of chromium into the atmosphere. The second most prevalent industrial use of chromium is for the production of refractory materials such as bricks for electric furnaces. This process releases chromium into the atmosphere in significant quantities as evidenced by a report that in the early 1970's nearly 10% of the chromium emissions in the U.S.A. were due to refractory processing [22]. The third most prevalent industrial use of chromium is the production of chemicals which are most commonly chromates and dichromates. The production of common chemicals such as sodium dichromate is a process which is well controlled and emissions from this process contribute less than 1% of the chromium released into the atmosphere [22]. Coal-fired power plants also contribute to the atmospheric levels of chromium due to fuel combustion contributing more than five times more chromium to the atmosphere than do commercial and residential furnaces [23]. Another source of atmospheric chromium is the result of cement production using kilns. Emission levels depend on the concentration of chromium in the limestone and the emission control equipment used. Other possible sources of atmospheric chromium emissions include: incinerator emissions, catalytic control emission devices, and asbestos breaklinings and other asbestos containing materials. Interestingly, chromium is one of the elements which is emitted into the atmosphere in equal amounts by natural (principally windblown dust) and anthropogenic (principally steelmaking) sources [1,24].

13.1.3.2. Chromium in water

Groundwater contaminated with chromium is nearly always the result of industrial processes. The most common industrial sources are the alloy industry, wood treatment, and chromium mining operations. Drinking water standards have been set to 0.05 μg/ml total chromium because of the toxic effects of Cr(VI) and the possibility of oxidation of Cr(III) to Cr(VI) [25,26].

Several documented cases of groundwater chromium contamination have been reported in the U.S.A. in the last half century. Most of these have occured in cases where there was shallow sand and gravel aquifers associated with industrial waste. In one instance waste from an aircraft plant where metals were plated, was a source of contamination. Water from mining and electroplating waste sites was also cited [1].

Industrial wastewaters are often treated to remove "heavy" metals such as chromium or in some cases, the effluents can be recycled in the industrial process or sometimes the waste can be used for another separate industrial process [27]. Biologic treatment such as the activated sludge treatment method is used for the treatment of industrial wastewater and also domestic sewage [1]. Chemical precipitation is the most common method for removing chromium from wastewater. This is achieved by reduction of hexavalent chromium to trivalent chromium and subsequent precipitation of the trivalent chromium. Electrochemical reduction is also used to reduce hexavalent chromium to trivalent chromium prior to precipitation as hydroxides. Sulfide precipitation is also used but it is considered a polishing step which should follow hydroxide precipitation. Other techniques include insoluble starch xanthate treatment, electrolytic ferrite treatment, cellulose xanthate treatment, and the use of ion-exchange resins, reverse osmosis, electrodialysis, solvent extraction, and liquid membranes [1]. This is by no means a complete list of all of the methods used to remove chromium from industrial wastewaters, but it clearly indicates the importance assigned to the treatment of water prior to its release into the environment.

13.1.4. Essentiality and toxicity

Chromium is an essential element for man and animals. The trivalent form is required to maintain normal glucose metabolism. Studies in more than five different countries have shown that some people could improve their glucose and/or fat metabolism by ingesting trivalent chromium compounds. The recommended safe and adequate daily dietary intake of chromium for humans is 50 to 200 μg/day. Chromium concentrations in the body fluids of normal healthy adults are usually less than 1 ng/ml. At this level, sensitive analytical methods are required and extreme caution must be taken to avoid sample contamination. Although this is possible with the technology available today, diagnosis of chromium deficiency by chemical or biochemical methods alone is not possible. As a result, a quantitative evaluation of the frequency of chromium deficiency in the human population is not currently possible [2].

Chromium toxicity is dependent on the chemical species with the following hexavalent compounds posing the most common risk: sodium chromate (Na_2CrO_4) and sodium dichromate ($Na_2Cr_2O_7 \cdot 2H_2O$), calcium chromate ($CaCrO_4 \cdot 2H_2O$), the zinc and lead pigment chromates previously mentioned, and chromic acid. Trivalent and elemental chromium species often encountered by industrial workers include ore ($FeOCr_2O_3$) and calcium chromite ($CaCr_2O_3$), chromium (III) oxide (Cr_2O_3), basic chromium sulfates, chromium metal and alloys [1].

The toxicology of chromium compounds has been reviewed by several groups including the US National Academy of Science and the International Agency for Research on Cancer [28,29]. When discussing toxicological effects it is very difficult to discern what oxidation state of chromium is responsible for the effects seen, although hexavalent chromium typically is what has been administered in studies where acute and chronic toxic effects have been recorded. Interpretation of toxicity study results is complicated by the fact that when administered, hexavalent chromium is reduced to trivalent chromium after penetration of membranes. In addition, the route of administration has complicated the interpretation of results from toxicity studies (e.g. hexavalent chromium orally administered is at least partially reduced to Cr(III) by acidic gastric juices) [30]. Additionally, the purity of chemical compounds used for these studies is often suspect, making a conclusive review of toxicity study data almost impossible. In any regard, the effects seen may be due mainly to the reduced trivalent chromium species rather than to the hexavalent species administered in such studies.

The chronic effects of chromium species (trivalent and hexavalent) reportedly include: skin and mucous membrane irritation, broncho-pulmonary effects, and systemic effects involving the kidneys, liver, gastrointestinal tract, and circulatory system [2]. Increased incidence of lung cancer has been reported by workers employed by the chromate producing industry in both the U.S.A. and Europe [31-35]. Increased risk of lung cancer associated with chromite exposure showed mortality rates which were 2 to 30 times higher than normal between 1945 and 1977 [34,36,37]. Recent data show a decreasing risk showing mortality rates approaching those of unexposed workers which is the result of technological advances and the development of an on-line processing of the chromium ore which has decreased the amount of hexavalent chromium compounds associated with the process [38]. Workers in the chromium pigment industry also are at increased risk of developing lung cancer. Langard and Vigander [39] studied Norwegian workers who were principally exposed to zinc chromate dust (up to 0.5 mg hexavalent chromium/m^3) and found a relative risk 44 times greater compared to the general population. A study of workers in a New Jersey pigment plant showed significant relative risks of 1.6 for workers employed for 10 or more years, and a relative risk of 1.9 for workers employed for at least 2 years. Painters who spray chrome pigment paints showed only a slightly increased relative risk as reported by Dalager [40] and automobile painters studied showed no statistically significant increased relative risk [41]. Welding fume particles have exhibited a positive response to the mammalian spot test indicating a mutagenic and possible carcinogenic potential [2]. The incidence of lung cancer in welders who had welded stainless steel for more than 5 years showed a relative risk of 4.4 compared to non-welders [42]. A correlation between nasal and sinonasal cancer and exposure to chromium, welding, soldering and dust was documented in 1983 [43]. An increased risk of stomach and pancreas cancers were documented by Sheffet et al. [44] while studying mortality in a pigment plant using lead and zinc chromates.

Because many individuals exposed to chromium compounds in an industrial setting are exposed to more than one form (hexavalent, trivalent, or metallic), it is difficult to draw any conclusions about increased incidence of cancer and the type of chromium compound. The IARC Working Group on the Evaluation of the Carcinogenic Risk of Chemicals to Humans [45] found that "there is sufficient evidence of respiratory carcinogenicity in men occupationally exposed during chromate production". Data on lung cancer risk in other chromium-associated occupations and for cancer at other sites are insufficient. The epidemiological data do not allow an evaluation of the relative contributions to carcinogenic risk of metallic chromium, trivalent chromium and hexavalent chromium or of soluble *versus* insoluble chromium compounds.

Chromium is generally accepted as being the second most common skin allergen in the general population, after nickel [1,46]. Chromate dermatitis cases typically involve the hands of individuals exposed to diverse occupational and environmental sources of chromium including: blueprints, primer paints, household chemicals and cleaners, cements, diesel engines utilizing anti-corrosive agents, upholstery dyes, leather tanning processes, welding fumes, and matches [2]. There is little evidence that the prevalence of contact dermatitis correlates with the dose but it is clear that chromium-containing compounds (particularly hexavalent compounds) in the workplace pose a potential hazard since many chromium compounds have a sensitizing capability. The chronicity of contact dermatitis is a function of the following: the ubiquity of chromium compounds in foods and objects which individuals touch, the persistence of chromium compounds in the skin, and the coexistence of other contact allergies (e.g. cobalt and nickel or rubber) [47,48].

13.2. SAMPLING PROCEDURES

Quantitation of chromium levels in environmental samples provides necessary information to scientists and medical professionals. Successful quantitation of low levels requires superior analytical expertise and requires stringent sampling procedures. Because chromium is ubiquitous in nature, care must be taken to avoid sample contamination during collection, preparation, and analysis. Clearly, quality control procedures must be invoked when doing chromium determinations. The use of reference materials with known chromium values serve as an invaluable aid to the accurate determination of low concentrations of chromium. In the WHO report on chromium [2] it was recently suggested that further development of analytical instrumentation and preanalytical processing techniques is needed to improve the current detection limit for chromium by an order of magnitudes. In this section, sample collection, handling, and preparation strategies are discussed and precautions to avoid sample contamination are highlighted.

References on p. 400

13.2.1. Sample collection

The primary concern relative to sample collection is that a representative subsample be obtained and that the sample is not subsequently contaminated. This requires that considerable thought and care must go into the sample collection phase. Kratochvil et al. [49] recently did an exhaustive review of the sampling literature highlighting the importance of an adequate sampling plan for the purpose of obtaining a representative sample. The ACS Committee on Environmental Improvement has made recommendations with regard to both sampling and analysis stating that an acceptable sampling program must include:

(1) a proper statistical design which takes into account the goals of the study and the associated uncertainties,

(2) instructions for sample collection, labeling, preservation and transport to the analytical facility, and

(3) training of personnel in the sampling techniques and procedures specified [50].

Sampling concerns were discussed generally in Chapters 2-5. The strategies for obtaining a representative sample for analysis were clearly outlined so the focus here will be on sample preparation and avoidance of sample contamination prior to analysis.

13.2.1.1. Tools and containers for sampling

Environmental samples routinely analyzed include air samples, water samples, rock and soil samples, and biological fluids and tissues. Air samples analyzed, like any gas sample, are microhomogenous so the number of samples required may vary with time and location [51]. Most atmospheric sampling is done with the ultimate goal of measuring pollutants or their effects in either a localized setting for industrial hygiene or monitoring on a more global scale for environmental concerns. Airborne chromium is primarily present as particulate matter. Particulates are clearly a health concern. A question exists as to whether respirable or inhalable particulates should be analyzed and appropriate samplers must be used. Ludwig and Robinson [52] discuss field sampling and site selection and Lynam [53] discusses sampling of automobile exhaust. Filtration is a technique which is commonly used to concentrate particulates. Filters used often include porous polymer membrane filters. Graphite electrodes have also been used as filters [49] and Benson and co-workers [49,54] developed high-purity microquartz filters which can be used up to temperatures of 500°C and which are insensitive to humidity and insoluble in most acids.

There are several national and international standards which address water sampling including ASTM D3370 [55] and ISO 5667 [56]. Reviews in the *Journal of the Water Pollution Control Federation* provide an overview of sampling procedures and difficulties. Owens et al. [57] reviewed four methods of sample collection and preservation and concluded that general cleanliness and the type of filtering apparatus had more of an effect on results than did the method of collection [49]. Many authors recommend minimizing processing done in the field. When sampling, the discrete depth sampling approach may be used or a water column sampler can be effectively utilized. Water samples are typically collected in acid cleaned polyethylene, polypropylene, or teflon bottles. Teflon coated GO-FLOP samplers are often used for seawater collection [49,58] as are Niskin and Hydro-Bios with the modified GO-FLO providing decreased possibility of contamination. Wires used to suspend samplers were evaluated to see which provided the least contamination. Plastic coated steel provided negligible contamination and Kevlar and stainless steel provided only slight contamination [49]. Precipitation in the form of rain or snow requires special sampling strategies and Aichinger [59] has described a fully automatic monitor to collect both rain and frozen precipitation.

Mineralogical sampling research serves as the basis for many of the general sampling theories in use today [60]. Kleeman has recommended that rock samples for analysis should pass a 120-mesh sieve and 0.5-g samples should be analyzed. Soil sample collection has been categorized by James and Dow [61] in three ways:

(1) microvariation (the variability between points in the soil separated by fractions of an inch),

(2) mesovariation (the variability between points separated by up to a few feet); and

(3) macrovariation (the variability between points separated on a scale larger than a few feet).

Microvariation could be due to the localized influence of a single fertilizer pellet while mesovariation would reflect the mechanical application of the pellets and macrovariation would reflect the natural soil properties [49]. Systematic sampling is recommended to identify macrovariation and is accomplished using a modified point sampling system where a cluster of samples is taken at each point. Equipment for soil sampling can be very simple or quite elaborate. Point samples collected manually may be obtained using a shovel, trowel, or spatula and can be stored in plastic bags or cups to prevent contamination. More elaborate soil sampling systems include the mechanical soil sampler developed by Ivancsics [62] which is tractor-drawn and equipped with disks containing spoons for sampling. Smith and co-workers [63] described a similar 3-wheeled self-propelled soil sampler. A remote-controlled sampler was designed for use on the Viking mission to Mars but the cost of such a system is prohibitive for most commercial applications [64].

Biological fluids and tissues are perhaps the most difficult samples to obtain without contamination because of the very low levels of chromium present in these materials. Biological tissues should be obtained by cutting with a titanium blade knife equipped with a Teflon or plastic handle. Veillon and Patterson [65] recommend immediate storage of tissues at $-20°C$ in clean plastic airtight containers or sealed bags and placing those containers in a large plastic bag containing several ice cubes to maintain 100% relative humidity. Biological fluids, especially serum, are difficult samples to obtain without contamination. Researchers in the Vitamin and Mineral Nutrition Research Laboratory at the U.S. Dept. of Agriculture in Beltsville, have established a method for collecting blood samples which avoids the conventional stainless-steel needles which so often are a significant source of contamination [66-68]. A summary of the recommendations relative to serum sample collection follows: Peripheral venipuncture blood samples are drawn using all plastic syringes (Safety Monovette, Sarstedt Inc., Princeton, NJ, U.S.A.) equipped with siliconized needles (butterflies) attached to a short length of small-bore polyvinylchloride tubing (Minicath, Seseret Medical Inc. Sandy, UT, U.S.A.). The siliconization process makes the needle sufficiently hydrophobic to prevent extensive contact of the serum with the metallic surface. Contamination control using this approach was evaluated by repeatedly drawing a low chromium serum solution through the needles with no detectable increase in the chromium concentration [66]. Urine sample collection for chromium analyses is more straightforward and involves collection of samples in standard disposable polypropylene specimen containers which have been acid cleaned.

Other sample types might include plant materials and animal tissues. Regardless of the sample type, care must be taken to ensure that samples do not become contaminated. All tools used for sample collection and storage should be teflon or polyethylene if possible to avoid potential contamination. In some instances metal tools may not be avoided, in these cases it is important to assess the potential sources of contamination. For low pH, liquid samples, chromium may be extracted into the sample. For neutral pH, powdered samples with low moisture content, the only likely source of chromium contamination might be due to abrasion and subsequent flaking off of metal material.

13.2.1.2. Cleaning of containers

Containers for collection and storage of samples should be made of an inert material which will neither contaminate the sample nor adsorb chromium onto the surface of the container, making determined chromium levels unreliably low. Materials preferred for trace element work are (in order of most preferable to least): Teflon (TFE, PFA, FEP, Tefzel), polyethylene, polypropylene, quartz, platinum, and borosilicate glass [69]. Containers should be free of occlusions within 1 mm of the surface and closures should be fabricated of similar polymers without cardboard liners or O-rings or gaskets. Several authors have reported procedures for cleaning sample containers ranging from boiling in concentrated nitric acid for a week to soaking at room temperature in a combination of dilute acids

TABLE 13.1

CLEANING OF CONTAINERS FOR CHROMIUM DETERMINATIONS

1.	Wipe the outside of the container with solvent (hexane, alcohol, or ketone) to remove surface dirt. Use a rinse bottle to rinse surface dirt from the inside of the container.
2.	Soak the container in a bath containing 1:1 dilution of Ultrex HCl and allow to stand for 5 days. When cleaning fluoropolymer containers which are very inert, it is best to heat to 80-100°C.
3.	Empty container and rinse with 18 MΩ ultrapure water and place in a bath containing a 1:1 dilution of sub-boiling distilled HNO_3 and allow to stand for several days. When cleaning fluoropolymer containers it is best to heat to 80-100°C.
4.	Empty container and rinse with 18 MΩ ultrapure water and then fill with ultrapure water and store until needed.
5.	Prior to use, empty, rinse with 18 MΩ ultrapure water and dry in a class 100 clean hood.

[69-72]. This author has found the procedures outlined in Table 13.1 to be very satisfactory for chromium determinations. Although polyethylene is a fairly inert material making it desirable for low level chromium sample containment, it can become yellowed and brittle after repeated exposure to high acid concentrations and may therefore, not always be appropriate.

13.2.2. Sample handling and storage

After the sample has been obtained, care must be taken to avoid contamination while manipulating the sample and while storing it. This is why sample handling in the field is kept to a minimum when the goal is the determination of an element as ubiquitous in nature as chromium.

13.2.2.1. Contamination control

The inability to control contamination can have a serious adverse effect on analytical results when determining elements at very low concentrations. Contamination is typically characterized by analyzing a blank which can be contaminated from any one of four sources including: the environment in which the analysis is performed, the reagents used in the analysis, the apparatus used for the determination, and the analyst [73]. The variability in the blank rather than the absolute value of the blank has a more serious effect on the accuracy of the analysis, requiring that a sufficiently large number of blank determinations must be done to provide a good estimate of the blank. Environmental contamination is a real problem with chromium because of the possibility of airborne particulate contamination. This problem can be minimized by use of HEPA filters which provide Class 100 clean air. Clearly, a complete Class 100 laboratory is desirable, but in its absence, a simple HEPA filtering unit can provide Class 100 clean air to a small work surface and many portable commercial units are now available making "portable" clean air possible. Reagents used in the analysis must be of the highest possible purity. Unfortunately, some commercial reagents are not satisfactory for chromium determinations as they are received. There are several ways to improve the commercial reagents including: isothermal distillation [74], double distillation from vycor, and sub-boiling distillation [71]. More recently, two commercial sources [National Institute of Standards and Technology (NIST), Gaithersburg, MD, U.S.A. and Seastar Chemicals, Seattle, WA, U.S.A.] of sub-boiling distilled acids have produced acids which we find to provide acceptably low chromium blanks. Typical chromium levels for sub-boiling distilled acids are less than 0.1 ng/ml chromium for nitric, hydrochloric, sulphuric, acetic, perchloric, and hydrofluoric acids and ammonia solution. Hydrogen peroxide is often used for wet ash digestion procedures in combination with HNO_3. Unfortunately, sufficiently high purity H_2O_2 is not routinely available for very low level chromium determinations. It is the author's experience that chromium contamination in 50% H_2O_2 (Fisher Scientific, Fair Lawn, NJ, U.S.A.) and 30% Ultex H_2O_2 (J.T. Baker, Phillipsburg, NJ, U.S.A.) is extremely variable from lot to lot requiring preliminary evaluation of each lot to check its suitability for use for chromium

References on p. 400

analyses. Another source of peroxide is Perone 50 (50% peroxide from DuPont, Wilming-ton, DE, U.S.A.) which was developed for the semiconductor industry. Chromium contamination in this product is also variable but this product is often used in the author's laboratory for chromium determinations. Ultrapure H_2O_2 can be prepared by a cryogenic preparation procedure which is expected to be set up at NIST. This will be an excellent source of high-purity H_2O_2 for the analytical community. The third source of contamination is the apparatus used. Section 13.2.1.2 outlined the preference of material types for sampling apparatus and containers. Materials which can provide significant amounts of chromium contamination include: rubber, stainless steel, and glass. These materials must be avoided if at all possible to minimize the likelihood of contamination. Other sources of contamination can include hot plates and muffle furnaces. Filter paper can also provide significant chromium impurities. Membrane filters provide less contamination than paper filters and cellulose acetate and polypropylene filters can be cleaned by acid leaching prior to use [73]. Other sources of contamination include pipet tips and autosampler cups which have not been acid washed. Gas cylinders which are in the laboratory should be covered with plastic bags to avoid flaking paint and rust from entering the laboratory atmosphere as dust. The fourth source of contamination is the analyst who can contami-nate samples by touching surfaces with which the sample will come into contact. Perspiration, cosmetics, lotions, and jewelry are all potential sources of contamination. Another major source of contamination is powdered gloves. Unpowdered disposable polyvinyl chloride gloves are preferable. Cross contamination from other samples being prepared in the laboratory is also a concern and requires that the analyst be keenly aware of the uses of glassware and take appropriate precautions to avoid cross contamination. The analyst also must take special precautions when handling ultrapure reagents to avoid contamination.

13.2.2.2. Sample homogenization

Sample may be homogenized by a variety of different means depending on the sample type and it's consistency. Samples which must be ground are prone to contamination. As Ingamells and Pitard [75] pointed out in their text, contamination during grinding can hardly be avoided. Any implement such as a mill or grinding container used for crushing and grinding must be evaluated to discern what contaminants are likely to be imparted into the sample prior to analysis. Steel mortars are undesirable when preparing samples for chromium analyses but if they are used they should be made of hardened plain-carb-on steel. For hard materials such as chromite, boron carbide mortars are useful. Wiley mills [70] have been used by some researchers and the chromium-plated 20 mesh screen is recommended because it has a harder surface than the stainless-steel screen, provid-ing less chromium contamination. Ball mills are useful for grinding some biological sample materials and the method of Ebdon et al. which utilizes grinding beads in polyethylene containers has been adopted by the author and used with glass beads and Teflon beads which provided relatively low chromium blanks [76] and particle sizes. If samples have to be sieved after grinding to obtain material with a particular particle size, it is best to use nylon or silk screens if possible. Brass screens may be a source of contamination but the solder to hold the screen in place may be a more significant contributor than the screen itself [75]. Any sampling equipment can be a possible source of cross-contamination if it is not properly cleaned between samples so the analyst must ensure that all of the homogenized sample is removed from the grinding device and that the equipment is effectively cleaned prior to future use.

Inhomogeneity is a concern for real world analyses and it is sometimes difficult to decide how to obtain a representative subsample from a whole non-uniform sample. Often the best approach is to homogenize the material and use an aliquot large enough to be compositionally representative of the sample. Homogenization methods for specific types of analyses have been reported in the literature. Difficulties arise when very large quantities of material are to be homogenized such as during the preparation of a large batch of quality control material. In those situations, if the material is powdered, rolling in a large plastic lined drum is recommended. If it is a liquid, a large volume can be mixed with a mechanical stirrer or magnetic stir bar providing that care is taken to avoid contamination from the stirrer.

Sample storage is another concern. The author typically stores perishable samples at low temperatures ($-40°C$) prior to analysis. When doing serum or urine analyses, Veillon [77] recommends lyophilization prior to storage. Although segregation may occur during freezing, the samples can be reconstituted or the whole sample can be ashed and analyzed. To maintain constant moisture in samples which are stored frozen, Veillon [77] recommends storing plastic sample containers in plastic bags which contain a few ice cubes so that they are in an approximately 100% humidity environment in the freezer. This avoids moisture loss through the plastic.

13.2.3. Sample digestion

An excellent general sample preparation reference is the text by Gorsuch which is entitled "The Destruction of Organic Matter" [78]. This text discusses the classical destructive sample dissolution procedures commonly used to prepare samples for analysis.

13.2.3.1. Wet ashing

The most common sample preparation methods are wet ashing and dry ashing. Wet ashing utilizes acids to decompose the sample and the most common acid combinations include: HF for dissolving silica, $HF-HNO_3$ to dissolve nickel-based alloys, $HF-HNO_3-HClO_4$ provides strong oxidizing power for many matrices, $HNO_3-H_2O_2$ is suitable for many biological materials. Ultrapure reagents (see section 13.2.2.1) are required to avoid contamination. Newer microwave dissolution technology reported by Kingston and Jassie [79] allows increased efficiency for acid decomposition because of increased pressure and should be considered if the necessary equipment is available. Recently Nakashima et al. [80] have reported a microwave method for the preparation of marine samples for flame atomic absorption spectrometry (AAS), graphite furnace AAS (GFAAS), and induc-

TABLE 13.2

WET ASH DIGESTION PROCEDURE (Nutrient Composition Laboratory)

1.	Weigh homogenized sample (0.5–1.0 g) into a silanized quartz test tube add 2 ml 18 MΩ water.
2.	Add 0.5–1.0 ml sub-boiling distilled nitric acid and heat samples at 80°C overnight.
3.	Add 1 ml ultrapure hydrogen peroxide, repeating until samples are clear.
4.	Allow samples to go to dryness.
5.	Dilute to a final volume of 3-10 ml.

tively coupled plasma atomic emission spectrometry (ICP-AES). Wet ashing is typically the method of choice for foods and powdered biological materials and a atypical wet ashing procedure used in the author's laboratory for chromium sample preparation is outlined in Table 13.2.

13.2.3.2. Dry ashing

Dry ashing is an alternative means of sample preparation and is accomplished by heating the sample in an open dish or crucible in air. Very often this is done in a muffle furnace located in a clean room. The stages of dry ashing include: sample drying, evaporation of volatile materials, and progressive oxidation of the non-volatile residue until all of the organic matter is destroyed. Ashing aids are sometimes added to samples to accelerate the oxidative process and/or prevent volatilization of the element of interest from the ash. Dry ashing offers the advantages of requiring few reagents, utilizing simple apparatus, lending itself to batch sample preparation and requiring limited operator attention. Table 13.3 outlines the dry ashing procedure which is routinely utilized in the author's laboratory for successful chromium determinations. Dry ashing is very often used for biological fluids. In some instances, unoxidized material remains after dry ashing overnight, in those cases the sample may be treated with hydrochloric or nitric acid, dried, and returned to the furnace to complete the ashing step.

TABLE 13.3

DRY ASH PREPARATION PROCEDURE (Nutrient Composition Laboratory)

1.	Pipet 2 ml of serum into a silanized quart test tube.
2.	Add 20 μl of 11% magnesium nitrate and gently vortex.
3.	Freeze samples at -20°C then freeze-dry overnight.
4.	Place samples in muffle furnace and heat: 100°C 1 hour; 150°C 1 hour; 200°C/hour; 250°C 1 hour; 480°C overnight.
5.	Dissolve ashed sera in 0.5 ml 5% sub-boiling distilled nitric acid.

13.2.3.3. Other techniques

Alternative sample preparation techniques include plasma ashing, fusions, dilution, and the direct analysis of solids. Plasma ashing is a low-temperature technique (typically < 150°C) which is quite time consuming. Volatile elements may still be lost using this technique, particularly if the sample is high in chloride. Fusions are another useful sample preparation technique. Lithium metaborate fusions are capable of dissolving nearly all minerals and are rapid, providing the analyst with solutions which are stable for extended periods of time [78]. Dilution of liquid samples prior to analysis is a technique which can be used for the analysis of beverages and water samples, depending on the level of chromium. Solids may be analyzed directly using analytical techniques such as GFAAS or neutron activation analysis, providing that an adequately homogeneous small sample is available. Alternatively , a slurry may be prepared using homogenized solid material and analyzed by GFAAS, flame AAS, or ICP-AES [81-83].

13.2.4. Preparation and storage of standard solutions

Stock chromium solutions (1000 mg/l) may be prepared using either of the following two methods.

(1) Weigh 1.000 g of high purity chromium metal (shot or flake) into a 1-liter volumetric flask and add 50 ml of high purity HCl. Heat to dissolve, then cool and dilute to one liter with 18 MΩ (high purity) water.

(2) Weigh 2.829 g of $K_2Cr_2O_7$ into a 1-liter volumetric flask and add 250 ml of 18 MΩ water to dissolve. Add 20 ml of HNO_3 and dilute to 1 liter [69].

Standard solutions should be stored in teflon or polyethylene bottles which have been acid cleaned (as discussed in section 13.2.1.2). All volumetric flasks and pipets used to prepare standards should be calibrated by weighing to ensure accuracy. Stock solutions should be weighed and the weight recorded prior to storage so that the standard may be checked for evaporative losses prior to subsequent use.

13.3 DETERMINATION PROCEDURES

Several analytical techniques are suitable for the determination of chromium at a variety of concentrations. The most popular techniques are flame and graphite furnace atomic absorption spectrometry (GFAAS) and inductively coupled plasma atomic emission spectrometry (ICP-AES). Table 13.4 lists the various analytical techniques which have been reported in the literature for the determination of chromium along with typical detection limits. Most of these techniques are not species specific unless the sample preparation step is specific to a particular form of chromium. The most sensitive methods include GFAAS, isotope dilution mass spectrometry and neutron activation analysis.

TABLE 13.4

ANALYTICAL METHODS FOR THE DETERMINATION OF CHROMIUM

Method	Detection limit (μg/ml)
Atomic absorption spectrometry [84]	
− Flame atomization	0.002
− Graphite furnace	0.00001[a]
Atomic emission spectrometry [85]	
− Inductively coupled plasma (ICP)	0.005
− ICP-mass spectrometry (ICP-MS)	0.00005
Isotope dilution mass spectrometry [86]	0.00005
Neutron activation analysis [87]	0.00001
(radiochemical separation)	
Colorimetric method (88)	0.02
Chromatographic methods	
− HPLC-AA [89]	0.06−0.12
− Reversed-phase HPLC [90,91]	0.004−0.010[b]
Polarography [92]	0.3
Miscellaneous	
− X-Ray fluorescence (93)	3.0
− Proton induced X-ray emission (PIXE) (94)	0.0003
− Atomic fluorescence (95)	0.01
− Electron spin resonance (ESR) (96)	0.007

[a] 20 μl sample volume.
[b] 10 μl sample volume.

13.3.1. Atomic absorption spectrometry

One of the most common spectroscopic techniques is atomic absorption spectrometry (AAS). It was first introduced in 1955 by Walsh and this technique revolutionized the field of trace element analyses. AAS is a popular technique because it is fairly fast and straightforward and the instrumentation is affordable. Flame atomization is typically used to determine chromium at concentrations of 0.1 μg/ml or greater while GFAAS offers sub ng/ml detection limits permitting very low level chromium determinations.

13.3.1.1. Flame atomic absorption spectrometry

Chromium is typically determined by flame AAS using an air-acetylene flame or a nitrous oxide-acetylene flame. Although chromium has a large number of resonance lines of similar sensitivity, the 357.9 nm line is most commonly used. At this line the characteristic concentration (the concentration required to provide an absorbance of 0.0044) is 0.04 μg/ml and the detection limit is 0.002 μg/ml [97]. The determination of chromium requires a fuel rich flame. The sensitivity is dependent on the air-acetylene ratio. Triple slot burners are sometimes used to improve stability [98]. Interferences in the flame AAS determination of chromium are possible. Chromium absorption is depressed in an air-acetylene flame in the presence of high concentrations of iron and nickel. The iron interference can be overcome by addition of 2% NH_4Cl to samples and standards. Aggett and O'Brien [99,100] have reported results from extensive studies where they have studied the formation of chromium atoms in air-acetylene flames concluding that the major cause of differences in the atom formation behavior of different compounds is due to the nature of the solid phase formed on desolvation and its subsequent thermal decomposition.

Flame AAS has been used to determine environmental levels of chromium in plants [101] and soils [102]. Jackson et al. [103] described a procedure for the determination of chromium in foods by flame AAS. More recently, a flame AAS method for the determination of chromium in tomato leaves following MIBK extraction was reported [104]. Cresser's review [105] on environmental analyses points out that solvent extraction often provides sufficient preconcentration such that flame AAS may be used rather than GFAAS as exemplified by recent reports of flame AAS determinations of chromium in plant materials and soil samples [106].

The Association of Official Analytical Chemists (AOAC) published a text called "Official Methods of Analysis" [107] which lists methods for a variety of analytes in specific matrices. U.S.A. and Canadian regulations and policy often stipulate the use of AOAC methods. The only chromium method listed in the 1984 centennial edition is a flame AAS method for the determination of chromium in water.

13.3.1.2. Graphite furnace atomic absorption spectrometry

One of the most readily accessible and sensitive analytical techniques for the determination of chromium is GFAAS. A review of chromium applications papers over the past decade convincingly shows that GFAAS is the single most widely used method for the determination of low levels of chromium. Chromium is most often determined at 357.9 nm using $Mg(NO_3)_2$ as a matrix modifier, an ashing temperature of 1650°C and an atomization temperature of 2500°C. This author [108] finds that optimum accuracy is obtained using peak area measurements, Zeeman background correction and platform atomization.

References on p. 400

Although GFAAS is a widely used and sensitive technique for the determination of chromium, it is also the technique most susceptible to matrix interference effects [109]. Urinary and serum chromium data prior to 1979 were obtained using GFAAS systems which had inadequate background correction capability resulting in measured chromium levels that were really measurements of non-specific background absorption rather than actual chromium concentrations [110]. Improved instrumentation with more intense background corrector lamps [111] or wavelength modulation [112] have overcome this problem and have permitted accurate determinations of normal urine and serum chromium levels at < 0.3 ng/ml and 0.16 ng/ml respectively [110]. More recently Baxter et al. [113] have analyzed the lack of success of deuterium background correctors for the determination of chromium in clinical samples and concluded that the interference is dependent on the gain setting of the photomultiplier tube and on the atomization temperature. They report that at temperatures in excess of 2400°C, the interference increases in severity and attribute this to emission caused by chromium together with potassium and sodium in the matrix. They recommend lowering the atomization temperature to 2400°C stating that conventional deuterium background correctors work adequately under these conditions and provide accurate determinations. Other approaches to overcoming problems with background correction include the use of a quartz-halogen source as described by Kayne et al. [111] or the use of Zeeman background correction [114].

GFAAS has been used extensively for the determination of chromium in environmental samples during the last decade. In 1980 Slavin [114] reported on the difficulty of determining chromium in the environment and in the workplace and highlighted the suitability of GFAAS for chromium determinations. Slavin noted that the GFAAS detection limit for chromium was 0.05 ng/ml (20 μl sample size) at that time. He also pointed out the benefit of using pyrolytically coated graphite tubes, platform atomization and Zeeman background correction. In 1983 Muzzarelli et al. [115] used GFAAS to determine chromium in human, cow's, and powdered milk using method of additions. GFAAS has also been used to determine chromium concentrations in plant tissues after wet ash sample digestion and concentration of Cr(III) by coprecipitation with iron [116,117]. The iron was then extracted into 4-methyl-2-pentanone from 9.6 M hydrochloric acid solution and discarded. Chromium was determined in the aqueous portion after treatment with hydrofluoric acid to dissolve silica. In 1986 Hoenig et al. [118] pointed out problems with chromium GFAAS determinations related to the irreproducible quality of the coating between different sets of graphite tubes. GFAAS has also been used to determine chromium levels in canned foods where researchers found that foods packaged in unlaquered cans contain higher levels of chromium and tin than foods packaged in laquered cans [119]. Analytical accuracy was verified by analyzing three National Bureau of Standards (NBS) [now known as the Institute for Standards and Technology (NIST)] standard reference materials with good success. Recently, Anderson et al. [120] reported chromium content data for several breakfast cereals concluding that chromium content of breakfast cereals varies widely and stating that a large potion of the chromium may arise from exogenous sources during the preparation and/or fortification of the cereal. Another paper by Anderson et al. [121] reported the results of a suplementation study done with turkeys which revealed that tissue chromium concentrations were not significantly affected by supplementation with up to 200 μg of Cr as chromium chloride per gram of diet. Another interesting application was reported where researchers analyzed slurry preparations of pine needles using a slurry preparation of NBS SRM 1575 for calibration [122]. Reported detection limits for chromium were 20 pg for both slurry preparations and wet ash digestion solutions. Analytical results from slurry analyses agreed well with results from dissolution methods.

Although literature reviews indicate that GFAAS is widely used for the determination of chromium in biological samples, marine materials, and geological samples, perhaps the largest number of recent GFAAS publications discuss the use of GFAAS for the determination of Cr in biological fluids (urine and serum). GFAAS methods for the determination of chromium in urine have been reported by Veillon et al. [123], Baxter et al. [124] and Halls and Fell [125]. The accuracy of the method of Veillon which used peak height measurements and method of additions, was verified by comparing GFAAS results to gas chromatography-mass spectrometry (GC-MS) data. This method was subsequently submitted to *Clinical Chemistry* as a proposed selected method [123]. Baxter et al. [124] reported an improved method in 1986 which permitted the direct analysis of urine samples using calibration against aqueous standards with probe atomization. More recently Halls and Fell [125] have reported a procedure which leads to decreased GFAAS cycle time to facilitate the large number of occupational exposure urine samples which come into the Glasgow Royal Infirmary. The GFAAS cycle time is decreased by omitting the char step. This leads to increased background but Zeeman effect background correction can compensate adequately for this. The final cycle times were 53 to 70 seconds and analytical accuracy was verified by comparing analytical data to data obtained using a more conventional furnace program. Serum chromium levels are also of particular interest to health professionals, both from the point of supplementation and also occupational exposure. Reported serum chromium levels have varied by more than 2000-fold over the last 25 years. Improved methodology and increased contamination control capabilities have led us to the current reported mean values of approximately 0.13 ng/ml [126]. The US Dept. of Agriculture has used GFAAS methods to determine chromium in urine for supplementation studies they have conducted [126]. Randall and Gibson [127] have likewise used GFAAS methods to determine chromium in urine and serum as indices of chromium status in tannery workers in Canada. Health hazards due to occupational exposure to Cr(III) have been ignored to date, because the rate of absorption (0.4%) of Cr(III) has been considered to be insignificant in industrial settings. This report [127], however, clearly showed increased serum and urine chromium levels for tannery workers which could not be attributed to Cr(VI) absorption. Similarly, Morris et al. [128] used GFAAS to determine chromium levels in plasma, urine and red blood cells in a group of stainless-steel welders with each sample type showing a significant increase in chromium level when stainless- steel welding rods are used.

Some of the most interesting applications using GFAAS for the determination of chromium in environmental samples have included the direct analysis of solid materials. Solids have either been placed directly into the graphite furnace for analysis or prepared as a slurry and introduced into the furnace [129]. In most instances quantitation is accomplished using aqueous standards or appropriate reference materials, peak area measurements, and the use of a L'vov platform.

13.3.2. Inductively coupled plasma spectrometry

The most prevalent atomic emission technique in use today is undoubtedly ICP-AES. It is an attractive technique because it can easily be automated and it provides detection limits within a factor of 2-3 of flame AAS. ICP-AES does not require a lamp source, it provides linear calibration ranges often covering 3-4 orders of magnitude of concentration, and it is particularly suited to the determination of refractory elements [130]. ICP-AES also offers the advantage of being an inherently simultaneous multielement technique. Because of the high plasma temperature, many consider ICP-AES to be more free of chemical interferences than flame AAS, but ICP-AES is prone to other types of interferences, the most notable of which are spectral interferences.

13.3.2.1. Inductively coupled plasma atomic emission spectrometry

ICP-AES has become one of the most widely used techniques of trace element analysis in the past decade [131]. Although ICP spectrometers are more difficult to operate than flame AAS systems, the inherent simultaneous multielement capability of the technique makes it attractive to analysts. Interferences such as line overlaps and baseline shift associated with ICP-AES are frequently compensated for on modern instruments. These instruments are very often computer controlled and are, for the most part, user friendly. ICP spectrometers may be either simultaneous multielement systems (quantometer) or sequential. Each has advantages. Clearly, the analyst has more flexibility with the sequential design and this is advantageous in laboratories where a variety of sample types and analyte concentrations are routinely encountered. With a sequential system the analyst is not limited to pre-selected wavelengths, the optimum conditions for the element of interest can be used for each sample type.

Ward et al. [132] reported a method for the determination of chromium and 22 other elements in agricultural and biological samples by ICP-AES. Samples were prepared using three different ashing procedures and the authors concluded that the two wet ash methods studied provide superior results to dry ashing. Typically chromium is determined at either 267.7 nm or 205.6 nm [133]. Kuennen et al. [134] reported a method for the determination of chromium and 13 other metals in foods using ICP-AES. Samples analyzed included lettuce, peanuts, potatoes, soybeans, wheat, corn, spinach, and beans. The accuracy of the method was verified by analyzing NBS Wheat Flour and Spinach Leaves. Matrix matched standards were used for quantitation. Goulden and Anthony [135] reported a method for the determination of trace metals in freshwater by evaporation and ICP-AES. Samples were preserved by adding HNO_3. A heated spray chamber was used to enhance the emission signal and the reported chromium detection limit was 0.27 ng/ml. Brezezinska and Van Loon [131,136] reported a method for the determination of 13 metals in particulates by ICP-AES. The samples (either membrane or glass filters) are dissolved in closed Teflon vessels placed inside a pressure cooker. McLaren et al. [137] reported ICP-AES results for the analysis of 14 elements in marine sediments concluding that incomplete sample dissolution was the cause of low chromium and titanium results as evidenced by spark source mass spectrography results from analysis of the residue. This highlights the importance of the sample pretreatment step prior to analysis by any method. Nakashima et al. [80] reported ICP-AES results form the analysis of a biological tissue and a marine sediment using microwave sample preparation prior to analysis. Excellent chromium data were obtained for the tissue sample but the sediment sample, again, was incompletely dissolved. Chromium is present in sediment in the form of chromite and dissolution is difficult unless heroic efforts are applied and multiple cycles of addition-evaporation of $HClO_4$–HNO_3 are made to the sample residues following the initial closed vessel decomposition [80,137]. Recently, Grant and Ellis [138] compared flame AAS and ICP-AES data obtained from the analysis of Rhode Island shellfish which are routinely tested for heavy metals because of the discharge of industrial and municipal wastes in the vicinity. The flame AAS detection limit for chromium was 0.03 μg/g and the ICP-AES detection limit was similarly 0.05 μg/g. Calibration with aqueous, non-matrix matched standards provided results comparable to results obtained using the method of standard additions for both flame AAS and ICP-AES. The authors concluded that the methods agree very favorably and note that ICP-AES may be used in lieu of flame AAS providing the analyst with a cost-effective method for accurately and precisely determining chromium and five other metals in shellfish while significantly reducing the instrumental analysis time.

13.3.2.2. Inductively coupled plasma mass spectrometry

Inductively coupled plasma mass spectrometry (ICP-MS) is a powerful technique. Interest in this technique is undoubtedly related to the excellent detection power of ICP-MS which provides detection limits 1–3 orders of magnitude lower than for ICP-AES. The ICP-MS technique also offers simultaneous multielement determinations with wide linear dynamic ranges and provides the ability to measure a wide array of elements. The ICP-MS technique offers the powerful capability of independently measuring multiple stable isotope concentrations. Stable isotope tracer experiments, isotopic "fingerprinting" and isotope dilution analyses can all be accomplished with this capability [139,140].

The determination of chromium by ICP-MS is not always possible. One common interference with ICP-MS determinations of chromium was reported by Beauchemin et al. [141] where chromium determinations were high suggesting an isobaric interference. Two polyatomic species were considered as possible interferents: $^{40}Ar^{12}C$ and $^{35}Cl^{16}OH$. The authors concluded that the (0.69%) chlorine concentration in the sample under consideration was sufficiently high to result in the presence of enough $^{35}Cl^{16}OH$ to seriously interfere with the determination of chromium. An interference which precludes the successful determination of chromium when performing isotope dilution experiments, is the isobaric interference on the ^{53}Cr spike due to $^{37}Cl^{16}O$. This interfering species is present in high chloride matrices and can be a problem when perchloric acid is used for the dissolution procedure if it is not completely removed during the evaporation process [142,143]. A third problem reported with isotope dilution ICP-MS determinations is due to incomplete equilibration of the isotopic spike in the sample [144].

13.3.3. Other techniques

In addition to AAS and AES, there are several other analytical techniques which are well suited to the determination of chromium in environmental samples. A partial list follows: neutron activation analysis, isotope dilution MS, colorimetric methods, chromatographic methods, and electrochemical methods. In this section, the suitability of these methods will be reviewed and representative applications will be discussed.

13.3.3.1. Neutron activation analysis

Neutron activation analysis (NAA) has been used to determine chromium in a variety of environmental samples [7,145]. NAA is based on the formation of radioactive isotopes after sample irradiation by neutrons from a nuclear reactor or other neutron source. The element of interest is identified and quantified by γ-spectroscopy. Low concentrations of chromium require long irradiation times at high flux. NAA is most often used as a reference method in inter-method comparisons because of its specificity and accuracy. The primary limitations are the requirement of a reactor and the cost of analysis. Since ^{32}P interferes with the ^{51}Cr peak, chromium often must be separated in NAA and this is done by distillation as the chromyl chloride [146]. The potential benefits of NAA include the multielement capability and the fact that non-destructive analyses can be performed.

13.3.3.2. Isotope dilution mass spectrometry

Isotope dilution mass spectrometry (IDMS) is based on the fact that many elements have two or more stable isotopes whose proportion in nature is constant. A known amount of a stable isotope is added to a sample and the ratio between the isotopes can then be measured by MS. The change in ratio of $^{50}Cr/^{52}Cr$ after addition of a ^{50}Cr spike gives an accurate estimate of the original concentration in the sample [7]. The advantage of this approach is that the total amount of chromium in the sample need not be measured, so 100% recovery is not necessary. Veillon et al. [147] used IDMS to determine Cr in pooled urine and serum. Trifluoroacetylacetonate (TFA) was used as a chelating agent and the $Cr(TFA)_3$ complex was isolated and measured by GC-MS. IDMS provides accurate and precise urinary chromium measurements. Accuracy was verified by running NBS SRM 1569 brewer's yeast as well as a pooled urine which was also analyzed by continuum source AAS. Jeandel and Minster [148] used IDMS to determine inorganic Cr(III) and total chromium in seawater and Dunstan and Carner [149] demonstrated the capability of this technique by successfully analyzing several biological reference materials. A significant limitation of the technique with regard to low level determinations is the magnitude and variability of the analytical blank and the potential for analyte loss by volatilization [149]. Although IDMS is a powerful technique, like NAA, specialized equipment is required and both methods are most often used as references in intermethod comparisons [150].

13.3.3.3. Colorimetric methods

The colorimetric determination of chromium is based on the formation of a complex resulting from the reaction of hexavalent chromium with 1,5-diphenyl carbazide. This complex may be measured colorimetrically at 540 nm. Other metals including iron and vanadium present in biological materials interfere with the colorimetric determination of chromium making this method less useful than alternative analytical techniques. In 1981 Bryson and Goodall [151] reported an improved colorimetric method where cerium(IV) was used to oxidize chromium completely at room temperature prior to complexation with diphenylcarbazide. More recently, Whitaker [152] has reported a flow injection method with colorimetric detection for the determination of total chromium. Chromium ions are oxidized in a flowing stream to Cr(VI) and then reacted with diphenylcarbazide with the product being measured spectrophotometrically in a 5-mm flow cell. Reported advantages of this approach include: decreased sample preparation time, high sample throughput (40 samples per hour) and good precision at low μg/ml levels.

13.3.3.4. Chromatographic methods

A variety of chromatographic techniques have been used to determine chromium over the past decade. Gas-liquid chromatography (GLC) requires that the chromium be converted into a volatile chelate, usually a fluorinated acetylacetonate. After separation on the GLC column, chromium can be detected by electron-capture detection, microwave-excited emission, atomic absorption or mass spectrometry (see Section 13.3.3.2). GLC combined with these detectors can provide good specificity and excellent detection limits (0.1 ng absolute) [153]. Potential problems include contamination from conventional stainless-steel needles used for sample injection, contamination from flame sealing reaction tubes, and impurities in the TFA is often a problem if it is not purified by distillation.

Chromium may also be determined using high-performance liquid chromatography (HPLC) which is particularly useful when analyzing inorganic compounds which are not suitable for the usual GLC methods because of low volatility or low thermal stability. Bond and Wallace [91] used reversed-phase HPLC based on formation, separation, and subsequent oxidation of dithiocarbamate complexes with electrochemical detection to determine Cr(III) and Cr(VI) as well as copper, nickel, and cobalt. An absolute detection limit of less than 1 ng was obtained. Willett and Knight [154] accurately determined chromium as chromium acetylacetonate in nitric acid digests of NBS SRM 1571 orchard leaves by reversed-phase HPLC on μBondapak C_{18} using 36% acetonitrile in water. Precisions were 3% R.S.D. for 6 injections and recoveries were 97-103%. In 1984 Lajunen et al. [90] studied the behaviour of Cr(III) chelates of 8-hydroxyquinoline on silica-gel, reversed-phase and size-exclusion columns using tetrahydrofuran-chloroform, methanol–

water or acetonitrile and tetrahydrofuran as mobile phases, respectively. Recently, Syty et al. [89] developed a method for the determination of Cr(III) and Cr(VI) in natural waters by ion-pairing HPLC with flame AAS as the detection method.

13.3.3.4. Miscellaneous methods

There are other methods which are used for the determination of chromium in environmental samples. These are not as commonly used for routine chromium determinations. One such method is spark source mass spectrometry (SSMS). This method is not recommended for high accuracy and precision analyses because the source is unstable and the method is highly dependent on sample homogeneity and matrix interferences [153]. Electrochemical methods of analysis are another source of chromium data. Korallus et al. [92] used polarography to determine the effect of ascorbic acid on the reduction of Cr(VI) to Cr(III) as an antidote for chromium poisoning. Interestingly, ascorbic acid is found to be very effective and this therapy prevents damage to the tubules associated with uraemia. Some electrochemical methods lack specificity and are prone to interferences but, with care, may provide detection of submicrogram amounts of chromium [153]. Energy dispersive X-ray fluorescence using a cobalt anode X-ray tube was reported by Potts et al. [93,155] for chromium determinations in geochemical samples. Proton-induced X-ray emission spectrometry (PIXE) is another method which has been used for the determination of chromium. PIXE provides considerably better detection limits than does conventional X-ray fluorescence spectrometry and may become an important tool in instances where non-destructive multielement analyses are required using microsamples. Sjogren [42] utilized PIXE to determine urine chromium levels as an indication of air exposure to stainless-steel welding fumes. Simonoff et al. [94] used PIXE to determine trace chromium in blood serum with a detection limit of approximately 0.3 ng/ml. Michel and co-workers [95] used a direct current plasma (DCP) as an excitation source for flame atomic fluorescence chromium determinations in some biological and botanical reference materials with a detection limit of 0.01 μg/ml.

13.3.4. Speciation

In the fields of biological, clinical, and environmental sciences, it is often crucially important to know the chemical forms as well as the total amount of an element in a sample since the chemical form determines the mode of interaction of the element in biological systems [131]. At the present time, although there is a large amount of total chromium data, there is not much information on the chemical forms of the chromium species present in complex samples. Although several analysts have developed analysis schemes which differentiate between Cr(VI) and Cr(III), little more has been done to develop methods for the identification of specific chromium species in environmental samples. Differentiation between Cr(VI) and Cr(III) is, however, important because of the potential toxicity of the hexavalent state compared with the trivalent state which is essential to man. Though methods based on solvent extraction, with or without prior oxidation, differentiate between these two oxidation states, few analytical data contain this important information [2].

Generally, when one is interested in species specific information chromatographic separation is possible if the compounds are relatively stable. Detection may then be accomplished using element selective techniques such as atomic absorption, fluorescence, or emission spectroscopy. Ion-exchange chromatography coupled with AAS has been used to determine Cr(III) and Cr(VI) in welding fumes [131]. Isozaki et al. [156] reported a method for the determination of Cr(III) and Cr(VI) using direct atomization of chromium from Chelex-100 chelating resin by graphite furnace AAS. Differentiation between different oxidation states is a step towards detailed species specific information. Research over the next decade will undoubtedly address the problem of obtaining detailed speciation information using gas chromatography and liquid chromatography coupled with suitable atomic spectrometric detectors. Availability of these data will significantly advance our understanding of the role of chromium in the natural and human environment.

13.3.5. Accuracy and reference materials

Although analytical methods and contamination control procedures have improved dramatically over the last decade, accurate low level chromium data are not easily obtained. Accurate and precise chromium analyses often require sophisticated procedures which need the full attention of a highly trained analytical chemist. Improved analytical capability may be linked to the increasing number of available reference materials which have certified chromium concentrations. The availability of a wide range of biological, geological, marine, and environmental materials has aided analysts in developing and validating their chromium methods and also facilitates day-to-day quality control evaluation of routine methods. Table 13.5 contains a list of 43 reference materials from 6 sources which have certified chromium concentrations. This is not a complete list of materials, but the identified materials clearly are of potential usefulness to the analyst analyzing samples of environmental interest. A review of the recommended values shows that there are few low level chromium materials, suggesting the need for the development of such materials to aid in validating methods and data at low chromium concentrations. In the future, differentiation between the levels of trivalent and hexavalent chromium will make reference materials even more useful to analysts. Eventually this will lead to the identification of specific species of the various forms of chromium in reference materials. This will require more research and will result only if there is some indication from the scientific community that it is important for the manufacturers of reference materials to provide such data.

References on p. 400

TABLE 13.5

REFERENCE MATERIALS WITH CERTIFIED TRACE CHROMIUM CONCENTRATIONS

Material	Certified concentration[a]	Source[b]
Botanical		
Citrus leaves SRM 1572	0.8 ± 0.2	NIST
Tomato leaves SRM 1573	4.5 ± 0.5	NIST
Pine needles SRM 1575	2.6 ± 0.2	NIST
Bowen's kale	0.369 ± 0.100	Bowen
Hay V-10	6.5 + 0.6 − 0.9	IAEA
Cotton cellulose V-9	0.11 ± 0.03	IAEA
Biological		
Brewer's yeast SRM 1569	2.12 ± 0.05	NIST
Non-fat milk powder SRM 1549	0.0026 ± 0.0007	NIST
Milk powder A-11	0.0177 ± 0.0038	IAEA
Mixed diet RM 8431	0.102 ± 0.0006	NIST
Human diet H-9	0.15 ± 0.04	IAEA
Animal muscle H-4	0.0091 ± 0.0022	IAEA
Biological fluids		
Human serum SRM 909	0.0913 + 0.0061 − 0.0050	NIST
Freeze dried urine - elevated SRM 2670	0.085 ± 0.006 μg/ml	NIST
Bovine serum SRM 1598	0.14 ± 0.08 ng/g	NIST
Coals		
Coal fly ash SRM 1633a	196 ± 6	ÑIST
Coal SRM 1635	2.5 ± 0.3	NIST

Table 13.5 (continued)

Material	Certified concentration[a]	Source[b]
Coal CRM-040	31.3 ± 2.0	BCR
Waters		
Water SRM 1643b	0.0186 ± 0.0004	NIST
Seawater NASS-2	0.175 ± 0.010 μg/l	NRCC
Seawater CASS-2	0.121 ± 0.006 μg/l	NRCC
Riverine water SLRS-1	0.36 ± 0.04 μg/l	NRCC
Fresh water W-4	9.9 + 0.6 μg/l − 0.9	IAEA
Sediments/Soils		
Estuarine sediment SRM 1646	76 ± 3	NIST
Estuarine sediment MESS-1	71 ± 11	NRCC
Estuarine sediment BCSS-1	123 ± 14	NRCC
Buffalo river sediment SRM 2704	135 ± 5	NIST
Sediment PACS-1	113 ± 8	NRCC
Lake sediment SL-1	104 ± 9	NRCC
Pond sediment CRM-2	75 ± 5	NIES
Marine sediment SD-N-1/2	149 ± 15	IAEA
Soil Soil-7	60 + 14 − 11	IAEA
Marine Tissues		
Lobster hepatopancreas TORT-1	2.4 ± 0.6	NRCC
Non-defatted lobster hepatopancreas LUTS-1	0.53 ± 0.08	NRCC
Dogfish liver DOLT-1	0.40 ± 0.07	NRCC

(Continued on p. 400)

References on p. 400

Table 13.5 (continued)

Material	Certified concentration[a]	Source[b]
Dogfish muscle DORM-1	3.60 ± 0.40	NRCC
Shrimp tissue MA-A-3/TM	1.11 ± 0.36	IAEA
Dried Copepoda MA-A-1/TM	1.1 ± 0.2	IAEA
Fish flesh MA-A-2/TM	1.3 ± 0.1	IAEA
Miscellaneous		
Urban particulate matter SRM 1648	403 ± 12	NIST
Wear metals in oil SRM 1085	298 ± 5	NIST
Uranium oxide SR-54	3.6 + 0.7	IAEA
Air filter Air-3/1	0.0050 ± 0.0003	IAEA

[a] Concentration are in μg/g, dry weight unless otherwise specified.

[b] Sources for reference materials include:
(1) IAEA, International Atomic Energy Aagency, Analytical Quality Control Services, Laboratory Seibersdorf, P.O. Box 100, A-1400 Vienna, Austria.
(2) NIST, National Institute for Standards and Technology, Office of Standard RM's, Room B311, Chemistry Bldg., Gaithersburg, MD 20899, U.S.A.
(3) BCR, Community Bureau of Reference, Commission of the European Communities, 200 Rue de la Loi, -1049 Brussels, Belgium.
(4) NRCC, National Research Council of Canada, Division of Chemistry, Ottawa, K1A OR6, Canada.
(5) Bowen, Dr. H.J.M. Bowen, Dept. of Chemistry, University of Reading, Whiteknights, P.O. Box 224, Reading RG6 2AD, U.K.
(6) NIES, National Institute for Environmental Studies, Japan Environment Agency, P.O. Yatabe, Tsukuba Ibaraki 300-21, Japan.

REFERENCES

1 N.O. Nriagu and E. Nieboer (Editors), *Chromium in the Natural and Human Environments*, Vol. 20, Wiley, New York, 1988, 571 pp.
2 *International Programme on Chemical Safety, Environmental Health Criteria: Chromium*, World Health Organization, Geneva, 1988, 197 pp.
3 B.E. Guthrie, W.R. Wolf and C. Veillon, *Anal. Chem.*, 50 (1978) 1900.
4 M. Fleischer, *Ann. NY Acad. Sci.*, 199 (1972) 6-16.
5 D.J. Lisk, *Adv. Agron.*, 24 (1972) 267-325.
6 E. Merian, *Toxicol. Environ. Chem.*, 8 (1984) 9-38.
7 S. Langard (Editor), *Biological and Environmental Aspects of Chromium*, Vol. 5, Elsevier, Amsterdam, 1982.

8 R.J. Lovett and F.G. Lee, *Environ. Sci. Technol.*, 10 (1976) 67-71.
9 K.S. Subramanian, C.L. Chakrabarti, J.E. Sueiras and I.S. Maines, *Anal. Chem.*, 50 (1978) 444-448.
10 H. Yamazaki, *Anal. Chim. Acta*, 113 (1980) 131-137.
11 R.E. Cranston and J.W. Murray, *Limnol. Oceanogr.*, 25 (1980) 1104-1112.
12 P. Benes and E. Steinnes, *Water Res.*, 9 (1975) 741-749.
13 H.J.M. Bowen, *Environmental Chemistry of the Elements*, Academic Press, London, 1979, 333 pp.
14 G.E. Batley and J.P. Matousek, *Anal. Chem.*, 52 (1980) 1570-1574.
15 J.F. Papp, *Chromium*, in: *U.S. Bureau of Mines, Minerals Yearbooks*, U.S. Government Printing Office, Washington, DC, 1983, p. 190.
16 J.A. Wolfe, *Mineral Resources: A World Review*, Chapman and Hall, New York, 1984, pp. 104-108.
17 J.H. DeYoung, M.P. Lee and B.R. Lipin, *International Strategic Minerals Inventory Summary Report - Chromium, U.S. Geological Survey Circular 930-B*, U.S. Dept. of the Interior, Washington, DC, 1984.
18 EPA, *Health Assessment Document for Chromium, Report No. EPA-600/8-83-014F*, Office of Health and Environmental Assessment, U.S. Environmental Protection Agency, Research Triangle Park, NC, 1984.
19 N.B. Pederson, in S. Langard (Editor), *Biological and Environmental Aspects of Chromium*, Elsevier, Amsterdam, 1982, pp. 249-275.
20 A. Yassi and E. Nieboer, in J.O. Nriagu and E. Nieboer (Editors), *Chromium in the Natural and Human Environment*, Wiley, New York, 1988, Ch. 17.
21 J.M. Pacyna, in J.O. Nriagu and C. Davidson (Editors), *Toxic Metals in the Atmosphere*, Wiley, New York, 1986, pp. 1-32.
22 GCA, *National Emissions Inventory of Sources and Emissions of Chromium*, U.S. National Technical Information Service, Springfield, VA, 1973, PB230-034.
23 J.M. Pacyna, *Trace Element Emission from Anthropogenic Sources in Europe, NILU Report 10/82*, The Norwegian Institute for Air Research, Lillestrom, 1982.
24 W.D. Balgord, *Science*, 180 (1973) 1168-1169.
25 Health and Welfare Canada, *Guidelines for Canadian Drinking Water Quality*, Ottawa, 1979.
26 US EPA, *National Interim Primary Drinking Water Regulations, Report No. EPA/570/9-76/003*, Washington, DC, 1976.
27 R.G.W. Laughlin, *Canadian Waste Exchange Program: Successes and Failures*, presented at the *National Conf. on Hazardous and Toxic Waste Management*, New Jersey Institute of Technology, Newark, NJ, 1980.
28 US NAS, *Chromium*, US National Academy of Sciences, Washington, DC, 1974.
29 IARC, *Chromium and Chromium Compounds*, in *Some Metals and Metallic Compounds*, International Agency for Research on Cancer, Monography on Evaluation of the Carcinogenic Risk of Chemicals to Humans, Vol. 23, Lyon, 1980.
30 S. Deflora and V. Boido, *Mutat. Res.*, 77 (1980) 307-315.
31 F.W. Taylor, *Am. J. Public Health*, 56 (1966) 218-229.
32 P.E. Enterline, *J. Occup. Med.*, 16 (1974) 523-526.
33 T.F. Manusco, in T.C. Hutchinson (Editor), *Proceedings of the International Conference on Heavy Metals in the Environment, Toronto, Canada, October 27-31, 1975*, Institute for Environmental Studies, Toronto, 1975, pp. 343-356.
34 M.R. Alderson, N.S. Rattan and L. Bidstrup, *Br. J. of Ind. Med.*, 38 (1981) 117-124.
35 P.L. Bidstrup and R.A.M. Case, *Br. J. Ind. Med.*, 13 (1956) 260-264.
36 W. Machle and F. Gregorius, *Public. Health Rep.*, 63 (1948) 1114-1127.
37 US PHS, *Mealth of Workers in Chromate Producing Industry*, US Public Health Service Publication No. 192, US Dept. of Health, Education, and Welfare, Washington, DC, 1953.
38 U. Korallus and N. Loenhoff, *Arbeitsmed. Sozialmed. Praventivmed.*, 16 (1981) 285-289 (in German).
39 S. Langard and T. Vigander, *Br. J. Ind. Med.*, 40 (1983) 71-74.

40 N.A. Dalager, T.J. Mason, J.F. Fraumeni, R. Hoover and W.W. Payne, *J. Occup. Med.*, 22 (1980) 25-29.
41 L. Chiazze, L.D. Ference and P.H. Wolfe, *J. Occup. Med.*, 22 (1980) 520-526.
42 B. Sjogren, *Scand. J. Work Environ. Health*, 6 (1980) 197-200.
43 S. Hernberg, P. Westerholm, K. Schultz-Larsen, R. Degerth, E. Kuosma, A. Englund, U. Engzell, H. Sand Hansen and P. Mutanen, *Scand. J. Work Environ. Health*, 9 (1983) 315-326.
44 A. Sheffet, J. Thind, A.M. Miller and D.B. Louria, *Arch. Environ. Health*, 37 (1982) 44-52.
45 *IARC*, in Lyons (Editor), *Some Metals and Metallic Compounds*, International Agency for Research on Cancer, IARC Monographs on the Evaluation of the Carcinogenic Risk of Chemicals to Humans, Vol. 23, 1980, pp. 205-323.
46 L. Polak, J.L. Turk and J.R. Frey, in P. Kallos, B.H. Waksman and A. DeWeck (Editors), *Progress in Allergy*, Basel, Karger, Vol. 17, 1970, pp. 145-226.
47 S. Fregert and H. Rorsman, *Arch. Dermatol.*, 90 (1964) 4-6.
48 S. Fregert, *Contact Dermatit*, 1 (1975) 96-107.
49 B. Kratochvil, D. Wallace and J.K. Taylor, *Anal. Chem.*, 56 (1984) 113R-129R.
50 *ACS Committee on Environmental Improvement*, 55 (1983) 2210-2218.
51 J.K. Taylor, *Quality Assurance of Chemical Measurements*, Lewis Publishers, Chelsea, MI, 1989, pp. 1-114.
52 F.L. Ludwing and E. Robinson, *J. Air Pollut. Control Assoc.*, 17 (1967) 664-669.
53 D.R. Lynam, *Abstr. Int. B.*, (Feb) 33 (1973) 369B.
54 A.L. Benson, P.L. Levins, A.A. Massucco, D.B. Sparrow and J.R. Valentine, *J. Air. Pollut. Control Assoc.*, 25 (1975) 274-277.
55 *ASTM D3370 Practices for Water Sampling*, Amer. Society Test. Mat., Philadelphia, PA, 1982.
56 *ISO 5667 Part 1 - Guidance on the Design of Sampling Programmes; Part 2 - Water Quality, Sampling, General Guide to Sampling Techniques*, Geneva, 1980.
57 J.W. Owens, E.S. Gladney and W.D. Purtymun, *Anal. Lett.*, 13 (1980) 253-260.
58 K.S. Subrmananian, J.C. Meranger, C.C. Wan and A. Corsini, *Int. J. Environ. Anal. Chem.*, 19 (1985) 261-272.
59 H.L. Aichinger, *Sci. Total Environ.*, 16 (1980) 279-283.
60 C.O. Ingamells and F.F. Pitard, *Applied Geochemical Analysis*, Wiley, New York, 1986, Ch. 1.
61 D.W. James, A.I. Dow, *Wash. Agric. Exp. Stn. Bull.*, 749 (1972) 1-11.
62 J. Ivancsics, *J. Commun. Soil Sci. Plant Anal.*, 11 (1980) 881-887.
63 D.B. Smith, A.J. Keaster, J.M. Cheshire and R.H. Ward, *J. Econ. Entomol.*, 74 (1981) 625-629.
64 J.A. Alic, *Digging into Mars*, Am. Soc. Engineering Educ., Washington, DC, 1982.
65 C. Veillon and K.Y. Patterson, in I.K. O'Neill, P. Schuler and L. Fishbein (Editors), *IARC Scientific Publications, No. 71*, International Agency for Research on Cancer, Lyon, 1986.
66 C. Veillon, K.Y. Patterson and N.A. Bryden, *Anal. Chim. Acta*, 136 (1982) 233-241.
67 R.A. Anderson, N.A. Bryden and M.M. Polansky, *Am. J. Clin. Nutrit.*, 41 (1985) 571-577.
68 J.A. Randall and R.S. Gibson, *Proc. Soc. Exp. Biol. Med.*, 185 (1987) 16-23.
69 M.S. Epstein, N.J. Miller-Ihli, H.M. Kingston and R. Watters, *Successful Atomic Spectrometric Determinations*, Society for Applied Spectroscopy Short Course Materials, Frederick, MD, 1987.
70 E.E. Cary and M. Rutzke, *J. Assoc. Off. Anal. Chem.*, 66 (1983) 850-852.
71 R.W. Dabeka, A. Nykytiuk, S.S. Berman and D.S. Russell, *Anal. Chem.*, 48 (1976) 1203-1206.
72 J.R. Moody and R.M. Lindstrom, *Anal. Chem.*, 49 (1977) 2264-2267.
73 T.J. Murphy, *Accuracy in Trace Analysis: Sampling, Sample Handling, and Analysis, Proceedings of the 7th IMR Symposium, Gaithersburg, MD*, October 7-11, 1974, NBS Special Publication 422, issued August 1976.
74 C. Veillon and D.C. Reamer, *Anal. Chem.*, 53 (1981) 549-550.

75 C.O. Ingamells and F.F. Pitard, *Applied Geochemical Analysis,* Wiley, New York, 1986, pp. 75-84.
76 N.J. Miller-Ihli, *Fresenius' J. Anal. Chem.,* 337 (1990) 271-274.
77 C. Veillon, *Anal. Chem.,* 58 (1986) 851A-866A.
78 T.T. Gorsuch, *The Destruction of Organic Matter,* Pergamon, Oxford, 1970.
79 H.M. Kingston and L.B. Jassie (Editors), *Introduction to Microwave Sample Preparation, Theory and Practice,* ACS, Washington, DC, 1988, 263 pp.
80 S. Nakashima, R.E. Sturgeon, S.N. Willie and S.S. Berman, *Analyst,* 113 (1988) 159-163.
81 N.J. Miller-Ihli, *J. Anal. At. Spectrom.,* 3 (1988) 73-81.
82 J. Willis, *Anal. Chem.,* 47 (1975) 1752.
83 L. Ebdon and J.R. Wilkinson, *J. Anal. At. Spectrom.,* 2 (1987) 39-44.
84 *Analytical Methods for Atomic Absorption Spectrophotometry,* Perkin-Elmer Corp., Norwalk, CT, 1982, pp. 12.3-12.4.
85 J. McLaren, personal communication.
86 C. Veillon, personal communication.
87 R. Greenburg, personal communication.
88 M. Bose, *Anal. Chim. Acta,* 10 (1954) 201-209.
89 A. Syty, R.G. Christensen and T.C. Rains, *J. Anal. At. Spectrom.,* 3 (1988) 193-197.
90 L.H.J. Lajunen, E. Eijarvi and T. Kenakkala, *Analyst,* 109 (1984) 699-701.
91 A.M. Bond and G.G. Wallace, *Anal. Chem.,* 54 (1982) 1706-1712.
92 U. Korallus, C. Harzdorf and J. Lewalter, *Int. Arch. Occup. Environ. Health,* 53 (1984) 247-256.
93 P.J. Potts, P.C. Webb and J.S. Watson, *J. Anal. At. Spectrom.,* 1 (1986) 467-471.
94 M. Simonoff, Y. Llabador, C. Hamon and G.N. Simonoff, *Anal. Chem.,* 56 (1984) 454-457.
95 M.S. Hendrick, P.A. Goliber and R. Michel, *J. Anal. At. Spectrom.,* 1 (1986) 45-50.
96 W.G. Bryson, D.P. Hubbard, B.M. Peake and J. Simpson, *Anal. Chim. Acta,* 116 (1980) 353-357.
97 B. Welz, *Atomic Absorption Spectrometry,* VCH Publishers, Weinheim, 1985, 506 pp.
98 S. Sprague and W. Slavin, *At. Absorpt. Newslett.,* 4 (1965) 293.
99 J. Aggett and G. O'Brien, *Analyst,* 106 (1981) 506-513.
100 J. Aggett and G. O'Brien, *Analyst,* 106 (1981) 497-505.
101 P.D. Parr and F.G. Taylor, *Environ. Exper. Botany,* 20 (1980) 157-160.
102 R.J. Bartlett and J.M. Kimble, *J. Environ. Qual.,* 12 (1983) 169-172.
103 F.J. Jackson, J.I. Read and B.L. Lucas, *Analyst,* 105 (1980) 359-370.
104 R. Farre, M.J. Lagarda and R. Montoro, *J. Assoc. Off. Anal. Chem.,* 69 (1986) 876.
105 M.S. Cresser, L. Ebdon and J.R. Dean, *J. Anal. At. Spectrom.,* 4 (1989) 1R-26R.
106 R. Ajlec, M. Cop and J. Stupar, *Analyst,* 113 (1988) 585.
107 S. Williams (Editor), *Official Methods of Analysis of the Association of Official Analytical Chemists,* Arlington, VA, 1984, Methods 33.089-33094, pp. 628-629.
108 N.J. Miller-Ihli, *A Method for Determining Cr in Complex Samples,* Eastern Analytical Symposium, Oct. 2-7, 1988, New York, NY, Paper No. 87.
109 C. Veillon, *Sci. Total Environ.,* 86 (1989) 65-68.
110 C. Veillon, K.Y. Patterson and N.A. Bryden, *Anal. Chim. Acta,* 164 (1984) 67-76.
111 F.J. Kayne, G. Komar, H. Laboda and R.E. Vanderlinde, *Clin. Chem.,* 24 (1978) 2151.
112 J.M. Harnly and T.C. O'Haver, *Anal. Chem.,* 49 (1977) 2187.
113 D.C. Baxter, D. Littlejohn, J.M. Ottaway, G.S. Fell and D.J. Halls, *J. Anal. At. Spectrom.,* 1 (1986) 135-139.
114 W. Slavin, *At. Spectrosc.,* 2 (1981) 8-12.
115 R.A.A. Muzzarelli, C.E. Eugeni, F. Tanfani, G. Caramia and D. Pezzola, *Milchwissenschaft,* 38 (1983) 453-457.
116 E.E. Cary and M. Rutzke, *J. Assoc. Off. Anal. Chem.,* 66 (1983) 850-852.
117 E.E. Cary, *J. Assoc. Off. Anal. Chem.,* 68 (1985) 495-498.
118 M. Hoenig, F. Dehairs and A.-M. deKersabiec, *J. Anal. At. Spectrom.,* 1 (1986) 449-452.

119 L. Jorhem and S. Slorach, *Food Additives Contam.*, 4 (1987) 309-316.
120 R.A. Anderson, N.A. Bryden and M.M. Polansky, *J. Food Comp.*, 1 (1988) 303-308.
121 R.A. Anderson, N.A. Bryden, M.M. Polansky and M.P. Richards, *J. Agric. Food Chem.*, 37 (1989) 131-134.
122 N. Carrion, Z.A. deBenzo, B. Moreno, A. Fernandez, E. Eljuri and D. Flores, *J. Anal. At. Spectrom.*, 3 (1988) 479-483.
123 C. Veillon, K.Y. Patterson and N.A. Bryden, *Clin. Chem.*, 28 (1982) 2309-2311.
124 D.C. Baxter, D. Littlejohn, J.M. Ottaway, G.S. Fell and D.J. Halls, *J. Anal. At. Spectrom.*, 1 (1986) 35-39.
125 D.J. Halls and G.S. Fell, *J. Anal. At. Spectrom.*, 3 (1988) 105-109.
126 R.A. Anderson, N.A. Bryden and M.M. Polansky, *Am. J. Clin. Nutrit.*, 41 (1985) 571-577.
127 J. Randall and R.S. Gibson, *Proc. Soc. Exptl. Biol. Med.*, 185 (1987) 16-23.
128 B.W. Morris, H. Griffiths, C.A. Hardisty and G.J. Kemp, *At. Spectrosc.*, 10 (1989) 1-3.
129 M. Cresser, J. Ebdon and J. Dean, *J. Anal. At. Spectrom.*, 4 (1989) 1R-26R.
130 W. Slavin, *Anal. Chem.*, 58 (1986) 589A-597A.
131 J.C. VanLoon, *Selected Methods of Trace Metal Analysis; Biological and Environmental Samples*, Vol. 80, Wiley, New York, 1985, 357 pp.
132 A.F. Ward, L.F. Marciello, L. Carrara and V.J. Luciano, *Spectrosc. Lett.*, 13 (1980) 803.
133 J.W. Jones, S.G. Capar and T.C. O'Haer, *Analyst*, 107 (1982) 353.
134 R.W. Kuennen, K.A. Wolnick and F.L. Fricke, *Anal. Chem.*, 54 (1982) 2146.
135 P.D. Goulden and D.H.J. Anthony, *Anal. Chem.*, 54 (1982) 1681.
136 A. Brezezinska and J.C. Van Loon, internal laboratory procedure, 1983.
137 J.W. McLaren, S.S. Berman, V.J. Boyko and D.S. Russell, *Anal. Chem.*, 53 (1981) 1802-1806.
138 W.A. Grant and P.C. Ellis, *J. Anal. At. Spectrom.*, 3 (1988) 815-820.
139 H.E. Taylor, *Spectroscopy*, 1 (1987) 20-22.
140 J.W. McLaren, A.P. Mykytiuk, S.N. Willie and S.S. Berman, *Anal. Chem.*, 57 (1985) 2907-2911.
141 D. Beauchemin, J.W. McLaren, S.N. Willie and S.S. Berman, *Anal. Chem.*, 60 (1988) 687-691.
142 J.W. McLaren, D. Beauchemin and S.S. Berman, *Anal. Chem.*, 59 (1987) 610-613.
143 D. Beauchemin, J.W. McLaren and S.S. Berman, *J. Anal. At. Spectrom.*, 3 (1988) 775-780.
144 D. Beauchemin, J.W. McLaren, A.P. Mykytiuk and S.S. Berman, *Anal. Chem.*, 59 (1987) 778-783.
145 H. Watanabe, *Soil Sci. Plant. Nutr.*, 30 (1984) 543-554.
146 J. Versieck, J. de Rudder and J. Hoste, in D. Shapcott and J. Hubert (Editors), *Developments in Nutrition and Metabolism.* Elsevier/North Holland Biomedical Press, Amsterdam, 1979, pp. 59-68.
147 C. Veillon, W.R. Wolf and B.E. Guthrie, *Anal. Chem.*, 51 (1979) 1022-1024.
148 C. Jeandel and J.-F. Minster, *Mar. Chem.*, 14 (1984) 347-364.
149 L.P. Dunstan and E.L. Garner, in D.D. Hemphill (Editor), *Trace Substances in Environmental Health*, University of Missouri, Columbia, MI, 1977, pp. 334-337.
150 J.M. Harnly, K.Y. Patterson, C. Veillon, W.R. Wolf, J. Marshall, D. Littlejohn, J.M. Ottaway, N.J. Miller-Ihli and T.C. O'Haver, *Anal. Chem.*, 55 (1983) 1417-1419.
151 W.G. Bryson and C.M. Goodall, *Anal. Chim. Acta*, 124 (1981) 391-401.
152 M.J. Whitaker, *Anal. Chim. Acta*, 174 (1985) 375-378.
153 A. Vercruysse (Editor), *Hazardous Metals in Human Toxicology, Techniques and Instrumentation in Analytical Chemistry*, Vol. 4, Elsevier, Amsterdam, 1984, 337 pp.
154 J.D. Willett and M.M. Knight, *J. Chromatogr.*, 237 (1982) 99-105.
155 P.J. Potts, P.C. Webb and J.S. Watson, *J. Anal. At. Spectrom.*, 1 (1987) 67-72.
156 A. Isozaki, K. Kumagai and S. Utsumi, *Anal. Chim. Acta*, 153 (1983) 15.

M. Stoeppler (Editor)/*Hazardous Metals in the Environment*
© 1992 Elsevier Science Publishers B.V. All rights reserved

Chapter 14

Nickel and Cobalt

Markus Stoeppler and Peter Ostapczuk

Research Center Juelich, Institute of Applied Physical Chemistry, P.O. Box 1913, D-W-5170 Juelich (Germany)

CONTENTS

14.1. Introduction 405
14.2. Environmental levels of nickel and cobalt 406
14.3. Sampling and sample pretreatment 408
14.4. Enrichment and separation 408
14.5. Analytical methods 410
    14.5.1. Overview 410
    14.5.2. Atomic absorption spectrometry 411
        14.5.2.1. Liquid samples 412
        14.5.2.2. Solid samples 414
    14.5.3. Atomic emission spectroscopy with plasmas 415
        14.5.3.1. Liquid and solid samples 416
    14.5.4. Inductively coupled plasma–mass spectrometry 417
    14.5.5. Liquid and solid samples 418
    14.5.6. Electrochemical methods 419
        14.5.6.1. Aqueous samples 420
        14.5.6.2. Other liquids and solid samples 422
    14.5.7. X-ray methods 424
    14.5.8. Nuclear methods 427
    14.5.9. Chromatographic methods 428
    14.5.10. Mass spectrometry 430
    14.5.11. Spectrophotometry and related techniques 431
14.6. Speciation 432
14.7. Quality control and reference materials 434
14.8. Conclusions and prospects 440
References 442

14.1. INTRODUCTION

Nickel (Ni) is widely distributed in the environment and is the twenty-fourth most abundant element in the earth's crust. Its atomic weight is 58.71; atomic number 28; density 8.9; melting point 1453°C; boiling point 2732°C. In metallic form it is silver-white and malleable. Its oxidation states are $+1$, $+2$ and $+3$. Important compounds are nickel oxide, nickel hydroxide, nickel sulfate, nickel chloride, nickel sulfate and nickel subsulfide. Nickel carbonyl [$Ni(CO)_4$] is a colorless, volatile liquid with a boiling point of 43°C.

Divalent nickel compounds are non-toxic for animals, plants and man at prevalent concentrations in natural waters, soils and food. In human beings adverse effects (dermatitis) from inorganic, water-soluble nickel compounds frequently occur following skin contact. Inhalation of these compounds mainly at the workplace can cause respiratory tract irritation and asthma. Inhalation of inorganic, water-insoluble nickel compounds in dusts or fumes can increase the risk of respiratory tract cancer. Nickel subsulfide is a potent carcinogen as has been shown in animal experiments. Because of its volatility and solubility in lipids nickel carbonyl is extremely toxic.

There are several comprehensive reviews about nickel chemistry, bio- and geochemistry [1–3] as well as toxicology and biological monitoring [3-7].

Cobalt (Co) is comparatively rare and is the thirty-second most abundant element in the earth's crust. Its atomic weight is 58.9; atomic number 27; density 8.9; melting point 1495°C, boiling point 2870°C. In metallic form it is steel-grey, shiny, hard, ductile and brittle, and magnetic. Its oxidation states are +2 and +3.

Important cobalt compounds are cobalt oxide, cobalt tetraoxide, cobalt chloride, cobalt sulfide and cobalt sulfate. Cobalt is a constituent of vitamin B12 and in this form essential for mammals, i.e. also for human beings. A lack of vitamin B12 causes pernicious anemia. On the one hand in various regions of Australia, New Zealand, Kenya, Florida and the Soviet Union cobalt levels in soils are very low which can result in extremely low cobalt contents in food. If this is the case, deficiency syndromes may occur, which has been especially observed in ruminants.

On the other hand, because of its industrial use, cobalt also poses a potential danger in occupational exposure primarily for metal workers. Exposure to cobalt-containing dust and fumes can cause adverse effects to lungs, heart and skin. Cobalt, like nickel, also belongs to the metals known as occupational carcinogens.

Cobalt chemistry, bio-geochemistry as well as toxicology and biological monitoring have recently been reviewed in great detail [7-11].

14.2. ENVIRONMENTAL LEVELS OF NICKEL AND COBALT

Typical prevalent contents for nickel and cobalt in air, precipitation, fresh- and seawater, plants, soils, sewage sludge, animal and human materials ranging from the low μg/kg to the high mg/kg level based on a number of recent compilations and original papers [3,5,8,10–33] are listed in Table 14.1. From this the significantly lower cobalt contents, except in cases of contents in human tissues due to implanted prostheses of cobalt–chromium alloys [34], are obvious and also generally pose greater difficulties for analysis.

TABLE 14.1.

TYPICAL LEVELS OF NICKEL AND COBALT IN BIOLOGICAL AND ENVIRONMENTAL
MATERIALS (see refs. 3, 5, 8, 10−33)

Material	Ni content		Co content		Unit
Air (non-polluted area)	0.5−4		0.0008−0.35		ng/m^3
Air (large cities)	9.5−20		1.9−80		ng/m^3
Dust (non-polluted area)	125		12		mg/kg
Dust (large cities)	≤225		≤ 20		mg/kg
Coal fly ash	2−115		5−70		mg/kg
River water	< 1−60		0.4−4		µg/l
Estuarine water	≤ 10		≤ 1		µg/l
Seawater, open sea	< 0.2−0.8		0.001−0.05		µg/l
Marine sediments	≤100		≤ 10		mg/l
Algae	≤ 60	(DW)	≤ 5	(DW)	mg/kg
Mussels (Mytilus) marine	≤ 10	(DW)	≤ 2	(DW)	mg/kg
Mussels, freshwater	≤ 15	(DW)	—		mg/kg
Oysters	≤ 8	(DW)	≤ 1	(DW)	mg/kg
Fish muscle	≤ 10	(DW)	≤ 0.05	(DW)	mg/kg
Soils (non-polluted)	≤ 70	(DW)	≤ 30	(DW)	mg/kg
Soils (polluted)	up to 2000 (DW) or even higher		—		mg/kg
Sewage sludge	≤300	(DW)	≤ 10	(DW)	mg/kg
Earthworm	≤ 20	(DW)	≤ 6	(DW)	mg/kg
Spruce shoots	≤ 10	(DW)	≤ 1	(DW)	mg/kg
Poplar leaves	≤ 7	(DW)	≤ 4	(DW)	mg/kg
Beech leaves	≤ 4	(DW)	≤ 0.2	(DW)	mg/kg
Potatoes	≤ 4	(DW)	≤ 2	(DW)	mg/kg
Carrots	≤ 10	(DW)	≤ 1	(DW)	mg/kg
Coffee	≤ 0.5	(DW)	≤ 0.2	(DW)	mg/kg
Coffee, instant	≤ 1	(DW)	≤ 0.3	(DW)	mg/kg
Tea leaves	≤ 10	(DW)	≤ 1	(DW)	mg/kg
Cocoa	≤ 10	(DW)	≤ 1.5	(DW)	mg/kg
Rice	≤ 0.3	(DW)	≤ 0.02	(DW)	
Wine	≤ 0.2		≤ 0.02		mg/kg
Meat, bovine	≤ 0.02	(DW)	≤ 0.02	(DW)	mg/kg
Liver, bovine	≤ 0.1	(DW)	≤ 1	(DW)	mg/kg
Herring gull eggs	≤ 0.08	(DW)	≤ 0.05	(DW)	mg/kg
egg shells	≤ 0.6		≤ 0.1		mg/kg
Human whole blood non-exposed	≤ 1		≤ 0.5		µg/kg
exposed	up to 10 and higher		—		µg/kg
Human urine	≤ 1		< 0.5		µg/l
Human ribs	≤ 0.5	(DW)	≤ 0.2	(DW)	mg/kg

(Continued on p. 408)

References on p. 442

TABLE 14.1 (continued)

Material	Ni content		Co content		Unit
Human lung tissue					
non-exposed	≤ 0.2	(DW)	≤ 0.05	(DW)	mg/kg
exposed	≤ 1	(DW)	≤ 0.1	(DW)	mg/kg

14.3. SAMPLING AND SAMPLE PRETREATMENT

Sampling of materials with nickel and cobalt contents above 0.2 mg/kg normally poses no problems if contamination is concerned. However, if meaningful results have to be obtained a reliable sampling strategy includes environmental, biological, toxicological and statistical considerations in relation to the respective analytical problem [2,3,5,8,11,35,36] as treated in some detail in the introductory part of this book. For lower nickel and cobalt levels contamination precautions in the laboratory are necessary because many tools are made of stainless steel. Thus favored materials for bottles, knives, pincers etc. are PTFE, other plastics, glass and quartz. Because of nickel and cobalt in laboratory dust laminar flow benches should be used if materials with low nickel and cobalt levels have to be treated, i.e. homogenized or prepared for decomposition.

In former years a strong influence from stainless-steel needles for blood sampling was reported on levels found during analysis [11]. However, a recent study with an improved and very sensitive graphite furnace atomic absorption spectrometric (GFAAS) method achieved identical results for nickel (and chromium) concentrations in serum irrespective of whether blood was collected with a stainless-steel needle or a PTFE *i.v.* cannula, provided the first 5 ml of blood was discarded [37].

Decomposition of all types of materials for the analysis of nickel and cobalt can be performed by common approaches, i.e. in open and closed systems for optical spectroscopy and other methods such as voltammetry [11,36,38]. A recently introduced very effective decomposition method for subsequent interference-free determination of nickel and cobalt by adsorptive (cathodic) stripping voltammetry is high pressure ashing with nitric acid in quartz vessels up to at least 300°C [39-41].

14.4. ENRICHMENT AND SEPARATION

Analyses of trace elements at very low levels in seawater and freshwater, body fluids and in complex environmental matrices frequently require preconcentration and/or selective separation. The reasons are either a lack of detection power of the applied methods or matrix interferences. Both is often the case for nickel and especially for cobalt because of its sometimes very low concentrations (see Table 14.1). Several techniques are available for this purpose and a number of examples will be given below.

The relatively simple solvent extraction procedure for several trace metals including nickel and cobalt from seawater using dialkyldithiocarbamates and heavy solvents (Freon, carbon tetrachloride) applying back extraction into an acid solution for determination by GFAAS first described by Danielsson et al. [42] is still in successful use in marine analytical chemistry [12,25,43,44]. Carbamates and oxines are frequently used for preconcentration of nickel and cobalt from body fluids as well [3,5,8,9,11,32,38]. There are in addition various other complexing agents in use for preconcentration by solvent extraction such as e.g. *meso*-tetra(*p*-sulfonatophenyl)porphyrine in ammoniacal solution and MIBK for direct analysis of the organic phase by GFAAS [45]. Other chelates are 2-hydroxy-1-naphthaldehyde 4-phenyl-3-thiosemicarbazide [46] and 2-hydroxy-1-napthtaloxime [47] in chloroform for nickel. Ultratrace amounts of cobalt (ng/l level) were extracted as nitrosonaphtal chelates into *m*-xylene for subsequent determination by photoacoustic spectroscopy [48]. High preconcentration factors for nickel and cobalt were achieved for GFAAS after solvent extraction with an alkylated oxine derivative [49]. Trace metals in ore concentrates, including nickel and cobalt, were separated by xanthate extraction for subsequent determination by flame AAS [50]. In another study a mixture of extractants containing Chelex-100, 4-ethyl-1,2,4-triazole and sodium or ammonium dithiocarbamate for the extraction of metals into chloroform for analysis by inductively coupled plasma–atomic emission spectrometry (ICP–AES) was used [51].

There are numerous papers describing the application of chelating resins for separation/preconcentration. Batch treatment with Chelex-100 resin was applied for subsequent ICP–AES determination, achieving a preconcentration factor of 100 [52]. A similar technique using a Chelex-100 column was recently described for shipboard pre-concentration of heavy metals, including cobalt and nickel. A concentration factor of 250 was reported [53]. Others used a combination of a column and a batch technique after preconcentration by solvent extraction [54], Chelex-100 columns in ammonium form [55], and a macroporous resin impregnated with 7-dodecenyl-8-quinolinol [56]. Cellulose collectors were also used for sorption of dithiocarbamate chelates [57] and Cellex CM ion exchangers for the determination of low nickel contents in water samples [58]. Also cellulose hyphan filters were reported to be useful for preconcentration of cobalt [59]. Oxine-silica gel columns were used for preconcentration prior to GFAAS determination of a number of metals, including nickel [60,61]. The same technique using silica-immobilized 8-hydroxychinoline was applied for subsequent multielement determination with ICP–mass spectrometry (MS) including cobalt in a riverine water reference material [62] and with GFAAS in seawater using a flow technique [63]. Cobalt was preconcentrated from aqueous matrices with N-(dithiocarboxy)sarcosine and Amberlite XAD-4 resin [64], XAD-resin was used for cobalt preconcentration by other authors as well [65].

Extraction-preconcentration procedures for trace metals from various aqueous matrices by flow systems with appropriate resins were reported by several authors (e.g. refs. 66–69).

The coprecipitation technique was used by a number of authors. Examples are the coprecipitation of trace metals from seawater and floating the precipitate with octadecylamine and 6 *M* HCl–MIBK–ethanol [70], coprecipitation with zirconium hydroxide

References on p. 442

[71,72], and collector precipitation with hexamethylene–ammonium–hexamethylene dithiocarbamate for trace metals in soils using a small amount of the iron present as a collector element [73]. Cobalt and other elements can be preconcentrated from fresh-water and seawater by coprecipitation with 1-(2-pyridylazo)-2-naphthol (PAN) prior to instrumental neutron activation analysis; the detection limit for cobalt being 40 ng/kg [74]. If gallium coprecipitation from 10 ml solution is performed a preconcentration factor of 200 times for subsequent microsample introduction into an ICP–AES could be achieved [75].

Other methods occasionally described for cobalt and nickel are preconcentration of cobalt on polyurethane foams loaded with PAN [76] chelation of heavy metals, including nickel and cobalt with 1,10-phenanthroline and floating with sodium lauryl sulfate [77]. A similar procedure by co-flotation on iron hydroxide using a colloidal gas aphron of sodium lauryl sulfate for subsequent AAS determination was described recently [78]. Another preconcentration method attaining a 550-fold enrichment uses liquid surfactant membranes (droplets of dilute hydrochloric acid in kerosene containing 2-ethylhexyl-phosphonic acid mono-2-ethylhexyl ester emulsion coated with sorbitan monooleate) for the determination of cobalt by GFAAS [79].

Electrochemical preconcentration using either mercury film electrodes [80] or a niobium wire working electrode [81] was also applied for subsequent multielement deter-minations by ICP–AES and ICP–MS including cobalt and nickel in different solid materials and waters, also seawater. An interesting approach also for future applications in com-bination with various determination methods is transformation of nickel into its gaseous carbonyl and thus separation/preconcentration for methods with optimal detection power [82–84].

14.5. ANALYTICAL METHODS

14.5.1. Overview

There are a couple of methods available that provide sufficient detection power for the direct determination of nickel and cobalt in numerous environmental and biological matrices.

In several cases, however, especially if ultratrace contents in fresh- and seawater and some biological materials have to be analyzed, various preconcentration/extraction procedures including some newly introduced approaches such as carbonyl genera-tion/preconcentration for nickel (see Section 14.4) as well as adsorptive stripping voltammetry are the only methods allowing at present determination down to the lowest natural levels.

In Table 14.2 typical detection limits for nickel and cobalt in commercial systems based on a recent compilation [85] are listed. The following sections deal with the analysis of nickel and cobalt in liquid and solid samples by optical spectroscopic methods (AAS and AES), ICP–MS, electroanalysis, X-ray spectrometry, chromatography and a few other less frequently applied methods.

TABLE 14.2.

TYPICAL RELATIVE DETECTION LIMITS FOR COBALT AND NICKEL FOR VARIOUS ANALYTICAL TECHNIQUES

Techniques: atomic spectroscopy (flame and furnace) voltammetry, inductively coupled plasma–mass spectrometry, total reflection x-ray fluorescence and instrumental neutron activation analysis for aqueous solutions or non-interfering matrix for NAA. Values are given in μg/l or μg/kg. The detection limit is defined as three times the standard deviation of the typical background noise or blank (3 s), tabulated values given as 2 s are converted to the more realistic 3 s values.

Element	ICP–AES[a]	Flame AAS	GFAAS[b]	Voltammetry[c]	ICP–MS[d]	TXRF[e]	INAA[f]	Radio-nuclide
Co	1.5 [3]	9	0.06	0.001	0.01	0.1	0.03	^{60m}Co
Ni	2 [4]	6	0.3	0.001	0.04	0.1	15	^{65}Ni

[a] Data for ICP–AES are based on determined values for the most sensitive lines with a 50 MHz conventional argon ICP at 15 pm bandwidth [86], data in parentheses are values reported by Jobin Yvon for their instruments JY 38 PLUS and JY 70 PLUS [87].

[b] Data for AAS are based on tabulated values [88] and a compilation for Zeeman GFAAS [89]. For graphite tube furnace a injected volume of 50 μl is assumed, absolute detection limits for Co are 3 pg, for Ni 15 pg (3 s) [89].

[c] Data from previous work performed by the late H.W. Nürnberg and co-workers [90].

[d] Data from a recent compilation [91].

[e] TXRF data have been provided by Michaelis [92] and are based for comparative purposes also on a sample volume of 50 μl.

[f] INAA data are based on a sample weight of 500 mg (non-interfering matrix) and the following irradiation and counting conditions: Thermal neutron flux 1×10^{13} $n \cdot cm^{-2} \cdot s^{-1}$, irradiation time 5 h (maximum), counting with a sample to detector distance of 2 cm; zero decay time before start of count and maximum counting time = 10 min.

14.5.2. Atomic absorption spectrometry

AAS doubtless is for both metals the predominantly applied method. Flame AAS and furnace AAS have already been treated in detail in several reviews (e.g. refs. 5,11,32,38,85,88,90,93). The determination of nickel by AAS is commonly performed at the 323.0 nm resonance line. However, there are also several other less sensitive lines that are useful if elevated contents have to be analyzed. The typical atomization temperature with maximal heating rate in the graphite tube (pyrotube) is 2300°C [11,88].

Cobalt is commonly determined at the 240.7 nm resonance line which is the most sensitive line but there is also a second line at 242.5 nm that is only about 20% less sensitive, however, showing a quite favorable signal-to-noise ratio the typical atomization

References on p. 442

temperature with maximal heating rate in the graphite tube (pyrotube) is 2200°C [93].

The remarkable improvement of AAS methodology during the last two decades has been of particular benefit for the analysis of both metals. This is mainly due to the commercial introduction of pyrolytically coated graphite tubes, L'vov platforms and matrix as well as analyte modifiers [94]. For nickel and cobalt also the successful use of molybdenum tubes with hydrogen as a matrix-modifying purge gas was reported [95]. This was supported in general by faster data evaluation with new and versatile computer systems with menu-driven software. Thus modern instruments now permit in many cases the direct analysis of nickel and cobalt in liquid and also in solid samples [38,93].

Although there is still no commercially available multielement AAS system with appropriate continuum light source, the possibility of also determining nickel and cobalt with remarkable figures of merit by this approach [96,97], recently also with photodiode array detection [98], is certainly worth being mentioned.

14.5.2.1. Liquid samples

The extraction and various other preconcentration procedures described in section 4 are mainly suited −and also applied− for subsequent flame and furnace AAS [42−79, 82−84]. Some examples of various matrices are described below.

Sturgeon et al. [99] determined sub μg/l levels of nickel in seawater by carbonyl generation using BH_4^- , transfer of the volatile nickel tetracarbonyl from a generation cell to a pre-heated graphite furnace. There it undergoes decomposition and deposition at > 500°C. Subsequent atomization at > 2700 K and calibration by identically treated reference solutions show an efficiency of approx. 80% and a detection limit of 4 ng/l based on a 10-ml test sample.

Grobenski et al. [100] determined a couple of elements directly in seawater under stabilized temperature platform furnace (STPF) conditions with magnesium nitrate as the matrix modifier with a detection limit of approx. 0.5 μg/l for nickel using peak area evaluation.

Boniforti et al. [101] performed an intercomparison of five methods for the determination of trace metals including nickel and cobalt in seawater using various preconcentration procedures. Best results for cobalt and nickel were obtained by preconcentration on Chelex-100 resin with final determination by ICP–AES and furnace AAS. The test of accuracy using the NRC NASS-reference seawater sample showed excellent agreement for cobalt and satisfactory agreement for nickel. Calmano et al. [102] described a procedure for the direct determination of nickel and other trace metals in the Elbe river estuary based on Zeeman background correction. Nickel analysis in seawater could only be performed using matrix modification and a detection limit reported to be 1 μg/l.

While in former graphite furnace work deuterium background correction subsequent to various preconcentration procedures, e.g. ion exchange for cobalt [103], was used for body fluids, the commercial introduction of instruments with Zeeman background correction permitted the increased use of simpler, often direct, methods [93]. Christensen et al. [104] described a rapid, simple and sensitive method for the direct determination of cobalt in whole blood and urine using Zeeman effect background correction after nitric acid precipitation of proteins at 60°C. This led to a detection limit of approx. 0.1 µg/l. The method was applied for the analysis of samples from unexposed and exposed workers. A similar method was used first by Sunderman et al. [105] and also successfully applied later by other workers [37] for the determination of nickel in whole blood. Nickel in serum was simply determined by Nixon et al. [27] with Zeeman effect background correction after dilution and addition of a surfactant. The reported values were below 0.2 µg/l for healthy persons.

Angerer et al. [32] and Heinrich and Angerer [106] described procedures for the determination of cobalt in urine and whole blood. For cobalt in urine a chelation/extraction procedure with hexamethylene–ammonium–hexamethylene dithiocarbamidate by the use of a xylene/diisopropyl ketone mixture prior to furnace-AAS was applied. A detection limit of about 0.1 µg/l was reported. Whole blood was diluted and directly injected into the graphite furnace attaining a detection limit of 2 µg/l which, however, only permits the analysis of blood samples of occupationally exposed workers. Direct GFAAS methods using STPF conditions for cobalt and nickel in urine, blood and serum were also used by Minoia et al. [29].

If, however, normal levels of cobalt in while blood have to be determined, even the most sophisticated direct procedures showed an insufficient sensitivity. For this purpose Andersen and Høgetveit [107] developed a method that is based on sample digestion in a mixture of nitric and perchloric acid. This is followed by chelation/extraction with ammonium pyrrolidine dithiocarbamate at pH 9 into methyl isobutyl ketone (MIBK) and determination in the graphite furnace. The attained detection limit of 0.06 µg/l plasma permitted the reliable determination of cobalt in the plasma of 32 non-exposed control persons with a range from 0.07 µg/l to 0.15 µg/l.

A method for the direct determination of cobalt in urine with Zeeman GFAAS using magnesium nitrate and nitric acid as matrix modifiers is described by Kimberly et al. [108]. A detection limit of 2.4 µg/l permitted the determination of potentially toxic levels of cobalt.

Andersen et al. [109] determined nickel in human plasma with a dilution step prior to quantification by Zeeman GFAAS. They reported a detection limit of 0.09 µg/l and a quantification limit of 0.31 µg/l and found a mean level of 0.65 µg/l for healthy, non-exposed persons.

Sunderman et al. [110] attained a detection limit of 0.5 µg/l in urine involving dilution of urine with dilute nitric acid and direct quantification with Zeeman-corrected GFAAS.

Matrix modification using palladium, recently suggested as approaching the utility of a "universal" matrix modifier in graphite furnace AAS [111], was tested by Sampson [112] for cobalt in plasma and urine. A detection limit of ≤ 0.15 µg/l was obtained even with deuterium background correction.

Determination of trace elements in biological fluids other than blood (urine, amniotic fluid, bile, breath, cerebrospinal fluid, hip-joint fluid, inter-ocular fluid, saliva, semen, sweat and synovial fluid) by GFAAS including cobalt and nickel was recently reviewed by Subramanian [113].

Nickel in oil, e.g. in corn oil, was determined by Ooms and van Pee [114] by dilution with MIBK (5 g oil to exactly 25 ml) and directly analyzed by GFAAS. The influence of the

References on p. 442

sample matrix was estimated by comparative measurements using the standard (analyte) addition method. Base line measurements with pure MIBK were also performed.

Analyses of cobalt and nickel in liquids with μg/l or even lower levels of nickel and cobalt always require effective cleaning procedures for all vessels used made of PTFE, other plastics, quartz or borosilicate glass. Blanks from airborne dust can be substantially reduced by the use of clean benches in appropriately equipped laboratories, thus current blank controls before and during analytical work are of paramount importance [115,116].

14.5.2.2. Solid samples

Analysis of cobalt and nickel in solid materials is usually performed after digestion but increasingly also directly by solid sampling or slurry techniques. Quantification is preferably performed by the graphite furnace but still the flame process can be used if the content of the analyte permits it. An example is the determination of copper and nickel in tea leaves after an official AOAC method of analysis [117]. Amounts of $\geq$ 3 g are wet ashed in nitric acid with the final addition of perchloric acid. Quantification is performed by flame AAS. Contents down to 50 μg/l in the analyte solution can be determined. The determination of nickel in mussel tissue during the U.S. mussel watch program was performed after HNO_3–$HClO_4$ digestion by flame AAS [17], flame and furnace were used for nickel and cobalt in polluted soils [28]. If the determination of nickel and cobalt in marine and human biological samples is concerned, this was mainly performed in the last decade after wet or pressurized decomposition in PTFE vessels with nitric acid by GFAAS using the standard addition technique [13,19,24].

Nickel and several other metals were determined by Nash et al. [118] in hydrogenated vegetable oils and fats by dispersion of the samples in 4-methyl-2-pentanone and subsequent determination in the graphite furnace. Nickel contents ranged from approx. 0.03 to 2.7 mg/kg.

Flame AAS is a method used still if nickel is to be determined in sewage sludges. For multielement analysis of trace metals at higher levels, however, flame AAS in many laboratories is being replaced more and more by ICP–AES (see Section 14.5.3). Alder and Batoren [119] and Alder et al. [120] have developed a method for the direct determination of nickel in human skin by GFAAS.

The nickel content of various foodstuffs was determined by Ellen et al. [121] using the graphite furnace. In most samples the nickel content was below 0.5 mg/kg.

Fishery products from Dutch coastal waters —sole, cod, plaice, herring, eel, pikeperch, shrimp and mussel— were analyzed for several trace elements by Vos et al. [122]. After dry ashing with magnesium nitrate as ashing aid, nickel was determined by GFAAS. For the different species analyzed the nickel contents were comparatively low and ranged from 0.03 mg/kg (sole) to 0.44 mg/kg (mussels) (fresh weight).

Sub-microgram amounts of nickel in plant materials were determined by Green and Asher [123]. After wet digestion of up to 0.2 g sample amounts with an HNO_3–$HClO_4$ mixture, nickel was extracted with ammonium–tetramethylene–dithiocarbamate with MIBK. Quantification was performed by GFAAS, accuracy checked with NIST reference material

SRM 1571 (Orchard leaves).

Borggaard et al. [124] determined cobalt in feedingstuffs — corn and grass — after wet ashing with nitric acid and hydrogen peroxides solvent extraction with 2-nitroso-1-naphtol in xylene and quantification by GFAAS. A detection limit of 1 μg/kg was achieved. A similar procedure for the determination of trace amounts of cobalt in feed grains and forage after dry ashing was described by Blanchflower et al. [125].

Alary et al. [126] published a simple and rapid procedure for the routine determination of cobalt in forage using dry or wet decomposition and quantification by Zeeman-corrected GFAAS with a detection limit of 20 μg/kg. The graphite furnace was also used for the analysis of nickel in various vegetables and contaminated soils by Frank et al. [15].

Carbonyl generation, already described above was adapted by Alary et al. [82] to the analysis of ultratrace amounts of nickel in a broad variety of materials ranging from biological and environmental to technical. The general procedure started with a decomposition appropriate for the respective matrix. Subsequently the analyte solution was allowed to react with sodium borohydride which reduced nickel to the metallic state and gaseous carbon monoxide for transformation before thermal decomposition of the carbonyl in a heated silica cell coupled with an AAS instrument. The detection limit was 10 ng in the aliquot used. Reliability of the method was proven by the analysis of samples with known nickel contents.

Ever since the early days of AAS the direct analysis of solids has been desired and attempted with flames and furnaces but was often hampered by interferences from the matrix which limited success. With the introduction of improved background correction systems, especially the Zeeman effect background correction, this could be realized for many elements and materials [127,128]. Although analysis of more volatile elements is common in this approach an increasing number of publications has also appeared for less volatile ones. Nickel was directly determined with acceptable accuracy in feedingstuffs [129], in polyethylene [130], in environmental reference materials [131] and in margarine [132].

Another possibility for direct analysis is the use of slurries. The advantage of this approach is that it can be applied with most commercial instruments and now also an automated system for sample introduction is commercially available. With this technique Hoenig et al. [133] performed automated sequential multielement slurry analysis using the same palladium modifier for six analytes, cobalt and nickel included. The reliability of the method could be confirmed by the analysis of certified soil and sediment reference materials. Since a distinct problem in solid sampling is calibration, Akatsuka and Atsuya [134] developed and applied synthetic reference materials by coprecipitation of eight metal ions, including cobalt and nickel. Detection limits reported were 12 μg/kg for cobalt and 43 μg/kg for nickel.

14.5.3. Atomic emission spectrometry with plasmas

Atomic emission spectrometry (AES) has drastically gained importance with the commercial introduction of plasmas as excitation sources from 1974. Most plasma-AES systems are equipped with an inductively coupled plasma and offer sequential as well as simultaneous multielement potential for up to seventy elements (refs. 135,136, see also Chapter 4). The plasma techniques also offer solid sampling approaches and the instrumentation commercially available at present has reached maturity as far as optical, technical and computer performance is concerned. Quite promising detection limits, particularly for refractory elements, compete favorably with flame AAS. Thus, doubtless ICP–AES systems, despite their relatively high running costs, are on the way to replacing flame AAS more and more if multielement determination is required.

From Table 14.2 it can be seen that the attainable detection limits for nickel and cobalt compared to flame AAS are to some extent superior as well. In ref. 135 applications of ICP–AES in various fields are reviewed, e.g. geology, agriculture and food, biological and clinical materials, and organics, also including solid sampling.

14.5.3.1. Liquid and solid samples

If multielement determinations have to be performed in samples with comparatively low levels of the elements to be analyzed, selective group preconcentration procedures as already mentioned have to be applied either directly in liquid or after appropriate dry or wet decomposition in solid samples. An example of decomposition − selective extraction and subsequent ICP–AES determination is multielement determination performed in Japanese ribs from the present time by Yoshinaga et al. [26]. The authors decomposed the rib material with nitric acid under pressure in PTFE vessels. For the determination of cadmium, cobalt, copper, nickel and lead these elements were extracted with ammonium pyrrolidine dithiocarbamate into chloroform, the solvent transferred into a PTFE beaker and treated with nitric acid after evaporation of the solvent to attain an appropriate solution for ICP–AES.

Optimization of ICP–AES conditions for various organic solvents and numerous metals in oils diluted with organic solvents and in solvents after separation by liquid chromatography and extraction/preconcentration, including nickel, was reported by Barrett and Pruszkowska [137].

The nickel tetracarbonyl method was applied by Drews et al. [84]. The authors adsorbed the carbonyl at liquid nitrogen temperature and determined nickel without any interference in a microwave-induced plasma. They attained an extremely low detection limit of 5 pg, corresponding to 5 ng/l in the analyte solution. The method was applied to seawater and urine without digestion and whole blood, blood serum and human hair after decomposition by a nitric acid–perchloric acid mixture.

Schramel and Xu Li-qiang [138] determined 14 elements including nickel in various botanical candidate reference materials by simultaneous ICP–AES using NIST SRM 1571 (Orchard leaves) as multielement reference after nitric acid pressure decomposition.

Dannecker et al. [139] discussed the analytical conditions for the determination of a number of metals, including cobalt and nickel in dust from industrial emissions after various wet decomposition procedures.

Taylor et al. [140] determined nickel and other trace metals for the certification campaign of the BCR reference material for city waste incineration ash. The obtained value was in quite good agreement with the certified one [141].

Brenner et al. [142] described a versatile technique for multielement ICP–AES analysis of geological samples. By combining mixed $HF–HClO_4–HNO_3–HCl$ acid NaOH and Na_2O_2 fusion techniques more than 30 elements including nickel and cobalt could be easily determined in large sample batches.

Karstensen and Lund [143] determined 19 elements, nickel and cobalt included, in BCR No. 176, city waste incineration ash by ICP–AES using two different decomposition methods: pressure decomposition with HNO_3 and HF and alkaline fusion.

Kanda and Taira [144] determined 17 elements in five reference sediments and soils, including nickel and cobalt, after a digestion procedure involving $HNO_3–HF$ and subsequent $HClO_4$ treatment in an open PTFE beaker. Final determination was performed with a computer-controlled rapid-scanning echelle monochromator.

Various decomposition procedures using a number of mixtures of $HNO_3–HF–HCl–HClO_4$ in closed (Parr bomb) and open systems with hot plate and microwave heating were applied by Que Hee and Boyle [145] for the analysis of environmental materials, nickel and cobalt included, by ICP–AES. Accuracy of the analyses was confirmed by simultaneously analyzing a couple of NIST certified reference materials.

The direct analysis of slurries by ICP–AES, including cobalt and nickel using finely powdered coal samples was described by Ebdon and Wilkinson [146] and Parry and Ebdon [147].

The flow injection method can be also used for calibration of an ICP–AES. Using the so-called generalized standard additions method 120 measurements per hour was attained recently by Giné et al. [148].

Another approach in solid sampling is coupling of a graphite furnace with an AES system (GFAES). Baxter and French [149] demonstrated the accurate determination of silver, manganese, nickel, lead and vanadium by this combination.

14.5.4. Inductively coupled plasma–mass spectrometry (ICP–MS)

ICP–MS is a comparatively new analytical approach. Since its commercial introduction in 1984 [150,151], despite very high investment costs, it is, although there are still some limitations to be overcome (refs. 152,153, see also Chapter 4), already used in a steadily increasing number of laboratories. This is because of its practically comprehensive element coverage and low detection limits that compete for a number of elements with that of the graphite furnace and are superior for many others. Especially for nickel and cobalt the attainable detection limits are somewhat superior to that for graphite furnaces (see Table 14.2). This makes ICP–MS the second sensitive commercially available instrumental analytical technique at present for nickel and cobalt. From this situation and the prospect of a further steady gain of promising instruments for routine use, several examples will be given below.

References on p. 442

14.5.5. Liquid and solid samples

ICP–MS is an ideal method for inland waters with low salt contents. Henshaw et al. [154] directly analyzed more than 250 water samples from lakes in the eastern United States by ICP–MS for 49 elements, including nickel and cobalt with averages below 1 μg/l. The system detection limits, evaluated by using field blanks, were less than 0.2 μg/l for most elements. NIST SRM 1643b certified water reference material was analyzed at the same time in order to check accuracy.

Since the ICP–MS cannot be used for liquids with comparatively high salt contents, e.g. seawater, if dilution is not applicable separation/preconcentration procedures are necessary for those samples. Plantz et al. [155] applied an on-line sample treatment for the separation of metal ions from alkali and alkaline earth elements prior to ICP–MS. They complexed V, Cr, Ni, Co, Cu, Mo, Pt, Hg and Bi with bis(carboxymethyl)dithiocarbamate, and adsorbed the complexes onto a polystyrene–divinylbenzene resin column at acidic conditions. The metal complexes are then removed from the column with a basic eluent. This technique was applied for the analysis of seawater and urine with typical detection limits in the range 8–80 ng/l.

Beauchemin et al. [156] analyzed seven metals, including nickel, by ICP–MS with the isotope dilution technique in an open ocean water reference material. This was performed after a 50-fold preconcentration of these metals by adsorption onto silica immobilized 8-hydroxyquinoline (I-8-HOQ) which separates the metals of concern from the bulk of the alkali and alkaline earth elements. Beauchemin and Berman [157] used a similar I-8-HOQ system but modified for flow injection analysis (FIA) technique achieving an on-line preconcentration improving the detection limits of several elements by a factor of 2–7 compared to ICP–MS alone.

Vanhoe et al. [158] described a method for the determination of seven trace elements (Fe, Co, Cu, Zn, Pb, Mo and Cs) in human serum by ICP–MS. The serum samples were only diluted with nitric acid and indium added as an internal standard. Results for iron, cobalt, copper and zinc had to be corrected for interferences from polyatomic ions by using a blank solution containing the same concentrations of sodium, sulfur, chlorine and calcium as human serum and important corrections were necessary for iron and cobalt. In order to check the accuracy a second-generation biological reference material was analyzed and showed good agreement with the certified values.

Ridout et al. [159] described the determination of 16 elements, including nickel and cobalt with ICP–MS in a marine reference material. The material (lobster hepatopancreas) was open-digested with nitric acid up to 160°C. Accurate results were obtained for most elements although some signal suppression was observed which was easily overcome by dilution. Corrections were made for interferences arising from chloride in the samples using isotope ratios of ^{35}Cl and ^{37}Cl.

McLaren et al. [160] recently reviewed various applications of IPC–MS with many details in marine analytical chemistry as far as the NRC coastal seawater CASS-2 and the non-defatted lobster hepatopancreas tissue LUTS-1 reference materials are concerned. The metals analyzed also include nickel and cobalt.

McLaren et al. [161] assessed the accuracy of analysis of geological materials by ICP–MS for ten trace elements (V, Mn, Co, Ni, Cu, Zn, As, Mo, Cd and Pb) in solutions of the NRC marine reference material BCSS-1. The material was decomposed in PTFE vessels under pressure with a mixture of nitric, perchloric and hydrofluoric acids and after evaporation to dryness re-dissolved in nitric acid to minimize interferences and appropriately diluted prior to analysis. Results obtained by external calibration with reference solutions showed some evidence of ionization suppression. Good accuracy was thus achieved by the method of standard additions for all elements except copper, for which an isobaric interference due to titanium oxide was observed.

Williams et al. [162] demonstrated the feasibility of solid sample introduction by slurry nebulization for ICP–MS by analyzing three certified soil reference materials and three industrial catalysts in comparison to an ICP–AES procedure. All samples contained nickel and cobalt. The finely ground samples —particle size below 3 μm— were dispersed in tetrasodium pyrophosphate solution and aspirated using a Babington-type nebulizer. Slurry concentrations of 0.05 g per 100 ml and 0.0001 g per 100 ml were used for ICP–MS and no cone blocking effects were observed. Calibration was performed with aqueous reference solutions. Agreement between values obtained by slurry nebulization and other analytical techniques or certified values was good for all elements investigated, except aluminium, for which ICP–MS results were approx. 10% low. The work thus confirmed a considerable potential on the part of ICP–MS for direct analysis of solids.

Hirata et al. [163] introduced powdered materials into an ICP–MS by using a spark dispersion-merging sample introduction technique for the direct multielement analysis of more than 40 elements, including nickel and cobalt in four geological standard rocks. The advantage of this method is that optimization of the operational settings can be carried out using an analyte solution in the same manner as conventional nebulization sample introduction.

14.5.6. Electrochemical methods

Electrochemical methods are among the classical trace analytical methods. Since their introduction around 1920 they have undergone a steady, sometimes slow, but during the last decade more pronounced progress. This is due to remarkable improvements by the introduction of new and more effective modes that enhance sensitivity, e.g. differential pulse, stripping and square wave voltammetry. Thus modern instruments provide for a number of metals the lowest detection limits of all instrumental analytical methods used at present coupled with a computer-controlled operation (refs. 164–168, see also Chapter 4). As far as nickel and cobalt are concerned the classical techniques were not sensitive enough for successful analysis of very low levels of cobalt and nickel by voltammetry because of the irreversibility of the Ni^{2+}/Ni and Co^{2+}/Co couples [169,170]. However, for elements that do not, or only with limitation, form amalgams another very promising approach, significantly broadening element coverage of electrochemical methods, is adsorptive (cathodic) stripping voltammetry [171]. The principle of this method is that metals forming a certain complex with comparatively large organic ligands can be accumulated by adsorption of the complex at the surface of, mainly, a hanging mercury drop electrode (HMDE) at a defined potential. The fundamental differences from the conventional technique of anodic stripping voltammetry (applied to the determination of cadmium, copper, lead, zinc etc.) are the adsorption step, and the potential scan in negative direction, thus the reduction (cathodic) current is measured. Usually the obtained signals are very sharp due to the single desorption mechanism. This cathodic stripping voltammetry approach hence is not limited to amalgam-forming elements.

Based on observations by several workers that dimethylglyoxime for nickel and cobalt [172–176] and 2-nitroso-1-naphtol especially for cobalt [177,178] significantly enhanced sensitivity a powerful new electroanalytical method was initiated that has led, applied to a number of elements not previously covered by voltammetry, to a remarkable extension of the number of elements accessible to electroanalytical methods with the additional advantage of extremely low detection limits [179,180]. The HMDE is still most frequently used for this technique. There are, however, also other electrodes in use, e.g. in the potentiometric stripping mode [181,182] described below in some detail.

14.5.6.1. Aqueous samples

Adsorptive stripping voltammetry is, due to its eminent sensitivity, an ideal method for the determination of ultratrace contents in natural waters. The necessary pretreatment in many cases simply consists of acidification or UV-irradiation [183] prior to voltammetric determination.

Pihlar et al. [176] determined nickel in seawater, tapwater, rainwater and wastewater and in various environmental and food samples by adsorptive stripping voltammetry. Determinations were performed in water for the "dissolved" metal after the conventional filtration procedure through 0.45 μm pore size membrane filters. Samples containing interfering matter (surfactants and other organic substances) were UV-irradiated in the presence of 0.01 M HCl and 0.03% H_2O_2. Determination of nickel is performed after neutralization, addition of a buffer and of 20 μl of a 0.1 M DMG solution to 20 ml of the sample. Under optimal conditions, i.e. optimized blanks, a detection limit of 1 ng/l was reported for aqueous solutions. Nickel levels range from < 1 μg/l in open seawater to ≤ 5 μg/l in rainwater [179].

Mart et al. [184] and Mart and Nürnberg [185] determined nickel and cobalt using the same methodological approach, in surface water along the tidal Elbe river and in the German Bight. In the tidal Elbe a mean of 4 μg/l dissolved and 5 μg/l total nickel and 0.1 μg dissolved and 0.2 μg total cobalt was found, whereas these values were on average 0.47 μg/l dissolved, 0.5 μg/l total and 0.011 μg/l dissolved, 0.013 μg/l total respectively for the open German Bight.

Gustavsson and Hansson [186] determined trace metals by graphite furnace and voltammetric methods in clean coastal Baltic Sea water samples. Nickel and cobalt were determined by the adsorptive stripping procedure using DMG described above [176]. The concentration of cobalt was too low for the graphite furnace (around 0.5 nM). The authors found that the voltammetric methods are more favorable for Zn, Cd, Pb, Cu, Ni and Co because they involved less handling of samples, were timesaving, avoiding the necessary solvent extraction for GFAAS, and required a relatively inexpensive instrumentation.

Braun and Metzger [187] used the mercury film electrode and DMG for determination of nickel. In water a detection limit of 20 ng/l in aqueous samples can be further reduced by the use of purer reagents.

Eskilsson et al. [188] determined nickel and cobalt in natural waters and biological material by reductive chronopotentiometric stripping analysis in a flow system without sample deoxygenation. The method of Pihlar et al. [176] was modified for chronopoten-tiometry to attain an automated system. Normally, flow potentiometric stripping is based on the potentiostatic reduction and simultaneous amalgamation of the trace metals to be analyzed on a freshly prepared mercury film electrode and the subsequent chemical re-oxidation of these metals in an appropriate matrix. During oxidation, the potential of the working electrode is monitored at constant time intervals and the time required for the re-oxidation gives the analytical signal. This had to be adapted to the properties of the metal chelates and the procedure finally applied consisted of plating the mercury film onto the glassy carbon electrode used, transfer of the solution containing the DMG complexes into the cell, potentiostatically adsorption of the complexes onto the electrode, reduction of nickel(III) and cobalt(II) in 5 M calcium chloride by means of constant current and simultaneously recording of potential *vs.* time behavior of the working electrode. The final step is removal of the mercury film and cleaning of the electrode prior to the next cycle. Analyte concentrations are evaluated by standard addition. Detection limits after one minute of potentiostatic adsorption were given as 9 and 11 ng/l for nickel(II) and cobalt(II) respectively.

Van den Berg [189] determined Fe, U, V, Cu, Zn, Mn, Mo, Ni and Co in water with adsorptive (cathodic) stripping. He applied the method of Pihlar [176] for nickel and cobalt and studied competition, i.e. suppression of the metal peak by natural organic ligands, and mentioned that this effect might be useful for speciation studies.

Newton and Van den Berg [190] evaluated the adsorptive stripping analysis of nickel, cobalt, copper and uranium in water using chronopotentiometry with continuous flow using an automated analyzer with a fast rate of data acquisition. The procedure was similar to that described by Eskilsson et al. [188]. The authors concluded that copper, uranium and nickel could be determined in the presence of dissolved oxygen but with significantly reduced sensitivity for nickel. In this case the detection limit, preceded by 60 s stirred adsorption was 0.6 nM for nickel (approx. 35 ng/l), cobalt could not be determined. In the absence of oxygen the detection limits were 0.1 nM for nickel and 0.1 nM for cobalt. A prolonged collection period can lower these values. This approach was successfully tested by analyzing nickel with continuous flow in water pumped on board a small vessel [190].

Donat and Bruland [191] developed a highly sensitive technique for the direct deter-mination of dissolved cobalt and nickel in seawater by adsorptive collection of cyclohexane-1,2-dione dioxime complexes. The authors performed detailed experiments to evaluate optimal conditions. Replicate analysis of seawater reference materials showed excellent agreement with certified values with an average RSD of 5%. Detection limits for cobalt and nickel depend upon reagent blanks. For 15 min adsorption periods 6 pM for cobalt and 0.45 nM for nickel were reported.

14.5.6.2. Other liquids and solid samples

If body fluids and solid materials have to be analyzed for nickel and cobalt by adsorptive stripping methods the applied decomposition procedures must be complete and blanks should be minimized if comparatively low contents have to be determined.

Golimowski et al. [174] used adsorptive stripping of nickel in wine using the linear sweep mode at a mercury drop electrode after wet or UV decomposition with nickel contents ranging from 30 to 70 μg/l. Under the selected working conditions the cobalt contents were below the detection limit of 10 μg/l analyte solution.

Pihlar et al. [176] digested biological materials including food either by open wet digestion with a mixture of nitric acid, sulfuric acid and hydrogen peroxide or by low temperature ashing. The values for nickel in food ranged from milk with 0.4 μg/l to 80 μg/kg in bread.

Meyer and Neeb [192] added weakly alkaline solutions of triethanolamine, ammonium chloride and dimethylglyoxime to ashes from the mineralization of various biological materials. If in these solutions adsorptive stripping voltammetry was applied the detection limit for cobalt was 0.05 μg/l. A 250-fold excess of nickel and a 25000-fold excess of zinc did not interfere.

Ostapczuk et al. [193] determined normal levels of nickel and cobalt in body fluids after open wet digestion with nitric and perchloric acid. The detection limit was determined by the digestion blank, i.e. 0.05 ng nickel and 0.005 ng Co per digestion.

Narres et al. [18] analyzed zinc, cadmium, lead, copper, nickel and cobalt in meat and organs of slaughter cattle after decomposition with a nitric acid–perchloric acid mixture. Nickel and cobalt were determined by adsorptive stripping voltammetry after the addition of dimethylglyoxime. Nickel values were generally low and ranged from < 1 to 16 μg/kg fresh weight, whereas cobalt, due to the occurrence of cobalt in vitamin B12, ranged from approx. 2 μg/kg in meat to approx. 265 μg/kg in liver.

Gammelgaard and Andersen [194] described a procedure for the determination of nickel in human nails by adsorptive stripping voltammetry. The samples were digested in a 10:1 mixture of nitric and sulfuric acids up to 170°C so that in the final analyte solution virtually no nitric acid was present. For determination DMG was applied and it was found that the only possible interferent was cobalt but the more negative potential for reduction of cobalt makes it only a problem when cobalt is present in much larger amounts than nickel. Since this is not so for nails and most other biological materials interference usually is no problem. Mean nickel values in finger- and toenails were found to be 2.14 ± 1.20 mg/kg and 0.78 ± 0.42 mg/kg respectively and the authors stated that the method used produced more reliable results for low nickel levels than the graphite furnace [194].

Eskilsson and Haraldsson [188] described in their paper the reductive chronopotentiometric stripping analysis of nickel and cobalt already mentioned above as well as the determination of nickel and cobalt in acid digests of sediments and biological material.

Subsamples of 1 g of these materials were digested in a 1:3 mixture of concentrated sulfuric and nitric acid with subsequent addition of 35% hydrogen peroxide. Blank samples were obtained by using the same digestion procedure with doubly distilled water instead of the sample. For the determination of cobalt sodium tetrahydroborate solution is added to the diluted acid digest in order to reduce the cobalt(III) formed during the decomposition procedure completely to cobalt(II) that only reacts with DMG. The authors also investigated the linear range dependent on the adsorption time, which in all electrochemical surface adsorption techniques is quite limited and thus linear only at a comparatively low surface coverage.

Adeloju et al. [195] developed a sequential multielement procedure for the determination of up to eight elements, i.e. selenium, cadmium, lead, copper, zinc, nickel and cobalt in the same solution. Arsenic was also determined under favorable conditions. Decomposition of the investigated materials was performed by mixtures of nitric acid and sulfuric acid until complete disappearance of nitric acid. A detailed scheme for the treatment of the analyte solution was presented including pH adjustment and DMG addition for the adsorptive stripping determination of nickel and cobalt. All procedures were carried out under clean air conditions. Preparations were made in a class 100 clean room and the analytical measurements in a class 1000 clean room. Further, the authors recommend the use of only high-purity reagents. At least three samples can be determined per day if seven elements are to be determined. Reliability of the procedure was demonstrated by the analysis of several certified reference materials.

Ostapczuk et al. [196–200] investigated the potential of square wave voltammetry, described first by Barker and Jenkins in 1952 (for details see Chapter 4), and now with modern commercial instrumentation easily accessible, for anodic and cathodic (i.e. adsorptive) stripping voltammetry with DMG. The authors observed that nickel can be determined with about forty-fold increased sensitivity which somewhat depends on the effectivity of the pretreatment procedure. The figure for cobalt is about twenty-fold (see Fig. 14.1). An additional advantage was the fast scan, which reduced the time needed for the determination of about 50% of that for the previously applied procedure. This methodological approach was applied for the determination of nickel and cobalt in numerous wines from recent vintages and human biological materials of exposed and unexposed collectives.

Gemmer-Čolos and Neeb [201] investigated the determination of cobalt and nickel and of other heavy metals in some biological materials, especially leguminous plants with special regard to their determination in lipid fractions. The interferences of the adsorptive stripping voltammetric determination of both elements by vanadium could be eliminated by proper adjustment of the initial wet digestion with sulfuric acid–hydrogen-peroxide. Data evaluation with a computed system (VA-processor) was very helpful in separating nickel and cobalt, especially if low amounts of cobalt have to be determined in the presence of a large excess of nickel.

Application of high-temperature high-pressure decomposition with nitric acid is an ultimate sample pretreatment for subsequent voltammetric procedures because of its

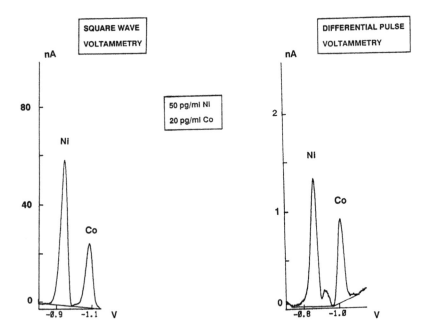

Fig. 14.1. Comparison of nickel and cobalt determination by adsorptive (cathodic) stripping voltammetry between the square wave and differential pulse mode.

property to decompose completely any organic compounds interfering in voltammetry [39,40].

Ostapczuk et al. [41] investigated the potential of this approach for the voltammetric determination of several trace metals but mentioned some problems if blanks for nickel and cobalt are concerned. The latter is, of course, only of importance if extremely low contents have to be determined. From experience gained recently, the use of extremely pure quartz vessels for decomposition with the HPA significantly decreased these influences.

14.5.7. X-ray methods

X-ray fluorescence is in many cases a useful multielement approach. The various modes of X-ray fluorescence induced by X-ray tubes, direct X-ray irradiation, gamma-radiation from radionuclide sources, but also by particles such as electrons, protons, α- and heavier particles and recent developments in this field, also including the comparatively new technique of total-reflection X-ray fluorescence are treated in some recent books and compilations (ref. 202–208, see also Chapter 4). Classical X-ray methods for the determination of nickel until 1980 are reviewed in detail by Stoeppler [11].

Several examples for the determination of nickel with conventional techniques are given first.

Rethfeld [209,210] determined heavy metals, including nickel, using a sequential X-ray spectrometer after drying, milling and disk preparation for the determination of heavy metals in foods and feedingstuffs of animal and vegetable origin, soil extracts, sewage sludges, partly after decomposition, and barn floor coating. Detection limits for nickel varied from 0.2 to 2.4 mg/kg depending on the matrix and the pretreatment procedure.

Talbot and Chang [211] determined 42 elements in four certified reference materials and cockle tissue by conventional XRF after fine grinding and preparation of pressed disks. Values for nickel and cobalt agreed fairly well with certified values, but were often below detection limits given as 4.5 mg/kg for cobalt and 1.5 mg/kg for nickel (corrected for 3 s) under the applied instrumental conditions.

Zeisler et al. [212] applied different non-destructive instrumental methods for multielement determination of up to 44 elements in marine bivalve tissue. The methods used were XRF, prompt gamma and neutron activation analysis. For the direct analysis of 21 elements including nickel and energy-dispersive XRF technique with "backscatter fundamental parameter" that provided information on total sample mass as well as on bulk sample composition was applied [213]. Sample preparation for analysis was done after storage of the marine samples at cryogenic temperatures ($< -150°C$) milling at the same temperature, followed by freeze drying and pelletizing with a KBr pellet press for the respective analytical techniques. Quality assurance was performed by analyzing a selection of appropriate certified reference materials (CRMs). Comparison of results for elements determined by more than one technique in sequence showed good agreement, as did results from the CRMs analyzed along with the samples. As far as the XRF procedure is concerned the authors mentioned that elemental concentrations at or below the 1 mg/kg-level could not be measured accurately or detected with this technique [212].

Coetzee and Lieser [59] investigated the elimination of interfering matrix constituents and preconcentration for the determination of cobalt in municipal and rain water samples and compared two procedures: (i) elution of iron from cellulose Hyphan filters used for the preconcentration of metal ions from solutions: cobalt remains on the filter and can be subsequently determined free from interference; (ii) selective prevention of the sorption of iron onto the filters by boiling the analyte solution in the presence of 1% H_2O_2 before starting the sorption process. The detection limit for cobalt was 0.4 μg/l in the presence of up to 1 mg/l iron.

A remarkable improvement of sensitivity is possible with total reflection X-ray fluorescence analysis (TRFA) because of a very effective suppression of background radiation [206–208]. In order to use the potential of this method in an optimized manner, i.e. as far as possible free from interfering elements, different procedures have been developed for various liquid and solid materials. In *liquids* after aliquotation and addition of an internal standard (often cobalt) several approaches are possible:

(i) direct measurement after evaporation of an appropriate aliquot on the sample carrier;

(ii) freeze drying, leaching of the residue with dilute nitric acid then proceeding as above;

(iii) either freeze drying and digestion of the dry residue with concentrated nitric acid followed by metal complexation with e.g. sodium dibenzyl dithiocarbamate or direct chelation. Subsequently the chelate is adsorbed on a chromatographic column, the complexes eluted with appropriate organic solvents.

In *solids* appropriate subsamples can be either directly analyzed, e.g. air dust samples collected on the sample carrier, or prepared after appropriate digestion, dilution, addition of standard solution or even preconcentration as described above for water samples [214].

The method was applied, using the different pretreatment versions described above by Michaelis et al. [215] for the analysis of urban air dust, including nickel and cobalt, by Stössel and Prange [216] for the determination of up to 20 elements in rainwater including nickel and cobalt with typical detection limits down to 5–20 ng/l for heavy metal traces, by Prange et al. [217,218] for up to 25 elements, including nickel and cobalt in seawater and in various stages if the water cycle, and by Michaelis and Prange [219] for the analysis of geological and environmental samples partly using cobalt as an internal standard. Michaelis [220] described multielement analysis of environmental samples by TXRF, NAA and ICP–AES and concluded that the use of several independent methods is the most promising approach to cope with a given analytical problem. Trace metal determination in whole blood, however, was, even after separation of iron, only possible for nickel with a detection limit around 2 μg/l which is not usually sufficient for unexposed and slightly occupationally exposed persons [221].

Gerwinski et al. [221] and Gerwinski and Goetz, [222] described the determination of a number of elements including nickel after various decomposition procedures in city waste incineration ash and in certified reference materials from BCR, Brussels. The results indicated in most cases recoveries from 94 to 104% for those elements for which certified values were available.

Mukhtar et al. [223] analyzed a range of sample types common to the water industry (soil, water, sewage sludge) for elemental composition by TXRF and ICP–AES. The authors came to the conclusion that TXRF offers lower detection limits than ICP–AES, 4–5 orders of magnitude range in calibration, a sample preparation precision of < 3% and a instrumental precision of < 1%. Using a very simple internal multielement standard solution consisting of the elements Se, V and Zn the obtained results compared favorably with those obtained by ICP–AES and with the available reference values.

Von Bohlen et al. [224] directly analyzed microgram amounts of different finely ground solid powders and smears of alloys with TXRF. The method was applicable for about 20 elements, including nickel. Only microgram quantities were necessary for the determination, the dynamic range of the method was about four orders of magnitude, and detection limits were in the mg/kg region. Results for certified reference materials are reported with relative standard deviations predominantly below 15%. Relative deviations from the certified values were of the same order of magnitude.

The analysis of small specimens using X-ray fluorescence induced by intense synchroton radiation is possible with an X-ray microprobe [205,206]. Giauque et al. [225] measured femtogram quantities of trace elements in biological materials, including nickel.

The beam spot size was less than 10 x 10 μm and the minimum detectable limits achieved varied from 3 to 70 fg for the elements Zn ($z = 30$) to K ($z = 19$).

14.5.8. Nuclear methods

Despite the remarkable progress in the development and application of many other modern instrumental analytical techniques, activation analysis, mainly with reactor neutrons, still remains a useful multielement method for fingerprint studies in environmental and biological materials and as an independent approach if analytical accuracy has to be confirmed [226,227]. As can be seen from Table 14.2 only the sensitivity for cobalt competes favorably with other trace analytical methods, therefore it was and still is more frequently used for cobalt determination whereas applications for nickel are limited to particular cases and higher contents; see examples below. For the development and control of analytical methods and other studies the use of radioactive isotopes (e.g. [63]Ni, half-life 92 years, [57]Co, half-life 270 days, [58]Co, half-life 70.78 days and ^{60}Co, half-life 5.7 years) is often very helpful. Application of [63]Ni was reviewed by Kasprzak and Sunderman [228], and radiotracers in general, also including cobalt, by Krivan [229].

A few examples will be given of nickel determination by activation analysis. Bem et al. [230] determined nickel in biological materials and seawater. Nickel was preconcentrated by coprecipitation with α-benzildioxime followed by neutron activation. Interference from copper, cobalt, manganese, chlorine and bromine that partially coprecipitate or were adsorbed onto the surface of the precipitate was removed by washing the chloroform extract of the dissolved precipitate with an appropriate solution. This solution contained citric acid and inactive carriers of the above mentioned elements and natrium except cobalt. Nickel was finally stripped into the aqueous phase with 2 M hydrochloric acid. This was followed by liquid scintillation counting of the [65]Ni formed (half-life 2.52 h) using a special scintillator mixture that was able to tolerate large volumes of water. The detection limit was 5 ng (absolute). Precision and accuracy were checked with four certified reference materials.

The relatively high sensitivity of neutron activation techniques for cobalt allowed some important contributions to normal levels of cobalt in biological materials, e.g. by Versieck et al. [234], for chromium and cobalt in human serum with a still acceptable value of approx. 0.1 μg/l for cobalt. Later, the same author and others determined cobalt and several other elements with INAA in a second-generation biological reference material (freeze-dried human serum) with an average of 3.6 μg/kg [235].

Chisela and Brätter [236] used epithermal reactor neutrons for the determination of trace elements via long-lived isotopes using a boron carbide filter. To demonstrate the applicability of the method results are given for human erythrocytes, plasma, urine and some certified biological reference materials for seven elements, cobalt included. The method is as accurate and precise as conventional thermal neutron activation but is faster. The detection limit for cobalt with this technique varied for different materials but was only slightly higher compared with thermal neutrons.

References on p. 442

Gerhardsson et al. [18] used NAA and AAS for their multielemental assay of tissues of deceased smelter workers and controls. A number of 23 elements, including cobalt were determined after irradiation of the samples with thermal neutrons after radiochemical separation.

Morselli et al. [237] determined 18 elements, including cobalt, in 22 dry atmospheric deposition samples. The results show characteristic groups of elements, i.e. anthropogenic, non-anthropogenic and toxic, and elements showing seasonal variations in an urban industrialized area in northern Italy.

Freitas and Martinho [238,239] used INAA and evaluation by the k_0 standardization method for multielement analysis, including cobalt, in certified reference materials and environmental samples (sediments). The results were compared with mostly non-certified reference values and published reference values obtained by other methods. Good agreement was found for most of the elements, however, for some elements large discrepancies are observed.

Zeisler et al. [212] determined many elements in bivalves including cobalt and also higher contents of nickel with INAA as well.

14.5.9. Chromatographic methods

Chromatographic methods consist of a combination of a compound specific separation technique with a compound (or element) specific detection system that makes this approach also well suited for speciation. As metals are not directly accessible to gas- and liquid chromatographic techniques they have to be transformed into appropriate organometallic compounds via chelation. Since the chelate used commonly reacts with several elements multielement determinations can also usually be achieved for low metal contents.

Meyer and Neeb [240] investigated the working conditions for the determination of cobalt and nickel in biological materials by chelate gas chromatography and adsorptive stripping voltammetry. After dry ashing the metals are transformed into their di(trifluoroethyl)dithiocarbamates (FDETC), taken up in a mixture of extremely pure saturated hydrocarbons. By using capillary columns and the on-column technique the detection limits for $Ni(FDETC)_2$ and $Co(FDETC)_2$ with the flame ionization detector were 2 ng absolute. Palladium chloride was initially added as an internal standard and good separation was achieved from other metals (iron, cadmium, copper and zinc). Accuracy was proved for both methods by the analysis of CRMs.

Schaller and Neeb [241] optimized the capillary gas chromatographic conditions for the determination of several trace metals including nickel and cobalt after extraction from aqueous solution with di(trifluoroethyl)dithiocarbamic acid. This was performed by coupling of capillaries and improvement of the injection technique as well as the (flame ionization and electron-capture) detectors. With this procedure, chromium(VI), cobalt, copper, nickel, zinc and cadmium could be simultaneously determined, using an ECD at the lower μg/kg or μg/l level. This method was applied by the same authors for the determination of cobalt in Rhine water [242].

Häring and Ballschmiter [243] described an high-performance liquid chromatographic (HPLC) procedure for the determination of copper, cobalt and nickel at 0.2 to 10 μg/l levels from aqueous solutions by chelation with diethyldithiocarbamate with preconcentration on a separate column. Separation is performed after direct transfer of the preconcentrated chelates to a reversed-phase phenyl column. Determination is performed by a UV/VIS spectrometer.

König et al. [244] determined nickel in hydrogenated fats by extracting nickel and iron with hydrochloric acid, and complexation with 1-hydroxy-2-pyridinthione (Na salt). The formed chelates are extracted with chloroform and separated by HPLC applying gradient elution. Nickel is then determined photometrically with an absolute detection limit of 0.3 ng nickel and a linear range up to 1 μg nickel.

Ichinoki et al. [245] determined mg/kg levels of cadmium, nickel, lead, zinc. cobalt, copper and bismuth in certified NIST reference materials Bovine liver and Oyster tissue. Subsamples of approx. 250 mg were ashed in a muffle furnace, the ash extracted with hydrochloric acid, the metals transformed into hexamethylenedithiocarbamate chelates and extracted into chloroform. After separation on a reversed-phase column, determination of the peaks was performed spectrophotometrically at 260 nm in a flow cell. Cadmium, nickel, lead, zinc and copper could be accurately determined from 0.5–850 mg/kg with a standard deviation around 7%.

Munder and Ballschmiter [246] investigated the chromatographic behavior of metal bis(ethoxyethyl)dithiocarbamates of 14 metals, including cobalt and nickel. After preconcentration from water (rivers and lakes) with an on-line system the chelates were separated by C_{18} reversed-phase liquid chromatography. A quaternary solvent mixture with admixture of a surfactant was used as eluent. Detection was performed by UV at 254 nm. The detection limits were 2 ng for cobalt and 3 ng for nickel.

Bond et al. [247] determined copper, nickel and lead in urine. Direct injection of freshly acidified and filtered (0.45 μm) urine samples were made onto a reversed-phase separation column with a guard column for sample clean-up. In the mobile phase a dithiocarbamate was included and the metal complexes were detected either electrochemically (copper and nickel) or spectrophotometrically (copper, nickel and lead). The procedure was shown to allow the determination of copper and nickel in urine of unexposed to occupationally exposed persons by electrochemical detection. Spectrophotometric detection, however, was not sensitive enough for direct determination of copper and nickel at lower concentration levels.

King and Fritz [248] determined cobalt, copper, mercury and nickel as bis(2-hydroxyethyl)dithiocarbamate (HEDC) complexes by HPLC. Ammonium-HEDC was used as a precolumn derivatizing reagent for the reversed-phase determination of cobalt(II), copper(II), mercury(II) and nickel(II). The metal–HEDC complexes were soluble in water, thus cobalt, copper and nickel were determined by direct injection of the aqueous solution onto a C_{18} column in the range of 5 μg/l to 10 mg/l with a precision of 1.5 to 3.2%; detection was at 405 nm by a variable-wavelength detector.

Baiocchi et al. [249] used 8-hydroxychinoline as a precolumn chelating agent for liquid chromatographic multielement determinations by conventional reverse-phase HPLC. With fixed wavelength UV absorption it could be shown that the determination of zinc(II), aluminium(III), cobalt(III) (oxidized prior to chelation in the presence of air) chromium(III), copper(II), gallium(III), indium(III) and iron(III) was feasible as 8-hydroxychinoline complexes at the μg/l level.

For gas chromatography and HPLC metals have to be transformed into metal chelates accessible to these techniques. Ion chromatography, however, initially predominantly applied for inorganic anion analysis is now also successfully applied to the direct separation and determination of inorganic cations [250,251]. Therefore a few examples from this still growing area in analytical chemistry will be given.

Rubin and Heberling [251], demonstrated the relatively simple enrichment possible with minimal blanks for transition metals with up to a factor of 10^3. This allowed the determination of cobalt at < 0.01 μg/l and nickel at 0.08 μg/l besides copper, iron(II), iron(III) and zinc in deionized water and a remarkable multielement capability and amenability to automation.

Yan and Schwedt [252] separated eleven heavy metals, nickel and cobalt included, and alkaline earth elements on a 200 mm long column in 24 min and performed the detection with the aid of a post-chromatographic derivatization with 4-(2-pyridylazo)resorcinol (PAR) or PAR–ZnEDTA solution. With PAR the sensitivity for zinc, nickel, cobalt, and iron(II/III) could be enhanced for subsequent UV/VIS spectrophotometry. After preconcentration from 300 ml water an absolute amount of 10.8 ng Ni was determined. Due to the very high enrichment factor the control of blanks is very important.

Joyce and Schein [253] demonstrated the separation/preconcentration of transition metals from complex matrices by pumping the analyte solution (e.g. 10 ml) through a chelation resin column which selectively retains transition metals but not alkali or alkaline earth metals or common anions. The transition metals concentrated on the column are then automatically eluted and determined. Quantification may be by ion chromatography or an atomic spectrometric method. Examples are given for the determination of iron(II), copper(II), nickel(II), zinc(II) and Mn(II) in seawater by ion chromatography with a value of 0.47 μg/l for nickel.

Yan and Worsfold [254] described an optimized flow-injection manifold for the chemiluminescence determination of cobalt(II), copper(II), iron(II) and chromium(III) by their catalytic effect on the luminol reaction. Detection limits are 0.0006, 0.08, 0.3 and 0.1 μg/l respectively. A procedure for the separation of cobalt, copper and iron on a low-capacity, silica-based cation-exchange column and post-column determination with the aid of the luminol reaction with detection limits of 0.01 μg/l for cobalt and 5 μg/l for copper was given.

14.5.10. Mass spectrometry

Nickel and cobalt can be determined by many classical and modern versions of mass spectrometry with typical detection limits down to the pg, in special cases, even the fg level [255–257].

Since only nickel has stable isotopes, it is accessible to the extremely sensitive, accurate and precise isotope dilution method that could significantly contribute to the determination of a number of toxicologically and environmentally important metals [255,258].

Mykytiuk et al. [259] determined trace concentrations of iron, cadmium, zinc, copper, nickel, lead and uranium by stable isotope dilution spark source mass spectrometry (IDMS), cobalt by internal standard. The seawater samples were preconcentrated on Chelex-100 and the concentrate was evaporated on a graphite or silver electrode. The results are compared with those obtained by GFAAS and ICP–AES. Results for two seawater samples 0.37 and 0.43 μg/l for nickel and 0.020 and 0.028 μg/l for cobalt were in acceptable agreement with GFAAS after solvent extraction and ion exchange.

Stukas and Wong [260] determined trace levels of copper, cadmium, lead, zinc, iron and nickel in seawater by IDMS. The authors applied thermal source IDMS together with ultra-clean room techniques after preconcentration by extraction with dithizone and by APDC/DDDC ion-exchange resins. Accuracy was tested by independent methods (voltammetry and graphite furnace AAS). Nickel values ranged from approx. 6 to approx. 10 nmol/l, i.e. from approx. 0.4 to approx. 0.6 μg/l.

Völkening and Heumann [261] used voltammetry and IDMS for the determination of a number of heavy metals in Antarctic snow. Nickel and other elements were analyzed by IDMS after adding spike solutions, for nickel ^{62}Ni, 200–300 ml subsamples were evaporated to a volume of about 10 ml under clean room conditions followed by cathodic deposition of the metals on platinum electrodes prior to further preparatory steps and IDMS measurements by thermal ionization. Nickel values in surface snow were in the range < 4.8 to 40 ng/kg.

14.5.11. Spectrophotometry and related techniques

As far as total elemental analysis is concerned UV/VIS spectrophotometry for cobalt and nickel is applied with decreasing frequency in conventional analytical chemistry and for the control of occupational exposure. This is despite the fact that in official analytical procedures spectrophotometric reference methods are still prescribed e.g. for the analysis of cobalt in fertilizers, plants and feedingstuff [262]. However, spectrophotometry is used with a growing tendency in metal-biochemistry and –as can be seen in Section 14.5.8– especially in chromatography if previous concentration and transformation steps and new approaches such as catalytic reactions and chemiluminescence are applied.

The use of catalytic spectrophotometric techniques might still be of benefit since they can lower detection limits for several orders of magnitude, compared to equilibrium spectrophotometry. This is especially the case for cobalt in various matrices as was reviewed by Müller [263].

Burns et al. [264] determined cobalt in the range of 0–10 μg spectrophotometrically at 625 nm after flow-injection extraction into chloroform of the ion associate ethylene-bis(triphenylphosphonium)tetrathiocyanatocobaltate(II) with an injection rate of 20/h. The calibration graph was linear up to 20 μg/ml and the detection limit was 0.23 μg/ml based on injection volumes of 500 μl.

Dawson and Lyle [265] extracted and determined cobalt spectrophotometrically with dithizone as cobalt(III) dithizonate with excess of dithizone in carbon tetrachloride. The calibration graph was linear for 1–10 μg of cobalt. Techniques to overcome most interferences, also from nickel, are described as well. The same authors [266] also described procedures for the spectrophotometric determination of iron and cobalt. Iron is determined as iron(II) with Ferrozine [disodium salt of 3-(2-pyridyl)-5,6-bis(4-phenylsulfonic acid)-1,2,4-triazine] and cobalt as the cobalt(III) dithizonate complex. The reduction to iron(II) prevented interference of iron(III) in the cobalt determination. Removal of interference by other metal ions was described.

Melgaredo et al. [267] developed a derivative spectrophotometric method for the simultaneous determination of microgram amounts of nickel, zinc and copper in an aqueous ethanolic medium with 2-(2-pyrideglazo)-2-naphtol (PAN). The method allowed the determination of nickel from 0.3 to 2.0 mg/l with good precision and accuracy.

Sakai et al. [268] used a luminol chemiluminescence detection/flow injection analysis technique coupled with ion chromatography for the selective determination of cobalt(II) at ng/l levels. When a 100 μl sample injection volume was used, the detection limit was 1 ng/l cobalt; the minimum detectable amount of cobalt was 100 fg. The calibration graph was linear above 10 ng/l and the linear dynamic range extended over six orders of magnitude. The RSD for ten replicate measurements of 30 ng/l cobalt was 3.8%.

Minggang et al. [269] examined the chemiluminescence of the reaction of tartaric acid with hydrogen peroxide in the presence of cobalt(II) in alkaline buffer media. The maximum emission wavelength was 460 nm. Foreign ions −iron(II), chromium(II) and manganese(II)− interfered when present in more than a 10-fold excess of cobalt(II). Several ions, however, could be tolerated when present in a greater excess. The concentration of linear responses ranged from $3.5 \cdot 10^{-9}$ to $2 \cdot 10^{-6}$ g/ml and the detection limit was $4 \cdot 10^{-11}$ g/ml. The procedure was applied for the determination of cobalt traces in human blood serum.

Jones et al. [270] separated cobalt by a simple cation-exchange system from other metal ions. Cobalt was detected by luminol chemiluminescence achieving a detection limit of 0.5 ng/l, using a 200 μl sample.

14.6. SPECIATION

Nickel and cobalt commonly do not form organometallic species, with the exception of nickel carbonyl which is an intermediate in the Mond process for nickel refining and also industrially used as a catalyst [3]. Speciation approaches for these metals e.g. in liquid and solid materials have to consider the differentiation in the liquid phase −ionic species at differing valency states, dissolved complexes− and metals adsorbed on colloidal particles and on distinct solid phases in soils and sediments determining bioavailability as recently described comprehensively (ref. 271 and also in Chapter 3). From this important and steadily growing field some typical examples of nickel and cobalt from work performed during the last decade will be given.

Tessier et al. [272] developed an analytical procedure that involved sequential chemical extractions for the partitioning of particulate trace metals (cadmium, cobalt, copper, nickel, lead, zinc, iron and manganese) into five fractions: exchangeably bound to carbonates, bound to Fe-Mn oxides, bound to organic matter and residual. The accuracy of this approach was verified and the developed scheme applied to river sediments. The results indicate only low amounts of nickel and cobalt in the exchangeable fraction and that bound to carbonates, while a significant fraction was found bound to Fe-Mn oxides. Relatively low was the amount of nickel and cobalt in the fraction bound to organic matter, while the residual fraction contained more than 50% of the total metal content for nickel and cobalt as well as for zinc, iron and manganese. This procedure was later successfully applied to the calculation of apparent overall equilibrium constants for the adsorption of some metals, nickel included, onto natural iron oxyhydroxides in oxidic lake sediments [273].

Hayase et al. [274] determined organically associated trace metals including nickel in estuarine seawater by chloroform extraction at pH 8 and pH 3. It was found that nickel, copper, molybdenum and manganese were extracted more at pH 8 than at pH 3, while cadmium and lead were not associated with the dissolved organic matter (DOM) at either pH 8 or 3. The metals extracted into chloroform at pH 8 were assumed to be associated with neutral or weakly basic DOM, while at pH 3 they could be associated with either the neutral (or weakly basic) DOM or two types of acidic DOM [274].

Nimmo et al. [275] used cathodic (adsorptive) stripping voltammetry to determine dissolved nickel, copper, vanadium and iron in seawater samples. It was shown that the concentration of the free metal ion and of the metal organic complexes could be computed from the CSV-labile and total dissolved metal concentrations. Total dissolved metal concentrations were obtained after UV-irradiation of acidified samples. In the studied area (Liverpool Bay) considerable portions of nickel (up to approx. 40%) were found to occur as stable organic compounds, whereas almost all copper (98–99%) occurred in less stable organic complexes [275].

Rudd et al. [276] applied a sequential extraction scheme for cadmium, copper, nickel, lead and zinc to sewage sludge. The largest cadmium and nickel fractions were extractable in 0.1 M sodium EDTA at pH 6.5, corresponding to the carbonates. Progressive acidification of liquid sludges mobilized nickel at all investigated pH values. Application of the sequential extraction scheme to the residues obtained following acidification indicated that nickel (and zinc) speciation was likely to change with decreasing pH.

Warren and Dudas [277] leached alkaline fly ash in a series of lysimeters with dilute sulfuric acid. Elements such as rubidium, caesium, lead, tantalum, titanium and hafnium were enriched in the highly leached portion of the residue sequence, suggesting association with the resistant internal silicium-rich glass matrix of ash particles. Between 50 and 80% of the total manganese, antimony, thorium, chromium, zinc, cobalt, scandium and rare earth elements were also retained in the highly leached ash residues.

References on p. 442

Wu and Wong [278] described a speciation procedure for a solid mixture of nickel species applicable to atmospheric monitoring of Ni^0, Ni^{2+} and NiO_x. The procedure involved magnetic extraction of metallic nickel to separate it from the water-soluble nickel salts and the insoluble oxides, followed by adsorptive (cathodic) stripping voltammetry to quantify the amount of nickel in each fraction. The separation scheme was validated by using calibration standards and by simulating air filter samples prepared with authentic nickel products. The recoveries for all three forms ranged from 94 to 105%.

Weber and Schwedt [279] investigated the speciation of nickel in coffee, tea and red wine. The scheme outlined for establishing the distribution of nickel relied on relatively simple techniques. The total amount of nickel (approx. 100 μg/l) is divided into different fractions by direct spectrophotometric determination, filtration and liquid–liquid extraction by ethyl acetate. The division of the latter was possible by means of thin-layer chromatography. Information about organic compounds obtained by chemical spray reactions, UV-visible and IR-spectroscopy, was compared with corresponding nickel determination. Characteristic nickel patterns dependent on chemical groups (e.g. phenolic and amine) were obtained.

Despite the fact that nickel carbonyl has a very short half-life, recent studies confirmed the appearance of traces of nickel carbonyl in the atmosphere near a busy traffic intersection, in town gas and in cigarette smoke [3]. In nickel refineries and the petrochemical industry, where nickel carbonyl is used as a catalyst, current monitoring of this very toxic compound is necessary. At the working place nickel carbonyl is quantitatively determined by portable chemiluminescent detectors developed by Stedman et al. [280,281] and further technically improved by Houpt et al. [282].

The principle of this detector is the conversion of nickel carbonyl by the action of ozone and carbon monoxide to an excited state of nickel oxide, which emits a photon during decay to the ground state. This radiation is determined with a chemiluminometer. The detection limit is approx. 1 μl in 1 l gas. More recently also other techniques for the determination of nickel carbonyl by decomposition in a quartz or graphite tube [82,83,283] or after preconcentration on chromosorb by microwave-induced plasma-AES [84] have been described.

14.7. QUALITY CONTROL AND REFERENCE MATERIALS

Quality control can be performed via different, usually simultaneously applied approaches: different and reference methods in intra- and interlaboratory comparisons and the use of appropriate reference materials.

Accuracy and comparability of nickel analysis in body fluids was significantly improved by the efforts of the IUPAC Subcommittee on Environmental and Occupational Toxicology of Nickel. The Subcommittee developed under the guidance of F.W. Sunderman Jr. a reference method for nickel in serum and urine by graphite furnace AAS [284]. It was subsequently applied in various interlaboratory comparisons [285–287].

TABLE 14.3.

REFERENCE MATERIALS WITH CERTIFIED RESPECTIVELY INFORMATION OR SIMILAR VALUES FOR COBALT CONCENTRATION

Concentrations are given on a dry weight basis and in mg/kg unless noted "liquid", than mg/l

Material	Code	Conc.	Error (confidence interval) (%)
I Biological materials, certified			
Human Serum	GHENT	0.0036	16
Milk Powder	IAEA-A-11	0.0045	17
Rice Flour	NIST-SRM-1568	0.02	50
Mixed Diet	NIST-RM-8431a	0.038	21
Single Cell Protein	BCR-CRM-274	0.039	7.7
Dogfish Muscle	NRCC-DORM-1	0.049	28
Bowen's Kale		0.0632	17
Fish Flesh	IAEA-MA-A-2/TM	0.08	12
Sargasso	NIES-CRM-09	0.12	8.3
Copepoda	IAEA-MA-A-1/TM	0.12	8.3
Hay Powder	IAEA-V-10	0.13	12
Human Hair	SHINR-HH	0.135	5.9
Dogfish Liver	NRCC-DOLT-1	0.157	23
Bovine Liver	NIST-SRM-1577a	0.21	24
Lobster Hepatopankreas	NRCC-TORT-1	0.42	12
Chlorella	NIES-CRM-03	0.87	5.7
II Biological materials, values only informative			
Rye Flour	IAEA-V-8	0.0025	
Milk Powder	NIST-SRM-1549	0.0041	
Wheat Flour	NIST-SRM-1567a	0.006	
Skim Milk Powder	BCR-CRM-151	0.006	
Skim Milk Powder	BCR-CRM-063	0.0062	
Skim Milk Powder	BCR-CRM-150	0.0064	
Rice Flour	NIES-CRM-10C	0.007	
Urine	NYCO-108	0.011	liquid
Citrus Leaves	NIST-SRM--1572	0.02	
Rice Flour	NIES-CRM-10B	0.02	
Rice Flour	NIES-CRM-10A	0.02	
Spruce Shoots	ESB-RM-F1	0.03	
Pine Needles	NIST-SRM-1575	0.1	
Tea Leaves	NIES-CRM-07	0.12	
Bovine Liver	CZIM-LIVER	0.37	
Oyster Tissue	NIST-SRM-1566	0.4	
Spruce Shoots	ESB-RM-F2	0.6	

(Continued on p. 436)

References on p. 442

TABLE 14. 3 (continued)

Material	Code	Conc.	Error (confidence interval) (%)	
III Environmental materials, certified				
Seawater	NRCC-NASS-2	0.000004	25	liquid
Seawater	NRCC-CAAS-2	0.000025	24	liquid
Estuarine Water	NRCC-SLEW-1	0.00046	15	liquid
Water	NIST-SRM-1643B	0.026	3.8	liquid
Fuel Oil	NIST-SRM-1634b	0.32	12	liquid
Bituminous Coal	NIST-SRM-1632b	2.29	7.4	
Apatite concentrate	ICHTJ-CVA-AC-1	2.72	10	
Vehicle Exhaust	NIES-CRM-08	3.3	9.1	
Soil	AMM-SO-1	3.9	7.7	
Coal (O.F.S.)	SABS-SARM-19	5.6	14	
Coal (Witbank)	SABS-SARM-18	6.7	13	
Coal	BCR-CRM-040	7.8	7.6	
Coal (Sasulbourg)	SABS-SARM-20	8.3	8.4	
Sewage Sludge	BCR-CRM-145	8.38	8.5	
Soil	IAEA-SOIL-7	8.9	9.6	
Sewage Sludge	BCR-CRM-114	9.06	6.6	
Estuarine Sediment	NIST-SRM-1646	10.5	12	
Marine Sediment	NRCC-MESS-1	10.8	17	
Marine Sediment	NRCC-BCSS-1	11.4	18	
Sewage Sludge	BCR-CRM-146	11.8	5.9	
River Sediment	NIST-SRM-2704	14.0	4.3	
Marine Sediment	NRCC-PACS-1	17.5	6.3	
Lake Sediment	IAEA-SL-1	19.8	7.6	
Coal Fly Ash	IRANT-ENO	26.1	4.8	
Pond Sediment	NIES-CRM-02	27	11	
City Waste	BCR-CRM-176	30.9	4.2	
Coal Fly Ash	ICHTJ-CTA-FFA-1	39.8	4.3	
Coal Fly Ash	IRANT-ECO	49.0	16	
Coal Fly Ash	IRANT-ECH	49.8	5.4	
Coal Fly Ash	IRANT-EOP	53.2	3.2	
Fly Ash	BCR-CRM-038	53.8	3.5	
Copper Plant Flue Dust	IRANT-KHK	313	7.0	
IV Environmental materials, values only informative				
Coal	NIST-SRM-1635	0.65		
Coking Coal	BCR-CRM-181	1.6		
Gas Coal	BCR-CRM-180	3.3		
Soil	BCR-CRM-142	7.9		
Steam Coal	BCR-CRM-182	8.8		
Steel Plant Flue Dust	IRANT-OK	9		
Soil	BCR-CRM-141	9.2		

TABLE 14. 3 (continued)

Material	Code	Conc.	Error (confidence interval) (%)
Soil	BCR-CRM-143	11.8	
Urban Particulate	NIST-SRM-1648	18	
Coal Fly Ash	NIST-SRM-1633a	46	

Further to these activities, interlaboratory cooperative studies using various independent analytical methods were conducted under the auspices of IUPAC, Commission on Toxicology by Christensen et al. [288] to establish concentrations of biologically important trace elements in commercially available lyophilized human serum, whole blood and urine. The main objective was to arrive at consensus concentration values in order to establish these materials as IUPAC proposed reference materials for chemical analysis. 16 elements and creatinine were characterized in urine, 10 elements in serum. Cobalt was only determined in urine with a consensus value of 100 $\mu g/l$, whereas the values for nickel in urine and serum were 40 and 3.2 $\mu g/l$ respectively.

Brix et al. [289] reported on the reproducibility in the determination of eight heavy metals in two marine plant (eelgrass) materials from the performance of an interlaboratory calibration with 14 participating laboratories. Nine laboratories analyzed nickel in the materials and confidence limits 1.0–3.1 mg/kg and 1.6–3.9 mg/kg respectively were found.

Kumpulainen and Paakki [290] reported on the analytical quality program employed within a FAO study on 14 trace elements in nationally representative staple foods of European countries. A number of certified biological reference materials were used to validate the analytical methods employed. The mean relative standard deviation was below 25% for nickel which was only analyzed in two materials, wheat flour and potato powder.

Byrne and Krasovec [291] reported on a very sensitive radiochemical neutron activation analytical method for nickel and cobalt in 10 reference materials from different sources. The given values either agreed quite well with certified or information values or are new for the investigated materials. Values for nickel ranged from approx. 0.011 to approx. 9.0 mg/kg, for cobalt from 0.0001 to 0.073 mg/kg.

Cornelis et al. [292] reported on the assistance of their laboratory in the certification of 31 environmental and food reference materials, issued by the Bureau of Reference Materials of the European Communities (BCR). Cobalt was with 80% one of the 10 most frequently certified elements in these materials and was determined by instrumental NAA and flame AAS. Nickel was less frequently analyzed and the only method applied for this element was ICP–AES.

TABLE 14.4

REFERENCE MATERIALS WITH CERTIFIED RESPECTIVELY INFORMATION OR SIMILAR VALUES FOR NICKEL IN INCREASING ORDER OF THE ELEMENTAL CONCENTRATION

Concentrations are given on a dry weight basis and in mg/kg unless noted "liquid", than mg/l.

Material	Code	Conc.	Error (confidence interval) (%)	
I Biological materials, certified				
Serum	KL-148-I	0.03	66	liquid
Serum	KL-148-II	0.05	40	
Cotton Cellulose	IAEA-V-9	0.09	61	
Wheat Flour	ARC/CL-WF	0.12	25.8	
Rice Flour	NIES-CRM-10A	0.19	15	
Potato Powder	ARC/CL-PP	0.193	22.3	
Urine	KL-142-II	0.2	37	liquid
Dogfish Liver	NRCC-DOLT-1	0.26	23	
Total Diet	ARC/CL-TD	0.271	14	
Rice Flour	NIES-CRM-10C	0.30	10	
Rice Flour	NIES-CRM-10B	0.39	10	
Fish	EPA-Fisch	0.54	51	
Citrus Leaves	NIST-SRM-1572	0.6	50	
Mixed Diet	NIST-RM-8431a	0.644	23	
Oyster Tissue	NIST-SRM-1566	1.03	18	
Fish Flesh	IAES-MA-A-2/TM	1.1	18	
Dogfish Muscle	NRCC-CORM-1	1.20	25	
Copepoda	IAEA-MA-A-1/TM	1.9	10	
Lobster hepatopankreas	NRC-TORT-1	2.3	13	
Human Hair	SHINR-HH	3.17	12	
Hay Powder	IAEA-V-10	4.0	14	
Tea Leaves	NIES-CRM-07	6.5	4.6	
II Biological materials, values only informative				
Human Serum	GHENT	0.0025	(best estimate)	
Serum	NYCO-105	0.0032		liquid
Skim Milk Powder	BCR-CRM-063	0.0112		
Urine	NYCO-108	0.04		
Milk Powder	IAEA-A-11	0.054		
Skim Milk Powder	BCR-CRM-151	0.056		
Skim Milk Powder	BCR-CRM-150	0.0615		
Urine (Normal)	NIST-SRM-2670	0.07		liquid
Rice Flour	NIST-SRM-1568	0.16		
Bovine Muscle	BCR-CRM-184	0.27		

TABLE 14.4 (continued)

Material	Code	Conc.	Error (confidence interval) (%)	
Urine (Spiked)	NIST-SRM-2670	0.3		liquid
Single Cell Protein	BCR-CRM-274	0.3		
Wholemeal Flour	BCR-CRM-189	0.38		
Pig Kidney	BCR-CRM-186	0.42		
Brown Bread	BCR-CRM-191	0.44		
Spruce Shoots	ESB-RM-F1	0.50		
Bowen's Kale		0.895		
Pine Needles	NIST-SRM-1575	3.5		
Spruce Shoots	ESB-RM-F2	6.4		

III Environmental materials, certified

Material	Code	Conc.	Error	
Seawater	NRCC-NASS-2	0.000257	10	liquid
Seawater	NRCC-CAAS-2	0.000298	12	liquid
Estuarine Water	NRCC-SLEW-A	0.000743	10	liquid
Water	NIST-SRM-1643b	0.049	6.1	liquid
Coal	NIST-SRM-1635	1.74	5.7	
Coal (Bituminous)	NIST-SRM-1632b	6.10	4.4	
Coal (Witbank)	SABS-SARM-18	10.8	6.5	
Coal (O.F.S.)	SABS-SARM-19	16	22	
Vehicle Exhaust	NIES-CRM-08	18.5	8.1	
Fly Ash	FISHER-SRS001	19.5	9.2	
Coal (Sasolburg)	SABS-SARM-20	25	6.0	
Coal	BCR-CRM-040	25.4	6.3	
Fuel Oil	NIST-SRM-1634b	28	7.1	
Soil	BCR-CRM-142	29.2	8.6	
Marine Sediment	NRCC-MESS-1	29.5	9.2	
Estuarine Sediment	NIST-SRM-1646	32	9.4	
Pond Sediment	NIES-CRM-02	40	7.5	
Sewage Sludge	BCR-CRM-145	41.4	5.8	
Estuarine Sediment	BCR-CRM-277	43.3	3.7	
Marine Sediment	NRCC-PACS-1	44.1	4.5	
River Sediment	NIST-SRM-2704	44.1	6.8	
Lake Sediment	IAEA-SL-1	44.9	17	
Steel Plant Flue Dust	IRANT-OK	49.8	10	
Marine Sediment	NRCC-BCSS-1	55.3	6.5	
Lake Sediment	BCR-CRM-280	73.6	3.5	
River Sediment	BCR-CRM-320	75.2	1.9	
Coal Fly Ash	IRANT-ENO	77.0	17	
Urban Particulates	NIST-SRM-1648	82	3.7	
Coal Fly Ash	IRANT-ECO	92	15	
Coal Fly Ash	ICHTJ-ETA-FFA-1	99	5.9	
Soil	BCR-CRM 143	99.5	5.5	
Coal Fly Ash	IRANT-EOP	108	15	

(Continued on p. 440)

TABLE 14.4 (continued)

Material	Code	Conc.	Error (confidence interval) (%)
Coal Fly Ash	IRANT-ECH	117	2.6
City Waste	BCR-CRM-176	123.5	3.4
Coal Fly Ash	NIST-SRM-1633a	127	3.1
Copper Plant Flue Dust	IRANT-KHK	188	10
Municipal Sludge	EPA-SLUDGE	194	7.7
Electroplating Sludge	FISHER-SRS010	195.7	6.7
Board Coating Sludge	FISHER-SRS009	347	6.3
Sewage Sludge	BCR-CRM 144	942	2.3
Incinerator Sludge	FISHER-SRS012	13088	8.3
Chrome Plating Sludge	FISHER-SRS011	41600	10

IV Environmental materials, values only informative

Cooking Coal	BCR-CRM-181	8.6	
Apatite Concentrate	ICH-CTA-AC-1	9	
Soil	AMM-SO-1	13	
Gas Coal	BCR-CRM-180	16	
Soil	IAE-SOIL-7	26	
Soil	BCR-CRM-141	30.9	
Steam Coal	BCR-CRM-182	39	
Fly Ash	BCR-CRM-038	194	

Alvarez [293] reported on six recently developed NIST food-related standard reference materials. Only in one material (Oyster tissue, SRM 1566a) was there a certified value given for nickel, whereas cobalt values, however not all certified, were given for five materials except for the Total Diet (SRM 1548) material. Using neutron activation analysis 32 major and trace elements were determined. The non-certified value for cobalt was given as 27.5 mg/kg [294].

Tables 14.3 and 14.4 summarize, based on a recent compilation by Toro et al. [295], reference materials commercially available at present from different suppliers with data (either certified or informative) for cobalt and nickel, Table 14.5 lists addresses and references of suppliers.

14.8. CONCLUSIONS AND PROSPECTS

In this chapter it was demonstrated that the instrumental analytical methods available at present, particularly graphite furnace AAS (GFAAS), voltammetry in the adsorptive (cathodic) stripping (ACS) mode, inductively coupled plasma–mass spectrometry (ICP–

TABLE 14.5.

ADDRESSES AND REFERENCES OF SUPPLIERS FOR THE CODES GIVEN IN TABLES 14.2 AND 14.4.

Most of these RMs can be purchased either directly from the suppliers or from Promochem, P.O. Box 1246, D-W-4230 Wesel, Germany

AMM-SO-1	Holynska et al. [296]
ARC/CL	Kumpulainen and Paaki [290]
BCR	Community Bureau of Reference, Belgium
Bowen's Kale	Bowen [297]
CZIM-Liver	Kucera et al. [298]
ESB-RM-F	Environmental Specimen Bank, KFA Juelich, Germany
Fisher	Fisher Scientific, Springfield, NJ, USA
Ghent	Versieck et al. [235]
IAEA	International Atomic Energy Agency, P.O. Box 100, A-1400 Vienna, Austria
ICHTJ-CTA	Commission of Trace Analysis of the Committee for Analytical Chemistry of the Polish Academy of Sciences and Institute of Nuclear Chemistry and Technology, Warszawa, Poland
IRANT	Institute of Radioecology and Applied Nuclear Techniques, Kosice, CSFR
KL	Kaulson Laboratories Inc, West Caldwell, NJ 07006, USA
NIES	National Institute of Science and Technology (formerly National Bureau of Standards) Washington, DC, USA
NRCC	National Research Council, Canada, Ottawa
NYCO	Nycomed Pharma, Oslo, Norway
SABS	South African Bureau of Standards, Pretoria, Republic of South Africa
SHINR	Shanghai Institute of Nuclear Research, Academia Sinica, Shanghai, People's Republic of China

MS) and for higher levels inductively coupled plasma–atomic emission spectrometry (ICP–AES) are excellent tools to cope with the broad concentration ranges of nickel and cobalt in biological and environmental materials, so that for nearly every task an appropriate and sometimes an additional reference method can be used. Preconcentration and selective preconcentration/separation approaches, e.g. the various solvent and column chelate extraction and ion chromatographic methods, recently also coupled with flow-injection techniques have enhanced the potential of less sensitive (GFAAS) or spectrometric and demiluminecence methods that are extremely prone to interferences. The introduction of high-pressure −or rather high-temperature− ashing and the commercial availability of square wave voltammetry have further improved accuracy and speed of adsorptive strip-ping techniques for both elements. As far as validation of methods is concerned, the impressive list of biological and environmental reference materials already available with

References on p. 442

442

values from the lowest to the highest natural contents is a promising basis for further reliable work.

Due to the efforts of many laboratories and suppliers around the world it can be expected that existing gaps in this area will be bridged in the near future, also by the introduction of an increasing number of second generation reference materials completely identical with natural materials prepared and stored at extremely low temperatures so that indefinitely long shelf lives can be achieved as was recently introduced for analytical tasks within environmental specimen banking [299,300], which is amply discussed in Chapter 3.

Despite the fact that some of the methods presently available reach absolute detection limits down to the pg level it can be forecast that particular problems with extremely small samples at surfaces, in solids as well as for speciation analysis in combination with chromatographic techniques, will require even lower routinely attainable detection limits. Since this will certainly be in conjunction with simultaneous or sequential fingerprint analysis for an increasing number of essential and non-essential elements the further growth and improvement of multielement techniques like total reflection and still more sensitive Sinchrotron excited X-ray modes and multielement mass spectrometric techniques making use of lasers can be expected [301,302]. Of the latter techniques certainly atomic absorption and atomic fluorescence might benefit from further lowering of detection limits and increase in reliability [303,304], which is also the case for furnace-AES combinations [305]. Finally furnace atomic non-thermal excitation spectrometry (FANES) a method currently being evaluated in detail should be mentioned as it might also be a promising technique in the near future for multielement analysis including nickel and cobalt [306].

REFERENCES

1 J.O. Niragu (Editor), *Nickel in the Environment*, Wiley, New York, Chichester, Brisbane, Toronto, 1980.
2 M. Anke and H.J. Schneider, *Nickel, 3. Spurenelement-Symposium, Jena, 1980*.
3 F.W. Sunderman, Jr. and A. Oskarsson, Nickel, in: E. Merian (Editor), *Metals and Their Compounds in the Environment*, VCH, Weinheim, 1991, pp. 1101–1126.
4 J.P. Rigaut, *Rapport preparatoire sur les critères de santé pour le nickel*, Doc. CCE/LUX/V/E/24/83, Luxembourg, 1983.
5 F.W. Sunderman, Jr., A. Aitio, L.G. Morgan and T. Norseth, *Toxicol. Ind. Health*, 2 (1986) 17–78.
6 T. Norseth, in L. Friberg, G.F. Nordberg and V.B. Vouk (Editors), *Handbook on the Toxicology of Metals, Vol. II*, Elsevier, Amsterdam, New York, Oxford, 1986, pp. 46–48.
7 H.F. Hildebrand and M. Champy (Editors), *Biocompatibility of Co-Cr-Ni-Alloys*, Plenum Press, New York and London, 1988.
8 G.N. Schrauzer, in E. Merian (Editor), *Metals and Their Compounds in the Environment*, VCH, Weinheim, 1991, pp. 879–892.
9 C.-G. Elinder and L. Friberg, in L. Friberg, G.F. Nordberg and V.B. Vouk (Editors), *Handbook on the Toxicology of Metals, Vol. II*, Elsevier, Amsterdam, New York, Oxford, 1986, pp. 211–232.
10 A. Ferioli, R. Roi and L. Alessio, in L. Alessio, A. Berlin, M. Boni, R. Roi (Editors), *Biological Indicators for the Assessment of Human Exposure to Industrial Chemicals*, EUR 11135 EN, 1987, pp. 49–61.

11 M. Stoeppler, in J.O. Nriagu (Editor), *Nickel in the Environment*, Wiley, New York, Chichester, Brisbane, Toronto, 1980, pp. 661–821.
12 K.W. Bruland, R.P. Franks, G.A. Knauer and J.H. Martin, *Anal. Chim. Acta*, 105 (1979) 233–245.
13 M. Stoeppler and H.W. Nürnberg, *Ecotoxicol. Environ. Safety*, 3 (1979) 335–351.
14 L.G. Danielsson, *Mar. Chem.*, 8 (1980) 199–215.
15 R. Frank, K.I. Stonefield, P. Suda and J.W. Potter, *Sci. Total Environ.*, 26 (1982) 41–65.
16 P.A. Yeats and J.A. Campbell, *Mar. Chem.*, 12 (1983) 43–58.
17 E.D. Goldberg, M. Koide, V. Hodge, A.R. Flegal and J. Martin, *Estuarine, Coastal, Shelf Sci.*, 16 (1983) 69–93.
18 H.-D. Narres, P. Valenta and H.-W. Nürnberg, *Z. Lebensm. Unters. Forsch.*, 179 (1984) 443–446.
19 S. Noriki, N. Ishimori, K. Harada and S. Tsunogai, *Mar. Chem.*, 17 (1985) 75–89.
20 L. Mart and H.W. Nürnberg, *Mar. Chem.*, 18 (1986) 197–213.
21 L. Brügmann, *Beitr. Meeresk.*, 55 (1986) 3–18.
22 T.D. Jickells, J.D. Burton, *Mar. Chem.*, 23 (1988) 131–144.
23 L. Gerhardsson, D. Brune, G.F. Nordberg and P.O. Wester, *Sci. Total Environ.*, 74 (1988) 97–110.
24 H.J. Raithel, K.H. Schaller, A. Reith, K.B. Svenes and H. Valentin, *Int. Arch. Occup. Environ. Health*, 60 (1988) 55–66.
25 K. Kremling and C. Pohl, *Mar. Chem.*, 27 (1989) 43–60.
26 J. Yoshinaga, T. Suzuki and M. Morita, *Sci. Total Environ.*, 79 (1989) 209–221.
27 D.E. Nixon, T.P. Moyer, D.P. Squillace and J.T. McCarthy, *Analyst*, 114 (1989) 1671–1674.
28 J. Einax, K. Oswald and K. Danzer, *Fresenius' J. Anal. Chem.*, 336 (1990) 394–399.
29 C. Minoia, E., Sabbioni, P. Apostoli, R. Pietra, L. Pozzoli, M. Gallorini, G. Nicolaou, L. Alessio and E. Capodaglio, *Sci. Total Environ.*, 95 (1990) 89–105.
30 P. Ostapczuk and M. Froning, *Nickel and Cobalt Contents in Materials of the German Environmental Specimen Bank*, to be published.
31 V. Krivan and K.P. Egger, *Fresenius' Z. Anal. Chem.*, 326 (1986) 41–49.
32 J. Angerer, R. Heinrich-Ramm and G. Lehnert, *Int. J. Environ. Anal. Chem.*, 35 (1989) 81–88.
33 R.M. Harrison and M.B. Chirgawi, *Sci. Total Environ.*, 83 (1989) 13–34.
34 R. Michel, M. Nolte, M. Reich and F. Löer, *Arch. Orthop. Trauma Surg.*, 110 (1991) 61–74.
35 G.V. Iyengar and W.E. Kollmer, *Trace Elem. Med.*, 3 (1986) 25–33.
36 A. Aitio and J. Järvisalo, *Pure Appl. Chem.*, 56 (1984) 549–566.
37 S. Bro, P.J. Jørgensen, J.M. Christensen and M. Hørder, *J. Trace Elem. Electrolytes Health Dis.*, 2 (1988) 31–35.
38 M. Stoeppler, in H.F. Hildebrand and M. Champy (Editors), *Biocompatibility of Co-Cr-Ni Alloys*, Plenum Press, New York, London, 1988, pp. 45–57.
39 G. Knapp, *Fresenius' Z. Anal. Chem.*, 317 (1984) 213–219.
40 G. Würfels, E. Jackwerth and M. Stoeppler, *Fresenius' Z. Anal. Chem.*, 329 (1987) 459–461.
41 P. Ostapczuk, M. Stoeppler and H.W. Dürbeck, *Fresenius' Z. Anal. Chem.*, 332 (1988) 662–665.
42 L.G. Danielsson, B. Magnusson and S. Westerlund, *Anal. Chim. Acta*, 98 (1978) 47–57.
43 R.F. Nolting, *Mar. Poll. Bull.*, 17 (1986) 113–117.
44 D. Chakraborti, F. Adams, W. van Mol and K.J. Irgolic, *Anal. Chim. Acta*, 196 (1987) 23–31.
45 A. Corsini, R. Di Frascia and O. Herman, *Talanta*, 32 (1985) 791–795.
46 S. Yamaguchi and K. Uesugi, *Nippon Kaisui Gakkaishi*, 39 (1986) 27–31.
47 K. Uesugi and S. Yamaguchi, *Nippon Kaisui Gakkaishi*, 37 (1984) 348–351.
48 T. Kitamori, K. Suzuki, T. Sawada, Y. Goshi and K. Motojima, *Anal. Chem.*, 58 (1986) 2275–2278.

444

49 V. Pavski, A. Corsini and S. Landsberger, *Talanta*, 36 (1989) 367–372.
50 E.M. Donaldson, *Talanta*, 36 (1989) 543–548.
51 B. Vos, *Fresenius' Z. Anal. Chem.*, 320 (1985) 556–561.
52 C.J. Cheng, T. Akagi and H. Haraguchi, *Anal. Chim. Acta*, 198 (1987) 173–181.
53 S.-C. Pai, T.-H. Fang, C.-T.A. Chen and K.-L. Jeng, *Mar. Chem.*, 29 (1990) 295–306.
54 R.E. Sturgeon, S.S. Berman, A. Desaulniers and D.S. Russel, *Talanta*, 27 (1980) 85–94.
55 G.R.W. Denton and C. Burdon-Jones, *Mar. Poll. Bull.*, 17 (1986) 96–98.
56 K. Isshiki, F. Tsuji and T. Kuwamoto, *Anal. Chem.*, 59 (1987) 2491–2495.
57 P. Burba and P.G. Willmer, *Fresenius' Z. Anal. Chem.*, 329 (1987) 539–545.
58 K. Pyrzynska, *Anal. Chim. Acta*, 238 (1990) 285–289.
59 P.P. Coetzee and K.H. Lieser, *Fresenius' Z. Anal. Chem.*, 323 (1986) 257–260.
60 R.E. Sturgeon, S.S. Berman, S.N. Willie and J.A.H. Desaulniers, *Anal. Chem.*, 53 (1981) 2337–2340.
61 P.A. Yates, *Sci. Total Environ.*, 72 (1988) 131–149.
62 D. Beauchemin, J.W. McLaren, A.P. Mykytiuk and S.S. Berman, *Anal. Chem.*, 59 (1987) 778–783.
63 S. Nakashima, R.E. Sturgeon, S.N. Willie and S.S. Berman, *Fresenius' Z. Anal. Chem.*, 330 (1988) 592–595.
64 Y. Sakai and N. Mori, *Talanta*, 33 (1986) 161–163.
65 T. Kiriyama and R. Kuroda, *Mikrochim. Acta*, 1 (1985) 405–410.
66 M. Bengtsson and G. Johansson, *Anal. Chim. Acta*, 158 (1984) 147–156.
67 Z. Fang, S. Xu and S. Zhang, *Anal. Chim. Acta*, 164 (1984) 41–50.
68 K. Bäckström and L-G. Danielsson, *Mar. Chem.*, 29 (1990) 33–46.
69 A. van Geen and E. Boyle, *Anal. Chem.*, 62 (1990) 1705–1709.
70 L. Cabezon, M. Caballero, R. Cela and J.A. Perez-Bustamante, *Talanta*, 31 (1984) 597–602.
71 K. Hiroshima, V. Mishima and T. Saki, *Sendi-shi Eisei Shikenshoho*, 13 (1983) 293–299.
72 S. Nakashima and M. Yagi, *Anal. Lett.*, 17 (1984) 1693–1703.
73 R. Eidecker and E. Jackwerth, *Fresenius' Z. Anal. Chem.*, 328 (1987) 469–474.
74 H. Bem and D.E. Ryan, *Anal. Chim. Acta*, 166 (1984) 189–197.
75 T. Akagi and H. Haraguchi, *Anal. Chem.*, 62 (1990) 81–85.
76 T. Braun and M.N. Abbas, *Anal. Chim. Acta*, 119 (1980) 113–119.
77 T. Hobo, K. Yamada and S. Suzuki, *Anal. Sci.*, 2 (1986) 361–364.
78 J.M. Diaz, M. Caballero, J.A. Pérez-Bustamente and R. Cela, *Analyst*, 115 (1990) 1201–1205.
79 T. Kumamaru, Y. Okamoto, M. Yamamoto, Y. Obata and K. Onizuka, *Anal. Chim. Acta*, 232 (1990) 389–391.
80 H. Matsusiewicz, J. Fish and T. Malinski, *Anal. Chem.*, 59 (1987) 2264–2269.
81 N.-S. Chong, M.L. Norton and J.L. Anderson, *Anal. Chem.*, 62 (1990) 1043–1050.
82 J. Alary, J. Vandaele, C. Escrient and R. Haran, *Talanta*, 33 (1986) 748–750.
83 R.E. Sturgeon, S.N. Willie and S.S. Berman, *J. Anal. Atom. Spectrom.*, 4 (1989) 443–446.
84 W. Drews, G. Weber and G. Tölg, *Anal. Chim. Acta*, 231 (1990) 265–271.
85 M. Stoeppler, in E. Merian (Editor), *Metals and Their Compounds in the Environment*, VCH, Weinheim, New York, Basel, Cambridge, 1991, pp. 105–206.
86 P.W.J.M. Boumans and J.J.A.M. Vrakking, *Spectrochim. Acta*, 42B (1987) 553–579.
87 Jobin Yvon, *ICP Spectrometer of ISA Jobin-Yvon*, Longjumeau Cedex, 1988.
88 B. Welz, *Atomic Absorption Spectrometry*, VCH Verlagsges Weinheim, Dearfield Beach, FL, Basel, 2nd ed., 1985.
89 W. Slavin, G.R. Carnrick, D.C. Manning and E. Pruszkowska, *At. Spectrosc.*, 4 (1983) 69–86.
90 M. Stoeppler and M.W. Nürnberg, in E. Merian (Editor), *Metalle in der Umwelt*, Verlag Chemie, Weinheim, Dearfield Beach, FL, Basel, 1984, pp. 45–104.
91 G.M. Hieftje and G.H. Vickers, *Anal. Chim. Acta*, 216 (1989) 1–24.
92 W. Michaelis, private communication, 1989.

93 W. Slavin, *Graphite Furnace AAS a Source Book*, Perkin-Elmer, Norwalk, CT, 1984.
94 E.J. Hinderberger, M.L. Kaiser and S.R. Koirtyohann, *At. Spectrosc.*, 2 (1981) 1–7.
95 K. Ohta and T. Mizuno, *Anal. Chim. Acta*, 217 (1989) 377–382.
96 J.M. Harnly, *Anal. Chem.*, 58 (1986) 933A–943A.
97 N.J. Miller-Ihli, *Talanta*, 37 (1990) 119–125.
98 B.T. Jones, B.W. Smith and J.D. Winefordner, *Anal. Chem.*, 61 (1989) 1670–1674.
99 R.E. Sturgeon, S.N. Willie and S.S. Berman, *J. Anal. At. Spectrom.*, 4 (1989) 443–446.
100 Z. Grobenski, R. Lehmann, R. Radziuk and U. Voellkopf, *The Determination of As, Cd, Cr, Mn, Mo and Ni in Sea Water Using Zeeman Graphite Furnace AAS*, Appl. Atom. Spectrom., No. 34E, Perkin-Elmer, 1984.
101 R. Boniforti, R. Perrazoli, P. Frigieri, D. Heltai and G. Queirazza, *Anal. Chim. Acta*, 163 (1984) 33–46.
102 W. Calmano, W. Alf and T. Schilling, *Fresenius' Z. Anal. Chem.*, 323 (1986) 865–868.
103 V.V. Lidums, *At. Abs. Newsl.*, 3 (1979) 71–72.
104 J.M. Christensen, S. Mikkelsen and A. Skov, *Chemical Toxicology and Clinical Chemistry of Metals*, IUPAC, 1983, pp. 68–68?
105 F.W. Sunderman, Jr., M.C. Crisostomo, M.C. Reid, S.M. Hopfer and S. Nomoto, *Ann. Clin. Lab. Sci.*, 14 (1984) 232–241.
106 R. Heinrich and J. Angerer, *Int. J. Environ. Anal. Chem.*, 16 (1984) 305–314.
107 I. Andersen and A.C. Høgetveit, Fresenius' Z. Anal. Chem., 318 (1984) 41–48.
108 M.M. Kimberley, G.G. Bailey and D.C. Paschal, *Analyst*, 112 (1987) 287–290.
109 I.R. Andersen, B. Gammelgaard and S. Reimert, *Analyst*, 111 (1986) 721–722.
110 F.W. Sunderman, S.M. Hopfer, M.C. Crisostomo and M. Stoeppler, *Ann. Clin. Lab. Sci.*, 16 (1986) 219–230.
111 G. Schlemmer and B. Welz, *Spectrochim. Acta, Part B*, 41 (1986) 1157–1165.
112 B. Sampson, *Anal. Proc.*, 25 (1988) 229–230.
113 K.S. Subramanian, *Prog. Analyt. Spectrosc.*, 11 (1988) 511–608.
114 R. Ooms and W. van Pee, *J. Am. Oil Chem. Soc.*, 60 (1983) 957–960.
115 Z. Zief and R. Speights, *Ultrapurity-Methods and Techniques*, Marcel Dekker, New York, 1972.
116 P. Tschöpel, L. Kotz, M. Schulz, M. Veber and G. Tölg, *Fresenius' Z. Anal. Chem.*, 302 (1980) 1–14.
117 AOAC (Asssoc of Official Anal. Chemists) *Official Methods of Analysis*, 15th ed., 1990, p. 242.
118 A.M. Nash, T.L. Mounts and W.F. Kwolek, *J. Am. Oil Chem. Soc.*, 60 (1983) 811–814.
119 J.F. Alder and M.C.C. Batoreu, *Anal. Chim. Acta*, 155 (1983) 199–207.
120 J.F. Alder, M.C.C. Batoreu, A.D. Pearse and R. Marks, *J. Anal. At. Spectrom.*, 1 (1986) 365–367.
121 G. Ellen, G. van den Bosch-Tibbesma and F.F. Douma, *Z. Lebensm. Unters. Forsch.*, 166 (1978) 145–147.
122 G. Vos, J.P.C. Hovens and P. Hagel, *Sci. Total Environ.*, 52 (1986) 25–40.
123 R.J. Green and C.J. Asher, *Analyst*, 109 (1984) 503–505.
124 O.K. Borggaard, H.E.M. Christensen and S.P. Lund, *Analyst*, 109 (1984) 1179–1182.
125 J.W. Blanchflower, A. Cannavan and D.G. Kennedy, *Analyst*, 115 (1990) 1323–1325.
126 J. Alary, P. Bourdon and J. Vandale, *Sci. Total Environ.*, 46 (1985) 181–190.
127 F.J. Langmyhr and G. Wibetoe, *Prog. Anal. At. Spectrosc.*, 8 (1985) 193–256.
128 F.J. Langmyhr, *Fresenius' Z. Anal. Chem.*, 322 (1985) 654–656.
129 W. Scholl, *Fresenius' Z. Anal. Chem.*, 322 (1985) 681–684.
130 A. Janssen, B. Brückner, K.-H. Grobecker and U. Kurfürst, *Fresenius' Z. Anal. Chem.*, 322 (1985) 713–716.
131 G. Schlemmer and B. Welz, *Fresenius' Z. Anal. Chem.*, 328 (1987) 405–409.
132 B. Krebs, *Fresenius' Z. Anal. Chem.*, 328 (1987) 388–389.
133 M. Hoenig, P. Regnier and R. Wollast, *J. Anal. At. Spectrom.*, 4 (1989) 631–634.
134 K. Akatsuka and I. Atsuya, *Anal. Chem.*, 61 (1989) 216–220.
135 P.W.J.M. Boumans, *Inductively Coupled Plasma Emission Spectroscopy* (2 Volumes), Wiley, New York, Chichester, Brisbane, Toronto, Singapore, 1987.

136 P. Schramel, in M. Stoeppler and R.F.M. Herber (Editors), *Trace Metal Analysis in Biological Systems,* Year Book Medical Publishers, Chicago, IL, in press.

137 P. Barrett and E. Pruszkowska, *Anal. Chem.,* 56 (1984) 1927–1930.

138 P. Schramel and Xu Li-qiang, *Fresenius' Z. Anal. Chem.,* 314 (1983) 671–677.

139 W. Dannecker, H. Berger, D. Mükker abd F. Meyberg in B. Welz (Editor), *Fortschritte in der Atomspektrometr. Spurenanalytik, Bd. 1,* Verlag Chemie, Weinheim, 1984, pp. 557–569.

140 P. Taylor, R. Dams and J. Hoste, *Anal. Lett.,* 18 (1985) 2361–2368.

141 B. Griepink and H. Muntau, *Fresenius' Z. Anal. Chem.,* 326 (1987) 414–418.

142 I.B. Brenner, Y. Lang, A. LeMarchand and P. Grosdaillon, *Int. Lab.,* 17 (1987) 18–33.

143 K.H. Karstensen and W. Lund, *J. Anal. At. Spectrom.,* 4 (1989) 357–359.

144 V. Kanda and M. Taira, *Anal. Chim. Acta,* 207 (1988) 269–281.

145 S.S. Que Hee and J.R. Boyle, *Anal. Chem.,* 60 (1988) 1033–1042.

146 L. Ebdon and J.R. Wilkinson, *J. Anal. At. Spectrom.,* 2 (1987) 325–328.

147 H.G.M. Parry and L. Ebdon, *Anal. Proc.,* 25 (1988) 69–71.

148 M.F. Giné, F.J. Krug, H.B. Filho, B.F. Dos Reis and E.A.G. Zagatto, *J. Anal. At. Spectrom.,* 3 (1988) 673–678.

149 D.C. Baxter and W. Frech, *Fresenius' Z. Anal. Chem.,* 328 (1987) 324–329.

150 R.S. Houk, *Anal. Chem.,* 58 (1986) 97A–105A.

151 G.M. Hieftje and G.H. Vickers, *Anal. Chim. Acta,* 216 (1989) 1–24.

152 G.H. Gillson, D.J. Douglas, J.E. Fulford, K.W. Halligan, S.D. Thanner, *Anal. Chem.,* 60 (1988) 1472–1474.

153 J.G. Williams and A.L. Gray, *Anal. Proc.,* 25 (1988) 385–388.

154 J.M. Henshaw, E.M. Heitmar and T.A. Hinners, *Anal. Chem.,* 61 (1989) 335–342.

155 M.R. Plantz, J.S. Fritz, S.G. Smith and R.S. Houk, *Anal. Chem.,* 61 (1989) 149–153.

156 D. Beauchemin, J.W. McLaren, A.P. Mykitiuk and S.S. Berman, *J. Anal. At. Spectrom.,* 3 (1988) 305–308.

157 D. Beauchemin and S.S. Berman, *Anal. Chem.,* 61 (1989) 1857–1862.

158 V. Vanhoe, C. Vandecasteele, J. Versieck and R. Dams, *Anal. Chem.,* 61 (1989) 1851–1857.

159 P.S. Ridout, H.R. Jones and J.G. Williams, *Analyst,* 113 (1988) 1383–1386.

160 J.W. McLaren, K.W.M. Siu, J.W. Lam, S.N. Willie, P.S. Maxwell, A. Pakpu, M. Koether and S.S. Berman, *Fresenius' J. Anal. Chem.,* 337 (1990) 721–728.

161 J.W. McLaren, D. Beauchemin and S.S. Berman, *J. Anal. At. Spectrom.,* 2 (1987) 277–281.

162 J.G. Williams, A.L. Gray, P. Norman and L. Ebdon, *J. Anal. At. Spectrom.,* 2 (1987) 469–472.

163 T. Hirata, T. Akagi and A. Masuda, *Analyst,* 115 (1990) 1329–1332.

164 A.M. Bond, *Modern Polarographic Methods in Analytic Chemistry,* Marcel Dekker, New York, 1980.

165 H.W. Nürnberg, *Pure Appl. Chem.,* 54 (1982) 853–878.

166 J. Wang, *Stripping Analysis,* VCH, Weinheim, Deerfield Beach, FL, Basel, 1985.

167 J. Wang, *Electroanalytical Techniques in Clinical Chemistry and Laboratory Medicine,* VCH, Weinheim, Deerfield Beach, FL, Basel, 1988.

168 G. Henze and R. Neeb, *Elektrochemische Analytik,* Springer, Berlin, Heidelberg, New York, Tokyo, 1986.

169 H. Grubitsch and J. Schukoff, *Fresenius' Z. Anal. Chem.,* 253 (1971) 201–205.

170 V. Gemmer-Colos, H. Tuss, D. Saur and R. Neeb, *Fresenius' Z. Anal. Chem.,* 307 (1981) 347–351.

171 J. Wang, *Int. Lab.,* 16 (1986) 11–12; 50–59.

172 P.J. Nangniot, *J. Electroanal. Chem.,* 14 (1967) 197–

173 G.V. Prokhorova, E.N. Vinogradova, I.S. Grebneva and E.V. Skobelnika, *Zh. Anal. Khim.,* 28 (1973) 123–126.

174 J. Golimowski, H.W. Nürnberg, P. Valenta, *Lebensmittelchem. Gerichtl. Chem.,* 34 (1980) 116–120.

175 C.J. Flora and E. Nieboer, *Anal. Chem.,* 52 (1980) 1013–1020.

447

176 B. Pihlar, P. Valenta and H.W. Nürnberg, *Fresenius' Z. Anal. Chem.*, 307 (1981) 337–346.
177 K.Z. Brainina, *Stripping Voltammetry in Chemical Analysis*, Wiley, New York, Toronto, 1972.
178 V. Gemmer-Čolos, G. Scollary and R. Neeb, *Fresenius' Z. Anal. Chem.*, 313 (1982) 412–413.
179 P. Ostapczuk and M. Stoeppler, *Chem. Ind.*, 4 (1990) 76–78.
180 J. Wang, *Fresenius' Z. Anal. Chem.*, 337 (1990) 508–511.
181 R.P. Baldwin, J.K. Christensen and L. Kryger, *Anal. Chem.*, 58 (1986) 1790–1798.
182 M. Trojanowicz and M. Matuszewski, *Talanta*, 36 (1989) 680–682.
183 W. Dorten, P. Valenta and H.W. Nürnberg, *Fresenius' Z. Anal. Chem.*, 317 (1984) 367–371.
184 L Mart, H.W. Nürnberg and H. Rützel, *Sci. Total Environ.*, 44 (1985) 35–49.
185 L Mart, H.W. Nürnberg, *Mar. Chem.*, 18 (1986) 197–213.
186 I. Gustavsson and L. Hansson, *Int. J. Environ. Anal. Chem.*, 17 (1984) 57–72.
187 H. Braun and M. Metzger, *Fresenius' Z. Anal. Chem.*, 318 (1984) 321–326.
188 H. Eskilsson and C. Haraldsson, *Anal. Chim. Acta*, 175 (1985) 79–88.
189 C.M.G. van den Berg, *Sci. Total Environ.*, 49 (1986) 89–99.
190 M.P. Newton and C.M.G. van den Berg, *Anal. Chim. Acta*, 199 (1987) 59–76.
191 J.R. Donat and K.W. Bruland, *Anal. Chem.*, 60 (1988) 240–244.
192 A. Meyer and R. Neeb, *Fresenius' Z. Anal. Chem.*, 315 (1983) 118–120.
193 P. Ostapczuk, P. Valenta, M. Stoeppler and H.W. Nürnberg, in S.S. Brown and J. Savory (Editors), *Chemical Toxicology and Clinical Chemistry of Metals*, Academic Press, London, New York, 1983, pp. 61–64.
194 B. Gammelgaard and J.R. Andersen, *Analyst*, 110 (1985) 1197–1199.
195 S.B. Adeloju, A.M. Bond and M.H. Briggs, *Anal. Chem.*, 57 (1985) 1386–1390.
196 P. Ostapczuk, M. Froning, M. Stoeppler and H.W. Nürnberg, in S.S. Brown and F.W. Sunderman (Editors), *Progress in Nickel Toxicology*, Blackwell, Oxford, 1985, pp. 129–132.
197 H.F. Hildebrand, B. Roumazeille, J. Decoulx, M.C. Herlant-Peers, P. Ostapczuk, M. Stoeppler and J.M. Mercier, in S.S. Brown and F.W. Sunderman (Editors), *Progress in Nickel Toxicology*, Blackwell, Oxford, 1985, pp. 169–172.
198 P. Ostapczuk, M. Apel, U. Bagschik, J. Golimowski, K. May, C. Mohl, M. Stoeppler, P. Valenta and H.W. Nürnberg, *Lebensmittelchem. Gerichtl. Chem.*, 39 (1985) 60–61.
199 M. Stoeppler, M. Apel, U. Bagschik, K. May, C. Mohl, P. Ostapczuk, R. Enkelmann and H. Eschnauer, *Lebensmittelchem. Gerichtl. Chem.*, 39 (1985) 60–61.
200 P. Ostapczuk, P. Valenta and H.W. Nürnbergm *J. Electronal. Chem.*, 214 (1986) 51–64.
201 V. Gemmer-Čolos and R. Neeb, *Fresenius' Z. Anal. Chem.*, 327 (1987) 547–551.
202 K.W. Liebhafsky, E.A. Schweikert and E.A. Meyers, in P.I. Elving (Editor), *Treatise on Analytical Chemistry*, Part I, Vol. 8, John, New York, 2nd ed., 1986, pp. 209–309.
203 C. Whiston, *X-Ray Methods*, Wiley, Chichester, 1987.
204 J.A. Helsen and B.A.R. Vrebos, *Int. Lab.*, 16 (1986) 66–71.
205 R. Klockenkämper, B. Raith, S. Divoux, B. Gonsior, S. Brüggerhoff and E. Jackwerth, *Fresenius' Z. Anal. Chem.*, 326 (1987) 105–117.
206 K.H. Janssens and F.C. Adams, *J. Anal. At. Spectrom.*, 4 (1989) 123–135.
207 W. Michaelis, A. Prange (Editors), *Totalreflexions-Röntgenfluoreszenzanalyse*, GKSS 86/E/61, GKSS-Forschnugszentrum Geesthacht, 1987.
208 A. Prange, *Spectrochim. Acta*, 44B (1989) 437–452.
209 H. Rethfeld, *Fresenius' Z. Anal. Chem.*, 310 (1982) 127–130.
210 H. Rethfeld, *Fresenius' Z. Anal. Chem.*, 324 (1986) 720–727.
211 V. Talbot and W.-J. Chang, *Sci Total Environ.*, 66 (1987) 213–223.
212 R. Zeisler, S.F. Stone and R.W. Sanders, *Anal. Chem.*, 60 (1988) 2760–2765.
213 R.W. Sanders, K.B. Olsen, W.C. Weimer, K.N. Nielson, *Anal. Chem.*, 55 (1983) 1911–1914.
214 A. Prange and H. Schwenke, *Adv. X-Ray Anal.*, 32 (1989) 209–218.

448

215 W. Michaelis, H. Böddeker, J. Knoth, H. Schwenke, GKSS 83/E/33, GKSS-Forschungszentrum Geesthacht, 1983.

216 R.P. Stössel and A. Prange, *Anal. Chem.*, 57 (1985) 2880–2885.

217 A. Prange, A. Knöchel and W. Michealis, *Anal. Chim. Acta*, 172 (1985) 79–100.

218 A. Prange, J. Knoth, R.P. Stössel, H. Büddeker and K. Kramer, *Anal. Chim. Acta*, 195 (1987) 275–287.

219 W. Michaelis and A. Prange, *Nucl. Geophys.*, 2 (1988) 231–245.

220 W. Michaelis, *Fresenius' Z. Anal. Chem.*, 324 (1986) 662–671.

221 W. Gerwinski, D. Goetz, S. Koelling and J. Kunze, *Fresenius' Z. Anal. Chem.*, 327 (1987) 293–296.

222 W. Gerwinski and D. Goetz, *Fresenius' Z. Anal. Chem.*, 327 (1987) 690–693.

223 S. Mukhar, S.J. Haswell, A.T. Ellis and D.T. Hawke, *Analyst*, 116 (1991) 333–338.

224 A. von Bohlen, R. Eller, R. Klockenkämper and G. Tölg, *Anal. Chem.*, 59 (1987) 2551–2555.

225 R.D. Giauque, A.C. Thompson, J.H. Underwood, Y. Wu, K.W. Jones and M.L. Rivers, *Anal. Chem.*, 60 (1988) 855–858.

226 G. Erdtmann and H. Petri, in P.J. Elving (Editor), *Treatise on Analytical Chemistry*, Part 1, Vol. 14, Wiley, New York, 2nd ed., 1986, pp. 419–643.

227 J. Hoste, in P.J. Elving (Editor), *Treatise on Analytical Chemistry*, Part 1, Vol. 14, Wiley, New York, 2nd ed., 1986, pp. 645–777.

228 K.S. Kasprzak and F.W. Sunderman, Jr., *Pure Appl. Chem.*, 51 (1979) 1375–1384.

229 V. Krivan, *Sci. Total Environ.*, 64 (1987) 21–40.

230 H. Bem, J. Holzbecher and D.E. Ryan, *Anal. Chim. Acta*, 152 (1983) 247–255.

231 C. Segebade, B.-F. Schmitt, H.U. Fusban and M. Kühl, *Fresenius' Z. Anal. Chem.*, 317 (1984) 413–421.

232 V. Krivan and K.P. Egger, *Fresenius' Z. Anal. Chem.*, 325 (1986) 41–49.

233 G.E.M. Hall, G.F. Bonham-Carter, A.I. Mac Laurin and S. Ballantyne, *Talanta*, 37 (1990) 135–155.

234 J. Versieck, J. Hostte, F. Barvier, H. Steyaert, J. De Rudder and H. Michels, *Clin. Chem.*, 24 (1978) 303–308.

235 J. Versieck, L. Vanballenberghe, A. de Kesel, J. Hoste, B. Wallaeys, J. Vandenhaute, N. Baeck, H. Steyaert, A.R. Byrne and F.W. Sunderman, Jr., *Anal. Chim. Acta*, 204 (1988) 63–75.

236 F. Chisela and P. Brätter, *Anal. Chim. Acta*, 188 (1986) 85–94.

237 L. Morselli, S. Zappoli, M. Gallorini and E. Rizzio, *Analyst*, 113 (1988) 1575–1578.

238 M.C. Freitas and E. Martinho, *Talanta*, 36 (198) 527–531.

239 M.C. Freitas and E. Martinho, Anal. Chim. Acta, 223 (1989) 287–292.

240 A. Meyer and R. Neeb, *Fresenius' Z. Anal. Chem.*, 321 (1985) 235–241.

241 H. Schaller and R. Neeb, *Fresenius' Z. Anal. Chem.*, 323 (1986) 473–476.

242 H. Schaller and R. Neeb, *Fresenius' Z. Anal. Chem.*, 327 (1987) 170–174.

243 N. Häring and K. Ballschmiter, *Talanta*, 27 (1980) 873–879.

244 K.H. König, D. Hollmann and B. Steinbrech, *Fresenius' Z. Anal. Chem.*, 317 (1984) 788–790.

245 S. Ichinoki, T. Morita and M. Yamazaki, *J. Liq. Chromatogr.*, 7 (1984) 2467–2482.

246 A. Munder and K. Ballschmiter, *Fresenius' Z. Anal. Chem.*, 323 (1986) 869–874.

247 A.M. Bond, R.W. Knight, J.B. Reust, D.J. Tucker and G.G. Wallace, *Anal. Chim. Acta*, 182 (1986) 47–59.

248 J.N. King and J.S. Fritz, *Anal. Chem.*, 59 (1987) 703–708.

249 C. Baiocchi, G. Saini, P. Berolo, G.P. Cartoni and G. Pettiti, *Analyst*, 113 (1988) 805–807.

250 J. Weiss, *Fresenius' Z. Anal. Chem.*, 327 (1987) 451–455.

251 R.B. Rubin and S.S. Heberling, *Int. Lab.*, 17 (1987) 54–60.

252 D. Yan and G. Schwedt, *Fresenius' Z. Anal. Chem.*, 327 (1987) 503–508.

253 R.J. Joyce and A. Schein, *Int. Lab.*, 20 (1990) 48–54.

254 B. Yan and P.J. Worsfold, *Anal. Chim. Acta*, 236 (1990) 287–292.

255 F. Adams, R. Gijbels and R. van Grieken (Editors), *Inorganic Mass Spectrometry*, Wiley, New York, 1988.

256 R.W. Odom, G. Lux, R.H. Fleming, P.K. Chu, I.C. Niemeyer and R.J. Blattner, *Anal. Chem.*, 60 (1988) 2070–2075.
257 K.G. Heumann (Editor), *Fresenius' Z. Anal. Chem.*, 331 (1988) 103–222.
258 K.G. Heumann, *Toxicol. Environ. Chem. Rev.*, 3 (1980) 111–129.
259 A.P. Mykytiuk, D.S.. Russell and R.E. Sturgeon, *Anal. Chem.*, 52 (1980) 1281–1283.
260 V.J. Stukas and C.S. Wong, in C.S. Wong, E. Boyle, K.W. Bruland, J.D. Burton and E.D. Goldberg (Editors), *Trace Metals in Sea Water*, Plenum, New York, 1983, pp. 513–536.
261 J. Völkening and K.G. Heumann, *Fresenius' Z. Anal. Chem.*, 331 (1988) 174–181.
262 K. Helrich (Editor), *Official Methods for Analysis of the Association of Official Analytical Chemists*, AOAC, Arlington, Virginia, 15th ed., 1990, pp. 30, 43–45 and 85–86.
263 H. Müller, *CRC Crit. Rev. Anal. Chem.*, 13 (1982) 313–372.
264 D.Th. Burns, N. Champalee and M. Harriott, *Anal. Chim. Acta*, 225 (1989) 123–128.
265 M.V. Dawson and S.J. Lyle, *Talanta*, 37 (1990) 443–446.
266 M.V. Dawson and S.J. Lyle, *Talanta*, 37 (1990) 1189–1191.
267 A.G. Melgarejo, A.G. Céspedes and J.M. Cano Pavon, *Analyst*, 114 (1989).
268 H. Sakai, T. Fujiwara, M. Yamamoto and T. Kumamaru, *Anal. Chim. Acta*, 221 (1989) 249–258.
269 L. Minggang, L. Xiachu and Y. Fang, *Talanta*, 37 (1990) 393–395.
270 P. Jones, T. Williams and L. Ebdon, *Anal. Chim. Acta*, 217 (1989) 157–163.
271 G.E. Batley (Editor), *Trace Element Speciation: Analytical Methods and Problems*, CRC Press, Boca Raton, FL, 1989.
272 A. Tessier, P.G.C. Campbell and M. Bisson, *Anal. Chem.*, 51 (1979) 844–851.
273 A. Tessier, F. Rapin and R. Carignan, *Geochim. Cosmochim. Acta*, 49 (1985) 183–194.
274 K. Hayase, K. Shitashima and H. Tsubota, *Talanta*, 33 (1986) 754–756.
275 M. Nimmo, C.M.G. van den Berg and J. Brown, *Estuarine, Coastal, Shelf Sci.*, 29 (1989) 57–74.
276 T. Rudd, D.L. Lake, I. Mehrotra, R.M. Sterritt, P.W.W. Kirk, J.A. Campbell and J.N. Lester, *Sci. Total Environ.*, 74 (1988) 149–175.
277 C.J. Warren and M.J. Dudas, *Sci. Total Environ.*, 76 (1988) 229–246.
278 T.-G. Wu and J.L. Wong, *Anal. Chim. Acta*, 235 (1990) 457–460.
279 G. Weber and G. Schwedt, *Anal. Chim. Acta*, 134 (1982) 81–92.
280 D.H. Stedman, D.A. Tammaro, D.K. Brauch and R. Pearson, Jr., *Anal. Chem.*, 51 (1979) 2340–2342.
281 D.H. Stedman, R. Pearson, Jr. and E.D. Yalvac, *Science*, 208 (1980) 1029–1031.
282 P.M. Houpt, A. van der Waal and F. Langweg, *Anal. Chim. Acta*, 136 (1982) 421–424.
283 L. Filkova, Report P-17-335-457-03, Institute of Hygiene and Epidemiology, Prague, 1985.
284 D.B. Adams, S.S. Brown, F.W. Sunderman Jr. and H. Zachariasen, *Clin. Chem.*, 24 (1978) 862–867.
285 S.S. Brown, A.D. DiMichiel and F.W. Sunderman Jr., in S.S. Brown and F.W. Sunderman (Editors), Academic Press, New York, 1980, pp. 167–170.
286 S.S. Brown, S. Nomoto, M. Stoeppler and F.W. Sunderman, Jr., *Pure Appl. Chem.*, 53 (1981) 773–781.
287 F.W. Sundermanm Jr., S.S. Brown, M. Stoeppler and D.B. Tonks, in H. Egan and T.S. West (Editors), *IUPAC Collaborative Interlaboratory Studies in Chemical Analysis*, Pergamon, Oxford, New York, 1983, pp. 25–35.
288 J.M. Christensen, M. Ihnat, M. Steoppler, Y. Thomassen, C. Veillon and M. Wolynetz, *Fresenius' Z. Anal. Chem.*, 326 (1987) 639–642.
289 H. Brix, J.E. Lyngby and H.-H. Schierup, *Mar. Chem.*, 12 (1983) 69–86.
290 J. Kumpulainen and M. Paakki, *Fresenius' Z. Anal. Chem.*, 326 (1987) 684–689.
291 A.R. Byrne and I. Krasovec, *Fresenius' Z. Anal. Chem.*, 332 (1988) 666–668.
292 R. Cornelis, S. Dyg, B. Griepink and R. Dams, *Fresenius' J. Anal. Chem.*, 338 (1990) 414–418.
293 R. Alvarez, *Fresenius' J. Anal. Chem.*, 338 (1990) 466–468.

294 M.M. Schantz, B.A. Brenner, Jr., S.N. Chesler, B.J. Koster, K.E. Hahn, S.F. Stone, W.R. Kelly, R. Zeisler and S.A. Wise, *Fresenius' J. Anal. Chem.*, 338 (1990) 501–514.

295 E.C. Toro, R.M. Parr, S.A. Clements, *Biological and Environmental Reference Materials for Trace Elements, Nuclides and Organic Microcontaminants*, IAEA/RL/128 (Rev. 1), IAEA, Vienna, 1990.

296 B. Holynska, J. Jasion, M. Lankosz, A. Markowicz, W. Baran, *Fresenius' Z. Anal. Chem.*, 332 (1988) 250–254.

297 H.J.M. Bowen, in W.R. Wolf (Editor), *Biological Reference Materials*, Wiley Interscience, New York, 1985, pp. 3–18.

298 J. Kucera, P. Mader, D. Mikolova, J. Cibulka, M. Polakowa, D. Kordik, *Report on the Interlaboratory Comparison of the Determination of the Contents of Trace Elements in Bovine Liver 12-02-01*, Report, Czechoslovak Institute of Metrology, Bratislava, November, 1989.

299 M. Stoeppler, F. Backhaus, J.D. Schladot, *Fresenius' Z. Anal. Chem.*, 326 (1987) 707–711.

300 J.D. Schladot, F.W. Backhaus, in S.S. Wise, R. Zeisler and G.M. Goldstein (Editors), *Progress in Environmental Specimen Banking*, U.S. Dept. of Commerce, NBS, Washington, DC, 1988, pp. 184–193.

301 G. Tölg, *Fresenius' Z. Anal. Chem.*, 331 (1988) 226–235.

302 G. Tölg, *GIT Fachz. Lab.*, 33 (1989) 971–978.

303 J.P. Dougherty, F.R. Preli, Jr. and R.G. Michel, *Talanta*, 36 (1989) 151–159.

304 J. Tilch, W. Süptiz, W. Gries and E. Zanger, in press.

305 R.E. Sturgeon, S.N. Willie, V. Luong and S.S. Berman, *Anal. Chem.*, 62 (1990) 2370–2376.

306 S. Geiss, J. Einax, J. Mohr and K. Danzer, *Fresenius' J. Anal. Chem.*, 338 (1990) 602–605.

M. Stoeppler (Editor)/*Hazardous Metals in the Environment*
© 1992 Elsevier Science Publishers B.V. All rights reserved

Chapter 15

Aluminium

Wolfgang Frech and Anders Cedergren

Department of Analytical Chemistry, University of Umeå, S-901 87 Umeå (Sweden)

CONTENTS

15.1. General introduction 451
15.2. Environmental levels of aluminium 452
15.3. Sampling and pretreatment procedures 455
    15.3.1. Blood, plasma and serum 455
    15.3.2. Biological tissues 460
    15.3.3. Plant materials 462
    15.3.4. Water 462
15.4. Methods for the determination of aluminium 463
15.5. Determination of aluminium by GFAAS 465
    15.5.1. Determination of aluminium in liquids, serum, blood, and plasma . . . 467
    15.5.2. Determination of aluminium in digested samples 469
    15.5.3. Direct determination of solid samples 469
15.6. Standard reference materials 470
15.7. Need for future research 470
References 470

15.1. GENERAL INTRODUCTION

Aluminium is one of the most common elements in the accessible part of the earth (see Table 15.1). In biological samples like tissues or blood, however, this element is normally present in the ppb to ppm range. This large difference gives rise to difficulties in connection with sampling, sample pretreatment, and the final determination. As a consequence many of the reported data concerning aluminium levels in biological systems suffer from systematic errors [1].

Aluminium is considered to be a non-essential element for human health and development, as no substantiating evidence has been found to indicate that it fulfils any vital biological function. For a long time aluminium was also thought to be totally biologically inactive, but this opinion has gradually changed and today great interest is focused on its metabolism and possible health effects [2,3]. The element has been implicated in the development of a number of neurological diseases such as Alzheimer's senile and pre-senile dementia, amyotropic lateral sclerosis, and Parkinson's dementia of Guam, as well as in dialysis encephalopathy [4-7]. The increased incidence of the osteodystrophic lesions

TABLE 15.1.

ELEMENTAL COMPOSITION OF THE ACCESSIBLE PART OF THE EARTH

Element	Weight (%)
O	49.4
Si	25.8
Al	7.5
Fe	4.7
Ca	3.4
Na	2.6
K	2.4
Mg	1.9
H	0.9

observed in dialysis patients has also been associated with aluminium toxicity [8,9]. A common characteristic of all these pathological conditions is the significantly increased level of the element in the grey matter of the brain [10].

Since human beings are widely exposed to aluminium through its prevalence in air, water, food, etc., and since it can be toxic after uptake in living organisms, it is important to increase knowledge about normal concentration levels and to establish the biological activity of its various forms [11].

15.2. ENVIRONMENTAL LEVELS OF ALUMINIUM

Aluminium is a major compound in a large number of minerals like feldspars, micas, and clays. However, at neutral pH the aqueous concentration of aluminium is low in

TABLE 15.2

ALUMINIUM CONCENTRATIONS FOUND IN WATER SAMPLES

	Typical level ($mg\ kg^{-1}$)	Ref.
Open seawater	<0.001	14,15
Lakes	0.01 – 0.5	16,17
Rivers	0.01 – 40	16,18
Tapwater	0.01 – 0.15[a]	19
Subsoil-water	0.01	

[a] Technical upper limit 0.15 $mg\ kg^{-1}$ (in Sweden) [21].

TABLE 15.3.

ALUMINIUM CONCENTRATIONS IN ANIMAL TISSUES AND RELATED MATERIALS

Matrix	Typical normal level ($mg\ kg^{-1}$ wet weight)		Ref.
Milk	0.02		23
Beef muscle	0.02 –	0.06	24
	0.14		25
Bovine liver	0.01 –	0.05	24
Rat brain	0.01 –	0.02	26
Rat bone	0.2 –	0.7	26,27
Beetle (*Carabus hortensis*)	60	– 120	28
Forest mouse liver (*Apodemus flavicollis*)	0.5 –	1	28
Flycatcher liver (*Ficedula hypoleuca*)	0.03 –	0.04	28
Human brain	0.4 –	2.2[a]	4,29,30
Human bone	2.4 –	3.9[a]	51
	0.6 –	1.7	52
	($mg\ l^{-1} \times 10^{-3}$)		
Human blood	< 1.6 –	5	31,32
Human plasma and serum	1.3		33
	2.0		34
	2.1		35
	2.8		3
	3.7		36
	< 4		37
	5.7		38
	5.9		39
	6.1		40,41
	6.2		42
	6.4		43
	6.5		44
	7.0		45
	7.3		46,47
	9.1		48
	9.2		49
	9.5		50

[a] Dry weight.

References on p. 470

subsoil water, lakes, rivers, and seawater, due to the formation of secondary hydroxo phases. At high and low pH levels the solubility increases substantially: this is also the case in the presence of complexing agents. For example, computer modelling (by means of which, among other things, equilibrium-distribution and predominance-area diagrams can be calculated) was used to show that the presence of oxalate in natural waters is of importance for the solubility, transportation, and speciation of aluminium [12]. At a concentration of 10^{-5} M, which is frequently found in soil solutions, the solubility of kaolinate, $Al_2(OH)_4Si_2O_5$ is increased by a factor of 8 at pH 5. This increase is mainly due to the formation of $(AlL)^+$ and $(AlL_2)^-$ where L stands for the oxalate ion. Calculations have also shown that the presence of ligands containing o-diphenolic groups [13] significantly increases the solubility of this clay mineral. This increase takes place between pH 5 and 9.

Tables 15.2, 15.3 and 15.4 show values for the aluminium found in different environmental matrices. As can be seen in Table 15.2 there is a great variation in the aluminium level found in lakes and rivers. The lowest values are found at neutral pH and a drastic increase takes place at lower pH values. For example, at a pH of 6 the aluminium level in Swedish lakes will be 0.2 – 0.3 mg kg^{-1} and at a pH of 4 values of 3 – 6 mg kg^{-1} are found. The aluminium levels in rivers are normally higher than in lakes. The very high value of 400 mg kg^{-1} has been observed in the delta of the Mekong river, at the onset of the monsoon rain period [21] during which time the pH drops to about 2.5. Values as high as 15 mg kg^{-1} have been registered in certain subsoil waters under strongly acidic conditions. It should be mentioned that the technical upper limit for aluminium in tap water is 0.15 mg kg^{-1} in Sweden. Occasionally relatively high aluminium values can be found in Swedish tap waters since aluminium sulphate is used in its purification.

Table 15.3 summarizes normal aluminium concentrations which have been found in animals and related materials. It should be pointed out that older literature values often suffer from systematic error, i.e. aluminium levels are in the order of tenths of mg kg^{-1}, which must be regarded as too high. The values for rats are from laboratory animals which have not been purposely exposed to aluminium. In particular, bone values increased drastically up to 18 mg kg^{-1} upon treating the animals with aluminium-containing antacids [26]. It was also found that the enhancement of aluminium in body tissues depend on the chemical form in which it was administered. Aluminium citrate gave rise to much higher levels than did aluminium hydroxide. The values for the beetle and the mouse are taken from a study of the role of acidification on metal uptake in forest areas in the south of Scandinavia. It was reported that the clay content of the ground was of utmost importance for the uptake of aluminium and iron. A high clay content efficiently counteracted a decrease in pH: however, in the pH-interval 4.5 to 4 the solubility of aluminium can change very quickly, so that large amounts can be dissolved.

Literature data reported for normal serum and plasma concentrations of aluminium vary considerably: the selection given in Table 15.3 includes only the lower values. Additional data which are equal to or greater than 0.01 mg l^{-1} may be regarded as unreliable in view of the fact that the vast majority of recently obtained results are much lower than this limiting value. For the lower reported values, it is not entirely clear whether the data represent the true aluminium contents of normal serum and whole blood or whether the

TABLE 15.4.

ALUMINIUM CONCENTRATIONS IN PLANTS AND RELATED MATERIALS

Matrix	Typical level (mg kg^{-1})		Ref.
Orange juice	0.1 –	0.6	19,53
Wine	0.3 –	2	19
Beer	0.06 –	1.2	19
Potatoes unpeeled, boiled	0.1		19
Tea, leaves	400	– 1000	20
Tea, steeped	0.9 –	5	19,20
Soya powder	1.7[a]		4
Beech buds	14 –	88[b]	54
Beech leaves	37 –	163	54
Beech seeds	6 –	13	54
Oxalis acetosella (blade)	140 –	310	28
Carex pilulifera (blade)	130 –	280	28

[a] Value established by W. Frech, unpublished results.
[b] The bud scale contained about 7 times more aluminium than the leaf primordia.

differences found should be ascribed to analytical errors rather than to biological variations.

A summary of aluminium contents in plant materials is given in Table 15.4. As can be seen, higher levels are found in plant materials than in animal materials. Some of the plant materials were analyzed without using any cleaning procedure (rinsing in water), in order to avoid losses of water-soluble ions like potassium (oxalis acetosella, corex pilulifera). This, however, results in systematic errors for aluminium. The values for leaf primordia from beech buds and from beech seeds are regarded as reliable since these parts of the plants are effectively protected from dust.

15.3. SAMPLING AND PRETREATMENT PROCEDURES

15.3.1. Blood, plasma and serum

Table 15.5 illustrates that different strategies must be applied depending on whether normal levels are to be dealt with, or increased levels, as such in the monitoring of uremic patients. With the very low aluminium concentrations in serum, systematic errors are easily introduced if the materials which can come into contact with the sample are not all carefully cleaned. Airborne contamination should also be avoided as far as possible. Table 15.6 gives a rough idea of the typical air fall-out of aluminium in different environments. Large variations are to be expected depending on various environmental factors like the

TABLE 15.5.

STRATEGIES FOR ALUMINIUM DETERMINATION IN BLOOD/SERUM/PLASMA SAMPLES

Objective	Normal level 5×10^{-3} mg l^{-1}	Increased level $0.02 - 1$ mg l^{-1}
Precautious against contamination	Rigorous	Adequate
Required sensitivity	0.5×10^{-3} mg l^{-1}	5×10^{-3} mg l^{-1}
Method	GFAAS GFAES	GFAAS ICP/DCP-AES ICP-MS GFAES

TABLE 15.6.

TYPICAL AIR FALL-OUT OF Al [55]

Location	μg m^{-2} day^{-1}
Outdoors	179
Analytical lab.	8.2
Clean-air lab.	0.9

climate, type of soil, building materials, etc. For samples with relatively large aluminium concentrations the mean error caused by airborne contamination will be negligible. Nevertheless, problems can arise since part of the airborne aluminium is typically concentrated in the form of a few relatively large particles which might be expected, statistically speaking, to contaminate perhaps one or two out of a hundred samples. Some laboratories cope with this problem by analyzing samples in triplicate, with the condition that consecutive aluminium signals from the same sample should not differ by more than 10%. A repeat analysis is considered if a greater difference is recorded [56]. A general rule should be to keep to a minimum the time of sample exposure to open air. Ericson and his co-workers [33] found very low normal values for aluminium, even when samples were taken in ordinary laboratory surroundings, and concluded that the major source of contamination is often the labware and containers used for sample storage. Table 15.7 shows their results for human blood serum. These workers used a thorough cleaning procedure for labware, including soaking in a detergent, followed by rinsing with distilled water and soaking for at least 2 h in a 100 ml/l nitric acid (reagent grade) bath, and then rinsing with

TABLE 15.7

ALUMINIUM IN SERUM OF NORMAL INDIVIDUALS (SAMPLING IN LABORATORY AIR) [33]

Sample No.	Al, μg ℓ^{-1}
1	2.1
2	1.2
3	1.0
4	1.6
5	2.1
6	1.2
7	2.3
8-15	1

TABLE 15.8

TRACE ELEMENTS LEACHED[a] FROM SAMPLING AND STORAGE DEVICES [33] (in mg ℓ^{-1} x 10^{-3})

Device	Cr	Mn	Se	Al	Zn	Cu
BD alcohol swabs	0.26	–	–	7.3	37	<5
Cliniswab IAA swabstick	3.0	4.6	1.5	510	<5	<5
Marion iodine swabstick	14.6	8.7	3.1	3200	384	287
Adsorbent cotton	1.0	–	–	113	64	24
Duoswab iodine scrub swabs	4.1	46	1.6	530	2140	25
S/P alcohol swabs	0.81	–	–	6.1	1020	<5
Falcon tube, black top	0	0	1.1	1.6	<5	<5
Falcon tube, blue top	0	0.02	1.5	1.8	<5	<5
Corning plastic container tube	0.1	0.13	1.0	1.8	27	<5
Vicra QUICK-CATH	1.8	0.67	0.8	11.8	143	<5
Jelco Cathlon IV	0.1	0.12	2.6	3.1	6	<5
BD Catheter 6743	0.7	0.22	2.5	4.6	9	<5
BD 20-ml syringe	0.1	0.33	1.6	2.0	14	<5
BD 30-ml syringe	0.2	0.11	1.9	2.0	46	<5
Monojet 6-ml syringe	0.04	0.04	0	3.1	3750	<5
Monojet 12-ml syringe	0.02	0	0	2.3	2090	<15
Stylex 10-ml syringe	0.15	–	–	13.8	3080	<5

[a] Leached overnight into 5 ml of 1 ml/h HNO_3.

References on p. 470

doubly-deionized water (DDW). The dried labware was stored in plastic bags until used. All labware was also rinsed again with DDW immediately before use. It should be mentioned that the plastic tubes used for sampling and storage were soaked in ultra-pure nitric acid solution (150 ml/l) overnight, rinsed with DDW, soaked overnight in Na_2EDTA solution (15 g/l) and again rinsed with DDW. Tubes were dried inside a laminar flow hood. Ericson et al. [33] also closely examined potential contamination factors like the choice of swabs, syringes, needles, and sampling tubes. Detailed results of this study are given in Table 15.8. It is important to realize that any specific product or procedure that ultimately has to be used in an aluminium determination by other workers should always be tested again for its contamination level. This is necessary since contamination of materials may change irregularly and so testing has to be performed at least each time a new batch of sampling aids is used. For example, a survey of the literature shows that for a certain brand of vacutainer tubes, used for the sampling of blood, contamination was found to be negligible in one case [43] or significant in another [53].

Blood is often taken with a separate syringe and samples are allowed to flow into, for example, a plastic tube. It has been recommended that a teflon catheter be used, through which 20–40 ml of blood should pass before the sample is taken. This volume is not, however, always acceptable and a more realistic waste volume might be 10 ml. It should be mentioned that rinsing is not possible when vacutainer tubes are used. Blood samples are typically stored in polypropylene since polyethylene is more often contaminated. As already mentioned the materials should be checked regularly for contamination. If heparin is added as an anticoagulant attention should be paid to its aluminium level. Table 15.9 shows aluminium levels in some commercial heparin preparations. From the most contaminated preparation a contribution larger than that of 0.01 mg l^{-1} blood is obtained if 10 μl of heparin per ml of blood is added. For the sampling of blood at normal levels, operators should wear gloves and hair-nets. Perspiration from the skin can also be a

TABLE 15.9.

ALUMINIUM CONCENTRATIONS IN COMMERCIAL HEPARIN PREPARATIONS [57]

Brand name Activity (μ/ml)	Aluminium concentration	
	ng Al/ml	ng Al/1000 units
Robins/1.000	962	962
Riker Lipo-Hepin/5.000	1.059	212
Upjohn/1.000	116	116
Sigma/1.000	69	69
Burns Biotek/1.000	57	57
Abbott Panheparin/5.000	248	50
ICN/1.000	18	18
Organon Liguaemin/20.000	32	1.6

source of contamination [58]. A widely employed method for the storage of blood, plasma, and serum is to freeze the sample below –20 °C. However, as has been pointed out [59], freezing may lead to protein-denaturation, and since part of the aluminium is bound to proteins an insoluble aluminium fraction may be obtained. For this reason one is recommended to store samples destined for aluminium analysis at +4 °C. It should be emphasized that contamination-free dissolution of blood or plasma in an acidic medium requires special equipment, such as a quartz vessel, special clean benches, and supra-pure acids.

TABLE 15.10.

COMMONLY USED DILUENTS FOR THE DETERMINATION OF ALUMINIUM IN SERUM, PLASMA, AND BLOOD BY GFAAS

Diluents	Ref.
H_2O	60
H_2O + HNO_3	59
H_2O + HNO_3 + $Mg(NO_3)_2$	38,44,61
H_2O, HNO_3, Triton X-100	52
H_2O, EDTA, NH_4OH, H_2SO_4	35

Often, blood, plasma, or serum samples are diluted with some reagents before determination. Table 15.10 gives the most commonly used diluents. A frequently-used procedure based on graphite furnace atomic absorption spectrometry (GFAAS), which has been discussed and evaluated by Slavin [61], is to mix 150 μl of serum with 150 μl of a reagent containing 0.4 g $Mg(NO_3)_2$ and 2 ml of HNO_3 per liter.

A digestion procedure for GFAAS determination using HNO_3 has been described [31], but the method is cumbersome when large sample throughput is required. Nevertheless, a wet-digestion procedure is useful for reference purposes since such a method minimizes losses of, e.g., endogenous aluminium: organic aluminium compounds are likely to be converted into inorganic aqua-complexes. The relatively large sample volume employed in this method is also advantageous in levelling out errors which may result from an uneven distribution of aluminium in frozen samples. In another application, flow injection analysis was used to shorten the analysis time [62]. Sample and reagent were injected, in parallel, into a double-injection valve. This double injector permitted synchronous merging of sample and reagent in a symmetrical system. After the sample and reagent had been mixed they flowed through a Pyrex coiled decomposition tube, located inside a microwave oven, where the heating encouraged mineralization of the blood. The authors pumped the mineralized sample directly into the nebulizer of a flame AAS instrument. It should be pointed out that flame AAS is not sensitive enough for the determination of aluminium at normal levels, which suggests that the system described might be advantageously connected to a more sensitive detector. In that case, however, the Pyrex-coil should be replaced by quartz to avoid contamination.

15.3.2. Biological tissues

The determination of trace elements in biological tissues includes, in most applications, a mechanical homogenization procedure which reduces the amount of sample needed for a representative analysis. If the homogeneity of the sample with respect to a certain element is adequate, however, the homogenization step can be omitted. Little is known about the sample amounts required to obtain representative samples for aluminium determinations, and no procedure for contamination-free homogenization has yet been published. As for the biological fluids discussed earlier, tissue samples should be prepared in a dust-free environment, and uncontaminated tools like acid-washed quartz

TABLE 15.11

SOME PROCEDURES SUGGESTED FOR THE DETERMINATION OF ALUMINIUM IN HUMAN AND ANIMAL TISSUES USING GFAAS

Type of tissue	Result (mg/kg) dry weight	Procedure	Ref.
Muscle Brain (human) Bone	1.2 1.3 2.4 – 3.9	Defatting with petroleum ether homogenization and dry ashing (400 °C brain) plus EDTA extraction	4
Brain (human)	1.3	Dry ashing at 600 °C Wet ashing $HNO_3/HClO_4$	29
Liver (human)	0.7 ± 0.3	Wet ashing $HClO_4/HNO_3/H_2SO_4$ at 200 °C	64
Bone (rat)	–	Freeze-drying, coarse fracturing, pulverization, low temp. ashing, dissolution in HCl	63
Bone (human)	0.5 – 5.0	Washed with methanol and defatted with chloroform, digestion in HNO_3 at 70 °C	52
Muscle (animal)	–	Dry ashing (500/600 °C), fusion with sodium carbonate – sodium borate, dissolution in HNO_3	25
Liver Bone (animal) Brain	–	Wet ashing in HNO_3	26,52
Liver (animal)	–	Dissolution in tetramethylammonium hydroxide and dilution with ethanol	65

knives must be used. Although several methods have been described for the determination of aluminium in serum, plasma, and blood, there are only a few procedures specifically concerned with its measurement in tissues. Table 15.11 summarizes some GFAAS-based procedures found in the literature.

Some authors recommend dry-ashing at 400 °C [4,51] followed by an EDTA-extraction or dry ashing at 600 °C [29] or 650 °C [30], and subsequent simple dissolution of the ash in nitric acid. However, dry ashing should only be used if dust can be excluded. Otherwise a systematic error may arise when correction is made for the blank, since the uptake of aluminium from the air and the vessels can vary, depending on whether or not a sample has been added. Quartz or porcelain should not be used since at temperatures above 500 °C aluminium oxide may react with these materials. The proposed wet-digestion methods include the use of mineral acids like nitric acid [26,34,52], a mixture of nitric and perchloric acid [29], and mixtures of nitric, perchloric acid and sulfuric acid [34]. Recently, Stevens [65] used hot aqueous tetramethylammonium hydroxide for the dissolution of soft tissues, followed by dilution with ethanol. He compared this technique with nitric acid-digestion and EDTA-extraction and found close agreement with only the former method. The lower values obtained by EDTA-extraction disagree with the findings of Alfrey et al. [4], who argued for the completeness of his EDTA-extraction method. One reason for the discrepancy may be the fact that Alfrey et al. [4] used much higher temperatures in the dry-ashing step.

As can be seen from Table 15.11, the values reported for control subjects are generally high, and of the order of 1 mg Al kg^{-1} of tissue (dry weight). For such high values contamination should not constitute as severe a problem as for blood and plasma at normal levels, provided that recommended precautions are taken. However, difficulties have been observed in aluminium determinations, even at higher concentrations. This is, for example, reflected in the large deviation in the results reported for animal muscle by the International Atomic Energy Agency [66], where aluminium values are included for orientation. The participating laboratories reported values for aluminium in the range 0.3 to 37 mg/kg^{-1}. Consequently, it has not been possible to check the accuracy of the higher reported aluminium values, even by reference to independent standard materials.

The results reported by Slanina et al. [26] (control, Spraque-Dawley rat brains with 0.01 - 0.02 mg kg^{-1} wet weight) are significantly lower than those obtained in tissue-sampling by other investigators, which again raises the question as to whether the higher values reported in the animal studies are subject to contamination errors. It should be mentioned that, by following the procedures recommended by D'Haese et al. [34] and Slanina et al. [26], it is possible to detect 50 and 2 μg, respectively of Al per kg tissue. In a procedure for the determination of aluminium in foods [25] the tissue was fused with a mixture of sodium carbonate-sodium borate after dry-ashing. These authors found that various forms of aluminium, such as aluminium oxide and aluminium silicate, could not be recovered from the samples unless fusion was applied. This means that the method of Slanina et al. [26] should not be employed for samples containing aluminium in certain inorganic forms. It should, however, be emphasized that the method based on a fusion procedure produces high blank values such that the detection limit is much increased.

References on p. 470

The accuracy of a dissolution procedure can be most easily tested using body fluids like blood or serum by comparing the results from dissolved samples with those obtained from direct determinations, which involve only a dilution step as discussed previously. For solid tissue samples it is possible to check the efficiency of the dissolution procedure by comparing the results from dissolved samples with those generated from the direct determination in the solid material using graphite furnace procedures. Such a comparison was made by Frech and Baxter [67], who found that aluminium in IAEA H-4 animal muscle and in lyophilized serum was sufficiently dissolved using nitric acid.

When solid materials are stored in acid-washed glass vessels contamination can occur and will increase after shaking. A lyophilized serum containing 0.02 mg of aluminium per kg increased its aluminium value to 0.4 ± 0.2 mg kg^{-1} when transferred to an acid-washed NBS glass container. After shaking for 1.5 h the aluminium value increased to 1.02 ± 0.50 mg kg^{-1} [67]. Some reference materials, for example NBS bovine liver, are delivered in glass containers. No certificate is given for aluminium but the material has been analyzed for this element by several laboratories. The published values have been used for comparison. However, in the light of the preceding results, such a procedure cannot be recommended for verifying an analytical method, since both the duration and manner in which the material has been transported and stored are likely to influence the aluminium concentration.

15.3.3. Plant materials

For plant materials, sampling is not as critical as for tissue and blood since the aluminium levels are generally relatively high (see Table 15.4). Nevertheless, one should take care when selecting tools for sampling and vessels for storage. A major contribution to the systematic error in analysing plant materials originates from dust.

15.3.4. Water

Since water is important for the aluminium levels in living organisms the procedures for sampling and storing may be of interest to the reader. According to international standards, polyethylene containers should be used for storing samples. As mentioned above, polypropylene may be less contaminated. Furthermore, solutions should be acidified with HNO_3 to a pH below 2. The filterable amount, and that adhering to suspended matter, may be determined from the same sample but, in this case, acidification should take place after filtration. Obviously, the aluminium-blank level in the filter must be tested before filtration of the sample in order to select a type of filter having sufficiently low background. Jardine et al. [68] found that polycarbonate membranes released comparatively small amounts of aluminium and that this material adsorbs aluminium only to a minor extent. Before determination of the samples, according to Swedish Standard SS 02 81 50, a digestion step (of a 40 ml sample with 10 ml HNO_3), including heating in an autoclave at 150 °C for 30 min, is recommended. This quantitation of aqueous aluminium has been discussed by Bloom and Erich [69].

15.4. METHODS FOR THE DETERMINATION OF ALUMINIUM

Approaches to trace aluminium determinations have, for the most part, involved the use of techniques which determine bulk levels. Only a few papers deal with methods for speciation of aluminium. For example, Kalik and Eichhorn [70] used ^{27}Al-NMR to elucidate the potential involvement of aluminium-complexes in Alzheimer's disease. Courtijn et al. [71] used a Chelex-100 ion-exchange column for the separation of different aluminium species dissolved in acidified waters in the Belgian Campine. Samples were filtered and the filtrates submitted to ion-exchange in order to separate aluminium, combined with humic acid, fluoride-complexed aluminium, and free or hydrolysed aluminium. For localization of aluminium in, for example, brain tissue a scanning electron microscope [72] has been used to detect aluminium deposits by energy-dispersive X-ray analysis. In spite of the fact that this technique is not particularly sensitive, Smith et al. [72] were able to locate aluminium deposits within the glomerular basement membrane of humans. An alternative approach providing better sensitivity is the use of laser-microprobe mass spectrometry. In this technique an intense laser beam is directed towards a specific organelle, and species vaporized from that region are then analyzed in a mass spectrometer [73]. This technique has been used to localize aluminium in the lysozomes of kupffer cells in liver tissue from patients on chronic heaemodialysis. Protein binding and aluminium speciation in blood serum were studied by Gonzalez et al. [75] using ultra-filtration and chromatographic techniques.

Analytical techniques for bulk aluminium determinations at trace-levels are: neutron activation, fluorimetry, inductively coupled plasma atomic emission spectrometry (ICP-AES, inductively coupled plasma mass spectrometry (ICP-MS), direct current plasma atomic emission spectrometry (DCP-AES), or graphite furnace atomizers and atomic absorption spectrometry (GFAAS).

Although neutron activation analysis is sufficiently sensitive for aluminium determination, relatively poor detection limits are obtained for biological samples containing silicon and phosphorus, both of which form ^{28}Al as a decomposition product upon radiation. The half-life of ^{28}Al is only 2.2 min and therefore instrumental neutron activation analysis (INAA) is difficult for samples containing large amounts of sodium chloride because of the intense γ-radioactivities of ^{24}Na ($t_{1/2}$ 15 h) and ^{38}Cl ($t_{1/2}$ 37 min). In conclusion, neutron activation analysis is not recommended for trace aluminium determinations in biological materials. Recently Kratochvil et al. [74] compared this technique with GFAAS for the biological reference material TORT-1 (lobster hepatopancreas containing 43 mg kg^{-1} aluminium) and found a very good agreement between these techniques. The INAA results were corrected for interferences from ^{28}Al produced by ^{38}P (n,d) ^{28}Al and ^{28}Si (n,p) ^{28}Al-reactions.

Fluorimetric methods are generally very sensitive, but in complex matrices the selectivity is low. This explains why very few papers are found in the literature. Nevertheless, Joannou and Piperaki [76] described an interesting method for serum samples, based on monitoring the rate of reaction of 2-hydro-1-naphthaldehyde-p-methoxybenzoyl hydrogen with aluminium ions. The emission of the resulting fluorescent metal-chelate formed is

References on p. 470

measured at 475 nm. Aluminium was measured in the supernatant of serum, after proteins were removed by precipitation with concentrated nitric acid. The detection limit for the method, defined as 3 times the standard deviation of the mean blank, was reported to be 0.13×10^{-3} mg l^{-1}.

Several authors [77-80] have reported on the use of ICP- or DCP-atomic emission spectrometry for the determination of aluminium at the 10^{-3} mg l^{-1} level. However, a major problem with using the argon-plasma technique is the intense and broad emission of calcium which increases the background and the detection limit for aluminium. Mauras and Allain [78] described an improved background-correction system which made it possible to reach a detection limit of 3×10^{-3} mg l^{-1} in serum. One drawback, however, with this technique is the relatively high cost and complexity of the instrumentation which will preclude its use in many routine laboratories. Taylor [81] used the more sensitive 167.08 nm line in the vacuum UV, where spectral interferences are relatively small. Modifications of the spectrometer had to be made to minimize the optical path length in air. The detection limit for aluminium in serum was 1.5×10^{-3} mg l^{-1}. Some investigators [80-82] have used an electrothermal vaporizer coupled to an ICP, which gives detection limits in serum around 2×10^{-3} mg l^{-1}. With conventional instrumentation the corresponding detection limit is around 6×10^{-3} mg l^{-1}. Sanz-Medel et al. [80] concluded that ICP-AES is better than GFAAS only for blood or serum aluminium levels above 0.03 mg kg^{-1}. In the light of the most recently published normal levels, additional preconcentration and matrix-separation procedures have been employed to facilitate the use of ICP-AES instrumentation in trace aluminium determinations [83]. It should be emphasized that such methods are relatively time consuming and increase the risk of sample contamination.

ICP-MS is a highly sensitive multi-element technique and can be used to investigate samples with unknown composition. With this technique, the above-mentioned spectral interferences observed with ICP-AES are circumvented. On the other hand, ICP-MS suffers from isobaric overlap and sample matrix-interferences. For example, Satzger [84] reports a signal enhancement for aluminium in the presence of magnesium and molecular nitrogen, due to a mass overlap. Hieftje and Vickers [85] have recently discussed the current status of the ICP-MS technique.

Baxter et al. [86] used graphite furnace AES with a constant temperature atomiser and reported a detection limit of 0.3×10^{-3} mg Al l^{-1} for samples of whole blood and cortex. Other emission techniques, like flame atomic spectrometry, are not sensitive enough for trace aluminium determinations.

The greatest success achieved by any of the techniques used for the determination of aluminium in biological specimens has been with GFAAS. This has become the method of choice because it offers the best combination of sensitivity, reliability, simplicity, and low cost. The reliability of the method has been improved substantially during the last few years as a consequence of the introduction of new types of pyro-coated high density graphite tubes in combination with the 'L'vov platform technique'. These improvements have been combined with other developments in the graphite furnace technique and incorporated in the so called STPF (stabilized temperature platform furnace) concept. This will be discussed in more detail below.

15.5. DETERMINATION OF ALUMINIUM BY GFAAS

This technique is based on the absorption of radiation by free atoms. This means that one pre-requisite for an interference-free determination is that the same fraction of aluminium in samples and standards is converted to free atoms. Freedom from interferences is always obtained if the analyte is completely converted to free atoms, independent of the matrix composition. (We assume here that the residence-time of atoms within the light-path is not influenced by the matrix.) Since the GFAAS technique is prone to interference effects, the method of standardization and the conditions for the formation of free aluminium atoms in the graphite tube are of fundamental importance for the accuracy of the results: these points will be discussed further.

L'vov and co-workers have estimated the extent of atom formation for a number of elements by comparing theoretically calculated with experimentally obtained characteristic masses using a platform-equipped Perkin-Elmer furnace [87]. For aluminium this ratio was found to be 0.69 which indicates that about 30% of the analyte was not converted to free atoms, provided the constants for calculating the characteristic mass are correct. When using a constant temperature furnace it was found that this ratio was 0.79 [88]. This can be explained by the higher gas-phase temperature obtained in the latter system leading to a more efficient dissociation of aluminium-containing molecules. In line with this, Gardiner et al. [89] observed a 20% reduction in the peak area when samples were atomized from the tube-wall instead of from the L'vov platform.

High temperature equilibrium calculations show [90] that even at relatively high temperatures significant amounts of gaseous aluminium oxides and hydroxides are formed. Recently, Styris and Redfield [91] monitored the gas-phase components of a graphite furnace at atmospheric pressure during the vaporization/atomization of an aluminium standard solution using mass spectrometry. Oxides and numerous carbides of aluminium were observed to appear before and during the formation of free aluminium atoms. The formation of aluminium carbides and cyanides in a graphite tube has also been discussed by L'vov [92] and in references cited therein. Styris and Redfield [91] further showed that, in the presence of nitric acid, gaseous aluminium cyanides were formed during atomization. They presented strong evidence that aluminium atoms are formed by thermal dissociation of $Al_2O_3(s)$. Mechanisms that explain the stabilization of aluminium by magnesium nitrate were shown to involve two temperature regions. During thermal pretreatment magnesium hydroxides are formed by hydration of magnesium oxide (the latter being formed during thermal pretreatment of magnesium nitrate). This prevents losses of aluminium as its hydroxides which were shown to occur between 1300 and 1550 K, in the absence of modifier. At temperature above 1500 K an oxidation-dissociation sequence, involving reduction of magnesium oxide, takes place. This results in retention of adsorbed aluminium oxide until the magnesium oxide is depleted, thereby delaying the atomization of aluminium to higher temperatures. Therefore, in the presence of the modifier, atomization takes place via thermal dissociation of the adsorbed aluminium oxides. Since aluminium forms stable gaseous cyanides [92], obtaining the optimum sensitivity requires that argon be used as the sheath gas instead of nitrogen [93].

References on p. 470

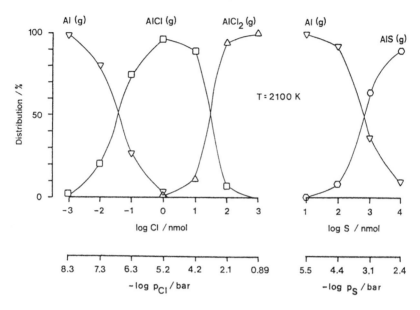

Fig. 15.1. Equilibrium distribution of aluminium species at 2100 K as a function of the partial presence of chlorine and sulphur, respectively (from ref. 90).

Many samples contain significant amounts of chlorine and sulphur. Therefore it is important to investigate the influence of these elements on the atomization behaviour of aluminium. The susceptibility to interference effects from chlorine and sulphur is indicated by the equilibrium calculations given in Fig. 15.1. As can be seen, as little as sub-nano-gram amounts of chlorine can give rise to significant signal depressions. Koirtyohann et al. [94] found that perchloric acid caused a 95% suppression of the aluminium signal. The acid was shown to be inter-chelated in the graphite tube, and the chlorine could not be removed at pretreatment temperatures below 1700 K. Residual chlorine caused a gas-phase interference effect, although this suppression could be reversed by the addition of excess ammonium carbonate or ammonium sulphate. Figure 15.2 shows thermal pre-treatment curves for 1 ng of aluminium atomized from a platform, in water and in the presence of various concentrations of copper chloride, using the above-mentioned modi-fier. As can be seen, the interferences from the copper chloride can be removed in this case by choosing a proper ashing temperature. Interestingly, the high stability of gaseous aluminium chloride has been used analytically in the determination of chlorine by graphite furnace molecular spectroscopy [95]. In contrast, the capacity for forming gaseous alumi-nium sulphide is very low. Experiments performed with 1 ng of aluminium in the presence of 2 to 100 μg of copper sulphate gave a recovery of 98–96% for an ashing temperature of 1970 K, using magnesium nitrate as modifier [90].

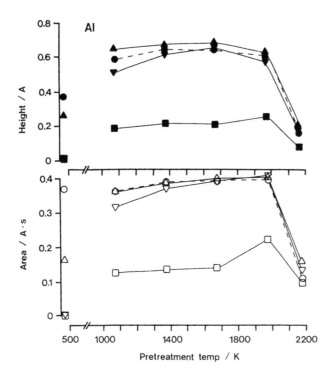

Fig. 15.2. Thermal pretreatment curves for 1 ng of aluminium (from platform) in water (○●) and in the presence of 0.1% (Δ ▲), 0.4% (∇ ▼) and 1% (□■) of copper chloride, respectively. The data points shown on the left of the figures were obtained without a modifier and without ashing.

15.5.1. Determination of aluminium in liquids, serum, blood, and plasma

As can be seen from Table 15.3, a very common application is the determination of aluminium in human serum and plasma. Some of the factors which can influence the precision, sensitivity, and accuracy of the determination were discussed by Gardiner and Stoeppler [56].

We will discuss two of the most frequently used procedures for the determination of aluminium in body fluids, both of which are based on the use of platform-equipped furnaces and peak area evaluation. Andersen and Reimert [52] propose dilution of the sample (plasma) 1+1 or 1+3 with a solution containing 10^{-3} M HNO_3 and 0.1% Triton X-100. A similar procedure was used by Oster [37]. For 1+1 diluted samples Gardiner and Stoeppler [56] reported problems during the pretreatment step. In this application no magnesium nitrate was added as modifier because it is endogenous to the samples. The only effect of the magnesium nitrate would be to facilitate the thermal decomposition of the organic sample compound. For tissue samples, which frequently need to be greatly diluted as a consequence of their intrinsically high aluminium concentrations (see

TABLE 15.12.

GRAPHITE FURNACE PROGRAM FOR THE DETERMINATION OF ALUMINIUM IN HUMAN SERUM ACCORDING TO ANDERSEN AND REIMERT [52]

Parameter	Step					
	1 Dry	2 Dry	3 Char	4 Atomize	5 Clean	6 Cool
Temperature ($^\circ$C)	120	200	1400	2500	2700	20
Ramp (s)	30	20	40	0^a	1	3
Hold (s)	20	10	30	7	2	5
Ar flow (ml min^{-1})	300	300	300	0	300	300

[a] Maximum power heating.

Table15.3), the magnesium nitrate present in the final solutions is probably not sufficient to act as a modifier. Table 15.2 shows the graphite furnace program used by Andersen and Reimert [52]. The long ramp times are necessary to avoid sputtering of the sample. The 396.2 nm line, which is more favourable when using the method of standard additions, was used since it gives calibration graphs with a larger linear range than does the more sensitive non-resonance line at 309.3 nm. It should be mentioned that when the deuterium-arc background-correction system is used the intensity of the continuum will be much lower at the longer wavelength, resulting in increased noise. When using the Zeeman-effect background-corrector such restrictions on the choice of wavelength do not exist. Since high pretreatment temperatures can be used for aluminium, the background absorbance emanating from biological samples is normally very low. Despite this fact, background correction is still mandatory for accurate determination at low levels.

Slavin [61] has critically discussed the conflicting recommendations which have appeared for the determination of aluminium in serum. He recommends a procedure based on the work of Bettinelli et al. [96], Leung and Henderson [44], and Brown et al. [38], in which the sample is diluted 1 + 1 with a solution containing 0.1% magnesium nitrate. Table 15.13 gives the procedures proposed by Slavin [61]. In this method oxygen is added to the sheath gas, during the pretreatment, to facilitate decomposition of the organic material. The main difference between the two procedures described here lies in the method of calibration. Slavin used a standard curve obtained from magnesium nitrate containing solutions whereas Andersen and Reimert [52] applied the method of standard additions (which explains why they used the longer wavelength). The question of standardization has been further discussed in an exchange of letters [97,98]. The main argument for a standard addition procedure is that it will give better accuracy because the sensitivity for aluminium is not sufficiently constant during the life-time of the tube. In practice, the analysis time is not increased since all samples must be run at least in duplicate in order to remove erroneous results stemming from airborne particulate contamination. Additions

TABLE 15.13.

GRAPHITE FURNACE PROGRAM FOR THE DETERMINATION OF ALUMINIUM IN HUMAN SERUM ACCORDING TO SLAVIN [61]

Parameter	Step						
	1 Dry	2 Char	3 Purge	4 Char	5 Atomize	6 Clean	7 Cool
Temperature (°C)	130	600	600	1700	2400	1600	20
Ramp (s)	1	30	1	15	0	1	1
Hold (s)	45	55	25	25	6	6	20
Ar flow (ml min^{-1})	300	–	300	300	0	300	300
O$_2$ flow (ml min^{-1})	–	50	–	–	–	–	–
Read (s)	–	–	–	–	0	–	–
Base line (s)	–	–	–	–	–7	–	–
Integration (s)	–	–	–	–	6	–	–

These conditions apply to the Zeeman 5000.

of standards to samples are then made during dilution to the duplicate. Slavin [97], on the other hand, argues that the method of standard addition is less precise than analyses against an analytical curve. Furthermore, all the errors associated with contamination and dilution play a greater role when the method of additions is used.

In general, both of the methods mentioned should give accurate results. We always recommend assessing the accuracy of the method used by either analyzing suitable standard reference materials or by recovery tests using the standard addition technique.

15.5.2. Determination of aluminium in digested samples

For digested tissues, including blood samples, the method recommended by Andersen and Reimert can be used [52].

15.5.3. Direct determination of solid samples

Results for the determination of aluminium in several biological materials by GFAAS and GFAES, using the solid sampling technique and a constant temperature furnace, have been given by Frech and Baxter [67]. These authors compared results with a dissolution procedure and found good agreement. With the solid sampling technique it was possible to calibrate against aqueous standards and to achieve a precision of better than 10% R.S.D. for NBS 1577a bovine liver, and IAEA H4 animal muscle, provided the sample mass was greater than 1.3 mg and the peak area evaluated. The solid sampling procedure was identified as a possible source of contamination. When the powdered sample came into contact with an acid-washed glass capillary, used for sample introduction, severe contamination occurred. Thomassen and Radziuk [99] used a commercially available slotted graphite tube with a solid sampling cup for the determination of aluminium in solid materials.

15.6. STANDARD REFERENCE MATERIALS

Only few standard reference materials for low aluminium levels in biological materials are available at present. A lyophilized human serum has been prepared [100] taking all possible precautions to avoid contamination. This material was analyzed for 16 elements including aluminium by selected laboratories. The mean value for aluminium was 1.8 x 10^{-3} mg l^{-1} based on the results from 7 laboratories using GFAAS and GFAES in combination with different sample preparation procedures, including the solid sampling technique. In view of the very low aluminium value the agreement between the laboratories was good. A bovine serum pool was collected and prepared by Veillon et al. [101] under carefully controlled conditions to minimize contamination with trace elements. In this material the aluminium concentration was determined by GFAAS and by inductively coupled plasma emission spectrometry. Unfortunately, only a recommended Al-value could be given (0.013 mg l^{-1}). The serum is available through the National Bureau of Standards' Office of Standard Reference Materials as Reference Material 8419. Nycomed AS, Norway have produced lyophilized human reference whole blood, serum, and urine, at several relatively high aluminium levels (0.05 – 0.16 mg l^{-1}). Because of the high aluminium concentrations, plasma atomic emission spectrometry could be used along with GFAAS for certification.

15.7. NEED FOR FUTURE RESEARCH

At present, no reference materials for biological tissues like liver, brain, etc. are available at the non-elevated levels of aluminium. The conditions for contamination-free preparation of such materials should be investigated. Furthermore, the development of non-spectrometric methods for trace aluminium determination is desirable to increase the reliability of the produced reference values. In order to establish the nature of possible toxicity of aluminium it is necessary to develop further the methods for speciation of aluminium compounds, and to increase the selectivity and sensitivity of methods for the characterization of aluminium distribution in, for example, tissue samples.

REFERENCES

1 A. Cedergren and W. Frech, *Pure Appl. Chem.*, 59 (1987) 221-228.
2 P.O. Ganrot, *Environ. Health Perspectives*, 65 (1986) 363-441.
3 J.R. Andersen, *Arch. Pharm. Chem., Sci. Ed.*, 15 (1987) 123-180.
4 A.C. Alfrey, G.R. LeGendre and W.D. Kaehny, *New Engl. J. Med.*, 294 (1976) 184-188.

5 D.R. Crapper Mc Lachlan and U. de Boni, *Neurotoxicol.*, 1 (1980) 3-16.
6 D.P. Perl, C.D. Gajdusek, R.M. Garruto, R.T. Yanagihara and C.J. Gibbs, *Science*, 217 (1982) 1053-1055.
7 Y. Yase, *Proc. 11th World Congress of Neurology*, Amsterdam, 1977, pp. 413-427.
8 H.A. Ellis, J.H. McCarthy and J. Herrington, *J. Clin. Pathol.*, 32 (1979) 832-844.
9 M.K. Ward, T.G. Feest, H.A. Ellis, F.S. Parkinson and D.N.S. Kerr, *Lancet, i* (1978) 841-845.
10 A.C. Alfrey, *Neurotoxicol.*, 1 (1980) 43-53.
11 C.T. Driscoll, Jr., J.P. Baker, J.J. Bisogni, Jr. and C.L. Schofield, *Nature (London)*, 284 (1980) 161-164.
12 S. Sjöberg and L.O. Öhman, *J. Chem. Soc. Dalton Trans.*, (1985) 2665-2669.
13 L.O. Öhman, *Ph.D. Thesis*, University of Umeå, Sweden, 1983.
14 M. Stoffyn, *Science*, 203 (1979) 651-653.
15 D.J. Hydes, *Science*, 205 (1979) 1260-1262.
16 W. Dickson and T. Schneider (Editors), *Proc. Studies in Environmental Science 30, Amsterdam, May 5-9, 1986*, Elsevier, Amsterdam, 1986, pp. 19-28.
17 H. Borg, *Water Res.*, 21 (1987) 65-72.
18 W. Dickson, *Vatten*, 39 (1983) 400-404.
19 O. Knoll, H. Lahl, J. Böckmann, H. Hennig and B. Unterhalt, *Trace Elements in Medicine*, 3 (1986) 172-175.
20 K.R. Koch, M.A.B. Pougnet and S. De Villiers, *Analyst (London)*, 114 (1989) 911-913.
21 Statens Livsmedelsverk, Uppsala, Sweden, SLV FS 1982:2.
22 L. Lundmark, personal communication.
23 P. Mattsson, J. Haegglund, H. Jönsson and B. v Hofsten, *Vår Föda*, 36 (1984) 414-417.
24 L. Jorhem, Statens Livsmedelsverk, Uppsala, Sweden, personal communication.
25 D.M. Sullivan, D.F. Kehoe and R.L. Smith, *J. Assoc. Off. Anal. Chem.*, 70 (1987) 118-120.
26 P. Slanina, Y. Falkeborn, W. Frech and A. Cedergren, *Food Chem. Toxic.*, 22 (1984) 391-397.
27 P. Slanina, W. Frech, A. Bernhardson, A. Cedergren and P. Mattsson, *Acta Pharmacol. Toxicol.*, 56 (1985) 331-336.
28 G. Tyler, E. Steinnes, L. Folkeson, E. Lobersli and E. Nyholm, *Final Report, MIL-3-project*, Nordisk Ministerråd, Oslo, 1985.
29 J.R. McDermott, I. Whitehill, *Anal. Chim. Acta*, 85 (1976) 195-198.
30 A.I. Arieff, J.D. Cooper, D. Armstrong and V.C. Lazarowitz, *Ann. Internal Med.*, 90 (1979) 741-747.
31 W. Frech, A. Cedergren, C. Cederberg and J. Vessman, *Clin. Chem.*, 28 (1982) 2259-2263.
32 P. Slanina, W. Frech, L-G. Ekström, L. Lööf, S. Slorach and A Cedergren, *Clin. Chem.*, 32 (1986) 539-541.
33 S.P. Ericson, K.L. McHalsky, B.E. Rabinow, K.G. Kronholm, C.S. Arceo, J.A. Weltzer and S.W. Ayd, *Clin. Chem.*, 32 (1986) 1350-1356.
34 P.C. D'Haese, F.L. Van de Vyver, F.A. de Wolff and M.E. de Broe, *Clin. Chem.*, 31 (1985) 24-29.
35 F.R. Alderman and H.J. Gitelman, *Clin. Chem.*, 26 (1980) 258-260.
36 N. de Baets, *Licentiaatsthesis*, University of Ghent, Ghent, 1978.
37 O. Oster, *Clin. Chim. Acta*, 114 (1981) 53-60.
38 A.A. Brown, P.J. Whiteside and W.J. Price, *Int. Clin. Prod.*, (1984) 16-24.
39 O. Guillard, K. Tiphanean, D. Reiss and A. Piriou, *Anal. Lett.*, 17 (1984) 1593-1605.
40 R.L. Bertholf, S. Brown, B.W. Renoe, M.R. Wills and J. Savory, *Clin. Chem.*, 29 (1983) 1087-1089.
41 M.R. Wills, C.S. Brown, R.L. Bertholf, R. Ross and J. Savory, *Clin. Chim. Acta*, 1145 (1985) 193-196.
42 S.N.E. Marsden, I.S. Parkinson, M.K. Ward, H.A. Ellis and D.N.S. Kerr, *Proc. Eur. Dial. Transplant Assoc.*, 16 (1979) 588-596.

43 M. Buratti, G. Caravelli, G. Calzaferri and A. Colombi, *Clin. Chim. Acta*, 141 (1984) 253-259.

44 F.Y. Leung and A.R. Henderson, *Clin. Chem.*, 28 (1982) 2139-2143.

45 W.D. Kaehny, A.C. Alfrey, R.E. Holman and W.J. Schorr, *Kidney Int.*, 12 (1977) 361-365.

46 I.S. Parkinson, M.K. Ward and D.N.S. Kerr, *Clin. Chim. Acta*, 125 (1982) 125-133.

47 F.A. De Wolff and B.G. Van der Voet, *Clin. Chim. Acta*, 160 (1986) 183-188.

48 E. Verhaeven, *Licentiaatsthesis*, University of Ghent, Belgium, 1981.

49 J.P. Clavel, M.C. Jaudon and A. Galli, *Ann. Biol. Clin.*, 40 (1982) 51.

50 V. Hudnik, E. Kozak and M. Marolt-Gomiscek, *Vestn. Slov. Kem. Drus.*, 30 (1983) 411-419.

51 G.R. Le Gendre and A.C. Alfrey, *Clin. Chem.*, 22/1 (1976) 53-56.

52 J.R. Andersen and S. Reimert, *Analyst (London)*, 111 (1986) 657-660.

53 J.L. Greger, *Food Technol.*, May (1985) 73-80.

54 M. Andersson, A-M. Balsberg-Påhlsson, G. Berlin, U. Falkengren-Grerup and G. Tyler, *Flora*, 183 (1989) 405-415.

55 R. Cornelis and P. Schutyser, in E. Quellhorst, K. Finke and C. Fuchs (Editors), *Trace Elements in Renal Insufficiency*, S. Karger, Basel, 1984, pp. 1-11.

56 P.E. Gardiner and M. Stoeppler, *J. Anal. At. Spectrom.*, 2 (1987) 401-404.

57 P.J. Kostyniak, *J. Anal. Toxicol.*, 7 (1983) 20-23.

58 M. Zief and J.W. Mitchell, in P.J. Elving and I.M. Kolthoff (Editors), *Contamination Control in Trace Element Analysis*, Wiley, New York, 1976, pp. 63-64.

59 H.P. Bertram, *Nieren-Hochdruck*, 10 (1981) 188-191.

60 F.J. Langmyhr and D.L. Tsalev, *Anal. Chim. Acta*, 92 (1977) 79-83.

61 W. Slavin, *J. Anal. At. Spectr.*, 1 (1986) 281-285.

62 M. Burguera, J.L. Burguera and O..M. Alarcon, *Anal. Chim. Acta*, 179 (1986) 351-357.

63 J. Smeyers-Verbeke and D. Verbeelen, *Clin. Chem.*, 31 (1985) 1172-1174.

64 D. Thornton, L. Liss and J. Lott, *Abstracts of papers in Conference on Aluminium Determinations in Biological Materials, Charlottesville, VA, 1983*, p. 1.

65 J. Stevens, *Clin. Chem.*, 30/5 (1984) 745-747.

66 Internat. Atomic Energy Agency, Analytical Quality Control Service, Lab./243, 1985-1986.

67 W. Frech and D. Baxter, *Z. Anal. Chem.*, 328 (1987) 400-404.

68 P.M. Jardine, L.W. Zelazny and A. Evans, Jr., *Soil Sci. Soc. Am. J.*, 50 (1986) 891.

69 P.R. Bloom and M.S. Erich, in G. Sposito (Editor), Environmental Chemistry of Aluminium. CRC Press, Boca Raton, Fl, 1989, Ch. 4.

70 S.J. Karlik and G.L. Eichhorn, *Abstracts of papers in Conference on Aluminium Determinations in Biological Materials, Charlottesville, VA*, 1983, p. 3.

71 E. Courtijn, C. Vandecasteele and R. Adams, *Sci. Total Environ.*, 90 (1990) 191-202.

72 P.S. Smith and J. McClure, *J. Clin. Pathol.*, 35 (1982) 1283-1293.

73 A.H. Verbrueken, F.L. Van de Vyer, R.E. Van Grieken, G.J. Paulus, W.J. Kisser, P. D'Haese and M.E. De Broe, *Clin. Chem.*, 30 (1984) 763-768.

74 B. Kratochvil, N. Motkosky, M.J.M. Duke and D. Ng, *Can. J. Chem.*, 65 (1987) 1047-1050.

75 E.B. González, J.P. Parajón, J.I.G. Alonzo and A. Sanz-Medel, *J. Anal. At. Spectrom.*, 4 (1989) 175-179.

76 P.C. Ioannou and E.A. Piperaki, *Clin. Chem.*, 32 (1986) 1481-1483.

77 P. Schramel, A. Wolf and B.J. Klose, *J. Clin. Chem. Clin. Biochem.*, 18 (1980) 591-593.

78 Y. Mauras and P. Allain, *Anal. Chem.*, 57 (1985) 1706-1709.

79 T. Nishima, *ICP Inf. Newsl.*, 8 (1982) 147-155.

80 A. Sanz-Medel, R.R. Roza, R.G. Alonso, A.N. Vallina and J. Cannata, *J. Anal. At. Spectrom.*, 2 (1987) 177-184.

81 G.A. Taylor, *Paper presented at SAC 86/3rd BNASS, Bristol, 20-26 July, 1986.*

82 S.E. Long, R.D. Snook and R.F. Browner, *Spectrochim. Acta Part B*, 40 (1985) 553-568.

83 H. Matusiewicz and R.M. Barnes, *Spectrochim. Acta Part B,* 39 (1984) 891-899.
84 R.D. Satzger, *Anal. Chem.,* 60 (1988) 2500-2504.
85 G.M. Hieftje and G.H. Vickers, *Anal. Chim. Acta,* 216 (1989) 1-24.
86 D.C. Baxter, W. Frech and E. Lundberg, *Analyst (London),* 110 (1985) 475-482.
87 B.V. L'vov, V.G. Nikolaev, E.A. Norman, L.K. Polzik and M. Mojica, *Spectrochim. Acta Part B,* 41 (1986) 1043-1053.
88 W. Frech and D.C. Baxter, *Spectrochim. Acta.,* 45B (1990) 867-886.
89 P.E. Gardiner, M. Stoeppler and H.W. Nurnberg, *Analyst (London),* 110 (1985) 611-617.
90 W. Frech, E. Lundberg and A. Cedergren, *Prog. Analyt. Atom. Spectrosc.,* 8 (1985) 257-370.
91 D.L. Styris and D.A. Redfield, *Anal. Chem.,* 59 (1987) 2891-2897.
92 B.V. L'vov, *J. Anal. Atom. Spectrom.,* 2 (1987) 95-104.
93 W. Welz, *Atomic Absorption Spectrometry,* Verlag Chemie, Weinheim, 2nd ed., 1985.
94 S.R. Koirtyohann, E.D. Glass and F.E. Lichte, *Applied Spectrosc.,* 35 (1981) 22-26.
95 K. Tsunoda, H. Haraguchi and K. Fuwa, *Spectrochim. Acta,* 35B (1980) 715-729.
96 M. Bettinelli, U. Baroni, F. Fontana and P. Poisetti, *Analyst (London),* 110 (1985) 19-22.
97 W. Slavin, *Spectrochim. Acta,* 42B (1987) 933-935.
98 J.R. Andersen, *Spectrochim. Acta,* 42B (1987) 929-932.
99 Y. Thomassen and B. Radziuk, *Poster presented at XXIV CSI, Garmisch Partenkirchen, 15-20 Sept. 1985.* Poster abstracts pp. 436-437.
100 J. Versieck, *J. Res. Natl. Bur. Stand.,* 91 (1986) 87-92.
101 C. Veillon, S.A. Lewis, K.Y. Patterson, W.R. Wolf, J.M. Harnly, J. Versieck, L. Vanballenberghe, R. Cornelis and T.C. O'Haver, *Anal. Chem.,* 57 (1985) 2106-2109.

M. Stoeppler (Editor)/*Hazardous Metals in the Environment*
© 1992 Elsevier Science Publishers B.V. All rights reserved

Chapter 16

Selenium[*]

Milan Ihnat

Land Resource Research Centre, Research Branch, Agriculture Canada, Ottawa, Ontario K1A 0C6 (Canada)

CONTENTS

16.1. Introduction 475
16.2. Analytical methodology reviews 478
    16.2.1. Analytical method nomenclature and codes 478
    16.2.2. Review articles 481
16.3. Analytical applications 483
16.4. Sampling, sample handling and storage 490
    16.4.1. Sampling 490
    16.4.2. Sample handling, sample and analyte preservation and storage 493
16.5. Sample treatment 495
16.6. Determination procedures and recommended methods 499
    16.6.1. Determination procedures 499
    16.6.2. Recommended methods 504
16.7. Conclusions 505
References 505

16.1. INTRODUCTION

This chapter summarizes the various analytical facets involved in the measurement of selenium in a wide variety of environmental materials. Based on a comprehensive review of the literature, it lists major reviews of analytical methodologies, and presents a compilation of method applications. Sampling and sample treatment and storage procedures specifically developed for this element are described and evaluated and constitute the main thrust of this report. A brief discussion of the various determinative techniques and procedures is presented; speciation and the analytical chemistry of organo selenium compounds is touched upon. A brief treatment of collaborative studies and reference or official methods closes the chapter.

---

[*] Contribution No. 90-82 from Land Resource Research Centre.

TABLE 16.1.

CONCENTRATIONS OF SELENIUM IN NATURAL MATERIALS

Ranges of mean or median concentrations of total selenium reported, reducing effects of extreme values and giving estimates of typical "normal" concentration ranges observed in various natural materials. The data may or may not include data from seleniferous regions or other extreme values. Information has been adapted from extensive, comprehensive compilations in the literature cited covering an extensive range of materials, specimens, samples and analytical data. Unless noted, coverage is worldwide. Refer to the original litarature for detailed listings and discussions.

Material	Se concentration (mg/kg)[a]	Ref.
Minerals	0.043–200 000	6
Rocks	0.0062–277	6
Soils (including seleniferous soils)	0.079–78	6
Coals and oils	0.17–3.36	6
Freshwater (rivers, streams, canals, lakes, reservoirs, groundwater, drinking water)[b]	0.000016–0.041	7
Freshwater (elevated levels, San Luis Drain Canal and Kesterson Reservoir, CA, USA)[b]	0.066–0.40	7
Freshwater sediments	0.20–29	7
Freshwater plankton and fish	0.079–28	7
Seawater	0.000025–0.00027	8
Marine sediments	0.02–9.6	8
Marine biological tissues[c]	0.02–780	8
Atmospheric particulate Se at urban and semi-rural locations[d]	0.27–11 ng/m^3	9
Atmospheric particulate Se at marine and remote continental locations	0.0048–0.58 ng/m^3	9
Precipitation (rain, snow)	0.0000047–0.0018	7,9
Foodstuffs (USA core foods)[e]	0.00068–11.4	3
Animal feed crops[f]	0.004–0.53	2
Animal feed ingredients and concentrates, premixes[f]	0.0005–6.2	2
Selenium accumulator plants (primary accumulators)	1–8512	2
Animal tissues[g]	0.002–41	4
Human tissues[h]	0.0005–8.7	5
Human tisues (elevated level in hair, high Se area with chronic selenosis in Peoples' Republic of China)[i]	3.7,32	5

[a] Concentrations are reported as mg/kg (mg/l for fluids or packed cells) unless otherwise noted (ng/m^3 for atmospheric materials), with as received, wet, dry and not reported bases combined.

[b] Dissolved total selenium.

[c] Wide range of plant and animal tissues.

d Excluding two high values of 116 and 120 ng/m^3.

e In-depth study selecting acceptable analytical data for the top approximately 100 foods (designated core foods) contributing to the selenium intake of the US population.

f Feed crops are typically those customarily consumed by the animal, growing on normal cultivated or uncultivated fields without the purposeful addition of selenium. Feed ingredients are processed feed and food crops and byproducts which together with concentrates and premixes constitute the mixed diet.

g Wide range of tissues and fluids from mainly swine, poultry, sheep and cattle in normal and seleniferous regions.

h Wide range of tissues and fluids from healthy and diseased subjects throughout the world.

i Examples of hair selenium concentrations of subjects in high selenium areas of the Peoples' Republic of China without evidence of selenosis (3.7 mg/kg, $n = 14$) and with chronic selenosis (32 mg/kg, $n = 65$).

Selenium, atomic number 34, atomic weight 78.96±0.03, is the third element in the oxygen group (VI) of the periodic table. Six stable naturally-occurring isotopes have mass numbers 74, 76, 77, 78, 80, and 82 with representative contributions of 0.9, 9.0, 7.6, 23.5, 49.6 and 9.4%, respectively. Nine radioactive isotopes have been reported with mass numbers 70, 72, 73, 75, 77, 79, 81, 83 and 84 with isotope 77 existing in both radioactive and stable variants. The element is a metalloid, existing in several allotropic forms. It occurs in chemical combination, widely dispersed in nature.

When studying the natural environment, one deals with a vast variety of materials encompassing rocks, minerals, soils, sediments, fossil fuels, plant and animal tissues, seawater, freshwater, precipitation, land and marine organisms, the atmosphere with its gaseous and particulate components, food, feedstuffs and human tissues. The occurrence and distribution of selenium in the environment and specific components have been reviewed in several major works. One of the most recent is a comprehensive and critical treatment in the volume "Occurrence and Distribution of Selenium" [1] which includes critical compilations of concentration data of endogenous selenium in natural materials. Therein are chapters dealing in-depth with plants and agricultural materials [2], foods [3], animal tissues [4], human tissues [5], geological materials and soils [6], freshwater systems [7], the marine environment [8] and the atmosphere [9]. Other reviews have appeared on selenium in soils [10], soils, plants and foods [11-14], freshwater [15], foods [16,17], and biogeochemical cycling [18-20]. Good detailed expositions on selenium in the agricultural context may be found in reports by Anderson et al. [21], Underwood [22], Moxon [23,24], and Rosenfeld and Beath [25] as well as in proceedings of conferences on selenium or trace elements [26-33].

Selenium is widely dispersed in nature, occurring generally in small amounts but over extreme concentration ranges in all phases, compartments and materials. The crustal abundance of it is 0.05 mg/kg, with a mean level in world soils of 0.4 mg/kg. The atmosphere contains about 1 ng/m^3 of particulate selenium and seawater has about 0.05 μg/l. In order to select and evaluate appropriate methods of analysis it is instructive to be cognizant of selenium concentrations anticipated in various environmental materials. For this purpose, Table 16.1 summarizes total selenium concentration ranges typically observed in a wide variety of natural materials. These values were adapted from the extensive compilations of literature data presented in ref. 1 and reflect ranges of mean or median concentrations to reduce effects of extreme values. Even so, total selenium concentrations vary by a factor exceeding 10^{10} from 0.0000047 mg/l in precipitation to 200 000 mg/kg in minerals, underlying the requirement for different approaches to selenium determination.

Selenium has a toxic-essential duality being necessary for animal and human well-being at low concentrations and leading to health impairment and death at higher toxic doses. Prior to the discovery in 1958 that it is a nutritional requirement, selenium's biological action was considered only acutely toxic and earlier reviews [34] cover only this facet [35-39]. During the intervening three decades research has emphasized selenium deficiency in animals and lately in humans [37,39-45] and a number of reviews of the biochemistry of selenium have appeared [25-27,37,40,45-54]. Selenium's toxic and essential biochemical roles thus impart to it a broadly-based environmental significance with the concomitant demands for varied analytical approaches.

16.2. ANALYTICAL METHODOLOGY REVIEWS

Both the basics and applications of the analytical chemistry of selenium have been subject to numerous reviews. A survey of some of those appearing during the past three decades, including books, chapters and review articles dealing with sampling, sample handling and decomposition and selenium measurement are recorded and discussed in this section.

16.2.1. Analytical method nomenclature and codes

In order to aid in the succinct and unambiguous presentation of methodological information in this section and throughout this chapter, this may be a good point at which to introduce terms used. In the context of this chapter, abbreviations for generic terms used for materials analyzed and analytical methods together with their definitions are as follows:

Materials

FOO   Foods, diets.

FEE   Feeds, feed crops, forages, roughages, animal diets, premixes, concentrates.

PLA   Plants, plant tissues, including fluids, crops, forestry materials and products.

ANI   Animals, animal tissues (including faeces, excluding fluids).

HUM   Human tissues, clinical specimens (including faeces, excluding fluids).

FLU   Animal and human fluids, fluid clinical specimens, blood, plasma, serum, urine, milk, lymph, sweat, saliva, other body fluids, blood cells.

BMA   Biological materials – any combinations of FOO, FEE, PLA, ANI, HUM, FLU.

SOI   Soil, fractions, inorganic and organic components, microbes, soil biota, humic/fulvic acids, amendments such as fertilizers, manure, sewage sludge and wastes.

GEO   Minerals, sediments, rocks, ores, coal, ash, fly ash, other mineralogical and geological materials, meteorites, lunar material.

WAT   Water, river water, fresh water, drinking water, seawater, rain, snow, ice, suspended sediments/particulates.

ATM   Atmosphere, air particulates, dry fallout, gases.

ORG   Other organic materials, organic compounds, oils, gasoline, petroleum products, proteins, enzymes.

WAS   Waste products (agricultural, residential, industrial), sewage, sludge, effluents, wastewater, refuse, manure, municipal waste, domestic waste, industrial waste, agricultural waste, chemical waste, pollutants.

Analytical methods

AAS   Atomic absorption spectrometry including flame atomic absorption, hydride generation and electrothermal atomization atomic absorption spectrometry with graphite furnace, carbon rod or tantalum strip.

EMI   Emission spectrometry (excluding plasma emission, and X-ray emission), emission spectrography, optical emission, flame emission, flame photometry, atomic emission, arc/spark emission, molecular emission, molecular emission cavity analysis (MECA), graphite furnace atomic emission.

FLR   Fluorescence spectrometry (excluding X-ray fluorescence) atomic fluorescence, flame atomic fluorescence, molecular fluorescence, fluorometry, spectrophotofluorometry, phosphorimetry.

ICP   Inductively-coupled plasma atomic emission spectrometry, direct current plasma atomic emission spectrometry.

CHR   Chromatography (all chromatographic techniques), ion-exchange chromatography, ligand-exchange chromatography, column chromatography, liquid-liquid chromatography, liquid-solid chromatography, affinity chromatography, gel-permeation chromatography, adsorption chromatography, high-performance liquid chromatography, gas chromatography, paper chromatography, thin-layer chromatography, high-performance thin-layer chromatography.

LAS   Light absorption spectrometry (molecular absorption), colorimetry, flame molecular absorption, reflectance spectrometry, turbidimetry, nephelometry.

MAS   Mass spectrometry (MS), inductively-coupled plasma MS, spark source MS, chemical ionization and electron impact MS, ion microprobe mass analysis, isotope dilution MS, gas chromatography-MS, liquid chromatography-MS.

NUC   Activation analysis, neutron activation analysis, instrumental neutron activation analysis (INAA), neutron activation analysis with radiochemical separation (RNNA), photon activation, charged particle activation, liquid scintillation counting.

References on p. 505

480

ELE	Electrochemical techniques, voltammetry, coulometry, electrogravimetry, electrodeposition, anodic stripping voltammetry, cathodic stripping voltammetry, polarography, ion specific electrode.
ELX	Electron spectrometry, X-ray spectrometry, X-ray photoelectron spectrometry = electron spectroscopy for chemical analysis, ultraviolet photoelectron spectrometry, particle induced X-ray emission, X-ray emission, X-ray fluorescence.
GRA	Gravimetry.
VOL	Volumetry, titrimetry.

Sample treatment/element isolation methods

SAM	Sampling, sample storage and preservation, drying, digestion, ashing, contamination, losses, material stability, solution stability.
SEP	Separation, isolation, concentration, precipitation, distillation, solvent extraction.

Abbreviations for specific determinative methods

INAA	Instrumental neutron activation analysis.
RNAA	Neutron activation analysis with radiochemical separation.
FAAS	Flame atomic absorption spectrometry.
EAAS	Electrothermal atomization atomic absorption spectrometry.
HAAS	Hydride generation atomic absorption spectrometry.
ICPAES	Inductively-coupled plasma atomic emission spectrometry.
HICPAES	Hydride generation inductively coupled plasma atomic emission spectrometry.
DCPAES	Direct current plasma atomic emission spectrometry.
IDMS	Isotope dilution mass spectrometry.
PIXE	Particle induced X-ray emission spectrometry.
F	Fluorometry (molecular).
GC	Gas chromatography.
LAS	Light absorption spectrometry.
XRF	X-ray fluorescence.
P	Polarography.
ASV	Anodic stripping voltammetry.
CSV	Cathodic stripping voltammetry.
HPLC	High-performance liquid chromatography.
HPTLC	High-performance thin-layer chromatography.
MECA	Microwave emission cavity analysis.
IC	Ion chromatography.
ICPMS	Inductively-coupled plasma mass spectrometry.
SSMS	Spark source mass spectrometry.
PAA	Photon activation analysis.
GC-MS	Gas chromatography-mass spectrometry.

16.2.2. Review articles

Table 16.2. summarizes more recent reviews bearing on the determination of selenium in environmental materials, covering a range of sample treatment and selenium measurement methods in a variety of natural matrices. References to earlier methods and reviews can be found therein. Although a few of these reviews are multielement, treating selenium as one of several elements, most [55,57-69, 71-73, 76-80, 82-84] are dedicated to this element. Three articles [57,72,84] review specific methods of analysis.

The determination of selenium at ultra-trace levels gas chromatography is reviewed by Dilli and Sutikno [57], Verlinden et al. [72] cover atomic absorption spectrometry and a technical report by Gunn [84] is a comprehensive, critical review of the determination of arsenic and selenium in raw and potable waters by hydride generation atomic absorption spectrometry. Methodological topics covered in the reviews include sampling, sample preservation and storage, sample decomposition and concentration, contamination considerations, selective separation, reduction of Se(VI) to Se(IV), data quality control and reference materials and determination methods such as gravimetry, volumetry (titrimetry), polarography, stripping voltammetry, light absorption spectrometry (colorimetry), fluorometry, emission spectrometry, X-ray fluorescence spectrometry, particle induced X-ray emission, instrumental and radiochemical separation neutron activation analysis, atomic absorption spectrometry (flame, electrothermal atomization, hydride generation), mass spectrometry (combined with gas chromatography, isotope dilution, spark source) and chromatography (ion, gas, high-performance liquid and thin-layer). Eleven reviews deal with one or more specific materials or classes of materials. Natural or potable water are the subjects of four reviews [59,60,71,84], air is treated in one [60], the class of biological materials is the subject in four articles [60,66,69,76], the determination of selenium in foods is covered in [70,74,82] and its determination and speciation in human urine are reviewed in [73]. Natural materials covered in all of the listed reviews include air particulates, animal and human tissues and fluids, feedstuffs, foods, plants and agriculture materials, water, soil and other geological materials and waste products. Major reviews on the analytical chemistry of selenium relevant to natural materials are [57,60,69,71,73,76-78,80,82,84] as well as the in-depth treatments of general analytical considerations by Green and Turley [55] and Nazarenko and Ermakov [79].

The following may be listed as critical reviews of selenium determination: refs. 55, 59, 60, 69, 71, 73, 76, 78, 79, 84, several discussing sources of error of given procedures, evaluating methods and suggesting analytical details. Green and Turley [55], Nazarenko and Ermakov [79] and Olson et al. [76] include specific analytical recommendations. The report by Gunn [84] formed the basis for the selection of a published procedure for further evaluation to yield a suitable method of monitoring arsenic and selenium in raw and potable waters.

References on p. 505

TABLE 16.2

ANALYTICAL METHODOLOGY REVIEWS

Review articles deal with the measurement of total selenium (main topic) as well as selenium species and compounds; major items covered are included here. Abbreviations are defined in text.

Materials[a]	Methods[b]	Year	Ref.
Org ani soi pla	sam sep gra vol ele las emi	1961	55
geo	sam nuc elx aas chr	1985	56
–	chr sam	1984	57
org bma	sam gra vol	1976	58
bma wat	sam flr aas	1984	59
atm bma wat	sam las aas chr nuc	1974	60
–	sam nuc mas flr chr aas	1989	61
–	chr mas nuc las flr ele	1989	62
fee pla ani flu	sam las flr aas chr ele mas nuc elx	1983	63
fee pla ani flu soi geo atm wat	sam sep gra vol ele las aas elx nuc	1974	64
soi geo pla ani fee foo wat atm	sam sep gra vol ele las flr aas chr nuc	1976	65
bma	sam sep nuc flr las ele aas	1967	66
org bma	sam sep gra vol ele las flr aas nuc	1973	67
–	sam sep gra vol	1962	68
bma	sam aas flr ele elx nuc	1984	69
foo	sam sep las flr las aas ele elx nuc chr	1977	70
wat	sam las flr aas elx, icp, nuc, mas, chr ele	1982	71
–	aas	1981	72
flu	sam las flr aas icp elx nuc chr ele vol	1984	73
foo	sam aas	1979	74
–	chr emi flr aas las	1984	75
bma	sam sep las flr aas chr ele vol gra mas nuc elx	1973	76
atm bma flu wat soi	sam nuc aas chr las ele elx	1981	77

TABLE 16.2 (continued)

Materials[a]	Methods[b]	Year	Ref.
wat atm bma	sam sep las flr aas icp emi elx nuc ele chr	1983	78
geo wat soi bma	gra vol las flr ele emi aas elx nuc sep	1972	79
hum flu wat	sam sep flr aas ele chr nuc emi elx	1988	80
–	chr sep	1974	81
foo	las flr chr elx nuc ele aas	1981	82
wat	aas chr	1981	83
wat	aas sam	1981	84
hum flu foo soi pla wat was geo	aas	1985	85

[a] The term bma is used when emphasis is not placed on specific materials but a variety of biological materials are alluded to; –: material not mentioned or of little relevance to thrust of article.

[b] The term sep is used when the separation of selenium receives pronounced treatment in the article. Catalytic methods utilizing spectrophotometry are included in las.

16.3. ANALYTICAL APPLICATIONS

In order to get a perspective of the analytical methodologies used for the measurement of selenium in natural materials, a literature survey was made. Those publications dealing with method development and/or application to actual materials as opposed to those dealing with theoretical and pure model solutions were retrieved and assessed. A total of 690 publications from among 4000 between 1960 and 1988 were selected. Although this is not a comprehensive survey covering all publications, method and materials, and is biased toward environmental materials, it should provide a representative picture of the applied analytical situation. The reports were perused and assigned key words pertaining to generic method, several specific methods, sample treatment, and classes of materials analyzed, as defined previously. In addition, information was gathered on whether the thrust was mainly method development or application or both and whether the method entailed separation of selenium from the matrix or solution prior to measurement. This information was sorted to provide an overview of analytical applications and trends as presented below.

A numerical summary of some information is provided in Table 16.3 and graphical presentations are depicted in Figs. 16.1-4. From Table 16.3 we observe that wet digestion

TABLE 16.3

SELENIUM ANALYTICAL PUBLICATIONS SUMMARY FOR 1960-1988

Based on publications retrieved and assessed by the author of this chapter.

Topic of publication	No. of publications	%
All publications	690	100.0
1 Wet digestion	394	57.1
Dry ashing	34	4.9
No sample treatment	182	26.4
2 Method development	226	32.8
Method application	385	55.8
Development & application	64	9.3
Selenium separation	458	66.4

(acid decomposition) is by far the preferred mode of sample treatment prior to determination of selenium, being favoured over dry ashing by a factor of almost 12. Analysis without any sort of treatment, save for drying and perhaps fine grinding, is the second most common sample treatment, used almost exclusively in INAA, XRF, and PIXE techniques, but also increasingly in the EAAS analysis of water and biological fluids. Method application is somewhat more dominant than method development according to this author's judgement of the publications. Subtotals within groups 1 and 2 do not add up to 100% due to non-reporting or insufficiently clear presentation in the publications. Separation of selenium from the digest was required in most instances.

Fig. 16.1 presents a numerical distribution of publications with general classes of analytical methods and sample treatment, specific methods and materials. In this and the other figures, publications have been grouped into the three decades 1960–69, 1970–79, 1980–88 (the latest year for which information was retrieved) as well as the overall period 1960–88. The general classes of methods reflect the dominant ones for selenium determination reported during this time period and include methods based on non-ICP emission spectrometry, gravimetry and volumetry. Sample treatment procedures are included here. Specific methods are subsets of the general categories: HAAS, EAAS (AAS); GC, HPLC (CHR); POL, SV (ELE); XRF, PIXE(ELX); INAA, RNAA(NUC). Materials mentioned in the reports have been assigned to the 11 general categories represented in Fig. 16.1C.

From Fig. 16.1A we see that analytical methods based on atomic absorption spectrometry, fluorometry and neutron activation analysis are generally the most popular as is the wet digestion procedure of sample treatment. A dramatic increase has occurred over 1960–88 in the application of atomic absorption-based procedures following development

DISTRIBUTION OF PUBLICATIONS

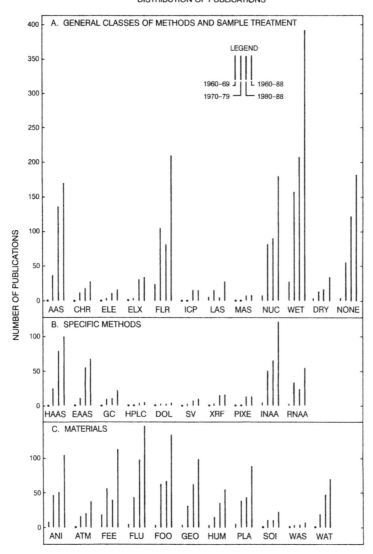

Fig. 16.1. Distribution of publications dealing with the determination of selenium by different methods in environmental materials grouped into time periods between 1960 and 1988. (A) General classes of analytical methods as well as sample treatments defined in the text. (B) Specific methods as subsets of some of the generic methods in (A). (C) Various classes of materials as defined in the text. Data represented by a point (•) at 0 ordinate indicate 0 or 1 publication. Some less common methods employed during this period such as emission (non-ICP), gravimetry and volumetry have been omitted.

in the early 1970's of hydride generation and electrothermal variants. Methods based on fluorometry passed a peak in 1970–79 but are still strongly represented in 1980–88, while usage of those based on neutron activation analysis, in use from at least the 1960's, continues to increase. No sample treatment and treatment by wet digestion have a large number of adherants with an increasing trend in both. A rise in usage of dry ashing is also evident but the numbers of publications reporting this procedure is relatively miniscule. The technique of light absorption spectrometry (LAS) shows a small number of reports passing through a maximum in 1970–79. Techniques based on chromatography, electrochemistry, electron spectroscopy, plasma emission spectrometry and mass spectrometry, not represented at all in 1960–69 (the latter three also not in 1970–79), all exhibit small but increasing numbers of publications. The reader may refer to Futekov et al. [86] for method trends based on analysis of literature citations in abstracts.

With respect to specific methods of analysis (Fig. 16.1B) those based on hydride generation and atomic spectrometry with flame or electrothermal decomposition, electrothermal atomization atomic absorption spectrometry, instrumental and radiochemical separation neutron activation analysis contain the most publications. The first three show increasing trends with time whereas the number of reports on RNAA usage peaked in the 1970–79 period. For the less prominent methods there are continuing increasing usage trends for especially gas chromatography, X-ray fluorescence and particle induced X-ray emission spectrometry.

Arranged by material (Fig. 16.1C) it is apparent that most materials reflect a good number of publications with the fewest number of papers reporting work on soil, waste and atmospheric particulate materials. With the exception of feeds, an increasing number of publications is evident for all materials.

A complementary perspective is offered by the presentation of this information expressed as a percentage in Fig. 16.2 and 3. In Fig. 16.2 it is evident that analytical procedures based on fluorometry, light absorption spectrometry and neutron activation analysis dominated in 1960-69, accounting for a total of in excess of 90% of all methods, with fluorometry by far the single most dominant at 60%. In 1970–79 fluorometry and neutron activation retained their dominance being joined by atomic absorption spectrometry with these three accounting for a total of almost 90% of all methods reported. The same three carry their dominance over to the last decade of the survey with atomic absorption spectrometry now leading the way. Other methods, notably chromatography and electrochemical ones have simultaneously gained some importance during this time for the measurement of selenium. With respect to sample treatment, the bulk of application always comes from wet digestion with, however, a significant trend toward those techniques requiring no sample destruction prior to determination.

Percentage distribution of publications plotted by material in Fig. 16.3 presents graphically the relationship among materials and trends with time. In 1960–69 feedstuffs was the most common material analyzed with animal and plant tissues second and third, respectively. No reports were retrieved dealing with atmospheric or soil samples. Within the next two decades somewhat more uniform distribution among materials resulted with foods being dominant in the period 1970–79 and animal and human fluids taking the lead in 1980–88.

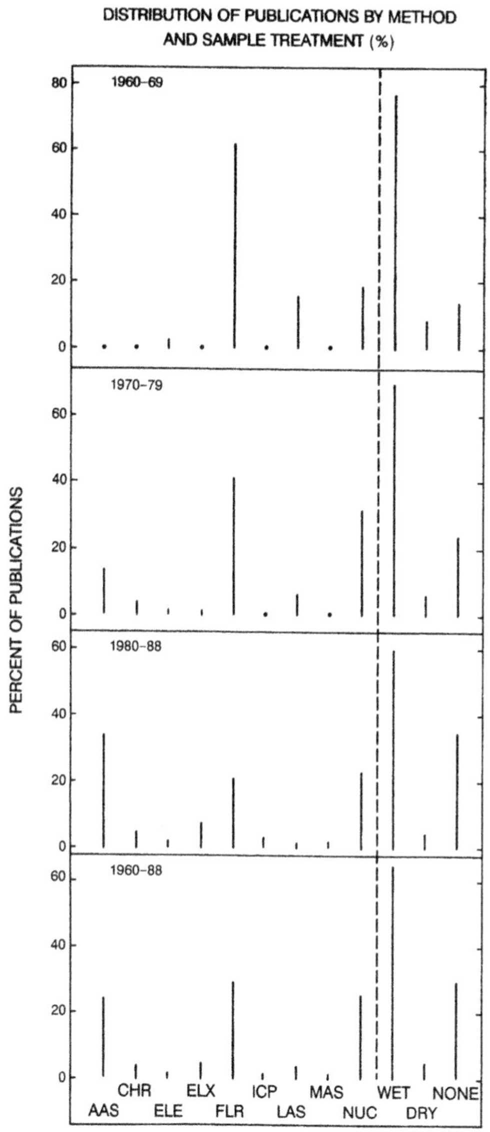

Fig. 16.2. Distribution (%) of analytical selenium publications by method and sample treatment grouped by time periods within 1960–1988. Points represent 0% or a very small value not plottable on the scale of the figure.

In order to get an impression of possible relationships between specific materials and analytical methods, the literature was sorted by all possible method/material keyword combinations. The percent rate of publication is plotted in Fig. 16.4 for each material as a function of analytical method for each decade for all materials and the overall time period.

488

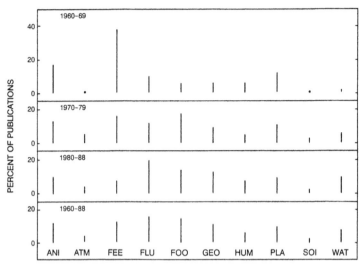

Fig. 16.3. Distribution (%) of analytical selenium publications by classes of materials grouped by time periods within 1960–1988.

Percentages total 100% within each material and period. This figure can easily be examined to provide information on (1) the relative emphasis on different methods for a given material, (2) the trend of (1) with time, (3) the relationship of a given method over different materials, (4) the trend of (3) with time, giving some striking observations.

Fluorometry is generally the most widely used method for most materials with the clear exception of atmospheric and geological materials, and human tissues. Neutron activation analysis has been and continues to be the method of choice of analysts dealing with atmospheric materials and human tissues while the three methods, atomic absorption, fluorometry and neutron activation are generally the most frequently utilized. Non-destructive techniques as neutron activation and electron spectroscopy (XRF and PIXE) are the two preferred techniques for the analysis of atmospheric particulates. Electron spectroscopic techniques are also applied to human tissue, soil and water samples while electrochemical techniques show, application to animal tissues and logically to water. Light absorption spectrometry shows most application to geological, plant and soil materials. With time, the trend is a decrease in fluorometric and spectrophotometric usage, a dramatic increase in the development and application of atomic absorption-based methods, a rather steady application of neutron activation analysis and small inroads being made by other methods. Method usage pertains not only to material applicability but reflects also method availability and analyst preference and experience.

DISTRIBUTION OF PUBLICATIONS BY
MATERIAL AND METHOD (%)

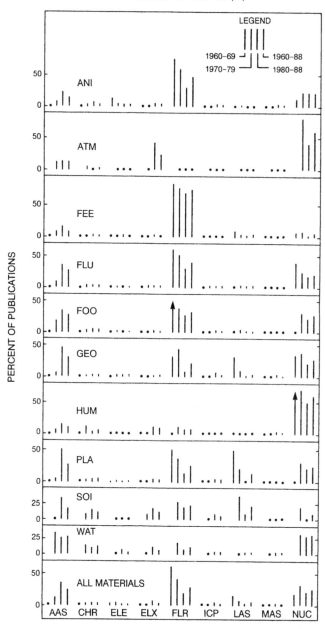

Fig. 16.4. Distribution (%) of analytical selenium publications by material and method grouped by time periods within 1960–1988. Within each material group, summation equals 100% within a given period. A blank indicates insufficient information (too few publications), a point represents 0–1%, and an arrow indicates a value off-scale.

16.4. SAMPLING, SAMPLE HANDLING AND STORAGE

Many steps make up the activity of chemical analysis. In the overall scheme of assigning an analytical result to a laboratory sample or ultimately to the population of samples or the environmental compartment of interest, consideration of all of the following steps is essential for the generation of acceptable information:

- presampling considerations and planning,
- sample collection (sampling),
- sample handling (including storage, preservation and transportation),
- reduction to laboratory sample,
- sample treatment (decomposition, preconcentration, separation),
- analyte measurement,
- data handling and
- interpretation.

Many sampling related steps occur early in the scheme, reflecting the effort necessary before actual determination of the analyte. Especially in the trace element field, analytical results are generally much more severely affected by the sampling and sample handling and preparation techniques than by the measurement. Designing a proper sampling plan and taking representative samples and maintaining their integrity are absolutely essential to guaranteeing correct and proper analytical information. Chapters 2–4 cover a general treatment of sampling and sample storage pertinent to inorganic trace analysis; a brief discussion of sampling more or less specific for selenium is presented in section 16.4.1 in this chapter. Items related to sample and analyte solution preservation and storage are included in section 16.4.2.

16.4.1. Sampling

Although little has appeared in the way of review articles regarding sampling specifically targetting selenium, several general reviews of relevance may be mentioned some of which touch upon this element. Kratochvil and Taylor [87] give a brief basic overview including the underlying statistics applied to the sampling of solid materials. Sampling methodology specific to soils is incorporated into the book on methods of soil analysis [88], while a detailed exposition of the theory and practice of sampling geological materials is presented by Ingamells and Pitard [89]. A critical review of contamination of clinical materials collected for trace element determinations is included in the review on trace elements in human body fluids and tissues by Versieck [90]. A thorough discussion of sampling and sample preparation methods for the analysis of trace elements in biological material is presented by Sansoni and Iyengar [91]. The literature on sampling and handling of environmental materials has been reviewed in ref. 92. Sampling and storage of natural water for trace metals has been critically and thoroughly reviewed by Sturgeon and Berman [93] and marine sampling has also been covered in-depth in a report by Erickson [94]. The collection of papers resulting from the 7th Materials Research Symposium published as a two-volume proceedings "Accuracy in Trace Analysis: Sampling, Sample Handling, Analysis " contains an extensive number of relevant contributions [95].

It is apparent from the literature and experience that unlike many other trace elements for whose determination materials must be sampled and handled with extreme care in order to avoid contamination, selenium does not pose such a problem. Only the usual, reasonable care is required to obtain acceptably accurate selenium levels. The usual good trace field and laboratory practices must be observed to prevent adsorption on vessels and pseudocontamination resulting from uncontrolled variations in time and sampling methods.

Selenium occurs in the environment in different oxidation states and compounds as well as distributed in various "phases" (e.g. true solution, colloidal dispersion and particulate in natural waters). Thus, for example, it has been determined in particulate and vapour phases in the atmosphere [9,96], as specific organo selenium compounds in the air [97], total selenium, Se(VI), Se(IV), Se(–II, 0) (colloidal), in freshwater and seawater [7,8] and organic compounds in seawater [8]. Sampling approaches for different materials which have been applied to selenium (often concurrently to other elements as well) are mentioned below.

With respect to the sampling of air, consideration should be paid to the occurrence of selenium as a variety of compounds in the gaseous or vapour state as well as the presence of the element in the particulate or aerosol phase in addition to measurement of total concentrations. Different sampling techniques capture different components to varying degrees. Active (as opposed to passive fallen dust collectors) atmospheric sampling devices to collect suspended particulate matter are of five types: cascade impactors using inertial particle size, cyclones based on centrifugation, impaction and filtration, liquid impingers using absorption, electrostatic precipitation and filter trapping by particle size. Although some trapping of vapour phase element will occur with most of these collectors, liquid impingers and cryogenic condensation traps are the proper ones for vapour phase collection.

Very few reports are in the literature dealing with air sampling for specifically measuring selenium. Two general reviews covering atmospheric sampling [98,99] may be mentioned with the former reviewing aerosol collection using the high volume (glass fiber filter) sampler, the dichotomous sampler (aerodynamic separation into two fine and coarse fractions) and the cascade impactor. The second review is an in-depth treatment of sampling of gases, vapours and particulates in air using for gases and vapours, whole air samplers, liquid impingers and bubblers, and solid sorbents, and for particulates, filters, impingers, cyclones and elutriators and impactors.

Several publications deal with the collection of atmospheric samples for selenium (and other elements) determinations in aerosols [100-102], as gaseous organic compounds [97,103,104], and in both vapour and particulate phases [96,105-108]. Identification and determination of alkyl selenide compounds was facilitated by their cryogenic trapping below –130°C [97,104] using the sampling system previously designed for organo lead compounds [109]. Hydrogen selenide was absorbed from the atmosphere in a liquid filled bubbler [103]. Kitto and Anderson [108] used a teflon filter for particle collection preceded

References on p. 505

by LiOH/glycerol-treated filters in a stacked filter arrangement to simultaneously collect particulate and gas samples. Mosher and Duce [96] used a system consisting of a Nuclepore filter for particle collection followed by activated charcoal filters or tubes to collect the vapour phase species. Vapour phase selenium dioxide, dimethylselenide, dimethyldiselenide and elemental selenium were not adsorbed by the filter but were completely retained on the charcoal adsorbents.

The types of sampling devices available for collecting natural water samples for trace analysis has been thoroughly reviewed by Sturgeon and Berman [93]. No single type is suitable for the various sampling techniques required for collecting water from the surface microlayer, surface and deepwater components. A variety of sampling devices are available: screens, drums, plates and a bubble interfacial microlayer sampler (BIMS) for the surface microlayer, sampling bottles and pumps for surface water, and a range of specially constructed bottle samplers with associated transport and collection mechanisms for deepwater sampling (Van Dorn, Niskin, transparent plastic Nansen, NIO bottles, Go-Flo, CIT sampler, Seakem sampler and NBS sampler). Reports including coverage of sampling for the determination of selenium in natural waters include those by Feder [110], Slawson, Jr. [111] and Rossmann and Barres [112]. Feder [110] conducted a reconnaissance geochemical survey of surface and ground waters of Missouri collecting water samples in acid-washed polyethylene bottles filled either by immersion in springs or from a faucet connected to well pumps. A 2-l sample was filtered through 0.45 μm filters then acidified with nitric acid. Slawson [111] investigated well water sampling by pumping and bailing and mentions the need to tailor the sampling protocol to the nature of the site. In a study of trace element concentrations in near-surface waters of the Great Lakes, Rossmann and Barres [112] filled 2-l acid-cleaned polyethylene bottles facilitated by a pressurized filtration tank. Filtration was through 0.5 μm Teflon filters by vacuum.

There are several publications on the collection of solid samples. They are of general relevance dealing mainly with geochemical and soil materials but touching on plant materials [88,89,113-117]. Several [88,89,113,115] treat planning and statistical aspects and sources of error necessary to consider before actually going out to collect the samples. Reference to determination of selenium in collected samples is made by Kubota [118], Henry and Kapadia [113], and Severson et al. [119]. Kubota [118] refers to large scale sampling programs undertaken since the 1930's to systematically study selenium in soils as a toxic factor to animals in the US southwest. The Bureau of Economic Geology report [113] summarizes sampling design and statistical model considerations for soil collections. The USGS report [119] considers experimentally a topic not considered sufficiently seriously by soil researchers, contamination of the sample by the sampling technique. A comparison is presented of analytical results for 30 elements, including selenium, in soils sampled in parallel by two sampling tools and associated techniques. The amount of contamination introduced by the USGS stainless-steel bucket-type Auger technique relative to the US Fish and Wildlife polyvinylchloride pipe-in-split barrel steel corer technique was not great enough to be measurable by the analytical techniques used.

In the biomedical and clinical field, considerations and planning early in the analytical scheme take on substantially more significance than in most other disciplines. Iyengar

[120] has thoroughly described the nature and effects of such "presampling" factors. He points out that although identification of analytical error arising from inappropriate methodology insufficiences in sampling and sample handling have received attention recently, an understanding of the impact of presampling factors on the entire analytical sequence is lacking and needed since the validity of the sample becomes questionable if the influence of presampling factors is neglected. Presampling factors may be defined as events associated with biological specimens *in situ* and before the arrival of the samples at the laboratory for analysis. A host of circumstances such as biological variation [genetic predisposition, long-term physiological influences (age, sex, geographical location, environment, diet, pregnancy, lactation), short-term physiological influences (circadian rhythms, recent meals, posture, stress), seasonal changes (physiologic and climatic)], postmortem changes (cell swelling, imbibition and autolysis), intrinsic errors (medication, hemolysis, subclinical conditions, medical restrictions) bear on the elemental makeup of the sample. The sampling of biological materials for trace element determinations has been detailed by Sansoni and Iyengar [91]. A critical review by Versieck [90] on trace elements, including selenium, in human body fluids and tissues treats sample contamination by collection devices and in this and other works [121,122] he points out the profound influences of sampling technique on trace element levels. An exposition of problems concerning sampling and sample preparation of platelets is documented by Kiem et al. [123].

Sampling requirements for the activity of specimen banking, the collection and storage of environmental specimens for current and retrospective analysis, are especially stringent and demanding. It is here that the concepts and practices of sample collection and long term storage take on a commanding role. Chapter 3 provides an introduction.

16.4.2. Sample handling, sample and analyte preservation and storage

Virtually every sample suffers to some extent the ravages of handling, preservation and storage prior to undergoing analysis in the laboratory. Steps within these activities include: separation of artifacts, drying, sterilization, preservation, grinding, homogenizing, subsampling, preparation of laboratory samples, transport and storage (containment). Some of the relevant variables with respect to sample integrity and analyte behaviour to be aware of are: adsorption, desorption, contamination, volatilization, photochemical or biological degradation, changes in speciation or oxidation state, temperature, time, nature of container and cleaning, agitation and phase separation (filtration, crystallization, extraction etc.). The systematics of handling, preservation and storage need to be considered.

Reviews or books which touch on these items are refs. 78,89,91-93,124-127 with the article by Raptis et al. [78] dealing specifically with selenium.

The relatively high volatility of selenium and its compounds necessitates consideration of losses of selenium by volatilization during sample handling and storage. Although earlier work with selenium accumulator plants suggested ambient temperature volatilization, several later studies on room temperature storage of forage crops and related plant materials [128-130] revealed, for up to about one year storage times, losses ranging from non-significant to 14% with most less than 5% compared to fresh or dried starting material.

Solid materials are usually dried for storage, grinding and for expression of analytical results on a dry matter basis. Lyophilization and oven drying procedures are used. Studies on losses by lyophilization of selenium from blood, components and solid tissues from the rat [131,132] and marine flora and fauna [133] indicated no significant losses and led to the suggestion that lyophilization is reliable for preparing such materials for total selenium determinations. Preconcentrations of natural water by lyophilization has been demonstrated acceptable for selenium as well as many other elements [134].

Effects of a range of drying conditions on selenium losses from a variety of materials have been reported. Collier and Parker-Sutton [135] summarize some literature and report no significant ($p = 0.01$) differences in the endogenous selenium contents of herbage dried overnight at 60°C and 100°C and for 32 h at 30°C. Using radioselenium, ^{75}Se, incorporated into forage and other plant species, Ehlig et al. [136] and Bisbjerg [129] observed on the average 98-99% recoveries after drying at 70°C for 30 h [136] and 18-230 h at 50°C as well as for an additional 24 h at 100°C [129]. In two other reports on agronomic plant species [128,130] no ^{75}Se was lost from rye grass drying for 4 days at 50°C but 4% was released after 4 days at 100°C [130]; in the case of mustard and radish no or small losses occurred below 50-60°C but greater (5–20%) losses were estimated at drying temperatures up to 100°C [128]. Radioselenium incorporated into oyster tissue showed only slight losses comparable to lyophilized controls when dried up to 105°C but showed appreciable losses when dried at 120°C [137,138]. Iyengar et al. [131] concluded that for ^{75}Se incorporated into rat blood and tissues, drying up to 105°C proved safe although statistically significant losses ($< 5\%$) were observed for drying at 120°C. Boiling, baking and broiling food preparation steps resulted in some appreciable losses depending on procedure and foodstuff [139]. Collection and partial characterization of volatile selenium compounds from alfalfa has been reported [140]. It appears that storage and drying considerations are only of moderate concern for this element and that storage of moderate temperature dried materials in tightly capped vessels is adequate. Correction of analytical results to a dry basis estimated by drying a separate subsample at some appropriate temperature less than 105°C is the usual way to proceed.

Selenium occurs in nature in a variety of compounds in oxidation states $+4$, $+6$ and -2. In order to estimate the element in these states steps must be taken to preserve oxidation and speciation during storage and processing. Some evidence has been presented [141] for decrease of selenide concentration in stored solid, milled, zinc selenide/silica mixtures. In order to prevent speciation changes in biogenic materials due to bacterial degradation, Cutter [142] immediately freezes such materials and later dries them at 40°C. Losses of selenium from oyster tissue by drying at 120°C has been explained by oxidation of the endogenous element to volatile SeO_2 [137]. Oxidation state changes are of even more concern in natural water samples. Acidification to pH 2 [143] or with 1 M HCl [144] of fresh or seawater and storage in pyrex or polyethylene containers was found to preserve both Se(IV) and Se(VI) satisfactorily whereas other work [145] indicated instability of Se(IV) stored in teflon containers with sulfuric acid.

Liquid sample storage entails consideration of sorption/desorption, contamination, and volatilization. Reviews have appeared covering the behaviour of various elements including selenium in model solutions [125], natural waters [78,92,93] and biological fluids [91]. Adsorption studies of Se(IV) from HNO_3, pH 3.8 and pH 7.0 solutions onto flint glass, pyrex and polyethylene surfaces for periods up to 15 days, showed losses from 1 to 8% increasing with increasing pH and between glass and polyethylene [146]. Insignificant losses of Se(IV) onto glass, polyethylene and polytetrafluorethylene were reported by Massée et al. [147]. Volatilization losses from polyethylene vials due to the formation of volatile compounds during irradiation for neutron activation analysis has been studied [148]. With respect to seawater, an early report [149] on 0.5 μg/l radioselenium in unacidified samples in pyrex showed no adsorption losses over six months. More recent work with fresh and seawater preserved to pH 2 or with 1 ml/l of 1 M HCl and stored in glass or polyethylene indicated good preservation of total selenium and its oxidation states for up to 4.5 months [143,144,150]. Storage of methyl species in air-tight containers was only good for one day before complete loss occurred [144]. Extensive tests by Cheam and Agemian [151] with 1 and 10 μg/l Se(IV) and Se(VI) in distilled water and harbour water in pyrex and polyethylene containers at pH 1.5, 5.4 and 7.2 showed preservation at pH 1.5 up to 4 months. These workers concluded that acidification to pH 1.5 in glass or polyethylene is satisfactory and although glass is better at pH 5.4 or 7.2, polyethylene containers are more convenient for handling and shipping. Storage of serum at −18°C for four years in cleaned polyethylene containers was satisfactory for selenium [152]. Hence preservation of aqueous samples with acid to pH <2 as for other trace elements is satisfactory for selenium.

The prospect of sample contamination with selenium from storage and handling equipment and the laboratory environment are nowhere near as severe as for other trace elements at the selenium levels usually encountered in natural materials. The usual good laboratory practices suffice. As the greatest concern will be contamination of the liquid sample or analytical solution during long term storage, reference to reviews and reports dealing with container cleaning will be of interest [92,134,145,147,148,153]. Most procedures, certainly the thorough one of Moody and Lindstrom [153] should be acceptable.

16.5. SAMPLE TREATMENT

For the majority of analytical procedures determining elemental composition, destruction of the sample matrix is a prerequisite to the measurement step. The element of interest must be recovered quantitatively without loss or contamination in a suitable form (speciation, oxidation state), free of the massive interfering original matrix or other components remaining after sample decomposition. Not one decomposition procedure works satisfactorily for all element-matrix combinations; hence an extremely wide variety of approaches are used. The major ones applied to the release of and measurement of total Se will be mentioned in this section.

Reviews of sample decomposition are available. Three books, one by Gorsuch "The Destruction of Organic Matter" [154], "A Handbook of Decomposition Methods in Analytical Chemistry" by Bock [155] and "Decomposition Techniques in Inorganic Analysis" by Dolezal et al. [156] are thorough general treatments as are other chapters [157-159] and a review by Tölg [160]. Several reviews dealing specifically with the determination of selenium [58-60,63-67,69,73,76-78,80], review sample decomposition procedures with seven of the listed references giving thorough reviews of sample treatments for this element.

A survey of sample treatments reported in the literature applied to environmentally-relevant materials prior to the measurement of selenium was made and is summarized in Table 16.4. Therein, matrix destruction methods are listed under four basic categories-wet digestion in open systems, wet digestion in closed vessels (pressure digestion), combustion and dry ashing or fusion. Matrices to which these treatments were directed are given, following the abbreviations defined in section 16.2 and several typical references are provided. As indicated previously, the volatility of selenium and its compounds has resulted in an overwhelming reliance on wet decomposition procedures as opposed to those involving dry ashing. Of the multitude of acid and other reagent mixtures reported in the literature, the most common ones are based on $HNO_3/HClO_4$, $HNO_3/HClO_4/H_2SO_4/H_2O_2$ and $HNO_3/HClO_4/H_2SO_4$. Many reports on the uses of these acid mixtures cite the development/adaptation researches of Olson et al. [174,180], Watkinson [171,172,282] and Levesque and Vendette [170] for $HNO_3/HClO_4$, Hoffman et al. [207] and Ihnat [196] for $HNO_3/HClO_4/H_2SO_4/H_2O_2$, and Fiorino et al. [194] for $HNO_3/HClO_4/H_2SO_4$. Oxidation with oxygen in a closed flask is often carried out according to procedures published by Allaway and Cary [257] and Gutenmann and Lisk (261].

Many research workers have conducted serious critical and in-depth investigations into sample decomposition techniques and procedures for selenium considering element losses, incomplete destruction of sample matrix and selenium compounds and conversion of selenium to oxidation states not amenable to detection by the analytical finish [171,175,178,180,193,196,264,266,283-291]. Gorsuch [283] did not include dry ashing techniques for selenium in his intensive study of the recovery of trace elements from various destruction techniques. The reason for this was virtually complete losses in the method of Middleton and Stuckey [292] using boiling to dryness with HNO_3. While this may have discouraged subsequent dry ashing studies, some have appeared. Studies involving dry ashing with or without incorporation of ashing aids $Mg(NO_3)_2$, $Mg(NO_3)_2/MgO$, $HNO_3/Mg(NO_3)_2$, indicated the possibility of selenium losses especially without ashing aids [130,155,224,277] but also good recoveries with ashing aids [224,271,275, 277,293].

Wet decomposition in open systems is by far the most widely used mode of matrix destruction for selenium determination and many researches have demonstrated the efficacy of acid digestions [162,171,175,178,180,193,194,196,206,216,224,283-285,287-291,294,295]. Nève and co-workers [285,288] carried out critical studies of some wet digestion methods for decomposition of biological materials for the determination of total selenium and Se(VI). They investigated digestion procedures of Shimoishi [296], Holak

[275] (dry ashing completion) and Ihnat [205] as well as other acid mixtures and observed that the $HNO_3/HClO_4/H_2SO_4/H_2O_2$ method of Ihnat [205] gave the best results for added and native selenium in a wide range of organic materials, and recommended it for total selenium. Of the three decomposition procedures, oxygen plasma ashing, dry ashing with $Mg(NO_3)_2$ and wet digestion with $HNO_3/H_3PO_4/H_2O_2$, Reamer and Veillon [224] found the latter the most satisfactory for speed, safety and efficiency. For the determination of exceedingly low ng/kg (parts per 10^{12}) quantities of the element, Piwonka et al. [294] advocate destruction with $HNO_3/HClO_4$. Unavailability of selenium to the final determination step occurs due to losses of the element by volatilization including that caused by sample charring, incomplete destruction of organo selenium compounds and oxidation to se(VI). The remedy to losses by sample charring is its avoidance by slow oxidation and/or by incorporation a larger volume of $HClO_4$ in the system $HNO_3/HClO_4/H_2SO_4/H_2O_2$ [196,283].

For proper reaction with the fluorescent reagents (e.g. diaminonaphthalene) in the fluorometric procedure, and with sodium tetrahydroborate in the hydride (H_2Se) generation AAS procedure, selenium must be present as Se(IV) in the solution. Release of the element from organic compounds and oxidation with acid reagents yields mixtures of oxidation states Se(IV) + Se(VI). Quantitative reduction of Se(VI) to the required tetravalent state Se(IV) is frequently accomplished by heating with HCl [132,144,162,171,174, 177,179,180,193,290,291,296-300] although H_2O_2 incorporated for digestion or specifically for reduction serves the purpose as well [196,207,209,301-305]. Incomplete decomposition of complex biological materials leaves chemical constituents including refactory organoselenium compounds which may be unsuitable for measurement of native selenium but added selenium as well and incomplete decomposition may be more serious than losses by volatilization [178,285,288]. Rigorous destruction with an acid mixture including $HClO_4$ is prescribed to yield suitable digests. Digestions based on $HNO_3/HClO_4$ [180] and $HNO_3/HClO_4/H_2SO_4$ with or without H_2O_2 [196,198] have been incorporated into recommended and official methods of analysis [179,193,209].

Wet decomposition in closed systems avoids losses by volatilization. Such digestions are often carried out using HNO_3 in PTFE, although uses of HNO_3 in combination with $HClO_4$, HF and other reagents have been reported and as well quartz or glassy carbon vessels have been described. A PTFE vessel within a stainless-steel encasement heated in an oven to 150°C is used in an official method [306,307] to digest foods with HNO_3 for the determination of Se together with other elements. Decomposition of human blood plasma [178] and biological materials [291] is effected with HNO_3 in a PTFE Perkin-Elmer Autoclave-3 system. It has, however, been reported [178,291,308,309] that organic residues remain following pressure decomposition with HNO_3; for proper selenium determinations an aliquot of the nitric acid solution is further treated with perchloric and sulfuric acids to obtain complete mineralization of the organic sample [291].

Combustion is carried out with oxygen either in a closed flask or in an open system whereby the volatilized selenium oxide is trapped externally. Closed flask combustion procedures are based on the Schöniger flask technique - developed by Schöniger [310] adapted and used for selenium determinations by several workers [256-258,261]. The

TABLE 16.4.

METHODS FOR MATRIX DESTRUCTION IN THE DETERMINATION OF SELENIUM IN BIOLOGICAL/ENVIRONMENTAL MATERIALS

Reagents[a]	Matrices[b]	No. of publ.[c]	Typical references[d]
Wet digestion in open systems			
HNO_3	*foo flu ani* geo hum pla was wat	11	161-166
HNO_3/H_2O_2	was wat	1	167
HNO_3/NH_4VO_3	pla	1	168
$HNO_3/HClO_4$	*fee flu ani pla foo* geo hum soi	127	169-180
$HNO_3/HClO_4/H_2O_2$	fee pla	2	181,182
$HNO_3/HClO_4/HF$	geo	6	183-188
$HNO_3/HClO_4/NaVO_3$	*fee* ani flu pla soi	4	189-192
$HNO_3/HClO_4/H_2SO_4$	*foo* ani atm flu geo hum wat	23	193-200
$HNO_3/HClO_4/H_2SO_4/H_2O_2$	*foo fee ani flu* pla hum geo	55	196,197,201-209
$HNO_3/HClO_4/H_2SO_4/Na_2MoO_4$	flu hum	1	210
$HNO_3/HClO_4/H_2SO_4/NH_4VO_3$	foo pla	2	211,212
$HNO_3/HClO_3/HClO_4$	ani geo pla	2	213,214
HNO_3/H_2SO_4	ani flu was wat	4	167,215-217
$HNO_3/H_2SO_4/HCl$	geo	1	218
$HNO_3/H_2SO_4/V_2O_5$	foo	1	219
HNO_3/H_3PO_4	geo	1	220
$HNO_3/H_3PO_4/H_2O_2$	ani flu foo hum pla	4	221-224
H_2SO_4/fuming HNO_3	geo org	1	225
H_2SO_4/H_2O_2	foo was wat	2	167,226
$H_2SO_4/HClO_4/Na_2MoO_4$	ani fee flu foo pla soi	10	210,227-233
$HClO_4$	foo geo	3	234-236
$HClO_4/H_2O_2$	fee	2	237,238
H_2O_2	wat	1	239
$HNO_3/HClO_4/KMnO_4/K_2S_2O_6/HF$	geo	1	240
$KMnO_4$	wat	1	241
$K_2S_2O_6$	was wat	3	167,242,243
$Na_2B_4O_7$	wat	1	143
Wet digestion in closed vessels (pressure digestion)			
HNO_3	ani foo hum pla was	5	244-247
HNO_3/HF	pla	1	248
$HNO_3/HClO_4$	flu	1	249
$HNO_3/Mg(NO_3)_2/HCl$	flu	1	250
$HNO_3/HCl/HF$	geo	3	251-253

TABLE 16.4 (continued)

Reagents[a]	Matrices[b]	No. of publ.[c]	Typical references[d]
Combustion			
O_2 in closed flask	*fee ani pla org* foo flu soi	19	254-261
O_2-open system volatilization	*geo* ani fee pla org soi	8	214,262-267
O_2-high frequency plasma	ani foo geo pla	3	224,236,268
Dry ashing or fusion			
$Mg(NO_3)_2$	ani fee foo pla	3	224,269,270
$Mg(NO_3)_2/MgO$	ani foo	2	271,272
$HNO_3/Mg(NO_3)_2$	ani fee flu foo geo hum pla	7	273-279
NaOH	geo	1	240
$NaOH/Na_2O_2/Fe_2O_3$	geo pla soi	1	280
$ZnO/MgO/Na_2CO_3$	geo	1	281

[a] Listing of reagents may not necessarily reflect actual sequence of usuage.

[b] Refer to text for abbreviation definitions. Matrices denoted in italics are most frequently cited in the literature; listing is generally in decreasing order of dominance.

[c] Total No. of references refer to the citation data base referred to in the text.

[d] Typical references are some of the more relevant ones for the more dominant matrices or all in the category if total No. was small.

oxygen flask technique entails an unfavourable sample to surface ratio; reports of both low and erratic [193] and good recoveries [311] have been reported. Dynamic open system volatilization as frequently reported [264,266,286,291,295], involves heating the sample in a stream of oxygen leading to rapid ashing and release of volatile SeO_2 which is trapped on a cold finger and subsequently dissolved in HNO_3. A wide variety of materials can be treated with good recovery of selenium. One other advantage is that the element is recovered free from interfering constituents.

16.6. DETERMINATION PROCEDURES AND RECOMMENDED METHODS

16.6.1. Determination procedures

The final major step in the analytical scheme is the actual determination or measurement of the analyte. This section will deal with determination procedures specific for selenium, mention official and recommended overall methods of analysis for total sele-

References on p. 505

TABLE 16.5.

DETERMINATION TECHNIQUES FOR SELENIUM IN BIOLOGICAL/ENVIRONMENTAL
MATERIALS

Technique	Matrices[a]	No. of publ.[b]	Typical references[c]
Fluorometry	*fee flu ani foo geo pla* hum soi was wat	209	170-172,174-176, 180,196,207,227, 257,284,312-314
Light absorption spectrometry (photometry)	*pla geo fee foo soi* flu	27	181,315-318
Hydride generation atomic absorption spectrometry	*foo geo pla ani fee flu wat* atm hum soi was	101	161,194,208,216, 225,243,319-325
Electrothermal atomization atomic absorption spectrometry	*flu wat geo foo* ami atm hum pla soi was	67	203,208,288,302, 326-334
Flame atomic absorption spectrometry	atm fee pla	4	237,335-337
Hydride generation atomic fluorescence spectrometry	flu geo	2	236,338
Instrumental neutron activation analysis	*hum flu foo atm ani wat pla geo* fee was	122	339-345
Radiochemical separation neutron activation analysis	*foo geo hum pla flu ani wat* atm fee soi	55	346-355
Gas chromatography	*foo flu hum pla wat* ani fee geo was atm	22	143,162,177,210, 277,279,356
High-performance liquid chromatography	*wat* flu soi	5	357-361
High-performance thin-layer chromatography	flu	1	362
X-ray fluorescence spectrometry	*atm wat* ani fee flu geo pla soi	18	363-368
Particle induced X-ray emission spectrometry	*flu hum* atm foo geo soi wat	14	369-374
Polarography	ani foo pla wat	5	195,375-377
Stripping voltammetry	ani fee foo geo wat	10	236,275,278,378-380

TABLE 16.5 (continued)

Technique	Matrices[a]	No. of publ.[b]	Typical references[c]
Inductively-coupled plasma atomic emission spectrometry	*geo flu* ani foo pla soi wat	15	185,249,381-385
Atomic mass spectrometry (SSMS, IDMS, ICPMS)	ani flu geo hum wat	6	166,222,386-389

[a] Refer to text for abbreviation definitions. Matrices denoted in italics are most frequently cited in the literature; listing is generally in decreasing order of dominance.

[b] Number of references refers to the citation data base referred to in the text.

[c] Typical of references are some of the more relevant ones for the more dominant matrices or almost all references in the category if the total number is small.

nium and touch upon speciation. Since determination procedures for selenium are covered quite thoroughly in the literature, this topic will be accorded only light review here.

For guidance to older historical methods of analysis reference may be made to publications cited in ref. 2. For the period covered in this chapter (1960–1988) reviews covering selenium include a book by Nazarenko and Ermakov [79], chapters in books or proceedings of symposia [33,55,61,62,64-68,76,80,159], a report [84] and articles in journals [57,58,60,71-73,77,78]. Verlinden et al. [72] review the determination of selenium by atomic absorption spectrometry, Dilli and Sutikno [57] review its determination at ultratrace levels by gas chromatography and Gunn [84] reviews the determination of arsenic and selenium in raw and potable waters by hydride generation atomic absorption spectrometry. Some review articles deal with specific materials; Gunn – raw and potable waters [84], Robberecht and Van Grieken – environmental waters [71], Shendrikar – air, water and biological materials [60], Bem and Raptis et al. – the environment and biological materials [77,78], Robberecht and Deelstra – urine [73], and Watkinson et al. [66], Olson et al. [76] – biological materials.

Using the literature data base referred to previously, a survey was made of determination procedures applied to environmentally-related materials and is summarized in Table 16.5. Specific measurement techniques are listed together with matrices to which these techniques were most commonly applied. The total number of publications from the population surveyed for this chapter are indicated as are some typical citations. During the three decade study period (1960–1988) fluorometry (molecular fluorescence) was the most popular measurement technique followed by instrumental neutron activation analysis, hydride generation atomic absorption spectrometry, electrothermal atomic absorption spectrometry and neutron activation analysis with radiochemical separation. A variety of other techniques based on photometry, chromatography, X-ray emission/fluo-

References on p. 505

TABLE 16.6.

DETECTION LIMITS OF SOME TECHNIQUES FOR SELENIUM DETERMINATION

Technique	Detection limit[a] (ng/ml or ng/g)	Ref.
Fluorometry	0.01–3, 0.5 ng, 10 ng	71,171,175,196
Light absorption spectrometry	10, 0–4000, 100–10 000, 5000–25 000, 300–10 000, 3000	77,390
Hydride generation atomic absorption spectrometry	0.25, 0.02–50, 0.005–3	71,84,208
Electrothermal atomization atomic absorption spectrometry	25, 0.02–36, 0.02–0.4	71,208
Flame atomic absorption spectrometry	50–1150, 82–3200	208
Instrumental neutron activation analysis	0.07–3, 0.5 ng, 0.5–100 ng	63,64,71
Neutron activation analysis with radiochemical separation	0.5, 8, 10, 20, 5 ng	67,346,349,350,353
Gas chromatography	0.02–100	57

[a] Detection limits have varying definitions in the literature and are often undefined. They refer either to the solution presented for determination or the original sample. Specifics of the analytical method and determination procedure conditions have profound effects on detection limit. Some data reflect extensive summaries.

rescence, electrochemistry and atomic emission and mass spectrometry round out the picture. The commonly reported materials are feed, fluid and animal tissues (fluorometry), human tissues, fluids and foods (INAA), food, geochemical materials and plants (HAAS), fluids, water and geochemical matrices (EAAS), and foods, geochemical materials and human tissues (RNAA). The reader is referred to the figures for more detailed relationships and trends. Table 16.6 presents a brief indication of detection limits of some of the more commonly applied techniques.

With the exception of instrumental procedures such as INAA, XRF and PIXE where the materials can be analyzed directly without treatment, most procedures require isolation of selenium to prevent interference by the other matrix constituents. The element is separated either directly from the material or most often from the solution resulting from decomposition. Most frequently used separation methods involve formation of H_2Se, extraction of selenium complexes (usually the piazselenoles), and chromatography. Reviews cover a variety of separation procedures [55,57,72,76,78,80,84]. Green and Turley [55] review thoroughly, methods used prior to the 1960's such as distillation of $SeBr_4$, coprecipation, ion exchange, cupferron extraction, electrolytic separation, and reductive precipitation. Gunn [84] and Dilli and Sutikno [57] cover gaseous hydride separation and piazselenole extractions in relation to atomic absorption and gas chromatographic determination steps, respectively.

Of the reviews on the technique of fluorometry, the earliest widely-used sufficiently detective technique for endogenous levels of selenium in natural materials [61,63,65-67,71,73,76,78,80], those of Watkinson [66], Olson et al. [76] and Robberecht and Van Grieken [71] are the most detailed. Olson et al. [76] conclude their article with description of three recommended methods, two based on this technique. The more recently developed technique of hydride generation atomic absorption spectrometry is reviewed in refs. 60, 61, 63, 65, 71-73, 77, 78, 80 and 84. Gunn [84] presents the most comprehensive review in a report dedicated to this technique applied to various waters. Verlinden et al. [72] and Robberecht and Van Grieken [71] also offer good treatment with the first of these devoted to atomic absorption spectrometric methods, the second concentrating on techniques for natural waters. Electrothermal atomization atomic absorption spectrometry is covered in refs. 63, 65, 71-73, 77, 78 and 80 with fairly in-depth coverage by Verlinden et al. [72], Robberecht and co-workers [71,73] and Raptis et al. [78]. Instrumental neutron activation analysis applications to selenium measurement are covered in [61,63,65-67,71,73,76-78,80], with relatively in-depth treatment by Olson et al. [76], Bem [77] and Watkinson [66]. Radiochemical separation neutron activation analysis is reviewed in refs. 64, 66, 67, 78 and 80. Reviews of other techniques are also available: light absorption spectrometry [55,60,63-67,71,73,76-78], gas chromatography [57,61,63,65,71,73,76-78,80], electrochemical techniques [55,63-67,71,76,78,80], X-ray fluorescence [63,64,71,73,76, 78,80], PIXE [78,80] and mass spectrometry [61,63,71,76].

The majority of selenium studies and determinations deal with the total concentrations of the element. Several older ones, however, have investigated volatile organoselenium compounds extractable from biological tissues and recent publications increasingly emphasize speciation characteristics of selenium, that is the distribution of the element among different oxidation states, compounds and phases. Speciation aspects are recently reviewed [6-9,62,71,73,78,84]. Chemical speciation in natural water is reviewed well by Robberecht and Van Grieken [71] who also summarize concentrations of Se(IV), Se(VI) and organoselenium compounds in drinking, fresh- and seawater, while chemical species in urine are discussed by Robberecht and Deelstra [73]. Tan and Rabenstein [62] discuss the analytical chemistry of organic and biochemical selenium covering selenoproteins such as glutathione peroxidase, selenoamino acids, and metabolites such as dimethylselenide, dimethyldiselenide, trimethylselenonium and hydrogen selenide. Berrow and Ure [6] review forms and distribution of selenium in soils, solubility and availability to plants. Distributions in freshwater systems of dissolved and colloidal selenium, Se(IV) and Se(VI) are treated by Cutter [7] and in seawater (including organic selenium) by Siu and Berman [8]. Particulate and vapor-phase selenium in the atmosphere is summarized by Mosher and Duce [9].

References on p. 505

16.6.2. Recommended methods

The overall method of chemical analysis as applied in the laboratory, encompasses sample decomposition analyte separation and measurement. For confident laboratory performance, it is advisable to follow a method validated and approved by an analytical agency or at least one recommended by researchers following in-depth critical studies. Fortunately for those in the selenium field, this elements' importance has led to a vast analytical research endeavour with among the countless methodology publications, a number of reports dealing with collaborative/interlaboratory studies, critical studies, and method recommendations. Collaborative studies designed to test specific methods have been conducted by Olson [180], Ihnat and co-workers [197,199,391], Holak [306] and Vaessen and Van Ooik [392]. Olson's [180] collaborative study of a fluorometric method for plants developed from that of Watkinson [282] and Ihnat's study of a similar fluorometric method [196,207] for foods were conducted under the auspices of the AOAC leading to official methods of analysis [179,209]. An extensive in-depth collaborative study by Ihnat and co-workers [199,391] of a hydride generation atomic absorption spectrometric method based on that developed by Ihnat and Miller [198] led to the recommendation that a hydride procedure in which free, unqualified choice of hydride generation equipment is permitted not be adopted as official. A later collaborative study by Holak [306] using a specific generator led to an official AOAC method [307]. Another study [392] tested the fluorometric method specifically for milk. Other interlaboratory studies in a different vien, intercomparing several different methods or for the purpose of assigning recommended concentration values for selenium in specific materials can be listed [193,393-398].

A number of methods manuals are available prescribing various approaches to the determination of selenium and the reader is directed to them to select an appropriate method applicable to his specific materials [76,179,209,305,307,399-408]. At least one of these, Official Methods of Analysis of the Association of Official Analytical Chemists subjects methods to the rigour of interlaboratory evaluation prior to acceptance; fluorometric methods for plants [179] and foods [209] and a hydride generation-atomic absorption spectrometric method for foods [307] are specified. Methods recommended and materials to which they apply are: fluorometry – biological materials [76], hydride generation atomic absorption spectrometry, light absorption spectrometry – water and wastewater [399], hydride generation atomic absorption spectrometry – urine and air [400], fluorometry – soil [401], electrothermal atomization and hydride generation atomic absorption spectrometry – water and wastes [402], hydride generation atomic absorption spectrometry – water and sediments [403], inductively coupled plasma atomic emission spectrometry – air and air particulates [404], fluorometry – water [405], fluorometry – foods [406], hydride generation atomic absorption spectrometry – human tissues, fluids and excreta [407], fluorometry – blood and other biological materials [305] and hydride generation atomic absorption spectrometry – sludges, soils and related materials [408].

16.7. CONCLUSIONS

Research on a multitude of aspects of selenium is well covered in an extensive, steadily increasing literature data base, which includes many reviews and critical studies/assessments to guide the research worker. Contamination during sampling, handling and laboratory manipulations is generally not as serious a concern as for other trace elements and the usual good careful work suffices for this element. Some care, however, is dictated in drying and matrix destruction to avoid losses of selenium by volatilization and/or conversion to non-reactive oxidation states. Many decomposition procedures have been described with those involving acid digestion using combinations of $HNO_3/HClO_4/H_2SO_4$ the most useful and common. A wide variety of determination techniques are applied to the measurement of selenium. The choice of method is based on detection capability, speed, and analyst preference. Research and development of measurement techniques and overall methods of analysis continues at a fast pace. Several official and recommended methods of analysis are available based on different procedure and material combinations. Initial studies related solely to selenium toxicity concerns have given way during the last three decades to an emphasis on the element's essentiality. As well, there is a trend to investigate speciation and bioavailability and antagonistic effects of other elements, impinging on selenium's physiological role in humans and animals.

REFERENCES

1 M. Ihnat (Editor), *Occurrence and Distribution of Selenium*, CRC Press, Boca Raton, FL, 1989.
2 M. Ihnat, in M. Ihnat (Editor), *Occurrence and Distribution of Selenium*, CRC Press, Boca Raton, FL, 1989, Ch. 5, pp. 33-105.
3 W.R. Wolf and A. Schubert, in M. Ihnat (Editor), *Occurrence and Distribution of Selenium*, CRC Press, Boca Raton, FL, 1989, Ch. 6, pp. 107-120.
4 R.C. Ewan, in M. Ihnat (Editor), *Occurrence and Distribution of Selenium*, CRC Press, Boca Raton, FL, 1989, Ch. 7, pp. 121-167.
5 Y. Thomassen and J. Aaseth, in M. Ihnat (Editor), *Occurrence and Distribution of Selenium*, CRC Press, Boca Raton, FL, 1989, Ch. 8, pp. 169-212.
6 M.L. Berrow and A.M. Ure, in M. Ihnat (Editor), *Occurrence and Distribution of Selenium*, CRC Press, Boca Raton, FL, 1989, Ch. 9, pp. 213-242.
7 G.A. Cutter, in M. Ihnat (Editor), *Occurrence and Distribution of Selenium*, CRC Press, Boca Raton, FL, 1989, Ch. 10, pp. 243-262.
8 K.W.M. Siu and S.S. Berman, in M. Ihnat (Editor), *Occurrence and Distribution of Selenium*, CRC Press, Boca Raton, FL, 1989, Ch. 11, pp. 263-293.
9 B.W. Mosher and R.A. Duce, in M. Ihnat (Editor), *Occurrence and Distribution of Selenium*, CRC Press, Boca Raton, FL, 1989, Ch. 12, pp. 295-325.
10 H.W. Lakin and D.F. Davidson, in O.H. Muth, J.E. Oldfield and P.H. Weswig (Editors), *Selenium in Biomedicine*, Avi, Westport, CT, 1967, Ch. 3, pp. 27-56.
11 A. Kabata-Pendias and H. Pendias, *Trace Elements in Soils and Plants*, CRC Press, Boca Raton, FL, 1984, Ch. 11, pp. 185-192.
12 R.J. Shamberger, *Sci. Total Environ.*, 17 (1981) 59-74.
13 R.J. Shamberger, *Biochemistry of Selenium*, Plenum Press, New York, 1983, Ch. 6, pp. 167-183.
14 D.E. Ullrey, in J.E. Spallholz, J.L. Martin and H.E. Ganther (Editors), *Selenium Biol. Med., (Proc. 2nd Int. Symp.)*, Avi, Westport, CT, 1981, Ch. 16, pp. 176-191.

15 P.V. Hodson, D.M. Whittle and D.J. Hallett, in J.O. Nriagu and M.S. Simmons (Editors), *Toxic Contaminants in the Great Lakes (Advances in Environmental Science and Technology, Vol. 14)*, Wiley, New York, 1984, Ch. 17, pp. 371-391.

16 W.H. Allaway, *Cornell Vet.*, 63 (1973) 151-170.

17 M.-T. Lo and E. Sandi, *J. Environ. Pathol. Toxicol.*, 4 (1980) 193-218.

18 J.O. Nriagu, in M. Ihnat (Editor), *Occurrence and Distribution of Selenium*, CRC Press, Boca Raton, FL, 1989, Ch. 13, pp. 327-340.

19 W.H. Allaway, E.E. Cary and C.F. Ehlig, in O.H. Muth, J.E. Oldfield and P.H. Weswig (Editors), *Selenium in Biomedicine*, Avi, Westport, CT, 1967, Ch. 17, pp. 273-296.

20 O.E. Olson, in O.H. Muth, J.E. Oldfield and P.H. Weswig (Editors), *Selenium in Biomedicine*, Avi, Westport, CT, 1967, Ch. 18, pp. 297-312.

21 M.S. Anderson, H.W. Lakin, K.C. Beeson, F.F. Smith and E. Thacker, *Selenium in Agriculture (U.S. Dept. Agric. Handbook, Vol. 200)*, 1961.

22 E.J. Underwood, in *Toxicants Occurring Naturally in Foods*, National Academy of Sciences, Washington, DC, 2nd ed., 1973, Ch. 3, pp. 43-87.

23 A.L. Moxon, in C.A. Lamb, O.G. Bentley and J.M. Beattie (Editors), *Trace Elem., Proc. Conf.*, Academic Press, New York, 1958, Ch. 12, pp. 175-192.

24 A.L. Moxon, in R.A. Zingaro and W.C. Cooper (Editors), *Selenium,* Van Nostrand Rheinhold, New York, 1974, Ch. 12, pp. 675-207.

25 I. Rosenfeld and O.A. Beath, *Selenium Geobotany Biochemistry, Toxicity and Nutrition*, Academic Press, New York, 1964.

26 O.H. Muth, J.E. Oldfield and P.H. Weswig (Editors), *Selenium in Biomedicine*, AVI Westport, CT, 1967.

27 J.E. Spallholz, J.L. Martin and H.E. Ganther (Editors), *Selenium Biol. Med., (Proc. 2nd Int. Symp.)*, AVI, Westport, CT, 1981.

28 D.J.D. Nicholas and A.R. Egan (Editors), *Trace Elements in Soil-Plant-Animal Systems*, Academic Press, New York, 1975.

29 C.F. Mills, I. Bremner, J.K. Chesters and J. Quarterman (Editors), *Trace Element Metabolism in Animals*, Livingstone, Edinburgh/London, 1970.

30 W.G. Hoekstra, J.W. Suttie, H.E. Ganther and W. Mertz (Editors), *Trace Element Metabolism in Animals-2*, University Park Press, Baltimore, MD, 1974.

31 M. Kirchgessner, D.A. Roth-Maier, H.-P. Roth, F.J. Schwarz and E. Weigand (Editors), *Trace Element Metabolism in Man and Animals-3*, Arbeitskreis fur Tierernahrungsforschung, Weihenstephan, 1978.

32 J.M. Gawthorne, J.McC. Howell and C.L. White (Editors), *Trace Element Metabolism in Man and Animals (TEMA- 4)*, Springer-Verlag, Berlin, 1982.

33 C.F. Mills, I. Bremner and J.K. Chesters (Editors), *Trace Elements in Man and Animals (TEMA-5)*, Commonwealth Agricultural Bureaux, Farnham Common, 1985.

34 A.L. Moxon and M. Rhian, *Physiol. Rev.*, 23 (1943) 305-337.

35 W.C. Cooper and J.R. Glover, in R.A. Zingaro and W.C. Cooper (Editors), *Selenium,* Van Nostrand Reinhold, New York, 1974, Ch. 11, pp. 654-674.

36 R.J. Shamberger, *Biochemistry of Selenium*, Plenum Press, New York, 1983, Ch. 7, pp. 185-206.

37 A.S. Prasad, *Trace Elements and Iron in Human Metabolism*, Plenum, New York, 1978, Ch. 9, pp. 215-250.

38 C.G. Wilber, *Clin. Toxicol.*, 17 (1980) 171-230.

39 G.-O. Yang, *Selenium Biol. Med. (Proc. 3rd Int. Symp.)*, 1987, pp. 9-32.

40 M.L. Scott, in C.L. Comar and F. Bronner (Editors), *Mineral Metabolism, An Advanced Treatise, Vol. 2, Part B*, Academic Press, New York, 1962, Ch. 37, pp. 543-558.

41 R.J. Shamberger, *Biochemistry of Selenium*, Plenum Press, New York, 1983, Ch. 2, pp. 31-58.

42 *Trace Elements in Human Nutrition*, WHO Tech. Rep. Ser., Vol. 532, 1973.

43 R.F. Burk, in A.S. Prasad and D. Oberleas (Editors), *Trace Elements in Human Health and Disease, Vol. 2, Essential and Toxic Elements*, Academic Press, New York, 1976, Ch. 30, pp. 105-133.

44 O.A. Levander, *Proc. N.Z. Workshop Trace Elem. N.Z.*, 1981, pp. 129-140.

45 G. Gissel-Nielsen, U.C. Gupta, M. Lamand and T. Westermarck, *Adv. Agron.*, 37 (1984) 397-460.
46 M.L. Scott, in D.L. Klayman and W.H.H. Gunther (Editors), *Organic Selenium Compounds: Their Chemistry and Biology*, Wiley-Interscience, Toronto, 1973, Ch. 13A, pp. 629-661.
47 H.E. Ganther, in R.A. Zingaro and W.C. Cooper (Editors), *Selenium*, Van Nostrand Reinhold, New York, 1974, Ch. 9, pp. 546-614.
48 B.-A. Gamboa Lewis, in J.O. Nriagu (Editors), *Environmental Biogeochemistry, Vol. 1, Carbon, Nitrogen, Phosphorous, Sulfur and Selenium Cycles*, Ann Arbor Science, Ann Arbor, MI, 1976, Ch. 26, pp. 389-409.
49 O.A. Levander, in A.S. Prasad and D. Oberleas (Editors), *Trace Elements in Human Health and Disease, Vol. 2, Essential and Toxic Elements*, Academic Press, New York, 1976, Ch. 31, pp. 135-163.
50 H.E. Ganther, D.G. Hafeman, R.A. Lawrence, R.E. Serfass and W.G. Hoekstra, in A.S. Prasad and D. Overleas (Editors), *Trace Elements in Human Health and Disease, Vol. 2*, Academic Press, New York, 1976, Ch. 32, pp. 165-234.
51 T.C. Stadtman, *Ann. Rev. Biochem.*, 49 (1980) 93-110.
52 R.J. Shamberger, *Biochemistry of Selenium*, Plenum Press, New York, 1983.
53 J.R. Marier and J.F. Jaworski, *Interactions of Selenium*, Natl. Res. Counc. Can. (Rep.) NRCC No. 20643 (1983).
54 A.T. Diplock, *CRC Crit. Rev. Toxicol.*, (1976) 271-329.
55 T.E. Green and M. Turley, *Treatise Anal. Chem.*, 7, Part 2, Sect. A, (1961) 137-205.
56 D.J. Swaine, *CRC Crit. Rev. Anal. Chem.*, 15 (1985) 315-346.
57 S. Dilli and I. Sutikno, *J. Chromatogr.*, 300 (1984) 265-302.
58 M.R. Masson, *Mikrochim. Acta*, I (1976) 419-439.
59 A.D. Campbell, *Pure Appl. Chem.*, 56 (1984) 645-651.
60 A.D. Shendrikar, *Sci. Total Environ.*, 3 (1974) 155-168.
61 S.A. Lewis and C. Veillon, in M. Ihnat (Editor), *Occurrence and Distribution of Selenium*, CRC Press, Boca Raton, FL, 1989, Ch. 2, pp. 3-14.
62 K.S. Tan and D.L. Rabenstein, in M. Ihnat (Editor), *Occurrence and Distribution of Selenium*, CRC Press, Boca Raton, FL, 1989, Ch. 3, pp. 15-24.
63 R.J. Shamberger, *Biochemistry of Selenium*, Plenum Press, New York, 1983, Ch. 10, pp. 311-327.
64 W.C. Cooper, in R.A. Zingaro and W.C. Cooper (Editors), *Selenium*, Van Nostrand Reinhold, New York, 1974, Ch. 10, pp. 615-653.
65 O.E. Olson, *Proc. Symp. Selenium-Tellurium Environ.*, 1976, pp. 67-84.
66 J.H. Watkinson, in O.H. Muth, J.E. Oldfield and P.H. Weswig (Editors), *Selenium in Biomedicine*, Avi, Westport, CT, 1967, Ch. 6, pp. 97-117.
67 J.F. Alicino and J.A. Kowald, in D.L. Klayman and W.H.H. Günther (Editors), *Organic Selenium Compounds: Their Chemistry and Biology*, Wiley-Interscience, New York, 1973, Ch. 17, pp. 1049-1081.
68 W.T. Elwell and H.C.J. Saint, *Compr. Anal. Chem.*, 1C (1962) 296-304.
69 G. Tölg, *Trace Elem. Anal. Chem. Med. Biol., Proc. 3rd Int. Workshop*, Walter de Gruyter, Berlin, 1984, pp. 95-125.
70 N.T. Crosby, *Analyst (London)*, 102 (1977) 225-268.
71 H. Robberecht and R. Van Grieken, *Talanta*, 29 (1982) 823-844.
72 M. Verlinden, H. Deelstra and E. Adriaenssens, *Talanta*, 28 (1981) 637-646.
73 H.J. Robberecht and H.A. Deelstra, *Talanta*, 31 (1984) 497-508.
74 F.L. Fricke, W.B. Robbins and J.A. Caruso, *Progr. Anal. At. Spectrosc.*, 2 (1979) 185-286.
75 L. Fishbein, *Int. J. Environ. Anal. Chem.*, 17 (1984) 113-170.
76 O.E. Olson, I.S. Palmer and E.I. Whitehead, *Methods Biochem. Anal.*, 21 (1973) 39-78.
77 E.M. Bem, *EHP Environ. Health Perspect.*, 37 (1981) 183-200.
78 S.E. Raptis, G. Kaiser and G. Tölg, *Fresenius' Z. Anal. Chem.*, 316 (1983) 105-123.
79 I.I. Nazarenko and A.H. Ermakov, *The Analytical Chemistry of Selenium and Tellurium*, Halsted Press, New York, 1972.

80 F. Alt and J. Messerschmidt, in H.A. McKenzie and L.E. Smythe (Editors), *Quantitative Trace Analysis of Biological Materials*, Elsevier, Amsterdam, 1988, Ch. 28, pp. 487-501.
81 J.A. Rodriguez-Vazquez, *Anal. Chim. Acta*, 73 (1974) 1-32.
82 H.-J. Hofsommer and H.J. Bielig, *Z. Lebensm.- Unters.-Forsch.*, 172 (1981) 32-43.
83 E. Schweiger, *Hydrochem. Hydrogeol. Mitt.*, 4 (1981) 247-262.
84 A.M. Gunn, *Tech Rep. WRC Medmenham, U.K.*, TR 169 (1981).
85 B. Welz, *Atomic Absorption Spectrometry*, VCH, Weinheim, 2nd edn., 1985.
86 L. Futekov, R. Paritschkova and H. Specker, *Fresenius' Z. Anal. Chem.*, 315 (1983) 342-344.
87 B. Kratochvil and J.K. Taylor, *Anal. Chem.*, 53 (1981) 924A-928A.
88 R.G. Petersen and L.D. Calvin, in A. Klute (Editor), *Methods of Soil Analysis Part 1*, Amer. Soc. Agronomy Inc. and Soil Sci. Soc. of Amer. Inc., Madison, WI, 1986, Ch. 2, pp. 33-51.
89 C.O. Ingamells and F.F. Pitard, *Applied Geochemical Analysis*, Wiley, New York, 1986.
90 J. Versieck, *CRC Crit. Rev. Clin. Lab. Sci.*, 22 (1985) 97-184.
91 B. Sansoni and V. Iyengar, *Spez. Ber. Kernforschungsanlage Juelich*, Jul-Spez.-13 (1978).
92 E.J. Maienthal and D.A. Becker, *Interface*, 5 (1976) 49-62.
93 R. Sturgeon and S.S. Berman, *CRC Crit. Rev. Anal. Chem.*, 18 (1987) 209-244.
94 P. Erickson, *Natl. Res. Counc. Can. (Rep.)*, NRCC No. 16472, 1977.
95 P.D. LaFleur (Editor), *Accuracy in Trace Analysis: Sampling, Sample Handling, Analysis*, Natl. Bur. Stand., Spec. Publ. (U.S.), No. 422, 1976, Vol. 1 and 2.
96 B.W. Mosher and R.A. Duce, *J. Geophys. Res.*, 88 (C11) (1983) 6761-6768.
97 S. Jiang, H. Robberecht and F. Adams, *Atmos. Environ.*, 17 (1983) 111-114.
98 R.W. Shaw and R.K. Stevens, *Ann. N.Y. Acad. Sci.*, 338 (1980) 13-25.
99 R.G. Melcher, T.L. Peters and H.W. Emmel, *Top. Curr. Chem.*, 134 (1986) 59-123.
100 B.L. Davis, L.R. Johnson, R.K. Stevens, W.J. Courtney and D.W. Safriet, *Atmos. Environ.*, 18 (1984) 771-782.
101 G.J. Keeler, W.W. Brachaczek, R.A. Gorse Jr., S.M. Japar and W.R. Pierson, *Atmos. Environ.*, 22 (1988) 1715-1720.
102 W.F. Gutknecht, M.B. Ranade, P.M. Grohse, A.S. Damle and P.M.P. Eller, *ACS Symp. Ser.*, 149 (1981) 95-107.
103 M. Kawamura and K. Matsumato, *Bunseki Kagaku*, 14 (1965) 789-795.
104 S. Jiang, W. DeJonghe and F. Adams, *Anal. Chim. Acta*, 136 (1982) 183-190.
105 Y. Hashimoto and J.W. Winchester, *Environ. Sci. Technol.*, 1 (1967) 338-340.
106 K.K.S. Pillay, C.C. Thomas Jr. and J.A. Sondel, *Environ. Sci. Technol.*, 5 (1971) 74-77.
107 A.D. Shendrikar and P.W. West, *Anal. Chim. Acta*, 89 (1977) 403-406.
108 M.E. Kitto and D.L. Anderson, *ACS Symp. Ser.*, 349 (1987) 84-92.
109 W.R.A. De Jonghe, D. Chakraborti and F.C. Adams, *Anal. Chem.*, 52 (1980) 1974-1977.
110 G.L. Feder, *Geol. Surv. Prof. Pap. (U.S.)*, 954-E (1979).
111 K.C. Slawson Jr., *14th Oil Shale Symp. Proc.*, (1981) 401-409.
112 R. Rossmann and J. Barres, *J. Great Lakes Res.*, 14 (1988) 188-204.
113 C.D. Henry and R.R. Kapadia, *Rep. Invest.-Univ. Tex. Austin*, Bur. Econ. Geol. No. 101, 1980.
114 E.I. Hamilton in B.E. Davies (Editor), *Applied Soil Trace Elements*, Wiley, New York, 1980, Ch. 2, pp. 21- 68.
115 J.J. Connor and A.T. Myers, *ASTM Spec. Tech. Publ.*, No. 540, 1973, pp. 30-36.
116 G.L. Feder, *Ann. N.Y. Acad. Sci.*, 199 (1972) 118-123.
117 H.L. Cannon, C.S.E. Papp and B.M. Anderson, *Ann. N.Y. Acad. Sci.*, 199 (1972) 124-136.
118 J. Kubota, *Ann. N.Y. Acad. Sci.*, 199 (1972) 105-117.
119 R.C. Severson, F. Paveglio, D.B. Hatfield and P. Briggs, *Open-File Rep.-U.S. Geol. Surv.*, No. 87-354, 1987.

120 G.V. Iyengar, *J. Res. Natl. Bur. Stand. (U.S.)*, 91 (1986) 67-74.
121 J. Versieck, *Trace Elem. Med.*, 1 (1984) 2-12.
122 J. Versieck and R. Cornelis, *Anal. Chim. Acta*, 116 (1980) 217-254.
123 J. Kiem, G.V. Iyengar, H. Borberg, K. Kasperek, M. Siegers, L.E. Feinendegen and R. Gross, *Nucl. Act. Tech. Life Sci., Proc. Int. Symp.*, (1979) 143-164.
124 D.E. Robertson, *Anal. Chem.*, 40 (1968) 1067-1072.
125 G. Tölg, *Talanta*, 19 (1972) 1489-1521.
126 M. Zief and R. Speights (Editors), *Ultrapurity, Methods and Techniques*, Marcel Dekker, New York, 1972.
127 D.E. Robertson, in P.D. LaFleur (Editor), *Accuracy in Trace Analysis: Sampling, Sample Handling, Analysis*, Natl. Bur. Stand., Spec. Publ. (U.S.), No. 422, 1976, pp. 805-836.
128 G. Gissel-Nielsen, *Plant Soil*, 32 (1970) 242-245.
129 B. Bisbjerg, *Risoe Natl. Lab.*, [Rep.] Risoe-R, No. 200 (1972).
130 T. Yläranta, *Ann. Agric. Fenn.*, 21 (1982) 84-90.
131 G.V. Iyenger, K. Kasperek and L.E. Feinendegen, *Sci. Total Environ.*, 10 (1978) 1-16.
132 E. Uchino, K. Jin, T. Tsuzuki and K. Inoue, *Analyst (London)*, 112 (1987) 291-293.
133 W.A. Maher, *Sci. Total Environ.*, 26 (1983) 173-181.
134 S.H. Harrison, P.D. LaFleur and W.H. Zoller, *Anal. Chem.*, 47 (1975) 1685-1688.
135 R.E. Collier and J. Parker-Sutton, *J. Sci. Food Agric.*, 27 (1976) 743-744.
136 C.F. Ehlig, W.H. Allaway, E.E. Cary and J. Kubota, *Agron. J.*, 60 (1968) 43-47.
137 H.O. Fourie and M. Peisach, *Analyst (London)*, 102 (1977) 193-200.
138 H.O. Fourie and M. Peisach, *Radiochem. Radioanal. Lett.*, 26 (1976) 277-290.
139 D.J. Higgs, V.C. Morris and O.A. Levander, *J. Agric. Food Chem.*, 20 (1972) 678-680.
140 C.J. Asher, C.S. Evans and C.M. Johnson, *Aust. J. Biol. Sci.*, 20 (1967) 737-748.
141 K. Hiraki and Y. Tamari, *Bull. Inst. Chem. Res. Kyoto Univ.*, 58 (1980) 228-234.
142 G.A. Cutter, *Anal. Chem.*, 57 (1985) 2951-2955.
143 C.I. Measures and J.D. Burton, *Anal. Chim. Acta*, 120 (1980) 177-186.
144 G.A. Cutter, *Anal. Chim. Acta*, 98 (1978) 59-66.
145 T.W. May and D.A. Kane, *Anal. Chim. Acta*, 161 (1984) 387-391.
146 A.D. Shendrikar and P.W. West, *Anal. Chim. Acta*, 74 (1975) 189-191.
147 R. Massée, F.J.M.J. Maessen and J.J.M. De Goeij, *Anal. Chim. Acta*, 127 (1981) 181-193.
148 K. Heydorn and E. Damsgaard, *Talanta*, 29 (1982) 1019-1024.
149 D.F. Schutz and K.K. Turekian, *Geochim. Cosmochim. Acta*, 29 (1965) 259-313.
150 H. Uchida, Y. Shimoishi and K. Tôei, *Environ. Sci. Technol.*, 14 (1980) 541-544.
151 V. Cheam and H. Agemian, *Anal. Chim. Acta*, 113 (1980) 237-245.
152 E.-L. Lakomaa, H. Mussalo-Rauhamaa and S. Salmela, *J. Trace Elem. Electrolytes Health Dis.*, 2 (1988) 37-41.
153 J.R. Moody and R.M. Lindstrom, *Anal. Chem.*, 49 (1977) 2264-2267.
154 T.T. Gorsuch, *The Destruction of Organic Matter*, Pergamon, Toronto, 1970.
155 R. Bock, *A Handbook of Decomposition Methods in Analytical Chemistry*, Wiley, Toronto, 1979.
156 J. Dolezal, P. Povondra and Z. Sulcek, *Decomposition Techniques in Inorganic Analysis*, Elsevier, New York, 1968.
157 H.H. Willard and C.L. Rulfs, in I.M. Kolthoff and P.J. Elving (Editors), *Treatise on Analytical Chemistry, Part 1, Vol. 2*, Interscience, New York, 1961, pp. 1027-1050.
158 E.C. Dunlop, in I.M. Kolthoff and P.J. Elving (Editors), *Treatise on Analytical Chemistry, Part 1, Vol. 2*, Interscience, New York, 1961, pp. 1051-1093.
159 T.T. Gorsuch, in I.M. Kolthoff and P.J. Elving (Editors), *Treatise on Analytical Chemistry, Part 2, Vol. 12*, Interscience, New York, 1965, pp. 295-372.
160 G. Tölg, *Pure Appl. Chem.*, 55 (1983) 1989-2006.
161 B. Welz, M. Melcher and G. Schlemmer, *Fresenius' Z. Anal. Chem.*, 316 (1983) 271-276.
162 C.J. Cappon and J.C. Smith, *J. Anal. Toxicol.*, 2 (1978) 114-120.
163 G.R. Sirota and J.F. Uthe, *Aquaculture*, 18 (1979) 41-44.

510

164 G.T.C. Shum, H.C. Freeman and J.F. Uthe, *J. Assoc. Off. Anal. Chem.*, 60 (1977) 1010-1014.
165 C.J. Cappon and J.C. Smith, *Arch. Environ. Contam. Toxicol.*, 10 (1981) 305-319.
166 P.S. Ridout, H.R. Jones and J.G. Williams, *Analyst (London)*, 113 (1988) 1383-1386.
167 D.D. Nygaard and J.H. Lowry, *Anal. Chem.*, 54 (1982) 803-807.
168 L.W. Anderson and L. Acs, *Environ. Sci. Technol.*, 8 (1974) 462-464.
169 R.D. Wauchope, *J. Agric. Food Chem.*, 26 (1978) 226-228.
170 M. Levesque and E.D. Vendette, *Can J. Soil Sci.*, 51 (1971) 85-93.
171 J.H. Watkinson, *Anal. Chem.*, 38 (1966) 92-97.
172 J.H. Watkinson, *Anal. Chim. Acta*, 105 (1979) 319-325.
173 O.E. Olson, *J. Assoc. Off. Anal. Chem.*, 56 (1973) 1073-1077.
174 O.E. Olson, I.S. Palmer and E.E. Cary, *J. Assoc. Off. Anal. Chem.*, 58 (1975) 117-121.
175 A.B. Grant, *N.Z.J. Sci.*, 6 (1963) 577-588.
176 C.Y. Chan, *Anal. Chim. Acta*, 82 (1976) 213-215.
177 H. Uchida, Y. Shimoishi and K. Tôei, *Analyst (London)*, 106 (1981) 757-762.
178 M. Verlinden, *Talanta*, 29 (1982) 875-882.
179 K. Helrich (Editor), *Official Methods of Analysis of the Association of Official Analytical Chemists, Vol. 1*, AOAC, Arlington, VA, 15th ed., 1990, sect. 969.06.
180 O.E. Olson, *J. Assoc. Off. Anal. Chem.*, 52 (1969) 627-634.
181 R. Duff and M. Chessin, *Nature (London)*, 208 (1965) 1001.
182 M.I.E. Long and B. Marshall, *Trop. Agric. (Guildford, U.K.)*, 50 (1973) 121-128.
183 W.A. Maher, *Anal. Lett.*, 16 (1983) 491-499.
184 C.C.Y. Chan, *Anal. Chem.*, 57 (1985) 1482-1485.
185 L. Halicz and G.M. Russell, *Analyst (London)*, 111 (1986) 15-18.
186 Y. Tamari, *Bunseki Kagaku*, 33 (1984) E115- E122.
187 N. Imai, S. Terashima and A. Ando, *Geostand. Newsl.*, 8 (1984) 39-41.
188 R.F. Sanzolou and T.T. Chao, *Geostand. Newsl.*, 11 (1987) 81-85.
189 P. Lindberg and S.O. Jacoboson, *Acta Vet. Scand.*, 11 (1970) 49-58.
190 H.E. Oksanen and M. Sandholm, *J. Sci. Agric. Finl.*, 42 (1970) 250-253.
191 H. Westermarck and P. Kurkela, *Proc. Int. Reindeer/Caribou Symp.*, 2nd (1980) 278-285.
192 P. Lindberg and S. Bingefors, *Acta Agric. Scand.*, 20 (1970) 133-136.
193 Analytical Methods Committee, *Analyst (London)*, 104 (1979) 778-787.
194 J.A. Fiorino, J.W. Jones and S.G. Capar, *Anal. Chem.*, 48 (1976) 120-125.
195 G.D. Christian, E.C. Knobloock and W.C. Purdy, *J. Assoc. Off. Anal. Chem.*, 48 (1965) 877-884.
196 M. Ihnat, *J. Assoc. Off. Anal. Chem.*, 57 (1974) 368-372.
197 M. Ihnat, *J. Assoc. Off. Anal. Chem.*, 57 (1974) 373-378.
198 M. Ihnat and H.J. Miller, *J. Assoc. Off. Anal. Chem.*, 60 (1977) 813-825.
199 M. Ihnat and H.J. Miller, *J. Assoc. Off. Anal. Chem.*, 60 (1977) 1414-1433.
200 H. Agemian and R. Thomson, *Analyst (London)*, 105 (1980) 902-907.
201 J. Nève, M. Hanocq and L. Molle, *J. Pharm. Belg.*, 35 (1980) 345-350.
202 V.C. Morris and O.A. Levander, *J. Nutr.*, 100 (1970) 1383-1388.
203 J. Kumpulainen, A.-M. Raittila, J. Lehto and P. Koivistoinen, *J. Assoc. Off. Anal. Chem.*, 66 (1983) 1129-1135.
204 R.J. Ferretti and O.A. Levander, *J. Agric. Food Chem.*, 24 (1976) 54-56.
205 M. Ihnat, *Anal. Chim. Acta*, 82 (1976) 293-309.
206 J. Nève and Hanocq, *Anal. Chim. Acta*, 93 (1977) 85-90.
207 I. Hoffman, R.J. Westerby and M. Hidiroglou, *J. Assoc. Off. Anal. Chem.*, 51 (1968) 1039-1042.
208 M. Ihnat, *J. Assoc. Off. Anal. Chem.*, 59 (1976) 911-922.
209 K. Helrich (Editor), *Official Methods of Analysis of the Association of Official Analytical Chemists, Vol. 1*, AOAC, Arlington, VA, 15th ed., 1990, sect. 974.15.
210 J.W. Young and G.D. Christian, *Anal. Chim. Acta*, 65 (1973) 127-138.
211 K.A. Wolnik, F.L. Fricke, S.G. Caper, G.L. Braude, M.W. Meyer, R.D. Satzger and R.W. Kuennen, *J. Agric. Food Chem.*, 31 (1983) 1244-1249.

212 M.H. Hahn, R.W. Kuennen, J.A. Caruso and F.L. Fricke, *J. Agric. Food Chem.*, 29 (1981) 792-796.
213 B. Irsch and K. Schäfer, *Fresenius' Z. Anal. Chem.*, 320 (1985) 37-40.
214 G. Kaiser and G. Tölg, *Fresenius' Z. Anal. Chem.*, 325 (1986) 32-40.
215 B. Lloyd, P. Holt and H.T. Delves, in P. Brätter and P. Schramel (Editors), *Trace Element Analytical Chemistry in Medicine and Biology*, Vol. 2, Walter de Gruyter, Berlin, 1983, pp. 1129-1141.
216 B. Lloyd, P. Holt and H.T. Delves, *Analyst (London)*, 107 (1982) 927-933.
217 K.I. Strausz, J.T. Purdham and O.P. Strausz, *Anal. Chem.*, 47 (1975) 2032-2034.
218 R. Bye and W. Lund, *Fresenius' Z. Anal. Chem.*, 313 (1982) 211-212.
219 E. Egaas and K. Julshamn, *At. Absorpt. Newsl.*, 17 (1978) 135-138.
220 N. Wells, *N.Z.J. Geol. Geophys.*, 10 (1967) 198-208.
221 C. Veillon, S.A. Lewis, K.Y. Patterson, W.R. Wolf, J.M. Harnly, J. Versieck, L. Vanballenberghe, R. Cornelis and T.C. O'Haver, *Anal. Chem.*, 57 (1985) 2106-2109.
222 D.C. Reamer and C. Veillon, *J. Nutr.*, 113 (1983) 786-792.
223 C.A. Swanson, D.C. Reamer, C. Veillon and O.A. Levander, *J. Nutr.*, 113 (1983) 793-799.
224 D.C. Reamer and C. Veillon, *Anal. Chem.*, 53 (1981) 1192-1195.
225 H.H. Walker, J.H. Runnels and R. Merryfield, *Anal. Chem.*, 48 (1976) 2056-2060.
226 K.W. Budna and G. Knapp, *Fresenius' Z. Anal. Chem.*, 294 (1979) 122-124.
227 R.C. Ewan, C.A. Bauman and A.L. Pope, *J. Agric. Food Chem.*, 16 (1968) 212-215.
228 R.C. Ewan, *J. Anim. Sci.*, 32 (1971) 883-887.
229 K.S. Dhillon, N.S. Randhawa and M.K. Sinha, *Ind. J. Dairy Sci.*, 30 (1977) 218-224.
230 M.C. Mondragon and W.G. Jaffé, *Arch. Latinoamer. Nutr.*, 26 (1976) 341-352.
231 S.A. Martinez, *Z. Lebensm. Unter.-Forsch.*, 144 (1970) 187-189.
232 L.M. Cummins, J.L. Martin, G.W. Maag and D.D. Maag, *Anal. Chem.*, 36 (1964) 382-384.
233 S. Xiao-quan, J. Long-zhu and N. Zhe-ming, *At. Spectrosc.*, 3 (1982) 41-44.
234 K. Lorenz in J.E. Spallholz, J.L. Martin and H.E. Ganther (Editors), *Selenium Biol. Med.*, *(Proc. 2nd Int. Symp.)*, Avi, Westport, CT, 1981, Ch. 47, pp. 449-453.
235 K. Lorenz, *Cereal Chem.*, 55 (1978) 287-294.
236 L. Ebdon and J.R. Wilkinson, *Anal. Chim. Acta*, 194 (1987) 177-187.
237 S. Ng, M. Munroe and W. McSharry, *J. Assoc. Off. Anal. Chem.*, 57 (1974) 1260-1264.
238 S. Ng and W. McSharry, *J. Assoc. Off. Anal. Chem.*, 58 (1975) 987-989.
239 Z. Stefanac, M. Tomaskovic and I. Bregovec, *Microchem. J.*, 16 (1971) 226-235.
240 V. Cheam and A.S.Y. Chau, *Analyst (London)*, 109 (1984) 775-779.
241 M. Lansford, E.M. McPherson and M.J. Fishman, *At. Absorpt. Newsl.*, 13 (1974) 103-105.
242 G. Pyen and M. Fishman, *At. Absorpt. Newsl.*, 17 (1978) 47-48.
243 P.D. Goulden and P. Brooksbank, *Anal. Chem.*, 46 (1974) 1431-1436.
244 K. Kolar and A. Widell, *Mitt. Geb. Lebensmittelunters, Hyg.*, 68 (1977) 259-266.
245 T. Stijve and E. Cardinale, *Mitt. Geb. Lebensmittelunters. Hyg.*, 65 (1974) 476-478.
246 A.M. Teilmann and J.C. Hansen, *Nord. Vet.- Med.*, 36 (1984) 49-56.
247 L. Kotz, G. Kaiser, P. Tschöpel and G. Tölg, *Fresenius' Z. Anal. Chem.*, 260 (1972) 207-209.
248 L. Kotz, G. Henze, G. Kaiser, S. Pahlke, M. Veber and G. Tölg, *Talanta*, 26 (1979) 681-691.
249 Z. Mianzhi and R.M. Barnes, *Appl. Spectrosc.*, 38 (1984) 635-644.
250 J. Petterson, L. Hansson, U. Ornemark and A. Olin, *Clin. Chem. (Winston-Salem, N.C.)*, 34 (1988) 1908-1910.
251 R.A. Nadkarni, *Anal. Chim. Acta*, 135 (1982) 363-368.
252 R.A. Nadkarni, *ACS Symp. Ser.*, 205 (1982) 147-162.
253 B. Bernas, *Anal. Chem.*, 40 (1968) 1682-1686.
254 J. Kubota, W.H. Allaway, D.L. Carter, E.E. Cary and V.A. Lazar, *J. Agric. Food Chem.*, 15 (1967) 448-453.
255 D.L. Carter, M.J. Brown, W.H. Allaway and E.E. Cary, *Agron. J.*, 60 (1968) 532-534.

512

256 T. Stijve, *Z. Lebensm. Unters.-Forsch.*, 164 (1977) 201-203.
257 W.H. Allaway and E.E. Cary, *Anal. Chem.*, 36 (1964) 1359-1362.
258 M.R. Masson, *Mikrochim. Acta*, 1 (1976) 399-411.
259 R. Engler and G. Tölg, *Fresenius' Z. Anal. Chem.*, 235 (1968) 151-159.
260 W. Ihn, G. Hesse and P. Neuland, *Mikrochim. Acta*, (1962) 628-633.
261 W.H. Gutenmann and D.J. Lisk, *J. Agric. Food. Chem.*, 9 (1961) 488-489.
262 J. Erzinger and H. Puchelt, *Geostand. Newsl.*, 4 (1980) 13-16.
263 H.W. Kolmer and S.E. Raptis, *Geostand. Newsl.*, 7 (1983) 315-318.
264 H.-B. Han, G. Kaiser and G. Tölg, *Anal. Chim. Acta*, 128 (1981) 9-21.
265 H. Heinrichs and H. Keitsch, *Anal. Chem.*, 54 (1982) 1211-1214.
266 A. Meyer, Ch. Hofer, G. Knapp and G. Tölg, *Fresenius' Z. Anal. Chem.*, 305 (1981) 1-10.
267 B. Morsches and G. Tölg, *Fresenius' Z. Anal. Chem.*, 219 (1966) 61-68.
268 S.E. Raptis, G. Knapp and A.P. Schalk, *Fresenius' Z. Anal. Chem.*, 316 (1983) 482-487.
269 L. Pavlik, J. Kalouskva, M. Vobecky, J. Dedina, J. Benes and J. Parizek, *Nucl. Act. Tech. Life Sci., (Proc. Int. Symp.)*, (1979) 213-223.
270 J. Sippola, *Ann. Agric. Fenn.*, 18 (1979) 182-187.
271 T.W. May, *J. Assoc. Off. Anal. Chem.*, 65 (1982) 1140-1145.
272 G.K.H. Tam and G. Lacroix, *J. Assoc. Off. Anal. Chem.*, 65 (1982) 647-650.
273 J.-P. Quinche, *Res. Suisse Vitic.,Arboric., Hortic.*, 11 (1979) 189-192.
274 J.-P. Quinche, *Schweiz. Landwirtsch. Forsch.*, 22 (1983) 137-144.
275 W. Holak, *J. Assoc. Off. Anal. Chem.*, 59 (1976) 650-654.
276 J.H. Hamence and D. Taylor, *J. Assoc. Public Anal.*, 18 (1980) 23-27.
277 C.F. Poole, N.J. Evans and D.G. Wibberly, *J. Chromatogr.*, 136 (1977) 73-83.
278 S.B. Adeloju, A.M. Bond and H.C. Hughes, *Anal. Chim. Acta*, 148 (1983) 59-69.
279 T.P. McCarthy, B. Brodie, J.A. Milner and R.F. Bevill, *J. Chromatogr.*, 225 (1981) 9-16.
280 T. Kojonen, *Suomen Kemistilehti*, B46 (1973) 133-138.
281 M.M. Schnepfe, *J. Res. U.S. Geol. Surv.*, 2 (1974) 631-636.
282 J.H. Watkinson, *Anal. Chem.*, 32 (1960) 981-983.
283 T.T. Gorsuch, *Analyst (London)*, 84 (1959) 135-173.
284 N.D. Michie, E.J. Dixon and N.G. Bunton, *J. Assoc. Off. Anal. Chem.*, 61 (1978) 48-51.
285 J. Nève, M. Hanocq and L. Molle, *Mikrochim. Acta*, 1 (1980) 259-269.
286 G. Knapp, S.E. Raptis, G. Kaiser, G. Tölg, P. Schramel and B. Schreiber, *Fresenius' Z. Anal. Chem.*, 308 (1981) 97-103.
287 H.J. Robberecht, R.E. Van Grieken, P.A. Van den Bosch, H. Deelstra and D. Vanden Berghe, *Talanta*, 29 (1982) 1025-1028.
288 J. Nève, M. Hanocq, L. Molle and G. Lefebvre, *Analyst (London)*, 107 (1982) 934-941.
289 H.F. Haas and V. Krivan, *Talanta*, 31 (1984) 307-309.
290 S.B. Adeloju, A.M. Bond and M.H. Briggs, *Anal. Chem.*, 56 (1984) 2397-2401.
291 B. Welz and M. Melcher, *Anal. Chem.*, 57 (1985) 427-431.
292 G. Middleton and R.E. Stuckey, *Analyst (London)*, 79 (1954) 138-142.
293 G.K.H. Tam and G. Lacroix, *J. Environ. Sci. Health*, B14 (1979) 515-524.
294 J. Piwonka, G. Kaiser and G. Tölg, *Fresenius' Z. Anal. Chem.*, 321 (1985) 225-234.
295 J. Raptis, G. Knapp, A. Meyer and G. Tölg, *Fresenius' Z. Anal. Chem.*, 300 (1980) 18-21.
296 Y. Shimoishi, *Analyst (London)*, 101 (1976) 298-305.
297 R. Massée, H.A. Van der Sloot and H.A. Das, *J. Radioanal. Chem.*, 35 (1977) 157-165.
298 K. Kurahasi, S. Inoue, S. Yonekura, Y. Shimoishi and K. Tôei, *Analyst (London)*, 105 (1980) 690-695.
299 S.B. Adeloju, A.M. Bond, M.H. Briggs and H.C. Hughes, *Anal. Chem.*, 55 (1983) 2076-2082.
300 S.P. Brimmer, W.R. Fawcett and K.A. Kulhavy, *Anal. Chem.*, 59 (1987) 1470-1471.

301 J. Nève, M. Hanocq and L. Molle, *Int. J. Environ. Anal. Chem.*, 8 (1980) 177-188.
302 J. Nève, M. Hanocq and L. Molle, *Anal. Chim. Acta*, 115 (1980) 133-141.
303 J. Nève, M. Hanocq and L. Molle, *Fresenius' Z. Anal. Chem.*, 308 (1981) 448-451.
304 J. Nève, M. Hanocq and L. Molle, in P. Brätter and P. Schramel (Editors), *Trace Element Analytical Chemistry in Medicine and Biology, Vol., 2*, Walter de Gruyter, Berlin, 1983, pp. 859-876.
305 H.A.M.G. Vaessen, A. van Ooik and P.L. Schuller, *IARC Sci. Publ.*, 8 (71) (1986) 477-485.
306 W. Holak, *J. Assoc. Off. Anal. Chem.*, 63 (1980) 485-495.
307 K. Helrich (Editor), *Official Methods of Analysis of the Association of Official Analytical Chemists, Vol. 1*, AOAC, Arlington, VA, 15th ed., 1990 sect. 986.15.
308 M. Würfels, E. Jackwerth and M. Stoeppler, *Anal. Chim. Acta*, 226 (1989) 17-30.
309 M. Würfels, E. Jackwerth and M. Stoeppler, *Anal. Chim. Acta*, 226 (1989) 1-16.
310 W. Schöniger, *Mikrokim. Acta*, (1955) 123-129.
311 H.J. Hofsommer and H.J. Bielig, *Deut. Lebensm. Rundsch.*, 76 (1980) 419-426.
312 F.B. Cousins, *Aust. J. Exp. Biol.*, 38 (1960) 11-16.
313 P. Lindberg, *Acta Vet. Scand.*, Suppl. 23 (1968).
314 R.J. Hall and P.L. Gupta, *Analyst (London)*, 94 (1969) 292-299.
315 W.J. Kelleher and M.J. Johnson, *Anal. Chem.*, 33 (1961) 1429-1432.
316 R.E. Stanton and A.J. McDonald, *Analyst (London)*, 90 (1965) 497-499.
317 I.H. Elsokkaro and A. Oien, *Acta Agric. Scand.*, 27 (1977) 283-288.
318 K.M. Holtzclaw, R.H. Neal, G. Sposito and S.J. Traina, *Soil Sci. Soc. Amer. J.*, 50 (1986) 75-78.
319 M. Verlinden, J. Baart and H. Deelstra, *Talanta*, 27 (1980) 633-639.
320 K.S. Subramanian, *Fresenius' Z. Anal. Chem.*, 305 (1981) 382-386.
321 H. Schnitzer and G. Lieck, *Landwirtsch. Forsch.*, 34 (1981) 1-7.
322 K. Schäfer and D. Behne, *Landwirtsch. Forsch.*, 37 (1984) 58-65.
323 C.H. Branch and D. Hutchison, *Analyst (London)*, 110 (1985) 163-167.
324 J. Agterdenbos, J.T. van Elteren, D. Bax and J.P. Ter Heege, *Spectrochim. Acta*, 41B (1986) 303-316.
325 N. Narasaki, *J. Anal. At. Spectrom.*, 3 (1988) 517-521.
326 K. Saeed, Y. Thomassen and F.J. Langmyhr, *Anal. Chim. Acta*, 110 (1979) 285-289.
327 K.S. Subramanian and J.C. Meranger, *Anal. Chim. Acta*, 124 (1981) 131-142.
328 G. Alfthan and J. Kumpulainen, *Anal. Chim. Acta*, 140 (1982) 221-227.
329 G.R. Carnick, D.C. Manning and W. Slavin, *Analyst (London)*, 108 (1983) 1297-1312.
330 B. Welz and G. Schlemmer, *J. Anal. At. Spectrom.*, 1 (1986) 119-124.
331 B. Welz, G. Schlemmer and J.R. Mudakavi, *J. Anal. At. Spectrom.*, 3 (1988) 93-97.
332 G. Morisi, M. Patriarca and A. Menotti, *Clin. Chem. (Winston-Salem, N.C.)*, 34 (1988) 127-130.
333 M.B. Knowles and K.G. Brodie, *J. Anal. At. Spectrom.*, 3 (1988) 511-516.
334 L. Ebdon and H.G.M. Parry, *J. Anal. At. Spectrom.*, 3 (1988) 131-134.
335 L. Futekov, R. Paritschkova and H. Specker, *Fresenius' Z. Anal. Chem.*, 306 (1981) 378-380.
336 C.M. Lau, A.M. Ure and T.S. West, *Anal. Chim. Acta*, 141 (1982) 213-224.
337 O.E. Olson, R.J. Emerick and I.S. Palmer, *At. Spectrosc.*, 4 (1983) 55-58.
338 L. Ebdon, J.R. Wilkinson and K.W. Jackson, *Anal. Chim. Acta*, 136 (1982) 191-199.
339 L. De Reu, R. Cornelis, J. Hoste and S. Ringoir, *Radiochem. Radioanal. Lett.*, 40 (1979) 51-60.
340 J.S. Morris, D.M. McKown, H.D. Anderson, M. May, P. Primm, M. Cordts, D. Gebhardt, S. Crowson and V. Spate, in J.E. Spallholz, J.L. Martin and H.E. Ganther (Editors), *Selenium Biol. Med., (Proc. 2nd Int. Symp.)*, (1981) 438-448.
341 E.P. Hamilton and A. Chatt, *J. Radioanal. Chem.*, 71 (1982) 29-45.
342 Y. Tanizaki and S. Nagatsuka, *Bull. Chem. Soc. Jpn.*, 56 (1983) 619-624.
343 J.R.W. Woittiez and B.J.T. Nieuwendijk, *J. Radioanal. Nucl. Chem.*, 110 (1987) 603-611.
344 A. Chatt, H.S. Dang, B.B. Fong, C.K. Jayawickreme, L.S. McDowell and D.L. Pegg., *J. Radioanal. Nucl. Chem.*, 124 (1988) 65-77.

345 R. Zeisler, R.R. Greenberg and S.F. Stone, *NBS Spec. Publ. (U.S.)*, 740 (1988) 82-90.
346 K. Heydorn and E. Damsgaard, *Talanta*, 20 (1973) 1-11.
347 E. Steinnes, *Anal. Chim. Acta*, 78 (1975) 307-315.
348 M. Gallorini, R.R. Greenberg and T.E. Gills, *Anal. Chem.*, 50 (1978) 1479-1481.
349 R.J. Rosenberg, *J. Radioanal. Chem.*, 50 (1979) 109-114.
350 D. Knab and E.S. Gladney, *Anal. Chem.*, 50 (1980) 825-828.
351 M. Gallorini, E. Orvini, A. Rolla and M. Burdisso, *Analyst (London)*, 106 (1981) 328-334.
352 E. Sabbioni, L. Goetz, A. Springer and R. Pietra, *Sci. Total Environ.*, 29 (1983) 213-227.
353 R. Zeisler, R.R. Greenberg and S.F. Stone, *J. Radioanal. Nucl. Chem.*, 124 (1988) 47-63.
354 A. Fajgelj, M. Dermelj, A.R. Byrne and P. Stegnar, *J. Radioanal. Nucl. Chem.*, 128 (1988) 93-102.
355 L. Xilei, D. Van Renterghem, R. Cornelis and L. Mees, *Anal. Chim. Acta*, 211 (1988) 231-241.
356 K. Tôei and Y. Shimoishi, *Talanta*, 28 (1981) 967-972.
357 G. Schwedt and A. Schwarz, *J. Chromatogr.*, 160 (1978) 309-312.
358 Y. Shibata, M. Morita and K. Fuwa, *Analyst (London)*, 110 (1985) 1259-1270.
359 T. Ishikawa and Y. Hashimoto, *Bunseki Kagaku*, 36 (1987) 542-546.
360 T. Ishikawa and Y. Hashimoto, *Bunseki Kagaku*, 36 (1988) 344-348.
361 D. Vezina and G. Bleau, *J. Chromatogr.*, 426 (1988) 385-391.
362 W. Funk, V. Dammann, T. Couturier, J. Schiller and L. Völker, *J. High Resolut. Chromatogr. Chromatogr. Commun.*, 9 (1986) 224-235.
363 A.T. Ellis, D.E. Leyden, W. Wegscheider, B.B. Jablonski and W.B. Bodnar, *Anal. Chim. Acta*, 142 (1982) 73-87.
364 H. Robberecht, R. Van Grieken, J. Shani and S. Barak, *Anal. Chim. Acta*, 136 (1982) 285-291.
365 F. Adams, P. Van Espen and W. Maenhaut, *Atmos. Environ.*, 17 (1983) 1521-1536.
366 H. Robberecht and R. Van Grieken, *Anal. Chim. Acta,* 147 (1983) 113-121.
367 A. Prange, J. Knoth, R.-P. Stöbel, H. Böddeker and K. Kramer, *Anal. Chim. Acta*, 195 (1987) 275-287.
368 M. Nagj, J. Injuk and V. Valkovic, *J. Radioanal. Nucl. Chem.*, 127 (1988) 243-252.
369 M. Berti, G. Buso, P. Colautti, G. Moschini, B.M. Stievano and C. Tregnaghi, *Anal. Chem.*, 49 (1977) 1313-1315.
370 W. Maenhaut, L. De Reu, U. Tomza and J. Versieck, *Anal. Chim. Acta*, 136 (1982) 301-309.
371 S. Landsberger, R.E. Jervis, G. Kajrys and S. Monano, *Int. J. Environ. Anal. Chem.*, 16 (1983) 95-130.
372 E. Clayton and E.K. Wooller, *Anal. Chem.*, 57 (1985) 1075-1079.
373 H. Duflou, W. Maenhaut and J. De Reuck, *Biol. Trace Elem. Res.*, 13 (1987) 1-17.
374 M. Simonoff, C. Hamon, P. Moretto, Y. Llabador and G. Simonoff, *Nucl. Instr. Met. Phys. Res.*, B31 (1988) 442-448.
375 G. Chauvaux, *Ann. Med. Vet.*, 117 (1973) 491-495.
376 A.G. Howard, M.R. Gray, A.J. Waters and A.R Oromiehie, *Anal. Chim. Acta*, 118 (1980) 87-91.
377 G.E. Batley, *Anal. Chim. Acta*, 187 (1986) 109-116.
378 M.W. Blades, J.A. Dalziel and C.M. Elson, *J. Assoc. Off. Anal. Chem.*, 59 (1976) 1234-1239.
379 R.S. Posey and R.W. Andrews, *Anal. Chim. Acta*, 124 (1981) 107-112.
380 Ph. Breyer and B.P. Gilbert, *Anal. Chim. Acta*, 201 (1987) 33-41.
381 M.H. Hahn, K.A. Wolnik, F.L. Fricke and J.A. Caruso, *Anal. Chem.*, 54 (1982) 1048-1052.
382 E. de Oliveria, J.W. McLaren and S.S. Berman, *anal. chem.*, 55 (1983) 2047-2050.
383 P. Fodor and R.M. Barnes, *Spectrochim. Acta*, 38B (1983) 229-243.
384 R.M. Barnes and P. Fodor, *Spectrohcim. Acta*, 38B (1983) 1191-1202.

385 M. Sugiyama, O. Fujino, S. Kihara and M. Matsui, *Anal. Chim. Acta*, 181 (1986) 159-168.
386 A Pilate and F. Adams, *Fresenius' Z. Anal. Chem.*, 309 (1981) 295-299.
387 D.C. Reamer and C. Veillon, *Anal. Chem.*, 53 (1981) 2166-2169.
388 A.R. Date and A.L. Gray, *Spectrohcim. Acta*, 40B (1985) 115-122.
389 A.M. Ure, J.R. Bacon, M.L. Berrow and J.J. Watt, *Geoderma*, 22 (1979) 1-23.
390 R.M. Parr, Y. Muramatsu and S.A. Clements, *Fresenius' Z. Anal. Chem.*, 326 (1987) 607-608.
391 M. Ihnat and B.K. Thompson, *J. Assoc. Off. Anal. Chem.*, 63 (1980) 814-839.
392 H.A.M.G. Vaessen and A. van Ooik, *Z. Lebensm. Unters. Forsch.*, 185 (1987) 468-471.
393 T.-S. Koh, T.H. Benson and G.J. Judson, *J. Assoc. Off. Anal. Chem.*, 63 (1980) 809-813.
394 J. Kumpulainen and P. Koivistoinen, *Kem- Kemi*, 8 (1981) 372-373.
395 B. Welz and M. Verlinden, *Acta Pharmacol. Toxicol.*, 59 Suppl. 7 (1986) 577-580.
396 M. Ihnat, M.S. Wolynetz, Y. Thomassen and M. Verlinden, *Pure Appl. Chem.*, 58 (1986) 1063-1076.
397 B. Welz, M.S. Wolynetz and M. Verlinden, *Pure Appl. Chem.*, 59 (1987) 927-936.
398 T.-S. Koh, *J. Assoc. Off. Anal. Chem.*, 70 (1987) 664-667.
399 *Stand. Methods. Exam. Water Wastewater*, Am. Publ. Health Assoc., 14th edn., 1976.
400 T.J. Kneip, *Health Lab. Sci.*, 14 (1977) 53-58.
401 J. Kubota and E.E. Cary, in A.L. Page (Editor), *Methods of Soil Analysis Part 2*, Amer. Soc. of Agronomy Inc., and Soil Sci. Soc. Amer., 1982, Ch. 27, pp. 485-500.
402 *Methods for Chemical Analysis of Water and Wastes*, U.S. Environ. Prot. Agency, Res. Dev., (Rep.) EPA- 600/4-79-020 (1983).
403 M.J. Fishman and L.C. Friedman (Editors), *Open File Rep. U.S. Geol. Surv.*, No. 85-495, 1985.
404 Int. Agency Res. Cancer Sci. Publ. No. 71, 8 (1986) 225-230.
405 R.J. Hall and P.J. Peterson, *Int. Agency Res. Cancer Sci. Publ.*, 71 (8) (1986) 355-361.
406 R.J. Hall and P.J. Peterson, *Int. Agency Res. Cancer Sci. Publ.*, 71 (8) (1986) 409-419.
407 K.S. Subramanian, *Int. Agency Res. Cancer Sci. Publ.*, 71 (8) (1986) 461-470.
408 *Methods Exam. Water Assoc. Mats.*, HMSO, London, 1987.

M. Stoeppler (Editor)/*Hazardous Metals in the Environment*
© 1992 Elsevier Science Publishers B.V. All rights reserved

Chapter 17

Quality assurance and validation of results

Bernard Griepink

Pelgrimsweg 3, B-1860 Meise (Belgium)

and

Markus Stoeppler

Research Center Juelich (KFA), Institute of Applied Physical Chemistry, P.O. Box 1913, D-W-5170 Juelich (Germany)

CONTENTS

17.1. Introduction 517
17.2. Relevant definitions (ISO) 519
17.3. Within laboratory quality measures 519
    17.3.1. Guidelines for quality assurance 520
    17.3.2. Statistical control 521
    17.3.3. Comparison with results of other methods 524
    17.3.4. Use of certified reference materials (CRMs) 525
    17.3.5. General 525
17.4 Intercomparisons 526
    17.4.1. Aims of intercomparisons 526
    17.4.2. Results of intercomparisons 528
    17.4.3. Effects of participation in intercomparisons 529
17.5. Certified reference materials (CRMs) 530
    17.5.1. Requirements 530
        17.5.1.1. Homogeneity 531
        17.5.1.2. Stability 531
        17.5.1.3. Similarity to the real sample 531
        17.5.1.4. Accuracy, uncertainty and traceability 532
    17.5.2. Preparation of candidate CRMs 532
    17.5.3. Availability of CRMs 533
References 533

17.1. INTRODUCTION

The amount of analyses performed in routine and research laboratories is immense as modern automated equipment allows a very high through-put. However, with the improvement of equipment and data handling, the quality of results has often been neglected.

This chapter provides means of ascertaining and assuring good quality of measurements. Most of the arguments and examples will be taken from environmental analysis but the conclusions hold in other fields as well. The measures to be taken within a laboratory, participation in intercomparisons and the application of certified reference materials will be dealt with.

Accuracy is a prerequisite of reliable analysis. Too many scientists have stated that a good reproducibility in time is sufficient in order to follow trends and demonstrate effects of actions by authorities to improve the quality of the environment or food. Such statements overlook modeling applications, theory development, etc. and ignore improvements in equipment and methodology. Moreover, results of different groups in different parts of the world are then incomparable.

It is already the case that results of the monitoring of various rivers and oceanic regions can hardly be compared over approximately 3–5 years. Older results obtained with a different method often show an unknown bias. This in turn means that a lot of correlation studies performed in the past and their conclusions should be ignored.

For achieving not only good reproducibility but also good accuracy, various measures are necessary. The tables and figures presented below indicate that good control of the quality of measurement has not yet been achieved.

A typical example (Table 17.1) illustrates the lack of accuracy that occurred in the determination of inorganic traces. Findings such as these are quite common and occur in many fields of analysis [1–9].

When results differ so much, they are not trustworthy. Poor performance by analytical laboratories creates economic losses: extra analyses, destruction of food and goods, court actions, etc. Estimated losses of this type for the F.R.G. were in the region of

TABLE 17.1

RESULTS OBTAINED IN THE DETERMINATION OF TRACE ELEMENTS IN PLANTS AND MILK POWDER

Element	Matrix	Found (mg/kg)		Quotient H/L	Most accurate value mg/kg
		Lowest (L)	Highest (H)		
Cd	Plant	0.050	6.654	133	0.10
	Milk powder	0.0004	4.500	11 250	0.0029
Hg	Plant	0.005	0.702	140	0.28
	Milk powder	0.0006	0.42	700	0.001
Pb	Plant	17.6	33.3	1.9	25.0
	Milk powder	0.068	5.5	81	0.105

2000 000 000 DM [10]. The cost in terms of quality of life cannot be estimated, but is certainly high.

17.2. RELEVANT DEFINITIONS (ISO)

Many terms used in the relatively young subdiscipline of analytical quality assurance and control have different definitions. This does not facilitate good communication between the scientists involved. Therefore the relevant ISO-definitions (taken from ISO-guides) are given below.

Quality: the totality of features and characteristics of a product, process or service that bear on its ability to satisfy STATED or IMPLIED needs.

Quality policy: the overall quality intentions and objectives of an organization as formally expressed by senior management.

Quality control: the operational technique and activities that are used to satisfy quality requirements.

Quality assurance: all those planned and systematic actions necessary to provide adequate confidence that a product, process or service will satisfy given quality requirements.

Certification system: a system that has its own rules of procedure and management for carrying out *certification of conformity*.

Laboratory accreditation: formal recognition that a *testing laboratory* is competent to carry out specific *tests* or specific types of *tests*.

Laboratory assessment: examination of a *testing laboratory* to evaluate its compliance with specific *laboratory accreditation criteria*.

17.3. WITHIN LABORATORY QUALITY MEASURES

Two basic parameters should be considered when discussing analytical results: accuracy ("absence of systematic errors") and uncertainty (coefficient of variation: confidence interval) as caused by random errors and random variations in the procedure.

In this context, accuracy is of primary importance. However, if the uncertainty in a result is too high, it cannot be used to reach any conclusion about the outcome of an experiment, to judge the quality of the environment or of food, nor can it be used for the diagnosis of a patient's illness. An unacceptably high uncertainty renders the result useless.

In the performance of an analysis, all basic principles of calibration, of elimination of sources of contamination and losses, and of correction for interferences, etc. should be followed [11–15].

A prerequisite for a good result is a correct calibration. Although calibration is so obvious and has been emphasized for as long as analytical chemistry has existed, experience has shown that, with the introduction of modern automated equipment, calibration has become a part of the analytical chain to which insufficient attention is being

References on p. 533

paid. It is perhaps necessary to stress once again that compounds of well known stoichiometry should be used, of which a possible water content is fixed. Careful consideration should be given to the choice of the method of calibration: using pure calibrant solutions, matrix-matched solutions or standard additions. In addition, internal standards should be used when appropriate. The chemicals used for the matrix-matched calibration should not contain the analyte to be determined.

Modern equipment often requires calibration with a similar matrix (e.g. solid sampling Zeeman Electrothermal Atomic Absorption Spectroscopy (ZETAAS)). In such a case, a calibration with more than one reference material to be verified by a calibration with a pure solution is recommended.

17.3.1. Guidelines for quality assurance [16–21]

Good Laboratory Practice (GLP) is one of the manifestations of the increased attention paid to the achievement of good quality results. GLP involves quality assurance (QA) although QA can and should be applied in cases where the formal GLP-system is not required. GLP is therefore the system with direct legal implications (e.g. (eco)toxicology). The need for a QA- or GLP-framework of guidelines to bring the quality of a laboratory to such a level that it meets predefined standards arises from the economic, scientific or political implications of laboratory results.

GLP-guidelines have been laid down in written form by OECD [22]; the principles of GLP as developed by the Food and Drug Administration (FDA) and the Environmental Protection Agency (EPA) were adopted by US law in 1979.

GLP describes how the laboratory should work, its organization and the way to produce valid results. Analytical methods are not mentioned. GLP-guidelines are implemented in a laboratory through a QA-program which is delineated in a QA-manual. This comprises a detailed description of the items listed below in order to assess and monitor the quality of the analytical results. The summary given below was compiled by Vijverberg and Cofino [17].

Test facilities, organization, and personnel
  – location of the laboratories and safety aspects
  – structure of the organization and allocation of responsibilities
  – education
  – management
Quality assessment
  – interlaboratory and intralaboratory testing programs
  – reference material
Statistical quality control
  – control charts
Apparatus, chemicals, reagents, and blank
  – apparatus
      preventive maintenance and maintenance
      calibration

cleaning glassware and non-glassware
- chemicals
    registration
    quality control checks
    rules for storage of waste
    handling and storage
- reagents
    standard solution
- blank

Sampling and storage
- sampling strategy
- way of sampling
- storage and preservation
- sample identification

Documentation
- study
- methods
- work-sheets
- notebook

Reporting of results
- not computerized
- computerized

Archiving of results.

In this chapter most attention will be paid to those aspects of the above listing where, e.g., the BCR (Community Bureau of Reference) and other organizations can be of help or play a role: inter- and intralaboratory testing programs, materials for use in setting up control charts, certified reference materials.

17.3.2. Statistical control

When a laboratory works at a constant level of high quality, fluctuations in the results become random and can be predicted statistically [23].

This implies in the first place that limits of determination and detection should be constant and well known [24]; rules for rounding off final results should be based on the performance of the method in the laboratory. Furthermore, if such a situation of absence of systematic fluctuations exists, normal statistics (e.g. regression analysis, *t*- and *F*-tests, analysis of variance etc.) can be applied [24] to study the results wherever necessary.

Besides these checks, control checks should be used as soon as the method is under control in the laboratory. A control chart provides a graphical way of interpreting the method's output in time, so that the repeatability of the results and the method's precision over a period of time can be evaluated at a glance. To do so, one or several standards of

References on p. 533

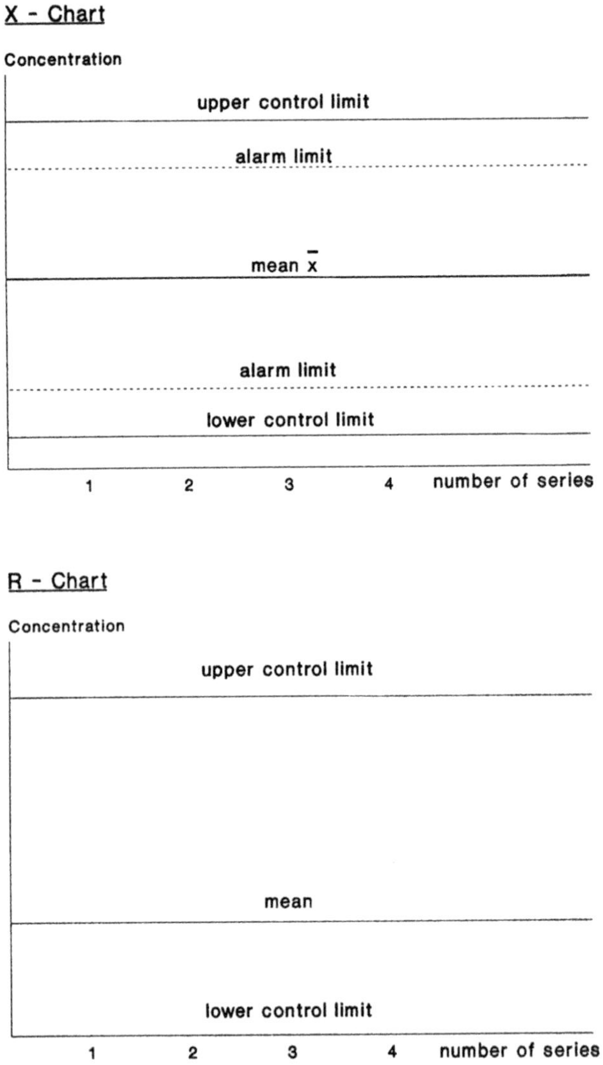

X - Chart

Concentration

| | 1 | 2 | 3 | 4 | number of series |

R - Chart

Concentration

Fig. 17.1. Example of X- and R-charts.

good homogeneity and stability should be analyzed with each batch of unknown materials. Some 5–10% of all samples should be used for this purpose [23].

In a SHEWHART control chart [25] the laboratory plots, for each standard analyzed at a time, the obtained value (X) or the difference between duplicate values (R). The X-chart additionally presents the lines corresponding to a risk of 5 resp. 1% that the results are

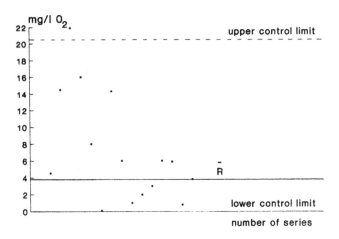

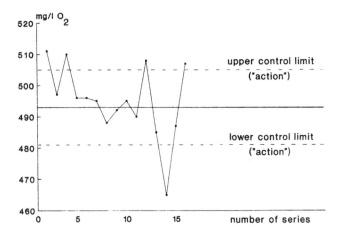

Fig. 17.2. R- and X-chart of a COD-determination [26].

not contained in the whole population of results. These lines are for "warning" and "action" resp. (Fig. 17.1). Fig. 17.2 presents an example of the determination of the chemical oxygen demand using dichromate (COD_{Cr}) of a control solution, taken from the quality handbook in the Nordisk Council of Ministers [26].

The results of a method are considered to be out of control if:

- the upper or lower control limit is exceeded ("action"),
- the same "alarm-line" is exceeded twice in succession,
- eleven successive measurements are on the same side of the line.

X- and R-chart systems are also being tested in which the weighing factor for each point of the curve decreases according to a geometric progression (e.g. let the value at moment i be X_i, then the value plotted in a chart for a moment i is: $(X_i + (X_{i-1}) \cdot 0.2 + (X_{i-2}) = 0.16 + \ldots)$. This allows an earlier detection of an upward or downward trend.

Cumulative sum ("cusum") charts are frequently used, particularly for early detection of a drift in the results. In these charts the cumulative sum of the differences between the found value and probable value are plotted against the result obtained at moment i:

$$s_i = \sum_{i=1}^{i} (X_i - X_{ref})$$

in which s_i is the cumulative sum, X_i is the measured value and X_{ref} is the reference value, which should be accurately known. Dewey and Hunt [27] presented a practical evaluation of the use of cusum-charts. It should be emphasized that good results can only be obtained if the value of X_{ref} is accurately known. A certified reference material of good quality should therefore be used.

The evaluation of a cusum-chart is done using the V-mask method [28–30].

17.3.3. Comparison with results of other methods

The use of a SHEWHART-chart enables the detection of whether or not a method is still in control; it is not, however, able to detect a systematic error which is present from the moment of introduction of the method in a laboratory. If possible, results should be verified by other methods.

All methods have their own particular source of error. For spectrometric or voltammetric techniques such a possible source of error can be the digestion of the sample, a step which is not necessary for instrumental neutron activation. The latter technique, however, may have errors such as shielding, insufficient separation of gamma-peaks, etc. which are not encountered in chemical analysis.

An independent method should therefore be used to verify the results of routine analysis. If the results of both methods are in good agreement, it can be concluded that the results of the routine analysis are not affected by a contribution of a systematic nature (e.g. insufficient digestion). This conclusion is strongest when the two methods differ widely, such as voltammetry and instrumental neutron activation. If the methods have similarities, such as a digestion step in the case of voltammetry and spectrometry, a comparison of the results would most likely lead to conclusions concerning the accuracy of the method of final determination, and not concerning the analytical result.

If the compared method is a method for which the sources of error and the effects of the various steps are well known, and if this method is relatively insensitive to human failures, it can be called a "Reference method".

The application of such a reference method, if possible, belongs to a good quality control system. However, good reference or independent methods do not always exist.

Furthermore, if the analyst using this reference method is not sufficiently experienced with the method, the application may even create errors. The results of a method are not only dependent on the method but also on the person using it.

Good reference methods for trace element determinations are: isotope dilution mass spectrometry (provided that the digestion takes place in a closed system after addition of the spike) and activation techniques.

17.3.4. Use of certified reference materials (CRMs)

Results can only be accurate and comparable worldwide if they are traceable. By definition, *traceability* of a measurement is achieved by an unbroken chain of calibrations connecting the measurement process to the fundamental units. In the vast majority of chemical analyses, the chain is broken because the sample is physically destroyed by dissolutions, calcination, etc. To ensure the full traceability it is necessary to demonstrate that no loss or contamination has occurred in the course of sample treatment.

Indeed, the physical calibration of all measuring instruments is not always of great importance for the chemist. The manner in which the calibration curve is established for the particular assays is of prime importance (it has to take into account all possible interferences) as well as the conditions of sample treatment.

The only possibility for any laboratory to ensure traceability in a simple manner is to verify the analytical procedure by means of a so-called matrix reference material certified in a reliable manner.

The laboratory which measures such a reference material by its own procedure and finds a value in disagreement with the certified value is thus warned that its measurement includes an error, of which the source must be identified.

Besides the aspect of achieving traceability, certified reference materials having well known and well documented properties should be used to:

- demonstrate the accuracy of results obtained in a laboratory,
- monitor the performance of the method (e.g. cusum charts),
- calibrate equipment which requires a calibrant similar to the matrix (e.g. optical emission spectrometry, X-ray fluorescence spectrometry),
- demonstrate equivalence between methods,
- detect errors in the application of standardized methods (e.g. ISO, ASTM, ECCLS, ...)

17.3.5. General

It is not sufficient that the laboratory applies the most modern equipment and techniques, nor that control charts and reference methods are used.

If the persons working with the equipment lack sufficient instruction and training, their results cannot be good.

The laboratory management should provide an infrastructure in which both equipment and personnel work in an optimal atmosphere, where the workers are motivated to perform their task in an excellent way and are not bothered with tasks which do not belong to their level of training, education and experience.

It is the human being and not the instrument who is, and should feel, responsible for the quality of the result.

17.4. INTERCOMPARISONS

When all necessary measures have been taken in the laboratory to achieve accurate results, the laboratory should demonstrate its accuracy in intercomparisons, which are also useful to detect systematic errors.

In general, besides the sampling error, three sources of error can be detected in all analyses:
(i) sample pretreatment (e.g. digestion, extraction, separation, preconcentration ...);
(ii) final measurements (e.g. calibration errors, spectral interferences, peak overlap, baseline and background corrections ...);
(iii) laboratory itself (e.g. training and educational level of workers, care applied to the work, awareness of pitfalls, management, clean bench facilities ...).

In the framework of a COST-program on the analysis of sewage sludge, Muntau estimated the uncertainty originating from the digestion method and from the final determination (AAS) method [31]. Fig. 17.3a and b give an illustration of the effects mentioned under (i) and (ii).

A relatively large laboratory, where various methods are applied by different analysts who are experienced in their techniques, may have been able to eliminate most errors due to (i) and (ii) in a continuously and carefully applied quality scheme. Other, smaller laboratories, or laboratories having perhaps two methods only, are not always fully in a position to assess the quality of their measurements with respect to error sources of these categories.

For both types of laboratories, however, intercomparisons to eliminate at least error source (iii) are necessary.

When laboratories participate in an intercomparison, different sample pretreatment methods and techniques of final determination are the focus of discussion, as well as the performance of these laboratories. If the results of such an intercomparison are in good and statistical agreement, the collaboratively obtained value is likely to be the best approximation of the truth.

17.4.1. Aims of intercomparisons

Before conducting an intercomparison the aims should be clearly defined. An intercomparison can be held:
(i) to detect the pitfalls of a commonly applied method and to ascertain its performance in practice;
(ii) to measure the quality of a laboratory or a part of a laboratory (e.g. audits for accredited laboratories);

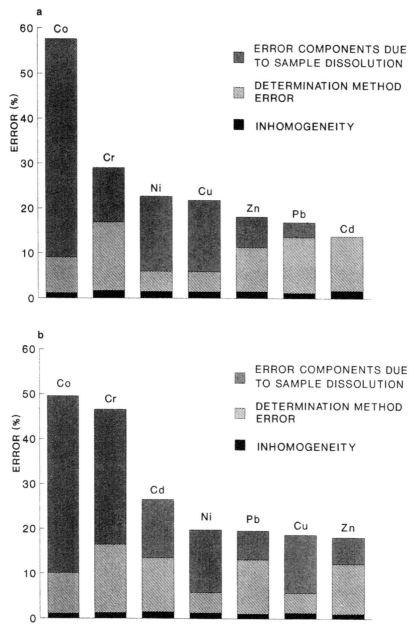

Fig. 17.3. (a) Error components contributing to total observed error (sample 1). (b) Error components contributing to total observed error (sample 2).

(iii) to improve the quality of a laboratory in collaborative work with mutual learning processes;

(iv) to certify the contents of a reference material.

In the ideal situation, where the results of all laboratories are under control, intercomparisons of types (ii) and (iv) will be held only. For the time being, however, types (i) and (iii) play an important role.

The organization of the collaborative trial, the sort of samples and the laboratories invited depend strongly on the aim of the exercise. The results submitted should be evaluated accordingly. The basic principles are given in norms and ISO-guidelines [32].

17.4.2. Results of intercomparisons

In his long career in the FDA, W. Horwitz (e.g. [33,34]) held many intercomparisons, especially on biotic or food-type materials. The laboratories (mostly American) in this quality system developed considerable skill over the years. Nevertheless, uncertainties were not reduced to zero. Horwitz found that the order of magnitude of uncertainties in an intercomparison is related to the concentration level. As a rule of thumb and a first estimate he found for many results in his intercomparisons amongst trained laboratories:

$$\text{C.V.} = 2^{(1-0.5 \log C)}$$

in which C.V. is the coefficient of variation (in %) and C the concentration of the analyte.

TABLE 17.2

COEFFICIENTS OF VARIATION IN SEDIMENT AND CABBAGE ANALYSIS OBTAINED IN AN INEXPERIENCED GROUP (A) AND A SELECTED (BCR) GROUP (B) OF LABORATORIES SHOWING GOOD PERFORMANCE OVER A LONGER PERIOD OF TIME

Element	Matrix	Found C.V. (%)		Expected (Horwitz) C.V. (%)	Rejected (%) Group A
		Group A[a]	Group B		
As	Sediment	22	9	15	—
Cd	Cabbage	9.4	4.5	10	50
	Sediment	56	12.6	15	28
Cu	Cabbage	14	5.5	15	26
	Sediment	2.7	2.3	3	35
Pb	Cabbage	14	8.4	20	64
	Sediment	18	5.0	8	53

[a] After rejection of obvious outliers (see last column)

This rule of Horwitz, however, is not fundamental. When participating laboratories have not been in quality exercises before, the results are likely to be worse than indicated; well trained and experienced laboratories with a good history in quality control, which can therefore be selected for certification purposes, may do better. A recent example (trace elements in a sediment and in red cabbage) illustrates this (Table 17.2).

17.4.3. Effects of participation in intercomparisons

Intercomparisons are a useful and important means to avoid systematic errors in laboratories, to assess the quality of work done in the laboratory, to motivate laboratory workers and to demonstrate the quality of a laboratory's result to its "customers" (e.g. public authorities).

In the following an example that was obtained in the BCR-program will be shown. Participants discussed thoroughly the reasons for discrepancy after an intercomparison, and changed and improved their methods in the light of the outcome of the discussion, before embarking on a new exercise with a different sample.

Table 17.3 presents results of four laboratories, A, B, C and D, participating in inter-comparisons of the same material in 1979 and 1982. The laboratories did not know that the same sample was used in both studies.

TABLE 17.3

SOME RESULTS OBTAINED BY FOUR MAJOR LABORATORIES IN EEC MEMBER COUNTRIES IN 1979 AND 1982 FOR THE SAME PLANT MATERIAL (contents in μg/g)

	Cd		Pb		Hg	
	1979	1982	1979	1982	1979	1982
A	0.106	0.106 ($\sigma = 0.006$)	25.2	25.7 ($\sigma = 1$)	0.250	0.334 ($\sigma = 0.04$)
B	0.114	0.118 ($\sigma = 0.006$)	25.5	26.9 ($\sigma = 0.5$)	0.289	0.254 ($\sigma = 0.02$)
C	0.161	0.114 ($\sigma = 0.006$)	29.4	27.4 ($\sigma = 1.2$)	0.288	0.296 ($\sigma = 0.0117$)
D	2.17	0.12 ($\sigma = 0.006$)	63.0	5.3 ($\sigma = 0.5$)	0.306	0.830 ($\sigma = 0.167$)
Certified value		0.10		35.0		0.28

Laboratories A and B are very well controlled and have produced good results over the years. Laboratory C has made an effort and was able to improve the quality of the determinations of cadmium and lead. Laboratory D has a good repeatability but continues to make wrong determinations.

References on p. 533

Experiences of various control schemes [e.g. AOAC (USA), VDLUFA (D), EPA (US), WRC (UK), TNO (NL) and others] all point in the same direction: participation in intercomparisons is a useful tool if the laboratory intends to improve the quality of work or to demonstrate this quality.

In intercomparisons for improving the quality of the participants' results, two similar samples with different contents should be studied wherever possible. The results of each laboratory can be presented in a so-called Youden plot [35,36]. The abscissa (x) is used for the first sample, the ordinate (y) for the second. Thus each laboratory is represented by one point. When the laboratory is situated in the vicinity of the point representing the most accurate values of the two samples, its performance is good. If the laboratory has good control but systematic errors occur (e.g. calibration), its position can be found close to the line $y = x$. Where the within laboratory consistency is poor, its results are in the lower right and upper left quadrant.

The evaluation becomes more difficult if the contents differ widely or if the matrix varies. However, in cases where such exercises were carried out on a regular basis, a definite improvement of the performance of the participants was indicated [37].

17.5. CERTIFIED REFERENCE MATERIALS (CRMs)

The use of certified reference materials has been mentioned previously in this chapter. This part will deal in more detail with the requirements for CRMs [48–50].

According to the IUPAC definition, the value of certified reference materials (NBS: "standard reference material") is "based on the consistent results obtained by using independent analytical techniques". The experience of BCR shows that this definition should be expanded to: "based on the consistent results obtained by using different analytical techniques which are likely to give accurate results for the particular matrix and which are applied by different independent laboratories". This also accounts for the important factor of the laboratory, illustrated in Sections 17.3 and 17.4.

CRMs can be categorized as follows:
(i) pure substances,
(ii) matrix materials,
(iii) CRMs for the calibration of relative methods (e.g. mass spectrometry of solid materials, XRF ...),
(iv) biological CRMs used for the measurement of biologically defined parameters (e.g. vitamin content in food).

CRMs should have the same traceability to the primary units as the transfer standards used for physical measurements.

17.5.1. Requirements

CRMs have to fulfill many requirements which in practice are often contradictory. Therefore compromises are necessary. A "fresh strawberries" CRM is impossible; ground and dried strawberry powder would be the best approximation.

17.5.1.1. Homogeneity

Standards in physics which are not destroyed when applied can be unique. CRMs in chemistry can be used once only, when taken out of their ampoule, bottle or other container. Therefore the amount of starting material should be high. This amount (e.g. 20–50 kg of dried and powderized material) must be homogeneous to assure that every sample out of this bulk should have the same composition. But even in a sample (e.g. a bottle with 40 g of the material) there should be homogeneity, as normally laboratories take sub-samples for analysis of about 100–1000 mg.

Homogeneity decreases with decreasing sub-sample size. The producer of a CRM therefore gives the minimum amount of sample for which the homogeneity is sufficient, i.e. for which the uncertainty caused by inhomogeneity does not exceed the uncertainty of the certified values.

Suitable techniques to study matrix homogeneity for trace elements are: XRF, INAA, ICP, etc. Their standard deviation should be small compared to the final uncertainty aimed at in the certification. In a homogeneity study all possible care is taken to obtain a good repeatibility; in this stage the accuracy of the techniques is of less importance.

The homogeneity of a sample is investigated by repeated analysis at different levels of intake and comparing the coefficients of variation of the results obtained with those expected from the analytical methods themselves [38,39]. This approach involves many individual repetitions, but various techniques of analysis [e.g. INAA, ICP, (ET)AAS] can be applied. Whenever a method is under full statistical control for a given matrix, the sampling constants method [40–43] can be followed.

Segregation is always possible if the CRM is not a pure compound. Matrix-CRMs and biological CRMs usually contain various phases of different particle size of different density. Segregation will even occur on a laboratory bench. Rehomogenization before taking the analytical sub-sample is required.

17.5.1.2. Stability

The user of a CRM should know the shelf life of his samples. The manufacturer of the CRM therefore carries out the necessary investigations to obtain the required information on the stability or even to establish an expiry.

Long-term tests are thus carried out at various temperatures and under different light conditions. Also, effects of short-term exposures to, e.g. high temperatures (simulating a transport situation) or moist air (simulating normal laboratory use) are studied.

In some cases a stability check is performed at regular intervals even when the materials are available.

17.5.1.3. Similarity with the real sample

When CRMs are used to verify the results of the analysis of a certain matrix, the sources of error encountered in the analysis of the CRM should be of the same or of a more difficult nature (e.g. same matrix, comparable contents of analyte and interfering elements) as in the real matrix. In such a case a good accuracy for the CRM may indicate a good accuracy for the unknown. This is often ignored, e.g. the matrix of NBS SRM 1632a ("coal") combusts more easily than most coal samples taken in a routine coal laboratory. Correct results for this SRM do not *ipso facto* imply correct results for any unknown coal.

17.5.1.4. Accuracy, uncertainty and traceability

The certified value should be the best approximation of the true value. Different methods of the best available accuracy should be applied by different outstanding laboratories. This is the only possibility to reduce systematic errors to such a level that they cannot be detected.

The uncertainty in the certified value should be sufficiently small to allow the CRM to be used for calibration or to be able to distinguish between a good and a bad result of the laboratory which uses the CRM for certification.

Traceability to the primary units is achieved by high standards in the certifying laboratories with regards to: calibration of balances, volumetric glassware and other tools of relevance to the value to be certified. Also the calibration of the technique of final determination should be done with traceable compounds, which means that the laboratories verify stoichiometry or composition (e.g. for calibration of trace elements, the contents in the stock solution can be verified by gravimetry or titrimetry).

17.5.2. Preparation of candidate CRMs

In preparing a candidate CRM the items mentioned above should be taken into full consideration.

For gases and liquids the achievement of homogeneity is not the most difficult problem; however, stability causes great concern.

Solid materials are difficult to homogenize. In addition the speciation of, e.g. trace elements, the binding of organic compounds to a matrix component and the concentrations of the components to be certified should not be changed considerably in the preparation.

Contamination is only allowed with those elements which will not be certified (e.g., in the grinding process contamination is unavoidable; a material to be certified for Cr or Ni cannot be ground in a stainless steel device ...).

If the speciation or the binding of organic pollutants is changed, the difficulties encountered in the real analysis no longer resemble the difficulties of the CRM. Careful drying (temperature, inert gas ...) is a prerequisite. A typical example is a fish material where most arsenic is present as arsenobetaine, which is difficult to digest. If in the course of the preparation the arsenobetaine is already partly oxidized (with possible risk of losses) the CRMs digestion is facilitated.

Drying, grinding, and sieving are common methods for homogenization. If the materials are not very stable, or prone to contamination, pilot studies are necessary [44–46].

The container material should not contaminate the sample but ensure the stability and allow an easy application of the material in laboratory practice [47].

17.5.3. Availability of CRMs

ISO regularly publishes surveys of available CRMs together with their producers. Materials are grouped in fields of application such as: geochemistry, physical chemistry, environment, food, biomedical measurements, etc.

The ISO-listing of 1982 gives some 300 manufacturers offering 200 different sets of materials. (ISO-Directory of Certified Reference Materials, 1st Ed. 1982, ISBN 9267 01027; to be ordered from: ISO-secretariat, Case Postale 56, CH-1211 Genève 20). The amount of CRMs available has rapidly increased since then.

A computerized information system developed by leading normalization and metrology institutes is now under final development and is expected to be introduced in normalization and QA-institutes in various countries soon [51].

REFERENCES

1 S.S. Berman and V.J. Boyko, *ICES 6th Round Intercalibration for Trace Metals in Seawater* JMG 6/TM/SW (1987).
2 K. Leichnitz, *Sicherheitsingenieur*, 7 (1986) 24.
3 A.R. Byrne, C. Camara-Rica, R. Cornelis, J.J.M. De Goey, G.V. Iyengar, G. Kirkbright, G. Knapp, R.M. Parr and M. Stoeppler, *Fresenius' Z. Anal. Chem.*, 326 (1987) 723–729.
4 K.S. Subramanian and M. Stoeppler, *Fresenius' Z. Anal. Chem.*, 328 (1986) 875–879.
5 H.P. Thier, *Mitt. Lebensm. Gerichtl. Chem.*, 30 (1976) 187.
6 C.J. Musial and J.F. Uthe, *J. Assoc. Off. Anal. Chem.*, 66 (1983) 22.
7 R. Boniforti, R. Ferraroli, P. Frigieri, D. Heltal and G. Queirazza, *Anal. Chim. Acta*, 162 (1984) 33–46.
8 R.F.M. Herber, M. Stoeppler and D.B. Tonks, *Fresenius' Z. Anal. Chem.*, 338 (1990) 269–286.
9 A. Knöchel and W. Petersen, *Fresenius' Z. Anal. Chem.*, 314 (1983) 105–113.
10 G. Tölg, personal communication, 1990.
11 P. Tschöpel and G. Tölg, *J. Trace Microprobe Techn.*, 1 (1982) 1.
12 K. Heydorn and E. Damsgaard, *Talanta*, 29 (1982) 1019–1024.
13 G. Tölg, Proc. Symp. *Anorganische Stoffe in der Toxikologie und Kriminalistik*, Verl. Dr. Dieter Helm, Heppenheim, 1983, p. 109.
14 B. Griepink, *Fresenius' Z. Anal. Chem.*, 317 (1984) 210–217.
15 B. Griepink, *Irish Chem. News*, (1986) 22.
16 ISO-guide 25, *General Requirements for the Technical Competence of Testing Laboratories*, 1982.
17 F.A.J.M. Vijverberg and W.P. Cofino, *ICES-report 6, Control Procedures: GLP and QA*, Copenhagen, 1987.
18 B. Kinsella and R. Willix, *Analytical Quality Assurance in the Marine Quality Assessment Programme*, Western Australian Institute of Technology, 1985.
19 F.M. Garfield, *Trends Anal. Chem.*, 4 (1985) 162.
20 J.K. Taylor, *Trends Anal. Chem.*, 3 (1984) 11.

534

21 A. Uldall, *Quality Assurance in Clinical Chemistry, Scand. J. Clin. Lab. Invest.*, 47 (Supplement 187) (1987).
22 OECD, *Principles of Good Laboratory Practice*, 1981.
23 J.K. Taylor, *Handbook for SRM-Users*, NBS Special Publication 260–100, 1985.
24 J. Vogelgesang, *Fresenius' Z. Anal. Chem.*, 328 (1987) 213–220.
25 SHEWHART, *Economic Control of Quality of the Manufactures Product*, D. Van Norstrand, New York, 1931, 1–501.
26 Nordisk Council of Ministers, *Handbog 1, Intern Kvalitetskontrol pa Vandlaboratorier*, Projekt No. 180.21-16, 1984-08-21/SAA, VKI-sag No. 44.216.
27 D.J. Dewey and D.T.E. Hunt, *The Use of Cumulative Sum Charts (Cusum Charts) in Analytical Quality Control*, Water Research Centre, Medmenham, Technical Report TR 174, 1972.
28 G. Kateman and F.W. Pijpers, *Quality Control in Analytical Chemistry*, Wiley, New York, 1981.
29 R.W.H. Edwards, *Ann. Clin. Biochem.*, 17 (1980) 205.
30 J.O. Westgard, T. Groth, T. Aronsson and C.H. De Verdier, *Clin. Chem.*, 23 (1977) 1981.
31 H. Muntau, (Joint Research Centre EEC, Ispra, I) personal communication.
32 ISO International Standard 5727 (1981).
33 W. Horwitz, L.R. Kamps and K.W. Boyer, *J. Assoc. Off. Anal. Chem.*, 63 (1980) 1344.
34 K.W. Boyer, W. Horwitz and R. Albert, *Anal. Chem.*, 57 (1985) 454–459.
35 W.J. Youden, *Ind. Qual. Contr.*, 15(11) (1959) 24–28.
36 W.J. Youden, *Statistical Techniques for Collaborative Tests*, Statistical Manual of the Assoc. Off. Anal. Chem., 1975, p. 1.
37 T. Aronsson and T. Groth, *Scand. J. Clin. Lab. Invest.*, 44 (suppl. 172) (1984) 51.
38 P. Schramel, J. Schmolck and H. Muntau, *J. Radioanal. Chem.*, 50 (1979) 179.
39 B. Griepink, E. Colinet, G. Guzzi, L. Haemers and H. Muntau, *Fresenius' Z. Anal. Chem.*, 315 (1983) 20–25.
40 C.O. Ingamells, *Talanta*, 21 (1974) 141.
41 C.O. Ingamells, *Talanta*, 23 (1976) 263.
42 B. Kratochvil, M.J.M. Duke and D. Ng, *Anal. Chem.*, 58 (1986) 102–108.
43 K. Heydorn and E. Damsgaard, *J. Radioanal. Nucl. Chem.*, 100 (1987) 539.
44 R.B. Lockman, *J. Assoc. Off. Anal. Chem.*, 63 (1980) 766.
45 B. Griepink, H. Muntau and E. Colinet, *Fresenius' Z. Anal. Chem.*, 315 (1983) 193–196.
46 P.J. Wagstaffe, B. Griepink, H. Hecht, H. Muntau and P. Schramel, *EUR-Report*, 10168 (1986).
47 J.R. Moody and R.M. Lindstrom, *Anal. Chem.*, 49 (1977) 2264.
48 B. Griepink and H. Marchandise, *Referenzmaterialien*, in: *Analytiker Taschenbuch, Band 6*, Springer Verlag, Berlin, 1986, pp. 1–16.
49 H. Marchandise and E. Colinet, *Fresenius' Z. Anal. Chem.*, 316 (1983) 669–672.
50 B. Griepink, *Fresenius' Z. Anal. Chem.*, 338 (1990) 360–362.
51 H. Klich and P. Caliste, *Fresenius' Z. Anal. Chem.*, 332 (1988) 552–555.

SUBJECT INDEX

Air, analytical methods for elements and organometallic compounds
- chromium 376
- lead 246
- mercury 263,267-273,277-280
- nickel 434
- selenium 491-492, 503

Alkyllead 245-246

Aluminium
- determination methods 463-470
- environmental levels 452-455
- reference materials 470
- sampling and pretreatment 455-462

Analysis of liquids other than water (mainly biological and beverages)
- aluminium 463,464,467-469
- arsenic 303,308,316,317,322,325, 327,329-334
- cadmium 186-188,196,199,202,204, 205,209
- chromium 390,391,394,396
- lead 245,246
- nickel and cobalt 412,413,415,416, 418,422,425-427,429
- selenium 500,501,504
- thallium 363,366

Analysis of solids
- aluminium 463,469,470
- arsenic 297-300,302,303,308,309, 311-313,315-320,324-332,334,335,337, 338
- cadmium 184-186,188-199,202-207,209
- chromium 389-396
- lead 244,250,251
- mercury 274-276,278-280
- nickel and cobalt 413-419,422,423, 425-429
- selenium 500,501,504
- thallium 361-366

Analytical methods, general
- atomic absorption spectrometry (AAS) 98-103

- atomic emission spectrometry (AES) 103-105
- atomic fluorescence spectrometry (AFS) 103
- chromatographic methods 117-118
- electrochemical methods 107-112
- inductively coupled plasma-mass spectrometry (ICP-MS) 106-107
- mass spectrometry (MS) 118-119
- nuclear methods 114-117
- spectrophotometry and related techniques 119-120
- X-ray fluorescence spectrometry (XRF) 112-114

Analytical methods, single metals
- aluminium 463-470
- arsenic 296-321
- cadmium 182-210
- chromium 387-396
- cobalt 410-432
- lead 232-243
- mercury 273-278
- nickel 410-432
- selenium 478-504
- thallium 361-368

Analytical tasks
- with extreme difficulties 5
- with great difficulties 5
- with moderate difficulties 3
- without particular difficulties 2
- with slightly increased difficulties 4

Arsenic
- alkylarsines 334
- arsenic compounds analysis 321-339
- arsenite and arsenate 292,322-334
- arsenobetaine 335-338
- arsenocholine 337-338
- compounds in the environment 292-294
- reference materials 339-340
- total arsenic analysis 299-321
- total arsenic and arsenic compounds 290-292

Atomic absorption spectrometry — cold vapor technique
- mercury 102,259,260,279–281

Atomic absorption spectrometry — hydride technique
- general 102
- arsenic 316-318
- lead 234
- selenium 500-502

Atomic absorption spectrometry with flame (FAAS)
- general 99,104-105
- arsenic 303-304
- cadmium 183-186
- chromium 388-389
- lead 232-234
- nickel and cobalt 411,413,416
- selenium 500
- thallium 364-365

Atomic absorption spectrometry with the graphite furnace (GFAAS)
- general 99-102
- aluminium 465-469
- arsenic 304-308
- cadmium 183-184,186-193
- chromium 389-392
- lead 233-234
- nickel and cobalt 411-415
- selenium 500-503
- thallium 365-366

Atomic absorption spectrometry with the graphite furnace — solids (solid sampling GFAAS)
- general 100-101
- aluminium 469-470
- arsenic 308
- cadmium 193-194
- chromium 391-392
- lead 250
- nickel and cobalt 414-415

Atomic emission spectroscopy (AES) with plasmas (Plasma AES)
- general 103-105,120-121
- aluminium 463-464
- arsenic 309-310
- cadmium 195-197

- chromium 392-393
- lead 233,235
- mercury 274-275,277
- nickel and cobalt 415-417
- selenium 501-502
- thallium 361-363

Atomic fluorescence spectrometry
- general 103,121
- arsenic 103,303
- cadmium 103,219
- chromium 396
- lead 103,235
- mercury 103,274,277-279
- selenium 103,500

Beverages, analysis, see: Analysis of liquids other than water

Biological materials, analysis, see also analysis of liquids other than water (mainly biological) and analysis of solids
- general 4-5

Cadmium
- analysis 182-210
- enrichment and separation 181-182
- in environmental and biological materials 179-181
- quality control and reference materials 213-217
- sampling and sample treatment 178,181
- speciation 210-213

Carbon furnace atomic emission spectrometry
- general 105

Chromatographic methods
- general 117-118,121
- aluminium 463
- arsenic 315,325-339
- cadmium 207-208,211-213
- chromium 395-396
- lead 239-243,245-246
- mercury 270-271,278-280
- nickel and cobalt 428-430
- selenium 500,502-503
- thallium 360-361

Chromium
- analytical procedures 387-396
- occurence, production, essentiality and toxicity 374-379
- quality control and reference materials 397-400
- sampling and sample treatment 379-387
- speciation 396-397

Cobalt (see nickel and cobalt)

Colorimetry (see spectrophotometry)

Coupled (hyphenated) methods see: Chromatographic methods

Detection limits — elements
- aluminium 101,105,106,111,114,464
- arsenic 101,105,106,111,114,297,300, 302,303,308,311-313,315,317-320,323, 325,332,333
- cadmium 101,105,106,111,114,116,185
- chromium 101,105,106,111,114,388
- lead 101,105,106,111,114,116,233
- mercury 101,105,106,111,114,116,274
- nickel and cobalt 101,105,106,111, 114,416,430
- selenium 101,105,106,111,114,464
- thallium 101,105,106,111,114,116,362ff

Detection limits — methods
- flame AAS 105
- graphite furnace AAS 101
- inductively coupled plasma-atomic emission 105
- inductively coupled plasma-mass spectrometry 106
- instrumental neutron activation analysis 116
- total reflection X-Ray analysis 114
- voltammetric methods 111

Diagram of a trace analytical procedure 3

Differential optical absorption spectrometry
- mercury 281

Dimethylmercury 258,266,271,279,280

Dimethylthallium 365,367

Electrochemical analysis (see also voltammetry and potentiometric stripping)
- general 107-112,121
- aluminium 111
- arsenic 310-312
- cadmium 200-204
- lead 237
- nickel and cobalt 419-424
- selenium 500
- thallium 366-368

Electrothermal atomic absorption spectrometry (ETAAS) (see: Atomic absorption spectrometry with the graphite furnace (GFAAS))

Environmental and biological levels and occurence of metals
- aluminium 452-455
- arsenic 292-294
- cadmium 179-181
- chromium 374-375
- cobalt 406-408
- lead 244
- mercury 261-264
- nickel 406-408
- selenium 476-478
- thallium 352-353

Environmental and technical materials (solid materials) analysis
- general 2-5

Environmental specimen banking
- coordination, cooperation and technical assistance 16,31
- historical aspects 20-26
- methods 32-43
- relationship to environmental monitoring, research and assessment 26-27
- role and application to the assessment of toxic and hazardous chemicals in the environment 27-30
- suitability of specimens 31-32

Essentiality
- arsenic 289-290
- cadmium 178

538

- chromium 377
- cobalt 406
- selenium 478

Flow injection technique 99,120,123, 195,199,208,210,211,218,317,318,325, 368,417,421,430-432

Graphite furnace-MS coupling (GFAAS-MS)
- thallium 364

Hydride techniques
- general 102
- arsenic interferences 320-321
  with atomic fluorescence 315
  with colorimetry 315
  with DC plasma AES 319
  with flame AAS 316
  with gaschromatography 315
  with graphite furnace 317
  with ICP-AES 319
  with microwave induced plasma AES 318
  with molecular emission 318
  with quartz tube AAS 316
- lead with AAS 234
- selenium with AAS 500-502

Hyphenated methods (see: chromatographic methods)

Inductively coupled plasma mass spectrometry
- general 106-107,121
- aluminium 106,463
- arsenic 106
- cadmium 197-200
- chromium 106,393-394
- lead 238-239
- mercury 106,277
- nickel and cobalt 106,416-419
- selenium 106,501
- thallium 106,364

In vivo X-ray fluorescence
- general 113
- cadmium 113,205-206
- lead 113

Lead
- analytical techniques 232-243
- applications
  alkyllead 245-246
  lead in air 246-248
  lead in sediments, soils and biological materials 250-251
  lead in water 248-249
  total lead 244-245
- in environmental materials 244

Liquids, see analysis of water and analysis of liquids other than water

Mass spectrometry
- general 118-119
- aluminium 463
- cadmium 208-210
- chromium 394-395
- lead 232
- nickel and cobalt 430-431
- selenium 500-503
- thallium 363-364

Mercury
- general 258-259
- environmental levels 260-264
- sampling procedures 264-273
- speciation 270-271,278-280
- total mercury 273-278
- toxicity 259-261

Methyl mercury
- general 117,258,259,263,271,278-280
- toxicity 259,260

Methylarsenic compounds 326-334

Neutron activation analysis, see Nuclear methods

Nickel and cobalt
- analytical methods 410-432
- enrichment and separation 408-410
- environmental levels 406-408
- quality control and reference materials 434-440
- sampling and sample pretreatment 408
- speciation 432-434

Nuclear methods
- general   114-117,121
- aluminium   116,463
- arsenic   116,296-300
- cadmium   116,206-207
- chromium   116,394
- lead   116
- mercury   116,274-275
- nickel and cobalt   116,416,427-428
- selenium   116,500-502
- thallium   116

Optical emission spectrometry (OES), see atomic emission spectrometry (AES)

Phenylmercury   259

Plasma AES (see Atomic emission spectroscopy with plasmas)

Polarography (see Voltammetry)

Potentiometric stripping analysis (PSA)
- general   111-112
- cadmium   200,204
- nickel and cobalt   421,422-423

Quality assurance
- general   3
- cadmium   213-217
- chromium   397-400
- definitions   519
- intercomparisons   526-530
- nickel and cobalt   434-440
- reference materials   525,530-533
- specimen banking   42-43
- within laboratory measures   519-526

Radiotracers
- general   115-117
- aluminium   116
- arsenic   116,313
- cadmium   116,206-207
- chromium   116
- lead   116
- mercury   116
- nickel and cobalt   116,427
- selenium   116
- thallium   116

Reference materials
- general   525-526,530-533
- aluminium   470
- arsenic   339-340
- cadmium   214-217
- chromium   397-400
- cobalt   435-437
- nickel   438-440
- preparation   532-533
- suppliers   217,400,441,533
- thallium   354

Sample collection
- aqueous samples   10
- biological fluids   12
- environmental and anthropogenic materials 13
- for mercury analysis   269-270
- for selenium analysis   492-493
- for speciation   14
- solid biological materials   12

Sample storage
- general   15,81
- for selenium   495
- in environmental specimen banks 16,40

Sample treatment
- general   73-75
- aluminium   459-460
- arsenic   295-296
- cadmium   181
- chromium   385-387
- decomposition procedures, overview 81-88
- lead   244
- mercury   274-275
- nickel and cobalt   408
- selenium   495-499
- thallium   357-361

Selenium
- analytical applications   483-489
- analytical methods   478-483
- analytical procedures   499-504
- sample treatment   495-499
- sampling, sample handling and storage 490-495

Separation and pre-concentration
- general   88-92

- aluminium 463
- arsenic 321-339
- cadmium 181-182
- chromium 395
- lead 239-243
- mercury 270-271
- nickel and cobalt 408-410
- thallium 357-361

Solid sampling GFAAS (see Atomic absorption spectrometry with the graphite furnace-solids)

Solids, see Analysis of solids

Speciation
- arsenic 157,321-339
- cadmium 157,210-213
- chromium 158,396-397
- cobalt 161,432-434
- copper 159
- general in sediments and soils 136-140
- lead 162,239-343
- mercury 278-280
- methods for sediments and soils 140-156,167-171
- nickel 162,432-434
- selenium 503
- zinc 163

Spectrophotometry and related techniques (Colorimetry)
- general 119-120,122
- aluminium 463-464
- arsenic 312-314
- cadmium 210
- chromium 395
- lead 235-236
- nickel and cobalt 431-432
- selenium 500-504
- thallium 368

Square wave voltammetry (see also: voltammetry)
- general 109
- cadmium 200,204
- nickel and cobalt 423,424

Thallium
- analytical methods 361-368
- decomposition 354-357

- occurence 352-353
- reference materials 354
- separation 357-361

Titration
- lead 236-237

Total reflection X-ray fluorescence spectrometry (TXRF)
- general 113-114,121
- arsenic 114,302
- cadmium 205
- chromium 114
- lead 114
- mercury 114
- nickel and cobalt 114,416,425-426
- selenium 114
- thallium 114

Toxicity
- aluminium 451-452
- arsenic 288-291
- cadmium 178
- chromium 374,375,377-379
- lead 231-232
- mercury 259-261
- nickel and cobalt 406
- selenium 478
- thallium 352

Voltammetry (and polarography)
- general 107-111
- aluminium 111
- arsenic 111,311-312
- cadmium 111,200-204
- chromium 111,396
- lead 111,237
- mercury 111
- nickel and cobalt 111,419-424
- selenium 111,500
- thallium 111,366-368

Water analysis
- aluminium 462
- arsenic 292,300,302,307,308,311-313, 316-319,321-326,328-334
- cadmium 186-187,195,198,200-202, 208,210-211
- chromium 377,380-381,392,394
- lead 248-249
- mercury 276-277,279-280

- nickel and cobalt   411-412,418,
  420-422,426-429,431,433
- selenium   482,483,492,494,495,500-504
- thallium   358,363,364,366,368

Wine, enological specimen bank
- general   49-53
- arsenic   70
- cadmium   53-56
- chromium   69
- cobalt   56-57
- copper   57-58
- lead   58-64
- mercury   64-67

- nickel   67
- zinc   67-70

X-Ray fluorescence spectrometry (XRF)
- general   112-114,121
- aluminium   463
- arsenic   301-303
- cadmium   204-206
- chromium   396
- lead   237-238,250-251
- nickel and cobalt   424-427
- selenium   500-502

Printed and bound by CPI Group (UK) Ltd, Croydon, CR0 4YY

03/10/2024

01040333-0005